Climate Variability and Climate Change Impacts on Land Surface, Hydrological Processes and Water Management

Climate Variability and Climate Change Impacts on Land Surface, Hydrological Processes and Water Management

Special Issue Editors

Yongqiang Zhang
Hongxia Li
Paolo Reggiani

MDPI • Basel • Beijing • Wuhan • Barcelona • Belgrade

Special Issue Editors

Yongqiang Zhang
Institute of Geographic Sciences and
Natural Resources Research,
Chinese Academy of Sciences
China

Hongxia Li
State Key Laboratory of Hydraulics and
Mountain River Engineering,
Sichuan University
China

Paolo Reggiani
Department of Civil Engineering,
University of Siegen
Germany

Editorial Office
MDPI
St. Alban-Anlage 66
4052 Basel, Switzerland

This is a reprint of articles from the Special Issue published online in the open access journal *Water* (ISSN 2073-4441) from 2018 to 2019 (available at: https://www.mdpi.com/si/water/climate_change_water)

For citation purposes, cite each article independently as indicated on the article page online and as indicated below:

LastName, A.A.; LastName, B.B.; LastName, C.C. Article Title. *Journal Name* **Year**, *Article Number*, Page Range.

ISBN 978-3-03921-507-2 (Pbk)
ISBN 978-3-03921-508-9 (PDF)

Contents

About the Special Issue Editors ix

Yongqiang Zhang, Hongxia Li and Paolo Reggiani
Climate Variability and Climate Change Impacts on Land Surface, Hydrological Processes and Water Management
Reprinted from: *Water* **2019**, *11*, 1492, doi:10.3390/w11071492 1

Venkataramana Sridhar, Hyunwoo Kang and Syed A. Ali
Human-Induced Alterations to Land Use and Climate and Their Responses for Hydrology and Water Management in the Mekong River Basin
Reprinted from: *Water* **2019**, *11*, 1307, doi:10.3390/w11061307 9

Qiaoling Guo, Yaoyao Han, Yunsong Yang, Guobin Fu and Jianlin Li
Quantifying the Impacts of Climate Change, Coal Mining and Soil and Water Conservation on Streamflow in a Coal Mining Concentrated Watershed on the Loess Plateau, China
Reprinted from: *Water* **2019**, *11*, 1054, doi:10.3390/w11051054 31

Tongho Ri, Jiping Jiang, Bellie Sivakumar and Tianrui Pang
A Statistical–Distributed Model of Average Annual Runoff for Water Resources Assessment in DPR Korea
Reprinted from: *Water* **2019**, *11*, 965, doi:10.3390/w11050965 46

Raphael Pousa, Marcos Heil Costa, Fernando Martins Pimenta, Vitor Cunha Fontes, Vinícius Fonseca Anício de Brito and Marina Castro
Climate Change and Intense Irrigation Growth in Western Bahia, Brazil: The Urgent Need for Hydroclimatic Monitoring
Reprinted from: *Water* **2019**, *11*, 933, doi:10.3390/w11050933 65

Lianzhou Wu, Tao Bai, Qiang Huang, Ming Zhang and Pengfei Mu
Influence of Power Operations of Cascade Hydropower Stations under Climate Change and Human Activities and Revised Optimal Operation Strategies: A Case Study in the Upper Han River, China
Reprinted from: *Water* **2019**, *11*, 895, doi:10.3390/w11050895 86

Pranesh Kumar Paul, Yongqiang Zhang, Ashok Mishra, Niranjan Panigrahy and Rajendra Singh
Comparative Study of Two State-of-the-Art Semi-Distributed Hydrological Models
Reprinted from: *Water* **2019**, *11*, 871, doi:10.3390/w11050871 110

Taereem Kim, Ju-Young Shin, Hanbeen Kim, Sunghun Kim and Jun-Haeng Heo
The Use of Large-Scale Climate Indices in Monthly Reservoir Inflow Forecasting and Its Application on Time Series and Artificial Intelligence Models
Reprinted from: *Water* **2019**, *11*, 374, doi:10.3390/w11020374 130

Xingxing Shang, Xiaohui Jiang, Ruining Jia and Chen Wei
Land Use and Climate Change Effects on Surface Runoff Variations in the Upper Heihe River Basin
Reprinted from: *Water* **2019**, *11*, 344, doi:10.3390/w11020344 155

Zhihong Yan, Shuqian Wang, Ding Ma, Bin Liu, Hong Lin and Su Li
Meteorological Factors Affecting Pan Evaporation in the Haihe River Basin, China
Reprinted from: *Water* **2019**, *11*, 317, doi:10.3390/w11020317 175

Yanjun Gao and Yongqiang Zhang
Effects of the Three Gorges Project on Runoff and Related Benefits of the Key Regions along
Main Branches of the Yangtze River
Reprinted from: *Water* **2019**, *11*, 269, doi:10.3390/w11020269 . . . . **193**

Mohammed Gedefaw, Denghua Yan, Hao Wang, Tianling Qin and Kun Wang
Analysis of the Recent Trends of Two Climate Parameters over Two Eco-Regions of Ethiopia
Reprinted from: *Water* **2019**, *11*, 161, doi:10.3390/w11010161 . . . . **209**

Narayan Nyaupane, Balbhadra Thakur, Ajay Kalra and Sajjad Ahmad
Evaluating Future Flood Scenarios Using CMIP5 Climate Projections
Reprinted from: *Water* **2018**, *10*, 1866, doi:10.3390/w10121866 . . . . **221**

Wenjia Deng, Jinxi Song, Hua Bai, Yi He, Miao Yu, Huiyuan Wang and Dandong Cheng
Analyzing the Impacts of Climate Variability and Land Surface Changes on the Annual
Water–Energy Balance in the Weihe River Basin of China
Reprinted from: *Water* **2018**, *10*, 1792, doi:10.3390/w10121792 . . . . **239**

Lei Tian, Jiming Jin, Pute Wu and Guo-yue Niu
Quantifying the Impact of Climate Change and Human Activities on Streamflow in a Semi-Arid
Watershed with the Budyko Equation Incorporating Dynamic Vegetation Information
Reprinted from: *Water* **2018**, *10*, 1781, doi:10.3390/w10121781 . . . . **254**

Qian Li, Tao Yang, Zhiming Qi and Lanhai Li
Spatiotemporal Variation of Snowfall to Precipitation Ratio and Its Implication on Water
Resources by a Regional Climate Model over Xinjiang, China
Reprinted from: *Water* **2018**, *10*, 1463, doi:10.3390/w10101463 . . . . **271**

**Batsuren Dorjsuren, Denghua Yan, Hao Wang, Sonomdagva Chonokhuu, Altanbold
Enkhbold, Xu Yiran, Abel Girma, Mohammed Gedefaw and Asaminew Abiyu**
Observed Trends of Climate and River Discharge in Mongolia's Selenga Sub-Basin of the
Lake Baikal Basin
Reprinted from: *Water* **2018**, *10*, 1436, doi:10.3390/w10101436 . . . . **284**

Jae Heon Cho and Jong Ho Lee
Multiple Linear Regression Models for Predicting Nonpoint-Source Pollutant Discharge from
a Highland Agricultural Region
Reprinted from: *Water* **2018**, *10*, 1156, doi:10.3390/w10091156 . . . . **302**

Jie Wu, Zhihui Wang, Zengchuan Dong, Qiuhong Tang, Xizhi Lv and Guotao Dong
Analysis of Natural Streamflow Variation and Its Influential Factors on the Yellow River from
1957 to 2010
Reprinted from: *Water* **2018**, *10*, 1155, doi:10.3390/w10091155 . . . . **319**

Changming Zhu, Xin Zhang and Qiaohua Huang
Four Decades of Estuarine Wetland Changes in the Yellow River Delta Based on Landsat
Observations Between 1973 and 2013
Reprinted from: *Water* **2018**, *10*, 933, doi:10.3390/w10070933 . . . . **337**

Qiuwen Zhou, Xu Zhou, Ya Luo and Mingyong Cai
The Effects of Litter Layer and Topsoil on Surface Runoff during Simulated Rainfall in Guizhou
Province, China: A Plot Scale Case Study
Reprinted from: *Water* **2018**, *10*, 915, doi:10.3390/w10070915 . . . . **362**

Xiaocong Liu, Zhonggen Wang, Yin Tang, Zehua Wu, Yuhan Guo and Yashan Cheng
Integrating Field Experiments with Modeling to Evaluate the Freshwater Availability at
Ungauged Sites: A Case Study of Pingtan Island (China)
Reprinted from: *Water* **2018**, *10*, 740, doi:10.3390/w10060740 . . . . **373**

Fupeng Li, Zhengtao Wang, Nengfang Chao and Qingyi Song
Assessing the Influence of the Three Gorges Dam on Hydrological Drought Using GRACE Data
Reprinted from: *Water* **2018**, *10*, 669, doi:10.3390/w10050669 . . . . **388**

Paula Cecilia Soto Rios, Tariq A. Deen, Nidhi Nagabhatla and Gustavo Ayala
Explaining Water Pricing through a Water Security Lens
Reprinted from: *Water* **2018**, *10*, 1173, doi:10.3390/w10091173 . . . . **405**

Zengchao Hao, Vijay P. Singh and Fanghua Hao
Compound Extremes in Hydroclimatology: A Review
Reprinted from: *Water* **2018**, *10*, 718, doi:10.3390/w10060718 . . . . **425**

About the Special Issue Editors

Yongqiang Zhang has 20 years of experience in hydrological modeling and remote sensing hydrology. He is currently Distinguished Professor at the Institute of Geographic Sciences and Natural Resources Research, Chinese Academy of Sciences. He worked in CSIRO Land and Water as Research Scientist, Senior Research Scientist, and Principal Research Scientist over the period 2006–2018. He is a prestigious Alexander von Humboldt Fellow and has won 12 professional awards, including the 2012 GN Alexander Medal. Zhang has around 150 publications, including journal papers published in various Nature journals (*Nature Climate Change, Nature Communications, Scientific Reports*) and other top journals of their research field (*Water Resources Research, Journal of Hydrology, Journal of Geophysical Research: Atmospheres, Remote Sensing of Environment, Global Change Biology*). He has >6700 Google Scholar citations, a h-index of 39, and >4000 ISI citations. His major research contributions include developing novel hydrological modeling approaches by using remote sensing techniques to significantly improve simulations and predictions of runoff (and streamflow and water availability) in gauged and ungauged catchments as well as development of the PML evapotranspiration model—a representative diagnostic model to evaluate IPCC global climate model simulations. Zhang is serving in various positions for six peer-reviewed journals, including as Associate Editor of *Journal of Hydrology* and *Journal of Geophysical Research: Atmospheres* as well as Editorial Board member of *Remote Sensing of Environment* and *Water*. He is a regular reviewer for Science and top hydrology/water resources journals (such as *Water Resources Research* and *Journal of Hydrology*). He serves on the PhD Advisory Committee at the University of Melbourne and University of Technology Sydney. He is an Assessor in numerous remote sensing hydrology projects for the European Union.

Hongxia Li is Associate Professor at Sichuan University. She has 10 years of experience in hydrological modeling, flood forecasting, and climate change impacts on water resources. Li has authored more than 30 publications, including journal papers published in *Journal of Hydrology, Water*, and *Advances in Meteorology*. She works on many scientific projects in the fields of flood forecasting, water resource allocation, climate change impacts on water resources, and water environment modeling and evaluation.

Paolo Reggiani holds the Chair of Water Resources Management and Climate Impact Research at the University of Siegen, Germany. In 1994, Reggiani graduated from the University of Trento, Italy, in Environmental Engineering with a MSc in Engineering. In 2000, he was awarded a PhD with Distinction in Environmental Engineering from the University of Western Australia. Between 1999 and 2000, Reggiani worked at CSIRO Land and Water in Floreat Park, WA, where he developed modeling approaches to assess dryland salinity in Australia. After his return to Europe in the late 2000s, he was a Marie Curie Fellow at the soil science lab Laboratoire des Transferts en Hydrologie et Environnement (LTHE) in Grenoble, France. In 2002, he was appointed Researcher and Scientific Consultant at the Dutch Institute Deltares. In 2014, he was appointed Full Professor at the University of Siegen. During his career, Reggiani has worked in various applied scientific and consulting projects in Asia, Africa, and Europe. An important focal point of his scientific work has been the interface between numerical weather prediction and hydrological forecasting, including the area of forecasting uncertainty as well as operational water management in data-poor areas. Reggiani has published numerous papers in different areas of hydrology, climate change impacts on water resources, flood forecasting, channel hydraulics, and soil science. He also acted as Coordinator of several projects, including the FP5 project "European Flood Forecasting System (EFFS). Reggiani is a member of several scientific boards on water resources issues. He frequently serves as an expert reviewer for scientific evaluations in the fields of catchment hydrology, flood forecasting, climate change impacts, geomorphology, and natural hazard risk assessment. He also participates in an advisory board to the Commission of Hydrology at the WMO.

Editorial

Climate Variability and Climate Change Impacts on Land Surface, Hydrological Processes and Water Management

Yongqiang Zhang [1],* , Hongxia Li [2] and Paolo Reggiani [3]

[1] Key Laboratory of Water Cycle and Related Land Surface Processes, Institute of Geographic Sciences and Natural Resources Research, The Chinese Academy of Sciences, Beijing 100101, China
[2] State Key Laboratory of Hydraulics and Mountain River Engineering, Sichuan University, Chengdu 610065, China
[3] Department of Civil Engineering, University of Siegen, 57068 Siegen, Germany
* Correspondence: yongqiang.zhang2014@gmail.com; Tel.: +86-10-64856515

Received: 16 July 2019; Accepted: 17 July 2019; Published: 18 July 2019

Abstract: During the last several decades, Earth´s climate has undergone significant changes due to anthropogenic global warming, and feedbacks to the water cycle. Therefore, persistent efforts are required to understand the hydrological processes and to engage in efficient water management strategies under changing environmental conditions. The twenty-four contributions in this Special Issue have broadly addressed the issues across four major research areas: (1) Climate and land-use change impacts on hydrological processes, (2) hydrological trends and causality analysis faced in hydrology, (3) hydrological model simulations and predictions, and (4) reviews on water prices and climate extremes. The substantial number of international contributions to the Special Issue indicates that climate change impacts on water resources analysis attracts global attention. Here, we give an introductory summary of the research questions addressed by the papers and point the attention of readers toward how the presented studies help gaining scientific knowledge and support policy makers.

Keywords: climate variability; climate change; land use change; hydrological processes; trends; water management; model; predictions

---

1. Introduction

It is commonly recognized that Earth´s atmosphere is subject to anthropogenic climate change due to enhanced greenhouse gas concentrations in the lower atmosphere. This development also influences hydrological processes across a range of spatial scales, reaching from the singular catchment to regional and global scales. To cope with these changes, it is necessary to implement efficient water management strategies at country, regional or global scale adaptation. To better grasp the mechanism and response to climate variability and climate change, it is crucial to stimulate multidisciplinary studies involving multiple cross-cutting disciplines such as hydrology, meteorology, remote sensing, ecology, engineering, and agriculture.

To address these challenges, continuing efforts need to be undertaken to gain insights on hydrological processes, and engage in more efficient water management strategies in a changing environment across spatial and temporal scales. This Special Issue of Water contributes toward this aim through broad research work on the hydrological consequences of climate and land use change and hydrological modeling approaches. We published twenty-four peer-reviewed papers, and grouped them into four categories (Table 1):

- Climate change and land use change impacts on hydrological processes;
- Trends and variation of hydrological variables, such as precipitation, runoff, actual evapotranspiration, and soil moisture;
- Hydrological modeling in simulating and predicting hydrological variables, such as precipitation, evapotranspiration and soil moisture in data-sparse regions, and
- Reviews on water prices and climate extremes

Table 1. Summary of 24 papers published in the special issue "Climate Variability and Climate Change Impacts on Land Surface, Hydrological Processes and Water Management" in Water Journal.

Categories	Authors	Title	Research Area	Research Fields
Climate change and land use change impacts on hydrological processes	Sridhar et al. [1]	Human-Induced Alterations to Land Use and Climate and Their Responses for Hydrology and Water Management in the Mekong River Basin	Mekong River Basin	Land use; Climate change; Water resources management; Hydrology model
	Guo et al. [2]	Quantifying the Impacts of Climate Change, Coal Mining and Soil and Water Conservation on Streamflow in a Coal Mining Concentrated Watershed on the Loess Plateau, China	Yulin	Climate change; Coal mining; Soil and water conservation
	Pousa et al. [3]	Climate Change and Intense Irrigation Growth in Western Bahia, Brazil: The Urgent Need for Hydroclimatic Monitoring	Western Bahia, Brazil	Climate change; Water security
	Shang et al. [4]	Land Use and Climate Change Effects on Surface Runoff Variations in the Upper Heihe River Basin	Upper Heihe River Basin	Climate change; Land use; scenario simulation; Hydrological simulation
	Gao and Zhang [5]	Effects of the Three Gorges Project on Runoff and Related Benefits of the Key Regions along Main Branches of the Yangtze River	Yangtze River	Runoff changes; Flood control
	Deng et al. [6]	Analyzing the Impacts of Climate Variability and Land Surface Changes on the Annual Water–Energy Balance in the Weihe River Basin of China	Weihe River Basin	Budyko; Climate variability; Land surface change
	Tian et al. [7]	Quantifying the Impact of Climate Change and Human Activities on Streamflow in a Semi-Arid Watershed with the Budyko Equation Incorporating Dynamic Vegetation Information	Wuding River Watershed	Budyko; Climate variability; Land surface change
	Wu et al. [8]	Analysis of Natural Streamflow Variation and Its Influential Factors on the Yellow River from 1957 to 2010	Yellow River	Streamflow variation; Intra-annual climate change
Hydrological trends and causality analysis	Gedefaw et al. [9]	Analysis of the Recent Trends of Two Climate Parameters over Two Eco-Regions of Ethiopia	Ethiopia	Trend analysis; Precipitation; Temperature
	Dorjsuren et al. [10]	Observed Trends of Climate and River Discharge in Mongolia's Selenga Sub-Basin of the Lake Baikal Basin	Selenga Sub-Basin of the Lake Baikal Basin	Precipitation; Temperature; River discharge
	Zhu et al. [11]	Four Decades of Estuarine Wetland Changes in the Yellow River Delta Based on Landsat Observations Between 1973 and 2013	Yellow River Delta	Estuarine wetlands; Spatiotemporal change analysis

Table 1. *Cont.*

Categories	Authors	Title	Research Area	Research Fields
	Li et al. [12]	Spatiotemporal Variation of Snowfall to Precipitation Ratio and Its Implication on Water Resources by a Regional Climate Model over Xinjiang, China	Xinjiang	Snowfall to precipitation ratio; WRF model
	Yan [13]	Meteorological Factors Affecting Pan Evaporation in the Haihe River Basin, China	Haihe River Basin	Evapotranspiration
	Li et al. [14]	Assessing the Influence of the Three Gorges Dam on Hydrological Drought Using GRACE Data	Yangtze River	Hydrological drought; Three Gorges Dam; GRACE
Hydrological model simulations and predictions	Kim et al. [15]	The Use of Large-Scale Climate Indices in Monthly Reservoir Inflow Forecasting and Its Application on Time Series and Artificial Intelligence Models	Han River basin in South Korea	Climate variability; Large-scale climate indices; Artificial intelligence model
	Wu et al. [16]	Influence of Power Operations of Cascade Hydropower Stations under Climate Change and Human Activities and Revised Optimal Operation Strategies: A Case Study in the Upper Han River, China	Upper Han River	Climate change; human activities; Power operation
	Paul et al. [17]	Comparative Study of Two State-of-the-Art Semi-Distributed Hydrological Models	Baitarani river basin in India	Grid-based; HRU-based
	Ri et al. [18]	A Statistical–Distributed Model of Average Annual Runoff for Water Resources Assessment in DPR Korea	DPR Korea	Runoff map; Hydrological model
	Cho and Lee [19]	Multiple Linear Regression Models for Predicting Nonpoint-Source Pollutant Discharge from a Highland Agricultural Region	Lake Soyang basin of South Korea	Diffuse pollutant discharge; Multiple regression model; Climate change
	Liu et al. [20]	Integrating Field Experiments with Modeling to Evaluate the Freshwater Availability at Ungauged Sites: A Case Study of Pingtan Island (China)	Pingtan Island in southeast China	Predictions in ungauged basins; Rainfall-runoff experiments; Distributed hydrological model
	Zhou et al. [21]	The Effects of Litter Layer and Topsoil on Surface Runoff during Simulated Rainfall in Guizhou Province, China: A Plot Scale Case Study	Guizhou province	Runoff; Simulated rainfall; Litter layer; Topsoil
	Nyaupane et al. [22]	Evaluating Future Flood Scenarios Using CMIP5 Climate Projections	Carson River in the desert of Nevada	Flood; Climate change; CMIP5
Review	Soto Rios et al. [23]	Explaining Water Pricing through a Water Security Lens	-	Water security; Water pricing; Sustainable water management
	Hao et al. [24]	Compound Extremes in Hydroclimatology: A Review	-	Compound extremes; Climate change; Multivariate distribution

2. Contributed Papers

2.1. Climate Change and Land Use Change Impacts on Hydrological Processes

There are eight papers published in this category. Sridhar et al. [1] evaluated human-induced alterations to land use and climate and their responses to hydrology and water management in the Mekong river basin. Authors used two hydrological models to evaluate the impacts of natural and climate-induced changes on water budget components, particularly streamflow. Model simulations show that wet season flows were increased by up to 10% and there was no significant change in dry season flows under natural conditions. Their results suggest an increasing trend in streamflow without the effect of dams, while the inclusion of a few major dams resulted in decreased river streamflow of 6% to 15%, possibly due to irrigation diversions and climate change. Guo et al. [2] quantified the impacts of climate change, coal mining, and soil and water conservation on streamflow in a coal mining concentrated watershed on the Loess Plateau, China. They found that relative to the baseline period, i.e., 1955–1978, the mean annual streamflow reduction in 1979–1996 was mainly affected by climate change, which was responsible for a decreased annual streamflow of 12.70 mm (70.95%). However, in a recent period of 1997–2013, the impact of coal mining on streamflow reduction was dominant, reaching 29.88 mm (54.24%). Pousa et al. [3] analyzed climate change and intense irrigation growth in western Bahia, Brazil and concluded that urgent management is required for hydroclimatic monitoring. They found that the irrigated area has increased over 150-fold in 30 years, and in the most irrigated regions, has increased by 90% in the last eight years only. Their findings suggest that a monitoring system in which the availability and demand of water resources for irrigation are actually measured and monitored is the safest path to provide water security to this region. Shang et al. [4] separated climate change impacts on surface runoff variations from land use impacts in the upper Heihe river basin. Authors found that in this region the contribution rate of climate change is 87.1%, while the contribution rate of land use change is only 12.9%. The climate change scenario simulation analysis shows that the change in runoff is positively correlated with the change in precipitation. The relationship with the change in temperature is more complicated, but the influence of precipitation change is stronger than the change in temperature. Under the economic development scenario of land use simulation, the runoff decreases, whereas under the historical trend and ecological protection scenario of land use simulation, the runoff increases. Gao and Zhang [5] analyzed the effects of the Three Gorges Project (TPG) on runoff and related benefits of the key regions along main branches of the Yangtze River. Their results show that the main benefits of TGP on flood control are remarkable in the reduction of disaster-affected population, the decrease of agricultural disaster-damaged area, and the decline of direct economic loss. Due to torrentially seasonal and non-seasonal precipitation, the sharp rebounds of three standards for Hubei and Anhui occurred in 2010 and 2016, and the percentage of agricultural damage area of five regions in the core and extended areas did not decline synchronously and performed irregularly. The five key regions along the main branches of the Yangtze River should establish a flood control system and promote the connectivity of infrastructures at different levels to meet the significant functions of TGP. Deng et al. [6] analyzed the impacts of climate variability and land surface changes on the annual water–energy balance in the Weihe river basin of China. Authors used the Budyko framework in which the catchment properties represent land surface changes, climate variability comprises precipitation (P) and potential evapotranspiration, and found that the contribution of land surface changes to runoff reduction in period I was less than that in period II, indicating that changes in human activity further decreased runoff. Tian et al. [7] used the Budyko framework incorporating dynamic vegetation information to quantify the impact of climate change and human activities on streamflow in Wuding river basin, a semi-arid basin within the Yellow River Basin. Their results show that climate change generated a dominant effect on the streamflow and decreased it by 72.4% in this basin. This climatic effect can be further explained with the drying trend of the Palmer severity drought index, which was calculated based only on climate change information. Wu et al. [8] analyzed natural streamflow variation and its influential factors on the Yellow River from 1957 to 2010. They found that the reduction of annual streamflow was mainly caused by a precipitation decline and a rise in temperature for all Yellow River regions before 2000, whereas the contribution of anthropogenic interference increased significantly—more than 45%, except for Tang-Tou region after 2000. In the humid Yellow River region, annual streamflow was more sensitive to annual precipitation than temperature, and the opposite situation was observed in the arid region.

2.2. Hydrological Trends and Causality Analysis

There are six papers published in this category. Gedefaw et al. [9] analyzed the recent trends of precipitation and temperature over two eco-regions of Ethiopia. Authors found that the effects of precipitation and temperature changes on water resources are significant after 1998 and the consistency in the precipitation and temperature trends over the two eco-regions confirms the robustness of the changes. Dorjsuren et al. [10] used observed data detecting trends of annual precipitation, air temperature, and river discharge at five selected stations in Mongolia's Selenga sub-basin of the Lake Baikal Basin. The observation results indicate that the average air temperature has significantly increased by 1.4 °C in the past 38 years and there exists a significantly decreasing trend in river discharge during that period. Zhu et al. [11] investigated estuarine wetland changes in the Yellow River Delta based on Landsat observations between 1973 and 2013. Their results show that natural wetlands are significantly decreased, meanwhile, the artificial wetlands are significantly increased. The main reason for wetland degradation in the Yellow River Delta is human activities such as urban construction, cropland expansion, and oil exploitation. Li et al. [12] investigated spatiotemporal variation of snowfall to precipitation ratio and its implication on water resources by a regional climate model over Xinjiang, China. Their results reveal that the snowfall is increased in the southern edge of the Tarim Basin, the Ili Valley, and the Altay Mountains, but decreased in the Tianshan Mountains and the Kunlun Mountains. However, the trends in snowfall/precipitation ratio are opposite in low-elevation regions and mountains of the study area. Yan et al. [13] attributed meteorological factors affecting pan evaporation in the Haihe River Basin (HRB). The average temperature, maximum temperature, and minimum temperature of the HRB increased, while precipitation, relative humidity, sunshine duration, wind speed and evaporation observed from pan exhibited a downward trend. Attribution analysis shows a significant reduction in sunshine duration, which was found to be the primary factor in the pan evaporation decrease, while declining wind speed was the secondary factor. Li et al. [14] assessed the influence of the Three Gorges Dam (TGD) on hydrological drought using GRACE remote sensing data. They proposed the dam influence index (DII) to assess the influence of the TGD on hydrological drought in the Yangtze River Basin (YRB) in China, and found that impoundments of the TGD between 2003 and 2008 slightly alleviated the hydrological drought in the upper sub-basin and significantly aggravated the hydrological drought in the middle and lower sub-basins, which is consistent with the Palmer drought severity index.

2.3. Hydrological Model Simulations and Predictions

There are eight papers published in this category. Kim et al. [15] used large-scale climate indices in monthly reservoir inflow forecasting for considering climate variability. They demonstrate that there exists potential to use climate indices in artificial intelligence models to improve the model performance, and the ARX-ANN and AR-RF models generally show the best performance among the employed models. Wu et al. [16] proposed an optimal operation model of cascade power stations based on the simulation model to generate single and joint optimal operation charts for future hydrological scenarios. Their modeling results show that under existing hydrological conditions, the modified single and joint operation charts would increase power generation by about 32 million and 47 million kWh for a case study carried out in the upper Han River, China. Paul et al. [17] developed a semi-distributed hydrological model (SHM) whose simulation appears to be superior in comparison to SWAT simulation in Baitarani River Basin in India for both calibration and validation periods. Furthermore, the SHM model is superior to the SWAT model in annual peak flow, monthly flow variability, and different flow percentiles. Differences in data interpolation techniques and physical processes of the models are identified as the probable reasons behind the differences among the models' outputs. Ri et al. [18] developed a statistical–distributed model of average annual runoff for water resources assessment in DPR Korea. The model was derived from 50 years' observations of 200 meteorological stations in DPRK, considering the influence of climatic factors. Based on the water balance equation and assumptions, the empirical relationship for runoff depth and impact factors was established and calibrated. Cho and Lee [19] used multiple linear regression (MLR) models for predicting nonpoint-source pollutant discharge from a highland agricultural region in South Korea. The explanatory variables used in the MLR models are the percentage of fields, sub-basin area, and mean slope of sub-basin as topographic parameters, and the number of preceding dry days, rainfall intensity, rainfall depth, and rainfall duration as rainfall parameters. The MLR models are good for simulating and predicting pollutant load except for total nitrogen. Liu et al. [20] integrated field experiments with modeling to evaluate the freshwater availability at ungauged sites in Pingtan Island, China. The simulation results indicate high heterogeneity and distinct seasonal dynamics in freshwater availability across the entire island. This is pioneering Prediction in Ungauged Basin (PUB) study for Chinese islands, which could provide reference for planning and management of freshwater in a water shortage area. Zhou et al. [21] conducted a plot scale study to investigate the effects of litter layer and topsoil on surface runoff during simulated rainfall. They investigated three kinds of plots: The thin litter layer with low soil bulk density type (T-L type), the thick litter layer with high soil bulk density type (T-H type), and the moderate litter depth and soil bulk density type (M type), and three artificial rainfall intensities (30 mm/h, 70 mm/h, 120 mm/h). The runoff volume was largest in the T-H type plot at different rainfall intensities and durations. Runoff in the M type plot had characteristics of both the T-L and T-H type plots. The runoff yielding speed was significantly higher and the runoff yielding time was significantly lower in the T-H type plot. Nyaupane et al. [22] evaluated future flood scenarios under CMIP5 climate projections for Carson River in the desert of Nevada. Altogether, 97 projections from 31 models with four emission scenarios were used to predict the future flood flow over 100 years using a best fit distribution. The developed floodplain map for the future streamflow indicated a larger inundation area compared with the current Federal Emergency Management Agency's flood inundation map, highlighting the importance of climate data in floodplain management studies.

2.4. Review

There are two papers published in this category. Soto Rios et al. [23] reviewed water pricing through a water security lens. This paper analyzed how water pricing can be used as a tool to enact the water security agenda, Three facets were reviewed for tackling water crises, including (i) economic aspects—the multiple processes through which water is conceptualized and priced, (ii) analysis of water pricing considering its effect in water consumption, and (iii) arguments for assessing the potential of water pricing as a tool to appraise water security. Hao et al. [24] reviewed compound extremes in hydroclimatology. This review covers different approaches for the statistical characterization and modeling of compound extremes in hydroclimatology, including the empirical approach, multivariate distribution, the indicator approach, quantile regression, and the Markov Chain model. Several key challenges in the statistical characterization and modeling of compound extremes include the limitation in the data availability to represent extremes and lack of flexibility in modeling asymmetric/tail dependences of multiple variables/events.

3. Conclusions

Over the last several decades, Earth's climate has experienced substantial changes because of global warming linked to increased anthropogenic atmospheric greenhouse gas concentrations. This process affects the hydrological cycle at different levels of observations, ranging from plot to catchment, regional and global scales. Enhancing our overall knowledge on this topic requires multi-disciplinary efforts to learn about hydrological processes and to engage in more efficient water management strategies under changing environmental conditions across those scales.

The research papers published in this Special Issue contribute significantly toward our understanding of the hydrological impacts of climate and land-use change as well as on hydrological modeling approaches in four main subject areas:

- Climate and land use change impacts on hydrological processes;
- Trends and variability of hydrological quantities, such as precipitation, runoff, actual evapotranspiration, and soil moisture;
- Hydrological modeling in simulating and predicting hydrological variables, such as precipitation, evapotranspiration and soil moisture in data-sparse regions; and
- Reviews on water prices and climate extremes

The twenty-four papers presented in this Special Issue reflect on the fact that climate change impact analysis on water resources is a very relevant, albeit challenging topic because of hydrological nonstationary under conditions of global change and the uncertainty related to model inputs, model parameterization, and model structure. The papers published in this issue can not only advance water sciences but support policy makers toward more sustainable and effective water management.

Author Contributions: Y.Q.Z. conceived and led the development of the Special Issue and this paper; H.X.L. and P.R. each contributed to the writing of this paper.

Funding: This study is supported by the CAS Pioneer Hundred Talent Program and IGSNRR Supporting Fund (YJRCPT2019-101).

Acknowledgments: As guest editors of this Special Issue, the authors acknowledge the journal editors, all authors submitting manuscripts to this Special Issue. Special thanks extend to referees who diligently reviewed all the submissions, which greatly improved quality of published papers.

Conflicts of Interest: The authors declare no conflict of interest.

References

1. Sridhar, V.; Kang, H.; Ali, S.A. Human-Induced Alterations to Land Use and Climate and Their Responses for Hydrology and Water Management in the Mekong River Basin. *Water* **2019**, *11*, 1307. [CrossRef]
2. Guo, Q.; Han, Y.; Yang, Y.; Fu, G.; Li, J. Quantifying the Impacts of Climate Change, Coal Mining and Soil and Water Conservation on Streamflow in a Coal Mining Concentrated Watershed on the Loess Plateau, China. *Water* **2019**, *11*, 1054. [CrossRef]
3. Pousa, R.; Costa, M.H.; Pimenta, F.M.; Fontes, V.C.; Brito, V.F.A.d.; Castro, M. Climate Change and Intense Irrigation Growth in Western Bahia, Brazil: The Urgent Need for Hydroclimatic Monitoring. *Water* **2019**, *11*, 933. [CrossRef]
4. Shang, X.; Jiang, X.; Jia, R.; Wei, C. Land Use and Climate Change Effects on Surface Runoff Variations in the Upper Heihe River Basin. *Water* **2019**, *11*, 344. [CrossRef]
5. Gao, Y.; Zhang, Y. Effects of the Three Gorges Project on Runoff and Related Benefits of the Key Regions along Main Branches of the Yangtze River. *Water* **2019**, *11*, 269. [CrossRef]
6. Deng, W.; Song, J.; Bai, H.; He, Y.; Yu, M.; Wang, H.; Cheng, D. Analyzing the Impacts of Climate Variability and Land Surface Changes on the Annual Water–Energy Balance in the Weihe River Basin of China. *Water* **2018**, *10*, 1792. [CrossRef]
7. Tian, L.; Jin, J.; Wu, P.; Niu, G.-Y. Quantifying the Impact of Climate Change and Human Activities on Streamflow in a Semi-Arid Watershed with the Budyko Equation Incorporating Dynamic Vegetation Information. *Water* **2018**, *10*, 1781. [CrossRef]
8. Wu, J.; Wang, Z.; Dong, Z.; Tang, Q.; Lv, X.; Dong, G. Analysis of Natural Streamflow Variation and Its Influential Factors on the Yellow River from 1957 to 2010. *Water* **2018**, *10*, 1155. [CrossRef]
9. Gedefaw, M.; Yan, D.; Wang, H.; Qin, T.; Wang, K. Analysis of the Recent Trends of Two Climate Parameters over Two Eco-Regions of Ethiopia. *Water* **2019**, *11*, 161. [CrossRef]
10. Dorjsuren, B.; Yan, D.; Wang, H.; Chonokhuu, S.; Enkhbold, A.; Yiran, X.; Girma, A.; Gedefaw, M.; Abiyu, A. Observed Trends of Climate and River Discharge in Mongolia's Selenga Sub-Basin of the Lake Baikal Basin. *Water* **2018**, *10*, 1436. [CrossRef]
11. Zhu, C.; Zhang, X.; Huang, Q. Four Decades of Estuarine Wetland Changes in the Yellow River Delta Based on Landsat Observations Between 1973 and 2013. *Water* **2018**, *10*, 933. [CrossRef]
12. Li, Q.; Yang, T.; Qi, Z.; Li, L. Spatiotemporal Variation of Snowfall to Precipitation Ratio and Its Implication on Water Resources by a Regional Climate Model over Xinjiang, China. *Water* **2018**, *10*, 1463. [CrossRef]
13. Yan, Z.; Wang, S.; Ma, D.; Liu, B.; Lin, H.; Li, S. Meteorological Factors Affecting Pan Evaporation in the Haihe River Basin, China. *Water* **2019**, *11*, 317. [CrossRef]
14. Li, F.; Wang, Z.; Chao, N.; Song, Q. Assessing the Influence of the Three Gorges Dam on Hydrological Drought Using GRACE Data. *Water* **2018**, *10*, 669. [CrossRef]
15. Kim, T.; Shin, J.-Y.; Kim, H.; Kim, S.; Heo, J.-H. The Use of Large-Scale Climate Indices in Monthly Reservoir Inflow Forecasting and Its Application on Time Series and Artificial Intelligence Models. *Water* **2019**, *11*, 374. [CrossRef]

16. Wu, L.; Bai, T.; Huang, Q.; Zhang, M.; Mu, P. Influence of Power Operations of Cascade Hydropower Stations under Climate Change and Human Activities and Revised Optimal Operation Strategies: A Case Study in the Upper Han River, China. *Water* **2019**, *11*, 895. [CrossRef]

17. Paul, P.K.; Zhang, Y.; Mishra, A.; Panigrahy, N.; Singh, R. Comparative Study of Two State-of-the-Art Semi-Distributed Hydrological Models. *Water* **2019**, *11*, 871. [CrossRef]

18. Ri, T.; Jiang, J.; Sivakumar, B.; Pang, T. A Statistical–Distributed Model of Average Annual Runoff for Water Resources Assessment in DPR Korea. *Water* **2019**, *11*, 965. [CrossRef]

19. Cho, J.H.; Lee, J.H. Multiple Linear Regression Models for Predicting Nonpoint-Source Pollutant Discharge from a Highland Agricultural Region. *Water* **2018**, *10*, 1156. [CrossRef]

20. Liu, X.; Wang, Z.; Tang, Y.; Wu, Z.; Guo, Y.; Cheng, Y. Integrating Field Experiments with Modeling to Evaluate the Freshwater Availability at Ungauged Sites: A Case Study of Pingtan Island (China). *Water* **2018**, *10*, 740. [CrossRef]

21. Zhou, Q.; Zhou, X.; Luo, Y.; Cai, M. The Effects of Litter Layer and Topsoil on Surface Runoff during Simulated Rainfall in Guizhou Province, China: A Plot Scale Case Study. *Water* **2018**, *10*, 915. [CrossRef]

22. Nyaupane, N.; Thakur, B.; Kalra, A.; Ahmad, S. Evaluating Future Flood Scenarios Using CMIP5 Climate Projections. *Water* **2018**, *10*, 1866. [CrossRef]

23. Soto Rios, P.C.; Deen, T.A.; Nagabhatla, N.; Ayala, G. Explaining Water Pricing through a Water Security Lens. *Water* **2018**, *10*, 1173. [CrossRef]

24. Hao, Z.; Singh, V.P.; Hao, F. Compound Extremes in Hydroclimatology: A Review. *Water* **2018**, *10*, 718. [CrossRef]

 *water*

Article

Human-Induced Alterations to Land Use and Climate and Their Responses for Hydrology and Water Management in the Mekong River Basin

Venkataramana Sridhar *, Hyunwoo Kang and Syed A. Ali

Department of Biological Systems Engineering, Virginia Polytechnic Institute and State University, Blacksburg, VA 24061, USA; hwkang@vt.edu (H.K.); syedaa@vt.edu (S.A.A.)
* Correspondence: vsri@vt.edu

Received: 26 April 2019; Accepted: 19 June 2019; Published: 25 June 2019

Abstract: The Mekong River Basin (MRB) is one of the significant river basins in the world. For political and economic reasons, it has remained mostly in its natural condition. However, with population increases and rapid industrial growth in the Mekong region, the river has recently become a hotbed of hydropower development projects. This study evaluated these changing hydrological conditions, primarily driven by climate as well as land use and land cover change between 1992 and 2015 and into the future. A 3% increase in croplands and a 1–2% decrease in grasslands, shrublands, and forests was evident in the basin. Similarly, an increase in temperature of 1–6 °C and in precipitation of 15% was projected for 2015–2099. These natural and climate-induced changes were incorporated into two hydrological models to evaluate impacts on water budget components, particularly streamflow. Wet season flows increased by up to 10%; no significant change in dry season flows under natural conditions was evident. Anomaly in streamflows due to climate change was present in the Chiang Saen and Luang Prabang, and the remaining flow stations showed up to a 5% increase. A coefficient of variation <1 suggested no major difference in flows between the pre- and post-development of hydropower projects. The results suggested an increasing trend in streamflow without the effect of dams, while the inclusion of a few major dams resulted in decreased river streamflow of 6% to 15% possibly due to irrigation diversions and climate change. However, these estimates fall within the range of uncertainties in natural climate variability and hydrological parameter estimations. This study offers insights into the relationship between biophysical and anthropogenic factors and highlights that management of the Mekong River is critical to optimally manage increased wet season flows and decreased dry season flows and handle irrigation diversions to meet the demand for food and energy production.

Keywords: hydrology; land cover; land use and climate change; water resources management; macro scale modeling

---

1. Introduction

The Mekong is one of the most important rivers in Asia. Its significance is evident from its geographical location, topographic variability, biodiversity, and large population of inhabitants in the basin. A cascade of dams, population increase, and climate change have also complicated hydrology and water resources management in the Mekong River Basin (MRB). The nexus of food–energy–water is highly pronounced as the basin relies on rice production and fisheries to feed the population. The conversion of lands from forests to agriculture, subsequent expansion and intensification of irrigation, and hydropower development projects have changed the characteristics of the MRB, in which the river previously flowed unhindered for most of its length [1,2]. The Tibetan Plateau in China—where the river originates at about 4000 m—and the downstream regions are going through natural and human-induced climatic changes and experiencing a general increase in precipitation and temperature in the 21st century, and this can affect the basin's hydrology. Low flow days are expected to decrease and flooding potential may also increase, and hence policies to mitigate the impacts are urgently needed [3,4].

Considerable implications of dam constructions, climate change, irrigation, and land use change to downstream ecosystems have resulted in numerous studies to predict floods, droughts, and sediment yield over the past two decades [5–7]. The next few paragraphs will cover some of these studies and identify the knowledge gaps that still exist. Specifically, the understanding of unintended consequences of dams require a comprehensive investigation of reservoir management [8]. Construction and initial filling of the upstream dams reduced the annual streamflow in wet seasons and increased the streamflow in dry seasons, resulting in a unique seasonal variation in the streamflow [9], and the dams had significant impacts on the low pulse duration. Besides, study authors [10,11] reported that construction of dams in the basin is expected to decrease total sediment transportation by 40%–80% over the whole basin, which would impact the river's morphology, aquatic biodiversity, ecosystem services, and agriculture.

By employing simulation models, many studies have projected the basin conditions, but the uncertainties in climate model projections are greater than those of the hydrological models; therefore, comparisons of different climate models and hydrological model outputs at a relatively high resolution are necessary to characterize these uncertainties [12,13]. Study authors [14,15] evaluated the climate change impacts on the hydrological characteristics of the Harvey River catchment in western Australia and the Richmond River catchment in eastern Australia using a rainfall-runoff model (HBV model) and climate model outputs from the Coupled Model Intercomparison Project 5 (CMIP5). The results suggested that there were decreases and increases in the mean annual flows due to the precipitation and temperature variabilities in the future. In another study [16], authors compared two different models (conceptual-HBV and distributed-BTOPMC) in several catchments in Australia and assessed the impacts of climate change on streamflow. Both models simulated a decrease in wet and dry season streamflow across the catchments. An evaluation of the water resource development scenarios over different future time periods' horizons by Piman et al. [17] reported reductions in the average wet season flows by 4%–14% and flow reversal to the Tonle Sap Lake by up to 16%. It predicted an increase in flooded areas by 5%–8% and in salinity intrusion areas in the Viet Nam Delta by up to 17% in the future. It was also reported that the small and nonlinear response of annual river discharge to progressive change in global mean temperature, the change in monthly river discharge varying from −16% to +55%—showed the greatest decrease in July–August and increase in May–June for natural flow only. The impacts of climate change for six catchments around the world, including the Mekong Basin, using a global hydrological model (GHM) and catchment-scale hydrological models (CHM) was performed by [18], and this study reported that substantial differences in the projected change of mean annual runoff between GHM and CHM were dependent on climate model outputs and did not evaluate the regulated flow impacted by the reservoirs. Finally, a semi-distributed hydrological model (SLURP) with the pattern-scaled GCM scenarios was used by [19] to assess the impact of climate change on the freshwater resources associated with GCM structure and climate change sensitivity in the Mekong River Basin.

The effect of land use land cover change (LULCC) impact on the water balance studied by Homdee et al. [20] using the soil and water assessment tool (SWAT) in the Chi River basin, Thailand, reported that land use changes impacted annual and seasonal water yield and evapotranspiration (ET). In addition, the conversion of forested area and agricultural lands affected the flow regimes in the basin. Replacing sugarcane with rice paddies resulted in clearly reduced water flows and increased ET by almost 5.0% during the dry season. Also, the increased conversion of rice paddies to farmland showed a significant effect on seasonal flows. Also, the results of this change showed a decrease in ET by 12.0% and an increase in water yield by 5.1% during the dry season. However, the implications of this study for the entire Mekong basin is not well understood. Another study [21] evaluated, the Mae Chaem River—which was subjected to land use change—by developing three plausible future forest-to-crop expansion scenarios and a scenario of crop-to-forest reversal based on the land cover transition from 1989 to 2000. In this study, the resulting hydrologic responses of the basin were simulated using the distributed hydrology soil vegetation model (DHSVM). The authors also reported that the expansion of highland crop fields affected annual and wet-season water yields compared with a similar expansion in the lowland–midland zone and that the downstream sections of the river were sensitive to irrigation diversion.

The effect of irrigation water abstraction on the streamflow, energy state, and fluxes was evaluated using a model simulation to predict changes in the Bowen Ratio, surface temperature, and water resources within the Mekong River Basin based on the variable infiltration capacity (VIC) macroscale hydrological model [22]. Their results revealed a significant decrease in the Bowen Ratio and surface temperature due to irrigation water withdrawal. The irrigation water withdrawals from runoff, river channels, and dams decreased the total monthly runoff by 32%. Study authors [1] identified the relative roles of precipitation and soil moisture in runoff variability in the Mekong River Basin and reported that simulated soil moisture plays an important role in determining the timing and amount of generated runoff.

However, while these studies reported the changing biophysical conditions of the basin, flow regimes, hydroclimatic extremes, and ecosystems, long-term simulation of the basin hydrology highlighting the role of land use and climate change as well as the effect of dams on the downstream flows have been limited. To our knowledge there is no study that compared SWAT and VIC simulations as well as with and without-reservoir effects. Given their differences in model structure and strengths in simulating global river basins, how they characterize the basin responses under changing conditions of land use and climate change needs a periodic reanalysis. Finally, two different hydrology models are implemented to understand how the major reservoirs play a role in modifying the peak flow in the wet season and low flows in the dry season. While management inputs are needed to precisely quantify the impoundment effects, sensitivity analysis of regulated and natural flows has the potential to know the role of human-induced changes to the flow regimes. Due to the range of predictions and uncertainties, it is imperative to evaluate the changing conditions in the basin in the multi-model framework in order to generate an updated assessment for policy decisions. Therefore, our objective is to evaluate two macroscale hydrological models in capturing basin responses and investigate the historical streamflow changes by explicitly considering the effect of dams and future projections of streamflow and other water budget components. We use both SWAT and VIC to evaluate the hydroclimatological behavior by including six major dams and two climate model projections combined with four global circulation models to characterize the peak flow regime shifts in the basin.

2. Materials and Methods

2.1. The Mekong River Basin

MRB covers an area of about 800,000 km^2 and the mainstem and its tributaries drain six countries: China, Myanmar, Thailand, Laos, Cambodia, and Vietnam. The basin is divided into seven sub-watersheds with flow stations and major dams as shown in Figure 1a. The upper reaches of the Mekong River flow through higher elevations in the Himalayan mountain ranges, through the steep terrain of Laos and Thailand and the lowlands of Cambodia, and into the delta in Vietnam before draining into the South China Sea. For both development and management of this transboundary river basin, a complex river basin agreement was formulated between the member countries and coordinated by the Mekong River Commission; however, rapid changes in this basin have necessitated a comprehensive understanding of conditions in a system modeling framework.

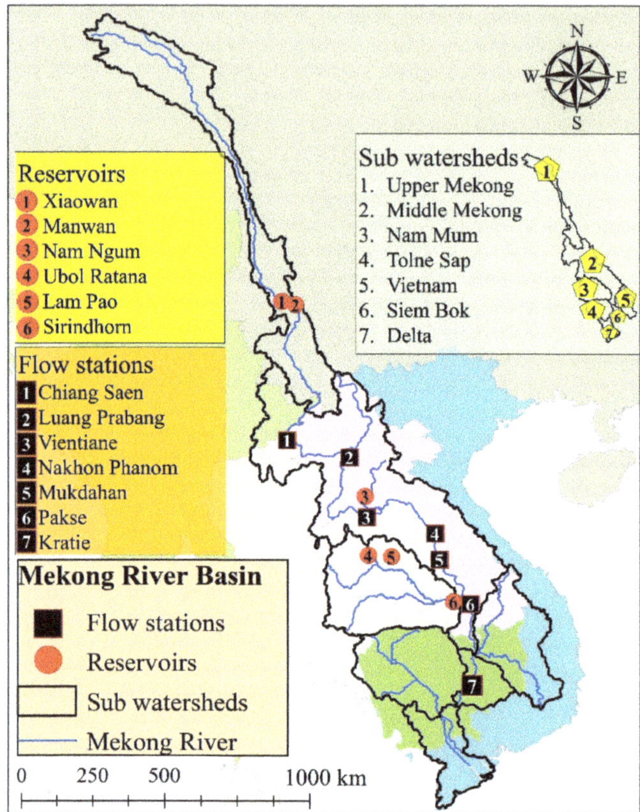

Figure 1. Location map of the Mekong River Basin with reservoirs (in red circles), flow stations (black boxes), and sub-watershed boundaries.

The mean annual discharge from the basin is approximately 15,000 km^3/year. The heterogeneous distribution of the precipitation follows an east–west gradient, with the mean annual value of 1200 mm. Nearly 70% of the annual precipitation in the MRB occurs during the monsoon season. However, the temperature and elevation variations follow a north–south gradient. The temperature in the MRB varies from 38 °C during March–April to 15 °C during November–February. Since conditions in the MRB are hot and humid with the glaciated portion for the upper region, the climate is classified as tropical monsoonal. The elevation drop of more than 4900 m in the MRB also affects the climate heterogeneity. A major portion of the MRB is covered with croplands (40%), followed by evergreen broadleaf forest (28%), closed shrublands (10.3%), and grasslands (9.3%). Irrigated wet season rice grown throughout the year and fishing (4.4 million tons per year) provide food security to more than 60 million people residing in the MRB. In addition, the hydropower potential of the MRB amounts to more than 88,000 MW, with only a small portion utilized. Hence, more than 450 dam projects are currently being planned/constructed by the member countries to take advantage of the hydropower capabilities of the MRB.

2.2. Hydrological Models

Both VIC and SWAT have been widely used in our previous studies in several basins around the world and both are currently incorporated in the Mekong River simulation studies. The physical diagrams of these models are available in published literature and websites [23,24]. The range of applications to evaluate water resource problems includes drought [25–28], water management [29,30], and climate impacts [31–34]. The VIC model is grid-based, whereas SWAT is a hydrologic response unit (HRU)-based model that is defined using soil, slope, and land-use data. The resolution of the VIC model varies depending on the availability of forcing data. In our study, we used 0.25° (about 25 km), while the SWAT model considered 1153 climate grids at the same resolution but subdivided into 2196 sub-watersheds. The VIC model was implemented to simulate the natural flows in the basin, whereas the SWAT model was used to simulate both natural and managed flows across selected reservoirs.

The SWAT model [35–37] is a river basin-scale, semi-distributed, and continuous model that generates hydrologic variables based on hydrologic response units (HRUs), which combine diverse land uses, soil types, and slopes. SWAT has been applied to various river basins around the globe to evaluate climate change impacts on streamflow [38–40], agricultural systems [41], and hydrologic extremes [25,26,42,43]. SWAT estimates several hydrologic components—such as surface runoff, baseflow, evapotranspiration (ET), and soil moisture—which are the primary variables for streamflow calculation (Equation (1)).

$$SW_t = SW_0 + \sum_{i=1}^{t} P_{day} - Q_{surf} - ET_a - W_{seep} - Q_{gw}, \tag{1}$$

where SW_t is the final soil water (mm) on day i, t is the time (days), SW_0 is the initial soil water on day i, P_{day} is the daily precipitation (mm), Q_{surf} is the surface runoff (mm), ET_a is the evapotranspiration (mm), W_{seep} is the water entering to the vadose zone from the soil layer (mm), and Q_{gw} is the return flow (mm).

The SWAT model needs a meteorological dataset (e.g., daily precipitation, maximum and minimum temperatures), digital elevation model (DEM), soil properties, and land use. For the historic simulation (1951–2015), a 0.25° resolution of the meteorological forcing dataset was applied [44,45]. The MRB was delineated as 2196 sub-watersheds to consider all climate grids (1153 grids). In addition, the Global Multi-Resolution Terrain Elevation Data 2010 (GMTED2010; 250-m resolution) [46] was applied, and the soil properties were obtained from the Food and Agriculture Organization of the United Nations dataset [47]. Finally, the Global Land Cover Characterization (GLCC) was used to determine land use [48].

The VIC model was also implemented to estimate the streamflow at the gage station locations for observed and projected future climates. The VIC is a semi-distributed, physically based hydrological model that solves water and energy balance for each grid separately at a designated daily time. The meteorological parameters for the execution of the model include precipitation from the APHRODITE dataset and minimum and maximum temperatures and wind speed from the Global Meteorological Forcing Dataset (GMFD) gridded dataset, available at 0.25° spatial and daily temporal resolution [44,45]. The vegetation texture—containing the land cover type, leaf area index, and albedo—was developed using the Advanced Very High Resolution Radiometer (AVHRR) at a 1 km spatial resolution. The soil class was taken from the United States Department of Agriculture (USDA) classification and pedo-transfer functions [49] applied to the Harmonized World Soil Database (HWSD) were combined to extract soil parameters.

The infiltration mechanism utilized in the Xinanjiang model [50] was adopted for use in the VIC model to generate the runoff from precipitation when it is higher than the available infiltration capacity. This scheme is commonly used in models that are used for flood forecasting, climate change studies, and water resource assessment in the humid and sub-humid regions of the world [51]. The model is capable of catchment response on any scale and can account for nonlinear spatial retention of soil moisture [52]. Also, the Xinanjiang model accounts for soil heterogeneity and assumes the variation of the infiltration capacity within an area [53]. In the VIC, the Xinanjiang formulation is assumed to hold for the upper soil layer only. The Xinanjiang model effectively assumes that runoff is generated by those areas for which precipitation, when added to soil moisture storage at the end of the previous time step, exceeds the storage capacity of the soil. When the precipitation is less than or equal to the available infiltration capacity, overland runoff is not generated. However, the soil moisture transfers from the upper soil layer to the lower soil layer for subsurface runoff generation using the Arno model conceptualization [54]. The top two layers of the three soil layers in the model respond to the rainfall, whereas the bottom layer corresponds to baseflow computed using the Arno model formulation [54]. The variable infiltration curve [55] governs the infiltration of water into the soil layer. The total ET is estimated using the Penman–Monteith approach and defined as the accumulation of evaporation from bare soil and canopy and transpiration from vegetation features. Since VIC is a unidimensional hydrological model, the fluxes are exchanged only in the vertical direction and the lateral movement in the subsurface layer is considered negligible. Moreover, the routing scheme developed by [56,57] is employed on the fluxes simulated by the VIC model for each grid to estimate the monthly streamflow at the gage station locations. The surface and subsurface fluxes of the grids were explicitly routed by the routing scheme using a unit hydrograph of a channel network, in which the node of the channel network represented each grid-cell of the VIC model.

The observed monthly streamflow from the seven gauging stations distributed across the basin—namely Chiang Saen, Luang Prabang, Nakhon Phanom, Vientiane, Mukdahan, Pakse, and Kratie—were used to calibrate and evaluate the VIC model. The VIC model has been used by the various studies for hydrological assessment of the MRB [8,22,58–60].

2.3. Choice of General Circulation Models

Figure 2 shows the distribution of wet/dry and cold/hot global circulation models (GCM) from 2 Representative Concentration Pathways (RCPs), 4.5 and 8.5, showing changing precipitation and temperature for 5 future periods F1 through F5 between 2006 and 2099. These models are GFDL-ESM2M, IPSL-CM5A-LR, MIROC-ESM-CHEM, and NorESM1-M. Each model was bias-corrected and statistically downscaled to 0.25° resolution by the Intersectoral Impact Model Intercomparison Project (ISI-MIP) [61]. These models exhibited a wide range of temperatures (1–6 °C) and precipitation changes (−5%–20%) in the basin and were widely used to predict climate change impacts on hydrology as well as in other basins [61–63]. Clearly, MIROC showed wetter and hotter conditions for the later part of the century, while GFDL and IPSL projected drier and cooler conditions through 2040.

2.4. Calibration and Simulation of Streamflows and Water Budget Components

The SWAT model was calibrated using the monthly streamflow and the SWAT calibration and uncertainty assessment tool (SWAT-CUP) [64] with 4 parameters (Table 1) at 7 stations. Similarly, VIC was also calibrated, and the results are shown in Table 2. As shown in Table 3, monthly calibration metrics of correlation coefficient (R^2) and Nash–Sutcliff (NS) efficiency were above 0.8 for both SWAT and VIC models. The parameters used to calibrate the VIC model included the variable infiltration curve parameter (b_i), the depth of the second and third soil layers (D), the fraction of maximum velocity of baseflow where non-linear baseflow begins (D_s), and the fraction of maximum soil moisture where non-linear baseflow occurs (W_s) with allowable ranges of 0.1–0.5, 0.1–1.5, 0–0.4, and 0.5–1.0 respectively. The calibration was carried out for the gage stations stepwise from upstream basins, with the exclusion of the regions already considered for the upstream station. The Nash–Sutcliffe efficiency coefficient [65] and coefficient of determination (R^2) between the monthly simulated and observed streamflows was used to evaluate the capability of the VIC model. This exercise was necessary to ensure that the model's parameters were able to characterize the hydrologic responses to changing environmental and bio-physical conditions.

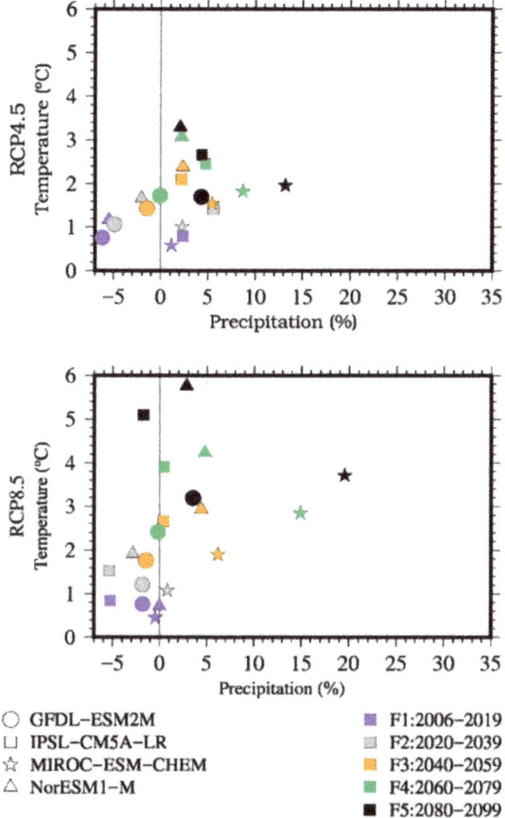

Figure 2. Choice of wet/dry and cold/hot global circulation models (GCM) from two representative concentration pathways (RCPs)—4.5 and 8.5—showing changing precipitation and temperature for five future periods F1 through F5 between 2006 and 2099.

Table 1. Description of the soil and water assessment tool (SWAT) model input parameters for the calibration.

Parameter	Description	Min	Max	Best Parameters
r_CN2.mgt	Curve number for moisture condition II	−0.2	0.2	0.06
v_ALPHA_BF.gw	Baseflow alpha factor	0	1	0.35
v_GW_DELAY.gw	Groundwater delay time	30	450	177
v_GWQMN.gw	Threshold water depth in shallow aquifer for back discharge	0	2000	1500

Notes: v_, denotes the default parameter is replaced by a given value; r_, means the existing parameter value is multiplied by (1 + a given value).

Table 2. Description of the variable infiltration capacity (VIC) model input parameters for the calibration.

S. No.	Parameter	Description	Allowable Range	
			Lower	Upper
1	b_i	variable infiltration curve parameter	0.1	0.5
2	D	the depth of soil layers	0.1	1.5

Table 2. *Cont.*

S. No.	Parameter	Description	Allowable Range	
			Lower	Upper
3	D_s	fraction of maximum velocity of baseflow where non-linear baseflow begins	0	0.4
4	W_s	fraction of maximum soil moisture where non-linear baseflow occurs	0.5	1

Table 3. Statistical indicators showing the hydrology model calibration and validation for the historical period between 1984 and 1992 in the Mekong River Basin.

Station	Calibration Period	Validation Period	Calibration				Validation			
			R^2		NS		R^2		NS	
			SWAT	VIC	SWAT	VIC	SWAT	VIC	SWAT	VIC
Chiang Saen	1984–1990	1991–1996	0.92	0.93	0.86	0.83	0.93	0.91	0.85	0.81
Luang Prabang	1984–1990	1991–1997	0.93	0.93	0.81	0.73	0.94	0.89	0.86	0.67
Vientiane	1984–1990	1991–1996	0.92	0.93	0.83	0.91	0.95	0.94	0.88	0.92
Nakhon Phanom	1984–1990	1991–1995	0.93	0.93	0.87	0.90	0.92	0.92	0.86	0.79
Mukdahan	1984–1990	1991–1995	0.93	0.94	0.89	0.86	0.93	0.94	0.88	0.83
Pakse	1984–1990	1991–1998	0.90	0.91	0.84	0.86	0.90	0.93	0.85	0.87
Kratie	1984–1990	1991–1998	0.90	0.90	0.85	0.85	0.91	0.93	0.86	0.86

2.5. Study Design

Our approach consisted of the following steps: monthly calibration of the hydrology models for the historic period, simulation of streamflows using the climate model outputs by dividing them into seven sub-basins with the outlets where the observations were available, evaluation of peak flows, assessment of flow changes in the context of reservoirs, and spatial mapping of temperature and precipitation anomalies and water budget components (ET and runoff). Monthly calibration of the hydrology models for the historic period was required in order to understand whether the models could capture the basin scale responses hydrologically and reliably so as to as extend to the other periods of interest [66,67]. Subsequent analysis was aimed to investigate if there were any differences in streamflows and peakflows considering the spatial and temporal variability of the forcings, land use and reservoir management. This sequential approach enabled us to understand and quantify the impact of spatial variability and shift in the flow regimes as shown in Figure 3. Finally, this study is framed to seek an answer for the suitability of these models for changing conditions in the future. To answer this question, we evaluated the differences between them in multiple variables, including peakflows, sub-basin scale hydrologic budgets and see whether they help us decide the suitability for decision making in the context of the sustainable management of the basin and food–energy–water nexus considering the data needs and resolution. For instance, food systems need high resolution, field scale data for decision making while hydropower and water would be decided based on the catchment scale runoff and inflows to the dam. This particular study offers insights based on 0.25°

forcings and a relatively high-resolution land use and soil properties with major reservoirs across the basin all of which can be integrated in a simple framework.

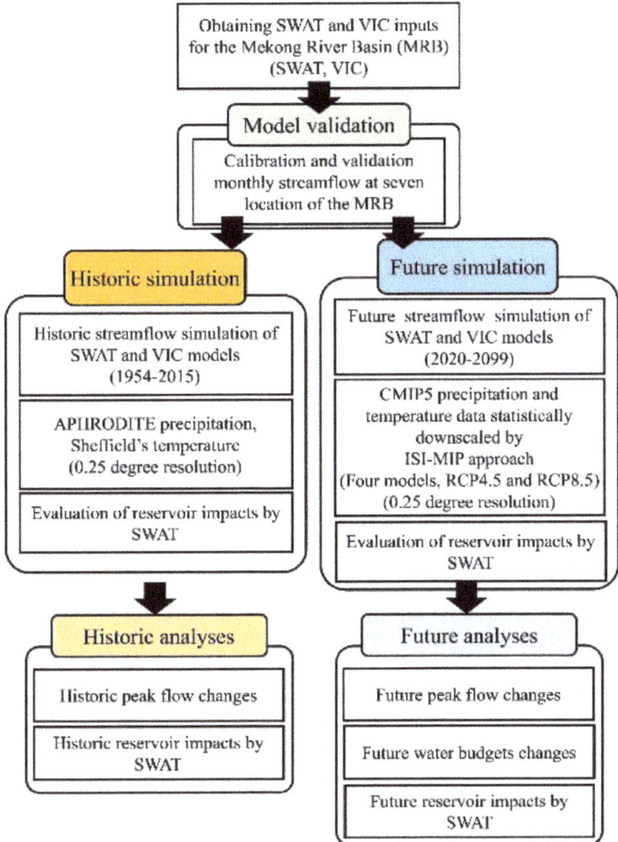

Figure 3. Flow diagram of the overall processes of hydrologic modeling and analyses. SWAT: soil and water assessment; VIC: variable infiltration capacity; APHRODTE: Asian Precipitation—Highly-Resolved Observational Data Integration Towards Evaluation; CMIP5: Coupled Model Intercomparison Project 5; ISI-MIP: Inter-Sectoral Impact Model Intercomparison Project; RCP: representative concentration pathway.

3. Results

3.1. Hydroclimatology of Streamflow

The annual hydrograph was primarily driven by the southwest monsoon in the basin and the typical flood hydrograph consisted of peak flows in the wet season (July–October) and relatively low flows in the dry season (January–May). Generally, the smooth hydrographs reflecting the size of the catchment were evident. Figure 4a–f show the long-term streamflow simulations by SWAT and VIC. The historical simulation period was between 1954 and 2015, and due to limited availability of observational data, a relatively short period between 1984 and 1990 was used for calibration and the remaining period from 1991 to 1996 for validation. The locations distributed across the entire lower Mekong from the upstream point in the basin—Chiang Sean to downstream at Kratie—demonstrated how the annual average streamflow gathered in magnitude from about 2000 m^3/s to 10,000 m^3/s.

Table 1 shows the list of four runoff-, base flow-, and groundwater-related parameters calibrated in SWAT. Similarly, the calibration parameters shown in Table 2 for VIC include the variable infiltration curve parameter, the depth of the second and third soil layers, the fraction of maximum velocity of baseflow where nonlinear baseflow begins, and the fraction of maximum soil moisture where nonlinear baseflow occurs.

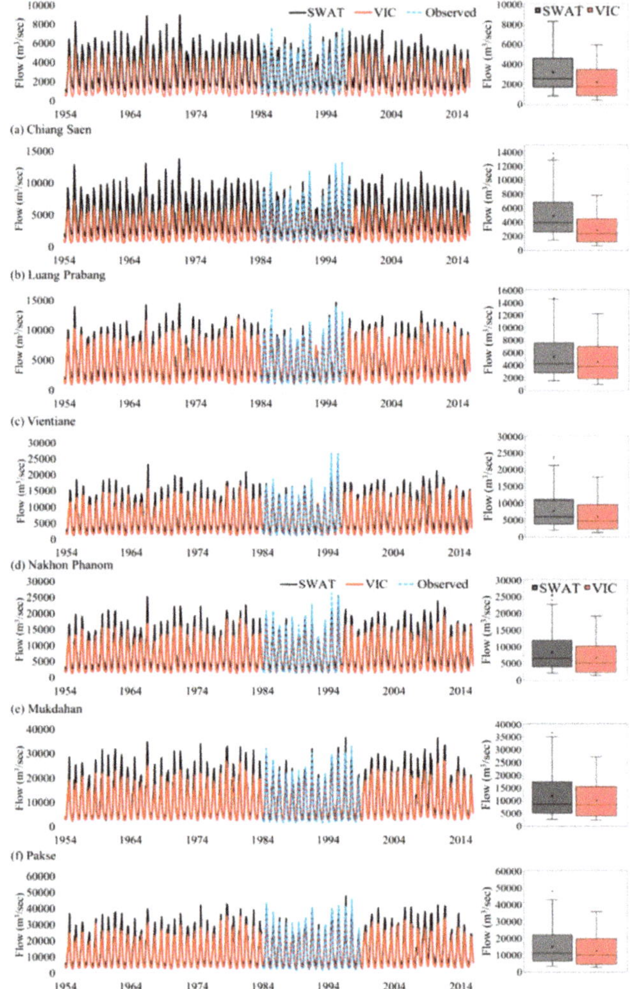

Figure 4. Long-term streamflow simulations (1954–2015) compared against observations (1984–1996) with two different time periods for calibration and validation by the SWAT and the VIC macro scale hydrological models. The box plot (on the right) shows the mean and spread of flows captured by SWAT and VIC.

Clearly, the multi-decadal simulations showed interannual variability in flows caused mostly by precipitation changes; however, the shifts in flows on annual scales were indistinguishable. The box plot (on the right) shows the mean and spread of annual streamflows captured by SWAT and VIC. While the mean values between these models across all seven stations were close, the spread was greater for SWAT. Also, the coefficient of variation (standard deviation/mean) computed for each of the gage locations showed a low value of less than 0.7 consistently for all the stations and they were comparable between the observed and simulated flows. This could be considered typical for a tropical river basin where variability was minimal.

3.2. Historical Peakflow Assessment

In order to understand the dynamics of drivers of change, particularly climate and land use, we evaluated the peakflow magnitudes simulated by SWAT and VIC. Our assessment of change in land use between 1992 and 2015 suggested a 3% increase in croplands and a 1–2% decrease in grasslands, shrublands and forests. Figure 5 shows the differences in peakflows between the two periods—1956–1965 and 2006–2015—to compare pre-development and post-development conditions in the basin. Other than a reduction in flows of 4–8% for Chiang Sean and less than 1% for Luang Prabang, all of the other flow stations indicated an increase of 8–11%. The decrease in flows in the upstream location can be attributed partly to climate change in the Tibetan Plateau. However, the tropical monsoon impacts on the lower portion of the basin were evident in the increased flows. These increased peakflows can result in flooding, and therefore impoundments of these flows can potentially reduce the risk of flooding in this basin, which is prone to seasonal flooding. This is further highlighted in Figure 5h, where the streamflow anomaly (%) for the seven locations between 1992 and 2015 decreased up to 4% for Chiang Saen and Luang Prabang. The remaining stations showed a positive anomaly of up to 5%. Noticeably, the differences between 1992 and 2015 in both SWAT and VIC showed no difference in anomaly, which suggested that natural flows between pre- and post-development of hydropower projects are not significant. In other words, while the flow alterations in the basin could not be attributed to land use changes in the basin, human-induced changes—such as irrigation diversions—and climate change can affect peakflows.

3.3. Projected Changes in Flows and Comparison of Models

Since the effect of climate change was evident with increased precipitation and temperature in the basin, it was considered appropriate to assess the climate change impacts on streamflow and other water budget components. Figure 6 shows the projected streamflows from the VIC and SWAT models between 2020 and 2099 for the same seven locations where calibration and validation of streamflows were performed for the historic periods. The results included the ensemble average of all four GCMs introduced in the earlier sections. The annual hydrographs resembled historical estimates of flows, with interannual variability and seasonal peaks. Most notably, the differences in SWAT and VIC were also similar to historic simulations, as SWAT produced more flows relative to VIC. While the hydrological model processes that caused the increased flow in SWAT are not discussed here in detail, the role of the calibration parameters that previously estimated higher flows could be substantial. Also, the irrigation extraction for croplands—whereby the streamflows remained mostly natural and hence the attenuation of flows was not obvious—was not explicitly considered.

Due to increased precipitation in the basin, as predicted by most of the GCMs, hydrological flow simulation had shown similar increases in peakflows, ranging from 10%–70% between RCP 4.5 and 8.5 scenarios. The hydrological model responses in the form of streamflow were directly proportional to increased precipitation, typical of a tropical basin. The substantial increases were also expected in the later part of the century across all flow stations between 2060 and 2099. Projected peakflow changes simulated by VIC and SWAT are shown in Figure 7. On the one hand, the reductions in dry season flows were not evident, and counterintuitively, on the other hand, the management of reservoirs and their releases can augment them.

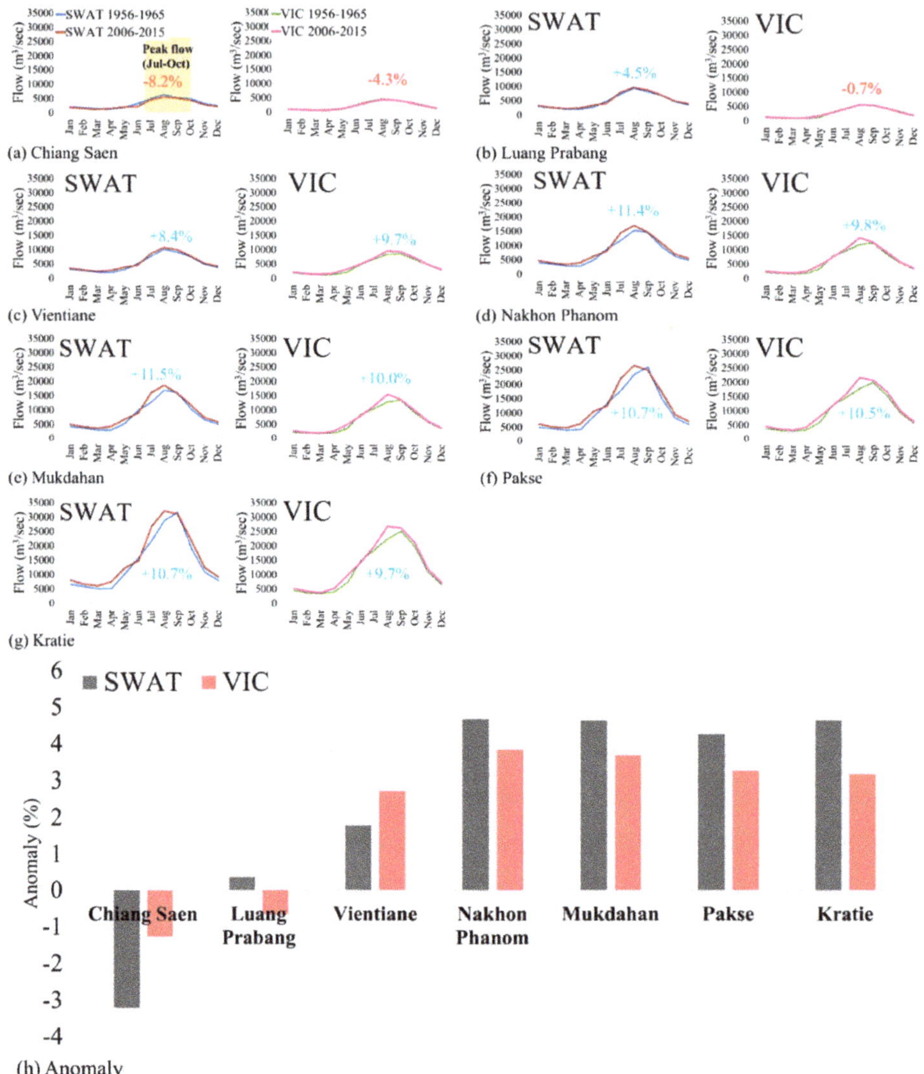

Figure 5. (**a–g**) Differences in peakflows between two periods; (**h**) streamflow anomaly (%) for seven locations between 1992 and 2015.

3.4. Projected Peakflow Estimation

Changes in peak flows were analyzed, and the shifts in peakflows with and without reservoirs simulated by the SWAT model for the period 1992–2015 (14 year average) are shown in Figure 8. A similar analysis for future climate projections from 2020–2099 (80 year average) from RCP 4.5 is shown in Figure 9. The results are similar for RCP 8.5, and the percentage changes in flows are shown in Table 4. In general, the reductions in flows in 2015 were lesser when compared to 1992 in simulations in which reservoirs were taken into account with certain parameters. These reductions, ranging from 3%–15%, can only be considered changes due to climate variability, as exact operation and releases of flows were not integrated into this analysis. When the simulations did not include reservoirs, both wet and dry season flows were higher for the same period. However, general reductions of up to

35% in dry season flows and 16% in wet season flows were identical in simulations that considered reservoirs in their analyses. This analysis emphasizes the importance of incorporating the actual reservoir operations to predict wet and dry season flows.

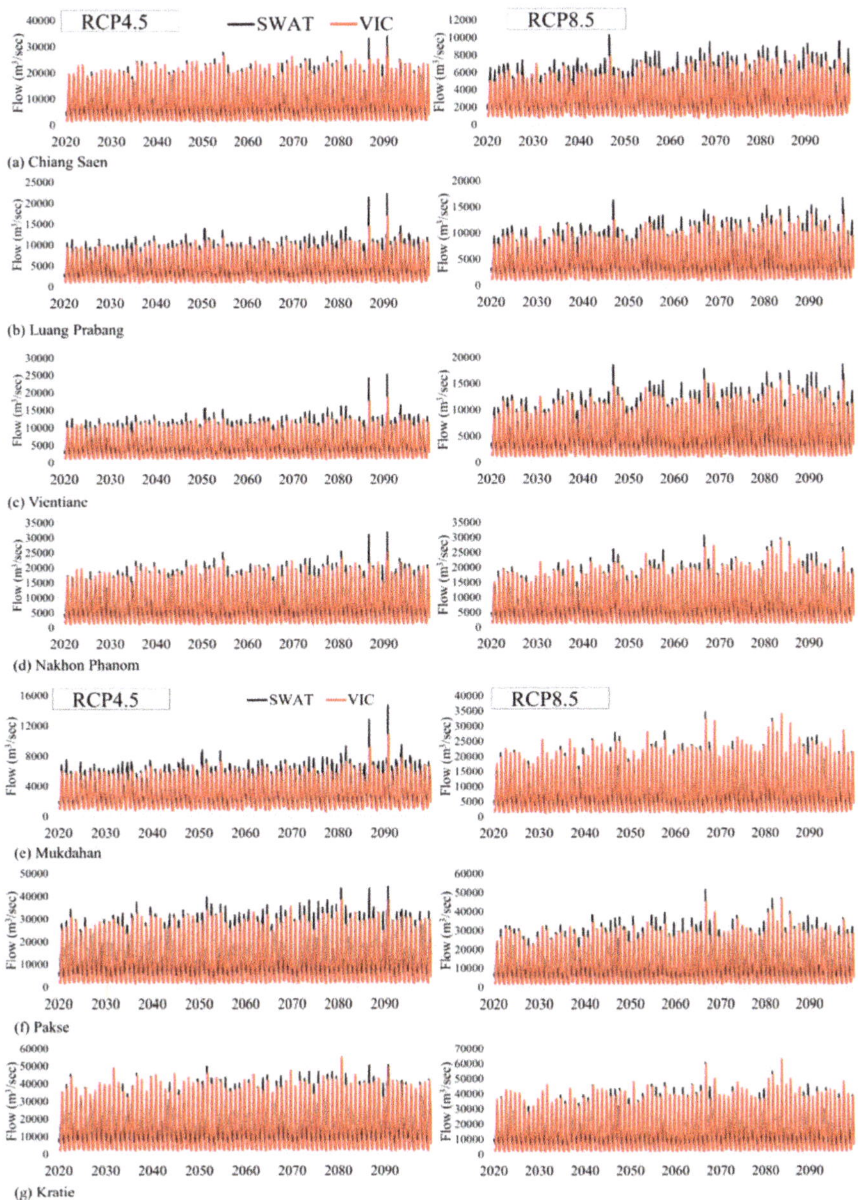

Figure 6. Projected streamflows from VIC and SWAT models between 2020 and 2099 for seven locations where calibration and validation of streamflows were performed for the historic periods.

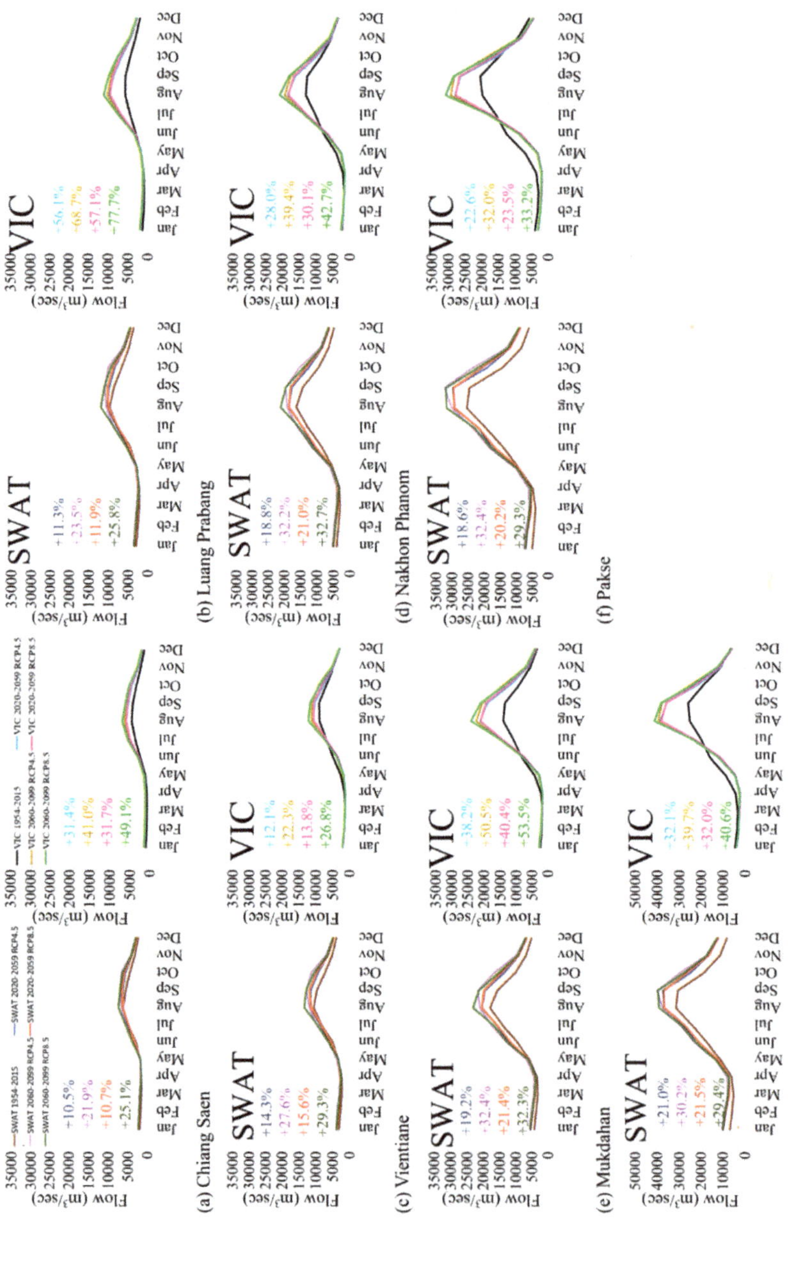

Figure 7. Projected peakflow changes between 2020 and 2099, simulated by VIC and SWAT.

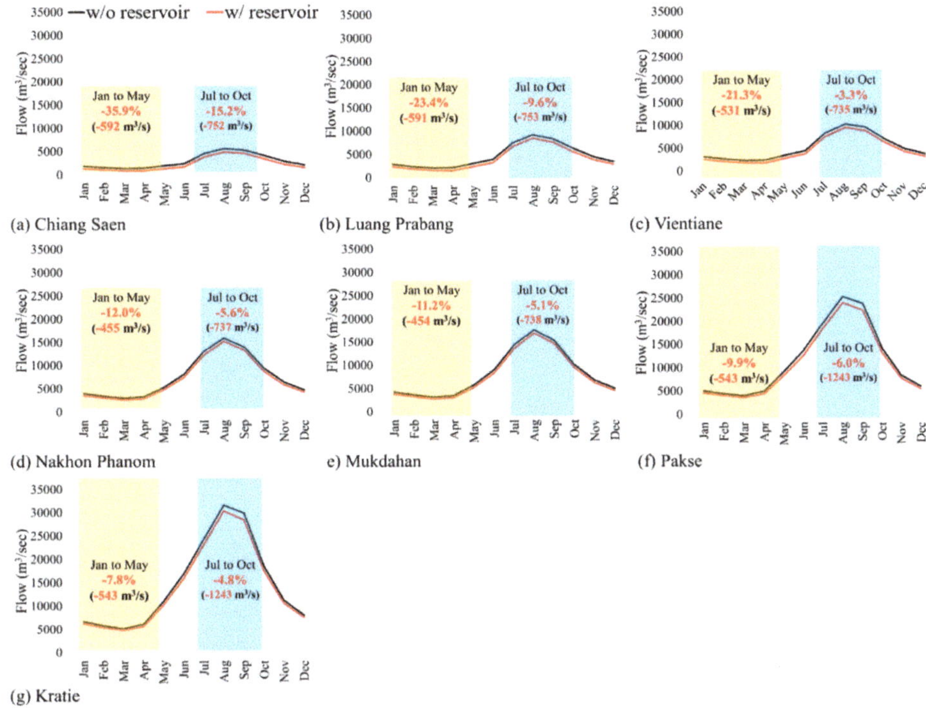

Figure 8. Shifts in peakflows with and without reservoirs, simulated by the SWAT model for the period between 1992 and 2015 (14 year average).

Table 4. Percentage change in dry- and wet-season flows projected by SWAT and VIC for RCP 4.5 and RCP 8.5 between 2020 and 2099.

Station	Season	SWAT				VIC			
		RCP4.5		RCP8.5		RCP4.5		RCP8.5	
		2020–2059	2060–2099	2020–2059	2060–2099	2020–2059	2060–2099	2020–2059	2060–2099
Chiang Saen	Wet	10.5	21.9	10.7	25.1	31.4	41.0	31.7	49.1
	Dry	17.6	25.7	14.7	21.3	33.6	41.3	32.2	36.6
Luang Prabang	Wet	11.3	23.5	11.9	25.8	56.1	68.7	57.1	77.7
	Dry	17.7	25.5	14.1	19.4	26.8	34.2	25.2	29.7
Vientiane	Wet	14.3	27.6	15.6	29.3	12.1	22.3	13.8	26.8
	Dry	20.4	28.6	16.5	22.0	−11.3	−5.3	−12.4	−8.6
Nakhon Phanom	Wet	18.8	32.2	21.0	32.7	28.0	39.4	30.1	42.7
	Dry	26.0	33.5	21.4	26.9	−8.7	−3.8	−10.8	−6.5
Mukdahan	Wet	19.2	32.4	21.4	32.3	38.2	50.5	40.4	53.5
	Dry	27.6	35.0	22.8	28.5	−3.3	1.9	-5.5	−0.7
Pakse	Wet	18.6	30.4	20.2	29.3	22.6	32.0	23.5	33.2
	Dry	31.7	39.0	26.3	33.4	−31.2	−27.0	−32.5	−28.2
Kratie	Wet	21.0	30.2	21.5	29.4	32.1	39.7	32.0	40.6
	Dry	37.2	42.8	31.4	39.4	−22.7	−19.0	−24.3	−18.8

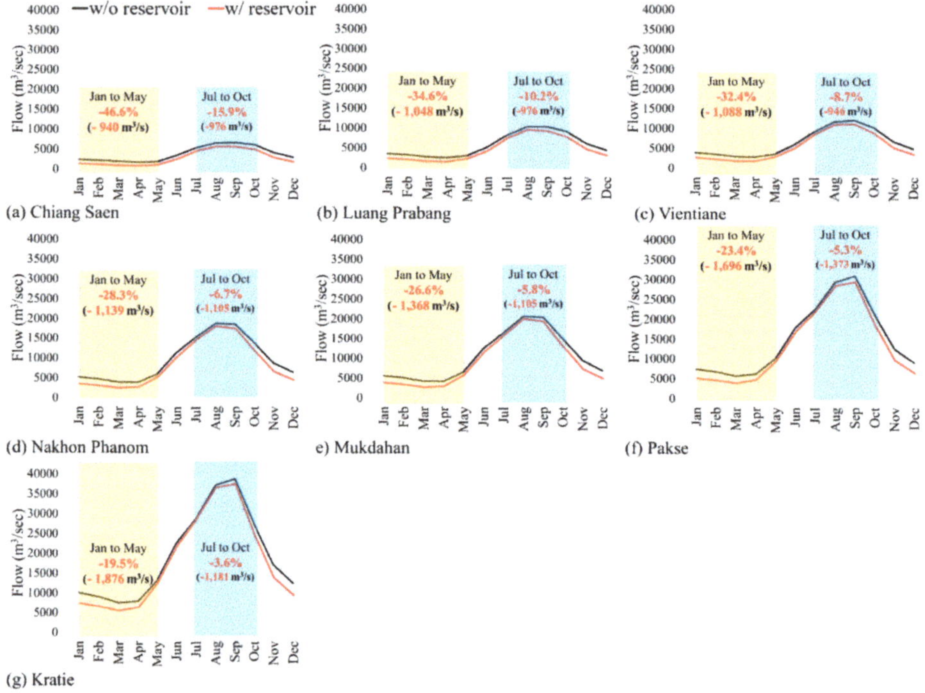

Figure 9. Shifts in peakflows (RCP 4.5) with and without reservoirs, simulated by the SWAT model for the period between 2020 and 2099 (80 year average).

3.5. Basin Scale Water Budget Analysis

Figure 10 presents the bar charts of ET and runoff changes for the historic and future periods from the SWAT and VIC models, and Figure 11 shows the spatial maps of changes in precipitation, temperature, ET, and runoff. The historic changes were the percentage alterations of the last 10 years (2006–2015) compared to the entire historic period (1954–2015), and the future changes were the percent increases of the future period (2020–2099) to the historic period. Both ET and runoff changes were computed for each sub-region and grid in the MRB. As in Figure 10, reductions in runoff of up to 6% in the historic period did not persist in other locations or into the future. Historic reductions were more pronounced in the upper portion. Both RCP 4.5 and RCP 8.5 projected increases in ET of 4–15% and runoff increases of up to 60%. In both historic and future periods, water budget changes were highly influenced by precipitation and temperature alterations in the MRB, as shown in Figure 11a–f. Relative to the past period, both RCPs projected increased precipitation between 10 and 60% across all of the sub-basins, with higher increases in the central and lower portions of the basin. However, temperature increases were notable in the upper and central sub-watersheds, ranging from 1–4 °C. This can potentially impact snowmelt-driven flow in the Tibetan Plateau before the monsoon season begins.

For the historic period, there were overall ET and runoff increases from the SWAT model for the entire MRB, except for the runoff in Upper Mekong. In the VIC model, there were also overall increases, but some regions showed ET and runoff decreases (e.g., Upper Mekong, Delta). The highest increase of runoff occurred in the middle of the basins from both SWAT and VIC (13.7% and 8.5% respectively; Figure 11j,p), and these increases were derived from the highest precipitation increase in the middle Mekong region during the last 10 years (Figure 11a). However, there were runoff decreases in the upper Mekong and Siem Bok (Figure 11j,p), and the precipitation decreases mainly derived from these in the corresponding areas (Figure 11a).

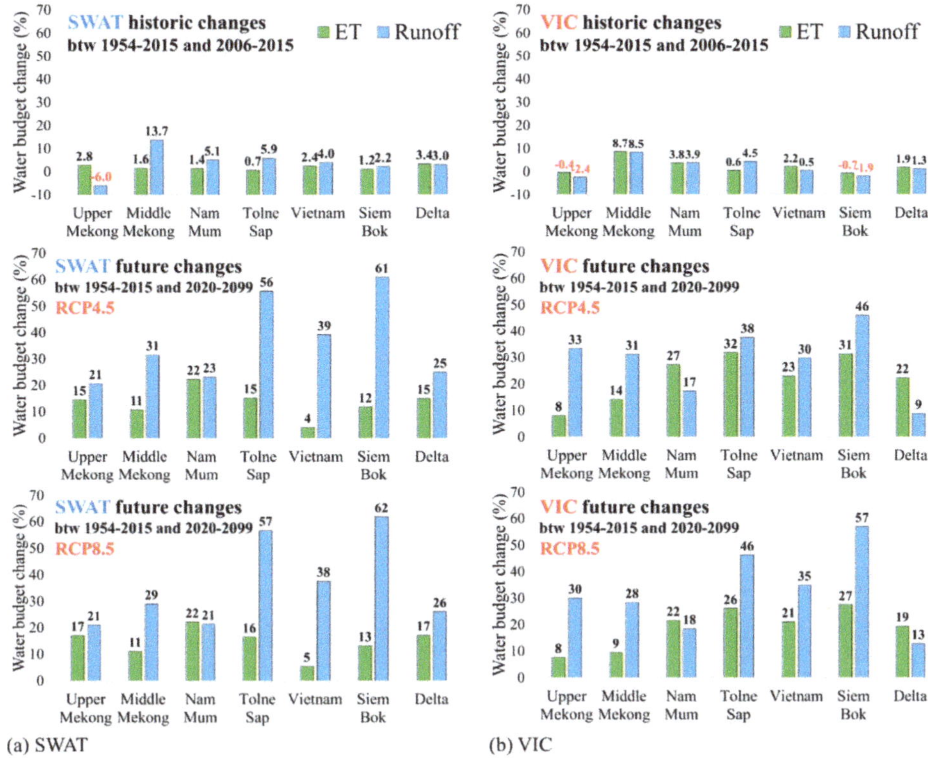

Figure 10. Water budget changes for historic and future periods for each sub-watershed; (**a**) SWAT; (**b**) VIC.

For the future period, there were ET and runoff increases for the entire regions and both models, and the precipitation and temperature increases were the main drivers of those changes. VIC projected decreased runoff in the delta in the future, and in general, estimates of the water budget from the two models were considerably different. The runoff estimations from the SWAT model were more sensitive to the precipitation increase. For instance, in the Tolne Sap region, where more than 40% precipitation increases occurred in the entire area, the runoff increases were 56% and 57% for RCP 4.5 and RCP 8.5 using the SWAT model, but 38% and 46% using the VIC model. In addition, the results of runoff increases in the Siem Bok showed similar results. These results were derived from the different runoff estimations between the two models. The SWAT model is based on the soil conservation service (SCS) curve number (CN) [68], while the VIC model uses the variable infiltration curve method [55]. The SCS CN method has been known to have a higher sensitivity for runoff estimation [69], and this could be credited for those discrepancies.

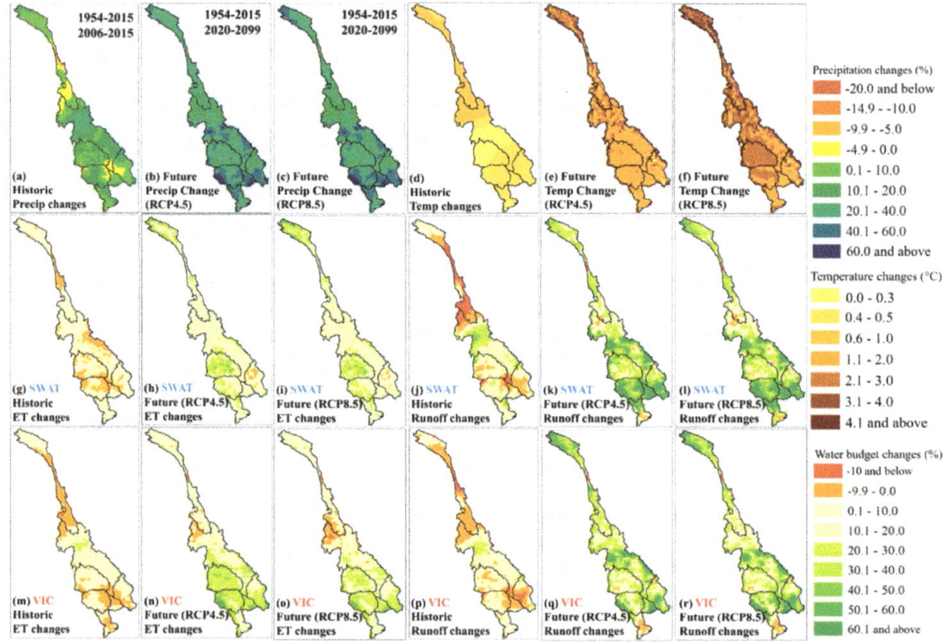

Figure 11. Spatial plot of precipitation, temperature, and water budget changes for historic and future periods. (**a–c**) Precipitation changes; (**d–f**) temperature changes; (**g–i**) evapotranspiration (ET) and runoff changes from the SWAT model; (**m–r**). ET and runoff changes from the VIC model.

4. Conclusions

The Mekong region is at the center of a multitude of changes, including hydropower development, land use, and climate change. While it is important to characterize the basin's responses to these changes at the local level, macroscale changes to climate and hydrology spanning all six basin countries at the sub-watershed domain is critical. This study evaluated historic streamflows and impacts of 21st century hydropower development on those flows through simulation. The limiting factor in considering basic scale changes arose from lack of information about the management of the currently operated dams. Hence, the initial simulation of flows could be used to identify the changes in dry- and wet-season flows driven primarily by both natural and ongoing anthropogenic-induced climatic changes while integrating reservoirs into the models to accommodate the inflows in a simplistic way. Several key insights were gained from this study. More broadly, the comparison of two hydrological model simulations highlighted that basin responses to peakflows for both historic and future periods were in close agreement despite the differences in the model formulations and both can serve as a predictive model for future water resources assessment with some improvements to field-scale crop water estimation. The increase in peakflow estimations due to increased precipitation in a changing climate was quantified for several locations and agreed with several previous studies. These increased peakflows are expected to be harnessed for both hydropower and irrigation water demand, and these new insights are useful for making policy decisions and developing operating procedures for water resources development projects. Additionally, the spatial variability in ET and runoff highlighted the need for a differential approach at the sub-basin level to sustain food and energy production in the context of drought and other anthropogenic-induced changes, including land use and population increase. Perhaps, irrigation water assessment by VIC was providing a more realistic estimates of ET and both hydrological models require improvements to simulate crop-specific, field scale estimates of water balance components.

Our findings from this investigation also suggested that the hydrological models SWAT and VIC were capable of predicting large-scale changes to the system by accounting for up to 90% variability under natural conditions. Relative changes in flows impacted by recent hydropower projects between 1992 and 2015 revealed that anomalies in peakflows during this period were less than 5%. This suggests that system-level changes were not identifiable due to modest land cover changes in croplands and forests unless storage and irrigation diversion were properly considered. Increased precipitation over several sub-watersheds also resulted in increased peakflows, as the monsoon season variability for multiple decades included only nominal changes.

Generally, climate models projected a wide range of temperature (1–6 °C) and precipitation changes (−5–20%) in the basin between 2020 and 2099. Corresponding increases in peakflows—ranging from 10–70% between RCP 4.5 and 8.5 scenarios—were expected to occur, leading to possible flooding and inundation unless the reservoir management for both peakflows and diversions for crop water requirements were optimally handled. Without reservoirs in the modeling assessment, both wet- and dry-season flows were higher, but general reductions of up to 35% in dry-season flows and 16% in wet-season flows were identical to simulations that considered reservoirs in their analysis. However, with expanded irrigated areas in the basins and increased peakflows, not only can conflicts be alleviated to manage dry season flows, but increased crop production and hydropower generation also become feasible.

Author Contributions: Conceptualization, V.S.; methodology, H.K. and S.A.A.; software, H.K. and S.A.A.; validation, V.S., H.K. and S.A.A.; formal analysis, V.S., H.K. and S.A.A.; investigation, V.S., H.K. and S.A.A.; resources, V.S.; data curation, V.S., H.K. and S.A.A.; writing—original draft preparation, V.S., H.K. and S.A.A.; writing—review and editing—V.S.; visualization, V.S., H.K. and S.A.A.; supervision, V.S.; project administration, V.S.; funding acquisition, V.S.

Funding: This work was supported in part by NASA under the award 80NSSC17K0259.

Acknowledgments: We thank Jiaguo Qi, William McConnell, Yadu Pokhrel, and David Hyndman from Michigan State University for their indirect assistance. We are grateful for the general discussions on the Mekong Basin with our partners from the Mekong River Commission, Asian Institute of Technology and the Vietnam National Mekong Committee.

Conflicts of Interest: The authors declare no conflict of interest.

References

1. Costa-Cabral, M.C.; Richey, J.E.; Goteti, G.; Lettenmaier, D.P.; Feldkötter, C.; Snidvongs, A. Landscape structure and use, climate, and water movement in the Mekong River basin. *Hydrol. Process.* **2008**, *22*, 1731–1746. [CrossRef]
2. Pokhrel, Y.; Burbano, M.; Roush, J.; Kang, H.; Sridhar, V.; Hyndman, D. A review of the integrated effects of changing climate, land use, and dams on Mekong river hydrology. *Water* **2018**, *10*, 266. [CrossRef]
3. Kiem, A.S.; Ishidaira, H.; Hapuarachchi, H.P.; Zhou, M.C.; Hirabayashi, Y.; Takeuchi, K. Future hydroclimatology of the Mekong River basin simulated using the high-resolution Japan Meteorological Agency (JMA) AGCM. *Hydrol. Process.* **2008**, *22*, 1382–1394. [CrossRef]
4. Thompson, J.R.; Green, A.J.; Kingston, D.G.; Gosling, S.N. Assessment of uncertainty in river flow projections for the Mekong River using multiple GCMs and hydrological models. *J. Hydrol.* **2013**, *486*, 1–30. [CrossRef]
5. Eastham, J.; Mpelasoka, F.; Mainuddin, M.; Ticehurst, C.; Dyce, P.; Hodgson, G.; Ali, R.; Kirby, M. *Mekong River Basin Water Resources Assessment: Impacts of Climate Change*; CSIRO: Canberra, Australia, 2008.
6. Shrestha, B.; Babel, M.S.; Maskey, S.; Van Griensven, A.; Uhlenbrook, S.; Green, A.; Akkharath, I. Impact of climate change on sediment yield in the Mekong River basin: A case study of the Nam Ou basin, Lao PDR. *Hydrol. Earth Syst. Sci.* **2013**, *17*, 1–20. [CrossRef]
7. Thilakarathne, M.; Sridhar, V. Characterization of future drought conditions in the Lower Mekong River Basin. *Weather Clim. Extrem.* **2017**, *17*, 47–58. [CrossRef]

8. Bonnema, M.; Hossain, F. Inferring reservoir operating patterns across the Mekong Basin using only space observations. *Water Resour. Res.* **2017**, *53*, 3791–3810. [CrossRef]

9. Li, D.; Long, D.; Zhao, J.; Lu, H.; Hong, Y. Observed changes in flow regimes in the Mekong river basin. *J. Hydrol.* **2017**, *551*, 217–232. [CrossRef]

10. Kummu, M.; Lu, X.; Wang, J.; Varis, O. Basin-wide sediment trapping efficiency of emerging reservoirs along the mekong. *Geomorphology* **2010**, *119*, 181–197. [CrossRef]

11. Wild, T.B.; Loucks, D.P. Managing flow, sediment, and hydropower regimes in the sre pok, se san, and se kong rivers of the mekong basin. *Water Resour. Res.* **2014**, *50*, 5141–5157. [CrossRef]

12. Perra, E.; Piras, M.; Deidda, R.; Paniconi, C.; Mascaro, G.; Vivoni, E.R.; Cau, P.; Marras, P.A.; Ludwig, R.; Meyer, S. Multimodel assessment of climate change-induced hydrologic impacts for a Mediterranean catchment. *Hydrol. Earth Syst. Sci.* **2018**, *22*, 4125–4143. [CrossRef]

13. Mendoza, P.A.; Clark, M.P.; Mizukami, N.; Newman, A.J.; Barlage, M.; Gutmann, E.D.; Rasmussen, R.M.; Rajagopalan, B.; Brekke, L.D.; Arnold, J.R. Effects of hydrologic model choice and calibration on the portrayal of climate change impacts. *J. Hydrometeorol.* **2015**, *16*, 762–780. [CrossRef]

14. Al-Safi, H.I.J.; Sarukkalige, P.R. The application of conceptual modelling to assess the impacts of future climate change on the hydrological response of the Harvey River catchment. *J. Hydro-Environ. Res* **2018**. [CrossRef]

15. Al-Safi, H.I.J.; Sarukkalige, P.R. Evaluation of the impacts of future hydrological changes on the sustainable water resources management of the Richmond River catchment. *J. Water Clim. Chang.* **2018**, *9*, 137–155. [CrossRef]

16. Al-Safi, H.I.J.; Kazemi, H.; Sarukkalige, P.R. Comparative study of conceptual versus distributed hydrologic modelling to evaluate the impact of climate change on future runoff in unregulated catchments. *J. Water Clim. Chang.* **2019**. [CrossRef]

17. Piman, T.; Lennaerts, T.; Southalack, P. Assessment of hydrological changes in the lower Mekong basin from basin-wide development scenarios. *Hydrol. Process.* **2013**, *27*, 2115–2125. [CrossRef]

18. Gosling, S.; Taylor, R.G.; Arnell, N.; Todd, M.C. A comparative analysis of projected impacts of climate change on river runoff from global and catchment-scale hydrological models. *Hydrol. Earth Syst. Sci.* **2011**, *15*, 279–294. [CrossRef]

19. Kingston, D.G.; Thompson, J.R.; Kite, G. Uncertainty in climate change projections of discharge for the Mekong River Basin. *Hydrol. Earth Syst. Sci.* **2011**, *15*, 1459–1471. [CrossRef]

20. Homdee, T.; Pongput, K.; Kanae, S. Impacts of land cover changes on hydrologic responses: A case study of chi river basin, Thailand. *J. Jpn. Soc. Civ. Eng. Ser. B1* **2011**, *67*, I31–I36. [CrossRef]

21. Thanapakpawin, P.; Richey, J.; Thomas, D.; Rodda, S.; Campbell, B.; Logsdon, M. Effects of landuse change on the hydrologic regime of the Mae Chaem river basin, NW Thailand. *J. Hydrol.* **2007**, *334*, 215–230. [CrossRef]

22. Tatsumi, K.; Yamashiki, Y. Effect of irrigation water withdrawals on water and energy balance in the Mekong River Basin using an improved VIC land surface model with fewer calibration parameters. *Agric. Water Manag.* **2015**, *159*, 92–106. [CrossRef]

23. Introduction to SWAT and the Instructional Videos. Available online: https://swat.tamu.edu/workshops/instructional-videos/ (accessed on 24 June 2019).

24. VIC Model Overview. Available online: https://vic.readthedocs.io/en/master/Overview/ModelOverview/ (accessed on 24 June 2019).

25. Kang, H.; Sridhar, V. Combined statistical and spatially distributed hydrological model for evaluating future drought indices in Virginia. *J. Hydrol. Reg. Stud.* **2017**, *12*, 253–272. [CrossRef]

26. Kang, H.; Sridhar, V. Assessment of Future Drought Conditions in the Chesapeake Bay Watershed. *JAWRA J. Am. Water Res. Assoc.* **2018**, *54*, 160–183. [CrossRef]

27. Kang, H.W.; Sridhar, V. Improved drought prediction using near real-time climate forecasts and simulated hydrologic conditions. *Sustainability* **2018**, *10*, 1799. [CrossRef]

28. Sehgal, V.; Sridhar, V. Watershed-scale retrospective drought analysis and seasonal forecasting using multi-layer, high-resolution simulated soil moisture for Southeastern U.S. *Weather Clim. Extrem.* **2019**, *23*, 100191. [CrossRef]

29. Hoekema, D.J.; Sridhar, V. A system dynamics model for conjunctive management of water resources in the Snake River basin. *JAWRA J. Am. Water Res. Assoc.* **2013**, *49*, 1327–1350. [CrossRef]

30. Sridhar, V.; Ali, S.; Modi, P.; Kang, H.; Quan, N.; Dat, N.D.; Kansal, M.D. Can we quantify the resilience of the Lower Mekong Basin in the face of dams and agricultural expansion? In Proceedings of the EWRI World Environmental & Water Resource Congress, Pittsburgh, PA, USA, 19–23 May 2019.

31. Kang, H.; Sridhar, V.; Mills, B.F.; Hession, W.C.; Ogejo, J.A. Economy-wide climate change impacts on green water droughts based on the hydrologic simulations. *Agric. Syst.* **2019**, *171*, 76–88. [CrossRef]

32. Sehgal, V.; Sridhar, V.; Juran, L.; Ogejo, J. Integrating Climate Forecasts with the Soil and Water Assessment Tool (SWAT) for High-Resolution Hydrologic Simulations and Forecasts in the Southeastern US. *Sustainability* **2018**, *10*, 3079. [CrossRef]

33. Jin, X.; Sridhar, V. Impacts of climate change on hydrology and water resources in the Boise and Spokane River Basins. *JAWRA J. Am. Water Res. Assoc.* **2012**, *48*, 197–220. [CrossRef]

34. Sridhar, V.; Jin, X.; Jaksa, W.T.A. Explaining the hydroclimatic variability and change in the Salmon River basin. *Clim. Dyn.* **2013**, *40*, 1921–1937. [CrossRef]

35. Arnold, J.G.; Srinivasan, R.; Muttiah, R.S.; Williams, J.R. Large area hydrologic modeling and assessment part I: Model development1. *JAWRA J. Am. Soc. Water Resour. Assoc.* **1998**, *34*, 73–89. [CrossRef]

36. Neitsch, S.L.; Arnold, J.G.; Kiniry, J.R.; Williams, J.R. *Soil and Water Assessment Tool Theoretical Documentation Version 2009*; Texas Water Resources Institute: Temple, TX, USA, 2011.

37. Arnold, J.G.; Moriasi, D.N.; Gassman, P.W.; Abbaspour, K.C.; White, M.J.; Srinivasan, R.; Santhi, C.; Harmel, R.D.; Van Griensven, A.; Van Liew, M.W.; et al. SWAT: Model use, calibration, and validation. *Trans. ASABE* **2012**, *55*, 1491–1508. [CrossRef]

38. Jha, M.; Arnold, J.G.; Gassman, P.W.; Giorgi, F.; Gu, R.R. Climate change sensitivity assessment on upper Mississippi River Basin streamflow using SWAT. *JAWRA J. Am. Water Resour. Assoc.* **2006**, *42*, 997–1015. [CrossRef]

39. Githui, F.; Gitau, W.; Mutua, F.; Bauwens, W. Climate change impact on SWAT simulated streamflow in western Kenya. *Int. J. Climatol.* **2009**, *29*, 1823–1834. [CrossRef]

40. Wu, Y.; Liu, S.; Abdul-Aziz, O.I. Hydrological effects of the increased CO2 and climate change in the Upper Mississippi River Basin using a modified SWAT. *Clim. Chang.* **2012**, *110*, 977–1003. [CrossRef]

41. Ficklin, D.L.; Luo, Y.; Luedeling, E.; Zhang, M. Climate change sensitivity assessment of a highly agricultural watershed using SWAT. *J. Hydrol.* **2009**, *374*, 16–29. [CrossRef]

42. Taye, M.T.; Ntegeka, V.; Ogiramoi, N.P.; Willems, P. Assessment of climate change impact on hydrological extremes in two source regions of the Nile River Basin. *Hydrol. Earth Syst. Sci.* **2011**, *15*, 209–222. [CrossRef]

43. Ashraf Vaghefi, S.; Mousavi, S.J.; Abbaspour, K.C.; Srinivasan, R.; Yang, H. Analyses of the impact of climate change on water resources components, drought and wheat yield in semiarid regions: Karkheh River Basin in Iran. *Hydrol. Process.* **2014**, *28*, 2018–2032. [CrossRef]

44. Yatagai, A.; Kamiguchi, K.; Arakawa, O.; Hamada, A.; Yasutomi, N.; Kitoh, A. APHRODITE: Constructing a long-term daily gridded precipitation dataset for Asia based on a dense network of rain gauges. *Bull. Am. Meteorol. Soc.* **2012**, *93*, 1401–1415. [CrossRef]

45. Sheffield, J.; Goteti, G.; Wood, E.F. Development of a 50-Year High-Resolution Global Dataset of Meteorological Forcings for Land Surface Modeling. *J. Clim.* **2006**, *19*, 3088–3111. [CrossRef]

46. Danielson, J.J.; Gesch, D.B. *Global Multi-Resolution Terrain Elevation Data 2010 (GMTED2010)*; No. 2011-1073; US Geological Survey (USGS): Sioux Falls, SD, USA, 2011.

47. FAO (Food and Agriculture Organization). *Digital Soil Map of the World and Derived Soil Properties*; Food and Agriculture Organization of the United Nations: Rome, Italy, 1995.

48. Global Land Cover Characterization (GLCC). Available online: https://www.usgs.gov/centers/eros/science/usgs-eros-archive-land-cover-products-global-land-cover-characterization-glcc?qt-science_center_objects=0#qt-science_center_objects (accessed on 24 June 2019).

49. Cosby, B.J.; Hornberger, G.M.; Clapp, R.B.; Ginn, T.R. A Statistical Exploration of the Relationships of Soil Moisture Characteristics to the Physical Properties of Soils. *Water Resour. Res.* **1984**, *20*, 682–690. [CrossRef]

50. Ren-Jun, Z. The Xinanjiang model applied in China. *J. Hydrol.* **1992**, *135*, 371–381. [CrossRef]

51. Ahirwar, A.; Jain, M.K.; Perumal, M. Performance of the Xinanjiang model. In *Hydrologic Modeling*; Springer: Singapore, Singapore, 2018; pp. 715–731.

52. Sahoo, B. The Xinanjiang model and its derivatives for modeling soil moisture variability in the land-surface schemes of the climate change models: An overview. In Proceedings of the International Conference on Hydrological Perspectives for Sustainable Development, New Delhi, India, 23–25 February 2005.

53. Liang, X.; Lettenmaier, D.P.; Wood, E.F.; Burges, S.J. A simple hydrologically based model of land surface water and energy fluxes for general circulation models. *J. Geophys. Res. Atmos.* **1994**, *99*, 14415–14428.

54. Franchini, M.; Pacciani, M. Comparative analysis of several conceptual rainfall-runoff models. *J. Hydrol.* **1991**, *122*, 161–219. [CrossRef]

55. Wood, E.F.; Lettenmaier, D.P.; Zartarian, V.G. A land-surface hydrology parameterization with subgrid variability for general circulation models. *J. Geophys. Res. Atmos.* **1992**, *97*, 2717–2728. [CrossRef]

56. Lohmann, D.; Nolte-Holube, R.; Raschke, E. A large-scale horizontal routing model to be coupled to land surface parametrization schemes. *Tellus A* **1996**, *48*, 708–721. [CrossRef]

57. Lohmann, D.; Raschke, E.; Nijssen, B.; Lettenmaier, D.P. Regional scale hydrology: II. Application of the VIC-2L model to the Weser River, Germany. *Hydrol. Sci. J.* **1998**, *43*, 143–158. [CrossRef]

58. Haddeland, I.; Lettenmaier, D.P.; Skaugen, T. Effects of irrigation on the water and energy balances of the Colorado and Mekong river basins. *J. Hydrol.* **2006**, *324*, 210–223. [CrossRef]

59. Västilä, K.; Kummu, M.; Sangmanee, C.; Chinvanno, S. Modelling climate change impacts on the flood pulse in the Lower Mekong floodplains. *J. Water Clim. Chang.* **2010**, *1*, 67–86. [CrossRef]

60. Zhou, T.; Nijssen, B.; Gao, H.; Lettenmaier, D.P. The contribution of reservoirs to global land surface water storage variations. *J. Hydrometeorol.* **2016**, *17*, 309–325. [CrossRef]

61. Hempel, S.; Frieler, K.; Warszawski, L.; Schewe, J.; Piontek, F. A trend-preserving bias correction–the ISI-MIP approach. *Earth Syst. Dyn.* **2013**, *4*, 219–236. [CrossRef]

62. Burhan, A.; Waheed, I.; Syed, A.A.B.; Rasul, G.; Shreshtha, A.B.; Shea, J.M. Generation of high-resolution gridded climate fields for the upper Indus River Basin by downscaling CMIP5 outputs. *J. Earth Sci. Clim. Chang.* **2015**, *6*, 1.

63. Ruan, Y.; Liu, Z.; Wang, R.; Yao, Z. Assessing the Performance of CMIP5 GCMs for Projection of Future Temperature Change over the Lower Mekong Basin. *Atmosphere* **2019**, *10*, 93. [CrossRef]

64. Abbaspour, K.C. *User Manual for SWAT-CUP: SWAT Calibration and Uncertainty Analysis Programs*; Swiss Federal Institute Aquatic Science and Technology: Dübendorf, Switzerland, 2011.

65. Nash, J.E.; Sutcliffe, J.V. River flow forecasting through conceptual models part I—A discussion of principles. *J. Hydrol.* **1970**, *10*, 282–290. [CrossRef]

66. Sridhar, V.; Hubbard, K.G.; Wedin, D.A. Assessment of soil moisture dynamics of the Nebraska Sandhills using long-term measurements and a hydrology model. *J. Irrig. Drain. Eng.* **2006**, *132*, 463–473. [CrossRef]

67. Sridhar, V.; Wedin, D.A. Hydrological behaviour of grasslands of the Sandhills of Nebraska: Water and energy-balance assessment from measurements, treatments, and modelling. *Ecohydrol. Ecosyst. Land Water Process Interact. Ecohydrogeomorphol.* **2009**, *2*, 195–212. [CrossRef]

68. USDA (U.S. Department of Agriculture). *Soil Conservation Service, National Engineering Handbook*; Hydrology (Section 4, Chapters 4–10); GPO: Washington, DC, USA, 1972.

69. Boughton, W.C. A review of the USDA SCS curve number method. *Soil Res.* **1989**, *27*, 511–523. [CrossRef]

Article

Quantifying the Impacts of Climate Change, Coal Mining and Soil and Water Conservation on Streamflow in a Coal Mining Concentrated Watershed on the Loess Plateau, China

Qiaoling Guo [1,2,3], Yaoyao Han [1], Yunsong Yang [4,*], Guobin Fu [5] and Jianlin Li [1,3]

[1] Institute of Resource and Environment, Henan Polytechnic University, Jiaozuo 454003, China; guoqiaoling@hpu.edu.cn (Q.G.); 15139179673@163.com (Y.H.); lijianlin@hpu.edu.cn (J.L.)
[2] School of Water Resources & Environmental Engineering, East China University of Technology, Nanchang 330000, China
[3] Collaborative Innovation Center of Coalbed Methane and Shale Gas for Central Plains Economic Region, Jiaozuo 454003, China
[4] Institute of Business Management, Henan Polytechnic University, Jiaozuo 454003, China
[5] CSIRO Land and Water, Private Bag 5, Wembley, WA 6913, Australia; GUOBIN.FU@CSIRO.AU
* Correspondence: yangyunsong@hpu.edu.cn or yys7401@163.com; Tel.: +1-553-899-2635

Received: 21 March 2019; Accepted: 28 April 2019; Published: 21 May 2019

Abstract: The streamflow has declined significantly in the coal mining concentrated watershed of the Loess Plateau, China, since the 1970s. Quantifying the impact of climate change, coal mining and soil and water conservation (SWC), which are mainly human activities, on streamflow is essential not only for understanding the mechanism of hydrological response, but also for water resource management in the catchment. In this study, the trend of annual streamflow series by Mann-Kendall test has been analyzed, and years showing abrupt changes have been detected using the cumulative anomaly curves and Pettitt test. The contribution of climate change, coal mining and SWC on streamflow has been separated with the monthly water-balance model (MWBM) and field investigation. The results showed: (1) The streamflow had an statistically significant downward trend during 1955–2013; (2) The two break points were in 1979 and 1996; (3) Relative to the baseline period, i.e., 1955–1978, the mean annual streamflow reduction in 1979–1996 was mainly affected by climate change, which was responsible for a decreased annual streamflow of 12.70 mm, for 70.95%, while coal mining and SWC resulted in a runoff reduction of 2.15 mm, 12.01% and 3.05mm, 17.04%, respectively; (4) In a recent period, i.e., 1997–2013, the impact of coal mining on streamflow reduction was dominant, reaching 29.88 mm, 54.24%. At the same time, the declining mean annual streamflow induced through climate change and SWC were 13.01 mm, 23.62% and 12.20 mm, 22.14%, respectively.

Keywords: streamflow reduction; climate change; coal mining; SWCM; coal mining concentrated watershed; the Loess Plateau

---

1. Introduction

Over the second half of the 20th century, the two factors which affected the change of catchment hydrology were climate change and human activities [1]. Plenty of studies have indicated that the streamflow of many rivers has changed due to climate change and anthropogenic activities [2,3], especially in arid and semi-arid regions. Climate change, for example, the redistribution of precipitation and temperature change, has affected hydrological systems and water resources [4,5]. Human activities, such as agricultural irrigation, cultivation, dam construction, reservoir operation, soil and water conservation (SWC), urbanization construction and coal mining could also affect hydrological processes, resulting in natural ecosystem and water resource problems [6,7].

The Loess Plateau, located in the middle reaches of the Yellow River, is the most severe soil and water loss region worldwide [8]. Most areas of the Loess Plateau comprises gully-hill dominated regions, with the most widely distributed loess on Earth. Intensive soil and water loss has resulted in water shortages, land productivity decline, and river ecosystem and environmental degradation. Soil from tributaries in middle reaches of the Yellow River is the major source of sediment in the lower reach of the Yellow River [9]. SWC are important measures for improving the ecosystem and the environment of the Loess Plateau. Since the 1970s, large-scale SWCs have been carried out by the Chinese government [10], which has brought about major changes to the runoff conditions and hydrological characteristics in tributaries of middle reaches of the Yellow River, and has had an impact on storm floods and river runoff [11,12]. The maximum runoff utilization rate of the SWC is 63%, which can significantly reduce the amount of water entering the river [13]. The Loess Plateau is rich in mineral resources [14], which plays a critical role on energy sources in China's economic development. The Shenfu-Dongsheng coalfield, accounting for 1/4 of the China's coal reserves, is located in the northern edge of Loess Plateau [15]. Subsidence and cracks formed by coal mining have changed the surface conditions, and thus, have altered runoff generation and confluence, leading to degradation of river ecological environment in the mining area [16,17].

The Kuye River Basin is located in a wind and water erosion interlaced area in the northern of the Loess Plateau [18]. It is a typical loess hilly landform [19], and is one of the most severely water and soil losing areas on the Loess Plateau. The Shenfu-Dongsheng coalfield is through the middle of the Kuye River Basin, which is the main source of sediment in the lower Yellow River. Su et al. analyzed precipitation and runoff changing trend in Kuye River Basin [20]. Guo et al. studied the trend of inner-annual runoff in Kuye River Basin [21]. Zhao et al. analyzed the flood characteristics and their changing trends in the Kuye River Basin [22]. Liu et al. and Bai et al. studied the impact of climate change and anthropogenic activities on runoff variation, indicating that SWC and coal mining had had an important impact on the runoff change in Kuye River Basin [23,24].

Three groups methods are used for assessing the effects of climate change and human activities on runoff variation: the paired catchments approach, the statistical method and hydrological modeling [25]. The paired catchments approach is usually considered in small experimental catchments; the statistical method can only analyze the impact of climate change and human activities on runoff variation, as it lacks a physical mechanism. Hydrological modeling is widely used to assess the effects of climate change and human activity on runoff variation. Wang et al. used a monthly water balance model to simulate the runoff of nine tributaries in the middle reaches of the Yellow River, achieving high simulation accuracy [26]. Xing et al. and Guo et al. simulated the runoff in Kuye River Basin by monthly water balance model and received satisfactory results [27,28]. Cheng et al. simulated daily and monthly discharges by SWAT model, and found that it is not effective [29]. Li et al. simulated the monthly runoff in Kuye River Basin using SWAT model, and the simulation effect was not satisfactory [30]. Considering that the monthly water balance model has a simple structure, few parameters and a good simulation effect, this method is used to separate the effects of climate change and human activities in this study.

In this work the objectives are: (1) Analysis of the annual streamflow variation since 1950s; and (2) Quantitative assessments of climate change, coal mining and SWC impacts on runoff decline in past 60 years. This work will provide a better understanding of the interactions between humans and nature, while also provide important insights into water-resource management in the Kuye River Basin.

2. Studied Watershed and Data

2.1. Studied Watershed

The Kuye River Basin is a first-tributary in the middle Yellow River. It has a main stream length of 242 km and a drainage area of 8706 km^2 [28]. There are two major tributaries (Wulanmulun River and Beiniuchuan River) in the upper reaches, and a large number of coal mine are along the two tributaries. The water system of the Kuye River Basin is shown in Figure 1. Affected by continental monsoons, the climate fluctuates dramatically throughout the year. The precipitation from June to September accounts for 75–81% of the annual total [31]. Rain storms usually take place in summer, especially in July and August. The drainage landform mainly consists of wind-dust region and hill-gully loess region [32]. The wind-dust region is relatively flat with hardly any vegetation, and the hill-gully loess region is covered by exposed soft bedrock.

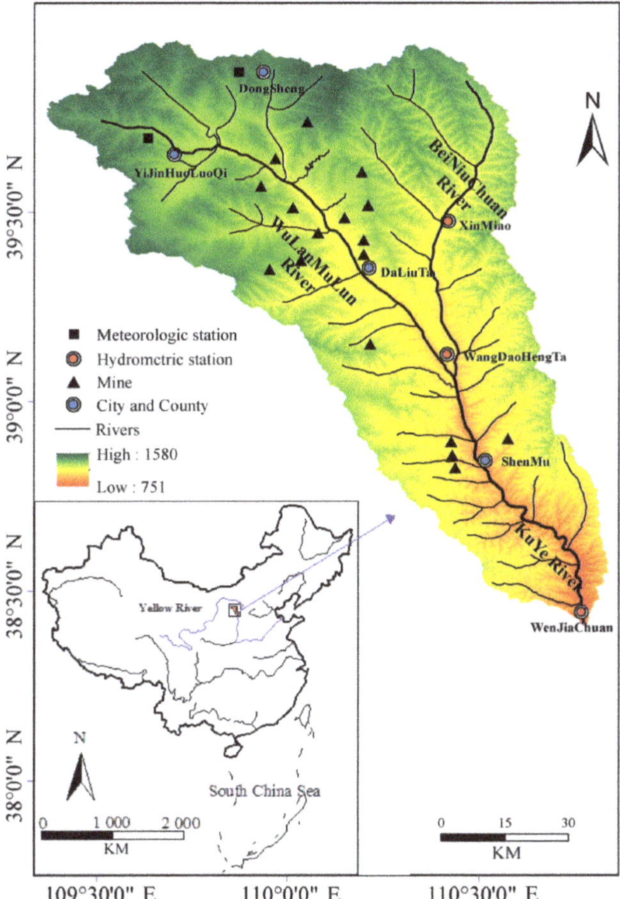

Figure 1. The sketch map of the Kuye River Watershed.

The coal resources are rich in the Kuye River Basin. There are 209 coal mines on both sides of the river; the coal seams are shallow, and the exploited mine areas reach 2482 km^2, which accounts for 28.51% of the basin areas. According to the survey data, the raw coal output was 6.25×10^6 t in 1991, 31.293×10^6 t in 2002, and 172.262×10^6 t in 2011, with the average annual growth of 7.905×10^6 t.

Along with the coal mining amount increasing, coal mining subsidence areas were 26.01 km^2 in 1991, 48.82 km^2 in 2000, and 266.15 km^2 in 2011, and the average annual growth reached to 11.435 km^2. The large surface cracks in the subsidence changed the surface characteristics, and impacted the streamflow (Figure 2).

Figure 2. The cracks in the coal mining subsidence.

Water and soil loss are serious in the Kuye River Basin. The SWC consists of terraces, grassland, afforestation, and the construction of sediment-trapping dams. Before 1979, the areas of SWC were 609.42 km^2. From 1980s to 1990s, the areas of SWC reached 1480.02 km^2 in the late 1990s [33]. By 2009, the areas of SWC were 3739.90 km^2, and the sediment-trapping dams were 1548. The increasing of SWC also transformed the surface characteristics, and impacted on runoff.

In Kuye River Basin, human influence on streamflow includes coal mining, SWC and the water abstraction from the river for domestic, irrigation, and industry uses. Among these influencing factors, coal mining and SWC are mainly impact factors, and the water abstraction from the river is only a small part [34,35]. Therefore, this study separated the impact of coal mining and SWC on runoff.

2.2. Data

The daily streamflow records of Wenjiachuan hydrological station which is the furthest downstream hydrological station (8645 km^2 at Wenjiachuan hydrological station) were used in this study. The daily streamflow data was provided by the Yellow River Conservancy Commission. The data of daily precipitation, average daily temperature was from two meteorological stations (Yijinhuoluo and Dongsheng), and four hydrological stations. The data of meteorological stations was obtained from the China Meteorological Administration, and the data of hydrological stations (Wangdaohengta, Xinmiao, Shenmu, and Wenjiachuan) was from the Yellow River Conservancy Commission. All the data were from 1955 to 2013. The monthly (year) streamflow, the monthly (year) precipitation, and the average monthly (year) temperature were calculated from daily discharge, daily rainfall, and average daily temperature.

2.3. Methods

First, the modified Mann–Kendall trend test was used to analyze the trends of annual streamflow [36]. Then, cumulative anomaly curves and the Pettitt test were applied to detect the abrupt change years of streamflow variables [37]. Finally, the contributions of climate change, coal mining and SWC on streamflow decreasing in the same period was analyzed using the monthly water-balance model (MWBM) and survey data. The brief introduction of MWBM, and the methods of distinguishing the impacts of climate change, coal mining and SWC on streamflow reduction are presented below.

2.3.1. Separating Climate Change, Coal Mining and SWC Impacts on the Streamflow

In an attempt to separate the effect of climate change, coal mining and SWC, we need to choose a baseline period. The streamflow difference between the baseline period and human induce period is the impact of climate change, coal mining and SWC, and is calculated as follows:

$$\Delta Q_{cm} + \Delta Q_{sb} + \Delta Q_c = Q_i - Q_b \tag{1}$$

where ΔQ_c, ΔQ_{cm} and ΔQ_{sb} are the contributions of climate change, coal mining and SWC on streamflow change, respectively; Q_i, Q_b are the observed streamflow during the human induced period and baseline periods, respectively.

$$\Delta Q_{cm} + \Delta Q_{sb} = Q_i - Q_m \tag{2}$$

$$\Delta Q_c = Q_m - Q_b \tag{3}$$

$$\Delta Q_{sb} = \sum_{i=1}^{4} \varepsilon_i A_i \tag{4}$$

where Q_m is the reconstructed streamflow by the monthly water-balance model (MWBM), ε_i is the water reduction coefficient, which use the research results of the reference [38], A_i is areas of terrace, afforestation, grassland and sediment-trapping dams.

The impact percentages from climate change (η_c), coal mining (η_{cm}) and SWC (η_{sb}), are stated as

$$\eta_c = \frac{\Delta Q_c}{\Delta Q_{cm} + \Delta Q_{sb} + \Delta Q_c} \times 100\% \tag{5}$$

$$\eta_{cm} = \frac{\Delta Q_{cm}}{\Delta Q_{cm} + \Delta Q_{sb} + \Delta Q_c} \times 100\% \tag{6}$$

$$\eta_{sb} = \frac{\Delta Q_{sb}}{\Delta Q_{cm} + \Delta Q_{sb} + \Delta Q_c} \times 100\% \tag{7}$$

The above equations were used to quantify the impacts of climate change, coal mining and SWC on streamflow variance in Kuye River Basin from 1955 to 2013. The next step is to reconstruct natural streamflow using the MWBM.

2.3.2. Brief Introduction of the MWBM

The MWBM was developed by the U.S. Geological Survey. It used a monthly accounting procedure based on the methodology, originally proposed by Thornthwaite [39]. Mean monthly temperature and monthly total precipitation are the input files to the model. The input parameters include runoff factor, soil-moisture-storage capacity, rain temperature threshold, maximum melt rate, direct runoff factor, snow temperature threshold and latitude of location. The individual components of the water balance included the amount of monthly precipitation (P) that is rain (P_{rain}) or snow (P_{rain}), direct runoff DRO, snow melt (SM), potential evapotranspiration (PET), soil-moisture storage withdrawal (STW), and runoff [40].

$$P_{snow} = P \times \left[\frac{T_{rain} - T}{T_{rain} - T_{snow}} \right] \tag{8}$$

$$P_{rain} = P - P_{snow} \tag{9}$$

$$DRO = P_{rain} \times drofrac \tag{10}$$

$$P_{remain} = P_{rain} - DRO \tag{11}$$

$$SMF = \frac{T - T_{snow}}{T_{rain} - T_{snow}} \times meltmax \tag{12}$$

$$SM = snostor \times SMF \tag{13}$$

$$PET = 13.97 \times d \times D^2 \times \frac{4.95 \times e^{0.062 \times T}}{100} \tag{14}$$

$$STW = ST_{i-1} - \left[abs(P_{total} - PET) \times (\frac{ST_{i-1}}{STC}) \right] \tag{15}$$

where P_{remain}, SMF, ST, SMF are remaining precipitation, snow melt fraction, soil-moisture storage, and the soil-moisture storage capacity, respectively.

The Nash Sutcliffe coefficient (NSE) and relative error (RE) were used to evaluate the performance of the model. NSE close to 1 and RE close to 0 are the good simulation result. The qualified conditions of simulation are that NSE is much more 0.6 and RE is less than 0.1 [41].

3. Results

3.1. Long-Term Variation of the Annual Streamflow Series

3.1.1. Trend Analysis of the Annual Streamflow Series

The annual streamflow in 1955–2013 at Wenjiachuan hydrological station was shown in Figure 3. The maximum annual streamflow was 13.706×10^8 m^3 and occurred in 1959. While, the minimum annual streamflow reached 1.187×10^8 m^3 in 2011. The mean annual streamflow from 1955 to 2013 was 5.187×10^8 m^3. Average annual streamflow in 1950s, 1960s, 1970s, 1980s, 1990s and the early 21st century were 6.827×10^8 m^3, 7.642×10^8 m^3, 6.867×10^8 m^3, 5.278×10^8 m^3, 4.226×10^8 m^3, and 1.919×10^8 m^3, respectively. Since the 1980s, average annual runoff began to decrease. At the beginning of the 21st century, average annual runoff decreased the most, which only accounted for 37% of mean annual streamflow. The annual streamflow has an obvious declining gradient of -0.113×10^8 m^3/year. When being analyzed using the modified M-K trend test, the annual streamflow during 1955–2013 presented a significant decreasing trend; the Zc reached the value of -2.325, and passed the 0.05 significance test.

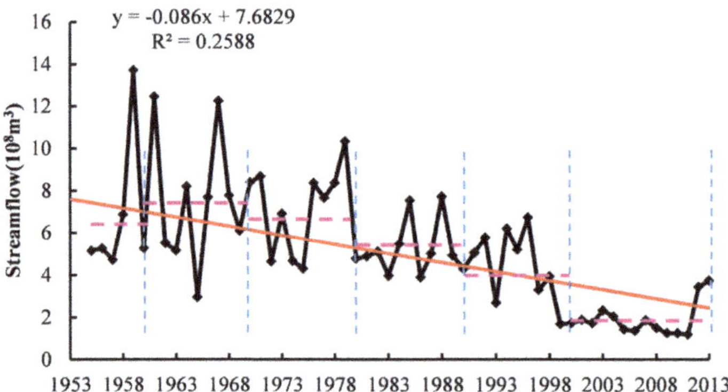

Figure 3. The variation trend of streamflow at Wenjiachuan hydrological station.

3.1.2. Abrupt Change Years of Annual Streamflow Series

To detect abrupt changes of annual streamflow change, cumulative anomaly curves and Pettitt test were used. Both Figures 4 and 5 show that the annual streamflow has an increasing trend before 1979, and that it then fluctuated slightly from 1979 to 1996, and finally, declined considerably after 1996. There were two significant change points in 1979 and 1996. According to the two change points, the annual streamflow series were divided into 3 stages which were 1955–1978, 1979–1996, and 1997–2013, respectively.

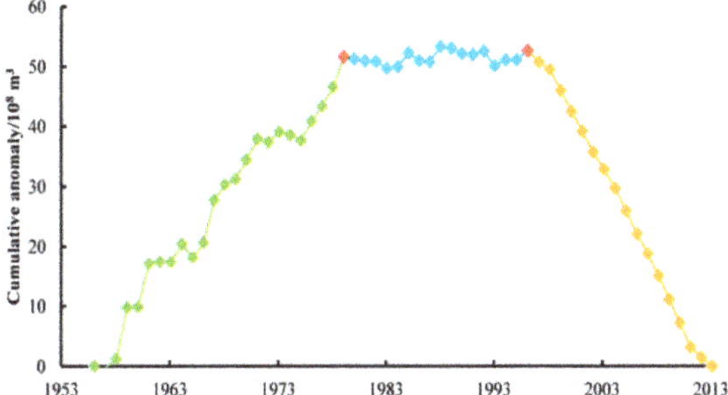

Figure 4. Cumulative anomaly curve of annual streamflow.

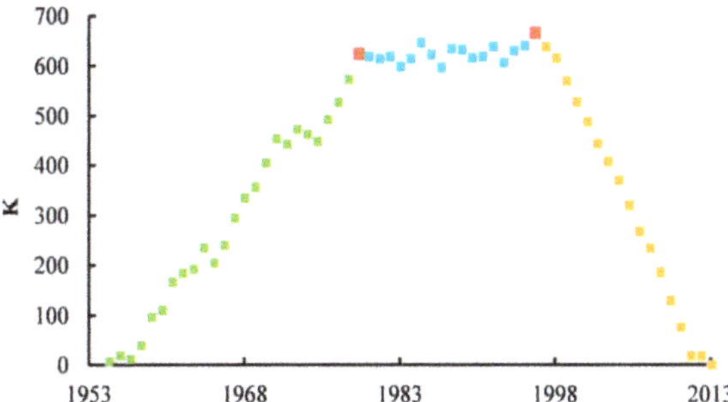

Figure 5. K value change calculated by Pettitt test.

3.2. Separating the Impacts of Climate Change and Human Activities by MWBM

3.2.1. Model Calibration and Verification

In this study, we took 1955–1978 as the baseline period. The observed climatic and streamflow data at the Wenjiachuan station in 1955–1970 was used for calibration, and the data from 1971 to 1978 were used for verification. Figure 6 show that the recorded and simulated data fit well. The points of the correlation between recorded and simulated streamflow are concentrated around the 1:1 line (Figure 7). In addition, The NSE in calibration period and verification period were 77.95% and 75.69%, respectively. Furthermore, the RE were 3.58% and 4.21%. Overall, the calibration and verification accuracies of the model were acceptable. The next step is investigating the effect of climate change and human activities in the human-induced periods by the MWBM.

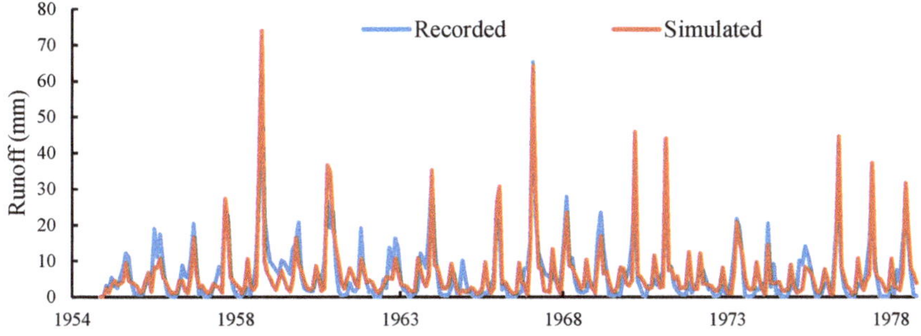

Figure 6. Monthly time series of recorded streamflow and simulated streamflow in 1955–1978 at Wenjiachuan station.

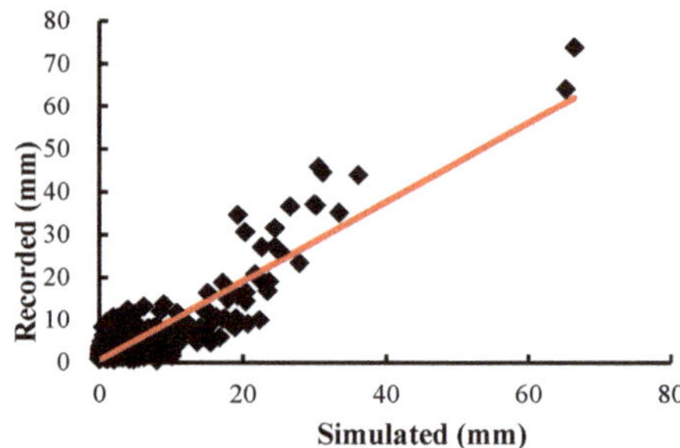

Figure 7. Correlation between simulated and recorded monthly streamflow in the baseline period.

3.2.2. Separating the Impacts of Climate Change and Human Activities

According to the abrupt change years, the streamflow series were divided into periods of the baseline period (1955–1978), human impacted (HIP) period (1979–1996), and human strong induced (HSIP) period (1997–2013), respectively. Given the recorded streamflow and reconstructed streamflow, the influence of climate change and human activities on streamflow in HIP and HSIP are summarized in Table 1.

Table 1. The contributions of climate change and human activities to annual streamflow reduction.

Periods	Recorded Streamflow (mm)	Reconstructed Streamflow (mm)	Total Change (%)		Impact by Climate Change		Impact by Human Activities	
			ΔQ (mm)	η (%)	ΔQ_c (mm)	η_c (%)	ΔQ_h (mm)	η_h (%)
1955–1978	81.02							
1979–1996	63.12	68.32	−17.90	22.09	−12.7	70.95	−5.20	29.05
1997–2013	25.93	68.01	−55.09	67.99	−13.01	23.62	−42.08	76.38
1979–2013	45.06	68.17	−35.96	44.38	−12.85	35.73	−23.11	64.27

During the period of 1979–2013, the recorded annual streamflow was obviously less than that of the baseline period. The absolute and relative values of total impacts of climate change and human activities on annual streamflow were −35.96 mm and 44.38%. Both climate change and human activities resulted in streamflow decreases compared to the baseline period. During 1979–1996, the climate change was the main factor that decreased streamflow with a contribution of 70.95% relative to the baseline period, while the reduction percentage due to human activities were only 29.05%. However, the contribution of climate variations to streamflow reduction dropped to 23.62%, corresponding that of human activities which ascended to 76.38% in 1997–2013. Human activity has become the major factor in reducing streamflow. Specific impacts of climate change on annual streamflow were −12.7 mm for HIP and −13.01 mm for HSIP, while the influence of human activities were from −5.20 mm in HIP to −42.08 mm in HSIP. On average, human activities and climate change were responsible for 64.27% and 35.73% of streamflow reduction, respectively.

3.3. Separating the Coal Mining and SWC Impacts on Streamflow Decreasing

The SWC in Kuye River Basin includes the construction of terraces, planting trees and afforestation, and building sediment-trapping dams. Water and soil loss is severe in the Kuye River Basin. Before 1979, the area of SWC was generally small, and less than 7% of the basin area. After 1979, a large number of SWC were carried out. Because there are not the information about the areas of the different SWC in every year. The analysis uses the areas in the survey years. The water reduction of SWC are calculated in HIP and HQIP by the water reduction coefficient and the areas of SWC in the representative years

(Table 2). The streamflow reduction caused by the SWC from −3.05 mm in HIP to −12.20 mm in HQIP. During 1979–2013, the average annual streamflow reduction was −7.57 mm.

Table 2. Cumulative area of SWC since the 1950s.

Year	Terrace		Afforestation		Grassland		Sediment-Trapping Dams	
	A (km²)	ε (m³/km²)	A (km²)	ε (m³/km²)	A (km²)	ε (m³/km²)	dam	ε (per dam/m³)
1959	5		27		22		0	
1969	33		97		52		2	
1979	66		415		110		8	
1989	67	44,600	1004	20,900	353	16,600	12	12,000
1996	99		1184		380		19	
2009	101		2652		938		1548	
2013	105		3555		638		2271	

Based on the influence of human activities and the SWC to annual streamflow reduction, respectively, we could calculate the contribution of coal mining to runoff reduction (Table 3). The average annual impact of the coal mining on streamflow were from −2.15 mm in HIP to −29.88 mm in HQIP. The percentage of the effect was increasing from 41.35% to 71.02%. However, the percentage of the effect for the SWC was decreasing from 58.65% to 28.99%. Seen from the impact quantity, the contribution of coal mining and SWC on streamflow decreasing showed an increasing trend. At the same time, the growth rate of the impact of coal mining was greater than that of the SWC. During 1979–2013, the influence of the SWC and the coal mining on streamflow was −7.57mm, 32.76% and −15.54 mm, 67.24%, respectively. Thus, the coal mining demonstrated the dominant influence on streamflow decline gradually.

Table 3. The contributions of the SWC and coal mining to annual streamflow reduction.

Period	Impact by Human Activities (mm)	Impact by SWC		Impact by Coal Mining	
		ΔQ_{sb} (mm)	η_{sb} (%)	ΔQ_{cm} (mm)	η_{cm} (%)
1979–1996	−5.20	−3.05	58.65	−2.15	41.35
1997–2013	−42.08	−12.20	28.99	−29.88	71.02
1979–2013	−23.11	−7.57	32.76	−15.54	67.24

3.4. Quantification Climate Change, Coal Mining and SWC Impacts on Streamflow Decreasing

Based on separating the impacts of climate change and human activities, and the contribution of coal mining and SWC among human activities, we could quantify the influence of climate change, coal mining and SWC on annual streamflow decline (Table 4). During 1979–1996, the annual runoff reduction was −17.90 mm induced by climate change, coal mining and SWC, among which the impacts of climate change was −12.70 mm, 70.95%; that of coal mining was −2.15 mm, 12.01%; and that of SWC was −3.05 mm, 17.04%, climate change was the main influencing factor. However, in 1997–2013, the impact of coal mining on annual streamflow was −29.88 mm, and that of climate change and SWC were −13.01 mm and −12.20 mm. The relative values of impacts were 54.24% for coal mining, 23.62% for climate change, and 22.14% for the SWC, respectively.

Table 4. The quantification of impacts of climate change, coal mining and SWC on streamflow decreasing.

Periods	Total Change (mm)	Impact by Climate Change		Impact by Coal Mining		Impact by SWC	
		ΔQ_c (mm)	η_c (%)	ΔQ_{cm} (mm)	η_{cm} (%)	ΔQ_{sb} (mm)	η_{sb} (%)
1979–1996	−17.90	−12.70	70.95	−2.15	12.01	−3.05	17.04
1997–2013	−55.09	−13.01	23.62	−29.88	54.24	−12.20	22.14
1979–2013	−35.96	−12.85	35.73	−15.54	43.22	−7.57	21.05

4. Discussion

4.1. Impact of Climate Change and Human Activities on Streamflow

In recent decades, the streamflow of many rivers around the world exhibited a decreasing trend because of climate change and human activity [42]. Over the past 40 years in the Central Rift Valley of Ethiopia, almost all rainfall indices have an increasing trend in the valley floor and a decreasing trend in the escarpment and highlands [43]. For responding to the effects of climate change, and simultaneously satisfy environmental, societal, and economic, the implementation of environment friendly techniques policies in Romania have been studied [44]. The research found that a 10% decrease in precipitation may cause a decrease in streamflow of between 19% in the tropical zone and 30% in the arid zone in Africa [45]. Climate change may contribute 26%–31% of streamflow decline relative to the base period in Beichuan river basin of China [46]. In northwest China, the streamflow reduction caused by climate change accounting for 14.3% [47]. For the tributaries in the middle reaches of the Yellow River basin, climate change accounted for more of the streamflow reduction in the Beiluo River and Yan River, while human activities has a greater effect on the streamflow reduction in other tributaries [48]. In this study, climate change along with human activity led to a decrease in streamflow in HIP to HSIP. However, the contribution of climate to streamflow reduction was in a relatively stable state, and the absolute amount of influence was −12.7 mm in HIP and −13.01 mm in HSIP. By contrast, the contributions of human activities to streamflow reduction between HIP and HSIP were significantly different and have an enhanced trend from −5.20 mm in HIP to −42.08 mm in HSIP. The two abrupt points of streamflow change were in 1979 and 1996, which were in consonance with the extensive SWC in the late 1970s (Table 2) and the massive coal mining in the late 1990s (Figure 9). During the HIP period, the contributions of climate change reached 70.95%, and was the main factor affecting streamflow reduction due to the low intensity of human activity. But, the total amount of streamflow decreasing significantly increased from −17.90 mm to −55.09mm to the HSIP, and 76.38% of the contributions was due to human activities, indicating that intensity of human activities has increased significantly since the end of 1990s.

4.2. Impacts of Coal Mining on Streamflow

The contribution of coal mining impacts on streamflow decline increased −27.73 mm from HIP to HSIP, and became the dominant factor affecting streamflow in HSIP. It is related to the dramatic increase in coal mining since the end of the 20th century (Figure 8). Raw coal production increased from 1102.58×10^4 t in 1996 to $17{,}262.21 \times 10^4$ t in 2011 which was 15 times that of 1996. A large number of coal mining produced significant impact on streamflow. The surface of mining areas are the aeolian sand of the Sarawusu Formation. The Quaternary Sarawusu group loose pore phreatic aquifer is significant for water supply, and the coal seam is under the Sarawusu group aquifer. After coal seam mining, the water conducting fissure extended to the surface (Figure 9) [49], which not only increased rainfall infiltration, but also lowered the phreatic level. The rapid loss of the phreatic and surface water inevitably led to streamflow reduction.

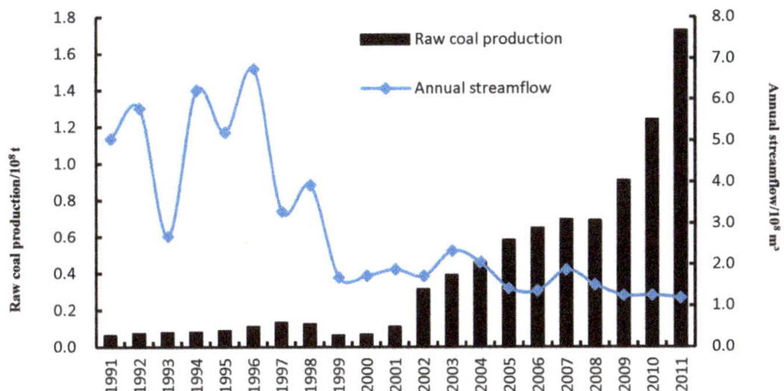

Figure 8. Relation of raw coal production and annual streamflow in Kuye River Basin.

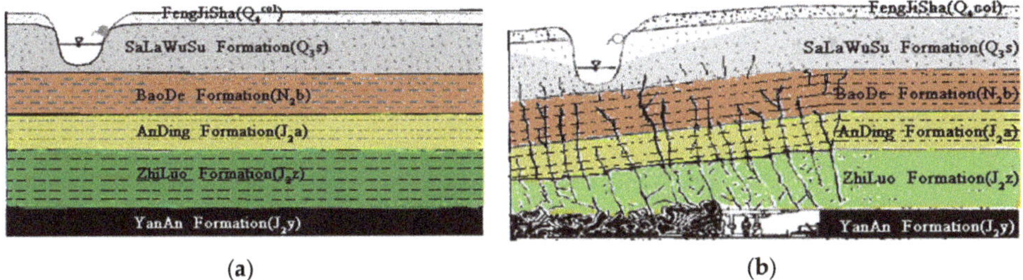

(a)   (b)

Figure 9. The coal mine aquifer in Kuye River Basin. (**a**) Before coal mining; (**b**) After coal mining.

4.3. Impacts of SWC on Streamflow

SWC techniques are widely used to alter soil and water processes and improve ecosystem environment. A study of sample plots from 22 countries indicated that afforestation, soil amendment and terraces may reduce annual streamflow by 55%, 48% and 44% respectively [50]. Both catch crops and weeds may enhance infiltration rates, delay and decrease the runoff discharge under single ring ponding conditions [51]. A cover of 50% of straw is able to delay the time to runoff initiation from 57 to 129 s, and mulching reduces the runoff coefficient from 65.6 to 50.5% in clementine plantations [52]. Surface runoff may be reduced by about 19% by the SWC in Ethiopian [53]. On the most severely eroded Loess Plateau in the world, large scale SWC were implemented, which induced streamflow decline [54]. The construction of terraces, planting trees and afforestation, and building sediment-trapping dams were the main measures of SWC in the watersheds which located on Loess Plateau. After building terracing, the topography of the basin has been changed, the rainfall infiltration has been greatly improved, and streamflow has been reduced. After planting trees and afforestation on bare hill slopes, a considerable proportion of rainfall can be intercepted by the canopy and evaporate into the atmosphere. Thus, the effective rainfall for runoff generation is reduced. The sediment-trapping dams are built in the ditch and channel of the Loess Plateau, and may intercept floods, and is an important measure to prevent water and soil loss. Although the percentage of the contribution of SWC impacts on streamflow reduction has decreased. In fact, its absolute amount of the contribution of SWC impacts on streamflow decline increased −9.15 mm, indicating the contributions of coal mining and SWC streamflow reduction were on an increasing trend, while the increase rate of coal mining was greater.

4.4. Uncertainties of Quantitative Assessment

The uncertainty of quantitative assessment came mainly from the following aspects. (1) Model parameters and input data may lead to uncertainty in the simulation process. For instance, in this study only mean monthly temperature and monthly total precipitation are as the climatic input data which should actually include sunshine, wind speed, evaporation and other factors. (2) We analyzed the contribution of climate change and human activities to streamflow variation with the assumption that the streamflow in the baseline period was not affected by human activities. However, in fact, the streamflow was also affected by human activities although with less intensity. (3) In this study, only coal mining and SWC were considered as elements of human activities affecting streamflow, and not considered water consumption for domestic, irrigation, and industry which also has the impact on streamflow. for example, Shenmu County, which accounts for more than 50% of the basin area, increased industrial water supply by nearly 10 times from 1980 to 2011; and the construction of the massive water landscape and urban grassland has increased evaporation loss and irrigation water consumption. So the quantitative assessment values of the contribution of coal mining and SWC were higher than the actual values. (4) It should also be note that the influence of climate change, coal mining and SWC on streamflow are not independent in theory and cannot be separated exactly. These factors interact with each other.

4.5. Prospects for Future Research

The researched watershed is located in the water and soil erosion zone of the Loess Plateau. The surface gully is vertical and horizontal, the terrain is broken, the loess is loose, the vegetation is scarce. At the same time, a large number of coal mines are distributed along the river, and the coal seams are shallow. Similar rivers in the middle reaches of the Yellow River include Wuding River, Tuwei River, etc. The commonality of these basins are: (1) the fragile ecological environment; (2) the main source of sediment in the Yellow River; (3) coal mining and SWC are the main human activities. The research conclusions are applicable to such watershed mentioned above. However, for the other coal mining concentrated watershed, the impact of coal mining on streamflow should been further studied. In addition, in this study, the water reduction of SWC are calculated by the water reduction coefficient and the areas of SWC in the representative years, and it made the accuracy of the results was affected to a certain extent. Future research should utilize hydrological models to effectively separate the effects of different types of human activity to streamflow.

4.6. Adaptive Strategies and Options

There are many watersheds in which coal mining and SWC are the main human activities on the Loess Plateau, for example, the Tuwei River Basin and the Wudin River Basin. Some previous studies indicate that the abrupt points of streamflow change were also in 1979 and 1996 for the Tuwei River and the Wudin River [55,56]. The contribution of climate change and human activities to streamflow reduction were 57.95% and 42.05% from 1997 to 1996, respectively. Nevertheless, the contribution of climate change dropped to 24.19%, and that of human activities ascended to 75.81% after 1996 in the Tuwei River basin. For the Wuding River, the contribution of climate change was 79.8% from the 1970s to the end of the 1990s, and human activity became the main factor affecting streamflow reduction to the 21st century [57]. The common features of these rivers are that they are located on the Loess Plateau and the ecological environment there is fragile. To protect water resources, local governments should adopt strategies such as strengthening water resources protection and popularizing water-preserving technology in coal mining, developing water saving irrigation technology, and reusing and recycling water resources in industry.

5. Conclusions

In this paper, we analyzed streamflow change trends, abrupt change years from 1955 to 2013, separated the contributions of climate change, coal mining and SWC impacts on streamflow decreasing. The main findings are as follows:

(1) The annual runoff presented a decreasing trend, and passed the 0.05 significance test during 1955–2013.The two significant change years was in 1979 and 1996.

(2) In the first impact period (1979–1996), climate change was the main factor for annual streamflow decreasing. Meanwhile, in the second impact period (1997–2013), coal mining was the dominant influence on streamflow decline.

(3) Compare two impact periods, the absolute value of climate change, coal mining and SWC impacts on streamflow reduction were all ascending, which indicated that the impacts of above three factors on streamflow decreasing were increasing. Meanwhile, the growth rate of coal mining impact on streamflow decline was greater than that of climate change and SWC.

(4) Quantifying the impacts of climate change, coal mining and SWC on streamflow decline by the MWBM and field investigation was reasonable and feasible.

Author Contributions: Q.G., Y.H. and Y.Y. conceived the study, analyzed the data and wrote the paper. G.F. and J.L. provided critical feedback on the manuscript.

Funding: This research work was supported by Key Research Project of colleges and universities of Henan Province in China (No.18A170006), Soft Science Research Project of Henan Province in China (No.182400410045), Soft Science Research of Key Research Project of colleges and universities of Henan Province in China (No.19A630013), Henan Provincial Natural Science Foundation Project of China (No.182300410155), The Innovation Scientists and Technicians Troop Construction Projects of Henan Province (No. CXTD2016053).

Conflicts of Interest: The authors declare no conflict of interest.

References

1. Jiang, C.; Xiong, L.H.; Wang, D.B.; Liu, P.; Guo, S.L.; Xu, C.Y. Separating the impacts of climate change and human activities on runoff using the Budyko-type equations with time-varying parameters. *J. Hydrol.* **2015**, *522*, 326–338. [CrossRef]

2. Bao, Z.X.; Zhang, J.Y.; Wang, G.Q.; Fu, G.B.; He, R.M.; Yan, X.L.; Jin, J.L.; Liu, Y.L.; Zhang, A.J. Attribution for decreasing streamflow of the Haihe River basin, northern China: Climate variability or human activities? *J. Hydrol.* **2012**, *460–461*, 117–129. [CrossRef]

3. Fu, G.B.; Chen, S.L.; Liu, C.M.; Shepard, D. Hydro-climatic trends of the Yellow River Basin for the last 50 years. *Clim. Chang.* **2004**, *65*, 149–178. [CrossRef]

4. Milliman, J.D.; Farnsworth, K.L.; Jones, P.D.; Xu, K.H.; Smith, L.C. Climatic and anthropogenic factors affecting river discharge to the global ocean, 1951–2000. *Glob. Planet. Chang.* **2008**, *62*, 187–194. [CrossRef]

5. Arrigoni, A.S.; Greenwood, M.C.; Moore, J.N. Relative impact of anthropogenic modifications versus climate change on the natural flow regimes of rivers in the northern Rocky Mountains, United States. *Water Resour. Res.* **2010**, *46*. [CrossRef]

6. Tu, A.G.; Xie, S.H.; Yu, Z.B.; Li, Y.; Nie, X.F. Long-term effect of soil and water conservation measures on runoff, sediment and their relationship in an orchard on sloping red soil of southern China. *PLoS ONE* **2018**, *13*, e0203669. [CrossRef] [PubMed]

7. Rossi, A.; Massei, N.; Laignel, B.; Sebag, D.; Copard, Y. The response of the Mississippi River to climate fluctuations and reservoir construction as indicated by wavelet analysis of streamflow and suspended sediment load, 1950–1975. *J. Hydrol.* **2009**, *377*, 237–244. [CrossRef]

8. Zhao, G.; Mu, X.; Wen, Z.; Wang, F.; Gao, P. Soil erosion, conservation, and eco-environment changes in the Loess Plateau of China. *Land Degrad. Dev.* **2013**, *24*, 499–510. [CrossRef]

9. Shi, H.; Shao, M.A. Soil and water loss from the Loess Plateau in China. *J. Arid Environ.* **2000**, *45*, 9–20. [CrossRef]

10. Chen, L.D.; Wei, W.; Fu, B.J.; Lv, Y.H. Soil and water conservation on the Loess Plateau in China: Review and perspective. *Prog. Phys. Geogr.* **2007**, *31*, 389–403. [CrossRef]

11. Xu, J.H.; Niu, Y.G. *Influence of Water Conservancy Project on Runoff and Sediment in Coarse Sand Area of the Middle Reaches of the Yellow River*; The Yellow River Water Conservancy: Zhengzhou, China, 2000; p. 6.

12. Liu, H.B.; Wang, G.Q.; Jian, H.R.; Wang, M.M. Research on the hydrological effects of soil and water conservation in the Qingjianhe River Basin of Loess Plateau. *J. Water Resour. Water Eng.* **2009**, *20*, 7–11.

13. Kang, L.L.; Wei, Y.C.; Zhang, S.L.; Liu, X.Q. Macroscopic analysis of runoff utilization for soil and water conservation in Loess Plateau. *J. Water Resour. Water Eng.* **2010**, *21*, 108–112.

14. Zhu, C.G. Soil and water conservation and ecological environment construction on the Loess Plateau. *Sci. Technol. Innov.* **2018**, *17*, 90–91.

15. Guo, Q.L.; Su, N.; Yang, Y.S.; Li, J.L.; Wang, X.Y. Using hydrological simulation to identify contribution of coal mining to runoff change in the Kuye river basin, China. *Water Resour.* **2017**, *44*, 586–594. [CrossRef]

16. Chiew, F.H.S.; Fu, G.B.; Post, D.A.; Zhang, Y.Q.; Wang, B.; Viney, N.R. Impact of coal resource development on streamflow characteristics: Influence of climate variability and climate change. *Water* **2018**, *10*, 1161. [CrossRef]

17. Pan, G. Impact Studies on the Mechanism and Quantitative Model of Coal Mining on River Runoff in Gujiao City of Shanxi Province. Master's Thesis, Zhengzhou University, Zhengzhou, China, 2015; p. 2.

18. Li, Q.Y.; Cai, Q.; Fang, H.Y. Contribution characteristics of wind erosion to the sediment yield in the Kuyehe River Watershed at time scales. *J. Nat. Resour.* **2011**, *26*, 674–682.

19. Luo, T.; Wang, W.L.; Wang, Z.; Jin, J. Experiment of water runoff and sediment yield on the disturbed lands in Shenfu-Dongsheng coalfield development and construction. *J. Northwest For. Univ.* **2011**, *26*, 59–63.

20. Su, H.; Kang, W.D.; Cao, Z.Z.; Zhu, L. Analysis on precipitation and runoff changing trend from 1954 to 2009 in Kuye River Basin. *Ground Water* **2013**, *35*, 14–17.

21. Guo, Q.L.; Xiong, X.Z.; Hao, B.; Bai, L. Variation trends of seasonal runoff distribution in Kuyehe basin over the past 50 years. *J. Arid Land Resour. Environ.* **2014**, *28*, 35–40.

22. Zhao, X.K.; Wang, S.J. Analysis on the flood characteristics and its change trend in the Kuye River Basin. *J. Arid Land Resour. Environ.* **2012**, *26*, 92–96.

23. Liu, E.J.; Zhang, X.P.; Zhang, J.J.; Lei, Y.N.; Xie, M.L. Variation of annual streamflow and the effect of human activity in the Kuye River during 1956 to 2005. *J. Nat. Resour.* **2013**, *28*, 1159–1168.

24. Bai, L.; Li, H.E.; He, H.M. Analysis on detection and attribution of runoff change in Kuye River Basin. *J. Hydroelectr. Eng.* **2015**, *34*, 15–22.

25. Li, Z.; Liu, W.Z.; Zhang, X.C.; Zheng, F.L. Impacts of land use change and climate variability on hydrology in an agricultural catchment on the Loess Plateau of China. *J. Hydrol.* **2009**, *377*, 35–42. [CrossRef]

26. Wang, M.L.; Guo, S.L. A primary analysis of runoff regime in the Yellow River Basin based on monthly water balance model. *Water Resour. Water Eng.* **1999**, *10*, 1–6.

27. Xing, X.P.; Zhang, W.Z.; Tang, F.F.; Liu, S.S.; Niu, H.J. Hydrological simulation of water balance model in the Kuyehe River catchment. *J. Water Resour. Water Eng.* **2012**, *23*, 73–78.

28. Guo, Q.L.; Han, Z.Y.; Yang, L.J.; Xiong, X.Z. Hydrological simulation of impacts of coal mining on surface runoff in Kuye River. *Adv. Sci. Technol. Water Resour.* **2015**, *35*, 19–22.

29. Cheng, L.; Xu, Z.X.; Luo, R.; Mi, Y.J. SWAT application in arid and semi-arid region: A case study in the Kuye River Basin. *Geogr. Res.* **2009**, *28*, 65–73.

30. Li, S.; Chen, Y.; Li, Z.J.; Yang, F.L. Study of coal mining disturbance to simulated monthly runoff values of Kuye River. *Yellow River* **2016**, *38*, 13–17.

31. Liang, K.; Liu, C.M.; Liu, X.M.; Song, X.F. Impacts of climate variability and human activity on streamflow decrease in a sediment concentrated region in the Middle Yellow River. *Stoch. Env. Res. Risk Assess* **2013**, *27*, 1741–1749. [CrossRef]

32. Sui, J.Y.; He, Y.; Liu, C. Changes in sediment transport in the Kuyeriver in the loess plateau in China. *Int. J. Sediment Res.* **2009**, *24*, 201–213. [CrossRef]

33. Wang, G.Q.; Zhang, J.J.; Li, Y.; Liu, C.S.; Bao, Z.X.; Jin, J.L. Analysis of runoff evolution and factor of driving force in Kuye river catchment. *J. Water Resour. Water Eng.* **2014**, *25*, 7–11.

34. Wu, X.J.; Li, H.E.; Dong, Y.; Liu, T.L. Quantitative identification of coal mining and other human activities on river runoff in northern Shaanxi region. *Acta Sci. Circumst.* **2014**, *34*, 772–780.

35. Wang, L.; Wei, S.P.; Wang, Q.J. Effect of coal exploitation on groundwater and vegetation in the Yushenfu coal mine. *J. China Coal Soc.* **2008**, *33*, 1408–1413.

36. Hamed, K.H.; Rao, A.R. A modified Mann-Kendall trend test for autocorrelated data. *J. Hydrol.* **1998**, *204*, 182–196. [CrossRef]

37. Pettitt, A.N. A nonparametric approach to the change-point problem. *Appl. Stat.* **1979**, *28*, 126–135. [CrossRef]

38. Ran, D.C.; Zuo, Z.G.; Wu, Y.H.; Li, X.M.; Li, Z.H. *Response of Water and Sediment to Human Activities Changes in the Middle Reaches of the Yellow River*; Science Press: Beijing, China, 2012; p. 125.

39. Thornthwaite, C.W. An approach toward a rational classification of climate. *Geogr. Rev.* **1948**, *38*, 55–94. [CrossRef]

40. McCabe, G.J.; Markstrom, S.L. *A Monthly Water-Balance Model Driven by a Graphical User Interface*; Open-File Report 2007-1088; U.S. Geological Survey: Reston, VA, USA, 2007; p. 6.

41. Jiang, X.H.; Gu, X.W.; He, H.M. The influence of coalmining on water resources in the Kuye River Basin. *J. Nat. Resour.* **2010**, *25*, 300–307.

42. Wu, L.H.; Wang, S.J.; Bai, X.Y.; Luo, W.J.; Tian, Y.C.; Zeng, C.; Luo, G.J.; He, S.Y. Quantitative assessment of the impacts of climate change and human activities on runoff change in a typical karst watershed, SW China. *Sci. Total Environ.* **2017**, *601–602*, 1449–1465. [CrossRef] [PubMed]

43. Alemayehu, M.; Woldeamlak, B.; Saskia, K.; Leo, S. Searching for evidence of changes in extreme rainfall indices in the Central Rift Valley of Ethiopia. *Theor. Appl. Climatol.* **2016**, 1–15. [CrossRef]

44. Rares, H.C.Z.; Saskia, K.; Zahra, K. The impact of political, socio-economic and cultural factors on implementing environment friendly techniques for sustainable land management and climate change mitigation in Romania. *Sci. Total Environ.* **2019**, *654*, 418–429.

45. Hasan, E.; Tarhule, A.; Kirstetter, P.; Clark, R.; Yang, H. Runoff sensitivity to climate change in the Nile River Basin. *J. Hydrol.* **2018**, *561*, 312–321. [CrossRef]

46. Wang, X.B.; He, K.N.; Dong, Z. Effects of climate change and human activities on runoff in the Beichuan River Basin in the northeastern Tibetan Plateau, China. *CATENA* **2019**, *176*, 81–93. [CrossRef]

47. Li, Y.Z.; Liu, C.M.; Yu, W.J.; Tian, D.; Bai, P. Response of streamflow to environmental changes: A Budyko-type analysis based on 144 river basins over China. *Sci. Total Environ.* **2019**, *664*, 824–833. [CrossRef]

48. Zhao, G.J.; Tian, P.; Mu, X.M.; Jiao, J.Y.; Wang, F.; Gao, P. Quantifying the impact of climate variability and human activities on streamflow in the middle reaches of the Yellow River basin, China. *J. Hydrol.* **2014**, *519*, 387–398. [CrossRef]

49. Lyu, X.; Wang, S.M.; Yang, Z.Y.; Bian, H.Y.; Liu, Y. Influence of coal mining on water resources: A case study in Kuye river basin. *Coal Geol. Explor.* **2014**, *42*, 54–57.

50. Xiong, M.Q.; Sun, R.H.; Chen, L.D. Effects of soil conservation techniques on water erosion control: A global analysis. *Sci. Total Environ.* **2018**, *645*, 753–760. [CrossRef]

51. Artemi, C.; Jesus, R.C.; Antonio, G.M.; Saskia, K. Hydrological and erosional impact and farmer's perception on catch crops and weeds in citrus organic farming in Canyoles river watershed, Eastern Spain. *Agric. Ecosyst. Environ.* **2018**, 1–11. [CrossRef]

52. Keesstra, S.D.; Rodrigo, C.J.; Novara, A.; Gimenez, M.A.; Pulido, M.; Di Prima, S.; Cerda, A. Straw mulch as a sustainable solution to decrease runoff and erosion in glyphosate-treated clementine plantations in Eastern Spain. An assessment using rainfall simulation experiments. *CATENA* **2018**. [CrossRef]

53. Melaku, N.D.; Renschler, G.S.; Flagler, J.; Bayu, W. Integrated impact assessment of soil and water conservation structures on runoff and sediment yield through measurements and modeling in the Northern Ethiopian highlands. *CATENA* **2018**, *169*, 140–150. [CrossRef]

54. Jiang, C.; Zhang, H.Y.; Wang, X.C.; Feng, Y.Q.; Labzovskii, L. Challenging the land degradation in China's Loess Plateau: Benefits, limitations, sustainability, and adaptive strategies of soil and water conservation. *Ecol. Eng.* **2019**, *127*, 135–150. [CrossRef]

55. Ren, Z.P.; Ma, Y.Y.; Wang, Y.S.; Xie, M.Y.; Li, P. Runoff changes and attribution analysis in tributaries of different geomorphic regions in Wuding River under ecological construction. *Acta Ecol. Sin.* **2019**, *39*, 1–10.

56. Li, Z.L. Effects of Environmental Changes on Runoff Situation in the Tuwei River Basin. Master's Thesis, Xi'an University of Technology, Xi'an, China, 2018.

57. Jiao, Y.; Lei, H.M.; Yang, D.W.; Yang, H.B. Attribution of discharge changes over Wuding River watershed using a distributed eco-hydrological model. *J. Hydroelectr. Eng.* **2017**, *36*, 34–44.

Article

A Statistical–Distributed Model of Average Annual Runoff for Water Resources Assessment in DPR Korea

Tongho Ri [1], Jiping Jiang [2,3,*], Bellie Sivakumar [4,5] and Tianrui Pang [3,6]

1   Department of Global Environmental Science, Kim Il Sung University, Pyongyang 999093, Korea; jjp_lab@sina.com
2   State Key Laboratory of Urban Water Resource and Environment, Harbin Institute of Technology, Harbin 150090, China
3   Shenzhen Municipal Engineering Lab of Environmental IoT Technologies, School of Environmental Science and Engineering, Southern University of Science and Technology, Shenzhen 518055, China; nicolaser941225@gmail.com
4   School of Civil and Environmental Engineering, University of New South Wales, Sydney, NSW 2052, Australia; s.bellie@unsw.edu.au
5   State Key Laboratory of Hydroscience and Engineering, Tsinghua University, Beijing 100084, China
6   School of Environment, Harbin Institute of Technology, Harbin 150090, China
*   Correspondence: jiangjp@sustech.edu.cn

Received: 26 March 2019; Accepted: 6 May 2019; Published: 8 May 2019

Abstract: Water resource management is critical for the economic development of the Democratic People's Republic of Korea (DPRK), where runoff plays a central role. However, long and continuous runoff data at required spatial and temporal scales are generally not available in many regions in DPRK, the same as in many countries around the world. A common practice to fill the gaps is to use some kind of interpolation or data-infilling methods. In this study, the gaps in annual runoff data were filled using a distributed runoff map. A novel statistical–distributed model of average annual runoff was derived from 50 years' observation on 200 meteorological observation stations in DPRK, considering the influence of climatic factors. Using principal component analysis, correlation analysis and residual error analysis, average annual precipitation, average annual precipitation intensity, average annual air temperature, and hot seasonal air temperature were selected as major factors affecting average annual runoff formation. Based on the water balance equation and assumptions, the empirical relationship for runoff depth and impact factors was established and calibrated. The proposed empirical model was successfully verified by 93 gauged stations. The cartography of the average annual runoff map was automatically implemented in ArcGIS. A case study on the Tumen River Basin illustrated the applicability of the proposed model. This model has been widely used for the development and management of water resources by water-related institutes and design agencies in DPRK. The limitation of the proposed model and future works are also discussed, especially the impacts of climate changes and topology changes and the combination with the physical process of runoff formation.

Keywords: average annual runoff; runoff map; hydrological model; GIS; DPR Korea

---

1. Introduction

Research regarding water resources estimation at regional and continental level contributes a lot in establishing water resources management policy [1–3], and water resources assessment is the first step for water resources development. Runoff plays a central role in water resources assessment. Generally, water resources are evaluated by means of average annual runoff [4,5], the mathematical expectation of multiyear observations of annual runoff, and can be described as runoff depth [4,6–12].

Usually, in hydrologic modeling, a water balance equation is used for the computation of average annual runoff, keeping in mind the factors affecting annual runoff [5,9,13–19].

Usually, average annual runoff is computed by means of multiyear data, while in some cases, from one-year data. However, long and continuous runoff data at required spatial and temporal scales are generally not available in many regions around the world, due to the costs involved in measurements, difficulty in accessing the locations of interest, and malfunctioning of the measurement devices, among others. A common practice to fill the gaps is to use some kind of interpolation or data-infilling methods. Many average annual runoff models have been reported for water resources development in regions which lack observational data [1,4,8,12,20–22]. The accuracy of the average annual runoff model mainly depends on the factors affecting annual runoff [23–27], which can be computed by principal component analysis and the factor analysis method [11,28,29].

Research has confirmed that the factors affecting runoff formation vary with locality [17]. It is well accepted that precipitation is the main meteorological factor for runoff formation [22]. The physical characteristics of the watershed underlying surface also play an important role, which includes land use, soil type, slope, vegetation, etc. [30–32]. Human activities like agriculture irrigation, urban water supply, and drainage, water division projects inevitably lead to changes in water resources.

Recently, lots of hydrological progress has been observed in modeling spatial variation of precipitation, infiltration, and evaporation with the advancement in 3S technology (GIS—geographic information system, RS—remote sensing, and GPS—global position system) [33] and computer information processing [5,34–38]. It is highly practiced in those regions for runoff estimation which lacks observational data, i.e., ungauged areas, for evaluating water resources.

Runoff maps are frequently used for highlighting spatiotemporal changes in water resources [10,20,39–42], while maps of meteorology factors are used for interpreting spatiotemporal changes in precipitation [13], evaporation [5], catchment classification, estimation of hydrologic response in ungauged catchments [10,12], etc.

In general, average annual runoff is computed by methods based on the geometric center of the river basin. Such methods, however, often fail to provide reliable runoff maps for small- and medium-sized rivers. Since hydrologic models take into account the local natural geographic, climatic, hydrologic characteristics and local data, such models are very useful for generating runoff maps in small and medium-sized rivers.

This study was rooted in the Democratic People's Republic (DPR) of Korea, where very limited hydrology studies were reported in the literature. Scientific water resources management is critical for the recent economic reformation and opening up of DPR Korea. Runoff-based water resources assessment on the whole nation is significant. However, due to the data limitation, advanced hydrology models with high resolution are not applicable in DPR Korea. An empirical statistical approach to produced distributed runoff results is preferred, especially for ungauged areas. Therefore, we attempted to address these issues in the present study. Making good use of 50 years of records in 200 meteorological stations, a statistical–distributed average annual runoff model was developed in this work.

The rest of this paper is organized as follows: In Section 2, the climate condition and the water balance relationship in DPR Korea is described. The factors affecting the average annual runoff are analyzed by principal component analysis (PCA) and explained. In Section 3, the new statistical annual runoff model is clearly described. Furthermore, model verification and cartography of the runoff map in GIS are presented. Section 4 presents the analysis and results for the Tumen River Basin. Sections 5 and 6 discuss and conclude the research.

2. Rainfall–Runoff Relationship and Runoff Impact Factors

2.1. The Precipitation and Temperature Characteristics in DPR Korea

DPR Korea has a temperate monsoon climate, four distinct seasons, average annual temperature varies from 8 °C to 12 °C, and average annual precipitation varies from 1000 to 1200 mm. Most of the precipitation falls in July–August (see Figures 1 and 2). Average annual monthly temperature and precipitation characteristics are shown in Table 1.

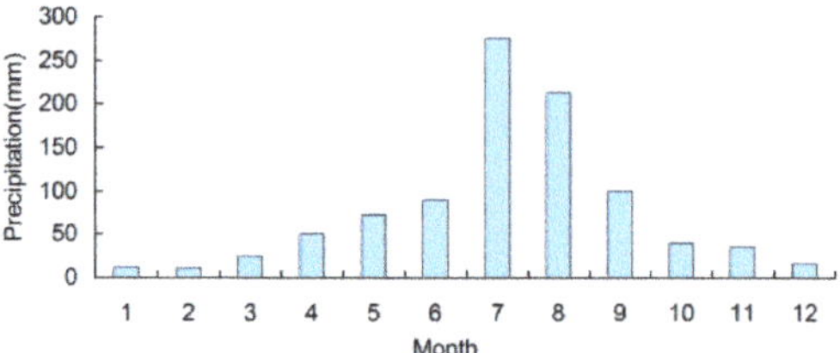

Figure 1. The monthly distribution of precipitation in the Democratic People's Republic (DPR) of Korea.

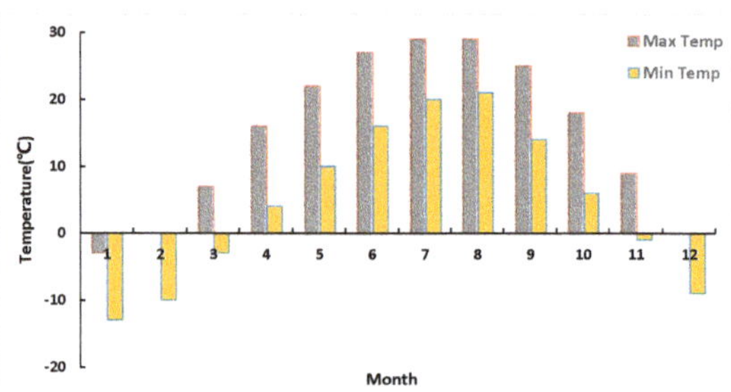

Figure 2. The monthly distribution of temperature in DPR Korea.

Table 1. The average annual temperature and precipitation in DPR Korea.

Month	Jan	Feb	Mar	Apr	May	Jun	Jul	Aug	Sep	Oct	Nov	Dec
Average daily maximum temperatures (°C)	−1	2	9	17	23	27	29	29	25	18	9	2
Average daily minimum temperature (°C)	−11	−8	−2	5	11	16	21	20	14	7	0	−7
Average precipitation amount (mm)	12	11	25	50	72	90	275	213	100	40	35	16
Average precipitation days (d)	5	4	5	7	8	9	14	11	7	6	7	6

2.2. Watershed Water Balance Relationship

Annual runoff is computed using the classical water balance Equation [43]:

$$Y = P - E \pm \Delta S \tag{1}$$

where Y is the annual runoff depth (mm); P is the annual amount of precipitation (mm); E is the annual amount of evaporation (mm); and ΔS is the change in water storage.

The change in basin water storage can be ignored, i.e., equal to zero, when considering multiyear average (see Equation (2)):

$$Y = P - E \tag{2}$$

Equation (2) can also be denoted by annual runoff coefficient $\varphi = Y/P$ and annual evaporation coefficient $\eta_E = E/P$.

$$\varphi = 1 - \eta \tag{3}$$

E can be computed indirectly by hydrometeorological observations using the values of P and Y. Calculating the values of E is challenging due to the influences from climatic factors, geographical factors, etc. [44]. Since soil type affects water exchanges with the atmosphere, Reder et al. evaluated the sensitivity of the annual average of runoff, precipitation, evaporation, and deep drainage to different soil types in China and argued the importance of clarifying the role of "geomorphological factors" [30]. Li et al. found that the effect of climate change on evapotranspiration was much more significant than the effect of land use and land cover changes in China [44].

Usually, the average annual runoff model is formulated by establishing a relationship between the parameters and elucidating the influence of factors affecting evaporation E or evaporation coefficient η_E.

In DPR Korea, the correlation coefficient between rainfall and runoff varies from 0.90 to 0.98, which makes the surface water consistent throughout the year. There are regional differences among river runoff, but 60% to 80% runoff flows occur in the summer.

2.3. Factors Affecting the Annual Runoff Formation by PCA

Time series of the main factors affecting runoff formation can be obtained using the principal component analysis method. The basic data matrix can be configured as follows:

$$Y(p \times n) = \begin{bmatrix} y_{11} & y_{12} & \cdots & y_{1n} \\ y_{21} & y_{22} & \cdots & y_{2n} \\ \vdots & \vdots & \cdots & \vdots \\ y_{p1} & y_{p2} & \cdots & y_{pn} \end{bmatrix}, \tag{4}$$

where y is annual runoff; p is the size of observational series; and n is the number of observation stations.

Then, the measured values can be represented as follows:

$$Y(t, x) = T_1(t)x_1(x) + T_2(t)x_2(x) + \cdots T_m(t)x_m(x) \tag{5}$$

Or:

$$Y(t, x) = \sum_{k=1}^{m} T_k(t)X_k(x) \tag{6}$$

where $T_k(t)$ is the coefficient related to the time; $X_k(x)$ is the function characterizing the distribution field; and m is the number of factors affecting average annual runoff formation, $k = 1, 2, \ldots, g, \ldots, m$.

Standardizing basis data and evolving obtains an $n \times n$ correlation matrix. It should satisfy a relationship of $m \le n$. In this case, the principal component analysis model can be obtained as follows:

$$(R - \lambda I)X = 0 \tag{7}$$

where R is correlation matrix and I is a unit matrix.

The characteristic equation is:

$$|R - \lambda I| = 0 \tag{8}$$

Time coefficient matrix is:

$$T_{gz} = \frac{\sum\limits_{z=1}^{m} Y_{tz}X_{gz}}{\sum\limits_{z=1}^{m} X_{gz}^2} \tag{9}$$

where $z = 1, 2, \ldots, m$.

The contribution rate is:

$$\beta = \frac{\lambda_k}{\sum\limits_{k=1}^{m} \lambda_k} \times 100\%, \tag{10}$$

where $\sum_{i=1}^{m} \lambda_i = m$ $(i = 1, 2, \ldots, m)$.

Eigenvalue $\lambda_k (k = 1 \sim m)$, eigenvector matrix $X(m \times n)$, and the time coefficient matrix can be computed using Equations (7)–(9) consecutively. We computed all the main components using the correlation coefficient between the time coefficient matrix and the factor variables.

The meteorological and hydrological observatory is in the countryside in DPR Korea (see Figures 3–5 for the spatial distribution of average annual precipitation and average annual temperature evaluated over the 1961–2010 period, respectively, by interpolation).

Figure 3. Network of meteorological stations (the northern area in DPR Korea).

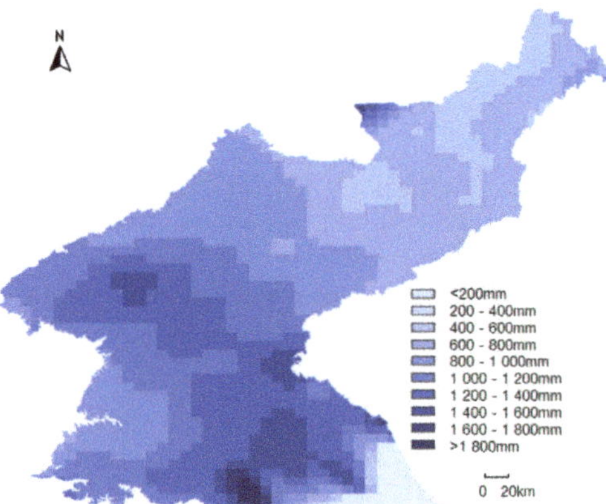

Figure 4. Spatial distribution of average annual precipitation evaluated over the 1961–2010 period (cell size: 10.0 km × 7.1 km).

Table 2 shows the contribution rate and the cumulative contribution rate for the components of factors affecting runoff, calculated using the hydrological observation station data in DPR Korea. The principal component analysis mainly highlights the first four factors. It is obvious that runoff depth in the mentioned regions in Table 2 can be computed using two factors, but for other river basins, at least four factors should be used. Next, the correlation coefficients between time coefficient s and major factors were computed.

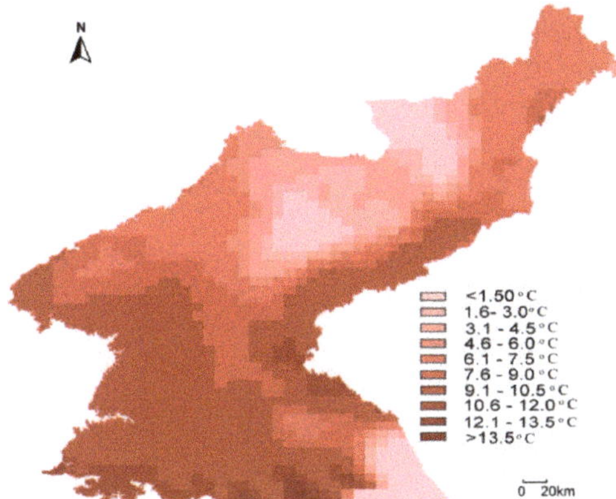

Figure 5. Spatial distribution of average annual temperature evaluated over the 1961–2010 period (cell size: 10.0 km × 7.1 km).

Table 2. The contribution rates for components (%).

No	Region	First Component	Second Component		Third Component		Fourth Component	
				Cumulative		*Cumulative*		*Cumulative*
1	Taedong River Bain	86.6	5.9	92.5	4.1	96.6	1.2	97.8
2	Chongchon River Basin	87.9	6.0	93.9	2.2	96.1	1.8	97.9
3	Ryesong River Basin, Rimjin River Basin	88.1	5.2	93.3	4.8	98.1	1.0	99.1
4	Abrok River Basin	74.9	7.2	82.1	6.4	88.5	3.8	92.3
5	East coast area	66.5	11.1	77.6	8.3	85.9	4.6	89.9
6	Whole DPR Korea	68.5	7.0	75.5	5.2	80.7	4.5	85.2

2.3.1. The Primary Factor: Average Annual Precipitation

Precipitation is obviously the major factor affecting runoff formation. The relationship between precipitation and time coefficient for the first component is computed. Variation in annual precipitation P is considered to be the first factor, and the time coefficient T_1 of the first principal component in the Taedong River Basin is shown in Figures 6 and 7. It is obvious from Figure 6 that the Taedong River Basin has a high correlation of 0.94 between $P(t)$ and $T_1(t)$, while for other river basins, it is 0.78 or more.

It is clear from Table 2 that in the whole area of DPR Korea, the first and second component contribute 68.5% and 7% to the river basin, respectively. It is difficult to identify and directly calculate the second, third, and fourth component. Therefore, they are computed using correlation and regression analysis.

In correlation analysis, the water balance equation is as follows:

$$\left. \begin{array}{l} h = P - z = P - E - u \\ z = E + u = P - h \\ H = h + u = P - E \end{array} \right\} \tag{11}$$

where h is the runoff depth of surface water; z is loss of precipitation; u is underground runoff depth; and H is whole runoff depth, including surface water and underground water.

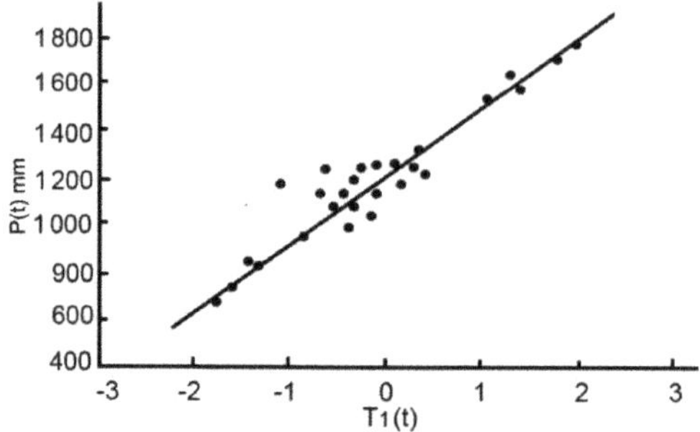

Figure 6. The relationship between $P(t)$ and $T_1(t)$ in the Taedong River Basin.

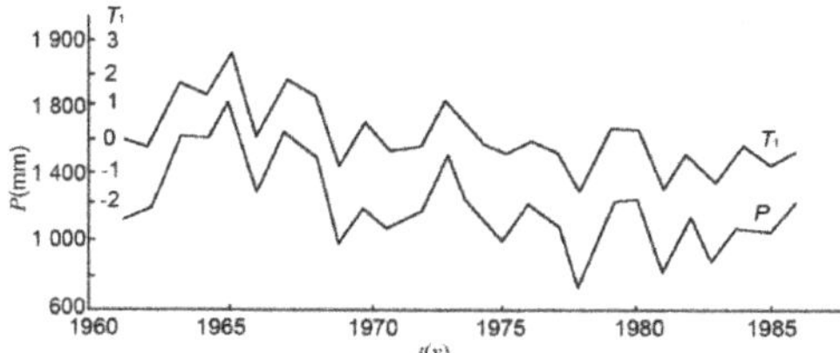

Figure 7. The change curves of $P(t)$ and $T_1(t)$ in the Taedong River Basin.

2.3.2. The Relationship between Average Annual Losses and Average Annual Air Temperature

The second factor can be computed indirectly using a water balance equation, considering annual loss as the effect of precipitation (the first factor) removed up to some level. Generally, most of the annual precipitation is lost by evaporation. Seepage losses are usually ignored while using the water balance relationship for the computation of average annual runoff.

As shown in Figure 8, high correlation exists between average annual precipitation losses z and average annual temperature T in Taedong River. The high correlation coefficient of 0.64 between z and T also denotes that groundwater movement is relatively active in the Taedong River Basin. It is worth noting that there is no linear relationship between z and T, as shown by the lower bound line (dotted line) of distributed dots. The average annual temperature is the primary factor, while the air temperature is the secondary factor affecting evaporation losses.

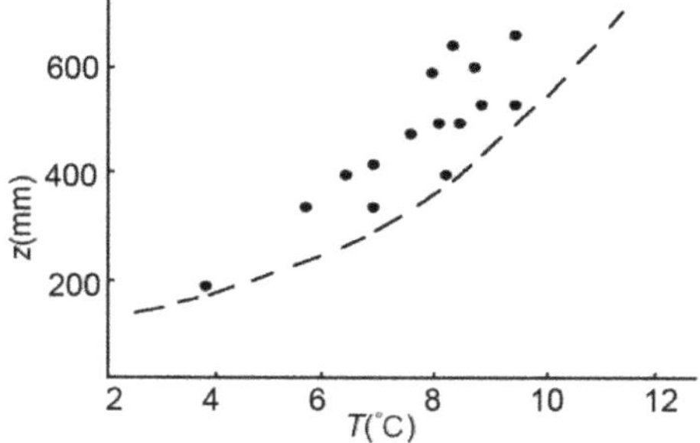

Figure 8. The relationship between the average annual losses z and the average annual temperature T (Taedong River).

2.3.3. The Relationship between Average Annual Losses and Average Annual Precipitation Intensity

Compute evaporation losses by primary and secondary factors affecting evaporation losses and then find out the residue for average annual loss as shown below:

$$\Delta z_1 = z - z' = z - f(P, T) \tag{12}$$

where z' is the calculated evaporation loss.

The residue Δz_1 is high in some areas (northern inland and northern area of the east coast of DPR Korea), as the remaining factors affecting evaporation losses were not considered.

Annual precipitation intensity is defined as:

$$I = \frac{P}{N_p} \tag{13}$$

The residual coefficient is defined as:

$$\eta_{\Delta z_1} = \frac{\Delta z_1}{P} \tag{14}$$

where N_p is the number of annual precipitation days.

The correlation coefficient between I and $\eta_{\Delta z_1}$ is -0.51, which denotes their inverse relationship (Figure 9). It was stated earlier that evaporation loss has a nonlinear relationship with precipitation intensity, which shows that the number of annual precipitation days (annual precipitation intensity) is one of the major factors affecting the annual runoff formation.

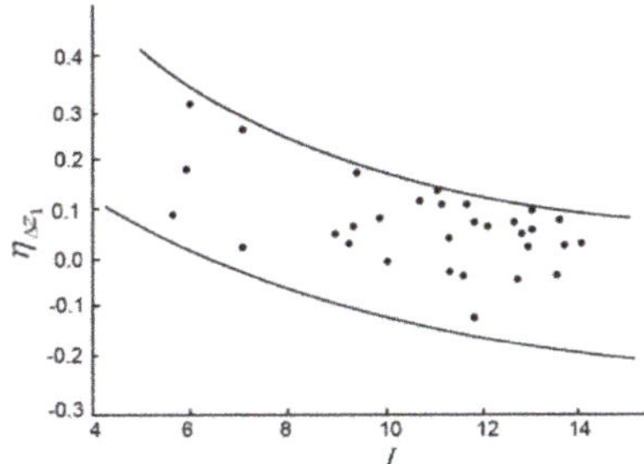

Figure 9. The relation between I and $\eta_{\Delta z_1}$ in DPR Korea.

2.3.4. The Relationship between Continuous Residue and Air Temperature of the Hot Season

Continuous residue (Δz_2) and its coefficient for annual losses can be written as follows:

$$\Delta z_2 = z - f(P, T, I) \tag{15}$$

$$\eta_{\Delta z_1} = \frac{\Delta z_2}{P} \tag{16}$$

It is obvious from the computation that the average range of variation in $\eta_{\Delta z_2}$ is less, compared to $\eta_{\Delta z_1}$, but some new factors affecting $\eta_{\Delta z_2}$ have a large value which cannot be ignored. The evaporation rate is higher in summer compared to winter owing to air temperature. Therefore, the average annual and air temperature of the summer should be considered. The difference in air temperature is given as follows:

$$\Delta t = T' - T \tag{17}$$

where T' is average air temperature from May to October and T is the average annual air temperature.

The relationship between the difference in air temperature (Δt) and the continuous residue coefficient ($\eta_{\Delta z_2}$) is shown in Figure 10. The correlation coefficient between Δt and $\eta_{\Delta z_2}$ is 0.74.

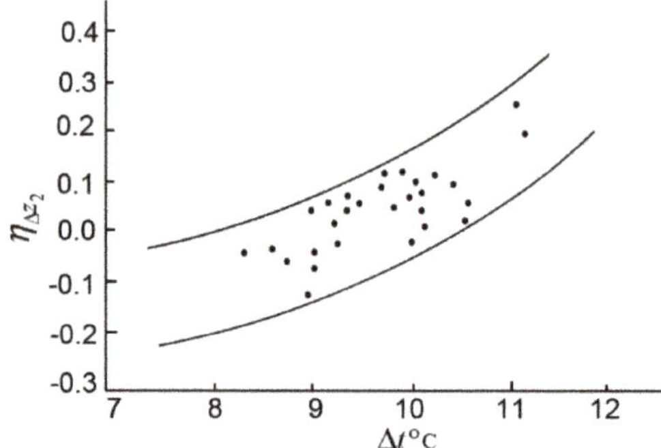

Figure 10. The relationship between Δt and $\eta_{\Delta z_2}$ in DPR Korea.

The air temperature of the hot season (or the difference of air temperature) is one of the main factors affecting annual runoff [4,45]. The difference (or range of variation) between the upper boundary and the lower boundary indicated by a broken line in Figures 9 and 10 is as follows:

$$\eta_{\Delta z_1}(I) = \left[\eta_{\Delta z_1}(I)\right]_{up} - \left[\eta_{\Delta z_1}(I)\right]_{down} \approx 0.30$$

$$\eta_{\Delta z_2}(\Delta t) = \left[\eta_{\Delta z_2}(\Delta t)\right]_{up} - \left[\eta_{\Delta z_2}(\Delta t)\right]_{down} \approx 0.20$$

The functional relationship of $\eta_{\Delta z_1}(I)$ and $\eta_{\Delta z_2}(\Delta t)$ is illuminated, as it causes 30% and 20% of the annual precipitation, which leads to the error reduction in annual runoff and/or annual losses computation. It is worth noting that the number of annual precipitation days, the air temperature of the hot season, average annual air temperature, and annual precipitation are the main factors in runoff formation. This can be expressed by the general function, which is given below:

$$h = f(P, T, I, \Delta t) \tag{18}$$

3. Development of an Empirical Average Annual Runoff Model

3.1. Model Development and Description

The general formula can be obtained from the water balance equation and runoff formation factors stated above:

$$\left.\begin{aligned} E &= f(P, T, I, \Delta t) \\ z &= f(P, T, I, \Delta t, \alpha) \end{aligned}\right\}. \tag{19}$$

Therefore:

$$\left.\begin{aligned} H &= P - E = f(P, T, I, \Delta t) \\ h &= P - z = f(P, T, I, \Delta t, \alpha) \end{aligned}\right\} \tag{20}$$

where α is a set of variables affecting the underground runoff.

Whole runoff coefficient φ_H and surface runoff coefficient φ_h can be written as follows:

$$\left.\begin{aligned} \varphi_H &= 1 - \eta_E = f(P, T, I, \Delta t) \\ \varphi_h &= 1 - \eta_z = f(P, T, I, \Delta t, \alpha) \end{aligned}\right\} \tag{21}$$

$E \leq z$ and $\eta_E \leq \eta_z$ because most of the losses (z) are caused by evaporation (E). Only E or η_E can be computed by the mathematical method using the observational data and the factors (P, T, and Δt) studied above.

Let us consider the relationship between the loss coefficient η_z, precipitation p, and air temperature t. Draw a curve for the west area of the river basin. The relationship is given as follows:

$$\eta_{Etp} = k_p \eta_{Et} \tag{22}$$

where:

$$k_P = \begin{cases} \left(\frac{P}{1000}\right)^{-1.16} (P > 1000 \text{ mm}) \\ \left(\frac{P}{1000}\right)^{-0.72} (P \leq 1000 \text{ mm}) \end{cases} \tag{23}$$

$$\eta_{Et} = \begin{cases} 0.218e^{0.109T} (T > 11\ ^\circ\text{C}) \\ 0.072e^{0.210t} (11^\circ\text{C} \geq T > 4\ ^\circ\text{C}) \\ 0.110e^{0.104t} (T \leq 4\ ^\circ\text{C}) \end{cases} \tag{24}$$

where η_{Etp} is the evaporation loss coefficient defined by the precipitation and air temperature, and k_P is the influence coefficient of annual precipitation.

As shown in Figure 11, some dots are remarkably deflected from the curve drawn by Equation (22). Quantities are computed to characterize the deflected degree for each region, which is given below:

$$\eta_{\Delta z} = \eta_z - \eta_{Etp} \tag{25}$$

$$k'_I = \frac{\eta_z}{\eta_{Etp}} \tag{26}$$

where $\eta_{\Delta z_1}$ is the first residual coefficient of losses, and k'_I is the proportionality coefficient which characterizes the deflected degree between η_z and η_{Etp}.

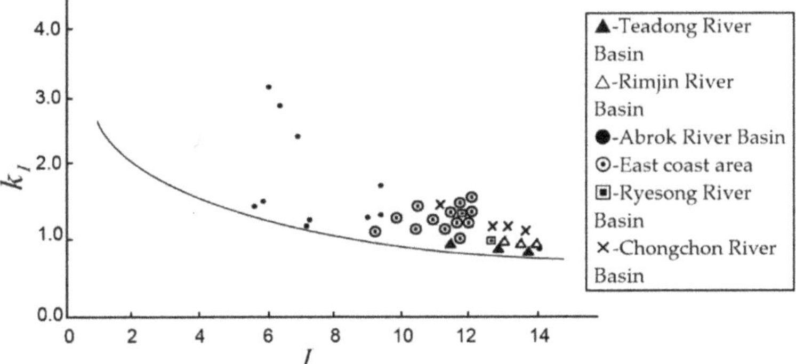

Figure 11. Scatter plot of k'_I vs. I and the relationship curve of $k_I = f(I)$.

The factor defining the first residue Δz_1 and the coefficient $\eta_{\Delta z_1}$ of annual precipitation intensity I have already been investigated. The relationship between k'_I and I is shown in Figure 11, where the $k_I = f(I)$ curve is drawn for the lower boundary condition, which estimates the underground runoff component in the future. Then, the equation of the relationship curve is given below:

$$k_I = 2.98I^{-0.39} \tag{27}$$

where k_I is the influence coefficient of annual precipitation intensity.

The deflection degree of dots from the curve can be computed as follows:

$$\eta_{EtpI} = k_I \eta_{Etp} \tag{28}$$

$$\eta_{\Delta zI} = \eta_z - \eta_{EtpI} \tag{29}$$

$$k'_{\Delta t} = \frac{\eta_z}{\eta_{EtpI}} \tag{30}$$

where η_{EtpI} is the evaporation coefficient calculated by the air temperature, precipitation, and number of annual precipitation days.

Continuous residue Δz_2 or coefficient $\eta_{\Delta z_2}$ is related to the difference between the average temperature T' of the hot season (May to October) and the average annual temperature T. Therefore, the relationship between coefficient $k'_{\Delta t}$ and Δt is analyzed as shown in Figure 12.

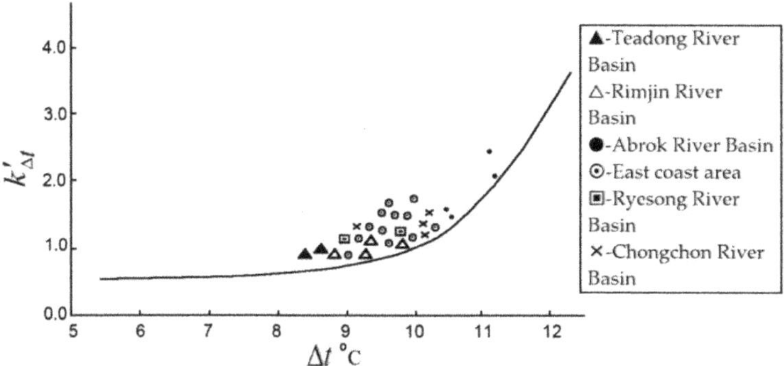

Figure 12. Scatter plot of $k'_{\Delta t}$ vs. Δt and the relationship curve of $k_{\Delta t} = f(\Delta t)$.

The lower bound line of the relationship defines the lost quantity due to the influence of factors $(P, T, I, \Delta t)$, as mentioned earlier.

The equation of the relationship curve can be written as follows:

$$k_{\Delta t} = \begin{cases} 0.089 e^{0.250\Delta t} & (\Delta t > 10\ ^\circ\mathrm{C}) \\ 0.1973 e^{0.1694\Delta t} & (\Delta t \le 10\ ^\circ\mathrm{C}) \end{cases} \tag{31}$$

The evaporation loss coefficient defined by P, T, I, and Δt can be represented as follows:

$$\eta_{EtpI\Delta t} = \eta_{Et} k_p k_I k_{\Delta t} \tag{32}$$

The third residue Δz_3 and coefficient $\eta_{\Delta z_3}$ can be calculated as follows:

$$\eta_{\Delta z_3} = \eta_z - \eta_{EtpI\Delta t} \tag{33}$$

$$\Delta z_3 = z - P\eta_{EtpI\Delta t} \tag{34}$$

It is worth noting that the values of Δz_3 and $\eta_{\Delta z_3}$ were found to be very small or close to zero except for some special regions. There was a small difference in the computed values for some regions (west coast zone, northern inland, and east coast zone), while a big difference for other regions (Taedong River Basin). It is reasonable to consider Δz_3 as the loss constituent by the underground runoff rather than evaporation loss by the climate factors. If Δz_3 is the loss component of the underground runoff, then the evaporation coefficient η_E and the evaporation can be represented as follows:

$$\eta_E = \eta_{Etp I \Delta t} = \eta_{Et} k_p k_I k_{\Delta t} \tag{35}$$

$$E = P \eta_{Et} k_p k_I k_{\Delta t} \tag{36}$$

Whole runoff coefficient φ_H and whole runoff depth can be represented as follows:

$$\varphi_H = 1 - \eta_{Et} k_p k_I k_{\Delta t} \tag{37}$$

$$H = P\left(1 - \eta_{Et} k_p k_I k_{\Delta t}\right) \tag{38}$$

Equations (37) and (38) are just the average annual runoff models.

This model has high accuracy as well as logical validity, universality, and objectivity that can be applied to all river basins in DPR Korea. Underground runoff can be computed from surface runoff using this model.

The average annual runoff model has some characteristics as follows. The first model considers all the influencing factors of average annual runoff formation and computes them. This model is based on factor analysis, which takes into account annual average precipitation P, air temperature t, annual precipitation intensity I, and hot season temperature t_0. This model is consistent and has logical validity.

This model is consistent with the physical nature of the natural phenomenon. If $P \rightarrow \infty$, then $k_p \rightarrow 0$, and finally $\eta_E \rightarrow 0$ and $\varphi \rightarrow 1$. Moreover, when precipitation is reduced to zero, i.e., $P \rightarrow 0$, precipitation P_a, called a lower limit precipitation as greater than zero at the moment that runoff reaches to zero, i.e., $Y = 0$ and/or $\phi = 0$, is present, i.e., $P_a > 0$, and P_a is determined by temperature T, precipitation intensity I, and temperature difference Δt.

This model demonstrates that when $P = 0$, not only $\eta_E = 1$ or $\varphi = 0$, but also $\eta_E = 1$ or $\varphi = 0$ subjected to $P = P_a > 0$ according to the values of T, I, and Δt. This model also considers the regional distribution characteristics of influencing factors. The annual runoff map of each factor influencing the annual runoff formation should be developed by a spatial interpolation tool before developing the annual runoff map. All those factors influencing the average annual runoff formation are computed through the Kriging interpolation method [46] (see Figures 4 and 5).

The interpolation accuracy of factor fields was high at a grid cell size 10×10 km, keeping in mind the density of the meteorological observation network in DPR Korea. The model using values of grid cell type is treated linearly values on the same grid cell. Therefore, grid cell size greatly affects the calculation accuracy of models.

This study, to correctly determine grid cell size, analyzed the spatial change of climatic factors according to the distance between observation stations for 200 stations, and the analysis results showed that values of all climatic factors linearly changed within 10~20 km. Additionally, we estimated the interpolation error by calculating values and observed values for grid cell sizes of 10 km, 15 km, and 20 km, respectively. The results as an example of average annual precipitation and average annual air temperature show that their relative errors are within 7~18% in a flat area and mountains for the case of more than 10 km, and within 0.2% for the case of 10 km. The study used a grid cell size of 10 km based on the above research.

Water resources can be exactly evaluated using the average annual runoff model based on the regional distribution of factors affecting the annual runoff formation. Therefore, an annual runoff map of grid cell type can be developed using this model.

3.2. Model Verification on Guauged Area

The proposed average annual runoff model was verified by relative errors between computed values and observed values of runoff depth in 93 gauged sites in DPR Korea. Table 3 shows the number of sites where the relative error is less than 5%; among the 93 sites, this applies to 83, meaning 89.25% of all sites, and all sites have a relative error of less than 10%. This denotes the usability of the empirical model in the whole area of DPR Korea.

Table 3. Relative errors of calculated runoff depth.

Statistics	Relative Error (%)		
	≤3.0	3.1~5.0	5.1~10.0
The number of sites	62	21	10
Rate occupied sites (%)	66.67	22.58	10.75
Accumulated number of sites	62	83	93
Accumulated rate of sites (%)	66.67	89.25	100

3.3. Cartography of Average Annual Runoff Map

Initially, 200 observation station data were interpolated from 1961 to 2010 using the Kriging interpolation technique for the factors (precipitation P, temperature, T, precipitation intensity I, and temperature difference Δt) affecting annual runoff formation using ArcGIS 9.2 (see Figures 4 and 5). Average annual runoff depth and average annual runoff modulus for the node points of the grid cells was computed using the average annual runoff model. Then, the values of the grid cell center were computed. Using Equation (38), the value of average annual runoff depth in each node points is computed as follows:

$$y_{i,i} = P_{i,j}\left(1 - \eta_{Et(i,j)} \times k_{p(i,j)} \times k_{I(i,j)} \times k_{\Delta t(i,j)}\right) \tag{39}$$

where:

$\eta_{Et(i,j)}$ is the influence coefficients of average annual temperature in grid cell i, j;

$k_{p(i,j)}$ is average annual precipitation in grid cell i, j;

$k_{I(i,j)}$ is average annual precipitation intensity in grid cell i, j;

$k_{\Delta t(i,j)}$ is the temperature of hot season affecting the annual evaporation in grid cell i, j.

A value of any grid cell center point is computed as the mean value of its four nodal points (Equation (40)).

$$Y_{i,j} = (y_a + y_b + y_c + y_d)/4 \tag{40}$$

The central values were updated, which leads to the completion of the average annual runoff map.

Water resources information can be easily obtained from the ungauged region using the runoff map developed by GIS spatial analysis tools. Average annual runoff depth can be computed in the upper basin of any river cross section using ArcGIS 9.2 and the runoff map of grid cell type using the weighted average method as follows:

$$Y = \frac{1}{n}\sum_i \sum_j Y_{ij}k_{ij} \tag{41}$$

where Y_{ij} is average annual runoff depth in a grid cell i, j; n is the number of grid cells included in the selected area; and $k_{i,j}$ is the area weight in grid cell i, j, namely:

$$k_{ij} = \frac{\Delta f_{ij}}{\Delta F} \tag{42}$$

where ΔF is the area of a grid cell having the size of 10 km × 10 km, i.e., $\Delta F = 100$ km^2; and Δf_{ij} (km^2) is the component area occupied on the grid cell.

$k_{ij} = 1.0$ and $k_{ij} \leq 1.0$ for completely and partially contained grid cells within the watershed boundary, respectively. Average annual runoff depth can be obtained directly from the average annual runoff map on the river basin, where the basin area is smaller than 100~200 km^2, i.e., $F \leq 100 \sim 200$ km^2. The abovementioned procedure is automatically executed by a spatial analysis tool in GIS.

4. Application on Tumen River Basin

4.1. The Natural Geographic Characteristics of Tumen River Basin

Tumen River is a boundary river passing through the border of DPR Korea, China, and Russia (see Figure 13). The area and length of the Tumen River Basin is 32,920.0 km^2 and 547.8 km, respectively. Tumen River is the second longest, and the basin area is the third largest in DPR Korea. The average annual precipitation of the river basin is below 600 mm. It is smaller than other basins because of the influence of the Hamgyong Mountain Range. Figures 14 and 15 show the map of average annual precipitation and the map of average annual temperature evaluated for 50 years in the Tumen River Basin.

Figure 13. Tumen River Basin (DPR Korea side).

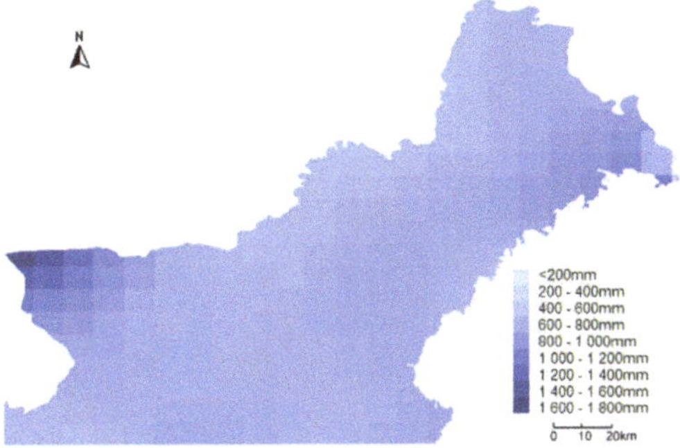

Figure 14. The map of average annual precipitation evaluated for 50 years in the Tumen River Basin (DPR Korea side).

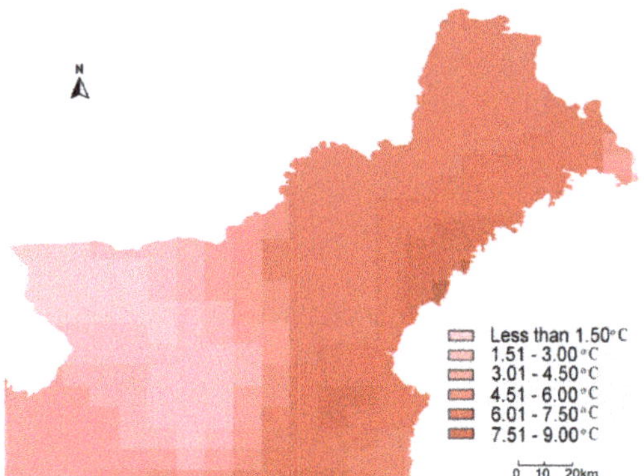

Figure 15. The map of average annual temperature evaluated for 50 years in the Tumen River Basin (DPR Korea side).

4.2. Water Resources of Tumen River Basin (DPR Korea Side)

The proposed average annual runoff model was applied to the Tumen River Basin (DPR Korea side) for water resources computation. Figures 16 and 17 show the map of average annual evaporation and average annual runoff depth of the Tumen River Basin (DPR Korea side), respectively. The mean value of average annual precipitation, average annual evaporation, average annual runoff depth, average annual runoff, and total water resources of the Tumen River Basin are approximately 604 mm, 254 mm, 350 mm, 11 $\ell/s \cdot km^2$, and 3,707,829 $\times 10^3$ m^3, respectively. It is evident from the results that great care is needed to protect water resources for ecological environment protection, establishment and implementation of the strategy for supports development because the water resources of the Tumen River Basin are running short due to exceeding evaporation, which is comparatively greater than in other river basins.

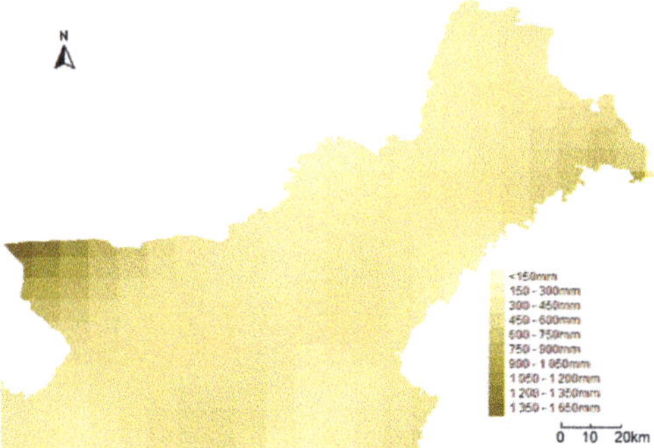

Figure 16. The map of average annual evaporation evaluated for 50 years in the Tumen River Basin (DPR Korea side).

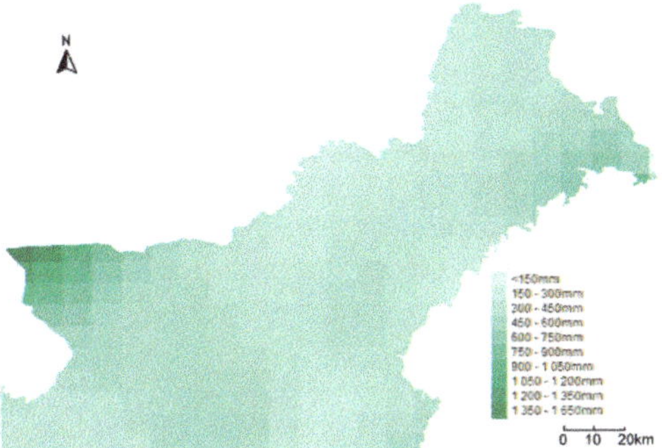

Figure 17. The map of average annual runoff depth evaluated for 50 years in the Tumen River Basin (DPR Korea side).

5. Discussion

At present, it seems we are more interested in modeling runoff at finer resolutions, i.e., daily and subdaily scales, rather than an annual scale. However, such a straightforward empirical model is also a useful management tool. An average annual runoff model was applied to assess the water resources of the river basin and can be used to determine average annual runoff in ungauged basins. This method is widely used for the development and management of water resources by the water-related institutes and the design agencies in DPR Korea.

Water resources can be easily computed in any region by a spatial analysis tool of GIS. The study of the distributed hydrologic model considering the physical, geographical, and meteorological factors influencing the runoff formation is one of the current trends in hydrological studies. Spatial variation in all those factors affecting runoff formation can be easily obtained by emerging technology such as 3S (GIS, RS, GPS). The distributed parameter model is more efficient than the lumped parameter model. This model develops a water resources map for the whole country in a very short time compared to the traditional models. The accuracy of the distributed model partly depends on the interpolation accuracy of the influencing factors in the model. In this study, variation in meteorological factors by topography was not considered. Due to a long length of 50 years of records in the case study of the Tumen River Basin, climate change may be taken into consideration in future work, and this is an important topic on recent water resource management studies [15,18,47–49]. The proposed statistical model can also be used for evaluation of climate change and human activity influence on the water resources of DPR Korea.

The formation of surface runoff–infiltration and Hortonian overland flow [36,50] is disregarded in the modeling process since the spatial–temporal scale in this study is much larger. The in situ process of runoff formation was not described in the model. It may be an interesting topic to be considered in future studies that focus on how to combine the infiltration and Hortonian flow process in our empirical model to improve the model accuracy.

DPR Korea is characterized by a combination of a continental climate and an oceanic climate. The calibrated model in the work can rationally be used on other countries or regions where runoff formation condition is the same as DPR Korea, and the proposed approach has universal applicability, since it is deduced from the basic water balance model. To apply to other countries or regions, it needs to consider different climate, runoff formation condition, and data availability in the local area, and the empirical relationship may differ. Hydrological models should be derived considering the meteorological, natural geographical conditions, and hydrological characteristics of the study area.

Comparison with more complex distributed hydrological models [51] like SWAT is able to support the verification of this new statistical model. However, due to data limitation, this is an issue at this stage.

6. Conclusions

This work developed an empirical hydrological model to provide fundamental information on the principal components of the water balance period in a predefined area over a selected time. It can also be used for future projections to some extent. PCA identified the four major factors contributing to the variation of average annual runoff in DPR Korea. The proposed average annual runoff model was composed of average annual precipitation, average precipitation intensity, average annual air temperature, and hot seasonal air temperature. It was proven through hydrologic data for 93 river basins of DPR Korea that the proposed statistical model can sufficiently reflect the physical nature of runoff formation in DPR Korea. Kriging interpolation tools of Arc GIS 9.2 were used to estimate the spatial distribution of factors affecting average annual runoff formation. A 10 km × 10 km grid cell size was used to interpolate factor fields, which was determined through analysis for spatial change of climate elements in 200 weather stations of DPR Korea. Average annual runoff depth for each grid cell can be easily computed through Equation (39) using the spatially distributed fields of factors affecting annual runoff formation. Further, in this study, a grid cell type runoff map was developed by the average annual runoff data of each grid cell and the proposed cartography of the average annual runoff map and use method. The case study on the Tumen River Basin demonstrates that this research work is highly significant for decision makers as it highlights variations in water resources, which are important for water resources development and management. The statistical–distributed hydrological model facilitates hydrologists in water resources assessment and information sharing in an ungauged area.

Author Contributions: T.R., conceptualization, data collection, formal analysis, software development, and writing original draft; J.J., conceptualization, writing original draft, review and editing the manuscript, supervision, methodology, and funding acquisition; S.B., imputing ideas, condensing the presentation of findings, and reviewing and editing the manuscript; T.P., reviewing and editing the manuscript and visualization.

Funding: This research was funded by the National Natural Science Foundation of China (Grant No. 51509061), the open research fund of State Key Laboratory of Urban Water Resource and Water Environment (MH20180645). Additional support was provided by the Southern University of Science and Technology (Grant No. G01296001).

Acknowledgments: We acknowledge the support of the National Natural Science Foundation of China and International Office in Harbin Institute of Technology to make this research possible. We wish to thank Afed Khan for language correction. We thank the anonymous reviewers for their thorough and thoughtful comments, which helped to improve this manuscript.

Conflicts of Interest: The authors declare no conflict of interest.

References

1. Kannan, N.; Santhi, C.; Arnold, J.G. Development of an automated procedure for estimation of the spatial variation of runoff in large river basins. *J. Hydrol.* **2008**, *359*, 1–15. [CrossRef]
2. Wang, Y.; Lei, X.; Liao, W.; Jiang, Y.; Huang, X.; Liu, J.; Song, X.; Wang, H. Monthly spatial distributed water resources assessment: A case study. *Comput. Geosci.* **2012**, *45*, 319–330. [CrossRef]
3. Xu, C.-Y.; Singh, V.P. Review on Regional Water Resources Assessment Models under Stationary and Changing Climate. *Water Res. Manag.* **2004**, *18*, 591–612. [CrossRef]
4. Hickel, K.; Zhang, L. Estimating the impact of rainfall seasonality on mean annual water balance using a top-down approach. *J. Hydrol.* **2006**, *331*, 409–424. [CrossRef]
5. Church, M.R.; Bishop, G.D.; Cassell, D.L. Maps of regional evapotranspiration and runoff/precipitation ratios in the northeast United States. *J. Hydrol.* **1995**, *168*, 283–298. [CrossRef]
6. Bishop, G.D.; Church, M.R. Automated approaches for regional runoff mapping in the northeastern United States. *J. Hydrol.* **1992**, *138*, 361–383. [CrossRef]
7. Bishop, G.D.; Church, M.R. Mapping long-term regional runoff in the eastern United States using automated approaches. *J. Hydrol.* **1995**, *169*, 189–207. [CrossRef]
8. Hickel, K.; Zhang, L. The Impact of Rainfall Seasonality on Mean Annual Water Balance in Catchments with Different Land Cover. In *CRC for Catchment Hydrology*; eWater: Canberra, Australia, 2003.

9. Milly, P.C.D. Climate, interseasonal storage of soil water, and the annual water balance. *Adv. Water Res.* **1994**, *17*, 19–24. [CrossRef]

10. Arnell, N.W. Grid mapping of river discharge. *J. Hydrol.* **1995**, *167*, 39–56. [CrossRef]

11. Menció, A.; Folch, A.; Mas-Pla, J. Identifying key parameters to differentiate groundwater flow systems using multifactorial analysis. *J. Hydrol.* **2012**, *472–473*, 301–313. [CrossRef]

12. Singh, R.; Archfield, S.A.; Wagener, T. Identifying dominant controls on hydrologic parameter transfer from gauged to ungauged catchments—A comparative hydrology approach. *J. Hydrol.* **2014**, *517*, 985–996. [CrossRef]

13. Christine, L.G.; John, D.A.; Scott, V.O. Mapping monthly precipitation, temperature, and solar radiation for Ireland with polynomial regression and a digital elevation model. *Clim. Res.* **1998**, *10*, 35–49.

14. Milly, P.C.D.; Eagleson, P.S. Effects of spatial variability on annual average water balance. *Water Resour. Res.* **1987**, *23*, 2135–2143. [CrossRef]

15. Daly, C.; Neilson, R.P.; Phillips, D.L. A Statistical-Topographic Model for Mapping Climatological Precipitation over Mountainous Terrain. *J. Appl. Meteorol.* **1994**, *33*, 140–158. [CrossRef]

16. McCabe, G.J.; Wolock, D.M. Temporal and spatial variability of the global water balance. *Clim. Chang.* **2013**, *120*, 375–387. [CrossRef]

17. Boughton, W.; Chiew, F. Estimating runoff in ungauged catchments from rainfall, PET and the AWBM model. *Environ. Model. Softw.* **2007**, *22*, 476–487. [CrossRef]

18. Arora, V.K. The use of the aridity index to assess climate change effect on annual runoff. *J. Hydrol.* **2002**, *265*, 164–177. [CrossRef]

19. Helfer, F.; Lemckert, C.; Zhang, H. Impacts of climate change on temperature and evaporation from a large reservoir in Australia. *J. Hydrol.* **2012**, *475*, 365–378. [CrossRef]

20. Krug, W.R.; Gebert, W.A.; Graczyk, D.J.; Stevens, D.L., Jr.; Rochelle, B.P.; Church, M.R. *Map of Mean Annual Runoff for the Northeastern, Southeastern, and Mid-Atlantic United States, Water Years, 1951–1980*; Water-Resources Investigations Report, 88-4094; U.S. Geological Survey: Reston, VA, USA, 1990; p. 15.

21. Zhou, X.; Zhang, Y.; Wang, Y.; Zhang, H.; Vaze, J.; Zhang, L.; Yang, Y.; Zhou, Y. Benchmarking global land surface models against the observed mean annual runoff from 150 large basins. *J. Hydrol.* **2012**, *470–471*, 269–279. [CrossRef]

22. David, M.W.; Gregory, J.M. Explaining spatial variability in mean annual runoff in the conterminous United States. *Clim. Res.* **1999**, *11*, 149–159.

23. Smith, M.B.; Koren, V.I.; Zhang, Z.; Reed, S.M.; Pan, J.-J.; Moreda, F. Runoff response to spatial variability in precipitation: An analysis of observed data. *J. Hydrol.* **2004**, *298*, 267–286. [CrossRef]

24. Kirkby, M.J.; Bracken, L.J.; Shannon, J. The influence of rainfall distribution and morphological factors on runoff delivery from dryland catchments in SE Spain. *CATENA* **2005**, *62*, 136–156. [CrossRef]

25. Ignatov, A.V.; Kichigina, N.V. Modeling elements and geographical factors of formation of the monthly runoff as exemplified by the Kuda river. *Geogr. Nat. Res.* **2010**, *31*, 396–402. [CrossRef]

26. Greenwood, A.J.B.; Benyon, R.G.; Lane, P.N.J. A method for assessing the hydrological impact of afforestation using regional mean annual data and empirical rainfall–runoff curves. *J. Hydrol.* **2011**, *411*, 49–65. [CrossRef]

27. Chang, H.; Johnson, G.; Hinkley, T.; Jung, I.-W. Spatial analysis of annual runoff ratios and their variability across the contiguous U.S. *J. Hydrol.* **2014**, *511*, 387–402. [CrossRef]

28. Westra, S.; Brown, C.; Lall, U.; Sharma, A. Modeling multivariable hydrological series: Principal component analysis or independent component analysis? *Water Resour. Res.* **2007**, *43*. [CrossRef]

29. Wotling, G.; Bouvier, C.; Danloux, J.; Fritsch, J.M. Regionalization of extreme precipitation distribution using the principal components of the topographical environment. *J. Hydrol.* **2000**, *233*, 86–101. [CrossRef]

30. Reder, A.; Rianna, G.; Vezzoli, R.; Mercogliano, P. Assessment of possible impacts of climate change on the hydrological regimes of different regions in China. *Adv. Clim. Chang. Res.* **2016**, *7*, 169–184. [CrossRef]

31. Beven, K.; Robert, E. Horton's perceptual model of infiltration processes. *Hydrol. Proc.* **2004**, *18*, 3447–3460. [CrossRef]

32. Dias, L.C.P.; Macedo, M.N.; Costa, M.H.; Coe, M.T.; Neill, C. Effects of land cover change on evapotranspiration and streamflow of small catchments in the Upper Xingu River Basin, Central Brazil. *J. Hydrol. Reg. Stud.* **2015**, *4*, 108–122. [CrossRef]

33. Yang, P.; Xia, J.; Zhan, C.; Qiao, Y.; Wang, Y. Monitoring the spatio-temporal changes of terrestrial water storage using GRACE data in the Tarim River basin between 2002 and 2015. *Sci. Total Environ.* **2017**, *595*, 218–228. [CrossRef] [PubMed]

34. Beven, K.; Binley, A. The future of distributed models: Model calibration and uncertainty prediction. *Hydrol. Proc.* **1992**, *6*, 279–298. [CrossRef]

35. Beven, K.J. *Rainfall-Runoff Modelling: The Primer*; Wiley: Hoboken, NJ, USA, 2011.

36. Beven, K. Infiltration excess at the Horton Hydrology Laboratory (or not?). *J. Hydrol.* **2004**, *293*, 219–234. [CrossRef]

37. De Jong van Lier, Q.; Sparovek, G.; Flanagan, D.C.; Bloem, E.M.; Schnug, E. Runoff mapping using WEPP erosion model and GIS tools. *Comput. Geosci.* **2005**, *31*, 1270–1276. [CrossRef]

38. Milewski, A.; Sultan, M.; Yan, E.; Becker, R.; Abdeldayem, A.; Soliman, F.; Gelil, K.A. A remote sensing solution for estimating runoff and recharge in arid environments. *J. Hydrol.* **2009**, *373*, 1–14. [CrossRef]

39. Cheng, Q.; Ko, C.; Yuan, Y.; Ge, Y.; Zhang, S. GIS modeling for predicting river runoff volume in ungauged drainages in the Greater Toronto Area, Canada. *Comput. Geosci.* **2006**, *32*, 1108–1119. [CrossRef]

40. Graczyk, D.J.; Gebert, W.A.; Krug, W.R.; Allord, G.J. *Maps of Runoff in the Northeastern Region and the Southern Blue Ridge Province of the United States During Selected Periods in 1983–1985*; Open-File Report, 87-106; U.S. Geological Survey: Reston, VA, USA, 1987; p. 12.

41. Sauquet, E. Mapping mean annual river discharges: Geostatistical developments for incorporating river network dependencies. *J. Hydrol.* **2006**, *331*, 300–314. [CrossRef]

42. Yan, Z.; Xia, J.; Gottschalk, L. Mapping runoff based on hydro-stochastic approach for the Huaihe River Basin, China. *J. Geogr. Sci.* **2011**, *21*, 441–457. [CrossRef]

43. Milly, P.C.D. Climate, soil water storage, and the average annual water balance. *Water Resour. Res.* **1994**, *30*, 2143–2156. [CrossRef]

44. Li, G.; Zhang, F.; Jing, Y.; Liu, Y.; Sun, G. Response of evapotranspiration to changes in land use and land cover and climate in China during 2001–2013. *Sci. Total Environ.* **2017**, *596–597*, 256–265. [CrossRef]

45. Caracciolo, D.; Deidda, R.; Viola, F. Analytical estimation of annual runoff distribution in ungauged seasonally dry basins based on a first order Taylor expansion of the Fu's equation. *Water Res.* **2017**, *109*, 320–332. [CrossRef]

46. Bostan, P.A.; Heuvelink, G.B.M.; Akyurek, S.Z. Comparison of regression and kriging techniques for mapping the average annual precipitation of Turkey. *Int. J. Appl. Earth Obs. Geoinf.* **2012**, *19*, 115–126. [CrossRef]

47. Li, L.; Zhang, L.; Xia, J.; Gippel, C.J.; Wang, R.; Zeng, S. Implications of Modelled Climate and Land Cover Changes on Runoff in the Middle Route of the South to North Water Transfer Project in China. *Water Res. Manag.* **2015**, *29*, 2563–2579. [CrossRef]

48. Chang, J.; Zhang, H.; Wang, Y.; Zhu, Y. Assessing the impact of climate variability and human activities on streamflow variation. *Hydrol. Earth Syst. Sci.* **2016**, *20*, 1547–1560. [CrossRef]

49. Chiew, F.H.S.; Teng, J.; Vaze, J.; Post, D.A.; Perraud, J.M.; Kirono, D.G.C.; Viney, N.R. Estimating climate change impact on runoff across southeast Australia: Method, results, and implications of the modeling method. *Water Resour. Res.* **2009**, *45*. [CrossRef]

50. Delfs, J.O.; Park, C.H.; Kolditz, O. A sensitivity analysis of Hortonian flow. *Water Res.* **2009**, *32*, 1386–1395. [CrossRef]

51. Devia, G.K.; Ganasri, B.P.; Dwarakish, G.S. A Review on Hydrological Models. *Aquat. Procedia* **2015**, *4*, 1001–1007. [CrossRef]

Article

Climate Change and Intense Irrigation Growth in Western Bahia, Brazil: The Urgent Need for Hydroclimatic Monitoring

Raphael Pousa *, Marcos Heil Costa, Fernando Martins Pimenta, Vitor Cunha Fontes, Vinícius Fonseca Anício de Brito and Marina Castro

Department of Agricultural Engineering, The Federal University of Viçosa, Viçosa MG 36570-900, Brazil; mhcosta@ufv.br (M.H.C.); fernando.m.pimenta@ufv.br (F.M.P.); vitor.fontes@ufv.br (V.C.F.); vfabrito@hotmail.com (V.F.A.d.B.); maricastrodasilva@gmail.com (M.C.)
* Correspondence: raphaelpousa@gmail.com; Tel.: +55-31-3899-1902

Received: 29 March 2019; Accepted: 30 April 2019; Published: 2 May 2019

Abstract: In Western Bahia, one of the most active agricultural frontiers of the world, cropland area and irrigated area are increasing at fast rates, and water conflicts have been happening at least since 2010. This study makes a hydroclimatic analysis of the water resources in Western Bahia, from both supply and demand viewpoints. Time series of precipitation for the period 1980–2015 and river discharge for the period 1978–2015 are analyzed, indicating a significant reduction of up to 12% in rainfall since the 1980s, and a reduction in river discharge in all stations studied, in both the rainy season and the dry season. Combined with that, irrigated area has increased over 150-fold in 30 years, and in the most irrigated regions, has increased by 90% in the last eight years only. Seven regions in Western Bahia have been identified where the potential for water use conflicts is critical. Moreover, the combination of reduced availability and increased demand of water resources indicates that, if current trends are maintained, conflicts over water may become more frequent in the next years or decades. A short-term alternative to avoid such conflicts is to largely avoid irrigation during the months with low discharge. However, a monitoring system in which the availability and demand of water resources for irrigation are actually measured and monitored, is the safest path to provide water security to this region.

Keywords: climate change; MATOPIBA agricultural frontier; water security; hydroclimatic analysis; water conflicts

---

1. Introduction

The relationship between water and conflict is an area of continued interest and debate in both the policy and water resources literature and in the popular press [1]. Conflicts arise by several socioeconomic, political, or biophysical causes, including proximity to the water source, government type, aridity, climate variability and change, and rapid population growth. The dispute becomes much more challenging when there are multiple causes for the conflict. This work provides a case study of a region where two factors, climate change and intense irrigation growth, contribute to increased friction on the use of water resources: Western Bahia, in Brazil.

The western part of the state of Bahia is one of the most active agricultural frontiers of the world, where land use transition started in 1985 [2]. Western Bahia (Figure 1) is part of a wider region called MATOPIBA (acronym formed by the states of Maranhão, Tocantins, Piauí, and Bahia), an agricultural frontier in the Cerrado biome in Brazil, and characterized by rapid changes in land cover and land use for cropland, especially soybean, and agricultural intensification through the adoption of new technologies, leading to high yields. In Western Bahia, cropland area has reached 2 million hectares in the 2016/2017 growing season, mainly soybeans, cotton, and maize [3].

A major difference between Western Bahia and the rest of MATOPIBA is that the impressive extensification has been followed by a no less impressive increase in irrigated area, which grew from 9 pivots in 1985 to 1550 center pivots in 2016 [4]. The region includes three river basins (Rio Grande, Rio Corrente, and the northern part of the Rio Carinhanha), all tributaries of the São Francisco River, and also sits on the top of the Urucuia aquifer [5], a vast geological formation that is connected to the rivers, and helps regulate their seasonality and interannual variability. Five small hydroelectric plants operate in the region, all on tributaries of the Rio Grande upstream of the town of Barreiras (Figure 1), with power ranging from 450 kW to 25 MW [6].

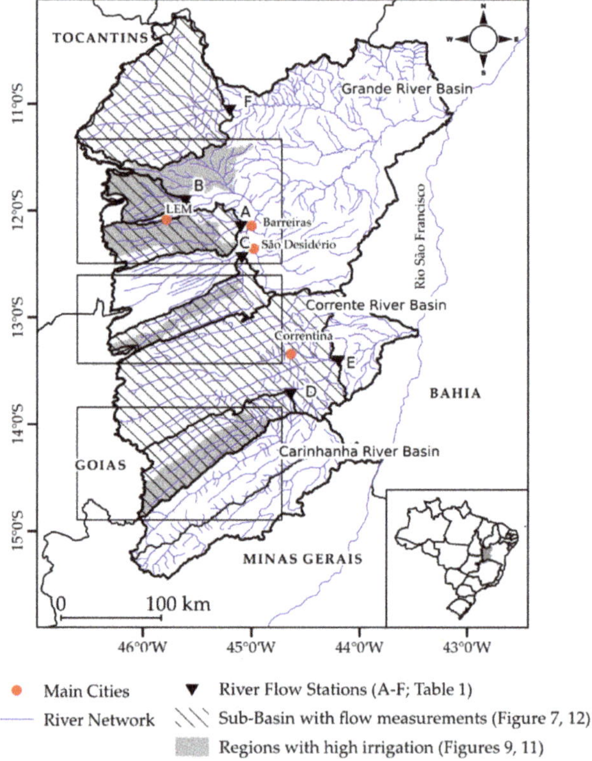

Figure 1. Study area representing the river networks, sub-basins with flow measurements analyzed in this study (hatched areas), regions with high irrigation (gray areas), and the location of main towns. LEM is the town of Luis Eduardo Magalhães. The rectangles represent the zoom areas detailed in Figure 5. River flow stations A–F are described in Table 1.

The long-term (1980–2015) precipitation, evapotranspiration, and runoff for the region are 1060 mm year^{-1}, 860 mm year^{-1}, and 200 mm year^{-1}, respectively. However, the region is located in the transition between the seasonally dry Cerrado biome to the west (annual precipitation >1200 mm year^{-1} and a six-month rainy season, from mid-October to mid-April) and the semi-arid Caatinga biome to the east (annual precipitation <800 mm year^{-1} and a four-month rainy season). Precipitation, evapotranspiration, and runoff are seasonal. Precipitation typically varies from 0–10 mm month^{-1} in the driest months (June, July, August) to about 150–200 mm month^{-1} in the rainiest months (December and January). Monthly evapotranspiration (from MODIS MOD16 product) typically varies between

20 mm month^{-1} in September and 85 mm month^{-1} in February. Despite the high seasonality in precipitation, the seasonal variability in runoff is relatively small, with maximum discharge about three times greater than the minimum discharge, which is an indication of the strong regulation provided by the Urucuia aquifer. Temperatures and solar radiation are high around the year and, with the aid of irrigation, would allow for year-round crops (five to six crop growing seasons in two years), limited only by phytosanitary regulations. These circumstances have contributed to the intense growth of irrigation in the region [7].

Conflicts over the use of water have become common in the region in the last decade, however few of them have been documented. Maybe the first documented conflict happened in 2010. The hydropower station *Sítio Grande* on the Rio das Fêmeas, a tributary of the Rio Grande, is the largest plant in the region and has a water permit of 12 m^3 s^{-1}, the largest water permit in the region, about 1/3 of the water rights granted in the Rio Grande basin [8]. Despite being the largest grant, this is a non-consumptive use of water, as the water is not withdrawn from the river, but instead it must be available at the river to flow through the turbines. The conflict happened during the initial filling of the lake, which interrupted the flow of the river for several days with environmental and social consequences.

Conflicts kept being reported informally through social networks, personal communications, etc. Another formal documentation happened in 2015, an El Niño year when the region experienced a severe drought (2015 annual rainfall was 674 mm, one of the lowest on record). On 11 December 2015, the Rio Corrente Basin Committee requested a temporary suspension on the concession of water use permits on the basin until further criteria for water permits on the basin are defined [9]. On 2 November 2017, the usually peaceful town of Correntina (population 32,000) made the national headlines [10], when about 500 people invaded one farm that received recent irrigation systems and destroyed a significant part of their facilities and equipment as a way of protesting against the appropriation of water by agribusiness. A week after, on 11 November, approximately 10,000 people marched peacefully through Correntina, in defense of the Rio Corrente and its tributaries [11].

Although the water use conflicts in the region are usually attributed to the immense growth rate of irrigation systems, climate variability may also play an important role. Being in the transition between the semi-arid and the seasonally dry tropical climate regions, Western Bahia may be a serious candidate for climate change. This study makes an hydroclimatic analysis of the water resources in Western Bahia, from both the supply and demand viewpoints.

2. Data and Methods

Data and methods used in this analysis are summarized in Figure 2. Long-term time series of precipitation and river discharge are analyzed to evaluate the availability of water resources for irrigation, while maps of irrigation areas and interviews with irrigators are produced to evaluate the demand of water resources. We conclude with recommendations to improve water management and avoid further water conflicts.

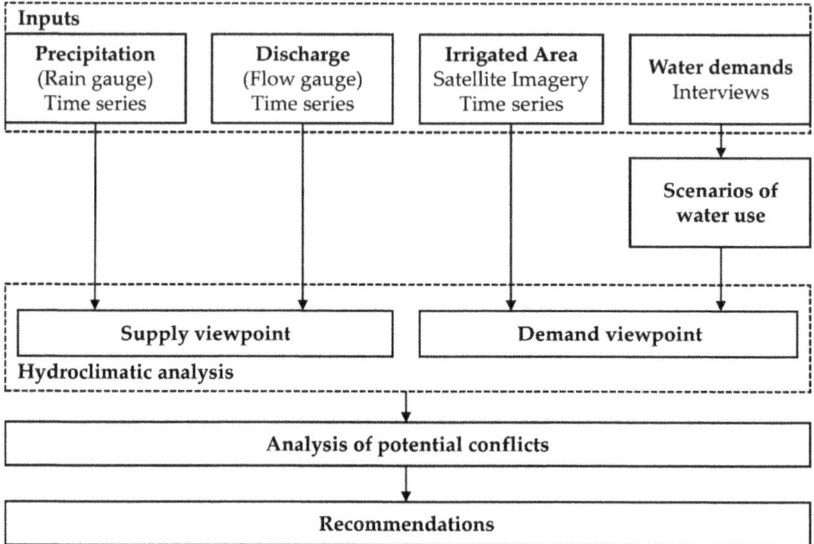

Figure 2. Flowchart illustrating the data and methods.

2.1. Precipitation and River Flow Data

Precipitation data in the region have been available through the INMET (*Instituto Nacional de Meteorologia*) and ANA (*Agência Nacional das Águas*) weather station and rain gauge networks since the 1930s, but the network was sparse and the time series were frequently interrupted. A somewhat dense network was available only in the late 1970s. To characterize regional patterns, the daily precipitation dataset of Xavier et al. [12] was used, which is available at a grid resolution of 0.25 × 0.25 (approximately 28 km × 28 km) for a 36-year period of 1980–2015. This dataset was assembled from the available rain gauge, conventional, and automatic weather stations. Original data were quality-controlled, and six different interpolation methods were tested (average of the five nearest data points; natural neighbor; thin plate spline; inverse distance weighting; angular distance weighting; and ordinary point kriging). The accuracy of the interpolation methods was evaluated by a cross-validation procedure, in which an observed data point was temporarily removed from the database, and then used to test the estimated value by each interpolation method at the location of the station. Angular distance weighting was considered the method with best skill [12,13]. To evaluate longer trends (before 1980), the station of Barreiras (WMO code 83236) was also used, which has nearly continuous data since 1961.

The daily river flow data (m^3/s) used are provided by ANA. The fluviometric data are available since the 1930s, with a low density of stations and significant gaps. From 1930s to 1970s, there are about 30% of gaps in the data series and some discontinued stations. Quantity and quality of data increased in the 1970s, with only 2% of gaps, and we initially selected 25 river flow stations with few gaps since 1978.

The granting of water use permits in the Grande and Corrente basins is an attribution of the State of Bahia, which uses the criterion that 80% of Q_{90} can be granted for human use (according to State Decree number 6296 of 21 March 1997). Q_{90} is the flow expected to be present in the river during at least 90% of the time, i.e., during 90% of the time series used in the calculation, there is a flow equal or greater than Q_{90} in the river.

To follow this criterion, our analyses of water availability are based on the flow duration curve of specific sections of the rivers. A flow duration curve is a cumulative frequency curve that shows the percent of time specified discharges were equaled or exceeded during a given period. Here, Q_{90} was calculated using the long-term series (LT Q_{90}), which is a more common criterion for granting water use permits, but we also calculate Q_{90} using two periods, to characterize hydroclimate change.

Although we analyzed data for 25 fluviometric stations, six stations were selected for a deeper analysis (Table 1, Figure 1). These stations are spread throughout the region and are representative of the regional variability. Moreover, four stations were chosen because they have dense irrigation systems upstream (A–D), while two of them (E,F) were chosen for their low irrigation density upstream.

The flow stations drain relatively large areas, and (with the exception of station B) may not be representative of the densest irrigated areas. Thus, seven ottobasins with the highest concentration of center pivots (represented by gray areas in Figure 1) were also analyzed. Ottobasins, or Otto-codified hydrographic basins, are areas of contribution of the stretches of the hydrographic network coded according to the topological system proposed by Otto Pfafstetter [14,15] and officially adopted by ANA to uniquely identify contribution areas in any watershed using a simple 10-base code. The system is hierarchical and recursive, and the higher number of digits in the ottobasin code implies a higher level of sub-division of a watershed.

Table 1. Selected river flow stations. The letters correspond to the labels of stations in Figure 1.

	ANA Station Code	River	Station Name	Municipality	Drainage Area (km²)	Station Coordinates
A	46543000	Rio de Ondas	Fazenda Redenção	Barreiras	5383.758	12°08′ S, 45°06′ W
B	46570000	Rio de Janeiro	Ponte Serafim	Barreiras	2522.118	11°54′ S, 45°36′ W
C	46415000	Rio Grande	Sítio Grande	São Desidério	4943.866	12°25′ S, 45°05′ W
D	45840000	Rio Formoso	Gatos	Jaborandi	7132.696	13°42′ S, 44°38′ W
E	45910001	Rio Corrente	Santa Maria da Vitória	Santana	29,643.660	13°24′ S, 44°12′ W
F	46790000	Rio Preto	Formosa do Rio Preto	Formosa do Rio Preto	14,326.870	11°03′ S, 45°12′ W

2.2. Statistical Tests

We applied four statistical analyses to detect changes in the rainfall time series. First, we applied the non-parametric Pettitt's test [16] for detecting changing points to the region-wide precipitation time series. This is a rank-based and distribution-free test for detecting a significant change in the mean of a time series and it is particularly useful when no hypothesis is required about the location of the changing point. The Pettitt test has been widely applied to detect changes in the observed hydroclimatic time series [17,18], and can only be applied to continuous time series. Considering a sequence of random variables $X_1, X_2, \dots, X_T$, which have a change point at $t = \tau$. As a result, $(X_1, X_2, \dots, X_\tau)$ have a common distribution function $F_1(X)$, but $(X_{\tau+1}, X_{\tau+2}, \dots, X_T)$ are distributed as $F_2(X)$, where $F_1(X) \neq F_2(X)$. The null hypothesis H_0 for this test is that the observations are independent and identically distributed (no change, or $\tau = T$), and is tested against the alternative hypothesis H_1: change (or $1 \leq \tau < T$); using the non-parametric statistic $K_T = \max|U_{t,T}|$ where:

$$U_{t,T} = \sum_{i=1}^{t} \sum_{j=t+1}^{T} \text{sign}(X_t - X_j).$$

The confidence level for a change-point is defined as

$$\rho = \exp\left(\frac{-6\,K_T^2}{T^3 + T^2}\right).$$

Second, we applied a classical Student's *t*-test to test the null hypothesis that the annual mean precipitation is not significantly different from one period to the other, where periods are divided at $t = \tau$, obtained by the Pettitt's test.

Third, we applied the Mann–Kendall test for trends in the time series, a non-parametric, distribution-free test that makes no assumptions of linearity or distribution of the values. This test has been recommended widely by the World Meteorological Organization for general trend analysis of time series [19]. Finally, we used box plots to evaluate the interannual variability of precipitation.

2.3. Irrigated Area

The irrigated area by center pivots was obtained by a four-step procedure. First, imagery from Landsat 5, 7, and 8 for the period 1990 to 2018 was processed using the Google Earth Engine cloud. The images were filtered using the median of the pixels for the dry period (April to September) and mosaicked for the study region, to produce a single region mosaic per year. Second, the filtered map was merged with the center pivots data from Landau et al. [20] and from the OpenStreetMaps project to obtain an initial pivots map of the region. Then, duplicated features and topology errors were removed from the dataset. Third, with the aid of the visible bands (RGB) and the normalized difference vegetation index (NDVI) from the generated mosaics, the center pivot features were digitized or erased according to the recognition in the images of each year. Finally, the annual center pivot geometries went through a trend and precision analysis of their positional components for positional accuracy validation, producing a final map without trends in center pivot sizes and with accuracy adequate to the scale of 1:150,000, compatible with the resolution of the Landsat images.

The resulting yearly maps for Western Bahia were further processed at the ottobasin scale, to select only the highly irrigated regions, i.e., ottobasins with at least 4% of their total area irrigated in 2018. A total of seven regions were selected (gray areas in Figure 1).

2.4. Calculations of Regional Water Demand for Irrigation

The regional water uptake for irrigation depends on (1) the effective area irrigated at some time (A_I), in km^2; (2) the reference evapotranspiration rate (ET_o), in mm/day; (3) the crop being irrigated and its stage of development, which are integrated into an adimensional "crop coefficient" K_c, that usually varies between 0.3 and 1.3; and (4) the efficiency of the system (ε), which for center pivots is typically around 0.8. The water uptake for irrigation (Q_I) in $m^3\ s^{-1}$ is the product of these four terms:

$$Q_I = \frac{A_I\ K_c\ ET_o}{86.4\ \varepsilon}.$$

Although we have mapped all center pivots in these regions, our estimates of irrigated area should be understood in terms of area with installed irrigation systems. These systems may be used fully, partially, or not at all, depending on the year and the season. Currently, there are no regionally consolidated data of the actual amount of irrigated area nor the crops planted per center pivot as a function of time.

To overcome this limitation, we interviewed 20 irrigators and one irrigation consultant (who consulted for several tens of irrigators). Interviews were conducted between July 2018 and October 2018, either in person or by phone. We asked questions about the frequency of irrigated crops a year, typical planting dates, crops planted, amount of irrigation applied, and the main reasons why they make their management decisions.

3. Results

3.1. Changes in Precipitation

Figure 3 shows the evolution of annual mean regional precipitation for the three basins. In addition, data for the Barreiras station are also shown. An analysis of Figure 3 indicates that two main characteristics of these time series stand out. First, annual mean precipitation presents strong interannual variability, ranging from ~600 to >1700 mm year^{-1}. In other words, individual precipitation years range from values typical of the semi-arid climate east of the region to values typical of the tropical seasonally dry climate west of the region. The interannual variability pattern is also consistent across the three basins, which indicates that it is large-scale driven. In addition, the regional pattern correlates well with the data at the Barreiras station, which allows for speculative interpretations in the period before 1980.

Second, basin-wide precipitation has not been greater than 1370 mm year^{-1} since 1992, while this level has been exceeded five times between 1980 and 1992, and another five times in the period from 1961 to 1979, if considering the Barreiras data. This is a major shift in the precipitation regime, that affects the regional decadal means. To be sure, we applied the non-parametric Pettitt's test for detecting changing points to the region-wide 1980–2015 time series, which confirms a changing point at $\tau = 13$ (1992), with $K_T = 87$ and significance level 4.6×10^{-9}.

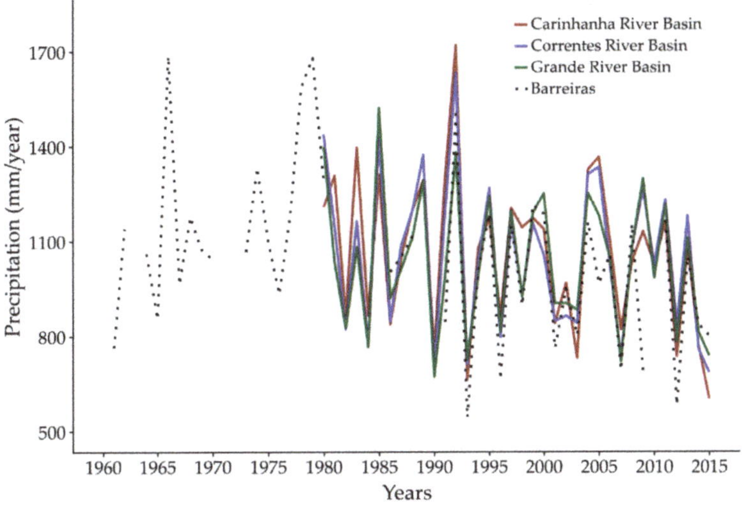

Figure 3. Annual mean precipitation for the three basins and historical precipitation for the Barreiras rain gauge.

Based on this evidence, the 36-year time series of rainfall and river flow were divided in two periods, namely P1 (1980–1992) and P2 (1993–2015), to test the hypothesis of precipitation change between the two periods. We also tested other divisions of the rainfall time series (two periods of 18 years, three periods of 12 years), but the division in P1 and P2 was the choice that yielded the highest significance in precipitation change.

Indeed, the regional patterns of mean precipitation for the two periods (P1, P2) show that isohyets are moving westward from P1 to P2 (Figure 4a,b), translating into a regional drying from P2 with respect to P1 (Figure 4c). Although the period of analysis is relatively small, significant precipitation change has been detected. Extremely likely ($\alpha = 0.05$) precipitation reduction, averaging −165 mm year^{-1} (−12% compared to P1 mean), appears in a core area in the west of the region. This core area is surrounded by very likely ($\alpha = 0.10$) precipitation changes (Figure 4c), and average drier conditions throughout nearly all the region (Figure 4c). Moreover, the Mann–Kendall trend test indicates precipitation trends consistent both in sign and in significance with the precipitation differences (Figure 4d).

Considering basin-wide averages, the significant reduction in precipitation has happened mainly in the months with higher precipitation (December and January) in the three basins (Figure 5, $\alpha = 0.05$). In addition, the interannual variability of precipitation, measured by both the interquartile difference and the range of variability, has also decreased in the three basins in P2, when compared to P1 (Figure 6).

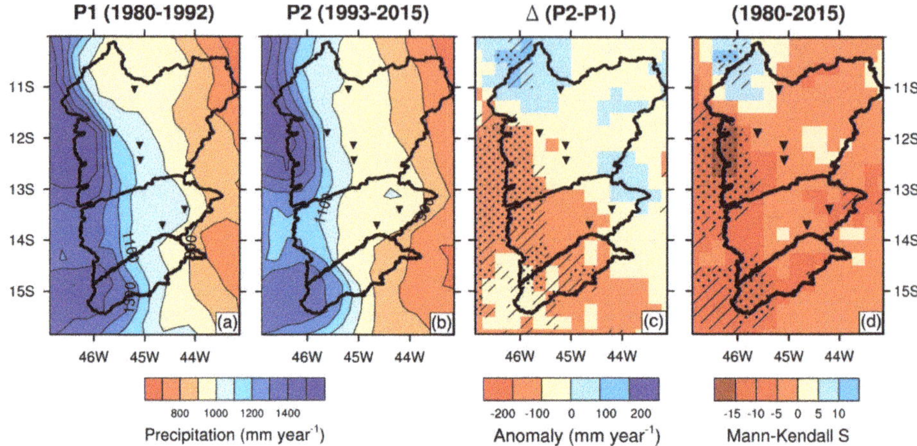

Figure 4. Average precipitation map for periods P1 (**a**), P2 (**b**) the difference between P2 and P1 (**c**), and the Mann–Kendall S statistic (negative values represent decreasing trends). The six selected river flow stations are also shown. Dotted areas represent differences significant at $\alpha = 0.05$, according to Student's t test (in **c**), or according to the Mann-Kendall test (in **d**), while shaded areas represent differences or trends significant at $\alpha = 0.10$.

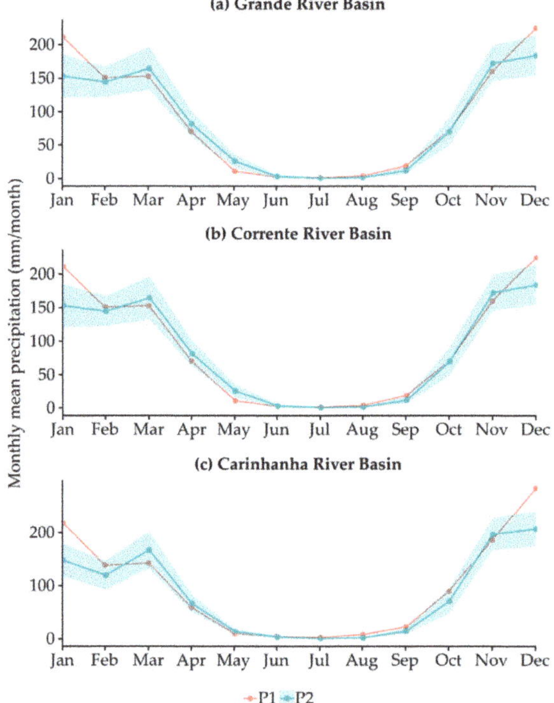

Figure 5. Monthly mean precipitation for two periods (P1 and P2) for the three basins. The shaded area in P2 is the confidence interval for the mean ($\alpha = 0.05$). Averages in P1 outside the shaded area are statistically different at this level of confidence.

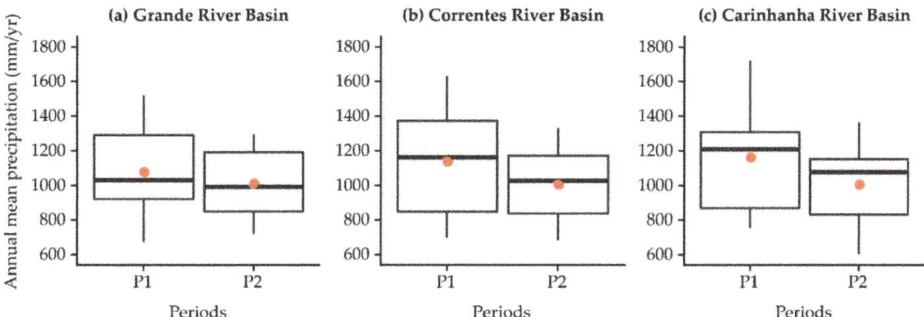

Figure 6. Boxplot displaying the median (thick lines), the lower and upper quartile (box), the mean (red dots), and the minimum and maximum of the distribution (whiskers) for annual values of precipitation in periods P1 and P2. Period P1, although much shorter than P2, has higher interannual variability.

Two main large-scale systems are likely the cause of the reduced rainfall in this region, both linked to higher interannual variability and more extreme drought years [21]. First, the warming of the tropical North Atlantic Ocean leads to a higher frequency of anomalously northward positions of the intertropical convergence zone (ITCZ). In addition, changes in the temperature of the Pacific manifested as extremes of the El Niño-southern oscillation (ENSO) are partially associated with extreme drought in the region [22,23].

We suggest that the strong interannual variability of precipitation is driven by the seasonal expansion of the subtropical high across northeast Brazil. In dry years (El Niño years and warmer North Atlantic years), it expands more to the west, reaching the Bahia/Tocantins border. In wet years (La Niña years and cooler North Atlantic years), it expands less, staying to the east of the São Francisco River.

3.2. Changes in River Flow

The flow duration curves for the six stations analyzed in this work are presented in Figure 7. These curves show the percent of time specified discharges were equaled or exceeded during each period (P1, P2). All panels show that discharge has been decreasing at all levels of probability.

The discharge of these rivers is heavily regulated by the Urucuia aquifer. Parallel flow duration curves, like Figure 7c–f, indicate that the decrease in discharge is mainly caused by the reduction in rainfall, and modulated by the aquifer. While a monitoring piezometer network of the aquifer was set up only in 2011, by 2015 it already shows groundwater level drawdown of up to 5 m [24].

On the other hand, when the decrease in discharge is smaller in the wet season (lower percentiles) and higher in the dry season (higher percentiles), which is the case of Figure 7b, this is an indication that withdrawal of water during the dry season may be playing a relevant role in the decrease of discharge. It is no coincidence that station B (also region R2 in Table 2) is the one among the six selected with the highest density of irrigation upstream (4.8% of the upstream area irrigated), while the other ones have less than 1% of their drainage area irrigated. We suggest that water withdrawal for irrigation is only detectable in fluviometric records when irrigation upstream is between 1% and 4% of the drainage area of the station.

In addition, it can also be verified in Figure 7 that nearly all discharge data recorded during P1 in these six stations are higher than Q_{90} of P2. This means that, even considering the long term (1978–2015), Q_{90} is mostly defined by the discharges observed in P2 only. This is confirmed by Figure 8, which clearly shows that most of the situations when daily discharge Q is smaller than the long-term Q_{90} (LT Q_{90}) happens after 2000. In an extreme case, station D in a very dry year like 2015, 72% of the days (263 days out of 365) had daily Q lower than the LT Q_{90}. Although this is a severe case, it is relatively common to find years when more than half of the days are below the LT Q_{90}.

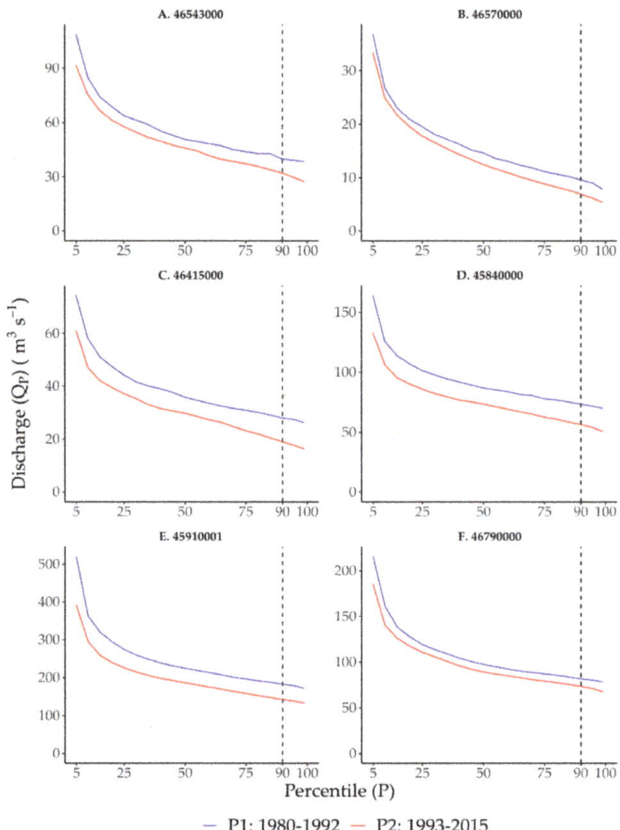

Figure 7. Flow duration curves for stations **A–F** for periods P1 and P2. Values in the x-axis are the probability that a given discharge (Q_p) is exceeded during that period. The intersection of the dashed line, representing the 90% percentile, and each curve, represents the Q_{90}.

Our interpretation of these data is that, even if considering its relatively short (38 years) duration, all river discharge time series are non-stationary. Discharge has been decreasing all over the spectrum and throughout the region, minimum discharges have been defined in the most recent years, and even when the most recent dry years have been considered for the definition of minimum discharges, the actual recent rate of occurrence of a relatively unlikely phenomenon like $Q < Q_{90}$ is four to seven times higher than the expected probability. Under these circumstances, probability discharges cannot be used to predict the distribution of future flows. In addition, although the results are not shown here, the above characteristics are consistent across 24 of the 25 stations analyzed, except the northwesternmost station (upstream of station F), only area where precipitation did not decrease.

Non-stationarity can be explained by several factors, such as changes in river basins by anthropogenic effects, climate change, and low-frequency climate variability [25]. Moreover, this does not seem to be the case of uncertainty dominating the distribution of extremes, as suggested by Serinaldi and Kilsby [26]. The entire flow duration curve has shifted down, not only the extreme values. This is very much consistent with the picture of rivers regulated by a decreasing-level aquifer, following a reduction of the aquifer recharge after reductions in precipitation, arguably caused by a strengthening of the South Atlantic subtropical high-pressure system associated with the warming of the North Atlantic and Central Pacific, the ultimate causes for the non-stationarity. This is our current reasoning for the attribution of causes, although it can only be verified through detailed hydrogeological modeling and large-scale climate dynamics studies, which are beyond the scope of this work.

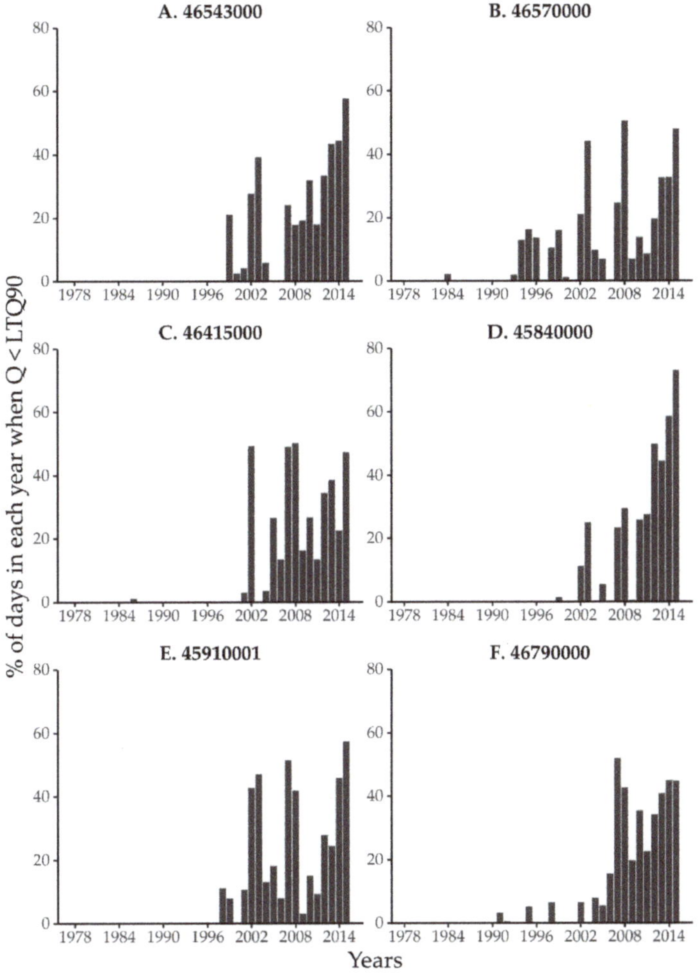

Figure 8. Percentage of days in each year when actual Q is below the long-term Q_{90}. Long-term Q_{90} is calculated for the period 1978–2015. The concentration of cases of Q < LT (long-term) Q_{90} after 2000 indicates drastic reductions in minimum discharges.

3.3. Trends in Irrigated Area and Water Uptake on the River Flows

To evaluate the effects of irrigation water uptake on the river flows, we choose seven regions with the highest concentration of irrigated area, where these effects are expected to be most significant. These seven regions have between 4.8% and 12.6% of the area of the ottobasin irrigated, while no other ottobasin in the region has more than 4%. The spatial evolution in irrigated area in these regions (R1 to R7) is shown in Figure 9. The total irrigated area in these seven regions was 662.4 km^2 in 2010, which increased to 1256.1 km^2 in 2018, a 90% increase in just eight years (Table 2). Figure 10 shows the temporal evolution of irrigated area for each region. Each region has a different pattern of growth, but all regions show a substantial increase in irrigation since the 1990s.

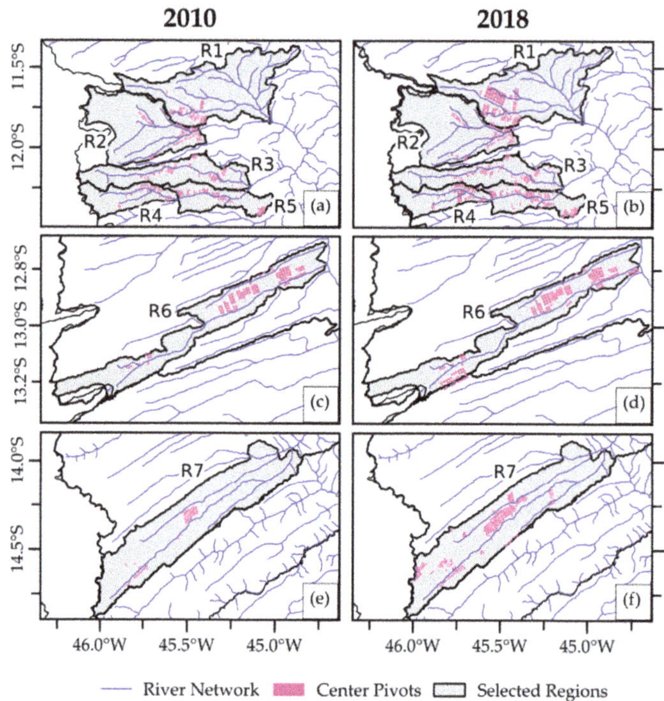

Figure 9. Evolution of irrigated area for selected regions for 2010 and 2018. Each center pivot ranges from 26.9 ha to 355 ha in area. The North–South scale of (**c**) and (**d**) is different than the same scale of (**a**), (**b**), (**e**), and (**f**).

Table 2. Selected regions (R1–7) where irrigated area is located in Figure 9, with corresponding ANA Ottobasin codes. The region R2 coincides with the drainage area of station B.

Region (Ri)	ANA Ottobasin Code	River	Total Area (km²)		Irrigated Area in 2018		
			Ottobasin	Region	Ottobasin (km²)	Region (km²)	% of Total Area
R1	76243	Rio Branco		3403.5	232.9	232.9	6.8%
R2	46570000 *	Rio de Janeiro		2522.1		122.2	4.8%
R3	762641	Rio Cabeceira de Pedras		1739.6		108.6	6.2%
R4	762691	Rio Borá		938.3		89.2	9.5%
R5	7626711	Rio de Ondas	778.64	1939.2	121.1	244.2	12.6%
	762661	Rio de Ondas mouth	222.33		33.9		
	762691	Rio Borá (upstream)	938.3				
R6	762891	Rio Grande	197.10	2075.2	42.0	194.7	9.4%
	76489	Rio Guará	295.04		11.1		
	762871	Rio Grande	361.45		42.1		
	76285	Rio Grande	789.94		37.2		
	76282	Vereda Passaginha	431.66		62.3		
R7	764271	Rio Pratudão	662.35	3865.0	14.3	264.2	6.8%
	76426	Riacho do Váu	702.94		115.0		
	764241	Rio Formoso	2499.73		134.9		

* Represents the fluviometric station code.

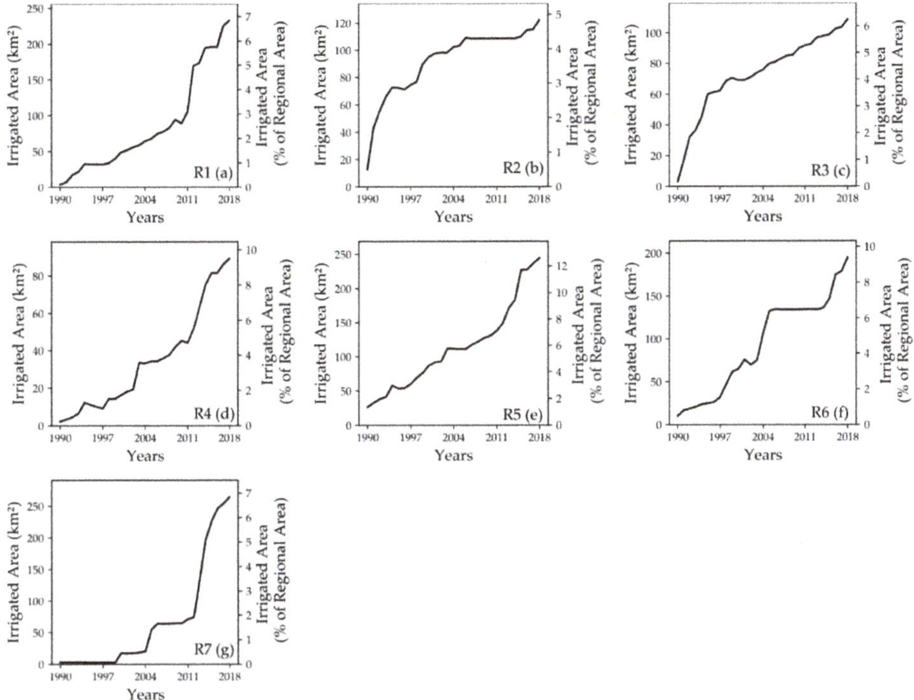

Figure 10. Evolution of irrigated area in selected regions, as defined in Figure 9 and Table 2.

With the area irrigated estimated and the areas with intense irrigation growth identified, in order to calculate water uptake, information on water demands per unit area are still needed (Section 2.4). From the interviews conducted, during the driest period of the year (September and October), when river discharge approaches Q_{90} values, water application rates (K_c ET_o/ε) could be as high as 10 mm day^{-1}, if the irrigator had a high consumption crop like maize at the peak of the cycle (K_c ~1.3); at the same time, when ET_o is very high, because cloudiness is low, incoming solar radiation is at a yearly maximum and relative humidity is low. However, this is a highly avoided situation by irrigators, because of the high costs of energy in September and October.

The electric energy fares in most of Brazil are flagged (green, yellow, or red) according to the actual costs to produce energy in the country. In 2017, on an annual mean, 63.8% of the electricity generated was hydroelectric, 17.2% was produced from fossil fuels sources, other renewables (biomass, solar, and wind) accounted for 17.6%, and nuclear participated with 1.3% [27]. Hydroelectricity is usually much cheaper than the other sources, and its availability is also seasonal. So, during the end of the rainy season, when reservoirs are at the highest level, proportion of hydroelectricity increases and costs decrease (green flag), while at the end of the dry season (September and October), when reservoirs are at the lowest level, proportion of fossil fuels (and costs) increase, leading to the red flag tariff (*Bandeira vermelha*). Although this is the general pattern, other factors, like interannual variability of rainfall, and any other significant changes in supply or demand may also affect the flagging, which is updated monthly.

If the irrigators generally avoid having high consumption of water during the driest months, what management do they usually practice at this time of the year? The results of our interviews indicate that the most common situations are either to have a crop at the end of the cycle when K_c is small (~0.65), or to have the crop cycle (and irrigation) finished by this time of the year.

We then consider two scenarios of regional water management for these seven regions, an aggressive one and a conservative one. In the aggressive scenario, maximum crop output is emphasized; regionally, irrigators would be planting year-round crops (either perennial crops like coffee, or sequential seasonal crops, like soy, cotton, maize, or beans); if sequential seasonal crops, one of the crops would be at the end of the cycle in September and October ($K_c = 0.65$), demanding on average 5 mm/day (150 mm/month) for all areas with irrigation systems installed; and irrigators plan to sow the next crop after the onset of rains in late October or November. The conservative scenario assumes that one-third of the irrigated area is cultivated with only two crops a year, and there is no irrigation during the driest months ($K_c = 0$); two-thirds of the irrigators still act aggressively, as in the previous scenario, some of them because they grow perennial crops and must irrigate year-round; all farmers still plan to sow the next crop after the onset of rains in late October or November, to minimize the costs of energy. Regional irrigation in this conservative scenario is the weighted average of the irrigation levels ($1/3 \times 0 + 2/3 \times 150$), or 100 mm/month.

Again, these are scenarios based on the declared experience of the local people. So far, there are no public yearbooks that document month-by-month variations in planted area, just snapshots of irrigated area that do not capture the quick growth of irrigation systems in the region, neither the seasonality, nor the timing of irrigated crops. In a future work, we plan to use remote sensing to estimate the actual amount of irrigated area and the irrigation period per center pivot as a function of time.

Water uptake for irrigation (Q_I) was estimated from the multiplication of the total area irrigated by the water application rates in the two irrigation management scenarios. Evolution of Q_I from 1990 to 2018 is shown in Figure 11, for regions R1 to R7. Discharge measurements are not available for these regions (except for R2, which will be analyzed again in Figure 12), so we use regionalized values of Q_{90} [28]. This technique is a downscale of discharge for drainage areas smaller than the available measurements, using empirical equations based on independent variables like area upstream or average precipitation upstream. These authors tested several empirical relationships, and the best skill low streamflow regionalization was obtained by basin-specific regression equations of Q_{90} against the upstream long-term annual rainfall minus an initial abstraction of 750 mm ($P_{eq750} = P - 750$) as independent variable [29].

Figure 11 plots 80% of long-term Q_{90} (the maximum discharge that could be granted for all human use, including irrigation). A water use conflict situation appears when the demanded water resources (blue or green lines) are higher than the availability of water resources (horizontal dashed red line). The aggressive scenario (blue line) implies conflicts in regions R1, R2, and R4. In the Rio de Janeiro region (R2), at least since 1997, these conflicts may have been sporadically occurring, depending on year-by-year decision to irrigate in the low flow months. This helps explain why the installation of center pivots has been halted between 2005 and 2013 (Figure 10b). Conflicts, however, may be avoided by a community decision to follow the more conservative scenario, which tolerates additional increases in irrigation area, as observed after 2015 (Figure 10b).

In R3 and R5, although the estimated demand of water resources has not yet reached the limit of 80% LT Q_{90}, they all show a quick increase in the use of water, and if the rates of irrigation growth continue to be high, water conflicts are imminent. In regions R6 and R7, although imminent conflicts may be in principle discarded, conflicts may still arise in the timescale of a decade or two, if irrigation growth rates remain high.

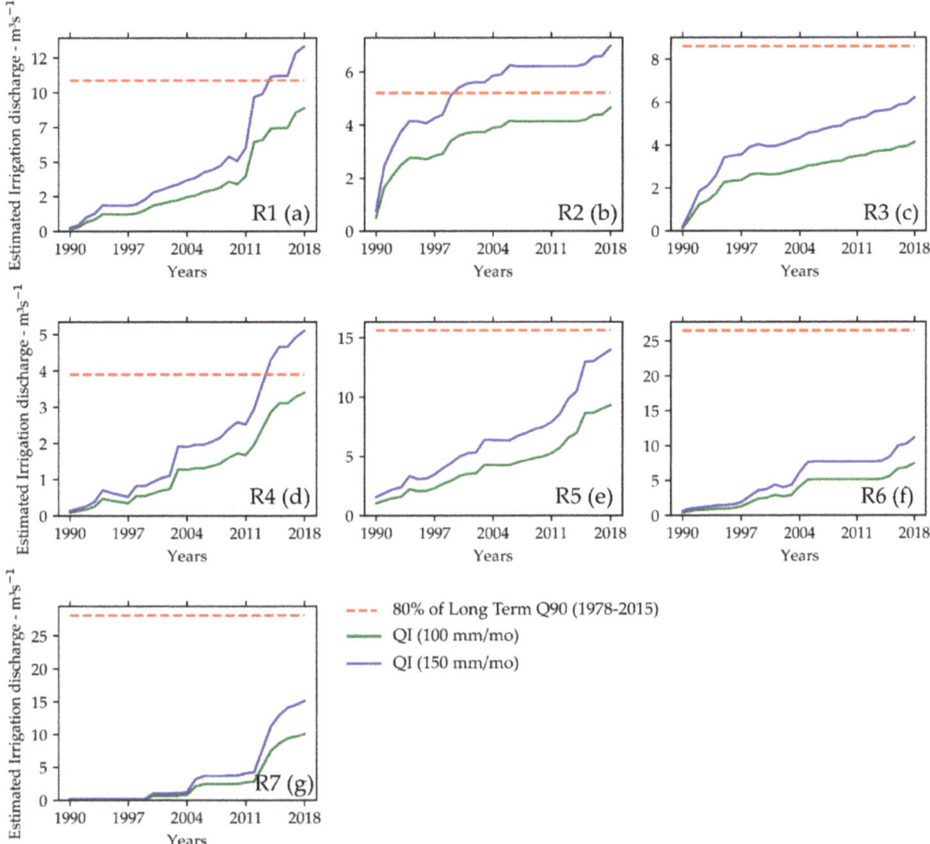

Figure 11. Scenarios of discharge uptake for irrigation in selected regions, assuming two water application rates (100 and 150 mm/month). 80% of long-term Q_{90} is the maximum river discharge that can be granted permission for human use [17].

The intense growth of irrigation systems (90% from 2010 to 2018, Table 2) is hardly the only concern for water users in Western Bahia. As described in Figure 7, the safe discharge for concession of water use permits, Q_{90}, has decreased everywhere in the basin. Q_{90} is mostly defined by the discharges observed in P2 only, which covers the period 1993–2015, so the use of updated hydrological information is crucial to minimize the hydroclimatic risks (Figure 8). In fact, Figure 12A–F does a similar water conflict analysis for the six selected hydrological stations. Of those, four (Figure 12A–D) have much irrigation upstream, while two of them (Figure 12E,F) have little irrigation upstream. A remarkable feature of Figure 12 is the decrease of Q_{90}, calculated only with data for P1 and only with data for P2 (black dashed lines).

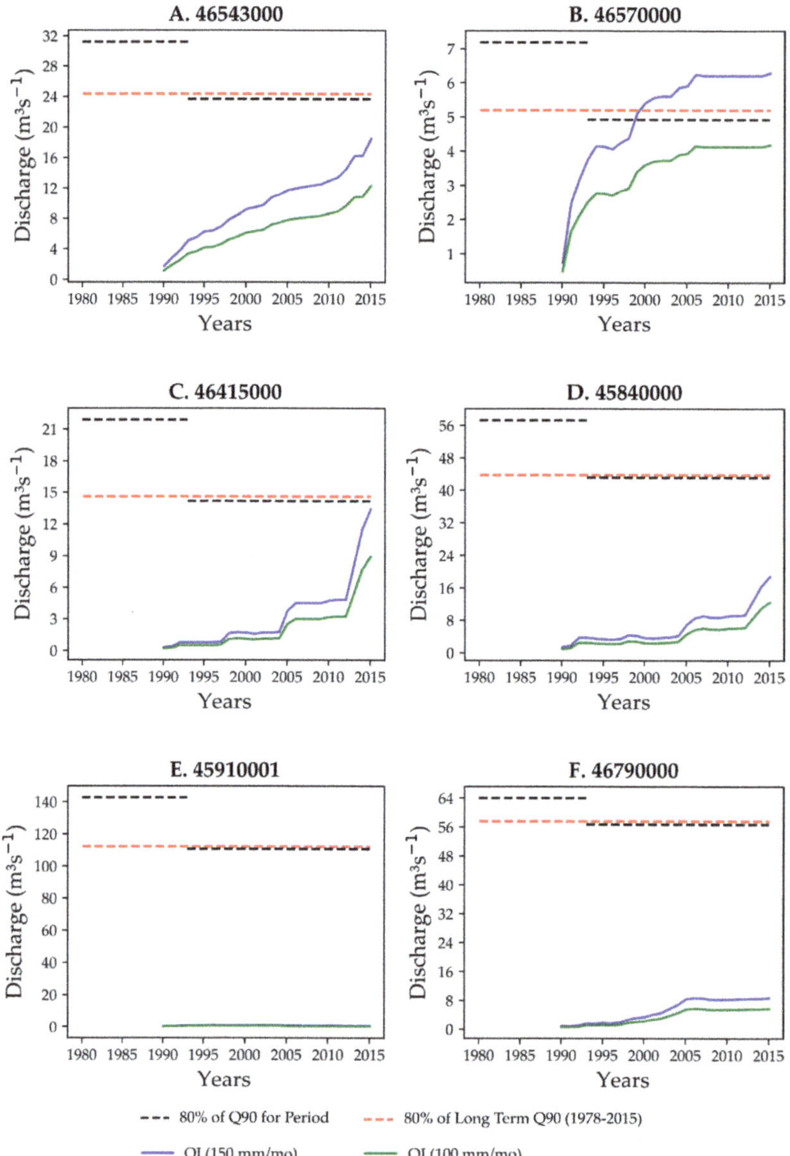

Figure 12. Evolution of discharge uptake for irrigation (QI) in the drainage area of each river flow station, assuming two water application rates (100 and 150 mm/month). 80% of long-term Q_{90} (dashed red line) is the maximum that can be allocated at that point. The black dashed lines represent the change in the availability of water resources from P1 to P2.

4. Discussion and Conclusions

4.1. Climate Change and Intense Iirrigation Growth: Increasing Water Stress

An analysis of Figure 12A–D indicates that conflicts of water use may arise much sooner if outdated hydroclimatic information is used to define water granting rights. For example, a hypothetical Q_{90}

defined using only pre-1992 data is between 15% and 60% higher than the Q_{90} calculated when the more recent years are considered (black dashed lines). Following the non-stationarity of the time series, the long-term Q_{90} (1978–2015) nearly coincides with the recent, shorter-term (P2) Q_{90} everywhere in the basin. In particular in Station B, which is coincident with R2, as early as the late 1990s, irrigation water demands, probably granted based on data of a wetter period, were no longer consistent with the decreasing availability of water resources. This inconsistency was probably understood in 2003, when discharge was low (below long-term Q_{90}) during 45% of the year (Figure 8b), leading to a halt in the installation of new center pivots in the area shortly after (Figure 10b). However, since 2015, irrigated area in the region has resumed its expansion, which is of much concern (Figure 10b). Water withdrawn for irrigation certainly affects the measurements, in particular in regions with a high density of irrigation systems like the sub-basin upstream of station B. This effect, however, is reduced in the other stations, which have a smaller density of irrigation systems.

At least seven sub-basins in Western Bahia are either in a state of conflict for the use of water or are moving rapidly towards it: Rio Branco, Rio de Janeiro, Rio Cabeceira de Pedras, Rio Borá, Rio de Ondas, Rio Grande (headwaters), and Rio Formoso. These sub-basins account for 17% of the area of Western Bahia. In these seven critical sub-basins, water conflicts are imminent, if irrigators actually irrigate in the driest months of the year, when discharge usually gets around or below Q_{90}. As a short-term alternative, conflicts can be avoided if irrigators largely avoid irrigation during these months. As shown in Figure 8, this is not restricted to a few months of the year. In many dry years, nearly half of the year daily discharges were below Q_{90}, with an extreme case in station D (Formoso river) in the very dry year of 2015, when daily Q was below Q_{90} during 72% of the year. To be sure, such Q_{90} already includes 2015 data.

Because of the declining rains and water resources, the water resource concession limits (80% of LT Q_{90}) may be reached with a much higher frequency than originally planned if outdated hydroclimate information is used. The combination of strong increase in demand of water for irrigation and the maintenance of low flows may bring much more critical consequences for water management in the region in the next years. Here, we discuss four different pathways to reduce water stress and increase water security: (i) Avoid irrigation during the low flow period; (ii) halt the installation of new irrigation systems; (iii) bet on a return to wet conditions; and (iv) invest in a hydroclimatic monitoring system.

4.2. Avoid Irrigation during the Low Flow Period

Avoiding irrigation during low-flow periods can be achieved by planting only two crops a year, one from November to February, and a second one from March to June. This is the most natural reaction to improve water security. This practice maximizes the use of rain during the six-month rainy season and cuts the use of irrigation to typically two months, reducing the water consumption not only because of the short irrigation period, but also because of the low ET rates at the end of the cycle. In addition to increasing water security, this practice also reduces production costs by avoiding the high costs of energy during the end of the dry season in Brazil, when additional energy tariffs (*bandeira vermelha*) are charged.

On the other hand, this practice has several drawbacks. Multiple cropping increases the revenue per plot, provides diversification of income, reduces pest pressure, and helps to maintain a more stable pool of farm labor, avoiding seasonal unemployment and the related social consequences [30]. It is also an important factor in the intensification of land use, which, if combined with additional conservation measures, reduces the pressure to expand cropland at the expense of natural ecosystems, possibly sparing land from deforestation. Moreover, this practice is not applicable on perennial crops.

4.3. Halt the Installation of New Irrigation Systems

Halting the installation of new irrigation systems, either through a ban of new water permits or by collective irrigators decision, is a short-term measure that vigorously attacks the problem from the viewpoint of the increasing demand. As stated earlier, this alternative has been requested by the Rio Corrente Basin Committee as a precautionary measure.

This option, however, does not resolve the conflicts of the regions already under a state of conflict, in particular in the Rio Grande basin, where the water demands are already too high in some places. It also does not address the problem of decreasing water availability. In addition, there are the economic consequences on jobs, tax revenue, and economic growth.

Although severe, it may be necessary in some regions with very high demands, in particular if the minimum discharges continue to decrease. A constant update of the low discharge values would be desired in this case.

4.4. Bet on A Return to Wet Conditions

The third alternative is to consider that current low precipitation period is not permanent and climate will return to the pre-1992 wet state.

Given the low skill of the current generation of interdecadal climate prediction models [31], it is hard to forecast whether these reducing precipitation trends will continue in the next decades. As said before, the location of Western Bahia in the transition between the semi-arid and the seasonally dry tropical climate regions makes it a serious candidate for climate change. CMIP5 simulations indicate a strengthening of the South Atlantic subtropical high, with a reduction of precipitation in the semi-arid of Northeast Brazil, and a possible expansion of the semi-arid climate over the region with a current seasonally dry climate [32]. This expansion, however, is somewhat uncertain given the relatively coarse resolution of the climate models involved in the CMIP5 ensemble (from 1.1° to 2.8°), when compared to the east–west dimension of the region (~2.5°). Despite the low skill of these models, the most likely scenario for the 21st century, as simulated by the CMIP5 ensemble, is an additional drying of the region, in particular in the months of September, October, and November, with very high agreement among models [21].

In addition, because the local rivers are connected to the Urucuia aquifer, pre-1992 river discharge levels may only be resumed after the aquifer previous levels of storage are restored, which may take from several years to several decades—our current understanding of the coupling of the aquifer and the rivers of the region is not sufficient to answer this question more precisely. Moreover, since the aquifer water level monitoring network was only set up in 2011, we do not know what the state of the aquifer was during the wet climate period.

In summary, climate models do not support a return to wet conditions—on the contrary. Even if climate models drying predictions do not materialize, it may take several decades to return to former conditions. Betting on this is a risky alternative.

4.5. Invest in A Hydroclimatic Monitoring System

The three alternatives discussed earlier may mitigate water conflicts, but all have their setbacks and risks. Western Bahia is an agriculture frontier under constant change, lacking a crucial element for management: Data. As stated earlier, the water resource management based on the long-term probability of hydroclimatic events requires at least a constant update of the low discharge values. But this does not seem to be sufficient. In the most water-stressed regions described above, a true management, in which the availability and demand of water resources for irrigation are actually measured and monitored, is the safest path to provide water security to this region. Such a monitoring system will allow a more confident and sustainable regional management of irrigated agriculture, maximizing the use of water resources, food production, and economic development, while reducing the risk of water conflicts.

This monitoring system should have three components: (1) Measurement and short term reporting of river discharge at key points in these basins, in particular in the sub-basins where the concentration of irrigation areas is higher; (2) a hydroclimatic forecast system to predict the availability of water resources at the period of lowest availability (September and October) several months in advance, in order to influence the irrigator decision to conduct an irrigated crop and when to plant this irrigated crop; and (3) monitoring and short-term reporting of the actual consumption of water for irrigation at the sub-basin scale. The latter can be done using either one of three possibilities: (i) Installation of hydrometers at each pumping station; (ii) correlation of water consumption with energy consumption, and monitor the latter; and (iii) monitor the actual evapotranspiration through operational remote sensing products, such as the MOD16 evapotranspiration product, correctly calibrated with field data for a reliable representation of reality. These measurements must be integrated at a monitoring center, which would periodically issue recommendations of how much area can be irrigated that year.

4.6. Final Remarks

It has been argued that water crises are mainly crises of governance [33]. Governance is a more inclusive concept than government itself, embracing the relationship between a society and its government. Governments mediate behavior through institutions, policies, laws, norms (like issuing water permits), and actions (like fiscalization and enforcement), but governance also relates to domestic activities, networks of influence, international market forces, the private sector, and civil society [34]. This concept has been incorporated by the Brazilian Policy on Water Resources (Law 9433 of 8 January 1997), which states that management of water resources must be decentralized with the participation of the government, users, and communities. The suggested monitoring system would provide the regional stakeholders (government agencies, agribusiness, and organized civil society) with the necessary data and decision-making tools to make key decisions.

One such key decision is the determination of how much area may be irrigated at each ottobasin in each year, a critical decision in this region with very high interannual variability of rainfall. The correct decision of how much area to be irrigated each year, according to the estimated availability of water resources, may contribute to avoid water use conflicts during the low water season, providing water security for both large irrigators and small farmers. If used wisely, it may also promote the long-term planning of the sustainable expansion of irrigation in the region, supporting the increase in the production of food, feed, and fiber, improving food security as well as water security.

Author Contributions: Conceptualization, methodology, project administration, and funding acquisition, M.H.C.; formal analysis, R.P.; writing—original draft preparation, review and editing, R.P. and M.H.C.; visualization, R.P. and F.M.P.; supporting data, V.C.F., V.F.A.d.B., and M.C.

Funding: This research was funded by PRODEAGRO (grant 011/2016) and CNPq (process 441210/2017-1). R.P. is supported through CAPES, Finance Code 001.

Conflicts of Interest: The authors declare no conflict of interest. The funders had no role in the design of the study, in the collection, analyses, or interpretation of data; in the writing of the manuscript, or in the decision to publish the results.

References

1. Yoffe, S.; Fiske, G.; Giordano, M.; Giordano, M.; Larson, K.; Stahl, K.; Wolf, T. Geography of international water conflict and cooperation: Data sets and applications. *Water Resour. Res.* **2004**, *40*, 1–12. [CrossRef]
2. Batistela, M.; Valladares, G.S. Farming expansion and land degradation in Western Bahia, Brazil. *Biota Neotrop.* **2009**, *9*, 61–76. [CrossRef]
3. AIBA (Associação de Agricultores e Irrigantes da Bahia). Anuário Agropecuário Oeste da Bahia—Safra 2015/2016. Available online: http://aiba.org.br/wp-content/uploads/2018/06/anuario-16-17.pdf (accessed on 15 July 2018).
4. ANA (Agência Nacional de Águas). Atlas de Irrigação—Uso da Água na Agricultura. 2017. Available online: http://arquivos.ana.gov.br/imprensa/publicacoes/AtlasIrrigacao-UsodaAguanaAgriculturaIrrigada.pdf (accessed on 17 December 2018).
5. ANA (Agência Nacional de Águas). Estudos hidrogeológicos na Bacia Hidrográfica do São Francisco—Sistema Aquífero Urucuia/Areado e Sistema Aquífero Bambuí. Comitê Bacia Hidrográfica do São Francisco. 2013. Available online: http://cbhsaofrancisco.org.br (accessed on 11 November 2018).

6. Web Map EPE—Sistema de Informações Geográficas do Setor Energético Brasileiro. Available online: https://gisepeprd.epe.gov.br/webmapepe/# (accessed on 17 April 2019).

7. Ministério do Desenvolvimento, Indústria e Comércio Exterior. Panorama Agroeconômico do Oeste da Bahia e Safra 2016/17. Available online: http://www.mdic.gov.br/images/REPOSITORIO/czpe/Eventos/ZPE_Agroneg%C3%B3cio/Panorama_do_agroneg%C3%B3cio_baiano_Aiba__Celestino_Zanella.pdf (accessed on 17 December 2018).

8. Almeida, W.A.; Moreira, M.C. Análise das outorgas da bacia do Rio Grande, Estado da Bahia. In Proceedings of the XLII Congresso Brasileiro de Engenharia Agrícola—CONBEA 2013, Campo Grande, Brazil, 27–31 July 2014.

9. Deliberação CBHRC 01/2015. Available online: https://www.conjur.com.br/dl/deliberacao-comite-bacia-corrente.pdf (accessed on 17 April 2019).

10. G1. Grupo invade fazendas e incendeia galpão em protesto no Oeste da Bahia. Available online: https://g1.globo.com/bahia/noticia/grupo-invade-fazendas-e-incendeia-galpao-em-protesto-no-oeste-da-bahia.ghtml (accessed on 17 April 2019).

11. Correio 24 horas. Guerra pela água em Correntina se arrasta desde 2015. Available online: https://www.correio24horas.com.br/noticia/nid/guerra-pela-agua-em-correntina-se-arrasta-desde-2015/ (accessed on 17 April 2019).

12. Xavier, A.C.; King, C.W.; Scanlon, B.R. Daily gridded meteorological variables in Brazil (1980–2013). *Int. J. Climatol.* **2016**, *36*, 2644–2659. [CrossRef]

13. Data Gridded Meteorological Data from 1980–2013 (and updated precipitation through 2015). Available online: http://careyking.com/data-downloads/ (accessed on 10 November 2018).

14. Pfafstetter, O. *Classificação de Bacias Hidrográficas*; Departamento Nacional de Obras de Saneamento: Rio de Janeiro, Brazil, 1989.

15. Verdin, K.L.; Verdin, J.P. A topological system for delineation and codification of the Earth's river basins. *J. Hydrol.* **1999**, *218*, 1–12. [CrossRef]

16. Pettitt, A.N. A non-parametric approach to the change-point problem. *Appl. Stat.* **1979**, *2*, 126–135. [CrossRef]

17. Verstraeten, G.; Poesen, J.; Demarée, G.; Salles, C. Long-term (105 years) variability in rain erosivity as derived from 10-min rainfall depth data for Ukkel (Brussels, Belgium): Implications for assessing soil erosion rates. *J. Geophys. Res.* **2006**, *111*, 1–11. [CrossRef]

18. Rybski, D.; Bunde, A.; Havlin, S.; von Stoch, H. Long-term persistence in climate and the detection problem. *Geophys. Res. Lett.* **2006**, *33*, 1–4. [CrossRef]

19. Mitchell, J.M., Jr.; Dzerdzeevskii, B.; Flohn, H.; Hofmeyr, W.L.; Lamb, H.H.; Rao, K.N.; Wallén, C.C. *Climatic Change*; WMO Technical Note No. 79; World Meteorological Organization: Geneva, Switzerland, 1966.

20. Landau, E.C.; Guimarães, D.P.; Souza, D.L. Concentração de áreas irrigadas por pivôs no Oeste da Bahia. In Proceedings of the Anais do Simpósio Regional de Geoprocessamento e Sensoriamento Remoto—GEONORDESTE 2014, Aracajú, Brazil, 18–21 November 2014.

21. Marengo, J.A.; Torres, R.R.; Alves, L.M. Drought in Northeast Brazil—Past, Present, and Future. *Theor. Appl. Climatol.* **2017**, *129*, 1189–1200. [CrossRef]

22. Kane, R.P. Prediction of Droughts in North-East Brazil: Role of ENSO and Use of Periodicities. *Int. J. Climatol.* **1997**, *17*, 655–665. [CrossRef]

23. Ambrizzi, T.; Souza, E.B.; Pulwarty, R.S. The Hadley and Walker Regional Circulations and Associated ENSO Impacts on the South American Seasonal Rainfall. In *The Hadley Circulation: Present, Past and Future*; Diaz, H.F., Bradley, R.S., Eds.; Kluwer Academic: Dordrecht, The Netherlands, 2004; Volume 21, pp. 203–235.

24. Marques, E.A.G.; Silva Júnior, G.C.; Illambwetsi, A.M.; Eger, G.Z.S.; Pousa, R.; Generoso, T.N.; Oliveira, J. Analysis of Groundwater Table and River Stage Fluctuations and their Relation to Rainfall and Water Use on Alto Grande Watershed, Northeastern Brazil. Unpublished work. 2018.

25. Bayazit, M. Nonstationary of hydrological records and recent trends in trend analysis: A state-of-the-art review. *Environ. Process.* **2015**, *2*, 247–542.

26. Serinaldi, F.; Kilsby, C.G. Stationarity is undead: Uncertainty dominates the distribution of extremes. *Adv. Water Res.* **2005**, *77*, 17–36. [CrossRef]

27. Ministério de Minas e Energia. Empresa de Pesquisa Energética, Brazilian Energy Balance. Available online: http://epe.gov.br/en/publications/publications/brazilian-energy-balance (accessed on 17 April 2019).

28. Oliveira, J.R.S.; Ribeiro, R.B.; Sousa, J.R.C.; Serrano, L.O.; Ramos, M.C.A.R.; Generoso, T.N.; Pruski, F.F. Hydrological Information System to quantify water availability (SIHBA). Unpublished work. 2019.

29. Pruski, F.F.; Rodriguez, R.D.G.; Nunes, A.A.; Pruski, P.L.; Singh, V.P. Low-flow estimates in regions of extrapolation of the regionalization equations: A new concept. *Eng. Agríc.* **2015**, *35*, 808–816. [CrossRef]

30. Richards, P.; Pellegrina, H.; VanWey, L.; Spera, S. Soybean development: The impact of a decade of agricultural change on urban and economic growth in Mato Grosso, Brazil. *PLoS ONE* **2015**, *10*, e0122510. [CrossRef] [PubMed]

31. Kirtman, B.; Power, S.B.; Adedoyin, J.A.; Boer, G.J.; Bojariu, R.; Camilloni, I.; Doblas-Reyes, F.J.; Fiore, A.M.; Kimoto, M.; Meehl, G.A.; et al. Near-term Climate Change: Projections and Predictability. In *Climate Change 2013: The Physical Science Basis. Contribution of Working Group I to the Fifth Assessment Report of the Intergovernmental Panel on Climate Change*; Stocker, T.F., Qin, D., Plattner, G.-K., Tignor, M., Allen, S.K., Boschung, J., Nauels, A., Xia, Y., Bex, V., Midgley, P.M., Eds.; Cambridge University Press: Cambridge, UK; New York, NY, USA, 2013; pp. 953–1028.

32. Magrin, G.O.; Marengo, J.A.; Boulanger, J.-P.; Buckeridge, M.S.; Castellanos, E.; Poveda, G.; Scarano, F.R.; Vicuña, S. Central and South America. In *Climate Change 2014: Impacts, Adaptation, and Vulnerability. Part B: Regional Aspects. Contribution of Working Group II to the Fifth Assessment Report of the Intergovernmental Panel on Climate Change*; Barros, V.R., Field, C.B., Dokken, D.J., Mastrandrea, M.D., Mach, K.J., Bilir, T.E., Chatterjee, M., Ebi, K.L., Estrada, Y.O., Genova, R.C., et al., Eds.; Cambridge University Press: Cambridge, UK; New York, NY, USA, 2014; pp. 1499–1566.

33. United Nations Educational, Scientific and Cultural Organization (UNESCO); World Water Assessment Programme. *Water, a Shared Responsibility*; The United Nations World Water Report 2; Berghahn Books: Paris, France; New York, NY, USA, 2006; pp. 43–86.

34. Rogers, P.; Hall, A.W. *Effective Water Governance*; Global Water Partnership Technical Committee Background Papers No. 7; Global Water Partnership: Stockholm, Sweden, 2003.

 *water*

Article

Influence of Power Operations of Cascade Hydropower Stations under Climate Change and Human Activities and Revised Optimal Operation Strategies: A Case Study in the Upper Han River, China

Lianzhou Wu, Tao Bai *, Qiang Huang, Ming Zhang and Pengfei Mu

State Key Laboratory Base of Eco-Hydraulic Engineering in Arid Area, Xi'an University of Technology, Xi'an 710048, China; wlzxaut@126.com (L.W.); wreshxaut@163.com (Q.H.); zhang_ming_1996@163.com (M.Z.); PengfeiMu@hotmail.com (P.M.)
* Correspondence: baitao@xaut.edu.cn; Tel.: +86-029-8231-2036

Received: 29 March 2019; Accepted: 24 April 2019; Published: 28 April 2019

Abstract: Climate change and human activities are two driving factors that affect the hydrological cycle of watersheds and water resource evolution. As a pivotal input to hydropower stations, changes in runoff processes may reduce the effectiveness of existing operation procedures. Therefore, it is important to analyze the influences of cascade hydropower stations under climate change and human activities and to propose revised optimal operation strategies. For the present study, three runoff series conditions including: Initial runoff, affected by only climate change, and affected by both climate change and human activities are examined by a simulation model to analyze the influence on power generation with four schemes. Additionally, an optimal operation model of cascade power stations is proposed based on the simulation model to generate single and joint optimal operation charts for future hydrological scenarios. The paper also proposes to change human activities based on optimizing operation rules to reduce its influence on downstream power stations. This procedure is theoretically applied and varied for three power stations in the upper Han River, China. The results show that the influence of climate change is greater than that of human activities in that power generation decreased by 17.95% and 12.83%, respectively, whereas combined, there is a reduction of 25.71%. Under existing hydrological conditions, the modified single and joint operation charts would increase power generation by about 32 million and 47 million kWh. Furthermore, after optimizing the upstream project, the abandoned water and power generation of these cascade power stations would reduce by 150 million m^3 and 5 million kWh, respectively. This study has practical significance for the efficient operation of cascade hydropower stations and is helpful for developing reservoir operation theory under changing environments.

Keywords: climate change; human activities; power operations; cascade joint operation chart; inter-basin water transfer project

---

1. Introduction

Climate change and human activities are two factors driving change in the hydrological cycle of watersheds and water resources in terms of hydrological response, energy structure, and the social economy, which have become the focus of current research in the field of global change [1,2]. Climate change leads to changes in atmospheric circulation including evaporation and precipitation conditions, through changes in rainfall distribution, and the evaporation and precipitation conditions in surface waters and soils [3,4]. Human activities including agricultural irrigation, water conservancy projects, and urbanization have directly affected the natural circulation of water resources in many regions [5,6]. These changes lead to decreases in runoff and increases in extreme events [7,8]. According to the IPCC assessment report, global climate has warmed over the past 100 years, and the climate change has seriously affected the streamflow regime [3]. Consequently, climate change and human activities have threatened the availability of water resources that is critical to human survival, especially in terms of energy structure and production [9,10].

According to the U.S. Climate Change Science Program (CCSP), a new clean energy structure that includes solar, wind, and hydro-power would be an effective solution to control CO_2 emissions caused by coal power generation [11]. Additionally, hydropower is also responsible for regulating the safety of power grid systems in future energy structures [12]. Take the current Chinese energy structure as an example, the hydropower installed capacity had exceeded 300 GW, about 50% of the total installed power capacity, up to 2015, and about half of total power generation was hydropower from 2000–2015 [13]. However, hydropower is vulnerable to climate change and human activities [14]. As an important input to the hydropower generation system, runoff and water distribution changes under the influences of climate change and human activities would directly impact the operation of power stations. Therefore, researching the influences of climate change and human activities on hydrological systems and establishing efficient coping strategies are of great significance for cascade power stations.

In recent years, scholars have carried out research on how climate change and human activities have affected the hydrological processes and the operation of hydropower stations. Viers considered it necessary to anticipate changing climatic and hydrological conditions for a similar period of time for operations of hydropower stations [15]. Harrison G P built a simulation model based on electricity systems to explore the sensitivity of power station operations to climate change [16]. Ehsani indicated that modifying reservoir operations and increasing the size and number of dams was necessary to offset the vulnerabilities of water resources to future climate uncertainties [17]. Ahmadi established the reservoir optimization scheduling model on the premise of considering future climate change, and the model coordinated the contradiction between the power generation guarantee rate and the vulne'rability of the reservoir [18]. Minville took the Peribonka basin water resources system in Canada as an example, and evaluated the impact of climate change scenarios on the adaptive scheduling results of the water resources system [19]. However, most previous research has focused on determining the characteristics and extent of change in hydropower systems under climate change [15–19]. There have been few studies on the combined effects of climate change and human activity coping strategies for power stations. Chang proposed an optimal adaptive operation chart for cascade hydropower system to increase power generation under changing environmental conditions providing novels methods for hydropower station operation under climate change [13].

Meanwhile, to balance the economic development of different regions, changing the distribution of water resources artificially has occurred more frequently [20]. For example, a number of Inter-Basin Water Transfer (referred to hereafter as "IBWT") projects have been built, so the runoff of the source area was changed [21]. The operation of original cascade hydropower stations under new hydrological conditions would be challenging for both water source areas and water intake areas [22,23]. Therefore, assessing the combined impact of climate change and human activities on the operations of hydropower stations and developing novel operational strategies to respond to changing conditions are more necessary.

Based on previous studies, the main objective of this study is to generate modified operation charts for the cascade power stations in the upper Han River to reduce the influence of climate change and human activities. This study analyzes the extent of impacts from climate change and human activities on power station operation both separately and combined. Then, the simulated reservoir operation charts are modified based on the traditional operation chart, and an optimal single and cascade operation charts are generated by the cuckoo algorithm based on the simulated chart. Finally, an optimal operation chart of the Hanjiang to Weihe River Valley Water Diversion Project (referred to hereafter as "the Project") is generated considering the downstream cascade power stations. This research would help quantify the impact of climate change and human activities on the operation of power stations, and provide reference value for the coping strategies for power stations.

2. Study Area and Data

2.1. The Upper Han River

The Han River is the largest tributary of the Yangtze River, China. The upper Han River is located before the Danjiangkou reservoir, with a length of 918 km and a drainage area of 95,200 km^2. The upper Han River is in an area with a subtropical humid climate. The annual rainfall distribution in this area is uneven and most runoff recharge is surface runoff from rainfall. The main flood season is from July–September; however, some small floods also occur from mid–late April [24].

The Han River occupies a prominent position in the social development of the Yangtze River Basin. Besides providing water to the provinces in the basin, the Han River is also the water source of some IBWT projects, such as the Project and the Mid-Line Project of the South-to-North Water Transfer Project. The Project being built is the only large-scale human activities in the upstream of the cascade power stations, like Figure 1 shows, the cascade power stations must be affected by the future operation of the Project [25]. Also, Chang has verified that the runoff in the upper Han River is mutated resulted from climate change [13]. Therefore, it is urgently needed to modify the operation charts for these hydropower stations to respond to future changing hydrological environments caused by climate change and human activities.

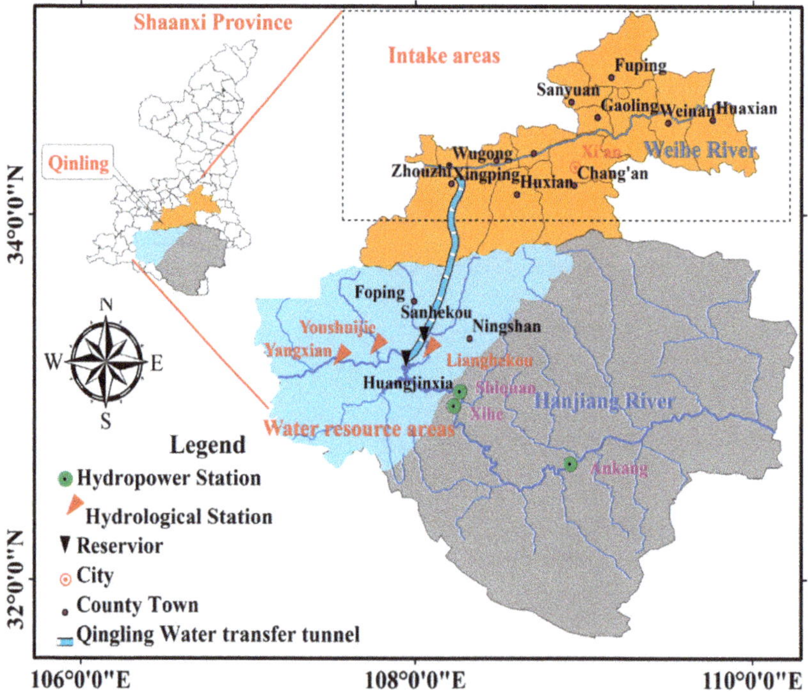

Figure 1. distribution map of power stations and reservoirs in the upper Han River, the blue area is water resource areas of the Project, the brown area is intake areas of the Project and the gray area is power stations in the downstream of the Project.

2.2. Cascade Hydropower Stations

There are three hydropower stations on the upper Han River, namely Shiquan, Xihe, and Ankang, which are the research objects of this paper. The characteristics of the cascade reservoirs are listed in Table 1.

Table 1. The characteristics of the cascade reservoirs of Shiquan, Xihe, and Ankang hydropower stations.

Index	Unit	Shiquan	Xihe	Ankang
Average annual discharge	m^3/s	308.3	378	621
Normal water level	m	410	362	330
Dead water level	m	400	360	305
Regulation storage	$10^8 m^3$	1.8	0.22	14.72
regulation performance	/	seasonal	Daily	incomplete yearly
Installed capacity	MW	225	180	852.5
Guaranteed output (Ng)	MW	32	21.8	175
Annual average power generation	10^8 kW·h	6.06	4.92	27.48
Maximum head	m	47.5	32.5	88
Minimum head	m	26.3	13	57
Maximum power flow	m^3/s	677.5	811	1500

These three power stations are an important part of the Northwest Power Grid, which is responsible for power generation, power grid peaking, and frequency modulation. In the upstream of the cascade hydropower station, the Project is being built, which will inevitably affect the operation of the cascade hydropower stations.

2.3. The Project

Uneven distribution of water resource is obvious in Shaanxi province, China, where the southern region has a large amount of water resource, and the central and northern regions is shortage. This situation is caused by weather conditions and the rapidity of the economic development and urbanization in the area coupled with a growing population and poor water resource management [26]. As a strategic project to improve the ecological environment and to upgrade industry, the Project is being developed in the upper Han River and will be in operation by 2025. The main task of the Project is to transfer multi-year average of 1.5 billion m^3 of water from the Han River to the Guanzhong region, including important cities, counties, and industrial parks. The Project consists of two water source areas connected by a water transfer tunnel. The Huangjinxia reservoir (HJX) in the main stream has abundant water with no regulation ability, and the Sanhekou reservoir (SHK) in a tributary has pluriennal regulation capacity with less water.

The reservoirs and power stations in the upper Han River are shown in Figure 1, the Project is located upstream of the Shiquan power station. The Shiquan reservoir has an annual average inflow of about of 10 billion m^3, it means that the amount of water transferred by the Project accounts for 15% of the inflow of the Shiquan reservoir.

2.4. Data Collection

In the present study, the monthly inflow data series for the three reservoirs were from 1954–2010 and were considered to be reasonable and representative. Information on the water transfer process of the Project from 1954 - 2010 was from the Yangtze River Water Resources Commission. The reservoir inflow data and the conventional hydroelectric operation charts (Figure 2) of Shiquan and Ankang were obtained from the hydropower plants. The operation chart consisted of four parts and the meaning and application of each is as follows:

(1) Part A refers to the guaranteed output area located between the upper and lower basic lines. If the water level at time *t* is in this part, then the hydropower station operates based on the guaranteed output.

(2) Part B refers to the increased output area located between the upper basic line and the anti-abandon water line. If the water level at time *t* is in this part, then the hydropower station should increase output based on the guaranteed output. In Shiquan and Ankang power station, the discount factor is 1.2, it means the hydropower station should operate based on 1.2 times guaranteed output.

(3) Part C refers to the decreased output area located between the lower basic line and the dead water line. If the water level at time *t* is in this part, then the hydropower station should decrease output based on the guaranteed output. In Shiquan and Ankang power station, the discount factor is 0.8, it means the hydropower station should operate based on 0.8 times guaranteed output.

(4) Part D refers to the flood control area. Once the water level is in this part, the reservoir should operate under the specified flood control rules.

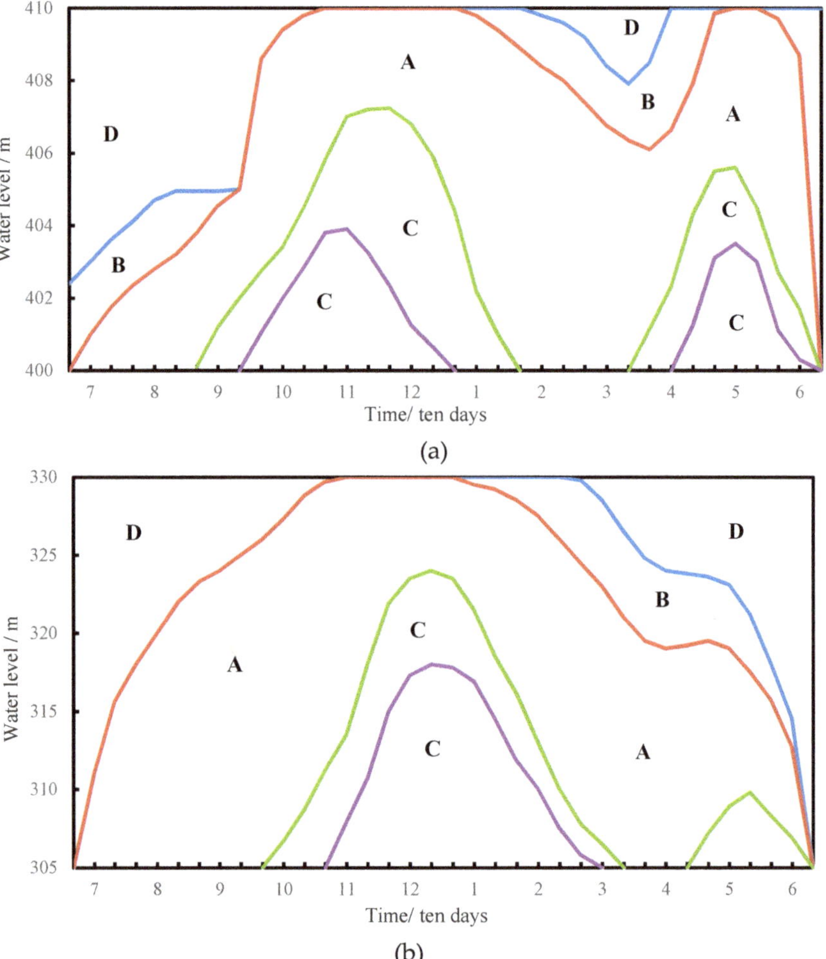

Figure 2. Conventional hydroelectric operation charts for Shiquan (**a**) and Ankang (**b**). The red line and green line are the upper basic line and lower basic line, respectively. The blue line and the purple line are the 1.2 *Ng* and 0.8 *Ng* line, respectively.

3. Materials and Methods

3.1. Variation of Runoff

The Mann-Kendall (MK) test is a nonparametric method for analyzing trends in time series and is recommended by the World Meteorological Organization [27,28]. Many scholars have used the MK test to analyze trends in precipitation, runoff, temperature, and water quality. The MK test is simple and easy to calculate and is applicable to data of non-normal distribution such as the data generally found in hydrology and meteorology studies. The detailed calculation process can be found in References [29,30].

3.2. Hydroelectric Operation Charts

Operation rules are an intuitive and practical way to guide the operation of reservoir, and the operation chart is a practical method for applying the rules in practice and so have been widely used in engineering operations [31,32]. The hydropower plant compares the conditions of the reservoir with the operation chart and accordingly stores or discharges water from the reservoir to meet the power generation requirements of the power system. For most hydropower plants, the hydroelectric operation charts were generated by historical runoff series and practical experience without hydrological forecasting [33]. However, under the combined influences of climate change and human activities, inflow runoff has changed since the traditional operation chart is designed according the initial runoff series. If the regulation capacity of a reservoir is limited compared with its inflow runoff, then runoff would largely determine the power generation of the hydropower stations. Especially if the guaranteed output area is too wide so that it further increases the difficulties in finding optimal global solutions. Therefore, exploring the coping strategies that are used to adjust conventional operation charts is of great urgency.

In recent years, much research regarding operation charts has been conducted, from which the methods of operation charts can be classified into three categories. One is the regular operation chart based on a simulation model with some manual corrections [34]. Second is the implicit stochastic optimal operation. The historical runoff series are input into the deterministic optimization model to obtain the optimal running solutions of the reservoir, and the operation rules are mined based on these solutions [35]. However, this operation chart is easily influenced by data mining methods and system errors. Third is to optimize the generalized operation chart directly [36]. The advantage being that the operation chart is optimized directly with less decision variables to avoid the "dimension disaster", and getting reasonable solutions by the long runoff series instead typical year. The optimization results can directly generate operation charts and can be analyzed and compared intuitively. Considering the existing conventional operation charts and basic rules, the third method was chosen for the present study. The four main parts are defined as follows:

(1) Part one: Generalize the initial operation charts, including the type and location of the selected water level line.

(2) Part two: Build a simulation model based on the basic rules and determine the objective functions.

(3) Part three: Choose the decision variables and an optimization method, and input the runoff series into the simulation model to calculate the objective function based on the generalized operation charts.

(4) Part four: Start evolution and iteration, and select the final operation chart corresponding to the best objective function result.

Reservoir operation methods have been greatly developed over the past 64 years. Little first applied dynamic programming and Markov chain methods to reservoir scheduling [37]. Evolutionary algorithms have also been widely used to optimize reservoir operations owing to their simple principles, easy implementation, parallel search capability, and global optimization ability, such as

Genetic Algorithm [38], Particle Swarm Optimization [39], Cuckoo Search Algorithm (CS) [40], and Differential Evolution Algorithm (DE) [41]. The CS algorithm was chosen for use in the present study owing to its superior search performance, fewer parameters, and robustness to obtain the optimal operation charts. CS mainly involves initializing the population, using Levi's flight to update the bird's nest position, and calculating fitness values; specific steps were shown in Reference [40].

3.3. Model Construction and Parameters

3.3.1. Simulation Model

The simulation model is constructed to analyze the influences of climate change and human activities singly and combined and to calculate the regular operation chart. This is also one part of the optimal model, which is used to obtain the optimal operation chart. The main purpose of establishing the simulation model is to generate the regular hydroelectric operation chart with the climate change and human activities data. The regular hydroelectric operation chart is drawn by the typical year method and co-output method. In this method, the dry season and wet season are determined first, then repeated to try out the power plant's output process until the output at time t is close to the basic output. The basic principle of the simulation model is water balance, and the main calculation processes are as follows:

Step one: Calculate the regulated flow during the dry season and determine the guaranteed output according to formulae (1)–(9).

$$Q_d = \frac{1}{T_d \cdot \left(\sum_{j=1}^{T_d} Q_i(j) + V_n \right)} \tag{1}$$

$$Q_w = \frac{1}{T_w \cdot \left(\sum_{j=1}^{T_w} Q_i(j) - V_n \right)} \tag{2}$$

$$V(t+1) = V(t) + (Q_i(t) - Q_o(t)) \cdot \Delta t \tag{3}$$

$$\overline{V}(t) = \frac{V(t+1) + V(t)}{2} \tag{4}$$

$$Z(t) = f_{vz}\left(\overline{V}(t) \right) \tag{5}$$

$$Z(t+1) = f_{qz}(Q_o(t)) \tag{6}$$

$$\overline{H}(t) = Z(t) - Z(t+1) - \Delta f \tag{7}$$

$$N'(t) = k \cdot Q_o(t) \cdot \overline{H}(t) \tag{8}$$

$$N_g = \frac{1}{T_d \cdot \sum_{j=1}^{T_d} N'(t)} \tag{9}$$

where Q_d and Q_w represent regulated flow during the dry season and wet season, respectively and Q_i and Q_o represent the inflow and outflow, respectively. T_d and T_w represent the length of the dry season and wet season, respectively. If the reservoir is in the dry season or wet season, $Q_o = Q_p$ or $Q_o = Q_f$, respectively. If the reservoir would not regulate the runoff, $Q_o = Q_i$. $V(t)$ and $\overline{V}(t)$ represent the reservoir storage and monthly average reservoir storage at t time, respectively. V_n represents the designed regulating reservoir storage, and Δt is the iteration step, which is 1 month. j represents the operation time point, which matches with runoff time series. $f_{vz}(\cdot)$ and $f_{qz}(\cdot)$ represent the functional relationship of $V \sim Z$ and $Q \sim Z$, respectively. $Z(t)$ and $Z(t+1)$ represent the water level at the beginning and end of time t, respectively. $\overline{H}(t)$ and Δf represent the water head for generation and head loss at time t, respectively. $N'(t)$ and N_g represent the output in the regulating period and the guaranteed output in the dry season, respectively.

Step two: Assume an initial power generation flow and calculate the initial reservoir storage based on the water balance formula and the upstream water level as follows:

$$V(t) = V(t+1) - (Q_i(t) - Q_o(t)) \cdot \Delta t \tag{10}$$

$\overline{V}(t)$, $Z(t+1)$, $Z(t)$, $\overline{H}(t)$, and $N'(t)$ are calculated as in formulae (4)–(8).

Step three: Compare the power plant output $N'(t)$ with N_g. Then adjust the outflow with Δq and return to step two:

$$\begin{cases} Q_o(t) = Q_o(t) - \Delta q, \ N'(t) > N_g \\ Q_o(t) = Q_o(t) + \Delta q, \ N'(t) < N_g \end{cases} \tag{11}$$

where Δq is the change in the outflow of the power plant according to the actual reservoir conditions.

Step four: If formula (12) is successfully calculated, then go ahead to step five, otherwise adjust the outflow and return to step two:

$$\left| N'(t) - N_g \right| < \delta \tag{12}$$

where, usually, $\delta = 0.01$, kW.

Step five: If formula (13) is successfully calculated, then stop, otherwise adjust the outflow and return to step two.

$$|Z_e - Z_{dead}| < \delta \tag{13}$$

where Z_e and Z_{dead} represent the water level at the end of whole period and the designed water dead level. The time trial ends when the above formula is satisfied; repeating all the steps until to the first period of the wet season.

3.3.2. Optimal Model of Cascade Hydropower Joint Operation

The main purpose of establishing an optimal model is to determine the optimized hydroelectric operation charts under the influences of climate change and human activities. However, with the development of reservoir operations over the past 60 years, researches on the cascade reservoir joint operations have been recognized by the public compared with single reservoir operation. Joint operation of cascade reservoirs with hydrological and hydraulic connections can obtain greater benefits than single reservoir operations. Additionally, in practice, cascade hydropower joint operations are one of bottlenecks to achieve more benefit in hydropower stations of the Han River. Therefore, an optimal model for cascade hydropower joint operations is established and maximizing power generation is the main objective function of the optimal model. The objective function and constraints of this model are as follows:

(1) Objective function

$$E = Max \sum_{t=1}^{T} \sum_{m=1}^{M} N_m(t) \cdot \Delta t \tag{14}$$

$$N_m(t) = k_m \cdot Q_o^m(t) \cdot \overline{h_m}(t) \tag{15}$$

where E is equal to the total power generation of the three power stations in operation series, 10^8 kWh. T and M represent the length of the operation cycle and the number of reservoirs, followed by Shiquan, Xihe, and Ankang. $\overline{h_m}(t)$ represents the water head of the m reservoir at time t, k_m represents the power coefficient of the m power station.

(2) Operational constraints

1) Water balance

$$V^m(t+1) - V^m(t) = \left(Q_i^m(t) - Q_o^m(t)\right) \cdot \Delta t \tag{16}$$

2) Water level

$$Z_{min}^m \leq Z^m(t) \leq Z_{max}^m(t) \tag{17}$$

3)   Maximum overflow

$$Q_o^m(t) \leq Q_{max}^m(t) \tag{18}$$

4)   Output of power station

$$N^m(t) \leq N_{ins}^m \tag{19}$$

$$N_{dry}^m(t) \leq N_g^m \tag{20}$$

5)   Operation lines are not allowed to be intersected in the operation chart optimization.

$$DZ_{k-1}^m(t) \leq DZ_k^m(t), \; t = 1, 2, \ldots, T, \; k = 1, 2, \ldots, K \tag{21}$$

where N_{ins}^m is the installed capacity of the m hydropower station; $DZ_k^m(t)$ is the value of the water level line of the operation chart of m hydropower station at time t and k is the number of the water level line.

3.3.3. Parameter Setting and Evaluation Indicators

According to optimal model determined in the present study, the value of the water level of the operation line of the operation chart is chosen as the decision variable, and the specific parameters are listed in Table 2.

Table 2. Parameters for the solution algorithm.

Parameters	CS Algorithm
Decision variable	Water level
Number of operation lines	4
Number of decision variables	48
Population size	400
Generation	5000
Discovery probability	0.25

In addition to the two indicators of power generation and guaranteed output, three other indicators are increased including the rate of water abandonment, the assurance rate of power generation, and the rate of water consumption to evaluate the optimized performance of the hydroelectric operation charts. The formulae are as follows:

1) The rate of water abandonment-P_a

$$P_a^m = \frac{1}{T \cdot \sum_{i=1}^{T} q_a^m(t) / Q_o^m(t)} \tag{22}$$

where $q_a^m(t)$ represents the discarded outflow at time t.

2) The assurance rate of power generation-P

$$P^m = \frac{f(N^m(t) > N_g^m)}{T} \cdot 100\% \tag{23}$$

where $f(N^m(t) > N_g^m)$ represents the number of the output of the power station is greater than the guaranteed output.

3) The rate of water consumption—P_w (m^3/k·Wh)

$$P_w = \frac{W}{E} \tag{24}$$

where W is equal to the water quantity used for power generation.

3.4. Calculation Schemes

In this study, the coping operation charts are the final objective for managers. Therefore, four operation scenarios are designed, including the initial runoff series, the influence of climate change, the influence of human activities, and combined influence of climate change and human activities.

Before the construction of the Project, there is no large-scale human development in the upper reaches of the cascade power station, so the Project represents human activity. Since the Project is not yet operational, if the existing observed runoff data mutates, it is caused by the influence of climate change. Runoff variation point (Y) is obtained by MK test. The runoff series before the variation point is the initial runoff series, and the data series after the variation point is the runoff series affected by the climate change. In other study of the research group, the simulated results of the Project are obtained. So the operation data of the Project is subtracted from the runoff data before and after the variation point respectively, and runoff data series affected by human activities only and affected by combined influence of climate change and human activities have been obtained.

Three kinds of operations charts as coping strategies are the final results in this paper. Chart 1 is the conventional corrections for traditional single reservoir operation chart. Data series of four operation scenarios are calculated in designed traditional operation charts. Chart 1 is generated from the results of the data series affected comprehensively by climate change and human activities. Chart 2 and Chart 3 is the optimal single and cascade reservoir operation chart, respectively. Both are generated by CS algorithm from the results of the data series affected comprehensively by climate change and human activities.

Six schemes are designed in terms of the operation cascade hydropower stations under climate change and human activities and its revised optimal operation strategies. The specific schemes are presented in Table 3. Correspondingly, the flow chart of calculation for coping operation charts are showed in Figure 3.

Table 3. Calculation schemes and the operation model.

Scheme	Operation Scenario	Operation Mode	Coping Strategy
1	**Initial natural runoff** (1954-Y)		
2	**Only climate change**: natural runoff (Y-2010)	Single reservoir operation in conventional operation charts	Chart 1
3	**Only human activities**: 1954-Y, natural runoff (1954-Y) minus transferred process (1954-Y)		
4	**Combined climate change and human activities**: natural runoff (Y-2010) minus transferred process (Y-2010)		
5	**Combined climate change and human activities**: natural runoff (Y-2010) minus transferred process (Y-2010)	Single reservoir optimization	Chart 2
6		Cascade reservoir optimization	Chart 3

There are four classes of operation chart including the conventional designed operation chart. The main reasons and purposes of the six schemes were as follows:

(1) To identify the influence of climate change and human activities on hydropower stations, the station was set to operate under four scenarios with the conventional operation chart. The power generation (E), guaranteed output (Ng), P_a, P, and P_w were used to quantitatively analyze the impacts.

(2) Chang. J. generates cascade joint operation charts for Shiquan, Xihe, and Ankang to respond to climate change in the upper Han River [13]. Also, the Project will run until 2030; therefore, calculating the operation charts after the Project has operated is one of the main tasks. Chang. J. indicates that optimal operation charts could lead to the generation of more power in the cascade stations [13]. Scheme five is therefore set in the optimal model.

(3) Before the Project operated, the operation chart in Reference [13] is applied. Once the Project has been operated, a cascade joint operation chart under climate change and human activities will be necessary. Therefore, scheme six is set in the cascade joint optimal model.

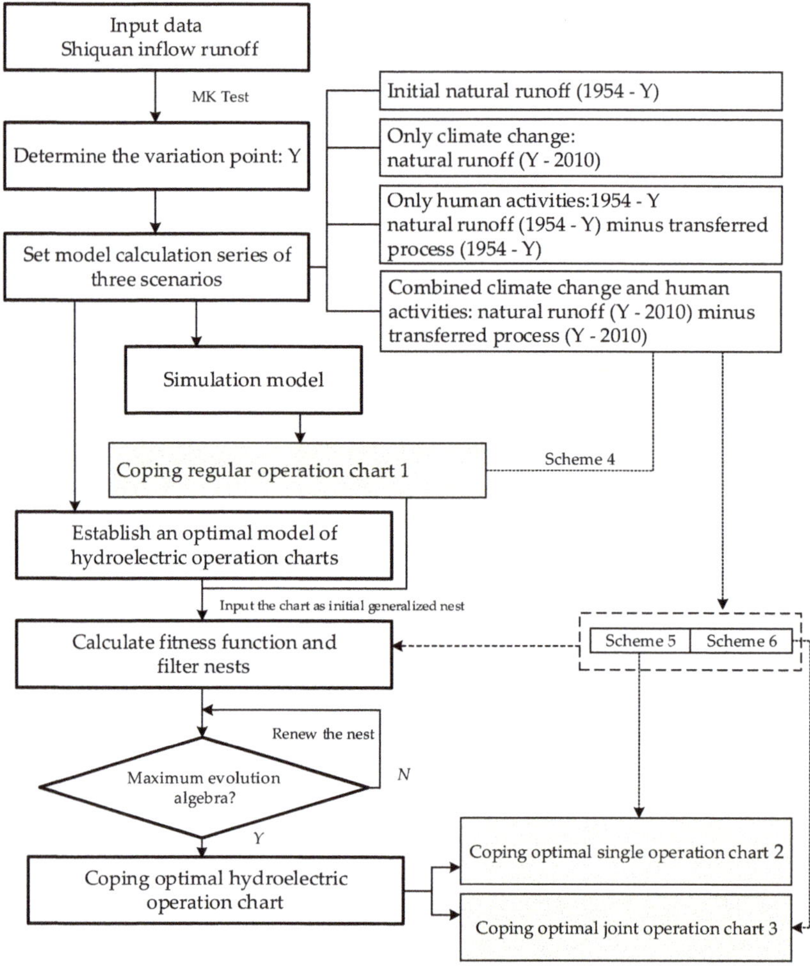

Figure 3. Flow chart for generating the optimal operation charts under climate change and human activities. See Section 3.2 in the main text for definition of the different parts. MK = Mann-Kendall test.

4. Results and Discussion

The results include analysis of runoff variation points, the influences of climate change and human activities on hydropower generation, and the coping regular and optimal operation charts. All operation charts are expanded based on the order of the schemes. Then we discuss the effects on cascade hydropower joint operations of the combined influence of climate change and human activities and generate three cascade hydropower joint operation charts.

4.1. Analysis of Runoff Variation Point

Only the runoff data series is used in the present study, therefore, the runoff variation point is acquired using the inflow runoff of Shiquan reservoir with the MK test (Figure 4). Then, we review the references to verify the existence of the variation point.

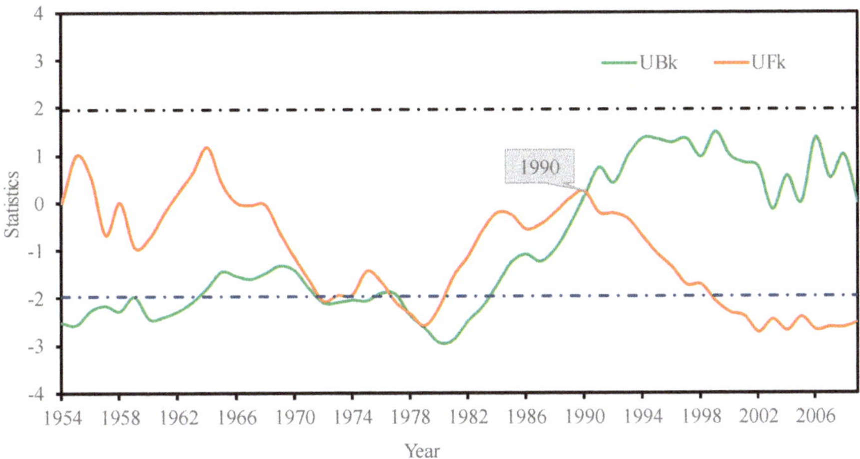

Figure 4. Mann-Kendall test results of inflow of Shiquan reservoir, UBk and UFk are time statistics.

In the MK test, the statistical variable is −2.5372, which indicates a decreasing trend for the runoff (Figure 4). The curves of UB_k and UF_k crossed in 1990, which may indicate that the runoff begins to change in this year.

He reports that increased in average temperature and decreased in precipitation leading to changes in the hydrological process from 1950–2005 in the upper Han River [42]. The climate become drier from 1980–2005. Similarly, Chang. J. also finds the runoff in the upper Han River has changed in 1990 [13]. Both of these studies report similar meteorological factors under climate change, and 1990 is regarded as the beginning of the observed variation. Therefore, 1990 is the variation point (Y) in this study, and the runoff data of four operation scenarios are listed as follows:

(1) Initial natural runoff (1954–1990)
(2) Only climate change: Natural runoff (1991–2010)
(3) Only human activities: Natural runoff (1954–1990) minus transferred process (1954–1990)
(4) Combined climate change and human activities: natural runoff (1991–2010) minus the transferred process (1991–2010)

4.2. Influences of Climate Change and Human Activities on Hydropower Operation

According to the results in Section 4.1 and the schemes in Table 3, the Operating results calculated in conventional operation charts are shown in Table 4 and Figure 5.

Table 4. Operating results of hydropower station under the influence of climate change and human activities (Scheme 1, Scheme 2, Scheme 3, and Scheme 4).

Scheme	1: Initial Runoff (1954~1989)					2: Climate Change (1990~2009)					3: Human Activities (1954~1989)					4: Climate Cange and Human Activities (1990~2009)				
Station	E	Ng	P (%)	Pw	Pa (%)	E	Ng	P (%)	Pw	Pa (%)	E	Ng	P (%)	Pw	Pa (%)	E	Ng	P (%)	Pw	Pa (%)
Shiquan	7.08	35	88	10.4	28.10	5.9	28	64	10.40	27.30	5.63	27	59	10.6	26.34	4.45	25	50	10.7	25.24
Xihe	4.95	23	91	14.3	22.42	4.75	20	80	14.3	22.11	3.91	18	75	14.8	21.34	3.71	16	65	15.1	20.37
Ankang	27.58	185	81	5.8	15.01	21.85	159	74	6.2	13.56	24.99	170	69	6.3	13.92	19.26	165	67	6.3	12.21
Cascade stations	39.61	243	87	10.17	21.84	32.5	207	73	10.30	20.99	34.53	215	68	10.57	20.53	27.42	206	61	10.70	19.27

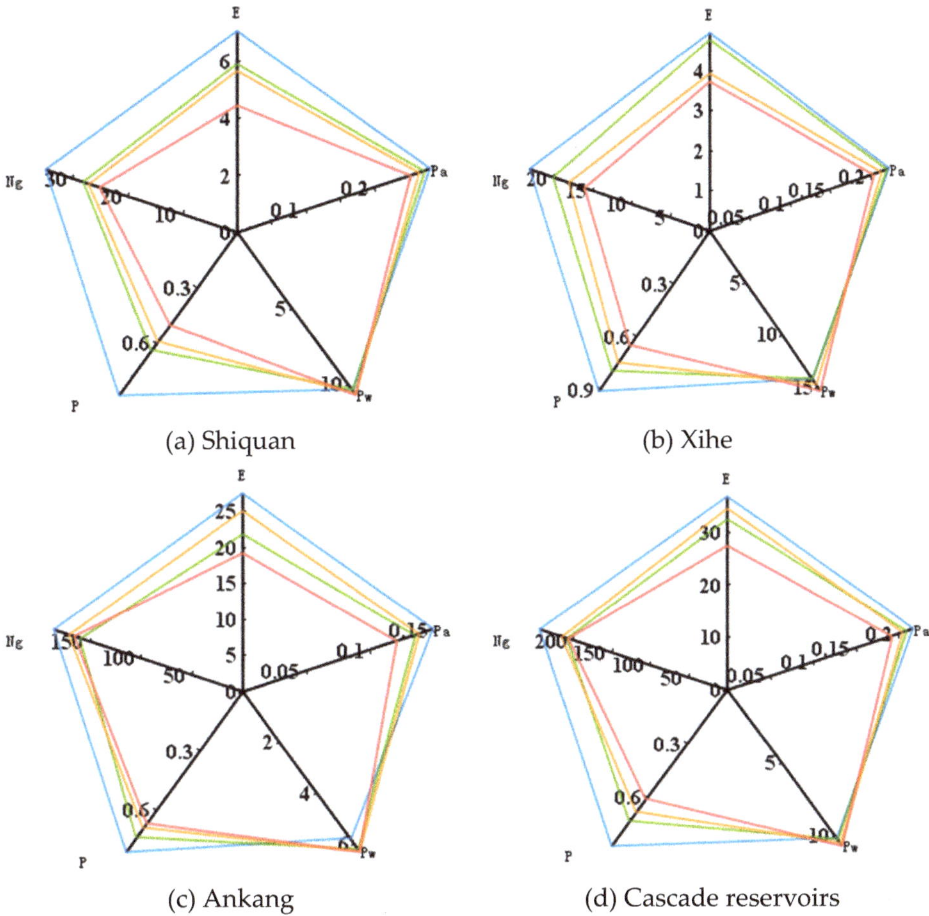

Figure 5. Operating parameters (E, N_g, P, P_w, and P_a) of the hydropower stations in the conventional operation chart under the four schemes. The blue, green, yellow, and red curve are in turn scheme 1 to 4.

The results of the present study show that both climate change and human activities affect the operation of the three hydropower stations under investigation. Only the power generation and guaranteed output of scheme 1 is found to reach or exceed the design value in the three power stations. Because the calculation series used in designing conventional operation charts is different from that used in the present study, scheme 1 is regarded as a reference standard of conventional operation charts rather than for designed values. Compared with scheme 1, the results indicate that:

(1) The results of schemes 2–4 are significantly worse than those of scheme 1. This indicates that the conventional operation chart is no longer suitable for the operation and development of the power stations under the varied hydrological situations. For example, the power generation of Shiquan under schemes 2–4 decrease by 1.18×10^8 kWh (16.67%), 1.45×10^8 kWh (20.48%), and 2.73×10^8 kWh (38.56%), respectively. Additionally, the same decreasing trend occurs in Xihe, which decreases by 0.2×10^8 kWh (4.04%), 0.3×10^8 kWh (6.06%), and 0.37×10^8 kWh (7.47%) and Ankang, which decreases by 5.73×10^8 kWh (20.78%), 2.59×10^8 kWh (9.39%), and 7.68×10^8 kWh (27.85%), respectively

(2) From the results of cascade reservoir operations, the influence of climate change is greater than that of human activities. For instance, power generation under schemes 2 and 3 decreases by 17.95%

and 12.83%, respectively. The reason for this is that the natural runoff after 1990 decreases by an annual average of about 2.5 billion m^3 in the upper Han River area.

(3) Based on the sensitivity of the power stations to these changes, Shiquan is the most affected followed by Ankang. The reason for this is that there is lower storage capacity in Shiquan than in Ankang, which means that Shiquan does not have sufficient capacity to save water and regulate the power head when runoff is reduced. This once again verifies the necessity to modify the operation charts under existing engineering and hydrological conditions.

4.3. Coping Hydropower Operation Charts under the Influence of Climate Change and Human Activities

Because the Project is planned to operate completely by 2030, hydropower operations of cascade power stations are currently mainly affected by climate change. Considering that [13] proposes adaptive operation charts and cascade joint optimal operation charts for the upper Han River, the main object of the present study is to develop the coping strategies to deal with the combined influence of climate change and human activities after 2030. As shown in Table 4, scheme 4 and 5 are used to generate the modified regular and optimal single hydropower operation charts, respectively. Scheme 6 is used to generate the modified optimal cascade hydropower operation chart. All the modified operation charts are listed in Figure 6.

Chart 1: Modified regular single reservoir operation charts

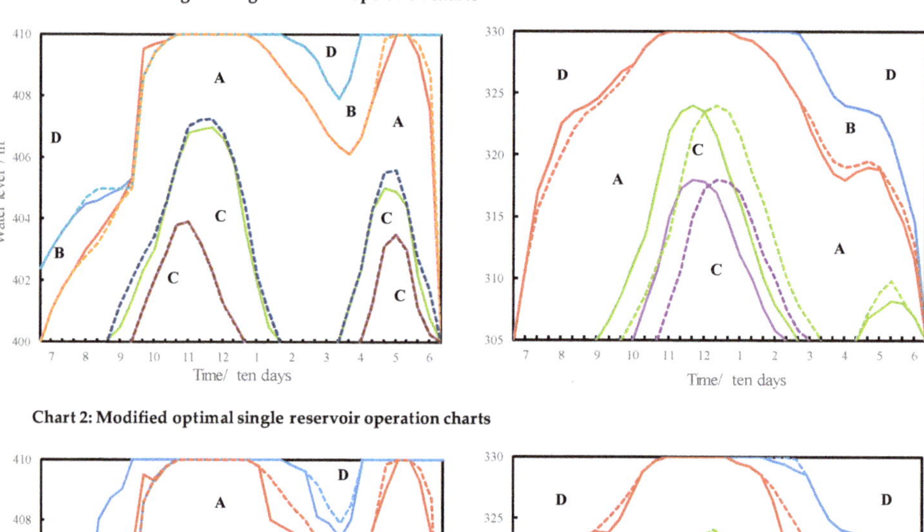

Chart 2: Modified optimal single reservoir operation charts

Figure 6. *Cont.*

Chart 3: Modified optimal cascade reservoirs operation charts

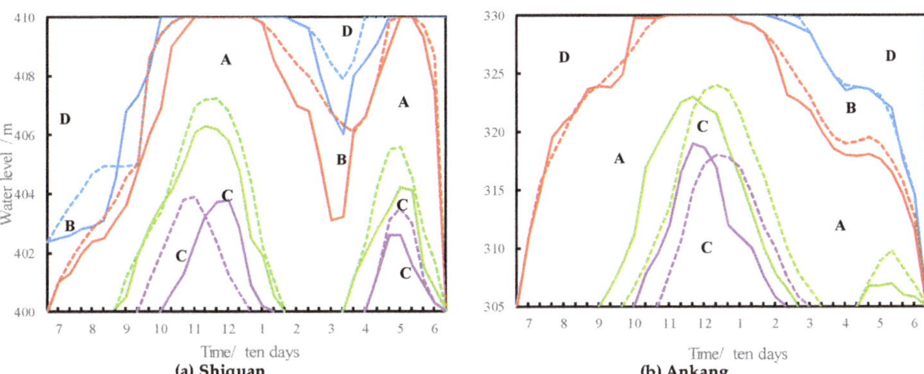

Figure 6. Modified hydropower operation charts for Shiquan (**a**) and Ankang (**b**) to address the effects of climate change and human activities. The solid and dashed lines indicate the modified and traditional operation lines, respectively. The red line and green line are the upper basic line and lower basic line, respectively. The blue line and the purple line are the 1.2 Ng and 0.8 Ng line, respectively.

4.3.1. Chart 1: Modified Regular Single Reservoir Operation Chart

Comparing Chart 1 in Figure 6 with the conventional operation charts (Figure 1), shows that the integral operation trend is similar to the traditional one. The upper and lower basic lines move slightly up and down. Specifically, the upper basic lines of both reservoirs move up a little during the main flood season and move down during the dry season. Furthermore, both lower basic lines move down during operation time, and the water storage period of the Ankang reservoir move forward for about 20 days. The main reason for this change is that reservoir inflow is decreased compared with the designed data series. During the main flood season, to improve power generation efficiency, the guaranteed output area is expanded and moves down, and the increased output area becomes smaller. This change would probably make the power stations work more in the guaranteed output area. Then, once the power plant is working in part A, it would be able to maintain a high water level and guarantee water demand during the dry season. Furthermore, in contrast to the previous conditions, the Ankang reservoir should store water in advance to raise the water level and avoid the power plant operating at a decreased output after the flooding season. In the dry season, the operation mode is different from during the wet season, in which the increased output area become larger making the power plant work in the increased output area, avoiding the abandoned water.

We assume that the cascade power stations are running under the Chart 1 (Figure 6) to verify their usefulness. The simulation results show that the effect of both operation charts could not reach the design value under the current runoff situation, but the modified chart is preferable to the conventional chart. Cascade power generation increases by about 12 million kWh, of which Ankang reservoir increases by about 9 million kWh. If the electricity price is calculated at RMB 0.25, then the increased value of the power generated is about RMB 3 million. Therefore, it is necessary and valuable to modify the operation charts for the changing hydrological environment, especially for reservoirs that are designed for power generation.

4.3.2. Chart 2: Modified Optimal Single Reservoir Operation Chart

Comparing the Chart 2 in Figures 6 and 7 shows that the overall trends at Shiquan and Ankang have not changed. After the calculation of the CS algorithm, part B of Shiquan increases to avoid abandoned water during the flood season because heavy rains often occur in the Han River basin, and rainfall is particularly heavy during the flood season. Additionally, the flooding season commonly lasts until October. Therefore, the hydropower plant should increase its outflow to improve power generation and avoid abandoned water. Part A of the Ankang reservoir increases to raise the water level, and the water level line of 0.8 N_p is slightly low compared with Figure 6b. All these changes are used to increase the power generation of the power stations.

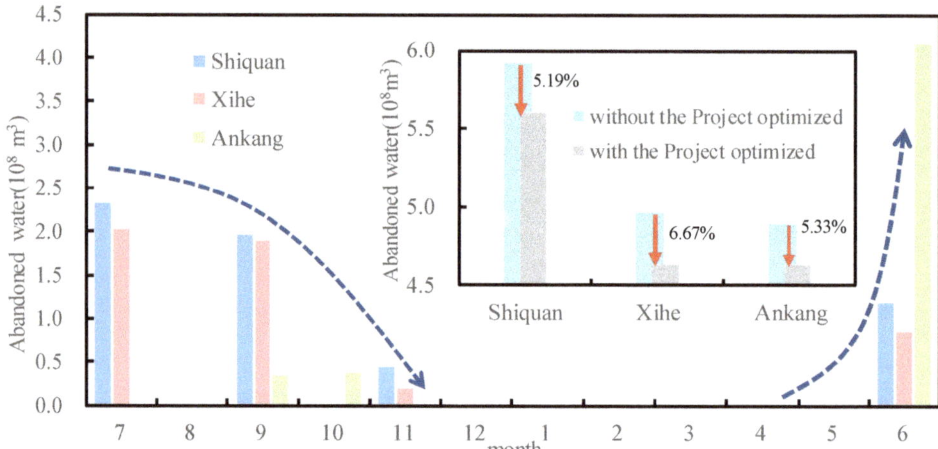

Figure 7. The monthly average change process of water abandoned of the Shiquan, Xihe, and Ankang reservoirs in condition of the Project optimized. The upper right part is the multi-year average change of abandoned water when the operation of Project is optimized or not.

Like in scheme 5, we assume that the cascade power stations are running under the Chart 2 to verify their usefulness. The simulation results show that the optimal chart outperforms the conventional chart. Cascade power generation increases by about 32 million kWh, of which Shiquan and Ankang reservoirs increases by about 9 and 23 million kWh, respectively. The increased value of the power generated is about RMB 8 million. Therefore, the single optimal operation chart (Chart 2) further improves the power generation of the power stations compared to the modified single regular operation chart (Chart 1).

4.3.3. Chart 3: Modified Optimal Cascade Reservoir Operation Chart

Comparing the Charts 2 and Chart 3, it can be seen that the operating areas of the Shiquan and Ankang reservoirs have not changed. In Chart 3, the upper basic lines moves down slightly during the dry season and the increasing output areas become lager. Similarly, under joint operation, the guaranteed output is increased to avoid abandoning water.

Like in scheme 5, we assume that the cascade power stations are running under the Chart 3 to verify their usefulness. The simulation results show that if the power stations run in a joint operation, the effects of human activities and climate change are greatly reduced. Cascade power generation increases by about 47 million kWh, of which Shiquan and Ankang reservoirs increases by about 12 and 35 million kWh, respectively. The increased value of the power generated is about RMB 11.25 million. These joint operation charts consider the effects of human activities and climate change that could be applied in the cascade hydropower stations in theory once the Project is finished in 2030. Climate change is a gradual process, and its effects on runoff are subtle. At present, runoff cannot be restored to its original state. However, it is possible to reduce the effect on runoff by making significant changes to human activities.

4.4. Optimal Operation of the Project

As the Project is based at the first cascade of the upper reaches of the Han River, its operation mode will directly affect the operation of the downstream power station group. [24] studies the joint optimization scheduling of the Huangjinxia (HJX) and Sanhekou (SHK) reservoirs of the Project to determine water supply, power generation, and energy consumption for the Project's own operation. While the modified optimal single reservoir and joint reservoirs operation charts of the cascade power stations increase power generation, abandoned water of reservoirs will always occur and reduce water energy efficiency. Therefore, based on the optimal models in Reference [24], preventing the occurrence of abandoned water at the downstream cascade power stations is as important as determining water supply and energy efficiency.

The results from schemes 5 and 6 show that abandoned water of cascade reservoirs often occur before and during the flood season (June–November). Therefore, the Project in upstream should increase the outflow constraint in the joint operation model during this period to avoid the cascade power stations generating excessive abandoned water

The operation process for the whole system includes two steps: (1) Obtaining reservoir outflow series for the joint optimal operation of the Project. These series are also taken as the inflow of the Shiquan reservoir. (2) Joint operation chart 3 is regarded as another fitness function. The model locates the series in which the corresponding power generation process resulted in higher levels of generated power coupled with lower levels of abandoned water. Then, following iterative optimization, an operation process for the Project is determined.

Figure 7 shows the monthly average change process of water abandoned of the cascade power stations in condition of the operation of the Project is optimized. When the Project operates in optimal situation, and the cascade power stations operate according the Chart 3, the operation results of the cascade power stations show: (1) The average annual of abandoned water would decrease by about 150 million m^3 and power generation would increase by 5 million kWh. (2) The Project would reduce the water level pressure for the downstream reservoirs before the flood season, and the abandoned water of Shiquan, Xihe, and Ankang power stations would decrease by 5.19%, 6.67%, and 5.33%, respectively. (3) The abandoned water from Shiquan and Xihe reservoirs always occurs at the same time, and the largest amount of abandoned water is July. These three reservoirs occur abandoned water in June and September at the same time, and Ankang reservoir has the largest abandoned water before the flood season (June).

The results for the joint operation model of the Project are suitable for the operation of the Project itself (data not shown in detail). Additionally, this section focuses on the operation rules of the Project, which are conducive to the operation of the downstream power stations.

Figure 8 shows that the differences between the two operation charts are mainly in terms of the hedging rule curve for abandoned water and the combined water supply. (1) The modified hedging rule curve for abandoned water shifts upward about 3 m from March–October compared with the initial curve. This part is defined as the operation area for preventing the abandonment of water. If the SHK water level at time t is in this location, then the SHK reservoir would be regarded as the first water resource to supply water and the HJX reservoir is the second. Otherwise, the HJX reservoir would still undertake the majority of the water supply task and reduce the probability that of water abandonment at the downstream power plants. (2) The modified hedging rule curve for abandoned water shifts down about 10 m from May–November compared with the initial curve. This part is defined as the guaranteed operation area of combined water supply. If the SHK water level at time t is in this location, then the HJX reservoir would be regarded as the first water resource. This change increased the probability of the HJX reservoir supply water for intake areas.

The water transferred scale of the Project is limited by the adjustable water volume which is approved by the government. Therefore, the basis of the Project operation chart change is in the range of adjustable water volume, what's more, HJX reservoir must take on more water supply tasks, and the SHK reservoir would mainly regulate the main stream runoff through the water supply. These measures have been validated in the model to mitigate the abandoned water of the downstream reservoirs. This also means that, if the effects of human activities are inevitable, then the optimal operation of the whole system could reduce the impact.

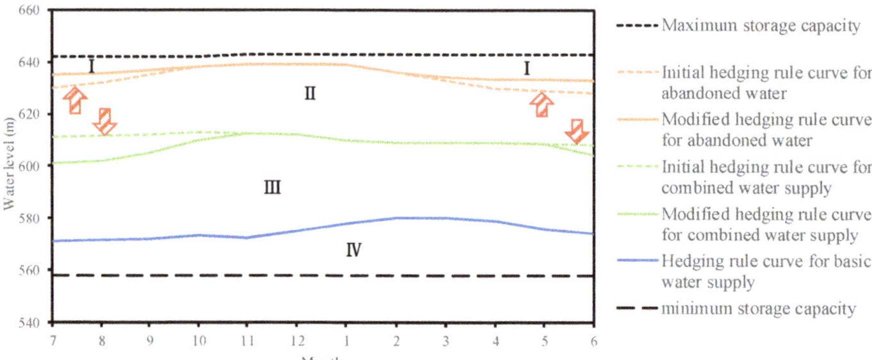

Figure 8. The modified operation chart for the Sanhekou reservoir of the Project. The original interpretation of the operation chart is listed in the supplementary material.

5. Conclusions

Many conventional operation rules for reservoir operations no longer apply to the current hydrological environment, which has been affected by climate change and human activities. Therefore, revised strategies are urgently needed for the optimized operation of cascade reservoirs. In the present study, the influence of climate change and human activities is analyzed and procedures are formulated to develop revised strategies. We consider three runoff series conditions including initial runoff, only affected by climate, and affected by both climate and human activities. A simulation model is applied to analyze the effects on power generation under four schemes, and a modified regular operation chart is generated through scheme 4. An optimal model for operation charts based on the simulation model is constructed to generate single and joint optimal operation charts for cascade power stations under the influence of climate change and human activities. We also attempt to change the influence of human activities by optimizing the rules of the Project to reduce its influence on power stations in the downstream. The primary conclusions are as follows:

(1) Both climate change and human activities affect the operation of cascade power stations. At the same time, the effect on power generation is greatest when climate change and human activities were combined, followed by climate change alone, and finally human activities alone. Compared with the initial condition, corresponding power generation decreases by 25.71%, 17.95%, and 12.83%, respectively. Furthermore, owing to geographical location and its own storage capacity, the Shiquan reservoir is the most sensitive to these changes.

(2) Three kinds of revised strategies for the cascade power stations are proposed herein, mainly by modifying existing operation charts. The three modified operation charts include a regular chart, an optimal single operation chart, and an optimal joint operation chart. Compared with the conventional chart, all three modified charts are preferable for the cascade power stations. The optimal joint operation chart shows better adaptability to the changes in runoff and the most evident increase in power generation (47 million kWh; RMB 11.25 million).

(3) Optimizing the upstream Project and slowing down its impact on the downstream power stations is another revised strategy proposed herein. If the Project works as in Figure 8, then the average annual abandoned water of the downstream power stations would decrease by about 150 million m^3, and the abandoned water of Shiquan, Xihe, and Ankang power stations would decrease by 5.19%,

6.67%, and 5.33%, respectively, which could increase power generation by 5 million kWh. The Project can also reduce the water level pressure before the flood season for the downstream power stations.

The present study has practical significance for the efficient operation of cascade hydropower stations and is informative for reservoir operation theory under changing environmental conditions. In future, studies should focus on power generation operations considering river ecology to solve the conflicting objectives of ecological benefit and power generation under changing environments.

6. Supplementary Material

6.1. The Multi-Objective Optimal Model for the Project

The multi-objective optimal model for the Project considers energy consumption, power generation, and water supply.

(1) Multi-objective function:

$$minF(x) = \left(E_{pump}, E_{power}, W \right) \tag{25}$$

Objective one: minimizing energy consumption

$$min \; E_{pump} = min \left[\sum_{t=1}^{T} \sum_{m=1}^{M} P_{pump}^{m}(t) \cdot \Delta t \right] \tag{26}$$

$$\sum_{m=1}^{M} P_{pump}^{m}(t) = \sum_{m=1}^{M} \frac{g \cdot q_{pump}^{m}(t)}{\eta_{pump}^{m}} \tag{27}$$

Objective two: maximizing power generation

$$max \; E_{power} = max \left[\sum_{t=1}^{T} \sum_{m=1}^{M} N_{power}^{m}(t) \cdot \Delta t \right] \tag{28}$$

$$\sum_{m=1}^{M} N_{power}^{m}(t) = \sum_{m=1}^{M} k \cdot Q_{power}^{m}(t) \cdot h(t) \tag{29}$$

Objective three: meeting water demand

$$W = \sum_{t=1}^{T} \sum_{m=1}^{M} Q_s(m,t) \cdot \Delta t \tag{30}$$

where Ω represents the set of optimal solutions for multi-objective operation models, E_{pump} represents the total energy consumption of two pump stations in an operation series, E_{power} represents the total power generation of two power stations in an operation series, W represents the transferred water quantity. T, M, and Δt represent the same as in the simulated operation model; $P_{pump}^{m}(t)$ represents the power from pump station m consumed in the period t, $q_{pump}^{m}(t)$ represents the water flow of pump station m transferred in the period t, η_{pump}^{m} represents the efficiency of pump station m, g represents gravity; $N_{power}^{m}(t)$ represents the power generation of power station m generated in the period t, $Q_{power}^{m}(t)$ represents the power flow of power station m used in the period t, $h(t)$ represents the water head of reservoir m in the period t, and k represents the power coefficient of power station m.

(2) Operational constraints

The operational constraints were as follows:

1) Water balance

$$V^m(t+1) - V^m(t) = \left[Q_I^m(t) - Q_O^m(t) - Q_S^m(t)\right] \cdot \Delta t \tag{31}$$

2) Water level

$$Z_{min}^2 \leq Z^2(t) \leq Z_{max}^2(t) \tag{32}$$

3) Transferable water quantity

$$\sum_{m=1}^{M} Q_S^m(t) \cdot \Delta t \leq W_{max}^{qty}(t) \tag{33}$$

4) Maximum overflow

$$Q_{power}^m(t) \leq Q_{max}^m \tag{34}$$

$$Q^{tunnel}(t) \leq Q_{max}^{tunnel} \tag{35}$$

5) Output of power station

$$N^m(t) \leq N_{installed}^{m,max} \tag{36}$$

$$N_{dry}^1(t) \geq N_{firm}^1 \tag{37}$$

6) Power of pump station

$$P^m(t) \leq P_{installed}^{m,max} \tag{38}$$

where $V^m(t)$ represents storage capacity of the m reservoir in t period (10^8 m^3); $Q_I^m(t)$, $Q_O^m(t)$, and $Q_S^m(t)$ represent the inflow runoff, outflow runoff, and water transferred flow of the reservoir m in period t, respectively (m^3/s); $Z^2(t)$ represents the water level of the SHK reservoir in period t, Z_{min}^2 represents the dead water level and $Z_{max}^2(t)$ represents the highest water level, including the flood control level during flooding season and the normal high water level during non-flooding seasons (m); $W_{max}^{qty}(t)$ represents the maximum transferable water quantity of the Han River in period t (10^8 m^3); $Q^m(t)$ represents the outflow of the power station m in period t, Q_{max}^m represents the maximum outflow of the power station m (m^3/s); $Q^{tunnel}(t)$ represents the average transferred flow in the Qinling tunnel in period t, Q_{max}^{tunnel} represents the maximum water transfer capability of the Qinling tunnel (m^3/s); $N^m(t)$ represents the output of power station m in period t, $N_{installed}^{m,max}$ represents the installed capacity of power station m, $N_{dry}^1(t)$ and N_{firm}^1 represent the output in the dry season and the firm power of HJX power station, respectively; $P^m(t)$ represents the power consumption of the pump station m in period t, $P_{installed}^{m,max}$ represents the installed capacity of pump station m. All variables were non-negative.

6.2. The Original Interpretation of the Operation Chart for the Project

The operation chart (Figure 8) includes four parts defined as follows:

(1) Part I

This part was defined as the operation area for preventing the occurrence of abandoned water. If the SHK water level at time t was in this location, then to save energy, the SHK reservoir would be regarded as the first water resource and its pump station would not work in this moment. At the same time, the HJX reservoir would be regarded as an auxiliary water resource. If the SHK reservoir is able to meet the water demands of the Guanzhong area, then the HJX pump station would not be needed, otherwise it would need to start supplying water.

(2) Part II

This part was defined as the guaranteed operation area for combined water supply. If the SHK water level at time t was in this location, then the HJX reservoir would be regarded as the first water resource and its pump station should supply the Guanzhong area and the SHK reservoir as much as possible. If the water in HJX is not sufficient, then the SHK reservoir would start to supply. In this part, the Project should meet the water demand of the Guanzhong area.

(3) Part III

This part was defined as the control operation area of combined water supply. In this part, the HJX and SHK reservoirs would supply the Guanzhong area together, and the HJX reservoir would not supply the SHK reservoir. The actual water supply of the Project would not meet the water demand because the actual water demand was applied according to the modified ratios set by the decision makers.

(4) Part IV

This part was defined as the minimum capacity water supply operation area. In this part, the HJX and SHK reservoirs both supply water according to their minimum capacity, and the HJX reservoir stops supplying the SHK reservoir.

Author Contributions: Conceptualization, T.B. and Q.H.; methodology, T.B. and L.W.; Data curation, M.Z.; investigation, P.M.; resources, T.B.; writing—original draft preparation, L.W.; writing—review and editing, T.B. and L.W.; supervision, Q.H.; project administration, Q.H.; funding acquisition, T.B. and Q.H.

Funding: This study is supported by the National Key R&D Program of China 2017YFC0405900), the National Department Public Benefit Research Foundation of Ministry of Water Resources (201501058), Planning Project of Science and Technology of Water Resources of Shaanxi (2017slkj-16, 2017slkj-27), the China Postdoctoral Science Foundation (2017M623332XB), the Postdoctoral Research Funding Project of Shaanxi Province (2017BSHYDZZ53), and the Basic Research Plan of Natural Science in Shaanxi Province (2018JQ5145).

Acknowledgments: The authors are indebted to the reviewers and editors for their valuable comments and suggestions.

Conflicts of Interest: The authors declare no conflict of interest.

References

1. Dey, P.; Mishra, A. Separating the Impacts of Climate Change and Human Activities on Streamflow: A Review of Methodologies and Critical Assumptions. *J. Hydrol.* **2017**, *548*, 278–290. [CrossRef]
2. Wang, W.; Shao, Q.; Yang, T.; Peng, S.; Xing, W.; Sun, F.; Luo, Y. Quantitative Assessment of the Impact of Climate Variability and Human Activities on Runoff Changes: A Case Study in Four Catchments of the Haihe River Basin, China. *Hydrol. Process.* **2013**, *27*, 1158–1174. [CrossRef]
3. Rind, D.; Rosenzweig, C.; Goldberg, R. Modelling the Hydrological Cycle in Assessments of Climate Change. *Nature* **1992**, *358*, 119–122. [CrossRef]
4. Williams, P.D.; Guilyardi, E.; Sutton, R.; Gregory, J.; Madec, G. A New Feedback on Climate Change from the Hydrological Cycle. *Geophys. Res. Lett.* **2007**, *34*, L08706. [CrossRef]
5. Jiang, S.; Ren, L.; Yong, B.; Singh, V.P.; Yang, X.; Yuan, F. Quantifying the Effects of Climate Variability and Human Activities on Runoff from the Laohahe Basin in Northern China Using Three Different Methods. *Hydrol. Process.* **2011**, *25*, 2492–2505. [CrossRef]
6. Wang, S.J.; Yan, M.; Yan, Y.X.; Shi, C.X.; He, L. Contributions of Climate Change and Human Activities to the Changes in Runoff Increment in Different Sections of the Yellow River. *Quat. Int.* **2012**, *282*, 66–77. [CrossRef]
7. Jeppesen, E.; Kronvang, B.; Meerhoff, M.; Søndergaard, M.; Hansen, K.M.; Andersen, H.E.; Lauridsen, T.L.; Liboriussen, L.; Beklioglu, M.; Ozen, A. Climate Change Effects On Runoff, Catchment Phosphorus Loading and Lake Ecological State, and Potential Adaptations. *J. Environ. Qual.* **2009**, *38*, 1930–1941. [CrossRef] [PubMed]
8. Bergström, S.; Carlsson, B.; Gardelin, M.; Lindström, G.; Pettersson, A.; Rummukainen, M. Climate Change Impacts on Runoff in Sweden‹ Assessments by Global Climate Models, Dynamical Downscaling and Hydrological Modelling. *Clim. Res.* **2001**, *16*, 101–112. [CrossRef]
9. Bacon, N.; Hoque, K. Climate Change, its Effects and Actions in the Light of the Fourth Assessment Report of the Ipcc. *Tájökológiai Lapok* **2009**, *8*, 241–268.
10. Qin, D.; Yong, L.; Chen, Z.; Ren, J.; Shen, Y. Latest Advances in Climate Change Sciences:Interpretation of the Synthesis Report of the Ipcc Fourth Assessment Report. *Adv. Clim. Chang. Res.* **2007**, *3*, 311–314.

11. Wise, M.; Calvin, K.; Thomson, A.; Clarke, L.; Bondlamberty, B.; Sands, R.; Smith, S.J.; Janetos, A.; Edmonds, J. Implications of Limiting CO_2 Concentrations for Land Use and Energy. *Science* **2009**, *324*, 1183–1186. [CrossRef]

12. Huang, H.; Zheng, Y. Present Situation and Future Prospect of Hydropower in China. *Renew. Sustain. Energy Rev.* **2009**, *13*, 1652–1656. [CrossRef]

13. Chang, J.; Wang, X.; Li, Y.; Wang, Y.; Zhang, H. Hydropower Plant Operation Rules Optimization Response to Climate Change. *Energy* **2018**, *160*, 886–897. [CrossRef]

14. Robinson, P.J. Climate Change and Hydropower Generation. *Int. J. Climatol.* **2015**, *17*, 983–996. [CrossRef]

15. Viers, J.H. Hydropower Relicensing and Climate Change. *J. Am. Water Resour. Assoc.* **2011**, *47*, 655–661. [CrossRef]

16. Harrison, G.P.; Whittington, H.W. Susceptibility of the Batoka Gorge hydroelectric scheme to climate change. *J. Hydrol.* **2002**, *264*, 230–241. [CrossRef]

17. Ehsani, N.; Vörösmarty, C.J.; Fekete, B.M.; Stakhiv, E.Z. Reservoir Operations under Climate Change: Storage Capacity Options to Mitigate Risk. *J. Hydrol.* **2017**, *555*, 435–446. [CrossRef]

18. Ahmadi, M.; Haddad, O.B.; Loáiciga, H.A. Adaptive Reservoir Operation Rules under Climatic Change. *Water Resour. Manag.* **2015**, *29*, 1247–1266. [CrossRef]

19. Minville, M.; Brissette, F.; Krau, S.; Leconte, R. Adaptation to Climate Change in the Management of a Canadian Water-Resources System Exploited for Hydropower. *Water Resour. Manag.* **2009**, *23*, 2965–2986. [CrossRef]

20. Akron, A.; Ghermandi, A.; Dayan, T.; Hershkovitz, Y. Interbasin Water Transfer for the Rehabilitation of a Transboundary Mediterranean Stream: An Economic Analysis. *J. Environ. Manag.* **2017**, *202*, 276–286. [CrossRef]

21. Guo, X.; Hu, T.; Tao, Z.; Lv, Y. Bilevel Model for Multi-Reservoir Operating Policy in Inter-Basin Water Transfer-Supply Project. *J. Hydrol.* **2012**, *424–425*, 252–263. [CrossRef]

22. Matete, M.; Hassan, R. Integrated Ecological Economics Accounting Approach to Evaluation of Inter-Basin Water Transfers: An Application to the Lesotho Highlands Water Project. *Ecol. Econ.* **2007**, *60*, 246–259. [CrossRef]

23. Li, Y.; Tang, C.; Wang, C.; Tian, W.; Pan, B.; Hua, L.; Lau, J.; Yu, Z.; Acharya, K. Assessing and Modeling Impacts of Different Inter-Basin Water Transfer Routes on Lake Taihu and the Yangtze River, China. *Ecol. Eng.* **2013**, *60*, 399–413. [CrossRef]

24. Wu, L.Z. Research on Multi-Objective Optimization Scheduling of the Water Transfer Project from Han to Wei. Master's Thesis, Xi'an University of Technology, Xi'an, China, 2017.

25. Kim, B.S.; Kim, B.K.; Kwon, H.H. Assessment of the Impact of Climate Change On the Flow Regime of the Han River Basin Using Indicators of Hydrologic Alteration. *Hydrol. Process.* **2015**, *25*, 691–704. [CrossRef]

26. Du, J.; Shi, C.X. Effects of Climatic Factors and Human Activities on Runoff of the Weihe River in Recent Decades. *Quat. Int.* **2012**, *282*, 58–65. [CrossRef]

27. Mann, H.B. Nonparametric Tests against Trend. *Econometrica* **1945**, *13*, 245–259. [CrossRef]

28. Kendall, M.G. Rank Correlation Methods. *Br. J. Psychol.* **1990**, *25*, 86–91. [CrossRef]

29. Yue, S.; Pilon, P.; Cavadias, G. Corrigendum to "Power of the Mann-Kendall and Spearman's Rho Tests for Detecting Monotonic Trends in Hydrological Series". *J. Hydrol.* **2002**, *264*, 262–263. [CrossRef]

30. Hamed, K.H. Trend Detection in Hydrologic Data: The Mann–Kendall Trend Test under the Scaling Hypothesis. *J. Hydrol.* **2008**, *349*, 350–363. [CrossRef]

31. Chang, F.J.; Lai, J.S.; Kao, L.S. Optimization of Operation Rule Curves and Flushing Schedule in a Reservoir. *Hydrol. Process.* **2010**, *17*, 1623–1640. [CrossRef]

32. Joeres, E.F.; Liebman, J.C.; Revelle, C.S. Operating Rules for Joint Operation of Raw Water Sources. *Water Resour. Res.* **1971**, *7*, 225–235. [CrossRef]

33. Wang, X.; Lei, X.H.; Jiang, Y.Z.; Wang, H. Reservoir Operation Chart Optimization Searching in Feasible Region Based On Genetic Algorithms. *J. Hydraul. Eng.* **2013**, *44*, 26–34. [CrossRef]

34. Long, L.N.; Madsen, H.; Dan, R. Simulation and Optimisation Modelling Approach for Operation of the Hoa Binh Reservoir, Vietnam. *J. Hydrol.* **2007**, *336*, 269–281.

35. Turgeon, A. Stochastic Optimization of Multireservoir Operation: The Optimal Reservoir Trajectory Approach. *Water Resour. Res.* **2007**, *43*, 225–236. [CrossRef]

36. Feng, M.; Liu, P.; Guo, S.; Shi, L.; Deng, C.; Ming, B. Deriving Adaptive Operating Rules of Hydropower Reservoirs Using Time-Varying Parameters Generated by the Enkf. *Water Resour. Res.* **2017**, *53*, 6885–6907. [CrossRef]
37. Little, J.D.C. The Use of Storage Water in a Hydroelectric System. *J. Oper. Res. Soc. Am.* **1955**, *3*, 187–197. [CrossRef]
38. Chen, L.; Chang, F.J. Applying a Real-Coded Multi-Population Genetic Algorithm to Multi-Reservoir Operation. *Hydrol. Process.* **2010**, *21*, 688–698. [CrossRef]
39. Celeste, A.B.; Billib, M. Evaluation of Stochastic Reservoir Operation Optimization Models. *Adv. Water Resour.* **2009**, *32*, 1429–1443. [CrossRef]
40. Ming, B.; Chang, J.; Huang, Q.; Wang, Y.; Huang, S. Optimal Operation of Multi-Reservoir System Based-On Cuckoo Search Algorithm. *Water Resour. Manag.* **2015**, *29*, 5671–5687. [CrossRef]
41. Reddy, M.J.; Kumar, D.N. Evolving Strategies for Crop Planning and Operation of Irrigation Reservoir System Using Multi-Objective Differential Evolution. *Irrig. Sci.* **2008**, *26*, 177–190. [CrossRef]
42. He, H.M.; Zhang, Q.F.; Jie, Z.; Jie, F.; Xie, X.P. Coupling Climate Change with Hydrological Dynamic in Qinling Mountains, China. *Clim. Chang.* **2009**, *94*, 409–427. [CrossRef]

 *water*

Article

Comparative Study of Two State-of-the-Art Semi-Distributed Hydrological Models

Pranesh Kumar Paul [1], **Yongqiang Zhang [2,*]**, **Ashok Mishra [3]**, **Niranjan Panigrahy [4]** and **Rajendra Singh [3]**

[1] IIT Khragpur, Kharagpur, West Bengal 721302, India; paulpranesh@iitkgp.ac.in
[2] Key Lab of Water Cycle and Related Land Surface Processes, Institute of Geographic Sciences and Natural Resources Research, Chinese Academy of Sciences, Beijing 100101, China
[3] AgFE Department, IIT Kharagpur, Kharagpur, West Bengal 721302, India; amishra@agfe.iitkgp.ernet.in (A.M.); rsingh@agfe.iitkgp.ernet.in (R.S.)
[4] Odisha University of Agriculture & Technology, Bhubaneswar-751003, Odisha, India; n.panigrahy@gmail.com
*   Correspondence: zhangyq@igsnrr.ac.cn; Tel.: +86-10-64856515

Received: 17 April 2019; Accepted: 21 April 2019; Published: 26 April 2019

Abstract: Performance of a newly developed semi-distributed (grid-based) hydrological model (satellite-based hydrological model (SHM)) has been compared with another semi-distributed soil and water assessment tool (SWAT)—a widely used hydrological response unit (HRU)-based hydrological model at a large scale (12,900 km^2) river basin for monthly streamflow simulation. The grid-based model has a grid cell size of 25 km^2, and the HRU-based model was set with an average HRU area of 25.2 km^2 to keep a balance between the discretization of the two models. Both the model setups are calibrated against the observed streamflow over the period 1977 to 1990 (with 1976 as the warm-up period) and validated over the period 1991 to 2004 by comparing simulated and observed hydrographs as well as using coefficient of determination (R^2), Nash–Sutcliffe efficiency (NSE), and percent bias (PBIAS) as statistical indices. Result of SHM simulation (NSE: 0.92 for calibration period; NSE: 0.92 for validation period) appears to be superior in comparison to SWAT simulation (NSE: 0.72 for calibration period; NSE: 0.50 for validation period) for both calibration and validation periods. The models' performances are also analyzed for annual peak flow, monthly flow variability, and for different flow percentiles. SHM has performed better in simulating annual peak flows and has reproduced the annual variability of observed streamflow for every month of the year. In addition, SHM estimates normal, moderately high, and high flows better than SWAT. Furthermore, total uncertainties of models' simulation have been analyzed using quantile regression technique and eventually quantified with scatter plots between P (measured data bracketed by the 95 percent predictive uncertainty (PPU) band) and R (the relative length of the 95PPU band with respect to the model simulated values)-values, for calibration and validation periods, for both the model simulations. The analysis confirms the superiority of SHM over its counterpart. Differences in data interpolation techniques and physical processes of the models are identified as the probable reasons behind the differences among the models' outputs.

Keywords: grid-based; HRU-based; SHM; SWAT; large scale basin

---

1. Introduction

Distributed hydrological models, with varying degree of complexity, are essential tools for modeling the spatial variability effects of basin characteristics and forcing variables (e.g., precipitation) on streamflow [1–4]. These models divide spatially heterogeneous space (basin or watersheds) into a number of near homogeneous units following various discretization schemes including: representative elementary area (REA) [5], grouped response unit (GRU) [6], representative elementary watersheds (REW) [7], hydro-landscape unit [8], triangular irregular network (TIN) [9], hydrological response unit (HRU) [10] and grid-based approaches [11]. Among these discretization schemes, HRU and square grid approaches are the most commonly used in hydrological modeling.

HRUs are formed by lumping individual areas of similar soil, topography, and land-use altogether within a sub-basin. However, there is no interaction between the HRUs, and these are routed individually to the sub-basin outlet [12,13]. Arnold et al. [13] studied the effect of HRU discretization on streamflow and concluded that many HRUs are too big to resolve into individual topographic positions since they occupy the landscape continuum from the divide up to valley bottom. They also identified that the impact of an upslope HRU management on a downslope HRU cannot be assessed. Furthermore, though the HRU-based approach is simple and computationally efficient, spatial information from high-resolution land-use or soil maps can be lost depending on the scale of the HRUs. On the other hand, grid-based discretization scheme uses aggregated spatial variations over each grid. The use of smaller HRUs, instead of grid cells, may yield similar results but incorporating raster data into the HRU based approach would require data transformation from simple grid geometry to a patchy geometry of irregular polygons. Therefore, a grid-based approach appears better to use to avoid the inconvenience.

To describe the basin topography accurately, the grid size is considered up to an acceptable range while keeping the trade-off between model simulation time and simulation accuracy to a minimum. Though, in theory, modeling with a finer grid cell resolution is expected to yield better results because of better-resolved model input data (e.g., rainfall, topography, land cover, etc.), it may not always happen [14]. Therefore, several studies have focused on examining the impact of grid cell size on model simulation results and model simulation time to find out the optimum resolution of grid cells for a particular modeling study. Finnerty et al. [15] illustrated the changes in water budget with continuous simulations at various spatial scales, ranging from 4 km × 4 km to 256 km × 256 km. Wood et al. [16] used a 1° × 1° gridded structure for modeling continental-scale basins. Kuo et al. [17] applied a variable-source-area hydrological model to grid sizes ranging from 10 to 600 m and observed increasing misrepresentation of the curvature of the landscape with increasing grid size. In modeling the 375,000 km^2 Senegal River basin, Andersen et al. [18] used grid cell resolution of 4 km × 4 km. Booij [19] compared three versions of Hydrologiska Byråns Vattenbalansavdelning (HBV) model [20] with different spatial resolutions in the Meuse river basin in Europe and found that the version with finer resolution reproduced a slightly improved average and extreme discharge behavior at the basin outlet in both calibration and validation periods. Recently, Haghnegahdar et al. [21] carried out a modeling study in a 2700 km^2 area with model grid cells of 15 km × 15 km resolution.

The effect of different spatial discretization schemes on streamflow simulation has been studied by researchers. For example, Abu El-Nasr et al. [22] assessed performances of fully distributed grid-based MIKE Systeme Hydrologique Europeen (SHE) and the semi-distributed HRU-based SWAT and showed that MIKE SHE can predict the overall variation of stream flow slightly better. There are more examples of studies investigated utility of different grid-based models and compared results with an HRU based SWAT model [23–30]. Arnold et al. [13] used a modified SWAT model, with landscape routing method, to compare modeling results, under four discretization methods: lumped, HRU, catena, and grid. The comparison showed that a high-resolution grid approach would include the impact of an upslope grid cell on a downslope grid cell and provide accurate spatial detailed output. Comparing SWAT model performances with HRU and grid-based structures, Pignotti et al. [31] concluded that the grid-based model under predicts streamflow from 5% to 50% with respect to the usual HRU-based model. Surfleet et al. [32] compared two HRU-based models namely the precipitation–runoff modeling system (PRMS) [33] and groundwater and surface-water flow (GSFLOW) with the grid-based variable infiltration capacity (VIC) model for future climate change analysis and concluded that the future changes can quantitatively be attributed not only to the scale of the models but also to the ability of models to represent hydrological processes. Findings of these various studies also pointed out that model simulation results also vary depending on several factors other than the spatial discretization scheme. These factors include the physiographic characteristics of the basin, seasonality of precipitation, season of the year, and dominating runoff producing mechanisms and, thus, emphasize the uncertainty of analysis of model simulation results for successful comparisons of different hydrological models in a particular study (e.g., [34,35]).

Keeping this in mind, this study aims at in-depth inter-comparison of simulation results of two state-of-the-art semi-distributed hydrological models, namely the satellite-based hydrological model (SHM) and soil and water assessment tool (SWAT), under similar discretization scale, and uncertainty related to the simulations [36,37] in a large scale (>1000 km^2) [13,36] sub-tropical river basin, namely Baitarani. The idea behind the similar discretization scale is to reduce the effect of different discretization schemes of the two models and analyze the effect of other factors on the streamflow simulation.

The remainder of this paper is organized as follows. The following section presents the description of the study basin and data used in the study. A description of the models along with sensitive parameters employed in the study is provided in Section 3. The methodologies of model setup, calibration, and validation procedure, as well as the consequent data analysis (including uncertainty analysis), are outlined in Section 4. The results are presented and discussed in Section 5. The final section, Section 6 provides conclusions.

2. Study Area and Data

The study has been performed in Baitarani river basin (12,900 km^2) in India which is bounded between 20°35′ N to 22°15′ N latitude and 85°10′ E to 87°03′ E longitude (Figure 1). It comes within the sub-tropical monsoon climate zone [38] and receives an annual rainfall of about 1450 mm (Annual Report, 2011-12, 2011). Almost 80% of the annual rainfall occurs during the four months of south-west monsoon season (June to September) that generates heavy flow and creates floods in lower reaches [39]. Daily temperature varies from 5 °C to 47.5 °C. The elevation of the basin ranges from 10 m to 750 m above mean sea level. Soils of this area vary from rich red loamy to gravely detritus.

For a consistent comparison of performances, the same datasets were used in SHM and SWAT models. Daily Rainfall and daily maximum and minimum temperature have been obtained from the India Meteorological Department (IMD), Pune at 1° × 1° resolution. Data have been interpolated to 5 km × 5 km resolution by using bi-linear interpolation technique to use as input into the SHM. Soil and land use land cover (LULC) maps were collected from the Food and Agriculture Organization (FAO) website (http://www.fao.org/soils-portal/soil-survey/soil-maps-and-databases/harmonized-world-soil-database-v12/en/) at 1 km × 1 km scale. The digital elevation model (DEM) of 30 m × 30 m resolution was taken from the Advanced Spaceborne Thermal Emission and Reflection Radiometer (ASTER) website (https://asterweb.jpl.nasa.gov/gdem.asp). All the static information (soil map, LULC map, and DEM) have been resampled into 5 km × 5 km resolution to use in the SHM. The weather database of SWAT is developed using the weather generator (WXGEN) model using the closest station scheme [40]. Observed streamflow data, at Anandpur gauging station (21.21° N, 86.12° E), were collected for the period of 1977 to 2004 from the Central Water Commission (CWC), Bhubaneswar, India.

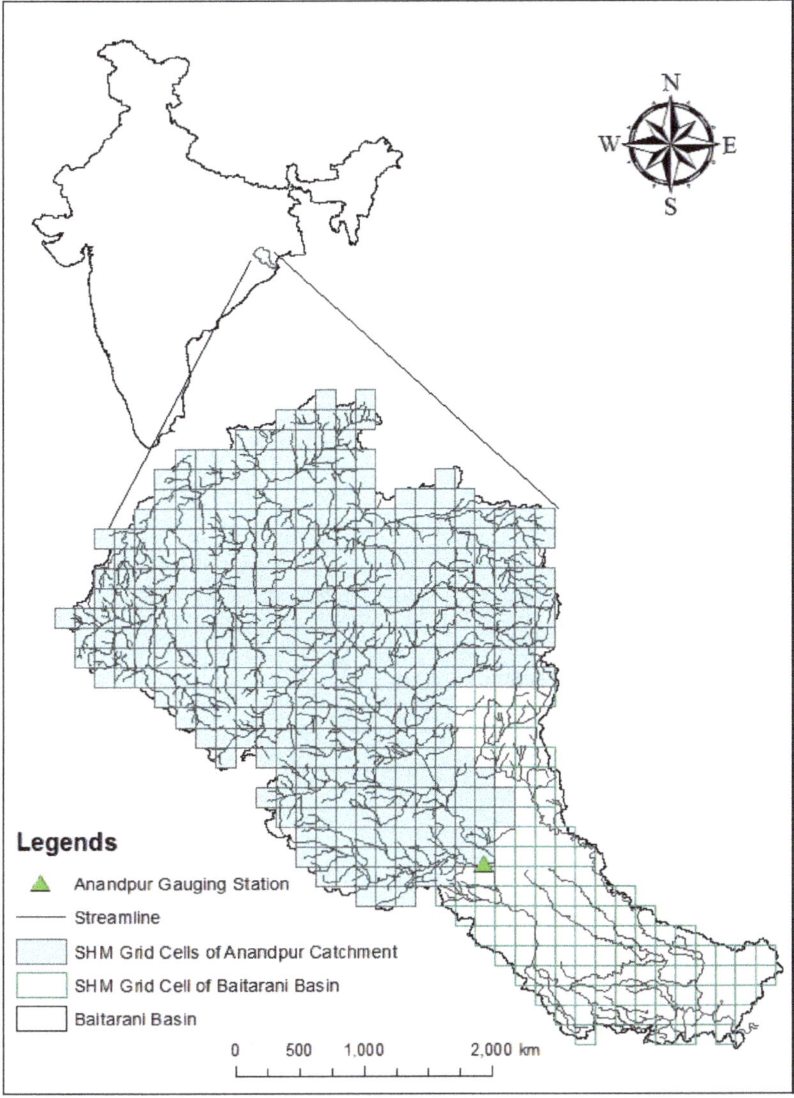

Figure 1. Index map of Baitarani river basin showing streamline and grid cells of SHM.

3. Comparative Discussion on SHM and SWAT

In this section, short descriptions of the SHM and SWAT are provided (Sections 3.1 and 3.2, respectively). Then the identified sensitive parameters of both the models, which have been used to calibrate the models, are discussed in Section 3.3.

3.1. Description of the SHM

The SHM works on 5 km × 5 km spatial grid resolution and properties at the center of a cell are assumed to be the properties of the cell. SHM has five modules: surface water (SW), forest (F), snowmelt (S), groundwater (GW), and routing (ROU). SHM grid cells corresponding to forest and snow land cover are modeled using the F and S modules, respectively; whereas other grid cells are modeled using the SW module.

In the SW module [41,42], the Soil Conservation Service (SCS) curve number (CN) method [43] is used to estimate the surface runoff along with the Hargreaves method [44] to estimate the potential evapotranspiration (PET). Soil moisture is estimated by using the water balance technique. The soil profile is considered as a single-layered zone of 300 mm, and moisture-holding and moisture-transmitting characteristics of the soil layer and underlying layer are considered to account for the soil moisture. Infiltrated water wets the soil layer, and excess water from the maximum capacity (saturation) contributes after percolation to GW module. The soil moisture is depleted by evapotranspiration, at a potential rate or actual rate, depending on soil moisture condition.

The F module serves, based on water balancing and the dynamics of the subsurface, to provide output in the form of runoff, soil moisture, evapotranspiration, and contribution to groundwater using the technique and parameters stated in [45]. Subsurface is reckoned on having soil matrix and macropores of main bypass and internal catchment types. The main bypass directly contributes to groundwater. Soil matrix is considered of having three layers, which are important with respect to water balance and change in soil moisture. After infiltration, the saturation of three layers gets started from the top in batch, and after complete saturation of the three layers, the excess water goes to groundwater. After a precipitation event, runoff generation occurs according to the antecedent moisture conditions in the subsurface.

The S module determines the snow density from snow albedo [46] for estimating snowmelt depth by using two different algorithms, viz., the temperature index algorithm and radiation-temperature index algorithm. Since the study area does not have any snow land cover; the S module is not considered in this study.

The GW module uses the contribution from SW, F, and S modules and generates baseflow following the water level variation process described in [47]. The resultant baseflow along with the surface runoff generated from other modules is routed up to the outlet as streamflow.

In SHM, a distributed routing technique [41], termed as time-variant spatially distributed direct hydrograph (SDDH) travel time method [48], was adopted. It requires the flow path, which is derived from DEM. The downstream cell, in the direction of the steepest descent, is defined from the DEM by the use of the flow direction geographic information system (GIS) function with a unique connection from each cell to the watershed outlet. This process produces a cell network to present the flow paths. The threshold number of upstream cells is set equal to two (based on trial and error) to delineate the channel network for the watershed. Any cell with a number of upstream draining cells equal to or greater than the threshold value is considered to be a channel cell, whereas others are considered as overland flow cells. The key point of this approach is the travel time estimation. SHM uses MySQL (open source software) as a relational database management system (RDBMS).

3.2. Description of SWAT

SWAT is used for simulation of the water cycle and its corresponding fluxes of energy and matter (e.g., sediment, nutrients, pesticides, and bacteria) as well as the impact of management practices on these fluxes at basin scale [49]. SWAT uses Microsoft Access as RDBMS. SWAT, however, first discretizes the watershed into a network of irregular sub-basins and then divides each sub-basin into HRUs. The model includes components for hydrology, sedimentation, crop growth, nutrients, and agricultural management [11]. A detailed description of all components of the model can be found in Arnold et al. [49] and Neitsch et al. [10].

In the present study, SWAT has been used with the Soil Conservation Service Curve Number (SCS-CN) method as a runoff generation technique along with the Hargreaves method to determine PET. SWAT calculates baseflow contribution to streamflow from groundwater depending on the water balance approach in a shallow aquifer [49]. In SWAT, runoff is first computed separately for each of the HRUs within the sub-basin and then routed through the stream network to obtain the total streamflow for the watershed. Since the study area does not have snow-covered land, the snow-melt runoff simulation procedure of SWAT is not discussed here.

3.3. Sensitive Parameters of Both the Models Used for Calibration

The total number of parameters of the two models varies in number for streamflow analysis. Three parameters of SHM and seven parameters of SWAT have been found sensitive for streamflow simulation (Table 1), in this study.

During calibration of SHM, parameters of SW and ROU modules have been changed manually (since an auto-calibration option is not available). For this purpose CN, Manning's roughness coefficient for overland cell (n_o), and Manning's roughness coefficient for channel cell (n_c) have been used as sensitive parameters [50]. The parameters of the F and GW modules have been set at their default values as recommended by the developers. The theoretical ranges of sensitive parameters are given in Table 1. CN is responsible for runoff generation in the SW module, and n_o and n_c affect the routing procedure of generated runoff and baseflow from a grid cell up to the outlet of a basin. Using calibrated values of the sensitive parameters, SHM simulates monthly streamflow at Anandpur gauging station of Baitarani basin.

For the SWAT model, seven sensitive parameters are identified (Table 1) for model calibration based on the analysis of parameter sensitivity using the Latin hypercube-one factor at a time (LH-OAT) method [51]. Curve number (Cn2) and baseflow recession constant (Alpha_bf) are responsible for runoff generation; delay time for aquifer recharge (Gw_delay) and threshold water level in a shallow aquifer for base flow (Gwqmn) are responsible for baseflow generation, and the soil evaporation compensation coefficient (Esco) is responsible for soil evaporation losses. Manning's n for the main channel (Ch_N2) and Effective hydraulic conductivity of soil (Ch_K2) are responsible for controlling river flow routing. Table 1 summarizes the sensitive parameters of both the models with corresponding hydrological processes, estimation methodology and their theoretical ranges. The table also focuses on the spatial variability of the sensitive parameters.

Table 1. Summary of the sensitive parameters, respective hydrological processes with methodology, theoretical range of their values and information on the spatial variation over the study area, used by SWAT and SHM to simulate hydrological components.

| | SWAT | | | | | SHM | | | | | |
List of Sensitive Parameters	Process	Method	Theoretical Range of Parameter Values	Calibrated Range of Parameter Values	Spatially Varied or Not	List of Sensitive Parameters	Process	Method	Theoretical Range of Parameter Values	Calibrated Range of Parameter Values	Spatially Varied or Not
Cn2	Runoff	SCS-CN [52]	0–100	50–100	Yes	CN	Runoff (SW)	SCS-CN [52]	0–100	30–100	Yes
Gwqmn	Baseflow	Threshold value based contribution from shallow aquifer storage	0–5000 (mm)	0–300	Yes						
Gw_delay			0–500 (day)	8	No						
Alpha_bf			0–1	0.5	Yes						
Ch_N2	River flow Routing	Variable storage/Muskingum	−0.01–0.03	0.03	No	n_o	Routing (ROU)	SDDH [48]	0.01–0.05	0.01–0.03	Yes
Ch_K2			0–500 (mm/h)	0.45	No	n_c			0.01–0.05	0.015	No
Esco	To compensate the soil evaporative demand with the depth of soil layers	Water balance	0–1	0.01–0.95	Yes						
	Total	7								3	

Enlargements: Cn2 (Curve number), Gwqmn (Threshold water level in shallow aquifer for base flow), Gw_delay (Delay time for aquifer recharge), Alpha_bf (Baseflow recession constant), Ch_N2 (Manning's n for main channel), Ch_K2 (Effective hydraulic conductivity of soil), Esco (Soil evaporation compensation coefficient), CN (Curve Number), n_o (Manning's roughness co-efficient for overland flow), n_c (Manning's roughness co-efficient for channel flow).

4. Methodology

4.1. Model Setup, Calibration, Validation

At first, both the models were setup with the same input data. For SHM setup, the Baitarani basin is represented by 498 grid cells of 25 km^2. The threshold values of LULC, soil, and slope were taken, respectively, 1%, 1%, and 2% for the development of HRUs in the SWAT model so that the average area of HRUs is around 25 km^2 and two discretization schemes come in a balanced scale. This assumption led to having 312 sub-basins and 511 HRUs in the Baitarani basin. However, the smallest HRU has an area of 1.9 km^2, and the largest HRU has an area of 52 km^2 with an average area of 25.2 km^2. Both the models were then calibrated (1977–1990) and validated (1991–2004) on a monthly basis. The performance evaluation of both the models has been done by comparing observed and simulated streamflows by using graphical interpretation and statistical indices, namely coefficient of determination (R^2), Nash–Sutcliffe efficiency (NSE), and percent bias (PBIAS) for the calibration and validation periods, separately. The used statistical analyses are discussed below.

4.1.1. Nash Sutcliffe Efficiency (NSE)

It is defined as one minus the sum of the absolute squared differences between observed and simulated values normalized by the variance of observed values [53]. It varies from $-\infty$ to 1, 1 being the perfect fit. It is chosen because of its extensive use in the field of hydrology, which facilitates comparison between different studies. However, it is highly sensitive to peak flows resulting in negligence of low flows.

$$NSE = 1 - \frac{\sum_{i=1}^{N}(Q_o - Q_{sim})^2}{\sum_{i=1}^{N}\left(Q_o - \overline{Qo}\right)^2} \tag{1}$$

where Q_o is the observed streamflow; Q_{sim} is the simulated streamflow; $\overline{Qo}$ is the average observed streamflow and N is the number of events in the time-series of streamflow.

4.1.2. Coefficient of Determination (R^2)

The coefficient of determination (R^2) describes the proportion of the total variance in the observed data that can be explained by a model. It ranges from 0 to 1, with higher values indicating better agreement, and is given by:

$$R^2 = \left[\frac{\sum_{i=1}^{n}\left(Q_o - \overline{Qo}\right)\left(Q_{sim} - \overline{Qsim}\right)}{\left\{\sum_{i=1}^{N}\left(Q_o - \overline{Qo}\right)\right\}^{0.5}\left\{\sum_{i=1}^{n}\left(Q_{sim} - \overline{Qsim}\right)\right\}^{0.5}}\right]^2 \tag{2}$$

where, $\overline{Qsim}$ is the average simulated value of streamflow.

4.1.3. Percent Bias (PBIAS)

It measures the average tendency of the simulated data to be larger or smaller than their observed counterparts. Its ideal value is 0. A positive value indicates model underestimation bias and a negative value indicates model overestimation bias.

$$PBIAS = \frac{\sum_{i=1}^{n}(Q_o - Q_{sim}) \times 100}{\sum_{i=1}^{n}Q_o} \tag{3}$$

4.2. Analysis of Results

After calibration and validation, the model-simulated streamflows were analyzed to compare model performances for both the periods with respect to annual peaks. Then, inter-annual variability of simulations of both the models for each month of the year, for the total period of analysis, were analyzed. Eventually, the capability of both models was compared using the five percentile series derived from observed data. Therefore, to understand the difference in the capability of the models to simulate different streamflow ranges in an improved manner, four percentile points of observed monthly streamflow, S5 (5th percentile), S25 (25th percentile), S75 (75th percentile), and S95 (95th percentile), were used to divide the overall flow range into five percentile series: low flows (<S5: <139.95 m^3/s), moderately low flows (S5–S25: 139.95 m^3/s to <349.4 m^3/s), normal flows (S25–S75: 349.4 m^3/s to <6590 m^3/s), moderately high flows (S75–S95: 6590 m^3/s to <20,760 m^3/s), and high flows (≥S95: ≥20,760 m^3/s). Finally, uncertainty analysis has been performed of the models.

Uncertainty Analysis

Using quantile regression, a stochastic approach [54], uncertainty from all sources was analyzed, as a whole and for monthly simulation of both the models at Anandpur gauging station for both the calibration and validation periods. The observed, simulated, and residual values of streamflow are linked with the following equation:

$$Q\,(t) = \hat{Q}(t) + e(t) \tag{4}$$

where $Q(t)$ is the observed daily streamflow, $\hat{Q}(t)$ is the simulated streamflow, and $e(t)$ is the residual.

The method assumes a functional relationship between residuals and estimates in the Gaussian domain, i.e., normalized quantile streamflow (NQS) and normalized quantile residual (NQR). A linear relation between NQS and NQR was also used in previous studies [55,56]. Hence, NQR may be expressed as:

$$NQR = a \times NQS + b \tag{5}$$

Different quantile regression lines may be obtained by minimizing the absolute bias by assigning different weights to positive and negative residuals in the Gaussian domain. Absolute bias can be considered for this purpose as an objective function (OF) which is expressed mathematically as:

$$OF \;=\; Min \sum \rho_\tau (mod[NQR - (a \times NQS + b)]) \tag{6}$$

where a is the slope, b is the intercept, and ρ_τ is the quantile regression function which pushes the regression line to the desired location.

To estimate the streamflow corresponding to a given confidence limit, the simulated streamflow is transformed to the Gaussian domain as NQS first, and then, the error in the Gaussian domain, NQR is estimated using the regression line (Equation (5)). The estimated error, NQR is transformed back to the original domain using the pre-estimated mean and standard deviation of the residual. Finally, the estimated residual is added to the daily simulated streamflow to obtain the streamflow which includes uncertainty. Regression lines were used to analyze uncertainty in the simulated streamflow for different confidence intervals. The slope and intercept of these lines are estimated by Equation (6) using the calibration period data. Furthermore, to verify the correctness of error models, the models were applied for both the calibration and validation periods.

Moreover, to have quantitative realization of uncertainty, P and R values have been calculated, and P vs. R plot has been generated for both the calibration and validation periods.

P-value represents the measured data bracketed by the 95 percent predictive uncertainty (PPU) band [57]. P-value has been determined by the following equation:

$$P - value = \frac{q_{in}}{N} \tag{7}$$

where, q_{in} are the total number of observed data points bracketed by the 95PPU band, N is the total number of observed data points.

R-value expresses the relative length of the 95PPU band with respect to the model simulated values [57]. R-value has been determined by the following equation:

$$R - value = \frac{\overline{d}_x}{\sigma_x} \tag{8}$$

where σ_x is the standard deviation of the model simulation x. $\overline{d}_x$ is the average distance between the upper and lower limit of the 95PPU band. $\overline{d}_x$ has been calculated using the following equation:

$$\overline{d}_x = \frac{1}{k} \sum_{l=1}^{k} \left(q_U - q_L\right)_l \tag{9}$$

where l is counter, k is the total number of simulated data points for streamflow q, q_U and q_L are the upper and lower limit of the 95PPU band.

Both the values vary between 0 and 1. P-value equal to 1 and R-value 0 represent the best model simulation with no uncertainty. In the P-Q plot, this point can be identified as the point of no uncertainty. Since to reach the point of no uncertainty is nearly impossible to achieve for any model simulation as a result of model uncertainties and measurement errors, the simulation nearest to the point may be considered as the simulation with the lowest uncertainty.

5. Results and Discussion

5.1. Calibration and Validation of the Models

Comparison between observed and models' simulated monthly streamflow are shown in Figure 2a for the calibration period and in Figure 2b for the validation period. Figures show good agreement among observed and simulated streamflow by both the models. However, SHM simulates the temporal pattern of observed streamflow relatively better in comparison to the SWAT model in both calibration and validation periods including the reproduction of peak flows. To strengthen this observation 1:1 scatter plots, between observed and models' simulations for the calibration and validation periods, have also been used (shown in Figure 2a,b). From the scatter plots it is evident that SWAT simulated streamflow deviates considerably from the observed streamflow with respect to the SHM simulated counterpart during both the calibration and validation periods. Moreover, scatter plots also depict that SWAT underestimates high flow more in comparison to SHM.

The goodness-of-fit statistics of both the models on monthly calibration and validation are shown in Table 2. Generally, if $R^2 > 0.6$, NSE > 0.5, and $-25\% \leq PBIAS \leq 25\%$, the model simulation results are judged as satisfactory [58,59]. Thus, both SHM and SWAT models have produced satisfactory model simulations for both the calibration and validation periods in the study area. However, the monthly streamflow simulated by SHM shows better fit with the observed monthly flow in comparison to the SWAT simulated streamflow during the calibration as well as validation periods. SHM shows similarity in results during both the calibration and validation periods with a slightly reduced PBIAS during validation than calibration period, thus, improvement in water balance dynamics. On the other hand, SWAT shows considerable deterioration in results during the validation period in comparison to the calibration period which is evident from the values of R^2 and NSE (Table 2). The results, thus, show improved performances of both SHM and SWAT simulations in comparison to previous studies performed at the Anandpur sub-basin [60–65].

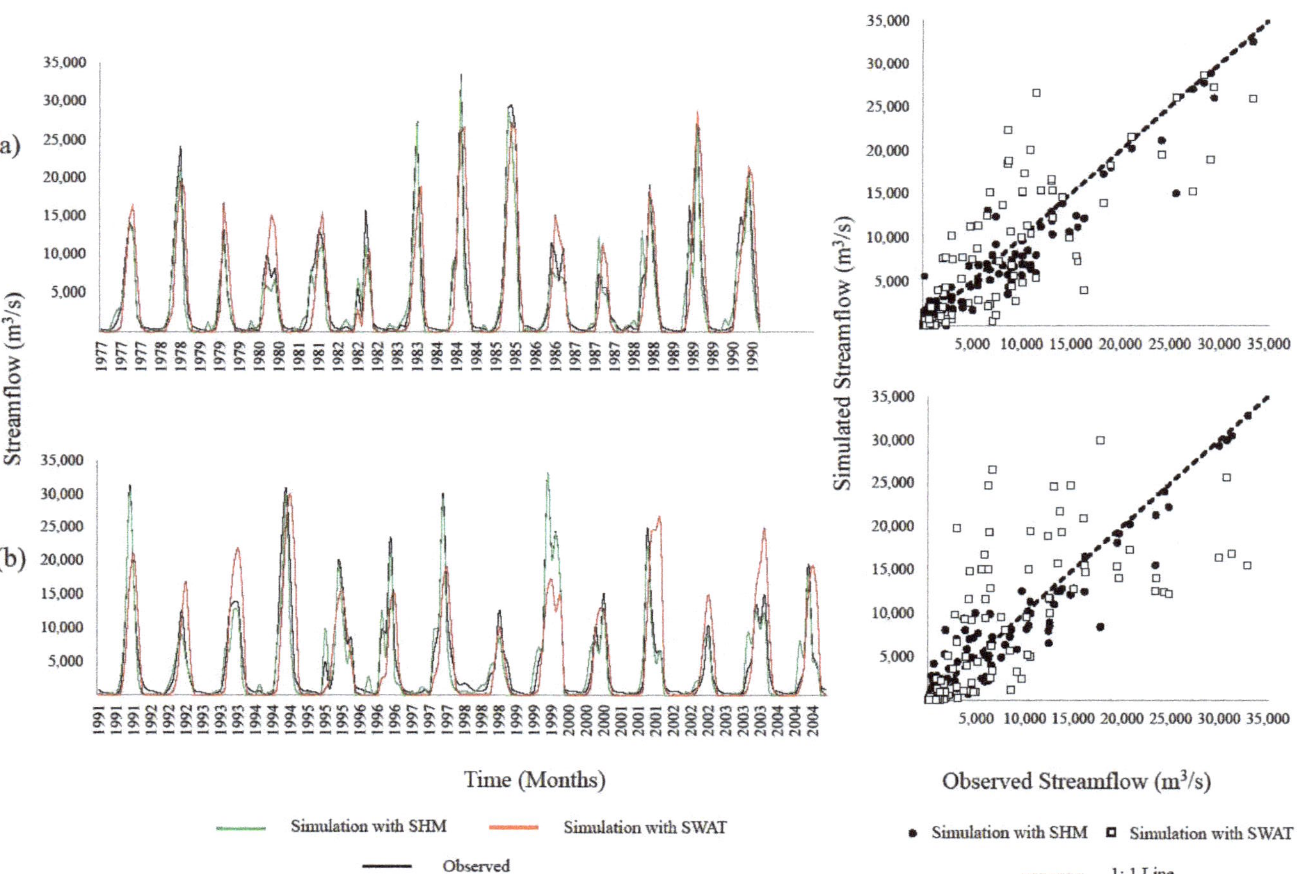

Figure 2. Comparison between observed and simulated monthly streamflow hydrographs and scatter plots during (**a**) calibration and (**b**) validation periods.

Table 2. Calibration and validation performances of the models at a monthly scale.

Period	Statistics	SHM	SWAT
	R^2	0.93	0.75
Calibration	NSE	0.92	0.72
	PBIAS	11.62	2.01
	R^2	0.93	0.58
Validation	NSE	0.92	0.50
	PBIAS	8.67	−1.4

5.2. Analysis to Compare Annual Peaks

To perform comparison of the annual peak simulation capabilities of both the models, observed and simulated annual peaks (from both the models) for the calibration (Figure 3a) and validation (Figure 3b) periods have been plotted against the 1:1 line. Figure 3 depicts that SHM reproduces annual peaks better than SWAT. Therefore, SHM can be a good option for streamflow simulation for extreme rainfall events as well as analyzing flooding possibility in the region. Findings are well comparable with the study performed by Baratti et al. [66] in which they estimated annual flood frequency for the same region.

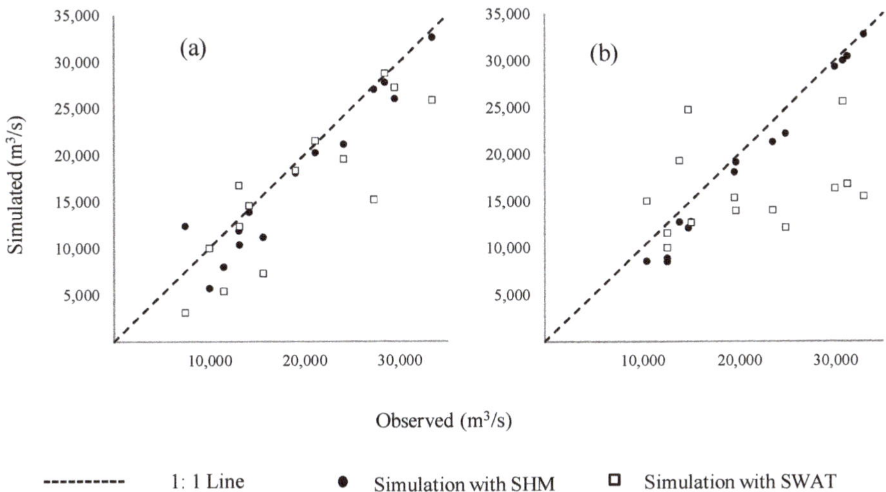

Figure 3. Comparison between observed and models simulated annual peaks during (**a**) calibration and (**b**) validation period.

5.3. Inter-Annual Variability of Model Simulations

To understand the difference in models' capabilities of producing inter-annual variability of monthly streamflow, comparison between observed and simulated monthly streamflow from both the models have been analyzed and are shown in Figure 4. From the figure, it is evident that SHM performs satisfactorily in simulating streamflow during the months of June to October (monsoon) season with the best simulation identified for the month of August throughout the analysis period. In addition, it is also evident that SHM reproduces observed streamflow better for all the months over the analysis period in comparison to SWAT streamflow.

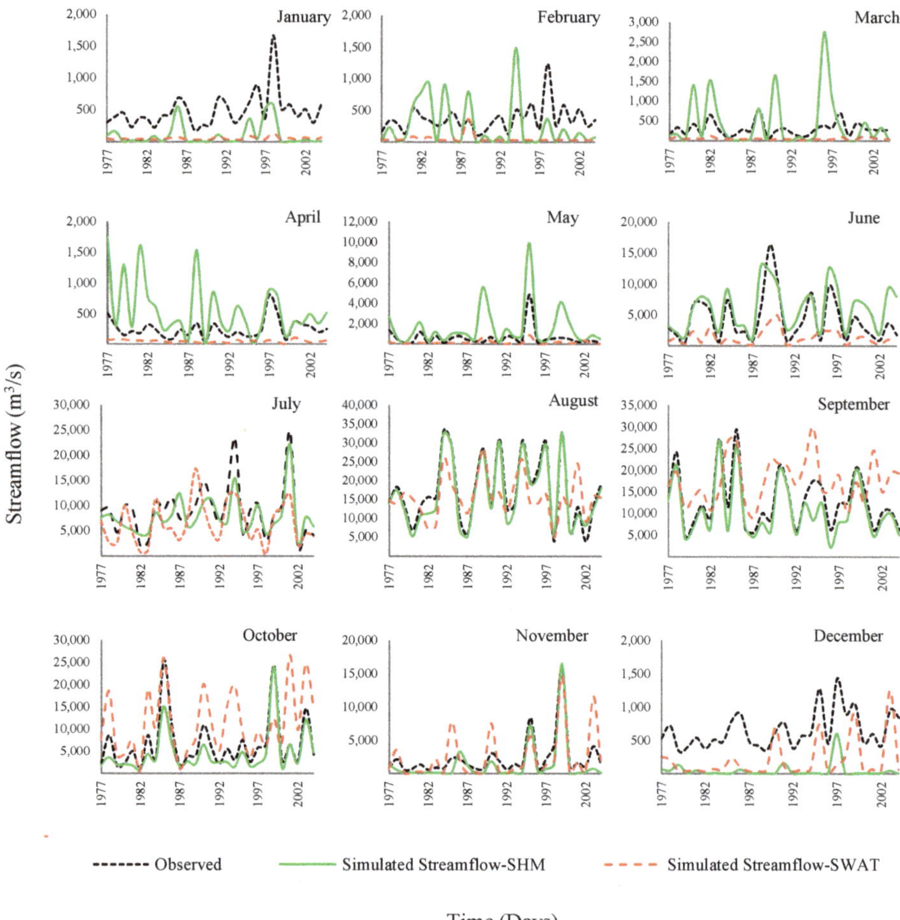

Figure 4. Comparison between observed and simulated monthly streamflow for each month of the year over the total period (calibration and validation) of analysis.

The differences in the results of two models for inter-annual variability on a monthly scale for the total period of analysis is mainly attributable to two reasons: different input data interpolation schemes and variation in modeling processes. First, meteorological data have been bi-linearly interpolated into 5 km × 5 km to run SHM and SWAT model and have used the meteorological data from the closest IMD grid to simulate monthly streamflow in a sub-basin instead of interpolation [67] (as stated earlier in Section 2). Different input data interpolation schemes and variation in spatial discretization create the difference in the spatial distribution of meteorological input for the models [30]. Second, apart from SCS-CN of the SW module, no modeling process of SHM matches with the SWAT model. However, the modeling process combination of SHM proved to be better in monsoonal months in comparison to SWAT; though modeling processes of both models may require improvement for the low rainfall months. In particular, water level variation approach of baseflow generation, in SHM, and water balance approach of baseflow calculation, in SWAT, may be compared with separate analysis for the non-monsoonal months, for the purpose. Furthermore, better calibration may also improve results for months with low rainfall.

5.4. Comparison of Model Simulations for Percentile Flows

The models' performance has also been analyzed in simulating streamflow of various magnitudes by considering five percentile classes (Section 4.2). The respective simulated flows by both models have been compared with that of observed streamflow by using scatter plots for the calibration (Figure 5a) and validation periods (Figure 5b). The performance of SWAT in simulating moderately low flows during the calibration period is better than simulating other streamflow percentiles during both the calibration and validation periods. SHM performs better for simulating normal, moderately high, and high flows during both the calibration and validation periods. Overall, both the models show an extremely poor performance in simulating low flows during both the periods and moderately low flows during the validation period.

The variation in percentile flow estimation of the models can also be attributable to different input data interpolation schemes. However, SCS-CN plays a major role in both models. Therefore, streamflow simulation may not be appropriate when the rainfall amount is small [30,49,68]. Similar results have been identified for the non-monsoonal months during analysis of inter-annual variability of the models, in the previous section. In addition, the different runoff generation technique of the F module and baseflow generation technique of the GW module of SHM (stated earlier in Section 3.1) with respect to techniques used in the SWAT model are also responsible for the different results of the months. In particular, the soil matrix and antecedent condition of the F module may play a role in the poor model simulation of SHM for low and moderate low flows. Moreover, the routing technique of SHM seems to be the reason behind the upper hand in simulating high flows in comparison to SWAT, by capturing the travel time of the streamflow in a better manner.

5.5. Uncertainty Analysis of Monthly Simulations

Figure 6a,c present the 95PPU uncertainty band for monthly simulation during the calibration period. Figure 6e,g present the 95PPU uncertainty band for monthly simulation during the validation period. Among them Figure 6a,e are for SHM simulations and Figure 6c,g are for SWAT simulations. In addition, Figure 6b,d present the scatter plot of NQR and NQS along with two regression lines: corresponding to upper and lower limits of 95% confidence interval (CI) and one corresponding to the median for the calibration period. Figure 6b,d are for the SHM and SWAT simulation, respectively. Figure 6f,h present the scatter plot of NQR and NQS along with two regression lines: corresponding to upper and lower limits of 95% confidence interval (CI) and one corresponding to the median for the validation period. Figure 6f,h are for SHM and SWAT simulation, respectively.

Figure 6a,c,e,g depict that most of the observed streamflow falls inside the defined bands, though the amount is higher for SHM simulations (Figure 6a,c). Moreover, from Figure 6a,c,e,g it is also evident that the width of 95PPU band is thinner for SHM simulations in comparison to SWAT simulations. Thus, it can be inferred that SHM has less uncertainty in model simulations in comparison to SWAT simulations. Figure 6b,d,f,h depict the relationship between residual and simulated streamflow in the Gaussian domain and confirm that the simulated streamflow is able to capture 95% of the observed streamflow during the calibration and validation periods for both the models. For estimating the collective uncertainty, Dogulu et al. [69] supported the use of the quantile regression (QR) technique due to its simplicity and linearity which has been used elaborately by Kumar et al. [70].

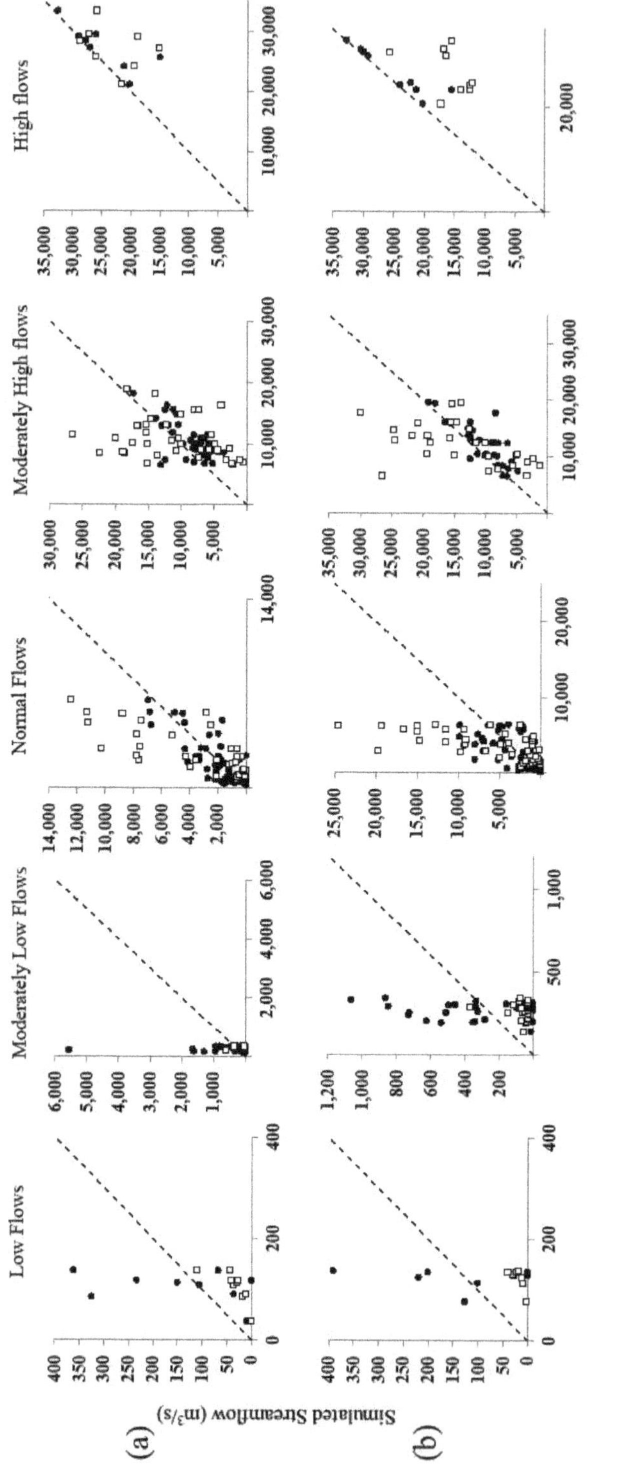

Figure 5. Comparison between observed and models' simulated five streamflow quantile ranges during (**a**) calibration period and (**b**) validation period.

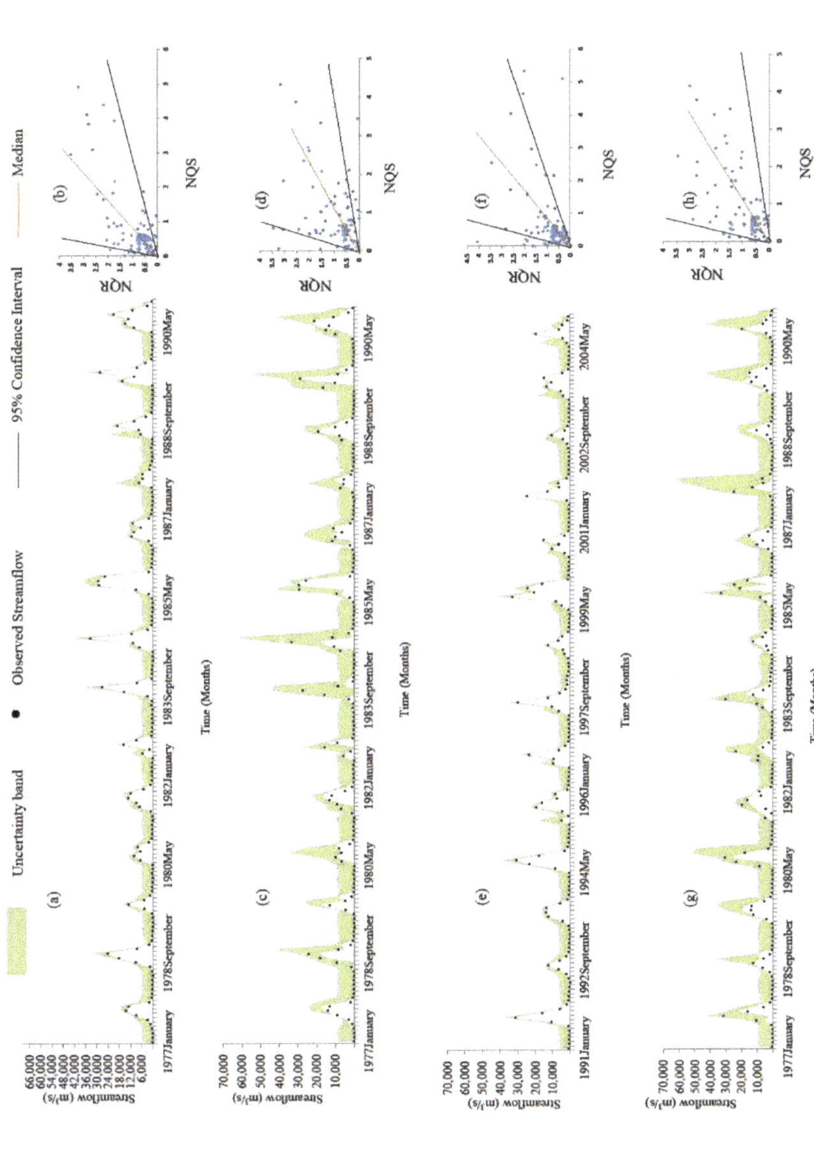

Figure 6. (**a**) Observed discharge and uncertainty band for SHM simulation for calibration period in normalized domain (**c**) Observed discharge and uncertainty band for SWAT simulation for calibration period in normalized domain (**e**) Observed discharge and uncertainty band for SHM simulation for validation period in normalized domain (**g**) Observed discharge and uncertainty band for SWAT simulation for validation period in normalized domain, (**b**) Error model of SHM simulation for calibration period, (**d**) Error model of SWAT simulation for calibration period, (**f**) Error model of SHM simulation for validation period, (**h**) Error model of SWAT simulation for validation period in normalized domain.

P and R values based on the uncertainty analysis results of SHM and SWAT simulations for the calibration and validation periods, respectively shown in Figure 7a,b, elaborate that SHM poses less uncertainty in monthly simulation than SWAT model.

Though uncertainties from all sources have been counted in the QR uncertainty analysis technique, spatial distributions of input data are different for the models due to different data interpolation techniques and model structures of the two models. Although the models' parameters take care of the modeling processes during calibration, the spatial variation of input data may affect the uncertainty of the models' simulation significantly. The results of the uncertainty analysis also represent this aspect and show that SHM represents the spatial variations of landscape characteristics and input data more accurately.

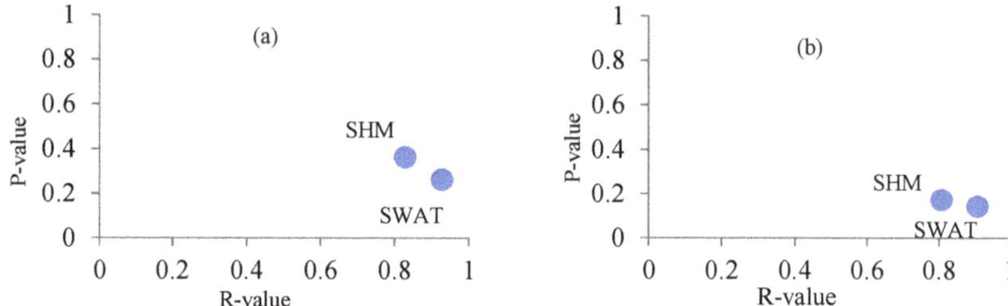

Figure 7. P-value vs. R-value for (**a**) calibration period and (**b**) validation period for the SHM and SWAT model simulations.

6. Conclusions

SHM and SWAT models were used to simulate the monthly streamflow at Anandpur gauging station of Baitarani basin for in-depth inter-comparison of the models' performances. The SWAT model was set to have an average size of the HRUs equal to 25.2 km^2, (nearly equal to the grid cell resolution of SHM, i.e., 25 km^2) so that the two discretization schemes were in similar scale. Results showed that although both SHM and SWAT have produced reasonable results, SHM performed better. To be more specific, SHM performed better in simulating annual peak flows, and reproduced the annual variability of observed streamflow for every month of the year. In addition, SHM estimates normal, moderately high, and high flows better than SWAT. Uncertainty analysis of simulated streamflow of both the models also supports the superiority of SHM model in comparison with SWAT model. Possible impacts of the model structure were also identified for the results.

In summary, SHM produced better results in comparison to SWAT at the monthly scale with proof of better model structure for the large research catchment. However, we cannot draw a conclusion that grid-based hydrological modeling is better than the HRU based. More researches should be carried out for comparing different discretization schemes for other Indian basins and other parts of the world.

Author Contributions: Conceptualization, Y.Z. and P.K.P.; methodology, P.K.P.; software, P.K.P.; validation, P.K.P.; formal analysis, P.K.P.; investigation, P.K.P.; resources, R.S.; data curation, P.K.P.; writing—original draft preparation, P.K.P.; writing—review and editing, Y.Z. and A.M.; visualization, Y.Z.; supervision, A.M., N.P., R.S. and Y.Z.; project administration, R.S.; funding acquisition, Y.Z.

Funding: This research was funded by Space Application Centre (SAC), Ahmedabad, grant number IIT/SRIC/AGFE/DWI/2013-14/124 and the APC was funded by the CAS Pioneer Hundred Talents Program.

Acknowledgments: This work is financially supported by the CAS Pioneer Hundred Talents Program, Space Application Centre (SAC), Ahmedabad (Grant no. IIT/SRIC/AGFE/DWI/2013-14/124). The technical support of programmer Partha Samanta in developing the model is acknowledged. We also acknowledge the technical support of R.P. Singh and P.K. Gupta from SAC, Ahmedabad, and our other project partners from NERIST, Itanagar, IIT Guwahati and IISc, Bangalore.

Conflicts of Interest: The authors declare no conflict of interest.

References

1. Liang, X.; Guo, J.; Leung, L.R. Assessment of the effects of spatial resolutions on daily water flux simulations. *J. Hydrol.* **2004**, *298*, 287–310. [CrossRef]
2. Kampf, S.K.; Burges, S.J. A framework for classifying and comparing distributed hillslope and catchment hydrologic models. *Water Resour. Res.* **2007**, *43*. [CrossRef]
3. Khakbaz, B.; Imam, B.; Hsu, K.; Sorooshian, S. From lumped to distributed via semi-distributed: Calibration strategies for semi-distributed hydrologic models. *J. Hydrol.* **2012**, *418–419*, 61–77. [CrossRef]
4. Smith, M.B.; Gupta, H.V. The Distributed Model Intercomparison Project (DMIP)—Phase 2 experiments in the Oklahoma region, USA. *J. Hydrol.* **2012**, *418–419*, 1–2. [CrossRef]
5. Wood, E.F.; Sivapalan, M.; Beven, K.; Band, L. Effects of spatial variability and scale with implications to hydrologic modeling. *J. Hydrol.* **1988**, *102*, 29–47. [CrossRef]
6. Kouwen, N.; Soulis, E.D.; Pietroniro, A.; Donald, J.; Harrington, R.A. Grouped response units for distributed hydrologic modelling. *J. Water Resour. Plan. Manag.* **1993**, *119*, 289–305. [CrossRef]
7. Reggiani, P.; Sivapalan, M.; Majid Hassanizadeh, S. A unifying framework for watershed thermodynamics: Balance equations for mass, momentum, energy and entropy, and the second law of thermodynamics. *Adv. Water Resour.* **1998**, *22*, 367–398. [CrossRef]
8. Winter, T.C. The concept of hydrologic landscapes. *J. Am. Water Resour. Assoc.* **2001**, *37*, 335–349. [CrossRef]
9. Vivoni, E.R.; Ivanov, V.Y.; Bras, R.L.; Entekhabi, D. Generation of triangulated irregular networks based on hydrological similarity. *J. Hydrol. Eng.* **2004**, *9*, 288–302. [CrossRef]
10. Neitsch, S.L.; Arnold, J.G.; Kiniry, J.R.; Srinivasan, R.; Williams, J.R. *Soil and Water Assessment Tool Theoretical Documentation Version 2009*; Texas Water Resources Institute Technical Report 406; Texas A & M University System, College Station: College Station, TX, USA, 2011.
11. Rathjens, H.; Oppelt, N. SWAT model calibration of a grid-based setup. *Adv. Geosci.* **2012**, *32*, 55–61. [CrossRef]
12. Gassman, P.W.; Reyes, M.R.; Green, C.H.; Arnold, J.G. The Soil and Water Assessment Tool: Historical development, applications, and future research directions. *Trans. ASAE* **2007**, *50*, 1211–1250. [CrossRef]
13. Arnold, J.G.; Allen, P.M.; Volk, M.; Williams, J.R.; Bosch, D.D. Assessment of different representations of spatial variability on SWAT model performance. *Trans. ASABE* **2010**, *53*, 1433–1443. [CrossRef]
14. Reed, S.; Koren, V.; Smith, M.; Zhang, Z.; Moreda, F.; Seo, D.J. Overall distributed model intercomparison project results. *J. Hydrol.* **2004**, *298*, 27–60. [CrossRef]
15. Finnerty, B.D.; Smith, M.B.; Seo, D.J.; Koren, V.; Moglen, G.E. Space-time scale sensitivity of the Sacramento model to radar-gage precipitation inputs. *J. Hydrol.* **1997**, *203*, 21–38. [CrossRef]
16. Wood, E.F.; Lettenmaier, D.; Liang, X.; Nijssen, B.; Wetzel, S.W. Hydrological modeling of continental-scale basins. *Annu. Rev. Earth Planet. Sci.* **1997**, *25*, 279–300. [CrossRef]
17. Kuo, W.-L.; Steenhuis, T.S.; McCulloch, C.E.; Mohler, C.L.; Weinstein, D.A.; DeGloria, S.D.; Swaney, D.P. Effect of grid size on runoff and soil moisture for a variable-source-area hydrology model. *Water Resour. Res.* **1999**, *35*, 3419–3428. [CrossRef]
18. Andersen, J.; Refsgaard, J.C.; Jensen, K.H. Distributed hydrological modelling of the Senegal River BasinModel construction and validation. *J. Hydrol.* **2001**, *247*, 200–214. [CrossRef]
19. Booij, M.J. Impact of climate change on river flooding assessed with different spatial model resolutions. *J. Hydrol.* **2005**, *303*, 176–198. [CrossRef]
20. Orth, R.; Staudinger, M.; Seneviratne, S.I.; Seibert, J.; Zappa, M. Does model performance improve with complexity? A case study with three hydrological models. *J. Hydrol.* **2015**, *523*, 147–159. [CrossRef]
21. Haghnegahdar, A.; Tolson, B.A.; Craig, J.R.; Paya, K.T. Assessing the performance of a semi-distributed hydrological model under various watershed discretization schemes. *Hydrol. Process.* **2015**, *29*, 4018–4031. [CrossRef]
22. Abu El-Nasr, A.; Arnold, J.G.; Feyen, J.; Berlamont, J. Modelling the hydrology of a catchment using a distributed and a semi-distributed model. *Hydrol. Process.* **2005**, *19*, 573–587. [CrossRef]
23. Singh, J.; Knapp, H.V.; Arnold, J.G.; Demissie, M. Hydrological modeling of the Iroquois river watershed using HSPF and SWAT. *J. Am. Water Resour. Assoc.* **2005**, *41*, 343–360. [CrossRef]
24. Im, S.; Brannan, K.M.; Mostaghimi, S.; Kim, S.M. Comparison of HSPF and SWAT models performance for runoff and sediment yield prediction. *J. Environ. Sci. Health Part A Toxic/Hazard. Subst. Environ. Eng.* **2007**, *42*, 1561–1570. [CrossRef]

25. Nasr, A.; Bruen, M.; Jordan, P.; Moles, R.; Kiely, G.; Byrne, P. A comparison of SWAT, HSPF and SHETRAN/GOPC for modelling phosphorus export from three catchments in Ireland. *Water Res.* **2007**, *41*, 1065–1073. [CrossRef]
26. Li, Z.; Xu, Z.; Li, Z. Performance of WASMOD and SWAT on hydrological simulation in Yingluoxia watershed in northwest of China. *Hydrol. Process.* **2011**, *25*, 2001–2008. [CrossRef]
27. Cornelissen, T.; Diekkrüger, B.; Giertz, S. A comparison of hydrological models for assessing the impact of land use and climate change on discharge in a tropical catchment. *J. Hydrol.* **2013**, *498*, 221–236. [CrossRef]
28. Xie, H.; Lian, Y. Uncertainty-based evaluation and comparison of SWAT and HSPF applications to the Illinois River Basin. *J. Hydrol.* **2013**, *481*, 119–131. [CrossRef]
29. Sommerlot, A.R.; Nejadhashemi, A.P.; Woznicki, S.A.; Giri, S.; Prohaska, M.D. Evaluating the capabilities of watershed-scale models in estimating sediment yield at field-scale. *J. Environ. Manag.* **2013**, *127*, 228–236. [CrossRef]
30. Zhang, L.; Jin, X.; He, C.; Zhang, B.; Zhang, X.; Li, J.; Zhao, C.; Tian, J.; DeMarchi, C. Comparison of SWAT and DLBRM for hydrological modeling of a mountainous watershed in arid northwest China. *J. Hydrol. Eng.* **2016**, *21*, 4016007. [CrossRef]
31. Pignotti, G.; Rathjens, H.; Cibin, R.; Chaubey, I.; Crawford, M. Comparative analysis of HRU and grid-based SWAT models. *Water* **2017**, *9*, 272. [CrossRef]
32. Surfleet, C.G.; Tullos, D.; Chang, H.; Jung, I.-W. Selection of hydrologic modeling approaches for climate change assessment: A comparison of model scale and structures. *J. Hydrol.* **2012**, *464–465*, 233–248. [CrossRef]
33. Flügel, W.-A. Combining GIS with regional hydrological modelling using hydrological response units (HRUs): An application from Germany. *Math. Comput. Simul.* **1997**, *43*, 297–304. [CrossRef]
34. Jiang, T.; Chen, Y.D.; Xu, C.Y.; Chen, X.; Chen, X.; Singh, V.P. Comparison of hydrological impacts of climate change simulated by six hydrological models in the Dongjiang Basin, South China. *J. Hydrol.* **2007**, *336*, 316–333. [CrossRef]
35. Najafi, M.R.; Moradkhani, H.; Jung, I.W. Assessing the uncertainties of hydrologic model selection in climate change impact studies. *Hydrol. Process.* **2011**, *25*, 2814–2826. [CrossRef]
36. Singh, V.P. *Computer Models of Watershed Hydrology*; Water Resources Publications, LLC: Littleton, CO, USA, 1995.
37. Haverkamp, S.; Srinivasan, R.; Frede, H.G.; Santhi, C. Subwatershed spatial analysis tool: Discretization of a distributed hydrologic model by statistical criteria. *J. Am. Water Resour. Assoc.* **2002**, *38*, 1723–1733. [CrossRef]
38. Dahm, R.J.; Singh, U.K.; Lal, M.; Marchand, M.; Sperna Weiland, F.C.; Singh, S.K.; Singh, M.P. Downscaling GCM data for climate change impact assessments on rainfall: A practical application for the Brahmani-Baitarani river basin. *Hydrol. Earth Syst. Sci. Discuss.* **2016**, *499*, 1–42. [CrossRef]
39. Nayak, P.C.; Sudheer, K.P.; Rangan, D.M.; Ramasastri, K.S. A neuro-fuzzy computing technique for modeling hydrological time series. *J. Hydrol.* **2004**, *291*, 52–66. [CrossRef]
40. Sharpley, A.N.; Williams, J.R. *EPIC—Erosion/Productivity Impact Calculator: 1. Model Documentation*; USDA Technical Bulletin No. 1768; USDA: Washington, DC, USA, 1990.
41. Paul, P.K.; Kumari, N.; Panigrahi, N.; Mishra, A.; Singh, R. Implementation of cell-to-cell routing scheme in a large scale conceptual hydrological model. *Environ. Model. Softw.* **2018**, *101*, 23–33. [CrossRef]
42. Paul, P.K.; Gaur, S.; Yadav, B.; Panigrahy, N.; Mishra, A.; Singh, R. Diagnosing credibility of a large-scale conceptual hydrological model in simulating streamflow. *J. Hydrol. Eng.* **2019**, *24*, 4019004. [CrossRef]
43. Chow, V.T.; Maidment, D.R.; Mays, L.W. *Applied Hydrology*; McGraw-Hill: New York, NY, USA, 2005; pp. 147–155.
44. Hargreaves, G.H.; Samani, Z.A. Reference crop evapotranspiration from ambient air temperature. *Am. Soc. Agric. Eng.* **1985**, *1*, 96–99. [CrossRef]
45. Das, P.; Islam, A.; Dutta, S.; Dubey, A.K.; Sarkar, R. Estimation of runoff curve numbers using a physically-based approach of preferential flow modelling. In *Hydrology in a Changing World: Environmental and Human Dimensions: Proceedings of the FRIEND-Water 2014*; IAHS Publication: Wallingford, Germany, 2014; Volume 363, pp. 443–448.
46. Smith, J.L.; Halverson, H.G. *Estimating Snowpack Density from Albedo Measurement*; Research Paper PSW-RP-136; U.S. Department of Agriculture, Forest Service, Pacific Southwest Forest and Range Experiment Station: Portland, OR, USA, 1979.

47. Sekhar, M.; Rasmi, S. Groundwater flow modeling of Gundal sub-basin in Kabini river basin, India. *Asian J. Water Environ. Pollut.* **2004**, *1*, 65–77. [CrossRef]

48. Du, J.; Xie, H.; Hu, Y.; Xu, Y.; Xu, C.Y. Development and testing of a new storm runoff routing approach based on time variant spatially distributed travel time method. *J. Hydrol.* **2009**, *369*, 44–54. [CrossRef]

49. Arnold, J.G.; Srinivasan, R.; Muttiah, R.S.; Williams, J.R. Large area hydrologic modeling and assessment Part I: Model development. *J. Am. Water Resour. Assoc.* **1998**, *34*, 73–89. [CrossRef]

50. Indian Institute of Technology Kharagpur. *Development of Conceptual Hydrological Model for Different Ecosystems of India*; Annual Report; Indian Institute of Technology Kharagpur: West Bengal, India, 2017; pp. 1–19.

51. Van Griensven, A.; Meixner, T.; Grunwald, S.; Bishop, T.; Diluzio, M.; Srinivasan, R. A global sensitivity analysis tool for the parameters of multi-variable catchment models. *J. Hydrol.* **2006**, *324*, 10–23. [CrossRef]

52. Mockus, V. Estimation of direct runoff from storm rainfall. In *SCS National Engineering Handbook*; U.S. Department of Agriculture: Washington, DC, USA, 1972; pp. 10.1–10.16.

53. Nash, J.E.; Sutcliffe, J.V. River flow forecasting through conceptual models part I—A discussion of principles. *J. Hydrol.* **1970**, *10*, 282–290. [CrossRef]

54. Koenker, R.; Bassett, G. Regression Quantiles. *Econometrica* **1978**, *46*, 33. [CrossRef]

55. Koenker, R.; Hallock, K.F. Quantile regression. *J. Econ. Perspect.* **2001**, *15*, 143–156. [CrossRef]

56. Weerts, A.H.; Winsemius, H.C.; Verkade, J.S. Estimation of predictive hydrological uncertainty using quantile regression: Examples from the National Flood Forecasting System (England and Wales). *Hydrol. Earth Syst. Sci.* **2011**, *15*, 255–265. [CrossRef]

57. Xue, C.; Chen, B.; Wu, H. Parameter uncertainty analysis of surface flow and sediment yield in the Huolin Basin, China. *J. Hydrol. Eng.* **2014**, *19*, 1224–1236. [CrossRef]

58. Moriasi, D.N.; Arnold, J.G.; Van Liew, M.W.; Bingner, R.L.; Harmel, R.D.; Veith, T.L. Model evaluation guidelines for systematic quantification of accuracy in watershed simulations. *Trans. ASABE* **2007**, *50*, 885–900. [CrossRef]

59. Moriasi, D.N.; Gitau, M.W.; Pai, N.; Daggupati, P. Hydrologic and water quality models: Performance measures and evaluation criteria. *Trans. ASABE* **2015**, *58*, 1763–1785. [CrossRef]

60. Mujumdar, P.P.; Ghosh, S. Modeling GCM and scenario uncertainty using a possibilistic approach: Application to the Mahanadi River, India. *Water Resour. Res.* **2008**, *44*, 1–15. [CrossRef]

61. Gosain, A.K.; Rao, S.; Arora, A. Climate change impact assessment of water resources of India. *Curr. Sci.* **2011**, *101*, 356–371.

62. Islam, A.; Sikka, A.K.; Saha, B.; Singh, A. Streamflow response to climate change in the Brahmani river basin, India. *Water Resour. Manag.* **2012**, *26*, 1409–1424. [CrossRef]

63. Mitra, S.; Mishra, A. Hydrologic response to climatic change in the Baitarni river basin. *J. Indian Water Resour. Soc.* **2014**, *34*, 10.

64. Paul, P.K.; Mishra, A. Streamflow assessment in changing monsoon climate in two neighbouring river basins of eastern India. *J. Indian Water Resour. Soc.* **2018**, *38*, 1–10.

65. Sindhu, K.; Durga Rao, K.H.V. Hydrological and hydrodynamic modeling for flood damage mitigation in Brahmani–Baitarani river basin, India. *Geocarto Int.* **2016**, *32*, 1004–1016. [CrossRef]

66. Baratti, E.; Montanari, A.; Castellarin, A.; Salinas, J.L.; Viglione, A.; Bezzi, A. Estimating the flood frequency distribution at seasonal and annual time scales. *Hydrol. Earth Syst. Sci.* **2012**, *16*, 4651–4660. [CrossRef]

67. Arnold, J.G.; Fohrer, N. SWAT2000: Current capabilities and research opportunities in applied watershed modelling. *Hydrol. Process.* **2005**, *19*, 563–572. [CrossRef]

68. Mishra, S.K.; Singh, V.P. *SCS-CN Method. Soil Conservation Service Curve Number (SCS-CN) Methodology*; Mishra, S.K., Singh, V.P., Eds.; Springer: Berlin, Germany, 2003; pp. 84–146.

69. Dogulu, N.; López López, P.; Solomatine, D.P.; Weerts, A.H.; Shrestha, D.L. Estimation of predictive hydrologic uncertainty using the quantile regression and UNEEC methods and their comparison on contrasting catchments. *Hydrol. Earth Syst. Sci.* **2015**, *19*, 3181–3201. [CrossRef]

70. Kumar, A.; Singh, R.; Jena, P.P.; Chatterjee, C.; Mishra, A. Identification of the best multi-model combination for simulating river discharge. *J. Hydrol.* **2015**, *525*, 313–325. [CrossRef]

Article

The Use of Large-Scale Climate Indices in Monthly Reservoir Inflow Forecasting and Its Application on Time Series and Artificial Intelligence Models

Taereem Kim [1], Ju-Young Shin [2], Hanbeen Kim [1], Sunghun Kim [1] and Jun-Haeng Heo [1,*]

[1] School of Civil and Environmental Engineering, Yonsei University, Seoul 03722, Korea;
   taereem@yonsei.ac.kr (T.K.); luckyboy89@yonsei.ac.kr (H.K.); sunghun@yonsei.ac.kr (S.K.)
[2] National Institute of Meteorological Science, Seogwipo 63568, Korea; hyjyshin@gmail.com
*   Correspondence: jhheo@yonsei.ac.kr; Tel.: +82-2-2123-2805

Received: 30 October 2018; Accepted: 13 February 2019; Published: 21 February 2019

Abstract: Climate variability is strongly influencing hydrological processes under complex weather conditions, and it should be considered to forecast reservoir inflow for efficient dam operation strategies. Large-scale climate indices can provide potential information about climate variability, as they usually have a direct or indirect correlation with hydrologic variables. This study aims to use large-scale climate indices in monthly reservoir inflow forecasting for considering climate variability. For this purpose, time series and artificial intelligence models, such as Seasonal AutoRegressive Integrated Moving Average (SARIMA), SARIMA with eXogenous variables (SARIMAX), Artificial Neural Network (ANN), Adaptive Neural-based Fuzzy Inference System (ANFIS), and Random Forest (RF) models were employed with two types of input variables, autoregressive variables (AR-) and a combination of autoregressive and exogenous variables (ARX-). Several statistical methods, including ensemble empirical mode decomposition (EEMD), were used to select the lagged climate indices. Finally, monthly reservoir inflow was forecasted by SARIMA, SARIMAX, AR-ANN, ARX-ANN, AR-ANFIS, ARX-ANFIS, AR-RF, and ARX-RF models. As a result, the use of climate indices in artificial intelligence models showed a potential to improve the model performance, and the ARX-ANN and AR-RF models generally showed the best performance among the employed models.

Keywords: Climate variability; Large-scale climate indices; Reservoir inflow forecasting; Ensemble empirical mode decomposition; Time series model; Artificial intelligence model

---

1. Introduction

Reservoir inflow forecasting is an essential task in dam operation and is strongly linked to water resource planning and management. Reservoir inflow forecasting has become increasingly complex and important due to changes in the frequency and magnitude of water-related disasters under climate change. To better understand the responses to climate change, a large number of models have been developed for more accurate and reliable inflow forecasting [1–9].

In the hydrological field, time series models are widely used to analyze the linear stochastic progress of observed time series and forecast future time series. Based on AutoRegressive Integrated Moving Average (ARIMA) family models proposed by [10], Seasonal ARIMA (SARIMA) and Seasonal ARIMA with eXogenous variables (SARIMAX) models have been widely applied to model hydrological time series considering seasonality [11–15]. Previous studies have successfully proved the applicability of the SARIMA model following the Box and Jenkins procedures, because of the simple mathematical structure, ideal representation of the statistical and correlation structures, and relatively small number of parameters [16]. In addition, hydrological variable forecasting has been performed using artificial intelligence models since the artificial intelligence technique began to be increasingly developed in the 1990s. The Artificial Neural Network (ANN) and Adaptive Neural-based Fuzzy Inference System (ANFIS) models have been frequently used and showed good performance in hydrological variable forecasting [1,17–23]. Since the ANN and ANFIS models consider both linear and nonlinear processes of the observed time series, they were suggested as alternatives to traditional time series models for the complex practice of hydrological variable forecasting. In addition to the above two classic artificial intelligence models, a new type of machine learning method, i.e., Random Forest (RF) model, has been recently introduced as a state-of-art artificial intelligence model in the hydrologic field. The RF model has produced more accurate and stable predictions with the additional advantage of handling nonlinear and non-Gaussian data series; therefore, it has been widely used in reservoir operations [24–26].

Many studies have focused on comparing the forecasting performances of time series and artificial intelligence models as numerous forecasting models begin to propose in recent decades [27–29]. Wang et al. [1] compared several artificial intelligence methods such as ANN, ANFIS, genetic programming and support vector machine models for monthly river flow discharges. They concluded that the best model differed depending on the evaluation criteria. Valipour et al. [4] compared AutoRegressive Moving Average (ARMA), ARIMA, and autoregressive ANN models for forecasting monthly inflow while increasing the number of parameters to improve accuracy. They concluded that the ARIMA model is more appropriate to forecast over 12 months while the autoregressive ANN model showed a better forecasting performance over five years. Emamgholizadeh et al. [30] compared the ANN and ANFIS models for forecasting the groundwater level at the Bastam Plain in Iran. The results showed that the ANFIS model leads to better performance than the ANN model. Li et al. [25] applied the RF model to compare the predictability of water level variations with various artificial intelligence models such as ANN, support vector regressions, and a linear model. They concluded that the RF model can be calibrated to provide information for water management and decision-making by providing efficient forecasting performance. They also stated that the lagged variables are important predictors, and the meteorological indices should be included in the future study.

Recently, large-scale climate indices have been employed in hydrological processes because it was proved that the climate indices can provide potential information about climate variability in the global climate system. Kashid et al. [31] predicted weekly rainfall using a genetic programming model with El Niño Southern Oscillation (ENSO) indices, Equatorial Indian Ocean Oscillation indices, outgoing longwave radiation, and lagged rainfall. They suggested that information about large-scale atmospheric circulation patterns can be successfully used for prediction of weekly rainfall with reasonable accuracy. Schepen et al. [32] provided evidence that lagged oceanic and atmospheric climate indices are potentially useful predictors of Australian seasonal rainfall by quantifying the pseudo-Bayes factor based on cross-validation predictive densities. Mekanik et al. [33] applied the ANN model and multiple linear regression analysis to forecast long-term rainfall using lagged ENSO and Indian Ocean Dipole (IOD) indices as potential predictors of spring rainfall. They suggested the use of combined lagged ENSO-IOD in ANN models can provide more reliable forecasting results, therefore, contributing significant positive impacts to water resource management. Abbot and Marohasy [34] evaluated the utility of climate indices in rainfall forecasting using the ANN model in Queensland, Australia. They focused on the selection of input variables including climate indices and concluded that the optimization of input variable selection could provide better performance in monthly rainfall forecasting. Li et al. [35] investigated teleconnections between large-scale ocean-atmosphere patterns as potential sources of nonstationarity in annual maximum flood series in the Wangkuai Reservoir watershed, China. They found that the North Pacific Oscillation (NPO), North Atlantic Oscillation (NAO), and Atlantic Oscillation (AO) indices all had significant correlations with flood peak.

To identify information on climate variability and trends due to the effects of climate change on hydro-climatic variables, a decomposition method named ensemble empirical mode decomposition (EEMD) was applied in recent studies. Wu et al. [36] suggested that empirical mode decomposition (EMD) can reveal intrinsic properties, i.e., trend and variability, in nonlinear and nonstationary data. Lee and Ouarda [37] predicted future precipitation and extreme hydrological variables by modeling a nonstationary oscillation process using the EMD process. Breaker and Ruzmaikin [38] used the EEMD to analyze a 154-year record of monthly sea level data. They identified that the extracted long-term trend modes contain variabilities on time scales consistent with the Pacific Decadal Oscillation (PDO). Shi et al. [39] applied EEMD to past temperature and precipitation data and found that the extracted low-frequency signals in temperature data significantly correlated with Northern Hemisphere temperatures. Castino et al. [40] analyzed the trend and oscillatory modes of river discharge in the southern Central Andes of northwestern Argentina using EEMD to find the statistically significant climate indices and time intervals. They determined the time intervals according to the mean period of the intrinsic mode functions (IMFs) extracted via EEMD and found the significant climate indices for each IMF. Kim et al. [41] suggested a procedure to select climate indices that affect long-term precipitation using EEMD to identify the relationship between long-term precipitation and climate indices. They found that the lagged NINO 1+2 and the Atlantic Multidecadal Oscillation (AMO) index should be preferentially considered as predictive indicators used to forecast monthly precipitation in South Korea. As these studies demonstrate, climate indices can be used as input variables because they provide predictive information on large-scale climate modes for long-term forecasting based on the global climate system under complex weather conditions. The EEMD can also be employed as an effective tool for reservoir inflow forecasting. Yu et al. [42] proposed three decomposition methods—Fourier transformation, EEMD, and singular spectrum analysis—for pre-processing, and used autoregressive input variables to support the vector regression model. They showed that this decomposition method is an effective method for reservoir inflow forecasting. Therefore, the use of large-scale climate indices and the EEMD for long-term forecasting is increased to utilize the predictive information provided by large-scale climate modes.

Two critical research questions for reservoir inflow forecasting are to identify highly-correlated climate indices, and to find a model which provides the best performance through applications. So far, many studies have a limited focus primarily on examining the relationship between climate indices and hydrologic variables. One of the important needs for today in monthly reservoir forecasting is not only identifying potential indicators, but also increasing their applicability for efficient water management strategies and operation. This study aims to apply and compare the model performance with highly correlated input variables including lagged inflow and lagged climate indices. For this purpose, time series and artificial intelligence models—i.e., SARIMA, SARIMA with eXogenous variables (SARIMAX), ANN, ANFIS, and RF models—were used to forecast the monthly reservoir inflow at the Soyanggang (SY), Chungju (CJ), and Goesan (GS) dams in South Korea. Two types of input variables, autoregressive variables (AR-) and a combination of autoregressive and exogenous variables (ARX-), were constructed to examine the impacts of climate indices for reservoir inflow. To select appropriate climate indices and construct input variables for monthly reservoir inflow, several statistical methods consisting of partial autocorrelation function (PACF), EEMD, cross-correlation analysis, and the backward elimination method were employed. The EEMD was applied to extract the oscillatory modes inherent in the reservoir inflow for selecting the climate indices with a significant correlation. Consequently, a total of eight models including time series and artificial intelligence models with two types of input variables (e.g., SARIMA, SARIMAX, AR-ANN, ARX-ANN, AR-ANFIS, ARX-ANFIS, AR-RF, ARX-RF) were constructed to forecast the monthly reservoir inflow. Model performance was compared using the correlation coefficient (r), root mean square error (RMSE), and Nash–Sutcliffe Efficiency (NSE).

The rest of this paper is organized as follows. Section 2 presents the employed forecasting models such as time series and artificial intelligence models. Section 3 presents data and study area, including monthly reservoir inflow and large-scale climate indices used in this study. Section 4 presents a process of input variable selection and explains the detailed methods. Application and results including model input variables, model parameters and setting, and the model performance are described in Section 5, and discussions are presented in Section 6. Section 7 concludes the paper.

2. Time Series and Artificial Intelligence Models

2.1. Seasonal AutoRegressive Integrated Moving Average (SARIMA) Model

The Box–Jenkins approach using the ARIMA family of models is widely used to forecast future values based on an observed time series. The ARIMA model parsimoniously interprets a stochastic process with autoregressive (AR) and moving average (MA) operators. If there is seasonality in the time series, the SARIMA model is more useful for modeling as it considers seasonality through a differencing procedure [10]. The SARIMA model is defined in Equation (1):

$$\phi(B)^P \Phi^s(B)^P \left(1 - B^d\right)\left(1 - B^D\right) y_t = \theta(B)^q \Theta^s(B)^Q \varepsilon_t \tag{1}$$

where y_t is a given time series, $\phi(B)^P$ and $\theta(B)^q$ are the non-seasonal AR and MA operators, and $\Phi^s(B)^P$ and $\Theta^s(B)^Q$ are the seasonal AR and MA operators with seasonal period, s, d and D are non-seasonal and seasonal differencing orders, ε_t represents white noise with zero mean and standard deviation σ_ε^2, and B is the backshift operator. The operators are defined in Equations (2)–(5), respectively.

$$\phi(B) = 1 - \phi_1 B - \phi_2 B^2 - \cdots - \phi_p B^p \tag{2}$$

$$\theta(B) = 1 - \theta_1 B - \theta_2 B^2 - \cdots - \theta_q B^q \tag{3}$$

$$\Phi^s(B) = 1 - \Phi_1 B^S - \Phi_2 B^{2S} - \cdots - \Phi_P B^{PS} \tag{4}$$

$$\Theta^s(B) = 1 - \Theta_1 B^S - \Theta_2 B^{2S} - \cdots - \Theta_Q B^{QS} \tag{5}$$

2.2. SARIMA with eXogenous Variables (SARIMAX) Model

The SARIMAX model is a multivariate time series model extended from the SARIMA model to consider the effect of the exogenous variables in a time series [43]. SARIMAX is useful in modeling a time series that has a dominating variable. The SARIMAX model is an advanced model of the ARIMA family because it can consider trends, seasonality, and exogenous variables. The SARIMAX model with exogenous variable (x_t) is defined in Equation (6):

$$\phi(B)^P \Phi^s(B)^P \left[\left(1 - B^d\right)\left(1 - B^D\right) y_t - c x_t\right] = \theta(B)^q \Theta^s(B)^Q \varepsilon_t \tag{6}$$

where c is the regression coefficient of the exogenous variable, and all other parameters are described in Section 2.1.

2.3. Artificial Neural Network (ANN) Model

The ANN model is a powerful machine learning technique that is designed to mimic the structure of the brain [44]. It has been widely applied in hydrology to improve the predictability of future hydrologic variables because it considers both linear and nonlinear structures. In general, the basic structure of the ANN model is three layers (input, hidden, and output) as shown in Figure 1.

Consider there are n number of input variables in the input node (x_i, $i = 1, 2, \ldots, n$), the p number of nodes in the hidden layer (z_j, $j = 1, 2, \ldots, p$), and the k number of output variables in the output node (y_m, $m = 1, 2, \ldots, k$). The ANN model can be described in Equations (7) and (8):

$$\hat{y}_m = f_y \left(\sum_{j=1}^{p} z_j W_{kj} + b_k \right) \tag{7}$$

$$z_j = f_z \left(\sum_{i=1}^{n} x_i W_{ji} + c_j \right) \tag{8}$$

where the weight parameters W_{kj} and W_{ji} indicate the strength of the connections between the nodes, b_k and c_j are the bias, f_y and f_z are the activation functions that are connected to each other with weight parameters.

The key purpose of the ANN model is to find the best weight parameters using a training algorithm. The backpropagation algorithm is most commonly used for ANN training by adjusting the weight parameters between the hidden and output layers to reduce the margin of error in the output. The optimal number of hidden nodes can be determined by trial-and-error approaches because there is no exact way to decide the number of hidden nodes. However, it was found that better results can be obtained when the number of hidden nodes is smaller than or equal to the number of input nodes [4,45]. In addition, there are several types of activation functions such as the sigmoid function, hyperbolic tangent function, and sign function that can learn nonlinear relationships between the input and output. The general process of ANN modeling is to construct a model which reduces errors in the training set and then applying this model to the test set.

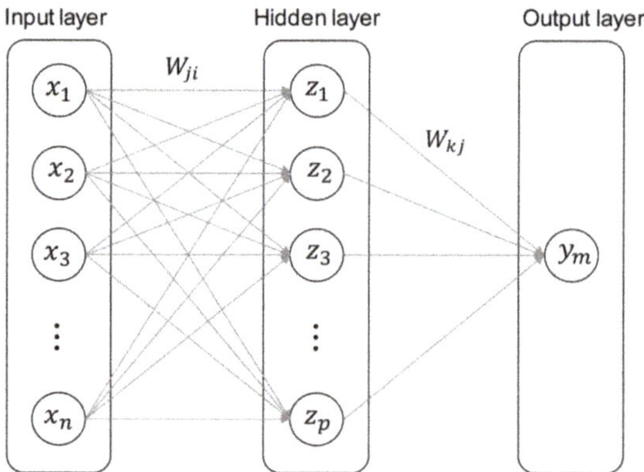

Figure 1. The basic structure of the Artificial Neural Network (ANN) model (input, hidden, and output layers).

2.4. Adaptive Neural-Based Fuzzy Inference System (ANFIS) Model

The ANFIS model is a multilayer feed-forward network that combines the ANN model and fuzzy logic based on the Takagi–Sugeno fuzzy inference system [46]. It is a powerful tool for hydrological forecasting because it integrates the advantages of both the ANN and fuzzy inference systems. The fuzzy reasoning system for a first-order Sugeno fuzzy model has the following two if-then rules:

- *Rule 1: if x is A_1 and y is B_1 then* $f_1 = p_1 x + q_1 y + r_1$
- *Rule 2: if x is A_2 and y is B_2 then* $f_2 = p_2 x + q_2 y + r_2$ where A_1, A_2 and B_1, B_2 are the membership functions of each input x and y, f_1 and f_2 are the output functions and p_1, q_1, r_1 and p_2, q_2, r_2 are linear parameters. The ANFIS model consists of five layers as shown in Figure 2.

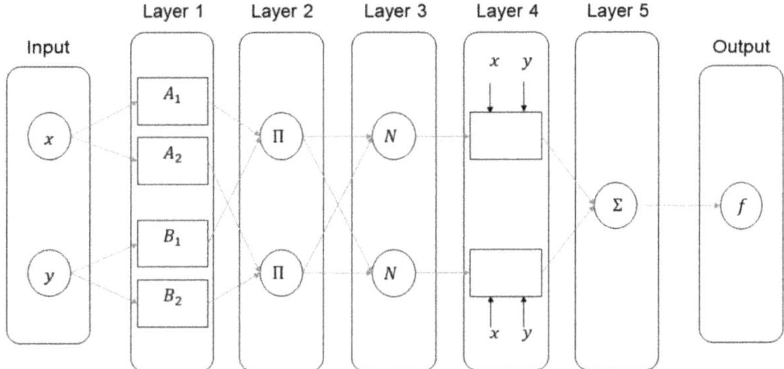

Figure 2. The basic structure of the Adaptive Neural-based Fuzzy Inference System (ANFIS) model.

Layer 1: The fuzzy membership grade for each node is generated by the fuzzy membership function (MF). An MF is an indicator function that defines how a point in the input space is mapped to a membership value between 0 to 1. The output of node i is defined in Equations (9) and (10):

$$Q_{1,i} = \mu_{A_i}(x), \ i = 1, 2 \tag{9}$$

$$Q_{1,i} = \mu_{B_{i-2}}(y), \ i = 3, 4. \tag{10}$$

where x and y are the input for node i, and μ_{A_i} (or $\mu_{B_{i-2}}$) is the degree of MF for a fuzzy set A_i (or B_{i-2}). The fuzzy set A_i (or B_{i-2}) is a linguistic label for a MF that could be given by appropriate functions such as Gaussian, generalized Bell shaped, trapezoidal shaped, and triangular shaped functions [1].

Layer 2: The incoming signals are multiplied to generate the output that represents the firing strength of the rules, as described in Equation (11).

$$Q_{2,i} = w_i = \mu_{A_i}(x) \times \mu_{B_i}(x), \ i = 1, 2 \tag{11}$$

Layer 3: The firming strength is normalized. The normalized firming strength for node i as described in Equation (12).

$$Q_{3,i} = \overline{w_i} = \frac{w_i}{(w_1 + w_2)}, \ i = 1, 2 \tag{12}$$

Layer 4: Consequent parameters $\{p_i, q_i, r_i\}$ at node i compute the contribution of the ith rule to the overall output. It can be described by Equation (13) using the output obtained from layer 3 $(\overline{w_i})$.

$$Q_{4,i} = \overline{w_i} f_i = \overline{w_i}(p_i x + q_i y + r_i) \tag{13}$$

Layer 5: Finally, the overall output is calculated from a single node by summation of all rules as described in Equation (14).

$$Q_{5,i} = \sum_i \overline{w_i} f_i = \frac{\sum_i w_i f_i}{\sum_i w_i} \tag{14}$$

In this study, the ANFIS model used hybrid learning that was a combination of descent gradient for precedents and least squares estimation for consequents. The normalized Gaussian MF and Bell-shaped

MF are popular, with the advantage of being smooth and nonzero at all points. The linguistic term of the normalized Gaussian MF and the Bell-shaped MF are given by Equations (15) and (16):

$$\mu_{A_i}(x) = \exp\left(-\frac{(x-c_i)^2}{2\sigma_i^2}\right) \tag{15}$$

$$\mu_{A_i}(x) = \frac{1}{1 + \left[\left(\frac{x-c_i}{a_i}\right)^2\right]b_i} \tag{16}$$

where σ_i, a_i, b_i, and c_i are the parameters of the membership function (i.e., c_i and σ_i are the center and width of the fuzzy set).

2.5. Random Forest (RF) Model

The RF model is a state-of-art machine learning technique which is a nonparametric white-box regression model [24]. For the regression, the RF model employs an ensemble-learning algorithm which operates by constructing a multitude of decision trees based on bootstrap samples from the training dataset. Unlike linear regression models, RF is the most robust technique for handling a combination of nonlinear interactions between the input variables and the output. The basic structure of the RF is shown in Figure 3.

The RF begins with many bootstrap samples that are randomly selected from the original input variables. A decision tree is built for each of the bootstrap samples. For each node of a decision tree, proper input variables are selected by binary partitioning. Finally, the forecasting result can be obtained by combining the results over all trees [47]. Therefore, three parameters need to be specified: (1) the number of trees; (2) number of variables in each node (default value is 5 for regression random forest); and (3) the number of input variables per node (default value is one third of the total number of variables).

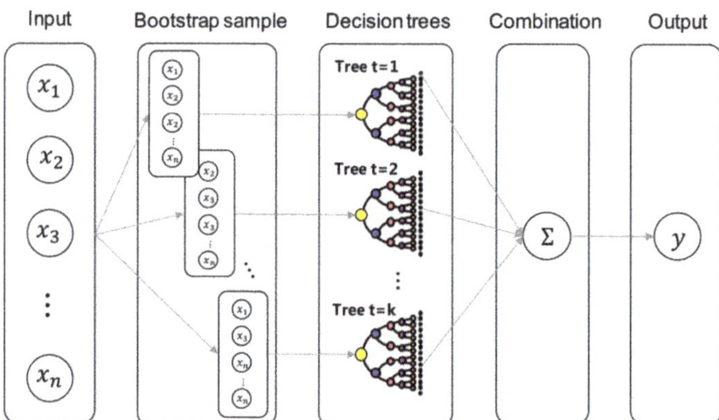

Figure 3. The basic structure of the Random Forest (RF) model.

3. Data and Study Area

Monthly reservoir inflow in the SY, CJ, and GS dams was employed in this study. The three dams are located in the Han River basin, and they are totally separate reservoirs on different branches of the Han River. The three dams have only natural reservoir inflow that is not affected by artificial control in their upstream areas. The Han River basin in South Korea plays an important role in supplying water to the capital, Seoul, and other metropolitan cities. Therefore, the three dams should be carefully operated to provide effective long-term water management.

The SY dam is a multi-objective dam located in the upstream section of the North Han River, which has a basin area of 2783 km². The CJ dam is the largest multi-objective dam in South Korea and is located on the South Han River, which has the basin area of 6705 km². The GS dam is a small hydropower dam located on the Dalcheon, a tributary of the South Han River, with a basin area of 671 km². There are seasonal characteristics to the inflow, with the monthly reservoir inflow concentrated during the flood season from June to September, since more than 70% of annual precipitation occurs in this season. The geographical locations of the three dams and basin areas are shown in Figure 4. Further detailed information about the three dams is presented in Table 1.

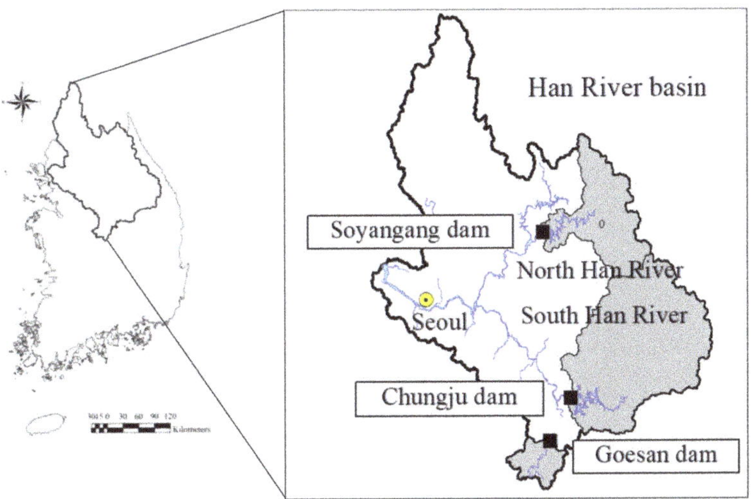

Figure 4. Locations of the Soyangang (SY), Chungju (CJ), and Goesan (GS) dams and their basins.

Table 1. Detailed information about the SY, CJ and GS dams.

Station	Type	Data Period	Basin Area (km²)	Volume (×10³ m³)	Water Supply Capacity (×10³ m³)	Mean Inflow (m³/s) Annual	Mean Inflow (m³/s) Seasonal (JJAS)
Soyangang dam	Multi-purpose (E.C.R.D[a])	January 1974–December 2016	2703	9600	1,900,000	68.28	221.99
Chungju dam	Multi-purpose (C.G.D[b])	January 1986–December 2016	6705	902	1,789,000	162.39	359.22
Goesan dam	Hydro-power (C.G.D[b])	January 1982–December 2016	671	19.6	5700	13.72	30.18

[a] Earth Core Rock Fill Dam. [b] Concrete Gravity Dam.

A total of 24 large-scale climate indices from the National Oceanic & Atmospheric Administration/Earth System Research Laboratory (NOAA/ESRL) were used in this study. The climate indices represent the atmospheric-oceanic circulation patterns, therefore, it can be possible to consider the impact of large-scale climate variability directly or indirectly on the monthly reservoir inflow series. Each index indicates an aspect of the geophysical system based on different measurements, and they are provided by the form of the monthly time series. Table 2 presents a list of the climate indices used in this study.

Table 2. List of the climate indices used in this study.

Climate Index	Classification	Climate Index	Classification
NINO 1+2 (NINO12)	ENSO/SST:Pacific	Tropical Northern Atlantic Index (TNA)	SST:Atlantic
NINO 3 (NINO3)	ENSO/SST:Pacific	Tropical Southern Atlantic Index (TSA)	SST:Atlantic
NINO 4 (NINO4)	ENSO/SST:Pacific	Carribbean SST Index (CAR)	SST:Atlantic
NINO 3.4 (NINO34)	ENSO/SST:Pacific	Pacific Decadal Oscillation (PDO)	Teleconnections
Bivariate ENSO Timeseries (BEST)	ENSO	Northern Oscillation Index (NOI)	Teleconnections
Multivariate ENSO Index (MEI)	ENSO	Pacific North American Index (PNA)	Teleconnections
Trans-Nino Index (TNI)	SST:Pacific	Western Pacific Index (WP)	Teleconnections
Western Hemisphere Warm Pool (WHWP)	SST:Pacific/SST:Atlantic	Eastern Atlantic/Western Russia (EAWR)	Teleconnections
Oceanic Nino Index (ONI)	SST:Pacific	North Atlantic Oscillation (NAO)	Teleconnections
Atlantic Multidecadal Oscillation (AMO)	SST:Atlantic	Southern Oscillation Index (SOI)	Atmosphere
Atlantic Meridional Mode (AMM)	SST:Atlantic	Quasi-Biennial Oscillation (QBO)	Atmosphere
North Tropical Atlantic SST Index (NTA)	SST:Atlantic	Artic Oscillation (AO)	Atmosphere

4. Input Variable Selection

The time series and artificial intelligence models were conducted by two types of input variables to compare the impacts of the climate indices. The first type of input variable only includes the autoregressive variables (AR-) such as the lagged inflow. The lag time was determined by the PACF. The second type of input variable includes the combination of autoregressive and exogenous variables (ARX-) composed of lagged inflow and lagged climate indices. To select the candidate climate indices for the second type of input variables, the EEMD method was employed to the monthly reservoir inflow series. A finite number of decomposed components, which are the IMFs, were extracted by the EEMD. Cross-correlation analysis was performed between the IMFs and the climate indices considering lag times from 1 to 12 months. The lagged climate indices with the highest correlation coefficient with each IMF were selected as the candidate climate indices. Therefore, the candidate variables include: (1) the lagged inflow obtained via PACF; and (2) the lagged climate indices obtained via EEMD. Finally, the backward elimination method was performed on the candidate variables to select the second type of input variables. Figure 5 shows the procedure followed to select the two types of input variables. Detailed methodologies are explained in the subsections.

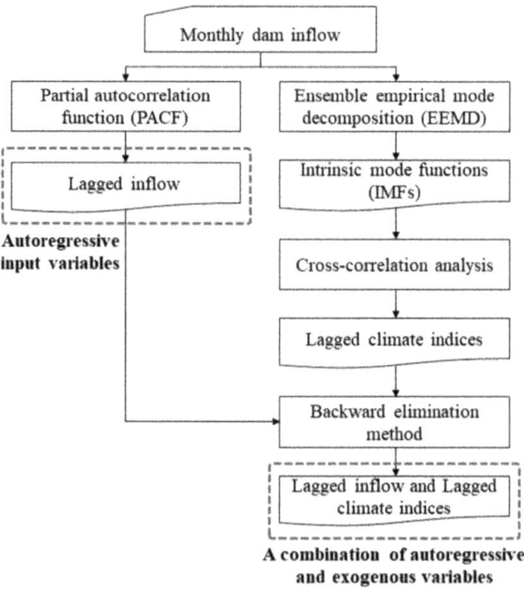

Figure 5. Procedure to select the two types of input variables: (1) Autoregressive variables (AR-); (2) a combination of autoregressive and exogenous variables (ARX-).

4.1. Partial Autocorrelation Function (PACF)

PACF is a statistic that measures the strength of relationships in a time series considering a time lag, after removing the presence of all other time lags. The range of PACF is between -1 and 1. In a PACF plot, it is able to identify the relationship between a time series and its lagged time series through the spikes and confidence intervals. For example, a significant spike at lag 12 in the PACF plot means that there is a strong correlation between a time series and the same time series with twelve intervals apart. PACF also provides useful information for selecting the autoregressive parameters in the time series models.

4.2. Ensemble Empirical Mode Decomposition (EEMD)

EMD has been recently introduced in the hydrology as an innovative analysis method to decompose statistically significant cycles and trends inherent in time series data [36,40,48]. EMD is a data-driven method that decomposes an original data series into a set of IMFs. IMFs indicate oscillatory modes that reflect the cycles and trends in the data series and should satisfy two conditions: (1) the number of extrema and zero crossings must either be equal or differ at most by one in the whole data series; (2) the mean value of the upper envelope defined by connecting all of the local maxima, and the lower envelope defined by connecting all of the local minima, must be zero at any point [47]. EEMD is a modified version of EMD proposed by Wu and Huang [49] to overcome the drawbacks of EMD due to the mode mixing problem. EEMD performs a sifting algorithm by adding an ensemble of white noise signals to the original data series, and the final result is obtained from the mean of the data series.

We will briefly introduce the sifting algorithm which is carried out to decompose a set of IMFs from an original data series $y(t), t = 1, 2, \ldots, n$ [36]. At first, the upper and lower envelopes are found by connecting the local maxima $(y_u(t))$ and local minima $y_l(t)$ using a cubic spline method. Second, the mean value between the local maxima and local minima is calculated, i.e., $y_{mean}(t) = (y_u(t) + y_l(t))/2$. Third, the $y_{mean}(t)$ is extracted from the original time series $y(t)$, i.e., $h(t) = y(t) - y_{mean}(t)$. If $h(t)$ satisfies the two conditions of IMFs, then $h(t)$ is the first IMF; else $h(t)$ is treated as $y(t)$ and the steps are repeated until $h(t)$ becomes the IMF. Fourth, a new data series is defined by extracting the IMF from $y(t)$ and repeating the steps until no more IMF can be extracted. The last IMF becomes the residue $r(t)$. Finally, $y(t)$ is defined as the sum of the IMFs and the residue as follows:

$$y(t) = \sum_{i=1}^{k} IMF_i(t) + r(t) \tag{17}$$

where k is the number of IMFs.

4.3. Cross-Correlation Analysis

Cross-correlation function is a measure of the strength of a linear correlation between two different time series considering a range of time lags. The range of the correlation coefficient (r) is between -1 and 1. If the two different time series are positively correlated, r is close to 1; if they are negatively correlated, r is close to -1; if they have no correlation, r is 0.

4.4. Backward Elimination Method

Backward elimination method is a kind of variable selection method beginning with all candidate input variables for a model. From the initially selected candidate input variables, the least significant variables are eliminated one by one until only one variable remains. The input variable showing the smallest contribution to the model performance is deleted at each step according to the model selection criteria. The relative importance of the input variable may be determined by removing the input variable [50]. The backward elimination method is useful for improving the model's performance with an iterative procedure, although it requires significant computational time.

5. Application and Results

5.1. Model Input Variables

To determine the autoregressive variables, the PACF was used to identify the lag time of the monthly reservoir inflow series. Figure 6 shows the PACF of the three dams with 95% confidence intervals. At all three dams, the spikes are mainly prominent at 1, 12, 24, and 36 months (lag1, lag12, lag24, lag36; y_{t-1}, y_{t-12}, y_{t-24}, y_{t-36}). Therefore, the first type of input variables for all the three dams includes the four cases of lagged inflows as shown in the second column of Table 3.

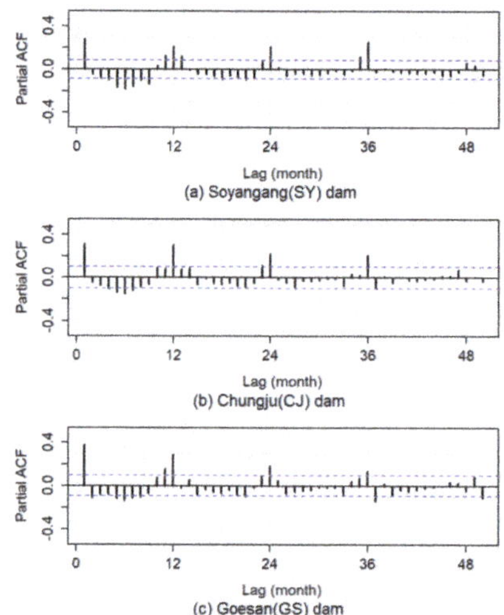

Figure 6. Partial Autocorrelation Function (PACF) of the three dams with 95% confidence intervals (dotted line): (**a**) SY dam, (**b**) CJ dam, (**c**) GS dam.

To determine the exogenous variables for the second type of input variables, the monthly reservoir inflow series was decomposed by EEMD to extract the IMFs. The IMFs indicate the different inherent frequencies of the reservoir inflow series. Lagged climate indices which have the highest correlation coefficient for each IMF were selected as the candidate exogenous variables. Figure 7 shows the IMFs (solid black line) of the three dams and the lagged climate indices which have the highest correlation coefficient for each IMF (blue dotted line). The reservoir inflow of the SY dam was decomposed into eight IMFs and a residue, and the reservoir inflow of the CJ and GS dams were decomposed into seven IMFs and a residue. For all three dams, the NINO12 index was mainly selected in the low-frequency IMF, the NTA and QBO indices were mainly selected in the middle frequency IMF, and the AMO index was mainly selected in the high-frequency IMF or residue. Table 4 shows the correlation coefficients (r) between the IMFs and the lagged climate indices in the SY, CJ, and GS dams. In the SY dam, the 4-month lagged NINO12 index had the highest correlation coefficient ($r = 0.76$) with the third IMF and the 6-month lagged AMO index has the second highest correlation coefficient ($r = -0.57$) with the eighth IMF. In the CJ dam, the 5-month lagged NINO12 index had the highest correlation coefficient ($r = 0.75$) with the third IMF, and the 12-month lagged AMO index had the second highest correlation coefficient ($r = 0.49$) with the seventh IMF. In the GS dam, the 5-month lagged NINO12 index had the highest and the second highest correlation coefficients ($r = 0.75$, $r = 0.45$) with the third and second IMFs, respectively. The 12-month lagged NTA index had the third highest correlation coefficient ($r = -0.40$) with the fifth IMF.

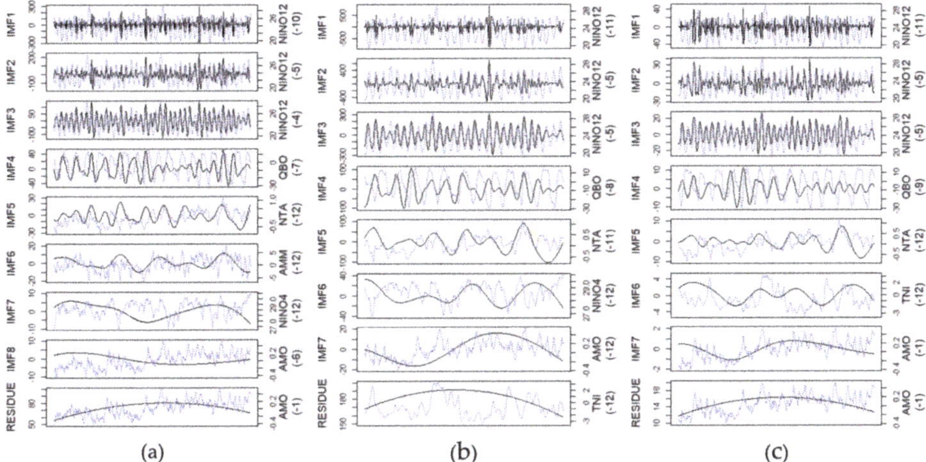

Figure 7. IMFs resulted from ensemble empirical mode decomposition (EEMD) (solid black line) and the lagged climate indices (blue dotted line) which have the highest correlation coefficient for each IMF. (**a**) SY dam, (**b**) CJ dam, (**c**) GS dam.

Table 3. The two types of input variables for the three dams.

Station	Autoregressive Variables (AR-)	A Combination of Autoregressive and Exogenous Variables (ARX-)
SY dam		Lag12, Lag36, NTA(12), AMO(6), NINO4(12), NINO12(10), AMM(12)
CJ dam	Lag1, Lag12, Lag24, Lag36	Lag36, TNI(12), AMO(12), NINO12(11), NTA(11), NINO12(5)
GS dam		Lag12, Lag36, NINO12(5), QBO(9), AMO(1)

To determine the second type of input variables for each dam, the backward elimination method was applied to the candidate variables, including the lagged inflows and the lagged climate indices. The backward elimination method results were shown in the third column in Table 3, which are the second type of input variables for each dam composed of a combination of autoregressive and exogenous variables. Therefore, a total of seven, six, and five input variables were finally selected for the SY, CJ, and GS dams. Notably, the 36-month lagged inflow, lagged NINO12 index, and lagged AMO index were identically selected as the second type of input variables at all the three dams.

Table 4. Correlation coefficients (r) between the intrinsic mode functions (IMFs) and the lagged climate indices in the SY, CJ, and GS dams.

	SY Dam				CJ Dam				GS Dam		
IMFs	**CI**	**Lag**	r	**IMFs**	**CI**	**Lag**	r	**IMFs**	**CI**	**Lag**	r
IMF1	NINO12	10	0.21	IMF1	NINO12	11	0.22	IMF1	NINO12	11	0.19
IMF2	NINO12	5	0.42	IMF2	NINO12	5	0.48	IMF2	NINO12	5	0.45
IMF3	NINO12	4	0.76	IMF3	NINO12	5	0.75	IMF3	NINO12	5	0.75
IMF4	QBO	7	−0.32	IMF4	QBO	8	−0.33	IMF4	QBO	9	−0.29
IMF5	NTA	12	−0.21	IMF5	NTA	11	−0.42	IMF5	NTA	12	−0.40
IMF6	AMM	12	−0.29	IMF6	NINO4	12	−0.38	IMF6	TNI	12	0.32
IMF7	NINO4	12	−0.15	IMF7	AMO	12	0.49	IMF7	AMO	1	0.30
IMF8	AMO	6	−0.57	RES	TNI	12	−0.21	RES	AMO	1	0.25
RES	AMO	1	0.47								

5.2. Model Parameters and Setting

A total of eight models including time series and artificial intelligence models with the two types of input variables (SARIMA, SARIMAX, AR-ANN, ARX-ANN, AR-ANFIS, ARX-ANFIS, AR-RF, and ARX-RF) were constructed. To calibrate and validate the time series models, the monthly reservoir inflow series was divided into train and test periods. The training period was set from the start of the data record to December 2012 and the test periods were set to the last four years (January 2013 to December 2016). In the artificial intelligence models, the monthly reservoir inflow series was divided into three parts, training, validation, and test periods to avoid overfitting problem. The training period was set from the start of the data record to December 2008 and the validation period was set to the next four years (January 2009 to December 2012). The test periods were set to the last four years (January 2013 to December 2016). Detailed descriptions of the model parameters and architectures are presented in the subsections.

5.2.1. SARIMA and SARIMAX Models

The SARIMA model is a conventional statistical model used to forecast future values. It is composed of autoregressive terms. The SARIMAX model, which is very similar to the SARIMA model, extends into the SARIMA model with an exogenous variable. Both the SARIMA and SARIMAX models consist of the parameters (p, d, q)(P, D, Q)[s]. In general, it is sufficient to consider up to two non-seasonal AR and MA parameters (p,q) and seasonal AR and MA parameters (P,Q) [51]. In this study, the SARIMA and SARIMAX models were composed of the non-seasonal AR and MA parameters (p,q) and the seasonal AR and MA parameters (P,Q) up to three (from 0 to 3) for more flexible model construction. Depending on the behavior of the reservoir inflow, non-seasonal and seasonal differencing orders were respectively set to 0 and 1 ($d = 0$, $D = 1$), and the seasonal periods were set to 12 ($s = 12$)) reflecting seasonality with 12-month intervals in the reservoir inflow (in Figure 5).

A total of 256 ($4 \times 1 \times 4 \times 4 \times 1 \times 4$) SARIMA models were respectively constructed for the SY, CJ, and GS dams. The number of constructed SARIMAX models was the number of SARIMA models multiplied by the number of combination autoregressive and exogenous variables in Table 3, because the SARIMAX model is able to consider one exogenous variable. The rest of the parameters were identical to the structure of the SARIMA model. A total of 1792 (256×7), 1536 (256×6), 1280 (256×5) SARIMAX models were respectively constructed for the SY, CJ, and GS dams. The best SARIMA and SARIMAX models were selected based on the minimum Akaike information criterion (AIC) value, which is commonly used for model selection. Table 5 shows the parameters of best SARIMA and SARIMAX models for the SY, CJ, and GS dams. For both model types, the best models were selected when the non-seasonal parameters had little effect ($p = 0$). In addition, the lagged inflow was selected as the exogenous variable for the SARIMAX models for all the three dams.

Table 5. Parameters of best SARIMA and SARIMAX models for the SY, CJ, and GS dams.

Station	SARIMA (p, d, q)(P, D, Q)[s]	SARIMAX (p, d, q)(P, D, Q)[s]
SY dam	SARIMA(0,0,0)(2,1,2)[12] $\Phi_1 -0.926$, $\Phi_2 -0.271$, $\Theta_1 -0.155$, $\Theta_2 -0.732$	SARIMAX(0,0,0)(3,1,1)[12]. $\times$ Lag12 $\Phi_1 -0.117$, $\Phi_2 -0.202$, $\Phi_3 0.148$, $\Theta_1 -0.954$, $c0.326$
CJ dam	SARIMA(0,0,0)(1,1,3)[12] $\Phi_1 -0.758$, $\Theta_1 -0.147$, $\Theta_2 -0.887$, $\Theta_3 0.148$	SARIMAX(0,0,0)(2,1,1)[12] $\times$ Lag36 $\Phi_1 -0.071$, $\Phi_2\ 0.052$, $\Theta_1 -0.949$, $c0.133$
GS dam	SARIMA(0,0,1)(1,1,1)[12] $\theta_1 0.171$, $\Phi_1 -0.066$, $\Theta_1 -0.888$	SARIMAX(0,0,0)(1,1,1)[12] $\times$ Lag12 $\Phi_1 0.152$, $\Theta_1 -0.937$, $c -0.254$

5.2.2. ANN Models

For the AR-ANN and ARX-ANN models, a number of hidden nodes ranging from 1 to 10 were considered in all three dams. The sigmoid activation function was used because of its superiority [4]. The optimal number of the hidden nodes for each dam was determined by considering the performance during the validation period. The correlation coefficient (r) and root mean square error (RMSE) of the AR-ANN and ARX-ANN models according to the number of the hidden nodes in the validation periods are shown in Figure 8. The optimal number of the hidden nodes was determined based on the highest r and the lowest RMSE. In cases of two criteria were different, the RMSE was preferred. Table 6 shows the optimal number of the hidden nodes of the AR-ANN and ARX-ANN models for the three dams.

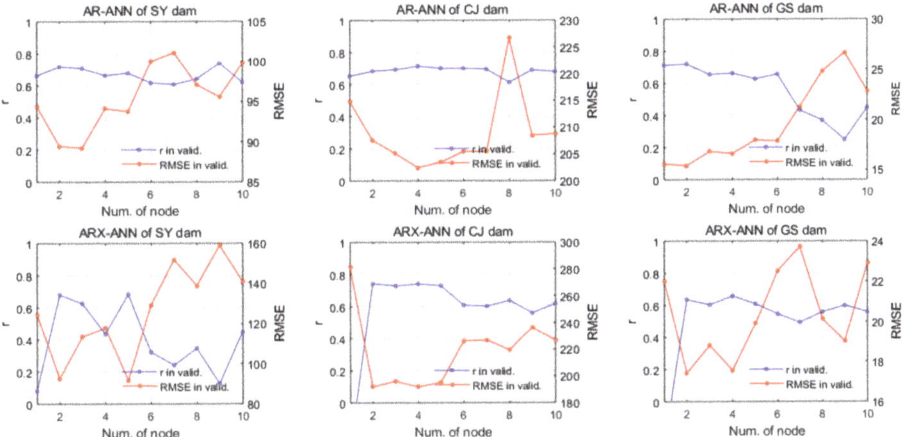

Figure 8. Correlation coefficients (r) and root mean square error (RMSE) of the AR-ANN and ARX-ANN models according to the number of nodes in the validation period for the SY, CJ, and GS dams. The blue line is the r-value in the validation period and the red line is the RMSE in the validation period.

Table 6. The optimal number of the hidden nodes of the AR-ANN and ARX-ANN models for the three dams.

Station	AR-ANN	ARX-ANN
SY dam	3	5
CJ dam	4	4
GS dam	2	2

5.2.3. ANFIS Models

For the AR-ANFIS and ARX-ANFIS models, two MFs were built on the normalized Gaussian MF ($c_i = -5$, $\sigma_i = 2$ and $c_i = +5$, $\sigma_i = 2$) and Bell-shaped MF (a = 4, b = 1, c = −5 and a = 4, b = 1, c = +5), and 20 epochs based on trial and error [52]. The optimal MF was determined by considering the performance during the validation period. The correlation coefficient (r) and root mean square error (RMSE) of the AR-ANFIS and ARX-ANFIS models according to the MF in the validation periods are shown in Figure 9. The optimal MF was determined based on the highest r and the lowest RMSE. In cases where two criteria were different, the RMSE was preferred. For the AR-ANFIS model, there were eight input MFs and 16 rules in the three dams because there are four cases of lagged inflow in the autoregressive input variable set. For the ARX-ANFIS model, the number of input MFs according to the number of input nodes was 14, 12, and 10 for the SY, CJ, and GS dams, respectively. Therefore, the number of rules were 128, 64, and 32. Table 7 shows the optimal MF, the number of input MFs and rules of the AR-ANFIS and ARX-ANFIS models for the three dams.

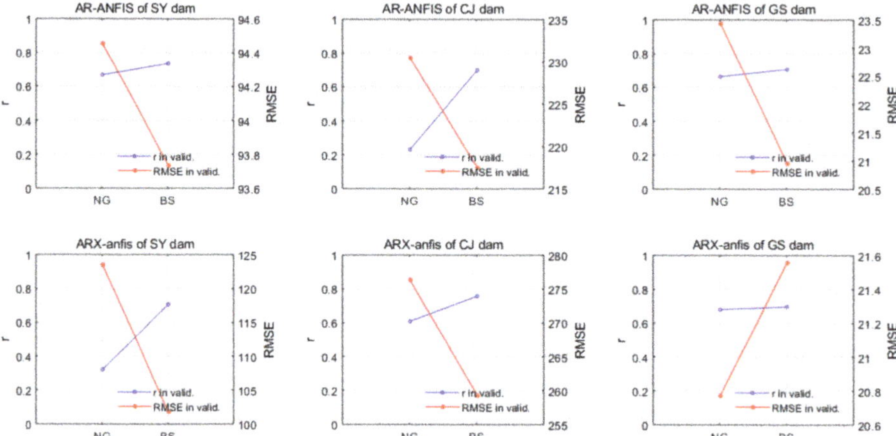

Figure 9. Correlation coefficients (*r*) and root mean square error (RMSE) of the AR-ANFIS and ARX-ANFIS models according to the MF in the validation period for the SY, CJ, and GS dams. The NS is normalized Gaussian MF and the BS is the Bell-shaped MF. The blue line is the r-value in the validation period and the red line is the RMSE in the validation period.

Table 7. A number of input MFs and rules of the AR-ANFIS and ARX-ANFIS models for the three dams.

Station	AR-ANFIS			ARX-ANFIS		
	Optimal MF	Number of Input MF (Layer2)	Number of Rules (Layer3)	Optimal MF	Number of Input MF (Layer2)	Number of Rules (Layer3)
SY dam	BS	8	16	BS	14	128
CJ dam	BS	8	16	BS	12	64
GS dam	BS	8	16	NG	10	32

5.2.4. RF Models

For the AR-RF and ARX-RF models, the number of trees ranging from 100 to 1000 (in 100 units) were considered in all the three dams. The optimal number of trees was determined by considering the performance during the validation period. The correlation coefficient (*r*) and root mean square error (RMSE) of the AR-RF and ARX-RF models according to the number of trees in the validation periods are shown in Figure 10. The optimal MF was determined based on the highest *r* and the lowest RMSE. In the case of two criteria were different, the RMSE was preferred. Table 8 shows the optimal number of trees of the AR-RF and ARX-RF models for the three dams.

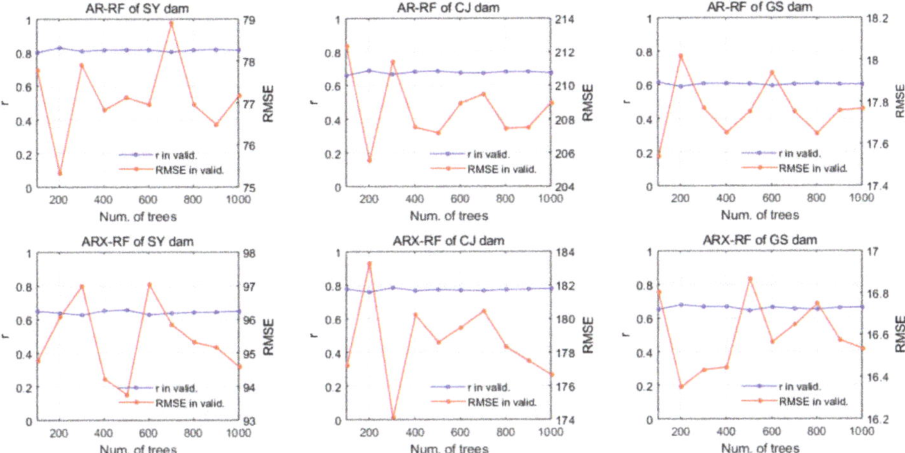

Figure 10. Correlation coefficients (r) and root mean square error (RMSE) of the AR-RF and ARX-RF models according to the number of trees in the validation period for the SY, CJ, and GS dams. The blue line is the r-value in the validation period and the red line is the RMSE in the validation period.

Table 8. The optimal number of trees of the AR-RF and ARX-RF models for the three dams.

Station	AR-RF	ARX-RF
SY dam	200	500
CJ dam	200	300
GS dam	100	200

5.3. Model Performance

Model performance can be evaluated based on different statistical measures. Two statistical criteria were used to estimate the model performance in this study. The correlation coefficient (r) is widely used to identify the linear relationship between observed and forecasted data. The value of r has a range of –1 to 1 that indicates either a negative or positive correlation. The root mean square error (RMSE) is used to measure the difference between observed and forecasted data. The value of RMSE indicates the magnitude of the error. The Nash–Sutcliffe Efficiency (NSE) is used to assess the predictive power of hydrological models. The value of NSE has a range from $-\infty$ to 1. The r, RMSE, NSE are defined in Equations (18), (19), and (20), respectively.

$$r = \frac{\sum_{t=1}^{T}\left(Y(t) - \overline{Y}(t)\right)\left(X(t) - \overline{X}(t)\right)}{\sqrt{\sum_{t=1}^{T}\left(Y(t) - \overline{Y}(t)\right)^2}\sqrt{\sum_{t=1}^{T}\left(X(t) - \overline{X}(t)\right)^2}} \tag{18}$$

$$RMSE = \sqrt{\frac{1}{T}\sum_{t=1}^{T}(Y(t) - X(t))^2} \tag{19}$$

$$NSE = 1 - \frac{\sum_{t=1}^{T}(Y(t) - X(t))^2}{\sum_{t=1}^{T}\left(X(t) - \overline{X}(t)\right)^2} \tag{20}$$

where T is data length, Y(t) and X(t) are the forecasted and observed data series, and $\overline{Y}$ (t) and $\overline{X}$ (t) are the mean of the forecasted and observed data series, respectively.

The calculated r, RMSE, and NSE during the training, validation, and test periods of SARIMA, SARIMAX, AR-ANN, ARX-ANN, AR-ANFIS, ARX-ANFIS, AR-RF, and ARX-RF models for the SY,

CJ, and GS dams were presented in Table 9. The underlined values indicate the best statistics among the employed models in the training, validation, and test periods. In the SY dam, the result shows that the ARX-RF and AR-RF models are very competitive and have good performance in the training and validation periods, respectively. On the other hand, in the test period, the AR-ANN or ARX-ANN models shows the best performance although it is slightly different according to the statistical criteria. In CJ dam, the ARX-RF model shows the best performance in the training and validation periods. On the other hand, the ARX-ANN model outperforms in the test period. In the GS dam, the ARX-RF model shows the best performance in the training period while the AR-ANN model shows the best performance in the validation period. The SARIMA and AR-RF models achieve the best performance in the test period depending on the statistical criteria. Overall, the ARX-RF model outperforms in the training period. Focusing on the test period, the ARX-ANN model generally provided better predictions, except for the GS dam. In summary, the time series model does not prove that the use of climate indices as an exogenous variable is more efficient. However, the use of climate indices generally tends to produce better performance in the ANN and RF models.

Table 9. Model performance of the SARIMA, SARIMAX, AR-ANN, ARX-ANN, AR-ANFIS, ARX-ANFIS, AR-RF, and ARX-RF models during the training, validation, and test periods for the SY, CJ, and GS dams.

Station	Model	r			RMSE			NSE		
		Train.	Vali.	Test	Train.	Vali.	Test	Train.	Vali.	Test
SY dam	SARIMA	0.65		0.53	84.17		83.69	0.42		−0.04
	SARIMAX	0.65		0.38	84.52		93.82	0.42		−0.31
	AR-ANN	0.66	0.70	0.58	82.89	89.18	<u>72.28</u>	0.44	0.49	<u>0.22</u>
	ARX-ANN	0.64	0.68	<u>0.63</u>	85.56	91.63	80.64	0.40	0.46	0.03
	AR-ANFIS	0.65	0.73	0.53	84.54	88.49	74.97	0.42	0.49	0.16
	ARX-ANFIS	0.61	0.71	0.41	87.93	89.53	86.96	0.37	0.48	−0.13
	AR-RF	0.94	<u>0.83</u>	0.47	42.61	<u>75.33</u>	77.78	<u>0.85</u>	<u>0.63</u>	0.10
	ARX-RF	<u>0.95</u>	0.66	0.50	<u>42.45</u>	93.76	75.71	<u>0.85</u>	0.43	0.15
CJ dam	SARIMA	0.65		0.57	203.35		179.26	0.41		−0.93
	SARIMAX	0.63		0.57	205.62		180.51	0.40		−0.96
	AR-ANN	0.64	0.71	0.52	204.71	202.42	162.32	0.40	0.48	−0.58
	ARX-ANN	0.61	0.74	<u>0.67</u>	209.49	191.73	<u>151.26</u>	0.37	0.53	<u>−0.37</u>
	AR-ANFIS	0.63	0.70	0.46	205.29	208.25	162.20	0.40	0.45	−0.58
	ARX-ANFIS	0.64	0.76	0.41	203.89	186.34	203.91	0.41	0.56	−1.50
	AR-RF	<u>0.93</u>	0.68	0.42	110.41	207.17	157.52	<u>0.83</u>	0.46	−0.49
	ARX-RF	<u>0.93</u>	<u>0.79</u>	0.40	<u>108.83</u>	<u>174.11</u>	169.53	<u>0.83</u>	<u>0.62</u>	−0.73
GS dam	SARIMA	0.65		<u>0.52</u>	17.44		15.55	0.42		−1.56
	SARIMAX	0.67		0.51	17.14		17.86	0.44		−2.38
	AR-ANN	0.62	<u>0.72</u>	0.37	17.88	<u>15.37</u>	13.67	0.39	<u>0.51</u>	−0.98
	ARX-ANN	0.69	0.63	0.51	16.48	17.41	14.21	0.48	0.37	−1.14
	AR-ANFIS	0.65	0.71	0.35	17.31	15.58	14.37	0.43	0.50	−1.19
	ARX-ANFIS	0.77	0.68	0.42	14.77	16.69	19.40	0.58	0.42	−2.99
	AR-RF	0.93	0.59	0.38	9.13	18.02	<u>13.45</u>	0.84	0.33	<u>−0.92</u>
	ARX-RF	<u>0.94</u>	0.68	0.46	<u>8.69</u>	16.35	16.76	<u>0.86</u>	0.45	−1.98

The reservoir inflow forecasting results from the SARIMA, SARIMAX, AR-ANN, ARX-ANN, AR-ANFIS, ARX-ANFIS, AR-RF, and ARX-RF models for the SY, CJ, and GS dams are shown in Figures 11–13, respectively. The training periods for the SY, CJ, and GS dams were respectively started in 1981, 1992, and 1987 considering the order of the time series models. In Figures 11–13, a part of forecasting results in the training period of the SARIMA and SARIMAX models were illustrated to the validation period. The general results of the time series models showed that they forecast only seasonal patterns even in both training, validation, and test periods. In addition, there were no significant differences between the SARIMA and SARIMAX models depending on the exogenous variables. The forecasted reservoir inflows of the SARIMA and SARIMAX models were very close, and were unable to forecast the extreme inflow during the flood season in both the training and test periods. This indicates the impact of the climate index was insufficient in the time series model. On the other hand, the exogenous variables have an impact on the performance of artificial intelligence models. The general results of the ARX-RF models showed that they forecast well not only seasonal patterns, but also peak flows during the flood season in the training and validation periods for all three dams. During the test period, the ARX-ANN model matched the seasonal patterns of inflow better than the other employed models, although it showed less accuracy in the volume. Although the ARX-ANN model showed lower forecasting performance in the training period than the ARX-RF model, the ARX-ANN model was able to make better forecasts of low and medium reservoir inflows than the ARX-RF model in the test period. By adding the validation period, we could avoid the overfitting problem. With regard to unsatisfactory performance in the test period captured by the NSE, this will be discussed further in the discussion section.

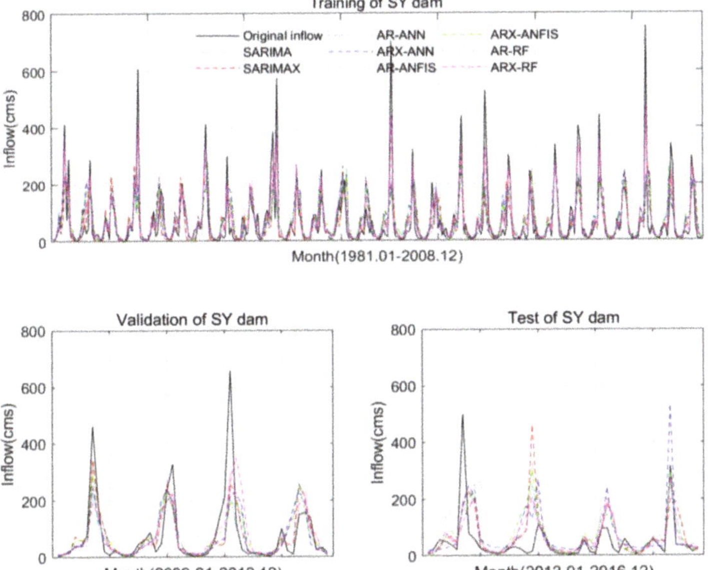

Figure 11. Forecasting results from the eight models in the SY dam during the training, validation, and test periods (a part of the forecasting results in the training period of the SARIMA and SARIMAX models were illustrated for the validation period).

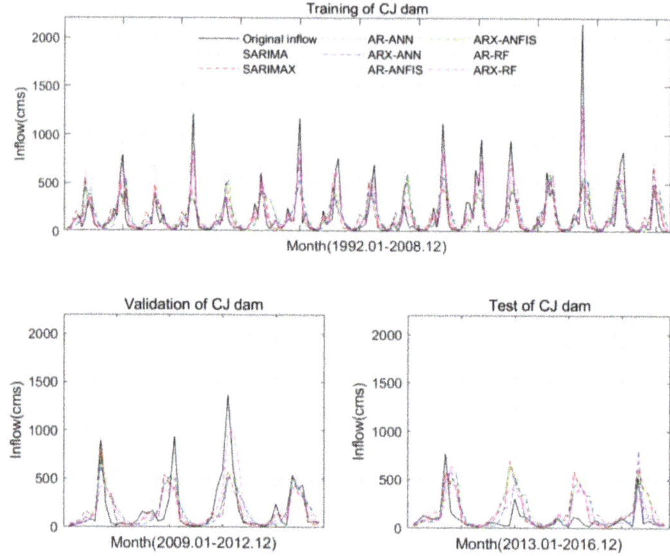

Figure 12. Forecasting results from the eight models in the CJ dam during the training, validation, and test periods (a part of the forecasting results in the training period of the SARIMA and SARIMAX models were illustrated for the validation period).

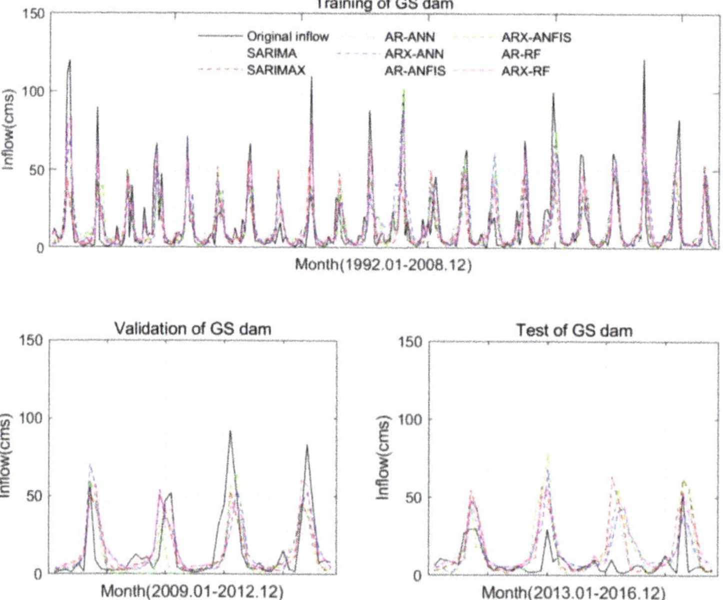

Figure 13. Forecasting results from the eight models in the GS dam during the training, validation, and test periods (a part of the forecasting results in the training period of the SARIMA and SARIMAX models were illustrated for the validation period).

6. Discussion

In this study, the statistical methods (e.g., PACF and EEMD) were employed to select the two types of input variables including lagged inflow and lagged climate indices. The PACF for all three dams mainly resulted in prominent spikes at 1, 12, 24, and 36 months. The EEMD showed that the third IMF had the highest correlation coefficient with the four or five-month lagged NINO12 index and residues, and the last IMF also had a considerably high correlation coefficient with the 1- or 12-month lagged AMO index. It was found that the EEMD can obtain the inherent climate variability and long-term trend in the reservoir inflow. Therefore, the statistical relationship between the reservoir inflow and the climate indices can be deeply examined through EEMD.

From the results of the backward elimination method, the 36-month lagged inflow, NINO12 index, and the AMO index were selected as a combination of autoregressive and exogenous variables for the three dams in Table 3. Consequently, we found that the three variables mainly affect reservoir inflow and play an important role as input variables. This finding supported the results obtained by Kim et al. [41] that identified the four-month lagged NINO12 and AMO indices as effective indicators for long-term precipitation in South Korea, as the precipitation is causally related to reservoir inflow. In addition, this result also agrees with previous studies [36,38,40,53,54] that showed the IMFs identified through EEMD include the cyclical variabilities and overall trends of time series data.

The comparison of model performance proved that the impact of the climate index was insufficient in the time series model while the exogenous variables have an impact on the performance of artificial intelligence models. The ARX-RF models generally forecasted well not only seasonal patterns, but also peak flows during the flood season in the training and validation periods for all the three dams. On the other hand, the ARX-ANN model generally showed better performance in the test period by matching the seasonal patterns of inflow although it showed less accuracy in the volume. However, the unsatisfactory performance in the test period captured by the NSE was due to severe weather conditions that rarely occurred in the observation period. To further understand the model performance under extreme weather conditions, the test period (2013–2016) was divided into each year, and the inflow characteristics were examined in each year. Figure 14(a) shows the percentage of the inflow rate of the year compared to the mean annual inflow in the 2013 to 2016 years for each dam. Figure 14(b) shows the percentage of the variation rate of the year compared to the mean annual variation in the 2013 to 2016 years for each dam. For example, the percentage of the inflow rate of 2013 year is calculated by (the mean annual inflow in 2013/the mean annual inflow in the training period) × 100(%). The percentage of the variation rate is calculated in the same way.

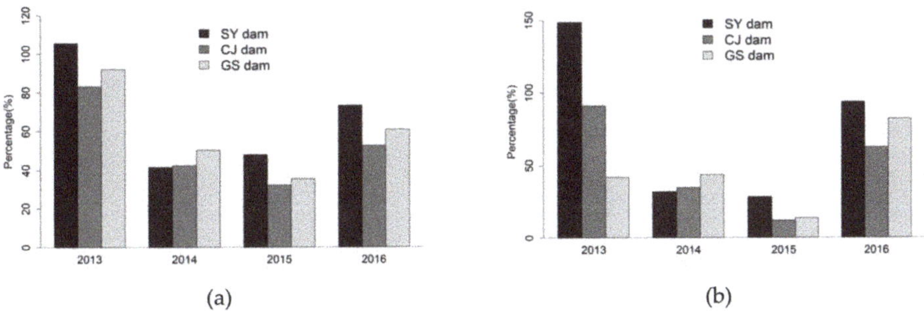

Figure 14. Percentages of (**a**) the inflow rate of the year compared to the mean annual inflow in the 2013 to 2016 years and (**b**) the variation rate of the year compared to the mean annual variation in the 2013 to 2016 years.

Notably, extreme weather conditions occurred during the test period. The dam inflow in 2013 approximated the mean annual inflow, however, there was a large difference in annual variation compared to the mean variation for the three dams. The annual inflow of the three dams was less than half the mean annual inflow in 2014 and decreased further in 2015 because a severe drought occurred in these two years. Due to the severe drought, the annual variation in inflow was also lower than average and there was minor difference in variation between the three dams. Inflow slightly increased in 2016 reaching half the mean annual inflow. In this study, the years of test period are defined as an ordinary year (2013), a drought year (2014), a severe drought year (2015), and a restored year (2016). Table 10 presents the model performances in the 2013 to 2016 years for the three dams.

In the ordinary year (2013), the time series models generally showed better performance than the artificial intelligence models. When the percentage of the annual variation was similar or smaller than the mean annual variation, such as was observed in GS dam, the AR-RF model generally showed the best forecasting accuracy based on both the RMSE and NSE. There were no significant differences according to the exogenous variables.

In the drought year (2014), the time series model performance was significantly poorer than during the ordinary year. The forecasting accuracy of the ARX-ANN and AR-RF models was generally better than other models based on the r and RMSE. The NSE captured poor performance at all dams because the mean inflows had been lowered due to drought effects. This result indicates that exogenous variables in the ANN model can improve forecasting accuracy during drought years, although there are limitations to forecasting the inflow volume.

In the severe drought year (2015), when the annual inflow was close to 50% of the mean annual inflow such as in the SY dam, all models showed improved forecasting accuracy. The ARX-ANFIS model showed the best forecasting accuracy based on the RMSE and NSE while the AR-RF model showed the best forecasting accuracy based on the r. However, when the annual inflow was lower than 40% of the mean annual inflow such as in the CJ and GS dams, the forecasting accuracy showed the worst performance, especially based on the NSE. This result shows a limitation of the use of exogenous variables in forecasting extreme weather conditions such as severe droughts, since the lagged climate indices mainly reflect the inherent climate variabilities and long-term trends.

In the restored year (2016), the ARX-RF and ARX-ANN models usually had the best forecasting accuracy for the SY and GS dams. Although the AR-ANN model showed better forecasting accuracy than the ARX-ANN model for the CJ dam based on the RMSE and NSE, the ARX-ANN model showed best forecasting accuracy based on the r. This result implies that the exogenous variables in the artificial intelligence models play an important role in forecasting accuracy under the restored climate conditions.

It was observed that the time series models have quite similarly forecasted inflow in the test period for the three dams. Mainly in the drought and severe drought years, the results generally described a limitation of the time series and artificial intelligence models, because the models are based on the observed data series. They only forecasted ordinary patterns of the observed inflow series. The r of the time series models decreased in the drought and severe drought years and slightly increased in the restored year because the time series models maintain seasonal patterns regardless of changing weather characteristics. The model performance in the artificial intelligence models was slightly different depending on the dam, period, climate conditions, and input variables. This also implies a difficulty to forecast events that have not been previously observed, although the model was trained very well. Among them, the AR-RF and ARX-ANN models generally showed better forecasting accuracy under drought conditions in our case study. Therefore, we can conclude that the use of artificial intelligence model and climate indices as exogenous variables has the potential to provide a suitable forecasting performance under the changing climatic conditions.

Reservoir inflow forecasting is still a challenge for reservoir managers all over the world. Reservoir managers should strive to operate the reservoir considering the various results from statistical, artificial intelligence, and dynamic models. In light of this decision-making, our study can be useful in making decisions for reservoir operation, because they showed good performance in forecasting inflow patterns by using lagged inflows and lagged climate indices. Finally, we also agree with Zhang et al. [55] that future studies should be ongoing by applying a new type of machine learning algorithm, i.e., deep learning, derived from the ANN model.

Table 10. Model performance of the SARIMA, SARIMAX, AR-ANN, ARX-ANN, AR-ANFIS, ARX-ANFIS, AR-RF, and ARX-RF models by year during test period for the three dams.

Station	Model	r				RMSE				NSE			
		2013	2014	2015	2016	2013	2014	2015	2016	2013	2014	2015	2016
SY dam	SARIMA	0.62	0.43	0.78	0.75	102.18	99.57	59.86	63.84	0.38	−9.02	−2.31	0.38
	SARIMAX	0.53	0.10	0.67	0.69	110.61	131.76	43.73	60.81	0.28	−16.54	−0.77	0.44
	AR-ANN	0.53	0.18	0.75	0.93	110.56	75.68	43.94	31.80	0.28	−4.79	−0.79	0.85
	ARX-ANN	0.50	0.58	0.71	0.99	114.18	80.57	49.75	63.31	0.23	−5.56	−1.29	0.39
	AR-ANFIS	0.50	0.23	0.80	0.90	112.66	80.56	44.31	36.59	0.25	−5.56	−0.82	0.80
	ARX-ANFIS	0.27	0.14	0.75	0.96	130.77	104.71	39.43	25.02	−0.01	−10.08	−0.44	0.91
	AR-RF	0.41	0.29	0.82	0.68	121.48	58.62	49.12	59.95	0.13	−2.47	−1.23	0.45
	ARX-RF	0.36	0.21	0.69	0.97	122.78	72.36	45.84	22.66	0.11	−4.29	−0.94	0.92
CJ dam	SARIMA	0.81	0.45	0.58	0.76	115.63	227.95	189.39	165.34	0.64	−7.96	−28.82	−0.52
	SARIMAX	0.75	0.40	0.59	0.78	140.97	221.25	204.75	139.94	0.47	−7.45	−33.85	−0.09
	AR-ANN	0.58	0.38	0.64	0.83	181.56	210.36	147.21	80.67	0.12	−6.63	−17.02	0.64
	ARX-ANN	0.62	0.83	0.42	0.99	169.29	137.67	190.67	86.89	0.23	−2.27	−29.22	0.58
	AR-ANFIS	0.54	0.30	0.68	0.76	182.45	214.87	131.65	91.90	0.11	−6.96	−13.41	0.53
	ARX-ANFIS	0.39	0.50	0.45	0.74	212.50	237.39	211.93	141.02	−0.21	−8.72	−36.34	−0.10
	AR-RF	0.43	0.36	0.66	0.78	182.52	167.42	172.71	89.85	0.11	−3.84	−23.80	0.55
	ARX-RF	0.35	0.47	0.61	0.69	230.87	162.34	153.13	108.92	−0.43	−3.55	−18.49	0.34
GS dam	SARIMA	0.87	0.47	0.05	0.62	9.52	16.66	19.43	14.87	−0.02	−4.11	−49.30	−0.16
	SARIMAX	0.88	0.48	0.10	0.62	9.32	18.26	22.81	18.30	0.02	−5.14	−68.31	−0.75
	AR-ANN	0.82	0.50	−0.21	0.01	10.10	15.14	13.00	15.70	−0.15	−3.22	−21.52	−0.29
	ARX-ANN	0.82	0.79	−0.39	0.71	8.89	15.67	18.74	11.49	0.11	−3.52	−45.80	0.31
	AR-ANFIS	0.79	0.44	−0.20	0.10	10.11	17.06	14.50	14.90	−0.15	−4.36	−27.00	−0.16
	ARX-ANFIS	0.77	0.58	−0.33	0.57	8.11	24.36	21.03	20.11	0.26	−9.93	−57.96	−1.12
	AR-RF	0.85	0.46	0.02	0.11	7.63	14.51	13.74	16.29	0.34	−2.88	−24.17	−0.39
	ARX-RF	0.80	0.61	-0.16	0.59	7.98	18.23	20.88	17.07	0.28	−5.12	−57.10	−0.52

7. Conclusions

In this study, the use of large-scale climatic indices as exogenous input variables to hydrologic forecasting models was considered to reflect climate variability due to climate change. To do this, a total of eight models including time series and artificial intelligence models (SARIMA, SARIMAX, AR-ANN, ARX-ANN, AR-ANFIS, ARX-ANFIS, AR-RF, and ARX-RF) were applied to monthly reservoir inflow forecasting. The results of this study led to the following conclusions:

(1) For input variable selection, the PACF and EEMD can be used to find lagged inflow and lagged climate indices that have a significant relationship with dam inflow. As a result, four lagged inflows (lag1, lag12, lag24, lag36) were selected as the autoregressive variables, and the 36-month lagged inflow, the lagged NINO12 index, and the lagged AMO index were commonly selected as the combinations of autoregressive and exogenous variables for the three dams. Therefore, the procedure for input variable selection using the PACF, EEMD, cross-correlation analysis, and backward elimination in this study is a suitable method for input variable selection.

(2) The ARX-RF model generally showed the highest forecasting accuracy in the training period, which proves that a combination of autoregressive and exogenous variables is useful for constructing an RF model for the three dams. In the test period, the ARX-ANN model generally showed the highest forecasting accuracy by capturing the seasonal patterns of reservoir inflow well, although there are limitations to its ability to accurately forecast inflow volume.

(3) The model performance in the test period (from 2013 to 2016) was examined according to the inflow characteristics of each year. The inflow of the three dams maintained the seasonal patterns in 2013. Drought occurred in 2014, and it worsened in 2015. The ordinary pattern was slightly restored in 2016. Although the model performance was not consistent in each year of the test period,

the ARX-ANN and the AR-RF models generally matched the seasonal patterns well, especially during the drought and restored years.

Based on this study, the results prove that the use of large-scale climate indices as exogenous variables has the potential to provide more efficient forecasting performance for water resource management and planning. Furthermore, there is a possibility to provide better forecasting results by using a state-of-art artificial intelligence models such as RF. Future studies should be required to identify the best forecasting models through many applications considering local weather conditions and inflow characteristics under the changing climate conditions for effective water management and planning.

Author Contributions: Conceptualization, T.K. and J.-Y.S.; Data curation, S.K.; Formal analysis, H.K.; Funding acquisition, J.-H.H.; Investigation, T.K. and H.K.; Methodology, T.K. and J.-Y.S.; Supervision, J.-H.H.; Visualization, H.K.; Writing–original draft, T.K.; Writing–review & editing, J.-H.H.

Funding: This research was supported by a grant(2017-MOIS31-001) from Fundamental Technology Development Program for Extreme Disaster Response funded by Korean Ministry of Interior and Safety(MOIS).

Acknowledgments: This research was conducted under the Development of long-term inflow forecasting technology and construction of water supply system (2nd year) supported by K-Water.

Conflicts of Interest: The authors declare no conflict of interest.

References

1. Wang, W.-C.; Chau, K.-W.; Cheng, C.-T.; Qiu, L. A comparison of performance of several artificial intelligence methods for forecasting monthly discharge time series. *J. Hydrol.* **2009**, *374*, 294–306. [CrossRef]
2. Wang, Y.; Guo, S.-L.; Chen, H.; Zhou, Y.-L. Comparative study of monthly inflow prediction methods for the Three Gorges Reservoir. Stoch. Environ. *Res. Risk Assess.* **2014**, *28*, 555–570. [CrossRef]
3. Wang, W.C.; Chau, K.W.; Xu, D.M.; Chen, X.Y. Improving forecasting accuracy of annual runoff time series using ARIMA based on EEMD decomposition. *Water Resour. Manag.* **2015**, *29*, 2655–2675. [CrossRef]
4. Valipour, M.; Banihabib, E.B.; Behbahani, S.M.R. Comparison of the ARMA, ARIMA, and the autoregressive artificial neural network models in forecasting the monthly inflow of Dez dam reservoir. *J. Hydrol.* **2013**, *476*, 433–441. [CrossRef]
5. Awan, J.A.; Bae, D.-H. Improving ANFIS Based Model for Long-term Dam Inflow Prediction by Incorporating Monthly Rainfall Forecasts. *Water Resour. Manag.* **2014**, *28*, 1185–1199. [CrossRef]
6. Kalteh, A.M. Wavelet Genetic Algorithm-Support Vector Regression (Wavelet GA-SVR) for Monthly Flow Forecasting. *Water Resour. Manag.* **2015**, *29*, 1283–1293. [CrossRef]
7. Kisi, O. Streamflow Forecasting and Estimation Using Least Square Support Vector Regression and Adaptive Neuro-Fuzzy Embedded Fuzzy c-means Clustering. *Water Resour. Manag.* **2015**, *29*, 5109–5127. [CrossRef]
8. Li, C.; Bai, Y.; Zeng, B. Deep Feature Learning Architectures for Daily Reservoir Inflow Forecasting. *Water Resour. Manag.* **2016**, *30*, 5145–5161. [CrossRef]
9. Moeeni, H.; Bonakdari, H.; Ebtehai, I. Integrated SARIMA with Neuro-Fuzzy Systems and Neural Networks for Monthly Inflow Prediction. *Water Resour. Manag.* **2017**, *31*, 2141–2156. [CrossRef]
10. Box, G.E.P.; Jenkins, G.M. *Time Series Analysis Forecasting and Control*; Holden-Day: San Francisco, CA, USA, 1970.
11. Carlson, R.F.; MacCormick, A.J.A.; Watts, D.G. Application of linear random models to four annual streamflow series. *Water Resour. Res.* **1970**, *6*, 1070–1078. [CrossRef]
12. Salas, J.D.; Tabios, G.Q.; Bartolini, P. Approaches to multivariate modeling of water resources time series. *J. Am. Water Resour. Assoc.* **1985**, *21*, 683–708. [CrossRef]
13. Burlando, P.; Rosso, R.; Cadavid, L.G.; Salas, J.D. Forecasting of short-term rainfall using ARMA models. *J. Hydrol.* **1993**, *144*, 193–211. [CrossRef]
14. Hipel, K.W.; McLeod, A.I. *Time Series Modelling of Water Resources and Environmental Systems*; Elsevier: Amsterdam, The Netherlands, 1994.
15. Mohan, S.; Vedula, S. Multiplicative seasonal ARIMA model for longterm forecasting of inflows. *Water Resour. Manag.* **1995**, *9*, 115–126. [CrossRef]
16. Papamichail, D.M.; Georgiou, P.E. Seasonal ARIMA inflow models for reservoir sizing. *J. Am. Water Resour. Assoc.* **2001**, *37*, 877–885. [CrossRef]

17. Campolo, M.; Soldati, A.; Andreussi, P. Artificial neural network approach to flood forecasting in the River Arno. *Hydrol. Sci. J.* **2003**, *48*, 381–398. [CrossRef]

18. Cheng, C.T.; Lin, J.Y.; Sun, Y.G.; Chau, K. *Long-Term Prediction of Discharges in Manwan Hydropower Using Adaptive-Network-Based Fuzzy Inference Systems Models*; Advances in Natural Computation; Springer: Berlin, Germany, 2005.

19. Jain, A.; Kumar, A.M. Hybrid neural network models for hydrologic time series forecasting. *Appl. Soft Comput.* **2007**, *7*, 585–592. [CrossRef]

20. Firat, M.; Gungör, M. Hydrological time-series modelling using an adaptive neuro-fuzzy inference system. *Hydrol. Process.* **2008**, *22*, 2122–2132. [CrossRef]

21. Trichakis, I.; Nikolos, I.; Karatzas, G.P. Comparison of bootstrap confidence intervals for an ANN model of a karstic aquifer response. *Hydrol. Process.* **2011**, *25*, 2827–2836. [CrossRef]

22. He, X.; Guan, H.; Zhang, X.; Simmons, C.T. A wavelet-based multiple linear regression model for forecasting monthly rainfall. *Int. J. Climatol.* **2014**, *34*, 1898–1912. [CrossRef]

23. Nawaz, N.; Harun, S.; Talei, A. Application of Adaptive Network-Based Fuzzy Inference System (ANFIS) for River Stage Prediction in a Tropical Catchment. *Appl. Mech. Mater.* **2015**, *735*, 195–199. [CrossRef]

24. Breiman, L. Random forest. *Mach. Learn.* **2001**, *45*, 5–32. [CrossRef]

25. Yang, T.; Gao, X.; Sorooshian, S.; Li, X. Simulating California reservoir operation using the classification and regression-tree algorithm combined with a shuffled cross-validation scheme. *Water Resour. Res.* **2016**, *52*, 1626–1651. [CrossRef]

26. Yang, T.; Asanjan, A.A.; Welles, E.; Gao, X.; Sorooshian, S.; Liu, X. Developing reservoir monthly inflow forecasts using artificial intelligence and climate phenomenon information. *Water Rosour. Res.* **2017**, *53*, 2786–2812. [CrossRef]

27. Ahmed, J.A.; Sarma, A.K. Artificial neural network model for synthetic streamflow generation. *Water Resour. Manag.* **2007**, *21*, 1015–1029. [CrossRef]

28. Noori, R.; Hoshyaripour, G.; Ashrafi, K.; Araabi, B.N. Uncertainty analysis of developed ANN and ANFIS models in prediction of carbon monoxide daily concentration. *Atmos. Environ.* **2010**, *44*, 476–482. [CrossRef]

29. Valipour, M. Long-term runoff study using SARIMA and ARIMA models in the United States. *Meteorol. Appl.* **2015**, *22*, 592–598. [CrossRef]

30. Emamgholizadeh, S.; Moslemi, K.; Karami, G. Prediction the groundwater level of Bastam Plain (Iran) by artificial neural network (ANN) and adaptive neuro fuzzy inference system (ANFIS). *Water Resour. Manag.* **2014**, *28*, 5433–5446. [CrossRef]

31. Kashid, S.S.; Ghosh, S.; Maity, R. Streamflow prediction using multi-site rainfall obtained from hydroclimatic teleconnection. *J. Hydrol.* **2010**, *395*, 23–38. [CrossRef]

32. Schepen, A.; Wang, Q.J.; Robertson, D. Evidence for using lagged climate indices to forecast Australian seasonal rainfall. *J. Clim.* **2012**, *25*, 1230–1246. [CrossRef]

33. Mekanik, F.; Imteaz, M.A.; Gato-Trinidad, S.; Elmahdi, A. Multiple regression and artificial neural network for long-term rainfall forecasting using large scale climate modes. *J. Hydrol.* **2013**, *503*, 11–21. [CrossRef]

34. Abbot, J.; Marohasy, J. Input selection and optimisation for monthly rainfall forecasting in Queensland, Australia, using artificial neural networks. *Atmos. Res.* **2014**, *138*, 166–178. [CrossRef]

35. Li, J.; Liu, X.; Chen, F. Evaluation of nonstationarity in annual maximum flood series and the associations with large-scale climate patterns and human activities. *Water Resour. Manag.* **2015**, *29*, 1653–1668. [CrossRef]

36. Wu, Z.; Huang, N.E.; Long, S.R.; Peng, C.K. On the trend, detrending, and variability of nonlinear and nonstationary time series. *Proc. Natl. Acad. Sci. USA* **2007**, *104*, 14889–14894. [CrossRef]

37. Lee, T.; Ouarda, T.B.M.J. Long-term prediction of precipitation and hydrologic extremes with nonstationary oscillation processes. *J. Geophys. Res. Atmos.* **2010**, *115*, D13107. [CrossRef]

38. Breaker, L.C.; Ruzmaikin, A. The 154-year record of sea level at San Francisco: Extracting the long-term trend, recent changes, and other tidbits. *Clim. Dyn.* **2011**, *36*, 545–559. [CrossRef]

39. Shi, F.; Yang, B.; Gunten, L.; Qin, C.; Wang, Z. Ensemble empirical mode decomposition for tree-ring climate reconstructions. *Theor. Appl. Climatol.* **2012**, *109*, 233–243. [CrossRef]

40. Castino, F.; Bookhagen, B.; Strecker, M.R. Oscillations and trends of river discharge in the southern Central Andes and linkages with climate variability. *J. Hydrol.* **2017**, *555*, 108–124. [CrossRef]

41. Kim, T.; Shin, J.-Y.; Kim, S.; Heo, J.-H. Identification of relationships between climate indices and long-term precipitation in South Korea using ensemble empirical mode decomposition. *J. Hydrol.* **2018**, *557*, 726–739. [CrossRef]

42. Yu, X.; Zhang, X.; Qin, H. A Data-driven Model Based on Fourier Transform and Support Vector Regression for Monthly Reservoir Inflow Forecasting. *J. Hydro-Environ. Res.* **2018**, *18*, 12–24. [CrossRef]

43. Jalalkamali, A.; Moradi, M.; Moradi, N. Application of several artificial intelligence models and ARIMAX model for forecasting drought using the Standardized Precipitation Index. *Int. J. Environ. Sci. Technol.* **2015**, *12*, 1201–1210. [CrossRef]

44. Yang, T.; Asanjan, A.A.; Faridzad, M.; Hayatbini, N.; Gao, X.; Sorooshian, S. An enhanced artificial neural network with a shuffled complex evolutionary global optimization with principal component analysis. *Inform. Sci.* **2017**, *418–419*, 302–316. [CrossRef]

45. Avarideh, F. Application of Hydro-Informatics in Sediment Transport. Ph.D. Thesis, Amirkabir University of Technology, Tehran, Iran, 2012.

46. Jang, J.S.R. ANFIS: Adaptive-Network-Based Fuzzy Inference System. *IEEE Trans. Syst. Man. Cybern.* **1993**, *23*, 665–685. [CrossRef]

47. Wang, L.; Zhou, X.; Zhu, X.; Dong, Z.; Guo, W. Estimation of biomass in wheat using random forest regression algorithm and remote sensing data. *Crop J.* **2016**, *4*, 212–219. [CrossRef]

48. Huang, N.E.; Shen, Z.; Long, S.R.; Wu, M.C.; Shih, H.H.; Zheng, Q.; Yen, N.C.; Tung, C.C.; Liu, H.H. The empirical mode decomposition method and the Hilbert spectrum for non-stationary time series analysis. *Proc. Math. Phys. Eng. Sci.* **1998**, *354*, 903–995. [CrossRef]

49. Wu, Z.; Huang, N.E. Ensemble empirical mode decomposition: A noise-assisted data analysis method. *Adv. Adapt. Data Anal.* **2009**, *1*, 1–41. [CrossRef]

50. May, R.; Dandy, G.; Maier, H. Review of Input Variable Selection Methods for Artificial Neural Networks. In *Artificial Neural Networks-Methodological Advances and Biomedical Applications*; Kenji, S., Ed.; InTechOpen: London, UK, 2011; ISBN 978-953-307-243-2.

51. Cryer, J.D.; Chan, K.S. *Time Series Analysis with Applications in R*, 2nd ed.; Springer: New York, NY, USA, 2008; ISBN 978-0-387-75959-3.

52. Sgurev, V.; Yager, R.R.; Kacprzyk, J.; Jotsov, V. *Innovative Issues in Intelligent Systems*; Springer: Cham, Switzerland, 2016; ISBN 978-3-319-27267-2.

53. Lee, T.; Ouarda, T.B.M.J. Stochastic simulation of nonstationary oscillation hydro-climatic processes using empirical mode decomposition. *Water Resour. Res.* **2012**, *48*, W02514. [CrossRef]

54. Kim, Y.; Cho, K. Sea level rise around Korea: Analysis of tide gauge station data with the ensemble empirical mode decomposition method. *J. Hydro-Environ. Res.* **2016**, *11*, 138–145. [CrossRef]

55. Zhang, D.; Lin, J.; Peng, Q.; Wang, D.; Yang, T.; Sorooshian, S.; Liu, X.; Zhuang, J. Modeling and simulating of reservoir operation using the artificial neural network, support vector regression, deep learning algorithm. *J. Hydrol.* **2018**, *565*, 720–736. [CrossRef]

Article

Land Use and Climate Change Effects on Surface Runoff Variations in the Upper Heihe River Basin

Xingxing Shang [1,2] , Xiaohui Jiang [1,2,*], Ruining Jia [2] and Chen Wei [2]

1   Shaanxi Key Laboratory of Earth Surface System and Environmental Carrying Capacity, College of Urban and Environmental Sciences, Northwest University, Xi'an 710127, China; shangXing_1028@163.com
2   Department of Environmental Engineering, College of Urban and Environmental Science, Northwest University, Xi'an 710127, China; jrn0505@163.com (R.J.); weichen_da@126.com (C.W.)
*   Correspondence: xhjiang@nwu.edu.cn; Tel.: +86-135-0382-1320

Received: 8 January 2019; Accepted: 13 February 2019; Published: 18 February 2019

Abstract: The runoff in the upper reaches of the Heihe River has been continuously abundant for more than a decade, and this has not happened previously in history. Quantitative analysis of runoff variation and its influencing factors are of great significance for the ecological protection of the basin. In this paper, the soil and water assessment tool model was used to simulate runoff in the study area, and the method of scenario simulation was used to quantitatively analyze the runoff response with respect to land use and climate change. According to the abruptness of the runoff sequence, the years before 2004 are categorized as belonging to the reference period, and after 2004 is categorized as the interference period. According to the analysis, compared with the reference period, the contribution rate of climate change is 87.15%, while the contribution rate of land use change is only 12.85%. The climate change scenario simulation analysis shows that the change in runoff is positively correlated with the change in precipitation. The relationship with the change in temperature is more complicated, but the influence of precipitation change is stronger than the change in temperature. According to the land use scenario simulation analysis, under the economic development scenario, the runoff decreased, whereas under the historical trend and ecological protection scenario, the runoff increased. Additionally, the runoff increased more under the ecological protection scenario.

Keywords: hydrological simulation; quantitative analysis; SWAT model; land use/cover change; climate change; scenario simulation

1. Introduction

Runoff is the product of the interaction between climate and land use change in a basin [1–3]. Climate change will directly change the spatial distribution and temporal variability of atmospheric precipitation and change the spatial configuration of runoff [4,5]. Changes in land use can directly lead to changes in the production and flow processes, which lead to changes in runoff [6,7]. To some extent, changes in land use also represent the impact of human activities on water resources [8]. Changes in runoff in a source area will directly affect life production in the middle and lower reaches [9]. This is crucial to revealing the characteristics of river basin runoff and its evolution against a background of land use change and climate change [10–12].

For arid and semiarid areas, where meteorological and hydrological monitoring data are scarce, it is particularly important to select appropriate methods to quantify the contribution rates of land use and climate change [13]. Many studies have been carried out on the impact of climate and land use change on the water resources of a basin [14–16]. The main methods used are long-term data comparative analysis, experimental comparative analysis and watershed hydrological simulation. The basin test method requires a long period of time and is difficult to implement, and this method is not suitable for large-scale watershed research [17]. The long-sequence data statistical method can be used to analyze hydro-meteorological data trends, but the spatial heterogeneity of the basin and the mechanism of land use and climate change on the hydrology of the basin cannot be considered. Large-scale watershed attribution analysis is also difficult [18]. Therefore, a semi-distributed hydrological model based on physical processes is selected in this paper to evaluate the hydrological response of climate variability and land use change and to further quantify the degree of impact. In the model setting, climate change and human activities are assumed to be independent factors that lead to changes in runoff [19]. The hydro-meteorological sequence is divided into reference stages and stages affected by land use change. Finally, the natural runoff during the impact of land use change is simulated, and the contribution of the two factors to runoff is calculated based on the water balance [20–22].

The main difficulties in this study are the determination of the mutation point and the contribution rate calculation. The determination of the mutation point uses statistical analysis methods, including the Mann-Kendall test method, wavelet analysis method, Pettitt test method, cumulative anomaly analysis method and so on [23–26]. The runoff in the reference period generally takes the measured runoff in the reference period of the basin as the reference value, and considers that the difference between the measured runoff and the reference value in the period of impact of land use change is caused by environmental changes. This difference consists of two parts: one is the climate change impact contribution, and the other is the contribution of land use change [27]. Using a hydrological model, according to the different periods of runoff mutation location, the meteorological data and land use data for different periods are combined to establish a real situation based on the combination of meteorological data and land use data before the mutation. In addition, the natural runoff is simulated under the influence of both climate and land use change using the meteorological data of the time period after mutation. The land use data are used for the pre-mutation time period to simulate the runoff under the influence of climate change alone using the time before the mutation. The meteorological data within the segment and the land use data during the post-mutation period simulate runoff under land use change alone [28–30].

Many researchers have conducted simulations of the upstream runoff for the Heihe River, but the quantitative analysis is relatively simple and uses the traditional mathematical statistics method [31]. Wang et al. used a wavelet analysis, wavelet neural network model and GIS spatial analysis for the Heihe River [32]. The analysis and prediction of watershed runoff showed that the increase in annual runoff has a causal relationship with the increase in upstream air temperature and precipitation. He et al. used the M-K test and cumulant slope change rate comparison method to calculate the contribution of climate change and human activities to runoff rate, and the researchers found that the upper reaches of the Heihe River are dominated by climate change, and the impact of human activities is small [33]. Other studies have been conducted in the Heihe River Basin. Wang et al. studied the impact of land use change on hydrological processes in the middle reaches of the Heihe River and found that human activities dominated the changes in runoff in the middle reaches of the Heihe River [34]. Zhang et al. studied the effects of irrigation on surface climate in the Heihe River Basin [35]. Although some scholars have conducted preliminary research on the runoff simulation of the SWAT model in the upper reaches of the Heihe River, it is necessary to conduct systematic research on the hydrological effects in the changing environment. Zhao et al. used the Hydrologiska Fyrans Vattenbalans model to study the corresponding effects of runoff on climate change in the Heihe River Basin [36]. He et al. used the Variable Infiltration Capacity model to analyze the uncertainty of runoff simulation in the upper reaches of the Heihe River [37].

In the study of the upper reaches of the Heihe River, the SWAT model is a relatively more used model, and it is more suitable for simulations with long time periods and continuous spatiotemporal runoff changes. It is convenient to use the spatial information provided by remote sensing and GIS to simulate the hydrological effects in many different scenarios. The application of the SWAT model in the upper reaches of the Heihe River began in the early 21st century. Liu et al. first applied the SWAT model to the Heihe River Basin [38]. In later studies, they carried out improvements in the SWAT model, including the study of the snowmelt module [39]. Due to the small number of meteorological stations in the Heihe River Basin, Zou, Meng et al conducted a coupling study of SWAT with other models to obtain a more accurate simulation [40,41]. Zhang, Luo and others used the SWAT model to simulate the runoff and evaporation of the Heihe River Basin [42,43]. They are analytical studies based on historical data. In this paper, the contribution rate of climate and land use change to surface runoff is separated based on hydrological model. And combined with the analysis of different climate and land use scenarios, on the one hand, it repeals the response of runoff to climate and land use change, on the other hand, it can make some predictions on the future changes in runoff under climate and land use scenarios.

In this paper, based on the hydrological model, we calculate the contribution rate of climate and land use change to surface runoff. Based on the scenario setting method of the model, the response of surface runoff to climate and land use change is studied. The possible scenarios are used to predict the runoff under future climate and land use conditions; it also provides a reference for the rational allocation of water resources in the basin. The main objectives of this paper are threefold: (1) determining the point of change of runoff based on long-term hydrological sequence; (2) quantitatively analyzing the contribution rate of climate and land use change to runoff impact; (3) through the scenario setting simulation method, studying the response of runoff to climate and land use change. The aim is to provide a reference for the rational allocation of living, production and ecological water in the basin.

2. Materials and Methods

2.1. Study Area and Data Sources

2.1.1. Study Area

The Heihe River Basin is the second largest inland river basin in China. Due to the arid and semiarid climate, water shortages are a major factor limiting the sustainable development of the socioeconomic and ecological environment in the region [44,45]. Recently, the grassland degradation trend in the Heihe River Basin has been obvious, and the ecological damage is serious [46]. To alleviate this series of problems, in August 2001, the State Council began to carry out comprehensive management of the Heihe River Basin and implement Heihe River water dispatching and integrated river basin management [47]. According to the upstream water supply situation, the difference in different annual water levels has led to an increased contradiction between water and water demand in the middle and lower reaches [48]. To achieve rational allocation of water resources, the upstream water supply trend must be understood. The upstream runoff of the Heihe River has been abundant for more than a decade, and such a history of runoff has never before been seen [49]. Therefore, it is necessary to analyze the causes of water abundance and combine these data into a model to study the contribution of climate change and land use factors and to develop and utilize water resources for the basin, which will provide a reference for rational planning.

The Heihe River originates in the Qilian Mountain region on the northern edge of the Tibetan Plateau. This area is a typical inland river basin in the arid region of northwest China, located in the middle of the Hexi Corridor [50]. The upstream area is attached to Qilian County in the Qinghai Province, and the basin area is approximately 10,000 km^2. This area is located in the central part of the Eurasian continent and is the site of the ancient Silk Road. The study area is far from the sea and the elevation ranges between 1600 m and 4800 m. Affected by the circulation of the westerly winds in the middle–high latitudes and the influence of polar cold air masses, the climate in the upper reaches is dry, and precipitation is scarce and concentrated [51]. The upper Heihe River is the main area of the Heihe River Basin. Surface runoff mainly comes from atmospheric precipitation and melting snow and ice. The runoff distribution throughout the year is basically the same as the precipitation process and during the high temperature season [52]. The runoff and precipitation are concentrated in summer and autumn, and the annual average precipitation exceeds 400 mm. This area is sparsely populated, and the main economic activities are forestry and animal husbandry. The low level of economic development has led to a limited level of water resource development and utilization [53]. The location of the study area is shown in Figure 1, as well as the distribution of the hydrology and meteorology stations.

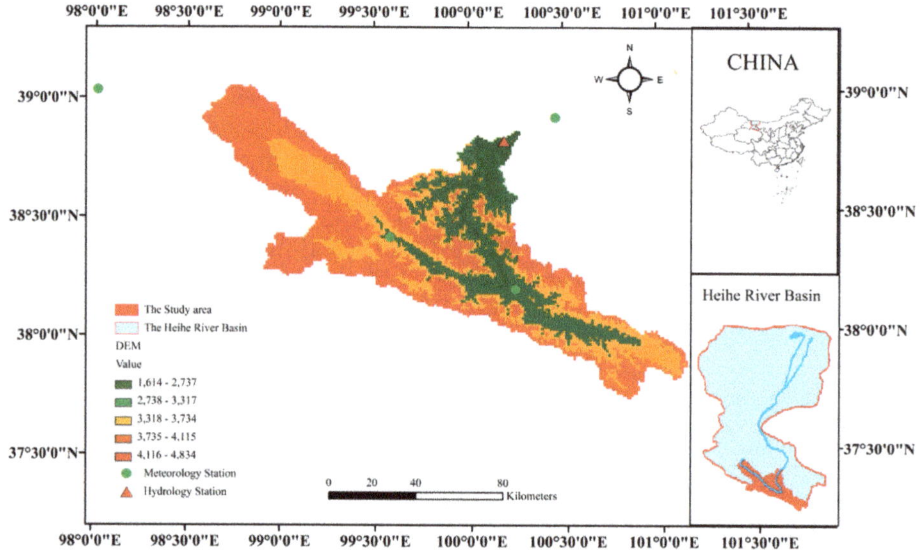

Figure 1. Upper reaches of the Heihe River Basin.

2.1.2. Data Sources

The data used in this paper are divided into two parts. The first part is used to analyze the water resource situation in the upper reaches of the Heihe River and the relationship between runoff and climate elements. The second part is the data needed for the soil and water assessment tool model (SWAT). The model data mainly include two parts: model input data and model calibration verification data. Model input data includes DEM data (1000 m × 1000 m), soil data (1:1,000,000), land use/cover data (1:100,000), and meteorological data. The model's calibration verification data is primarily the runoff and flow of the hydrological station. The data are based on ARCGIS 9.3 unified projection processing, in which all spatial data are converted into a unified projection with a spatial reference of Beijing_1954_GK_Zone_17N. DEM data and land use/cover data are from the Cold and Arid Regions Science Data Center. According to the standard, land use data are divided into six categories: forestland, grassland, water area, cultivated land, unused land, and urban construction land, as shown in detail in Table 1 and Figure 2. Soil data are from the HWSD-World Harmony Soil Database and include 8 soil classes, 14 soil classes, and 24 subcategories. See Figure 3 for details. The time series of climate data were selected from January 1980 to December 2008 and include daily precipitation, maximum and minimum temperature, humidity and wind speed. Meteorological data and hydrological data were downloaded from the Heihe River Bureau for model calibration and evaluation and include the Qilian station, Zhangye station, Yeniugou station, Tuole station, and Yingluoxia station, seeing Tables 2 and 3 for details.

Table 1. Land use/cover type.

Land Use Type	Land Use Secondary Classification	SWAT Code
urban construction land	urban resident land, rural resident land	URLD
forestland	forestland, shrub land, woodland, other woodland	FRST
grassland	high-covered grassland, medium-covered grassland, and low-covered grassland	HAY
water area	lakes, swamps, glacial snow, canals, beaches	WATR
unused land	sand, Gobi, swamp, bare rock	BALD
arable land	mountain drylands, plain dryland	AGRL

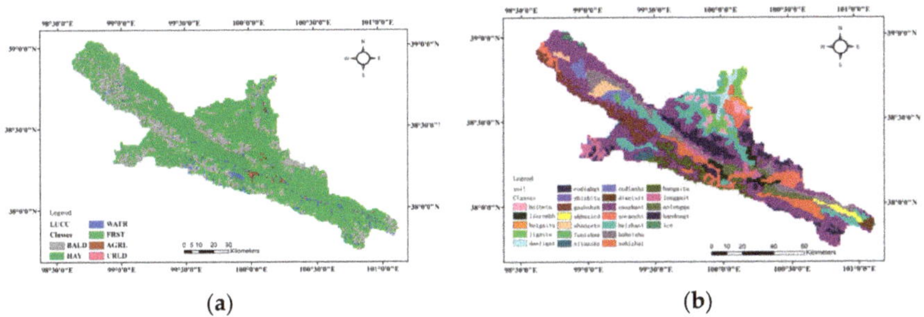

Figure 2. (**a**) Distribution of land use/cover types; (**b**) Distribution of soil types.

Table 2. Meteorological station information.

Station Name	Longitude/°	Latitude/°	Elevation/m
Tuole station	98.42	38.8	3367
Yeniugou station	99.58	38.42	3320
Qilian station	100.25	38.18	2787.4
Zhangye station	100.43	38.93	1482.7

Table 3. Hydrological station information.

Station Name	Longitude/°	Latitude/°	Elevation/m
Yingluoxia station	100.18	38.82	1700

2.2. Methodology

2.2.1. Mann-Kendall Trend Test

The Mann-Kendall test was performed on the mutation point. The Mann-Kendall test method is a nonparametric method. The M-K test is simple to calculate, and the results are not disturbed by a few outliers in the time series such as temperature, rainfall, and runoff. At the same time, it is not necessary for the sample to follow a certain distribution, and it can clearly indicate the start time of the sudden change of the time series of temperature, rainfall, runoff, and the like [54].

It is assumed that time series X_t, such as temperature, rainfall, runoff, etc., are composed of n randomly independent and identically distributed samples, such as $X = x1, x2, \ldots, xn$.

First, construct an order column:

$$s_k = \sum_{i=1}^{k} r_i, k = 2, 3, \ldots, n \tag{1}$$

when $x_i > x_j$, r_i has a value of 1, otherwise it has a value of 0.

The order column s_k is the cumulative number of times the value of the stat r_i at time i is greater than the value at time j.

Based on the assumption that the sample $X = \{x_1, x_2, \ldots, x_n,\}$ is a random independent and identical distribution, the normalized order column s_k is defined as the statistic UF_k.

$$UF_k = \frac{|s_k - E(s_k)|}{\sqrt{Var(s_k)}} \tag{2}$$

where $UF_1 = 0$, and $E(s_k)$, $Var(s_k)$ are the mean and variance, respectively, of the cumulative number s_k.

The above process is repeated in time series x in reverse order $x_n, x_{n-1}, \ldots, x_1$, making $UB_k = -UF_k, k = n, n-1, \ldots, 1, UB_1 = 0$.

Given the significance level $\alpha = 0.05$, the critical value $U_{0.05} = \pm 1.96$, the two statistical sequence curves of UF_k and UB_k, and the two critical values are plotted on the same graph for analysis.

If one of the statistics UF_k or UB_k is positive, it can be judged that time series such as temperature, rainfall, and runoff have an upward trend. In addition, when the absolute values of the statistics UF_k or UB_k exceed the threshold of the significance level, it can be further judged that the rising or falling trend of the time series such as temperature, rainfall, runoff, etc. is significant; in particular, the statistics UF_k or UB_k are at the level of significance. When the critical values intersect, the intersection point is the sudden change point, and the corresponding time is the sudden change time of temperature, rainfall and runoff [55].

2.2.2. Cumulative Anomaly Method

The anomaly is a commonly used statistic that indicates that the runoff deviates from the normal situation [56]. The difference between a certain value and the average value of a series of values is the anomaly; that is, $x_i - \bar{x}$. Any runoff series can be transformed into a sequence with an average value of 0 after anomaly processing. The cumulative anomaly is a statistical method for judging the trend of discrete data points by curve. The calculation process involves first calculating the anomaly value of annual runoff and then accumulating values year-by-year according to the time series to obtain the variation process of the cumulative anomaly value with time. The trend of discrete data points can be visually judged by the curve. The cumulative anomaly for a sequence X at a certain time t is expressed as follows:

$$X_t = \sum_{i=1}^{t}(x_i - \bar{x}), \quad t = 1, 2, \ldots, n \tag{3}$$

$$\bar{x} = \frac{1}{n}\sum_{t=1}^{n} x_t \tag{4}$$

The runoff cumulative anomaly curve can be used to characterize the abundance of runoff change. When the curve changes downward, this indicates that the runoff enters the dry season. An upward change indicates that the runoff enters the wet season, and a horizontal change indicates that the runoff enters the flat period.

2.2.3. Soil and Water Assessment Tool (SWAT Model)

The SWAT model is a typical distributed hydrological model based on the GIS platform, which was developed by the United States Department of Agriculture [57]. The model can predict the trend and impact of runoff changes under different land use patterns, soil conditions, and river basin management conditions in large watersheds [58]. The data required for the SWAT model include topography, soil, land use/cover, weather, hydrology, etc., and different databases can be selected depending on the purpose of the study [59]. The SWAT-simulated watershed hydrological process is divided into the land phase of the hydrological cycle and the convergence phase of the hydrological cycle. The entire water circulation system follows the law of water balance, and the formula is as follows:

$$SW_t = SW_0 + \sum_{i=1}^{t}\left(R_{day} - Q_{surf} - E_a - W_{seep} - Q_{gw}\right) \tag{5}$$

where SW_t is the final soil moisture content, mm; SW_0 is the initial soil moisture content of the i-th day, mm; t is the time, d; R_{day} is the precipitation of the i-th day, mm; Q_{surf} is the surface runoff of day i, mm; E_a indicates the amount of evapotranspiration on day i, mm; W_{seep} indicates the amount of water entering the vale zone from the soil profile on day i, mm; and Q_{gw} indicates the return flow amount on day i, mm.

The runoff simulation in the SWAT model is the SCS runoff curve method based on daily precipitation data and the Green&Ampt infiltration method based on time precipitation data. The SCS runoff curve number model links soil type, runoff, land use and management measures to provide a basis for estimating runoff under various land uses and soil types. According to the collected precipitation data, this paper selects the SCS runoff curve method to simulate the runoff.

The SWAT model has the following basic assumptions: The ratio between the actual water storage amount F and the maximum water storage capacity S is equal to the ratio of the runoff Q to the difference between the rainfall P and the initial loss I_a; a linear relationship between I_a and S. Its rainfall-runoff relationship expression is as follows:

$$\frac{F}{S} = \frac{Q}{P - I_a} \tag{6}$$

According to the water balance, it can be obtained that:

$$F = P - I_a - Q \tag{7}$$

Therefore, Equation (6) can be derived as follows:

$$Q = \frac{(P - I_a)^2}{S + P - I_a} \tag{8}$$

I_a is affected by factors such as land use, farming methods, irrigation conditions, canopy interception, etc. It has a certain proportional relationship with the maximum possible permeability S. Based on the analysis of a large number of long-term experimental results, the SWAT model provides that the most suitable scale factor for I_a and S is 0.2:

$$I_a = 0.2S \tag{9}$$

S is closely related to the underlying surface factors such as land use type, soil type and slope. The model can introduce CN to better determine S. The formula is as follows:

$$S = \frac{25400}{CN} - 254 \tag{10}$$

CN is a dimensionless parameter. The CN value reflects a comprehensive parameter of the characteristics of the pre-rainfall watershed. It is a combination of factors such as soil moisture, slope, land use type and soil type.

The principle of runoff simulation of the SWAT model is as follows: when the rainfall reaches the ground, the water infiltration rate is larger due to the dryness of the surface soil. The continuous rainfall process causes the soil moisture to increase, which leads to the decrease of water infiltration rate. When the rainfall intensity is greater than the infiltration rate, the filling begins. Once the surface is filled, the surface runoff will be generated. The hydrological simulation of the SWAT model is based on the water balance equation [38].

2.2.4. Parameter Sensitivity Analysis and Model Calibration and Validation

For the calibration of the model, the Yingluoxia hydrological station is selected, which controls the upstream outlets and has a strong representativeness, which helps to improve the accuracy of the model. First, the annual scale simulation is performed, and then the monthly scale simulation is performed. SWAT-CUP software was used for the calibration of the model. The SUFI-2 algorithm was used for iterative calculation. According to previous research experience, the parameters were selected for the LH-OAT sensitivity analysis. First, the initial range of the model is determined, and then multiple operations are performed until the optimal value of the parameter is determined. The model was evaluated using the decision coefficient R^2 and the model efficiency coefficient NSE to achieve good results:

$$R^2 = \frac{\left[\sum_{i=1}^{n}\left(Q_i^{obs} - Q_{mean}\right)\left(Q_i^{sim} - Q_{smean}\right)\right]^2}{\sum_{i=1}^{n}\left(Q_i^{obs} - Q_{mean}\right)^2 \sum_{i=1}^{n}\left(Q_i^{sim} - Q_{smean}\right)^2} \tag{11}$$

$$NSE = 1 - \left[\frac{\sum_{i=1}^{n}\left(Q_i^{obs} - Q_i^{sim}\right)^2}{\sum_{i=1}^{n}\left(\left(Q_i^{obs} - Q_{mean}\right)\right)^2}\right] \tag{12}$$

where Q_i^{obs} is the observed streamflow, Q_i^{sim} is the simulated streamflow, Q_{smean} and Q_{mean} are the average simulated and observed streamflow values, respectively, and n is the simulation number. The range of R^2 is 0~1, and the closer this value to 1, the better the simulation effect. For NSE, greater than 0.5 indicates that the simulation result is acceptable, and the NSE is between 0.5–0.65, indicating suitable simulation results [60].

2.2.5. Contribution Rate Calculation

The semi-distributed hydrological model, the SWAT model, is used to calculate the contribution rate of climate change and human activities to runoff effects. The reference period and the interference period are accurately divided according to the abrupt position of the runoff, and then the meteorological, hydrological and land use data rates of the reference period are used to determine the hydrological model parameters. The period before the runoff mutation point is the reference period, and the period after the runoff mutation point is the interference period. To analyze the contribution of the calculation of land use and climate change, the following scenarios are used for analysis; see Table 4. Based on scenario 1, scenario 3 is compared with the common impacts of land use change on runoff, scenario 2 is compared with the impact of climate change on runoff, and finally, the impacts of land use and climate change on runoff during different periods are quantitatively analyzed in the upper reaches of the Heihe River.

Table 4. Scenarios for quantitative attribution analysis.

Scenarios	Land Use/Cover Data	Meteorological Data
1	1980s	1980–2003
2	1980s	2004–2008
3	2000s	1980–2003
4	2000s	2004–2008

Q1, Q2, Q3, and Q4 are the average annual runoffs simulated under scenarios 1, 2, 3, and 4, respectively. In addition, the following formula is used to complete the calculation of the contribution rate of climate change and land use change.

$$\alpha_c = \frac{Q_3 - Q_1}{Q_4 - Q_1} \times 100\% \tag{13}$$

$$\alpha_h = \frac{Q_2 - Q_1}{Q_4 - Q_1} \times 100\% \tag{14}$$

2.2.6. Scenario Setting and Model Analysis

To further explore the impact of climate change on runoff in the upper reaches of the Heihe River, the range of possible future variabilities in climate change, precipitation and temperature changes were given. The following scenarios are determined: The existing precipitation conditions remain unchanged, precipitation is increased by 10% and 20%, and the precipitation is reduced by 10% and 20%, which gives a total of 5 scenario options. Additionally, the existing temperature is maintained, reduced by 0.5 °C, 1 °C, 1.5 °C and 2 °C, and increased by 0.5 °C, 1 °C, 1.5 °C, and 2 °C, for a total of 9 options.

To further explore the impact of land use on the runoff in the upper reaches of the Heihe River, according to the scenario analysis of the Western Data Center future trend of land use in the Heihe River, the ecological protection trend, the economic development trend and historical trend are adopted. (http://westdc.westgis.ac.cn/) Based on the historical development trend and existing problems of the Heihe River Basin, this dataset uses the Dyna-CLUE model to simulate land use development scenarios in the Heihe River Basin in 2020 and 2030.

3. Results and Discussion

3.1. Parameter Sensitivity Analysis and Model Calibration and Validation

Through the sensitivity analysis of the SWAT model, 14 parameters with higher sensitivity were selected to calibrate and verify the model (Table 5). The initial values and the range of the parameters can refer to the existing research, which can save time for parameter adjustment and improve efficiency. The period of 1980–1984 was used as the model's warm-up period, 1985–1998 was used as the model's calibration period, and 1999–2008 was used as the validation period of the model. SWAT-CUP was used to calibrate the model parameters, and the monthly streamflow at the upstream outlet at the Yingluoxia station was calibrated and adjusted. The monthly streamflow R^2 and ENS at the Yingluoxia station during the calibration period were 0.75 and 0.65, respectively, and the verification periods were 0.71 and 0.63. The SWAT model is suitable for the upper reaches of the Heihe River. The simulation results are shown in Figure 3 and Table 6.

Table 5. Parameter sensitivity analysis.

Parameter	Sensitive	Value Range	Fitted Value
CN2.mgt	2	(35,98)	29.532
Ch_K2.ret	5	(−0.01,500)	21.402
Ch_N2.ret	12	(0,0.2)	0.098
ESCO.hru	1	(0,1)	0.815
EPCO.hru	14	(0,1)	0.188
CANMX.hru	10	(0,100)	6.634
SOL_Z.sol	4	(0,3500)	0.187
SOL_K.sol	6	(0,2000)	−0.248
SOL_AWC.sol	3	(0,1)	0.358
GWQMN.gw	8	(0,5000)	3398.502
GW_Delay.gw	13	(0,500)	17,795
REVAPMN.gw	11	(0,500)	0.225
GW_REVAP.gw	7	(0.02,0.2)	0.02
ALPHA_BF.gw	9	(0,1)	0.008

CN2: Moisture condition SCS curve number; Ch_K2: River effective water transfer coefficient; Ch_N2: Manning's "n" value for the main channel; ECSO: Soil evaporation compensation factor; EPCO: Plant transpiration compensation coefficient; CANMX: Maximum canopy storage; SOL_Z: Depth from soil surface to bottom of layer; SOL_K: Saturated hydraulic conductivity; SOL_AWC: Available water capacity of the soil layer; GWQMN: Threshold water level in shallow aquifer for base flow; GW_Delay: Groundwater delay coefficient; REVAPMN: Threshold water level in in shallow aquifer for "revap"; GW_REVAP: Groundwater "revap" coefficient; ALPHA_BF: Baseflow recession constant.

Table 6. Calibration and validation results of SWAT model for monthly streamflow.

Period	Measured Average m^3/s	Simulated Average m^3/s	Nash-Suttcliffe	R^2
1985–1998	51.65	58.52	0.65	0.75
1999–2008	66.08	57.62	0.63	0.71

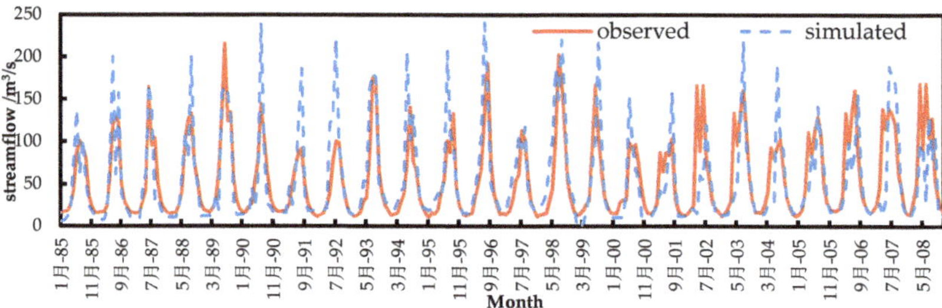

Figure 3. Calibration and validation results of SWAT model for monthly streamflow.

3.2. Trends in Annual Runoff

Figure 4 shows the linear analysis of the runoff from 1958 to 2017 and the 5-year moving average curve of the hydrological station at the water outlet of the Heihe source area. In the past 60 years, the annual runoff of the Yingluoxia shows an increasing trend, on the whole. The annual runoff at the Yingluoxia station reached its maximum in 2017, at 23.31×10^8 m^3, while the minimum appeared in 1971 at 10.32×10^8 m^3, with a tendency to change of 0.93×10^8 m$^3 \cdot 10$ a^{-1}. After entering the 21st century, runoff is generally large, showing a fluctuating rising trend.

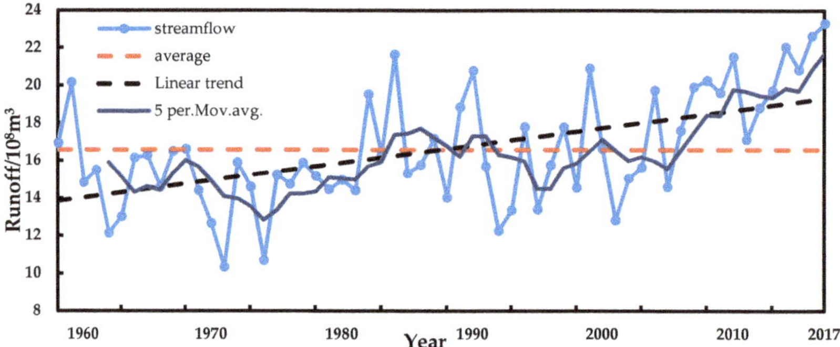

Figure 4. Trend analysis of runoff at Yingluoxia station.

We use a combination of two methods to identify the point of abrupt changes in the runoff sequence. In the M-K curve, the UF_k and UB_k graphs are plotted. If the value of UF_k or UB_k is greater than 0, this indicates that the sequence is on an upward trend, and less than 0 indicates a downward trend. When these values exceed the critical line, this indicates a significant increase or decrease. The range exceeding the critical line is determined as the time zone in which the mutation occurs. If there is an intersection between the curves of UF_k and UB_k, and the intersection is between the critical lines, then the moment corresponding to the intersection is the time when the mutation starts.

As shown in Figure 5a, in the 1980s, the UF_k value began to be greater than 0 and was always greater than 0, indicating that the runoff sequence of Yingluoxia station began to rise from the 1980s. In 2008, the UF_k value was greater than 1.96, indicating that the trend of increasing the runoff was significant based on a significance level test of 0.05. Also, an intersection of the curves appeared in 2004. As seen in the cumulative anomaly curve of Figure 5b, the cumulative anomaly change process can be roughly divided into three stages: from 1958 to 1980, the annual runoff showed a decreasing trend; from 1981 to 2003, the annual runoff showed a relatively gradual fluctuation trend; after 2004, runoff showed a clear upward trend. The mutations may be caused by symptoms of climate change, such as increased precipitation. In the 21st century, the global climate is warming, and the climate in northwestern China is warm and humid, but there is a significant hysteresis effect, which may lead to sudden changes of runoff appearing in 2004. Combining the two methods, the mutation point was set to 2004. Therefore, in the follow-up study, we divided the study period into a reference period pre-2004 and an interference period after 2004.

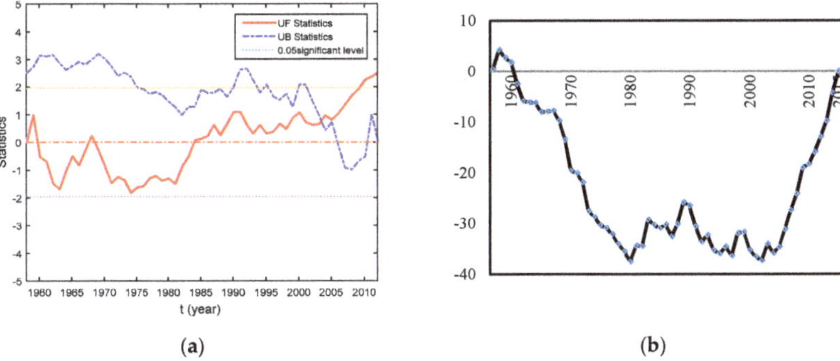

(a)   (b)

Figure 5. (a) Change trends of the M-K test; (b) Cumulative anomaly of the annual runoff in Yingluoxia station.

3.3. Precipitation and Runoff Correlation Analysis

The Double Mass Curve (DMC) is a common method for testing the consistency of relationships between two parameters and their changes. The DMC is the relationship between the continuous cumulative value of one variable and the continuous cumulative value of another variable plotted in the Cartesian coordinate system. It can be used to test the consistency of hydro-meteorological elements. The DMC of the runoff and precipitation in the upper reaches of the Heihe River was plotted to test the correlation between the two factors. The M-K test and the cumulative anomaly curve analysis of the annual runoff of the Yingluoxia hydrological station have been well verified in the precipitation-runoff DMC. As seen in Figure 6, the precipitation-runoff DMC is roughly divided into three phases, 1958–1979, 1980–2003, and 2004–2014.

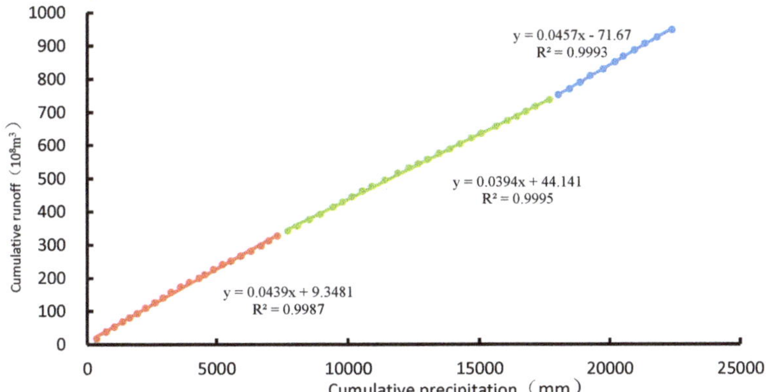

Figure 6. The precipitation-runoff DMC.

Figure 7 shows the correlation analysis between runoff and precipitation changes. From the annual scale, the runoff changes are consistent with the precipitation changes, and there is a clear correlation. In addition, the correlation passed the significance test.

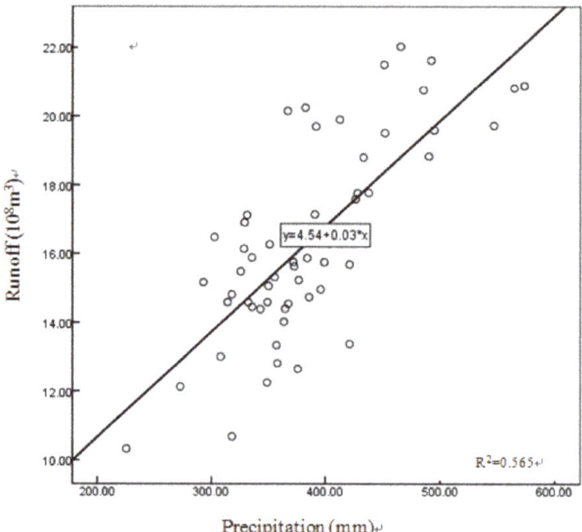

Figure 7. The correlation analysis between precipitation and runoff.

3.4. Contribution of Land Use and Climate Change to Runoff Variation

The different meteorological data and land use data for the designated natural period and interference period are combined and the runoff contribution rate of land use and climate change in the Heihe River Basin is calculated according to different scenarios. In the upper reaches of the Heihe River, the contribution rate of climate change runoff is much greater than the contribution rate of human activities.

The simulation results are shown in Table 7. The simulated annual runoff is 622.67 m^3/s in the reference period and 710.90 m^3/s in the interference period. Compared with the reference period, the total runoff increased during the interference period by 88.23 m^3/s. Among the scenarios, the increase caused by climate change is 76.89 m^3/s, and the increase caused by human activities is 11.34 m^3/s. The contribution rate of climate change is 87.15%, while the contribution rate of human activities is only 12.85%.

He et al. calculated the contribution rate of climate change and human activities to runoff in the upper reaches of the Heihe River by the elastic coefficient method, and found that the contribution rate of climate change is greater than that of human activities, but because of the analysis based on statistical characteristics, the land use type and soil are ignored the type and other physical mechanisms [32]. Lin et al. identified the effects of climate and land use change on runoff and evapotranspiration through hydrological model separation. The role of climate change is found to be much greater than land use change [50]. Meng et al. used the SWAT model to study the Aksu River in the northwest inland area and found that climate factors and human activities were responsible for 92.28% and 7.72% of the variability, respectively [61]. These are consistent with the results of this paper.

Table 7. Simulated response of human activities and climate change in Heihe River Basin.

Time Interval	Period	Simulated Annual Runoff m³/s	Total Increase m³/s	Effect of Human Activities on Runoff		Effect of Climate Change on Runoff	
Reference period	1980–2003	622.67	——	——	——	——	——
Interference period	2004–2008	710.9	88.23	Increase m³/s	Proportion	Increase m³/s	Proportion
				11.34	12.85%	76.89	87.15%

3.5. Scenario Simulation of Climate Change and Land Use Cover Change

3.5.1. Climate Change Factor

The SWAT model was used to analyze the combined scenarios of different meteorological data. According to Table 8, the influence of temperature on runoff is complicated, and the increase or decrease in runoff caused by temperature cannot be determined. When the temperature is lowered, evaporation is reduced, which leads to an increase in surface runoff. However, when the temperature rises, the change of surface runoff presents uncertainty due to the conflict effect caused by evaporation and snowmelt runoff. An increase in temperature causes an increase in evaporation, resulting in a decrease in surface runoff. At the same time, an increase in temperature will also lead to an increase in glacial snowmelt runoff, which will increase surface runoff. In the case of maintaining the precipitation in the upper reaches of the Heihe River, the runoff also changed with the change in temperature, but all scenarios showed an increasing trend; as the temperature increases, the increase in surface runoff has been alleviated. The effect of rainfall on runoff is positive. With the temperature of the upper reaches of the Heihe River remains unchanged, the runoff increases with increasing rainfall, and vice versa.

Table 8. Relative variation of mean annual runoff for different scenarios/%.

Temperature	Precipitation				
	−20%	−10%	0	+10%	+20%
−2 °C	−11.87	−4.45	17.39	23.21	38.65
−1.5 °C	−15.85	−6.76	13.13	18.86	32.97
−1 °C	−20.69	−7.70	9.11	11.24	21.75
−0.5 °C	−22.45	−15.33	4.35	8.96	16.50
0	−30.63	−20.85	0	7.45	14.69
+0.5 °C	−21.35	−9.64	9.2	10.52	22.49
+1 °C	−23.44	−12.17	6.29	12.48	25.86
+1.5 °C	−23.96	−12.96	5.36	16.63	26.52
+2 °C	−25.79	−15.59	4.43	18.72	30.03

Table 9 shows the response of runoff to lower temperatures and higher precipitation. As shown in Table 9, when $\Delta T = 0$, it means the response of the runoff to the increase of precipitation when the temperature is constant. It can be seen that the increase in precipitation increases the runoff, and the more the precipitation increases, the more the runoff increases. When $\Delta P = 0$, it means the response of the runoff to the temperature decreases when the precipitation is constant. It can be seen that the lower the temperature, the more the runoff increases, and the more the temperature is lowered, the more the runoff increases. When both ΔT and ΔP are not zero, it means that the temperature decreases and the precipitation increases. It can be found that when the temperature decreases and the precipitation increases, the runoff increases the most.

Table 9. Response of runoff to temperature reduction and precipitation increase.

Temperature Change	Precipitation Change		
	ΔP=0	ΔP=10%	ΔP=20%
ΔT $= 0$	0	7.45	14.69
ΔT $= -0.5$	4.35	8.96	16.50
ΔT $= -1.0$	9.11	11.24	21.75
ΔT $= -1.5$	13.13	18.86	32.97
ΔP $= -2.0$	17.39	23.21	38.65

Table 10 shows the response of runoff to both elevated temperature and precipitation. It can be seen from Table 10 that when ΔP = 0, when the precipitation is constant, the runoff will increase when the temperature rises. When both ΔT and ΔP are not zero, it indicates the response of the runoff when the temperature and precipitation increase simultaneously. When the increase of precipitation reaches 10% or more, the runoff increases, indicating that the effect of precipitation on runoff is more significant. Due to the large area of glaciers in the upper reaches of the Heihe River, when the temperature rises, the increase of glacial snowmelt will also lead to an increase in runoff.

Table 10. Response of runoff to simultaneous increase in temperature and precipitation.

Temperature Change	Precipitation Change		
	ΔP=0	ΔP=10%	ΔP=20%
ΔT $= 0$	0	7.45	14.69
ΔT $= 0.5$	9.2	10.52	22.49
ΔT $= 1.0$	6.29	12.48	25.86
ΔT $= 1.5$	5.36	16.63	26.52
ΔT $= 2.0$	4.43	18.72	30.03

Table 11 shows the response of runoff to simultaneous decrease in temperature and precipitation. From Table 11, when ΔT = 0, it means that when the temperature is constant, the runoff decreases when the precipitation decreases. When both ΔT and ΔP are not zero, it indicates the response of the runoff when both temperature and precipitation decrease. It can be seen that the decrease in temperature and the decrease in precipitation have the opposite effect on runoff. When the precipitation is constant, the temperature decreases, the runoff increases, and the lower the temperature, the larger the runoff. When the temperature is constant, the decrease in precipitation will result in a decrease in runoff, and the lower the temperature, the lower the runoff.

Table 11. Response of runoff to simultaneous decrease in temperature and precipitation.

Temperature Change	Precipitation Change		
	ΔP=-20%	ΔP=-10%	ΔP=0
ΔT $= 0$	-30.63	-20.85	0
ΔT $= -0.5$	-22.45	-15.33	4.35
ΔT $= -1.0$	-20.69	-7.70	9.11
ΔT $= -1.5$	-15.85	-6.76	13.13
ΔT $= -2.0$	-11.87	-4.45	17.39

Table 12 shows the response of runoff to elevated temperatures and reduced precipitation. It can be seen from Table 12 that when ΔT = 0, and the temperature is constant, when the precipitation decreases, the runoff will decrease, and the more the precipitation decreases, the smaller the runoff. When both ΔT and ΔP are not zero, it indicates the response of the runoff when the temperature rises and precipitation decreases. It can be seen that both the increase in temperature and the decrease in precipitation can reduce the runoff.

Table 12. Response of runoff to temperature increase and precipitation decrease.

Temperature Change	Precipitation Change		
	$\Delta P=-20\%$	$\Delta P=-10\%$	$\Delta P=0$
$\Delta T = 0$	-20.69	-7.70	0
$\Delta P = 0.5$	-21.35	-9.64	9.2
$\Delta P = 1.0$	-23.44	-12.17	6.29
$\Delta P = 1.5$	-23.96	-12.96	5.36
$\Delta P = 2.0$	-25.79	-15.59	4.43

3.5.2. Land Use Change Factor

According to the historical development trend and existing problems in the Heihe River Basin, the Dyna-CLUE model was used to simulate the land use scenarios of the 2020 and 2030 (Figure 8). Considering the actual land use setting in 2000 as a basic scenario, the impact of future land use changes on runoff was analyzed. The dataset was provided by the Heihe Plan Science Data Center, National Natural Science Foundation of China.

Under natural scenarios, land-use change evolves according to existing trends. The ecological protection scenario is to strictly limit the land use type to occupy land for forest land, grassland and water land, strictly implementing the measures of returning farmland to forests and grasslands. Under circumstances of economic development, with the development of the social economy and the increase of the urban population, the demand for industry, residential and public land is urgent, leading to the continuous expansion of the urban scale, leading in turn to an increase in urban construction land and cultivated land.

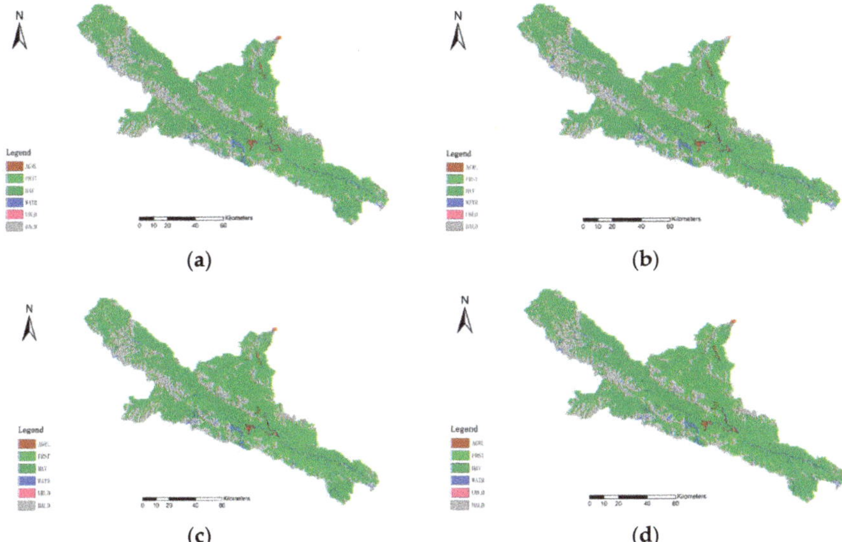

Figure 8. *Cont.*

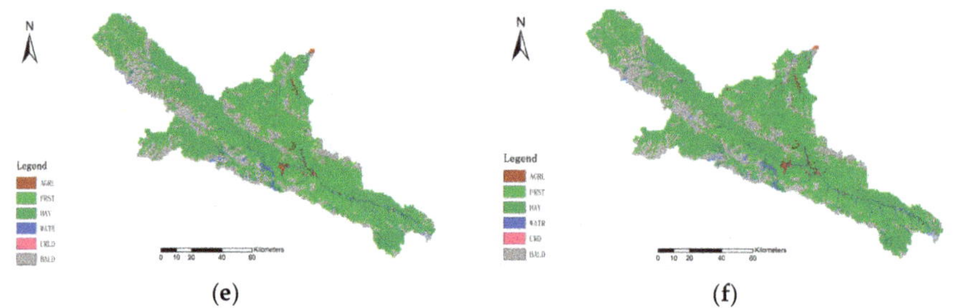

Figure 8. Distribution of land use types under different simulation scenarios. (**a**) Natural growth scenario in 2020, (**b**) Natural growth scenario in 2030, (**c**) Ecological protection scenario in 2020, (**d**) Ecological protection scenario in 2030, (**e**) Economic development scenario in 2020, (**f**) Economic development scenario in 2030.

It can be seen from Table 13 that under the future scenario, land use areas of various types will increase or decrease to different degrees. Under natural growth, from 2020 to 2030, the cultivated land will increase, and the forest and grassland area will decrease. Under the protection situation, from 2020 to 2030, the area of cultivated land, forest land and grassland will increase, and the area of bare land will decrease. Under the economic development situation, from 2020 to 2030, the area of cultivated land will increase, the area of bare land will increase, and the area of forest land and grassland will decrease.

Table 13. Area ratio of different land use scenarios.

Type of Landuse	Natural Growth Scenario		Ecological Protection Scenario		Economic Development	
	2020	2030	2020	2030	2020	2030
AGRL	12.009%	12.569%	11.565%	11.778%	12.889%	13.764%
FRST	12.930%	12.786%	13.060%	13.102%	12.680%	12.413%
HAY	35.344%	34.824%	35.753%	35.848%	34.533%	33.728%
WATR	2.295%	2.294%	2.294%	2.293%	2.290%	2.290%
URLD	0.983%	0.986%	0.959%	0.972%	1.037%	1.091%
BALD	36.439%	36.541%	36.369%	36.007%	36.570%	36.714%

The land use scenario simulation results are shown in Table 14. In 2020, under the natural growth scenario and ecological protection scenario, runoff showed an increasing trend, but the degree of increase was different between the scenarios, while under economic development, runoff showed a decreasing trend. Under the natural growth scenario, the area of grassland decreased, the area of cultivated land and bare land increased, and by 2030, the amount of runoff increased. Under the ecological protection scenario, the area of arable land and bare land decreased, and the area of forestland and grassland increased. The increase in runoff was larger than that under the natural growth scenario, which shows that the increase in forestland and grassland has a greater impact on the increase in runoff than that of cultivated land or bare land. In future planning, the area of ecological land such as forest and grass can be appropriately increased.

In the future land use scenario, the SWAT model is used to simulate, and the results show that the runoff increases or decreases in different situations. In the case of natural growth and ecological protection, runoff shows an increasing trend. However, in the case of ecological protection, the increase in runoff is more significant. Under the economic development situation, the runoff is reduced due to the large reduction in grassland area. It can be seen that the construction of ecological construction land such as forest land and grassland can, to a certain extent, improve the regional microclimate, improve the soil environment, reduce the surface temperature, reduce the direct evaporation of water, and have a positive effect on the surface runoff.

Table 14. Change of runoff under land use/cover scenarios.

	Land Use/Cover Change Scenarios	Change Rate of Runoff
	(a)	0.27%
2020	(c)	0.34%
	(e)	−0.10%
	(b)	1.57%
2030	(d)	3.74%
	(f)	−0.22%

4. Conclusions

By analyzing the impact of climate and land use change on runoff in this paper, a qualitative analysis is turned into a quantitative analysis for the development and utilization of water resources in the Heihe River Basin, providing a reference base. In this paper, the runoff change trend at the Yingluoxia station in the upper reaches of the Heihe River over the past 60 years is analyzed and combined with the SWAT model to simulate runoff. We combine different scenarios of climate and land use change to simulate surface runoff. The results show the following:

(1)   The annual runoff in the upper reaches of the Heihe River is increasing. The long-term runoff sequence was mutated in 2004. Therefore, the runoff sequence was divided into the reference period before 2004 and the interference period after 2004.

(2)   The SWAT model has good applicability in runoff simulation of the Yingluoxia hydrological station. It can be used for contribution rate calculations and scenario simulations.

(3)   The contribution rates of climate change and land use to watershed runoff are very different. In the upper reaches of the Heihe River, the contribution rate of climate change to runoff change is 87.15%, while human activities contribute only 12.85%.

(4)   According to different temperature and rainfall scenarios, the simulation analysis shows that decreased temperature causes increased surface runoff. However, when the temperature rises, the change of surface runoff presents uncertainty due to the conflict effect caused by evaporation and snowmelt runoff. While an increase in rainfall will lead to an increase in runoff, but the amount of increase will differ. It can be found that when the precipitation increases and the temperature decreases, the runoff increases the most. When the temperature increases and the precipitation decreases, the amount of runoff reduction is most significant.

(5)   Land use in different scenarios has different effects on runoff. Both natural and ecological conservation trends lead to increased runoff, but the increase in runoff is greater under the ecological protection scenario, while under economic development, runoff showed a decreasing trend. The increase of forest land and grassland area caused the increase of surface runoff. It can improve the regional climate to a certain extent and have a positive effect on surface runoff. Also, the cultivated land has a negative contribution to soil and water conservation and has a negative effect on the occurrence of surface runoff.

Although climate change plays a key role in the runoff changes in the UHRB, the impact of human activities cannot be ignored. Against the background of climate change, according to the water resource utilization management objectives, watershed management measures can be adjusted to realize the rational layout of land use and then change the river basin runoff trend, which provides a reference for effective scientific planning in river basins. In the future watershed management process, while meeting upstream water demand, the land use structure can be adjusted and rationally distributed, thereby increasing the upstream water output and ensuring water use in the lower reaches of the basin. In this paper, research regarding a certain contribution rate was conducted, but there are still many factors that have not been considered. For example, climate change can be combined with evapotranspiration, and other human activities can also be added to considerations for further research.

Author Contributions: Methodology, X.S., C.W.; software, X.S.; validation, X.S.; formal analysis, X.S.; investigation, X.S., R.J.; writing—original draft preparation, X.S.; writing—review and editing, X.S.; supervision, X.J.; project administration, X.J. All authors have read and approved the final manuscript.

Funding: This research was supported by the National Natural Science Fund Major Research Plan (91325201), the National Natural Science Fund (51779209), and the National key research and development plan (2017YFC0404303).

Conflicts of Interest: The authors declare no conflict of interest.

References

1. Huntington, T.G. Evidence for intensification of the global water cycle: Review and synthesis. *J. Hydrol.* **2006**, *319*, 83–95. [CrossRef]
2. Wang, R.; Kalin, L.; Kuang, W.; Tian, H. Individual and combined effects of land use/cover and climate change on Wolf Bay watershed streamflow in southern Alabama. *Hydrol. Process.* **2014**, *28*, 5530–5546. [CrossRef]
3. Liu, D.D.; Chen, X.H.; Lian, Y.Q. Impacts of climate change and human activities on surface runoff in the Dongjiang River basin of China. *Hydrol. Process.* **2010**, *24*, 1487–1495. [CrossRef]
4. Xia, J.; Liu, C.Z.; Ren, G.Y. Opportunity and Challenge of the Climate Change Impact on the Water Resource of China. *Adv. Earth Sci.* **2011**, *26*, 1–12.
5. Zhang, L.; Srinivasan, R.; Bai, Z.K. Analysis of streamflow response to climate variability and land use change in the Loess Plateau region of China. *CATENA* **2017**, *154*, 1–11. [CrossRef]
6. Birkinshaw, S.J.; Guerreiro, S.B.; Nicholson, A.; Liang, Q.; Quinn, P.; Zhang, L.; He, B.; Yin, J.; Fowler, H.J. Climate change impacts on Yangtze River discharge at the Three Gorges Dam. *Hydrol. Earth Syst. Sci.* **2017**, *21*, 1911–1927. [CrossRef]
7. Li, Z.; Liu, W.Z.; Zhang, X.C.; Zheng, F. Impacts of land use change and climate variability on hydrology in an agricultural catchment on the Loess Plateau of China. *J. Hydrol.* **2009**, *337*, 35–42. [CrossRef]
8. Liu, J.Y.; Zhang, Q.; Deng, X.Y.; Ci, H.; Cheng, X.H. Quantitative analysis the influences of climate change and human activities on hydrological processes in Poyang Basin. *J. Lake Sci.* **2016**, *28*, 432–443.
9. Bao, Z.X.; Zhang, J.Y.; Wang, G.Q.; Fu, G.; He, R.; Yan, X.; Jin, J.; Liu, Y.; Zhang, A. Attribution for decreasing streamflow of the Haihe River basin, northern China: Climate variability or human activities? *J. Hydrol.* **2012**, *460/461*, 117–129. [CrossRef]
10. Kibria, K.N.; Ahiablame, L.; Hay, C.; Djira, G. Streamflow Trends and Responses to Climate Variability and Land Cover Change in South Dakota. *Hydrology* **2016**, *3*, 2. [CrossRef]
11. Ma, L.; Liu, T.X.; Ma, L.; Sun, M.; Ding, T.; Xin, X.H. The effect of climate change and human activities on the runoff in the upper and middle reaches of the Liaohe River, Inner Mongolia. *J. Glaciol. Geocryol.* **2017**, *3*, 470–479.
12. Gao, C.; Ruan, T. The influence of climate change and human activities on runoff in the middle reaches of the Huaihe River Basin, China. *J. Geogr. Sci.* **2018**, *28*, 79–92. [CrossRef]
13. Gao, L.M.; Zhang, Y.N. Spatio-temporal variation of hydrological drought under climate change during the period 1960–2013 in the Hexi Corridor, China. *J. Arid. Land* **2016**, *8*, 157–171. [CrossRef]
14. Li, L.J.; Zhang, L.; Wang, H.; Meng, X. Assessing the impact of climate variability and human activities on streamflow from the Wuding river basin in China. *Hydrol. Process.* **2007**, *21*, 3485–3491. [CrossRef]
15. Zhang, Y.; Guan, D.; Jin, C.; Wang, Z. Impacts of climate change and land use change on runoff of forest catchment in northeast China. *Hydrol. Process.* **2014**, *28*, 186–196. [CrossRef]
16. Zhang, A.J.; Wang, B.D.; Cao, M.M. Influence research of climate change and human activities on runoff contribution. *Northeast Water Conserv. Hydropower* **2012**, *1*, 6–11.
17. Yin, Z.L.; Xiao, H.L.; Zou, S.; Lu, Z.X. Process of the Research on Hydrological Simulation in the Mainstream of the Heihe River, Qilian Mountains. *J. Glaciol. Geocryol.* **2013**, *35*, 438–466.
18. Tao, H.; Bai, Y.G.; Mao, W.Y. Observed and Projected Climate Changes in Tarim River Basin. *J. Glaciol. Geocryol.* **2011**, *35*, 738–743.
19. Hu, S.S.; Zheng, H.X.; Liu, C.M.; Yu, J.J.; Wang, Z.G. Assessing the Impacts of Climate Variability and Human Activities on Streamflow in the Water Source Area of Baiyangdian Lake. *Acta Geogr. Sin.* **2012**, *67*, 62–70. [CrossRef]

20. Ma, C.K.; Sun, L.; Liu, S.Y.; Shao, M.; Luo, Y. Impact of climate change on the streamflow in the glacierized Chu River Basin, Central Asia. *J. Arid Land* **2015**, *7*, 501–513. [CrossRef]

21. Wu, F.; Zhan, J.; Su, H.; Yan, H.; Ma, E. Scenario-based impact assessment of land use/cover and climate changes on watershed hydrology in Heihe River Basin of northwest China. *Adv. Meteorol.* **2014**, *2015*, 11. [CrossRef]

22. An, M.; Zhang, B.; Sun, L.; Zhang, T.; Yang, B.; Wang, D.; Zhang, C. Quantitative analysis of dynamic change of land use and its influencing factor in upper reaches of the Heihe river. *J. Glaciol. Geocryol.* **2013**, *35*, 355–363.

23. Wang, Y.; Ding, Y.J.; Ye, B.S.; Liu, F.J.; Wang, J.; Wang, J. Contributions of climate and human activities to changes in runoff of the Yellow and Yangtze rivers from 1950 to 2008. *Sci. China Earth Sci.* **2012**, *56*, 1398–1412. [CrossRef]

24. Li, Y.G.; He, D.; Li, X.; Zhang, Y.; Yang, L. Contributions of Climate Variability and Human Activities to Runoff Changes in the Upper Catchment of the Red River Basin, China. *Water* **2016**, *8*, 414. [CrossRef]

25. Li, F.P.; Zhang, G.X.; Xu, Y.J. Separating the impacts of climate variation and human activities on runoff in the Songhua River Basin, Northeast China. *Water* **2014**, *6*, 3320–3338. [CrossRef]

26. Wang, G.Q.; Zhang, J.Y.; Liu, J.F.; Yan, J. Quantitative assessment for climate change and human activities impact on river runoff. *China Water Resour.* **2008**, *2*, 55–58.

27. Rahman, K.; da Silva, A.G.; Tejeda, E.M.; Gobiet, A.; Beniston, M.; Lehmann, A. An independent and combined effect analysis of land use and climate change in the upper Rhone River watershed, Switzerland. *Appl. Geogr.* **2015**, *63*, 264–272. [CrossRef]

28. Zhao, Y.; Yu, X.X.; Zheng, J.K.; Wu, Q.Y. Quantitative effects of climate variations and land-use changes on annual streamflow in Chaobai River Basin. *Trans. Chin. Soc. Agric. Eng.* **2012**, *28*, 252–260.

29. Li, Z.; Liu, W.Z.; Zheng, F.L.; Hu, H.C. The impacts of climate change and human activities on river flow in the Loess Tableland of China. *Acta Ecol. Sin.* **2010**, *30*, 2379–2386.

30. Hu, X.L. Analysis on Runoff Variation Regularity and Regional Water Resources Optimum Allocation in the Heihe River Basin. *Hydrology* **2003**, *23*, 32–37.

31. Wang, J.; Meng, J.J. Research on runoff variations based on wavelet analysis and wavelet neural network model: A case study of the Heihe River drainage basin (1944–2005). *J. Geogr. Sci.* **2007**, *3*, 327–338. [CrossRef]

32. He, X.Q.; Zhang, B.; Sun, L.W.; Jin, S.L.; Zhao, Y.F.; An, M.L. Contribution rates of climate change and human activity on the runoff in upper and middle reaches of Heihe River basin. *Chinese J. Ecol.* **2012**, *31*, 2884–2890.

33. Sang, Y.F.; Wang, Z.G.; Liu, C.M.; Yu, J. The impact of changing environments on the runoff regimes of the arid Heihe River basin, China. *Theor. Appl. Climatol.* **2014**, *115*, 187–195. [CrossRef]

34. Wang, G.; Liu, J.; Kubota, J.; Ma, X. Effects of land-use changes on hydrological processes in the middle basin of the Heihe River, northwest China. *Hydrol. Process.* **2007**, *21*, 1370–1382. [CrossRef]

35. Zhang, X.; Xiong, Z.; Tang, Q. Modeled effects of irrigation on surface climate in the Heihe River Basin, Northwest China. *J. Geophys. Res. Atmos.* **2017**, *122*, 7881–7895. [CrossRef]

36. Zhao, N.; Zeng, X.F.; Liu, H. Response of runoff to climate change in Heihe River Basin based on HBV model. *Yangtze River* **2018**, *49*, 34–38.

37. He, R.; Pang, B.; Zhang, L.Y.; Shi, R. Uncertainty analysis of VIC model on GLUE method. *J. Beijing Normal Univ. (Nat. Sci.)* **2015**, *50*, 576–580.

38. Wang, Z.G.; Liu, C.M.; Huang, Y.B. The Theory of SWAT Model and its Application in Heihe Basin. *Prog. Geog.* **2003**, *22*, 79–86.

39. Huang, Q.H.; Zhang, W.C. Improvement and Application of GIS-based distributed SWAT Hydrological Modeling on High Altitude, Cold, Semi-arid Catchment of Heihe River Basin, China. *J. Nanjing Normal Univ. (Nat. Sci.)* **2004**, *28*, 21–26.

40. Ruan, H.W.; Zou, S.B.; Lu, Z.X.; Yang, D.W.; Xiong, Z. Coupling SWAT and RIEMS to simulate mountainous runoff in the upper reaches of the Heihe River basin. *J. Glaciol. Geocryol.* **2017**, *4*, 384–394.

41. Meng, X.Y.; Shi, C.X.; Liu, S.Y.; Lei, X.H.; Liu, Z.H.; Ji, X.N. CMADS Datasets and Its Application in Watershed Hydrological Simulation: A Case Study of the Heihe River Basin. *Pearl River* **2016**, *37*, 1–19.

42. Zhang, L.; Nan, Z.T.; Xu, Y. Hydrological Impacts of Land Use Changeand Climate Variability in the Headwater Region of the Heihe River Basin, Northwest China. *PLOS ONE* **2016**, *11*. [CrossRef]

43. Luo, K.S.; Tao, F.L.; Deng, X.Z.; Moiwo, J.P. Changes in potential evapotranspiration and surface runoff in1981–2010 and the driving factors in Upper Heihe River Basin in Northwest China. *Hydrol. Process.* **2017**, *31*, 90–103. [CrossRef]

44. Lu, Z.Y.; Xiao, H.L.; Wei, Y.P.; Zou, S.B.; Ren, J. Advances in the study on the human-water-ecology evolution in the past two thousand years in Heihe River Basin. *Adv. Earth Sci.* **2015**, *30*, 396–406.

45. Jiang, X.H.; Liu, C.M. The influence of water regulation on vegetation in the lower Heihe River. *Geogr. Sci.* **2010**, *20*, 701–711. [CrossRef]

46. Liu, X.L.; Dong, G.T.; Zhao, M.J.; Jiang, X.H.; Fan, Z.J.; Yin, H.J.; Guo, X.W. Evaluation of the Heihe River Water Diversion Scheme Adaptability. *Yellow River* **2017**, *39*, 65–69.

47. Cheng, J.Z.; Lu, Z.X.; Zou, S.B.; Yin, Z.L. Variation of the runoff in the upper and middle reaches of the main Heihe River and its causes. *J. Glaciol. Geocryol.* **2017**, *39*, 123–129.

48. Li, B.; Li, C.Y.; Liu, J.Y.; Hu, S. Decreased streamflow in the Yellow River Basin, China: Climate change or human-induced? *Water* **2017**, *9*, 116. [CrossRef]

49. Wang, S.J.; Yan, Y.X.; Yan, M. Contributions of precipitation and human activities to the runoff change of the Huangfuchuan drainage basin: Application of comparative method of the slope changing ratio of cumulative quantity. *Acta Geog. Sin.* **2012**, *67*, 388–397.

50. Dang, S.Z.; Liu, C.M.; Wang, Z.G.; Wu, M.Y. Analyses on Temporal Variations of Snowmelt Runoff Time in the Upper Reaches of Heihe River and Its Climate Causes. *J. Glaciol. Geocryol.* **2013**, *34*, 920–926.

51. Bai, L.M.; Hou, H.Y.; Yang, L.F. Analysis on change trend of runoff in Yingluoxia station of Heihe River. *Yellow River* **2013**, *35*, 23–25.

52. Zhang, X.F. Climate change and its driving effect on the runoff in the Heihe basin. *Resour. Environ. Yangtze Basin* **2014**, *23*, 542–548.

53. Cao, H.; Huang, Q.; Chang, J.X. Analysis on variation characteristics on the runoff series of the Heihe River. *J. Water Resour. Water Eng.* **2008**, *19*, 69–72.

54. Sun, Y.Y.; Mu, X.M.; Gao, P. Temporal Change in Runoff in the Upper and Middle Reaches of Yellow River. *Res. Soil Water Conser.* **2017**, *24*, 59–64.

55. Mo, G.Y. *Runoff Response to the Climate and Land Use Change in the Longtan Basin*; Guangxi University: Nanning, China, 2017; pp. 23–25.

56. Yang, L.S.; Feng, Q.; Yin, Z.L.; Wen, X.H.; Si, J.H.; Deo, R.C. Identifying separate impacts of climate and land use/cover change on hydrological processes in upper stream of Heihe River, Northwest China. *Hydrol. Process.* **2017**, *31*, 1100–1112. [CrossRef]

57. Luo, Y.; Sophocleous, M. Two-way coupling of unsaturated-saturated flow by integrating the SWAT and MODFLOW models with application in an irrigation district in arid region of West China. *J. Arid Land* **2011**, *3*, 164–173. [CrossRef]

58. Yu, W.J.; Nan, Z.T.; Zhao, Y.B.; Li, S. Improvement of snowmelt implementation in the SWAT hydrologic model. *Acta Ecol. Sin.* **2013**, *33*, 6992–7001.

59. Ficklin, D.L.; Barnhart, B.L. SWAT hydrologic model parameter uncertainty and its implications for hydro-climatic projections in snowmelt-dependent watersheds. *J. Hydrol.* **2014**, *519*, 2081–2090. [CrossRef]

60. Van Liew, M.W.; Garbrecht, J. Hydrologic simulation of the Little Washita River experimental watershed using SWAT. *J. Am. Water Resour. Assoc.* **2003**, *39*, 413–426. [CrossRef]

61. Meng, F.H.; Liu, T.; Huang, Y.; Luo, M.; Bao, A.M.; Hou, D.W. Quantitative detection and attribution of runoff variations in the Aksu River Basin. *Water* **2016**, *8*, 338. [CrossRef]

Article

Meteorological Factors Affecting Pan Evaporation in the Haihe River Basin, China

Zhihong Yan [1], Shuqian Wang [1,*], Ding Ma [2], Bin Liu [1,*], Hong Lin [1] and Su Li [1]

[1] School of Water Conservancy and Hydroelectric Power, Hebei University of Engineering, Handan 056021, China; yanzhihong0526@126.com (Z.Y.); Red-Lin@outlook.com (H.L.); susan627530@163.com (S.L.)
[2] Hebei Hydrology and Water Resources Survey Bureau, Shijiazhuang 050000, China; martin19880118@163.com
* Correspondence: wsq9681@sina.com (S.W.); liubin820104@163.com (B.L.); Tel.: +86-310-312-3077 (S.W.); +86-310-312-3702 (B.L.)

Received: 16 December 2018; Accepted: 10 February 2019; Published: 13 February 2019

Abstract: Pan evaporation (E_{pan}) is an important indicator of regional evaporation intensity and degree of drought. However, although more evaporation is expected under rising temperatures, the reverse trend has been observed in many parts of the world, known as the "pan evaporation paradox". In this paper, the Haihe River Basin (HRB) is divided into six sub-regions using the Canopy and k-means (The process for partitioning an N-dimensional population into k sets on the basis of a sample is called "k-means") to cluster 44 meteorological stations in the area. The interannual and seasonal trends and the significance of eight meteorological indicators, including average temperature, maximum temperature, minimum temperature, precipitation, relative humidity, sunshine duration, wind speed, and E_{pan}, were analyzed for 1961 to 2010 using the trend-free pre-whitening Mann-Kendall (TFPW-MK) test. Then, the correlation between meteorological elements and E_{pan} was analyzed using the Spearman correlation coefficient. Results show that the average temperature, maximum temperature, and minimum temperature of the HRB increased, while precipitation, relative humidity, sunshine duration, wind speed and E_{pan} exhibited a downward trend. The minimum temperature rose 2 and 1.5 times faster than the maximum temperature and average temperature, respectively. A significant reduction in sunshine duration was found to be the primary factor in the E_{pan} decrease, while declining wind speed was the secondary factor.

Keywords: evapotranspiration; Pan evaporation; TFPW-MK; Haihe River Basin

---

1. Introduction

Global warming has become an indisputable fact [1]. Temperature records indicate that the earth has warmed by approximately 0.6 °C during the 20th century [2]. This increase in global temperature has significantly impacted the natural environment, ecosystem, and social economy [3], and has led to a series of changes in hydrological factors, such as precipitation, evaporation, water infiltration, soil moisture, river runoff, and groundwater flow, all of which affect the global hydrological cycle. This, in turn, causes temporal and spatial redistribution of water resources, and thereby threatens water security, food security, social security, and national security [4,5]. As a key component in the hydrological cycle, evapotranspiration is associated with water balance and water exchange, as well as surface energy balance; hence, of all components of the water cycle, evapotranspiration is the factor most directly affected by climate change [6]. Therefore, analyzing the climate sensitivity of evapotranspiration has important theoretical and practical implications for understanding the impact of climate change on the hydrological cycle [7].

Evapotranspiration is the process of water transport from the earth's surface to the atmosphere [8]. As a core process of the climate system, evapotranspiration closely links the hydrological cycle, energy budget, and carbon cycle [9]. Pan evaporation (E_{pan}) [10] is the most universal and simplest way to measure evapotranspiration, which is often used to indicate the humidity level of a given regional climate [11]. Although E_{pan} cannot directly represent the evaporation of the water surface, it has a close correlation with water surface evaporation. Therefore, it has remained an important reference indicator in the assessment of water resources, water resources planning, and the design of irrigation systems, to name a few examples [12]. As the global temperature rises, E_{pan} should theoretically gradually increase. However, in reality, only certain regions in the world have an E_{pan} value that is consistent with theoretical expectations, and the majority of the world's regions have been found to have declining E_{pan} values. This phenomenon is called the "pan evaporation paradox" [3]. Specifically, countries such as Spain [13], Iran [14], Israel [15], and Brazil [16] have been found to have increasing E_{pan} values, and countries such as the former Soviet Union, the United States [17,18], New Zealand [19], China [20–23], Thailand [24], India [12], Nigeria [25], and Australia [26,27] have been found to have declining E_{pan} values. Correctly interpreting the overall declining trend of E_{pan} in the context of rising global temperatures and uncovering the main meteorological factors that affect the reduction of E_{pan} is of great importance to accurately predict future hydrological cycles.

Many scholars have studied the temporal and spatial changes of E_{pan} at global and regional scales, as well as the causes of such changes. According to their findings, the causes of E_{pan} reduction can be categorized as follows. (1) An increase in humidity in the surrounding environment of the evaporation pan: Brutsaert and Parlange ascertained that the decrease in E_{pan} value was due to an increase in the volume of evaporation from the land surface, considering the difference between evaporation from the land surface and the evaporation volume observed through the evaporation pan [28]. Zuo et al. employed observational data from 62 conventional meteorological stations with solar-radiation observation equipment in China to analyze in detail the relationship between E_{pan} and corresponding environmental factors, as well as the environmental factors' responses to global climate change. The researchers discovered that E_{pan} was most correlated to atmospheric relative humidity [20]. (2) Changes in precipitation: Tebakari et al. [24] analyzed the temporal and spatial variation of E_{pan} in Thailand from 1982 to 2000 and concluded that both E_{pan} and precipitation showed a declining trend. This conclusion was inconsistent with findings from the United States, where E_{pan} was found to be decreasing while precipitation was increasing [29]. Jaswal et al. utilized evaporation and rainfall data from 1971 to 2000 from 58 stations that were evenly distributed in India to analyze the overall correlation between evaporation and rainfall in a year, as well as their correlation in winter, summer, monsoon season, and post-monsoon season. The results showed that, in southern India, the evaporation trend had a complementary relationship with rainfall during the same period [30]. (3) A decrease in the diurnal temperature range: Peterson et al. compared data from both the United States and the former Soviet Union from 1950 to 1990 and found a steady decline in E_{pan} values in all investigated regions (except Central Asia), as well as a decline in diurnal temperature range. E_{pan} and diurnal temperature range were thus clearly correlated. Therefore, the researchers concluded that the reduction in the diurnal temperature range, caused by an increase in cloud cover, consequently caused the reduction in E_{pan} [17]. (4) A reduction in solar radiation: Roderick and Farquhar found that E_{pan} values observed in many parts of the world over the past 50 years showed a clear downward trend and asserted that such a decline was caused by the reduction in overall solar radiation resulting from an increase in cloud cover and aerosol concentrations [3]. (5) A reduction in wind speed: Burn and Hesch conducted a trend analysis on the evaporation data of 48 sites in the Canadian Prairies over three analysis periods and concluded that wind speed has a substantial influence on the decreasing trend of evaporation, while vapor-pressure deficit has a significant influence on the increasing trend of evaporation [31]. Hoffman et al. studied the changes in E_{pan}, rainfall, wind speed, temperature, and vapor-pressure deficit from 1974 to 2005 taken from 20 climate stations in the Cape Floristic Region (CFR), South Africa, and suggested that the reduction in E_{pan} was likely due to a reduction in wind speed [32]. Yang and Yang analyzed the daily E_{pan}, temperature, wind speed, solar radiation, and relative humidity of 54 meteorological stations in China for 1961 to 2001 and concluded that the reduction in E_{pan} in the majority of regions in China is due to a decrease in wind speed [33]. (6) The comprehensive impact of meteorological elements: Roderick and Farquhar analyzed data from Australia for 1970 to 2002 and found that E_{pan} values showed a downward trend. The results showed that such a change might be related to a decrease in solar radiation, wind speed, and diurnal temperature range [26]. Sheng examined E_{pan} data and other meteorological factors from 468 meteorological stations in China, measured simultaneously from 1957 to 2001, and found that the main influential factors of E_{pan} were solar radiation, diurnal temperature range, and wind speed, while the influence of humidity was the weakest factor [21]. Liu et al. investigated data for 1955 to 2001 taken from 671 sites in China. The results revealed an overall decline in E_{pan}. In addition, diurnal temperature range and wind speed were found to have the greatest correlation with such a decline [22]. Based on the aforementioned studies, the causes of the reduction of E_{pan} appear to be very complicated. Owing to the location, climate, atmospheric differences, and even the differences in the length of the data series, the conclusions of these studies are inconsistent. Therefore, identifying the impact of various meteorological variables on E_{pan} trends is critical to quantifying the impact of global warming.

The HRB is located in a region with a warm semi-arid climate and a continental monsoon climate. This area is sensitive to climate change and is a region with a fragile ecological environment. Owing to the area's dense population and rapid economic development, as well as its status as one of China's major wheat producers, the contradiction between water supply and water demand is prominent in the area. Water shortages have become a major factor restricting sustainable economic and social development in the HRB [34]. E_{pan} in the HRB generally exhibits a decreasing trend [35–37], which is largely consistent with E_{pan} trends in other regions of China [38–48]. However, scholars have differing views on the causes of the E_{pan} trend in the HRB. Zheng et al. analyzed the effects of temperature, wind speed, solar radiation, and atmospheric pressure on E_{pan} in the HRB for 1957 to 2001 and concluded that wind speed is the main factor leading to the decrease of E_{pan} in the region [49]. Hao et al. selected eight meteorological elements from 34 climate stations for 1958 to 2011 in order to analyze the spatial and temporal variations in the HRB. The results showed that the potential evapotranspiration in the region was negatively correlated with relative humidity and was positively correlated with diurnal temperature range [50]. Guo and Ren examined data observed from the evaporation pans of 117 meteorological stations for 1956 and 2000 and analyzed the changes in evaporation in the Huang-Huai-Hai River Basin. The findings showed that the direct climatic cause of the decrease in evaporation may be a reduction in sunshine duration and solar radiation. In addition, a reduction in wind speed and diurnal temperature range may also play an important role [51]. In summary, though most of the papers consider that the decrease of wind speed is a main factor causing the E_{pan} declining, but different literatures have different conclusions on the influence of other meteorological factors on the E_{pan} decreasing. So, this study aimed to analyze the trend of changes in E_{pan} in the HRB using data collected by 44 meteorological stations for 1961 to 2010, and to explore the temporal and spatial variation laws of E_{pan}, as well as the main driving forces of declining E_{pan} trend in the region.

2. Materials and Methods

The HRB is located between 112–120° E longitude and 35–43° N latitude, with the Bohai Sea to the east, the Yellow River to the south, the Yunzhong and Taiyue Mountains to the west, and the Inner Mongolia Plateau to the north. The total area of the HRB is 320,600 km^2, accounting for 3.3% of the total area of the country. The HRB spans eight provinces: Beijing, Tianjin, Hebei, Shanxi, Shandong, Henan, Inner Mongolia, and Liaoning. It is a political and cultural center and an economically developed region of strategic significance in China. The HRB has two major rivers: the Hai River and the Luan River. The Hai River, which is the main water system for the area, consists of the Ji Canal River, Chaobai River, North Canal, Yongding River, Daqing River, Ziya River, and Zhangwei River, as well as plains rivers, such as the Tuhai River and Majia River, each of which enters the sea individually. The Luan River includes itself and the rivers along the eastern coast of Hebei Province. The annual average temperature range of the basin is between 1.5 °C and 14 °C, the annual average relative humidity is between 50% and 70%, and the average annual precipitation is 539 mm (semi-arid climate). The annual average land-surface evaporation is 470 mm, and the water surface evaporation is 1100 mm. The geographical location and topographic distribution of the study area are shown in Figure 1.

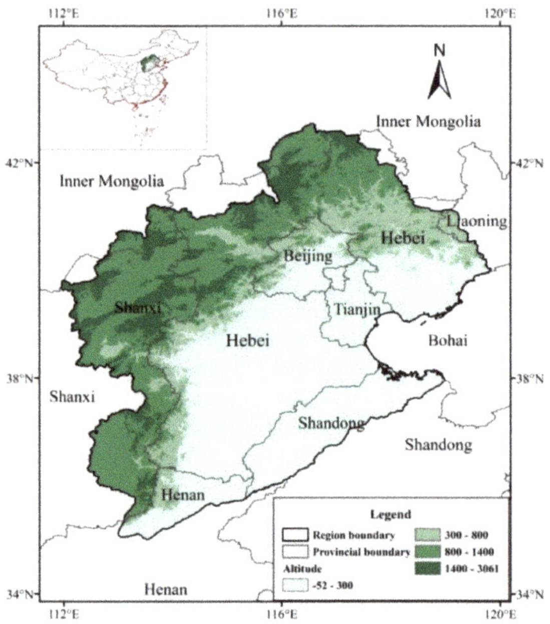

Figure 1. Location and topography of the HRB.

Meteorological data from the HRB and 55 meteorological stations in the surrounding area, provided by the National Meteorological Center of the China Meteorological Administration, were used in this study, including the daily average temperature, highest, and lowest temperatures, average relative humidity, sunshine duration, wind speed, precipitation, and daily evaporation from an evaporation pan 20 cm in diameter. In terms of missing data, the following rules were respected: when the daily data for five or more days were missing for a specific meteorological element in a month, the data of the entire month were considered missing; when the data for one or more months were missing, the data of the entire year were considered invalid. The time series of the data was from 1961 to 2010, and the length of the time series was 50 years. After excluding the station data that did not satisfy the time series requirements, data for 44 stations were retained.

In order to better analyze the seasonal changes of the elements, the data were divided into spring (March to May), summer (June to August), fall (September to November), and winter (December to February). The annual temperature (average, maximum, and minimum), annual average relative humidity, annual average sunshine duration, and annual average wind speed of each station were calculated based on the mean of the daily data. The annual E_{pan} and precipitation were calculated by summing the daily data. The same methods were applied to obtain the seasonal data for each element.

2.1. Canopy and k-means Clustering

In order to explore patterns in the spatial and temporal variations of the HRB climate, as well as geomorphological differences in the region and the climatic characteristics, this study selected eight representative indicators as references to categorize the HRB into several sub-regions. The indicators included geodetic coordinates (X and Y values), elevation, average temperature, precipitation, relative humidity, sunshine duration, and wind speed. The Canopy and *k*-means clustering technique was adopted. This method performs clustering in two stages. In Stage 1, the Canopy clustering algorithm is used to calculate the similarity of the objects and to categorize similar objects in the same subset (canopy). In Stage 2, the *k*-means clustering algorithm [52] is used to cluster the points in each canopy. Once Stage 1 is complete, the algorithm only needs to accurately cluster the points in each canopy, which greatly reduces the time spent on the accurate calculation of all data points that was performed in a traditional clustering algorithm. In addition, the number of canopies obtained in Stage 1 can be used as the K value in K-means clustering, which may minimize the irrational selection of the K value to a certain degree. The Canopy and *k*-means clustering technique not only greatly reduces the calculation of distance between points, the result is also more accurate when compared to general clustering methods [53,54].

MATLAB software (MATLAB 9.0, R2016a, MathWorks, Natick, MA, USA) was used to train the algorithm. The clustering results of the stations and their spatial distribution are presented in Figure 2. According to the clustering results, the HRB could be divided into six sub-regions (Figure 3). Detailed information for each sub-region is shown in Table 1.

Table 1. Basic information for the sub-regions of the HRB.

Sub-Regions	Climate Type	Number of Meteorological Stations	Ratio of Meteorological Stations (%)
I	Temperate monsoon climate	7	16
II	Temperate continental climate	7	16
III		7	16
IV		5	12
V	Temperate monsoon climate	9	20
VI		9	20

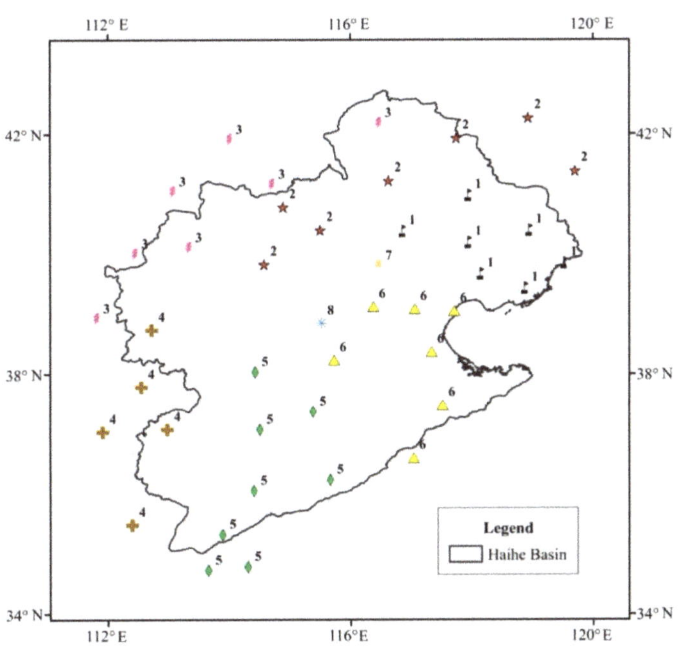

Figure 2. The Canopy and *k*-means clustering results and spatial distribution of the stations in the HRB.

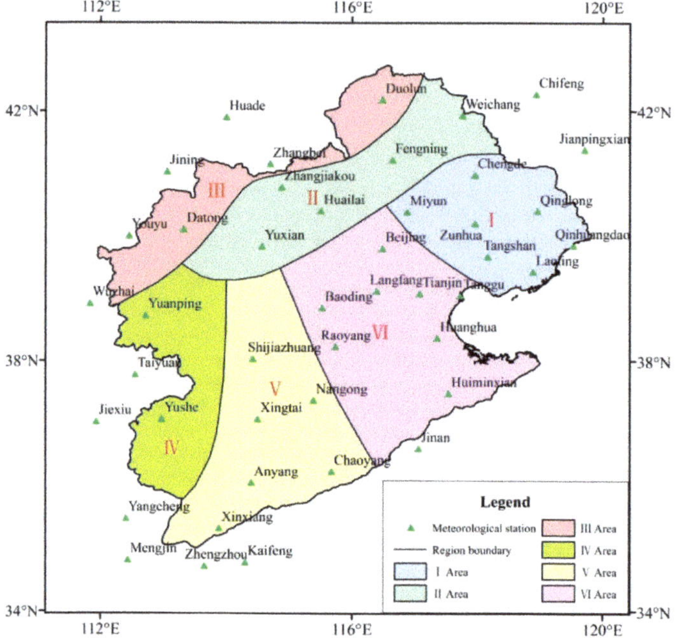

Figure 3. The sub-region categorization of the HRB.

2.2. Trend-Free Pre-Whitening Mann-Kendall Test (TFPW-MK)

The Mann-Kendall (M-K) test [55,56] is a non-parametric statistical method. Compared to parametric statistical methods, the M-K test does not require samples to follow a certain distribution, the results are not subject to interference from a few outliers, and the method is simple and efficient in calculations. Therefore, it is commonly used to detect trends in a series of values. For that reason, the M-K test is suitable for examining the trend of the hydrological variables in this study [57–59]. Assuming X_1, X_2, $\cdots$, X_n is a time series, n is the length of the time series; then, the M-K method defines the statistical variable S as follows:

$$S = \sum_{k=1}^{n-1} \sum_{j=k+1}^{n} sgn\left(x_j - x_k\right) \tag{1}$$

$$sgn\left(x_j - x_k\right) = \begin{cases} +1 & if\left(x_j - x_k\right) > 0 \\ 0 & if\left(x_j - x_k\right) = 0 \\ -1 & if\left(x_j - x_k\right) < 0 \end{cases} \tag{2}$$

where x_j and x_k are the measured values of years j and k, respectively; and k, $j \leq n$ and $k \neq j$.

When the number of samples is greater than 10, Z is calculated as follows:

$$Z = \begin{cases} \dfrac{S-1}{\sqrt{Var(S)}} & S > 0 \\ 0 & S = 0 \\ \dfrac{S+1}{\sqrt{Var(S)}} & S < 0 \end{cases} \tag{3}$$

$$Var(S) = \left[n(n-1)(2n+5) - \sum_{t} t(t-1)(2t+5)\right]/18 \tag{4}$$

where, Z is a normally distributed statistic, and *Var(S)* is the variance. If the Z value is positive, the data shows an increasing trend; if the Z value is negative, the data shows a decreasing trend [60]. Given the level of significance α, if $|Z| \geq Z_{1-\alpha/2}$, the null hypothesis is rejected, and the trend of the time series data (increasing or decreasing) is statistically significant at α.

The existence of serial correlation increases the probability that the M-K test will detect a significant trend [61,62]. The meteorological and hydrological data are mostly skewed and do not follow the same distribution, and there may be autocorrelation. Thus, in this paper, the TFPW method proposed by Yue et al. [62] is used to limit the influence of serial correlation; then, the significance of the time series is assessed by the M-K test.

The TFPW-MK steps are as follows:

Step 1. Use the Theil–Sen estimator(TSA) [63–66] to estimate the slope β of a trend in sample data. The slope of a trend is estimated using the TSA [63–66], and it is estimated as follows:

$$\beta = Median\left(\frac{X_j - X_i}{j - i}\right), \ \forall i < j \tag{5}$$

where β is the estimate of the slope of the trend, and X_i is the ith observation. The slope determined by the TSA is a robust estimate of the magnitude of a trend. Since the publication of Hirsch et al. [67], the TSA has been popularly employed to identify the slope of trends in hydrological time series [68–70].

Step 2. If $\beta = 0$, there is no need to continue trend analysis; if $\beta \neq 0$, it is assumed to be linear, and the sample data are detrended as:

$$Y_t = X_t - T_t = X_t - \beta t \tag{6}$$

Step 3. The lag-1 serial correlation coefficient r_1 of Y_t is calculated using Equation (7), and then the autocorrelation is removed by Equation (9).

$$r_1 = \frac{\frac{1}{n-1} \sum_{t=1}^{n-1} [X_t - E(X_t)][X_{t+1} - E(X_t)]}{\frac{1}{n} \sum_{t=1}^{n} [X_t - E(X_t)]^2} \tag{7}$$

$$E(X_t) = \frac{1}{n} \sum_{t=1}^{n} X_t \tag{8}$$

$$Y_t' = Y_t - r_1 Y_{t-1} \tag{9}$$

Step 4. The identified trend T_t and the residual Y_t'' are blended by

$$Y_t'' = Y_t' + T_t = Y_t' + \beta t \tag{10}$$

Step 5. Verify the significance of trend of the blended series using the MK test.

2.3. Spearman Correlation Coefficient

The Spearman correlation coefficient [71] is a nonparametric test method independent of distribution and can be used as an indicator to measure the relationship between two variables. If there are no repeated values in the data, the Spearman correlation coefficient is +1 or 1 when two variables are monotonously correlated. For a sample of size n, the n raw scores X_i, Y_i are converted to ranks x_i, y_i, and the Spearman correlation coefficient ρ can be calculated as [72,73].

$$\rho = 1 - \frac{6\sum d_i^2}{n(n^2 - 1)}, \tag{11}$$

where $d_i = x_i - y_i$ is the difference between the two ranks of each observation. The correlation degree between X_i, Y_i can be used according to the grading standards of ρ shown in Table 2 [74].

Table 2. Grading table of Spearman correlation coefficient (ρ).

Grading Standards	Correlation Degree
$\rho = 0$	no correlation
$0 < \lvert\rho\rvert \leq 0.19$	very week
$0.20 \leq \lvert\rho\rvert \leq 0.39$	weak
$0.40 \leq \lvert\rho\rvert \leq 0.59$	moderate
$0.60 \leq \lvert\rho\rvert \leq 0.79$	strong
$0.80 \leq \lvert\rho\rvert \leq 1.00$	very strong
1.00	monotonic correlation

3. Results

3.1. Trend and Significance Analysis

This study adopted the M-K test to analyze E_{pan}, temperature (average, maximum, and minimum), precipitation, relative humidity, sunshine duration, and wind speed of the HRB. The TFPW method was used to eliminate the trends and autocorrelation of meteorological sequence data before the M-K Test was applied. If the value of Z was positive, the data exhibited an upward trend, while if the value of Z was negative, the data exhibited a downward trend. The threshold of the significance level was defined as $\alpha = 0.05$. If the change trend of a given meteorological variable was found to be significant at this level, then $\lvert Z\rvert > Z_{\frac{\alpha}{2}} = 1.96$ [75]. The results of significance test of the interannual variations are shown in Figure 4, and that of the seasonal variations are shown in Table 3.

For TFPW-MK test detrending, the TSA method is used to calculate the magnitude of the trend of the meteorological variables. The rates of the meteorological elements of each sub-region for 1961 to 2010 are shown in Table 4.

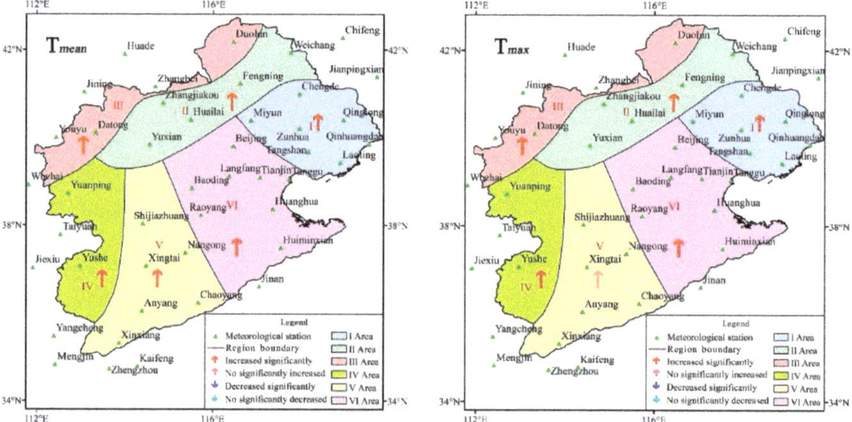

Figure 4. *Cont.*

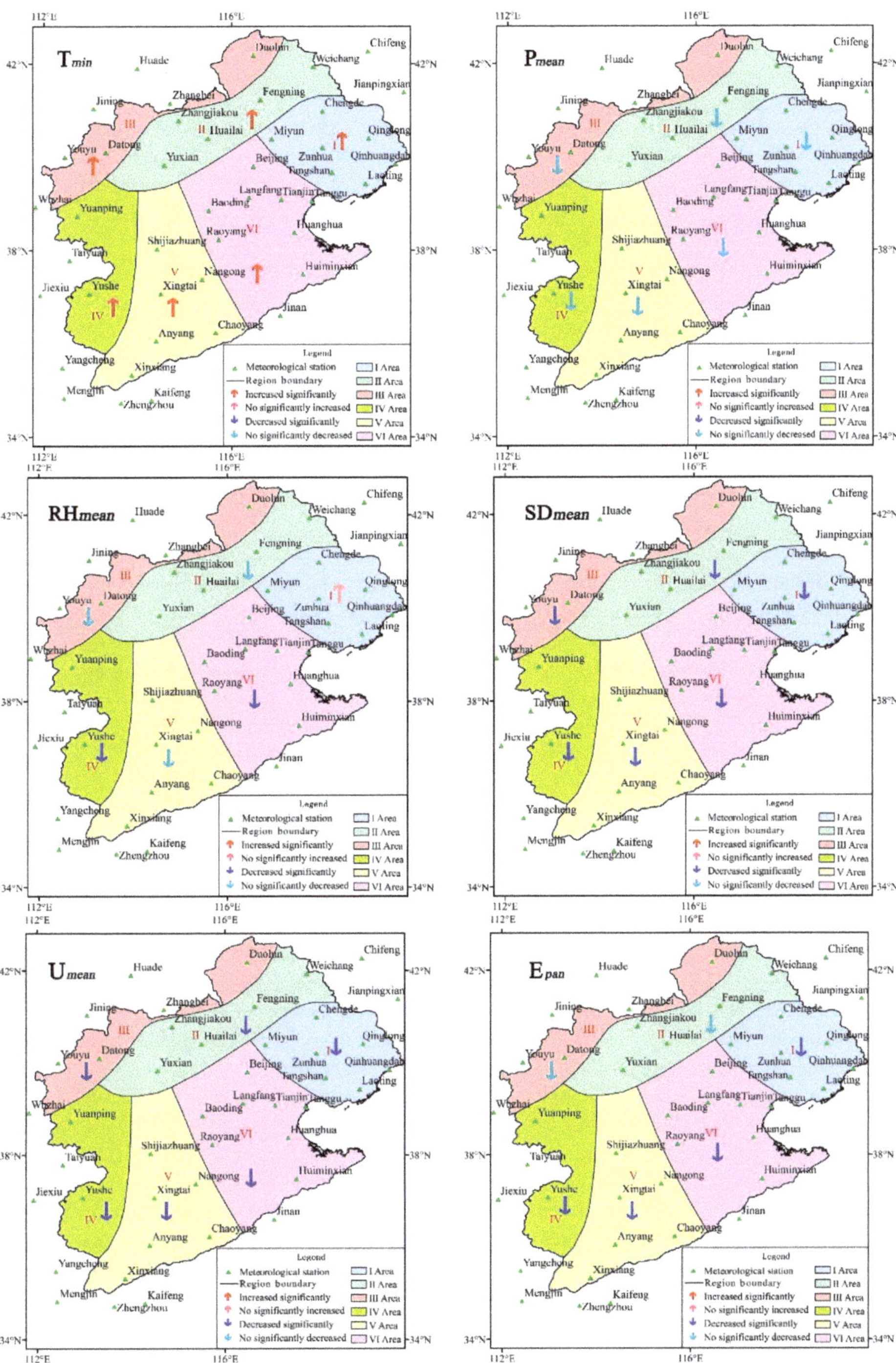

Figure 4. Results of significance test on the interannual variations of the meteorological elements.

As shown in Figure 4, the interannual average temperatures (T_{mean}) in sub-regions I to VI presented a significant upward trend. With the exception of sub-region V, the average maximum temperatures (T_{max}) of all sub-regions also increased significantly. The significance level of the trend of the average minimum temperatures (T_{min}) was consistent with that of the T_{mean}. The T_{mean} in spring and winter in sub-regions I to VI increased significantly; in summer, it increased significantly only in sub-regions II and III. In fall, it increased significantly in the majority of sub-regions (except for sub-region I). Only the significance levels for T_{max} in winter of sub-regions I to VI were consistent with those of T_{max}. The T_{min} of sub-regions I to VI in all four seasons significantly increased. However, it can be seen from Table 2 that the T_{min} rose more rapidly, followed by the T_{mean}, with the T_{max} rising the most slowly.

The interannual variations of average precipitation (P_{mean}) showed a downward trend of the six sub-regions: the decline rates were 22.50, 5.02, 4.07, 23.49, 5.35 and 19.71 mm/10a, respectively. However, the trend is not significant. The P_{mean} in spring showed an upward trend; only the trends of sub-regions II and VI were significant. In summer, it was found to be declining, except for sub-region V; the declining trends of sub-regions I, II, and VI were significant. The P_{mean} in fall showed an upward trend, except for sub-regions IV and V; The P_{mean} in winter showed an upward trend except for sub-region VI. The trends were not statistically significant in fall and in winter. As the decline of precipitation in summer offsets the increase of precipitation in spring, the interannual P_{mean} of the region showed a general downward trend.

The interannual variations of average relative humidity (RH_{mean}) of all sub-regions showed a downward trend, except for sub-region I; however only the trends of sub-regions IV and VI were statistically significant. The change rates of sub-regions I to VI were 0.20%/10a, −0.35%/10a, −0.30%/10a, −0.71%/10a, −0.56%/10a, and −0.78%/10a, respectively. As the increase in the RH_{mean} in fall and winter was larger than the sum of the reduction of the RH_{mean} in spring and summer, the RH_{mean} in sub-region I was found to be increasing. The RH_{mean} in spring, summer, and fall of sub-regions I to VI was found to be declining, except for sub-region I in fall and sub-region V in summer; however, the trends of the majority of sub-regions were not significant. The changes in the RH_{mean} of winter were not consistent across sub-regions I to VI, and none of the analyzed trends were significant.

The interannual variations of the average sunshine duration (SD_{mean}) in sub-regions I to VI showed a significant downward trend; the decline rates were 0.27, 0.13, 0.17, 0.29, 0.31, and 0.32 h/10a, respectively. In addition, the significance values for the changes in the SD_{mean} per season in sub-regions I to VI were consistent with those of the interannual variation.

The interannual average wind speed (U_{mean}) of sub-regions I to VI had significantly decreased; the rates were 0.16, 0.15, 0.28, 0.05, 0.17, and 0.16 m/(s·10a), respectively. The seasonal variation of the U_{mean} was showed a downward trend in all sub-regions, except for sub-region VI in summer. The U_{mean} in summer and fall in sub-region IV was not significant, while the declining trends of the U_{mean} per season of the other sub-regions were found to be significant.

The interannual E_{pan} of sub-regions I to VI was found to be decreasing over the research period at speeds of 55.2, 11.7, 24.4, 30.7, 82.6, and 65.7 mm/10a, respectively. However, the trends of E_{pan} for all sub-regions were significant except for sub-regions II and III. The E_{pan} of sub-region II presented an upward trend in summer, fall, and winter, while the E_{pan} of other sub-regions showed a downward trend in all seasons. Apart from sub-region IV, the E_{pan} of other regions was significantly decreased in spring. The significance test results in summer aligned with those of the interannual variations. The changes of E_{pan} in fall and winter in the majority of sub-regions were not statistically significant.

Table 3. Z-Value of the Mann-Kendall test on the meteorological variables ($\alpha = 0.05$).

Time	Sub-Regions	T_{mean}	T_{max}	T_{min}	P_{mean}	RH_{mean}	SD_{mean}	U_{mean}	E_{pan}
Spring	I	3.04 *	2.24 *	4.58 *	1.24	−0.15	−3.71 *	−7.04 *	−4.00 *
	II	3.58 *	2.46 *	5.14 *	2.06 *	−0.75	−2.63 *	−5.69 *	−2.61 *
	III	3.38 *	1.91	3.86 *	1.02	−0.90	−2.48 *	−7.29 *	−3.16 *
	IV	3.60 *	2.79 *	2.89 *	0.50	−2.43 *	−2.19 *	−3.41 *	−1.29
	V	3.25 *	0.84	5.59 *	0.90	−0.49	−2.31 *	−6.17 *	−3.45 *
	VI	3.86 *	2.24 *	5.84 *	1.97 *	−1.59	−2.86 *	−6.57 *	−3.28 *
Summer	I	1.77	1.44	3.28 *	−2.73 *	−1.00	−4.70 *	−5.24 *	−2.07 *
	II	3.18 *	2.59 *	5.15 *	−2.28 *	−2.07 *	−2.86 *	−2.58 *	0.10
	III	3.06 *	2.34 *	4.95 *	−1.92	−1.96	−2.99 *	−6.31 *	−0.17
	IV	1.39	1.19	3.03 *	−1.57	−0.94	−5.50 *	0.55	−2.26 *
	V	0.50	−1.37	3.65 *	0.18	0.15	−5.77 *	−5.89 *	−4.62 *
	VI	1.86	0.64	4.33 *	−2.19 *	−2.21 *	−5.62 *	−5.67 *	−3.25 *
Autumn	I	1.66	1.67	3.15 *	0.72	0.67	−5.02 *	−6.27 *	−3.31 *
	II	3.20 *	2.14 *	4.90 *	1.20	−1.69	−3.76 *	−4.55 *	0.07
	III	2.94 *	2.33 *	3.76 *	0.90	−0.03	−3.70 *	−6.37 *	−0.70
	IV	2.71 *	2.88 *	2.36 *	−1.04	−1.84	−3.40 *	−1.32	−0.20
	V	3.09 *	1.77	3.75 *	−1.52	−2.64 *	−3.58 *	−6.14 *	−1.41
	VI	2.63 *	1.82	4.08 *	0.25	−2.68 *	−4.82 *	−5.29 *	−2.43 *
Winter	I	4.08 *	2.58 *	5.25 *	0.22	1.79	−4.45 *	−6.79 *	−2.98 *
	II	4.12 *	2.79 *	5.42 *	0.87	0.25	−3.38 *	−5.52 *	2.07 *
	III	3.83 *	2.99 *	4.37 *	0.07	0.12	−4.10 *	−5.94 *	−0.52
	IV	4.55 *	2.59 *	5.40 *	0.85	0.64	−4.94 *	−4.55 *	−1.51
	V	4.13 *	1.20	5.81 *	0.42	−0.49	−4.30 *	−5.92 *	−2.23 *
	VI	4.40 *	2.46 *	5.60 *	−1.16	−0.85	−3.93 *	−6.27 *	−1.78

* Trends statistically significant at the 95% confidence level.

Table 4. Climate tendency rates of the meteorological elements per sub-region from 1961 to 2010.

Time	Sub-regions	T_{mean} (°C/10a)	T_{max} (°C/10a)	T_{min} (°C/10a)	P_{mean} (mm/10a)	RH_{mean} (%/10a)	SD_{mean} (h/10a)	U_{mean} (m/s/10a)	E_{pan} (mm/10a)
Interannual	I	0.3	0.2	0.4	−22.50	0.20	−0.27	−0.16	−55.2
	II	0.4	0.3	0.5	−5.02	−0.35	−0.13	−0.15	−11.7
	III	0.4	0.3	0.5	−4.07	−0.30	−0.17	−0.28	−24.4
	IV	0.3	0.3	0.3	−23.49	−0.71	−0.29	−0.05	−30.7
	V	0.3	0.1	0.4	−5.35	−0.56	−0.31	−0.17	−82.6
	VI	0.3	0.2	0.5	−19.71	−0.78	−0.32	−0.16	−65.7
Spring	I	0.3	0.3	0.4	3.84	−0.10	−0.24	−0.21	−21.8
	II	0.4	0.3	0.6	4.03	−0.31	−0.13	−0.19	−14.2
	III	0.3	0.2	0.4	2.40	−0.40	−0.17	−0.36	−18.1
	IV	0.3	0.3	0.3	1.66	−1.38	−0.17	−0.06	−12.7
	V	0.3	0.1	0.5	3.65	−0.34	−0.17	−0.19	−32.7
	VI	0.4	0.3	0.6	5.50	−0.90	−0.21	−0.23	−27.7
Summer	I	0.1	0.1	0.2	−33.14	−0.25	−0.39	−0.08	−13.7
	II	0.3	0.3	0.4	−13.95	−0.82	−0.14	−0.04	0.5
	III	0.3	0.2	0.3	−9.60	−0.79	−0.20	−0.17	−1.4
	IV	0.1	0.1	0.2	−13.00	−0.34	−0.45	0.02	−17.9
	V	0.0	−0.1	0.2	1.07	0.05	−0.52	−0.13	−40.0
	VI	0.1	0.1	0.3	−26.22	−0.80	−0.50	−0.10	−24.7
Autumn	I	0.1	0.1	0.2	2.45	0.17	−0.27	−0.12	−8.7
	II	0.3	0.2	0.5	2.89	−0.50	−0.11	−0.10	0.3
	III	0.3	0.2	0.4	2.81	−0.03	−0.16	−0.25	−2.8
	IV	0.2	0.3	0.2	−6.67	−1.09	−0.26	−0.03	−0.8
	V	0.2	0.2	0.4	−9.48	−1.42	−0.28	−0.14	−5.9
	VI	0.2	0.2	0.4	1.67	−1.11	−0.31	−0.12	−9.0
Winter	I	0.5	0.4	0.6	0.16	0.81	−0.20	−0.19	−4.5
	II	0.6	0.4	0.8	0.28	0.08	−0.09	−0.23	3.8
	III	0.6	0.5	0.7	0.04	0.10	−0.17	−0.35	−1.0
	IV	0.5	0.4	0.6	0.71	0.41	−0.32	−0.12	−4.2
	V	0.5	0.2	0.7	0.45	−0.35	−0.35	−0.19	−8.0
	VI	0.6	0.3	0.7	−0.87	−0.49	−0.26	−0.19	−3.6

3.2. Sensitivity Analysis

Changes in meteorological variables led to temporal and spatial fluctuations in E_{pan}, and their roles appeared to be different in different sub-regions. The influence of each meteorological element on changes in E_{pan} depended on two factors: the sensitivity of the meteorological elements toward E_{pan} and the change trend and corresponding significance level of the meteorological element. Therefore, it was necessary to analyze the meteorological elements that drive changes in E_{pan} for each sub-region, to qualitatively evaluate the contribution of each meteorological element on the changes of E_{pan}. The Spearman correlation coefficient was applied to qualitatively analyze the effects of T_{max}, T_{mean}, T_{min}, P_{mean}, RH_{mean}, SD_{mean}, and U_{mean} on E_{pan}. The results are shown in Table 5.

The sensitivity factors that caused changes in E_{pan} in the HRB were different. It can be seen from Table 5 that there were significant correlations between E_{pan} and T_{min}, P_{mean}, RH_{mean}, SD_{mean}, and U_{mean} in sub-region I. Although the correlations between E_{pan} and P_{mean} and RH_{mean} were significant, the trends of P_{mean} and RH_{mean} were not significant. Therefore, the meteorological elements that affected the interannual changes in E_{pan} of sub-region I were T_{min}, SD_{mean}, and U_{mean}. E_{pan} was found to have a negative correlation with T_{min}, which does not align with what is commonly expected. E_{pan} and SD_{mean} and U_{mean} were positively correlated. The significant decline in SD_{mean} and U_{mean} directly led to the significant decline of E_{pan}. Based on the above analysis, the significant decrease in E_{pan} in sub-region I was mainly due to the significant reduction in the SD_{mean} and U_{mean}. The Spearman correlation coefficient between E_{pan} and SD_{mean} was 0.75, indicating that the correlation is strong (see Table 2). The Spearman correlation coefficient between E_{pan} and U_{mean} was 0.46, indicating a moderate correlation. Thus, the primary factor affecting E_{pan} decline in sub-region I was SD_{mean}, followed by U_{mean}. According to the same analysis method, the factors responsible for E_{pan} decline in each sub-region were also analyzed. The primary factor affecting the decline of E_{pan} in sub-regions I, III, IV, V, and VI was SD_{mean}, followed by U_{mean}, while the reasons for the changes in E_{pan} in sub-regions II was the significant reduction in the SD_{mean}.

In addition, the contributing factors to E_{pan} in sub-regions I to VI for each season were also analyzed. In spring, the decrease in E_{pan} in sub-regions I to VI was primarily caused by a significant decrease in SD_{mean}, followed by a decline in U_{mean}. In summer, significant decreases in SD_{mean} and U_{mean} were the primary and secondary driving factors of E_{pan} in sub-regions I, V, and VI, respectively; the decrease in E_{pan} for sub-regions III and IV was mainly due to the significant decrease in SD_{mean}. Moreover, the change trend of E_{pan} in sub-region II was found to be the opposite to that for other sub-regions, which could be due to a significant increase in temperature, as well as a significant drop in P_{mean} and RH_{mean}. In fall, the decline in E_{pan} in sub-regions I, III, V, and VI was attributed to a significant reduction in SD_{mean} and U_{mean}, while the decline in E_{pan} in sub-region IV was caused by a significant reduction in SD_{mean}. In addition, the change trend of E_{pan} in sub-region II was the opposite of that for other sub-regions, which was mainly caused by a significant increase in T_{max} and T_{mean}. In winter, apart from sub-region II, where a significant increase in temperature resulted in an upward trend in E_{pan}, the reduction of E_{pan} in all sub-regions was caused by a significant reduction in SD_{mean} and U_{mean}.

In summary, the primary factor responsible for the decline in E_{pan} in the HRB was a reduction in sunshine duration, followed by a reduction in wind speed. The factors responsible for the reduction in E_{pan} in each sub-region were consistent with the overall reduction in E_{pan} of the HRB. However, the correlation between U_{mean} and E_{pan} was not significant in sub-region II, where the only influential factor was the decrease of sunshine duration.

Table 5. The Spearman correlation coefficients between Meteorological Elements and E_{pan} per Sub-Region.

Time	Sub-Regions	T_{mean}	T_{max}	T_{min}	P_{mean}	RH_{mean}	SD_{mean}	U_{mean}
Interannual	I	0.01	0.22	−0.35 *	−0.46 *	−0.68 *	0.75 *	0.46 *
	II	0.26	0.40 *	0.00	−0.64 *	−0.57 *	0.51 *	0.12
	III	0.05	0.15	−0.11	−0.52 *	−0.55 *	0.44 *	0.43 *
	IV	0.14	0.29 *	−0.13	−0.31 *	−0.47 *	0.61 *	0.59 *
	V	−0.06	0.37 *	−0.48 *	−0.38 *	−0.33 *	0.83 *	0.73 *
	VI	−0.16	0.19	−0.45 *	−0.40 *	−0.22	0.76 *	0.62 *
Spring	I	0.18	0.33 *	−0.20	−0.68 *	−0.71 *	0.84 *	0.55 *
	II	0.13	0.30 *	−0.19	−0.69 *	−0.66 *	0.68 *	0.48 *
	III	0.09	0.27	−0.20	−0.50 *	−0.67 *	0.62 *	0.61 *
	IV	0.45 *	0.65 *	0.09	−0.73 *	−0.73 *	0.86 *	0.68 *
	V	0.46 *	0.72 *	−0.05	−0.66 *	−0.72 *	0.84 *	0.60 *
	VI	0.22	0.46 *	−0.18	−0.78 *	−0.63 *	0.84 *	0.65 *
Summer	I	0.63 *	0.71 *	0.12	−0.42 *	−0.81 *	0.72 *	0.43 *
	II	0.68 *	0.79 *	0.30 *	−0.64 *	−0.83 *	0.64 *	0.34 *
	III	0.59 *	0.74 *	0.15	−0.66 *	−0.89 *	0.58 *	0.12
	IV	0.61 *	0.71 *	0.08	−0.48 *	−0.78 *	0.59 *	0.33 *
	V	0.52 *	0.76 *	−0.09	−0.50 *	−0.69 *	0.82 *	0.71 *
	VI	0.48 *	0.66 *	−0.03	−0.36 *	−0.58 *	0.78 *	0.53 *
Autumn	I	0.02	0.17	−0.32 *	−0.53 *	−0.70 *	0.70 *	0.53
	II	0.40 *	0.49 *	0.16	−0.60 *	−0.61 *	0.43 *	0.09
	III	0.19	0.37 *	−0.09	−0.56 *	−0.81 *	0.62 *	0.30 *
	IV	0.14	0.59 *	−0.35 *	−0.65 *	−0.83 *	0.76 *	0.59 *
	V	0.11	0.53 *	−0.39 *	−0.54 *	−0.68 *	0.81 *	0.35 *
	VI	−0.10	0.24	−0.44 *	−0.49 *	−0.53 *	0.77 *	0.55 *
Winter	I	0.16	0.40 *	−0.01	−0.60 *	−0.62 *	0.68 *	0.36 *
	II	0.68 *	0.72 *	0.62 *	−0.36 *	−0.46 *	0.35 *	−0.01
	III	0.49 *	0.55 *	0.44 *	−0.37 *	−0.74 *	0.31 *	0.29 *
	IV	0.26	0.57 *	0.01	−0.51 *	−0.71 *	0.70 *	0.57 *
	V	0.25	0.63 *	−0.11	−0.54 *	−0.72 *	0.75 *	0.47 *
	VI	0.21	0.47 *	0.00	−0.38 *	−0.61 *	0.66 *	0.43 *

* Trends statistically significant at the 95% confidence level.

4. Discussion

As noted above, when global temperature increases, the overall temperature of the HRB increases. However, there are differences in the spatial and temporal distribution. From 1961 to 2010, the lowest temperature increased approximately two times faster than the highest temperature, and approximately 1.5 times faster than the average temperature. This result is consistent with the results from an analysis of variations in annual temperatures by Zheng et al. [49] in the HRB for 1957 to 2001, but it is different from the results obtained by Salinger and Griffiths [76], which indicated that the lowest temperature rose approximately three times faster than the highest temperature globally from 1951 to 1998. Meanwhile, the average temperature, highest temperature, and lowest temperature in spring and winter increased much more quickly than the corresponding values in summer and autumn, for both the HRB and every sub-region. The temperatures in winter increased most significantly, three times faster than those in summer, and two times faster than those in autumn. Additionally, temperatures decreased to some extent in some areas of the HRB. For example, unlike the highest temperature in other sub-regions, which increased, the highest temperature in area V decreased.

The "evaporation paradox" also exists in the HRB. With respect to both the whole area and the sub-regions in the HRB, except for the slight increase in E_{pan} in sub-region II in autumn and winter, E_{pan} decreased, and this decline mostly occurred in spring and summer. This result agrees with the conclusions of Zheng et al. [49] and Liu et al. [77]. However, it contradicts with the findings in Liu et. al. [78] that from 1992 to 2007, E_{pan} significantly increased in North China, including the HRB. The correlation between the general decrease in E_{pan} and the temperature variation was weak, suggesting that the increase in temperature was not directly related to the decrease in E_{pan}. The main factors responsible for the decline in E_{pan} in the HRB included decreases in sunshine duration, which was the main factor, and in wind speed. This result differed from the conclusions of Zheng et al. [49], which stated that the decrease in wind speed is the main factor responsible for the decrease in E_{pan}. In addition, this result does not agree with the conclusions from Liu et al. [79] that in the semi-humid/semi-arid region of China (including the HRB), decreases in diurnal temperature range, sunshine duration and wind speed were found to be the main factors contributing to the pan evaporation declines. Liu et al. [78] concluded that wind speed and solar radiation are the main factors that led to the decline in pan evaporation in North China, which differs from our findings. However, it is generally accepted that wind speed is one of the main driving factors of the decrease in E_{pan} in the HRB. This statement is consistent with the conclusion that wind speed is one of the main factors driving the decrease in E_{pan} in areas such as the Canadian Prairies [31], the Cape Floristic Region in South Africa [32], and Australia [26].

The factors attributed to the decrease in sunshine duration and wind speed also vary among studies [36,80–82]. For sunshine duration, multiple studies have concluded that this decrease may be related to the increase in aerosols and other air pollutants [3,83]. In other studies, it is argued that the decrease may be related to the increase in cloud cover. In addition, several studies have reported a correlation between a decrease in sunshine duration and urbanization [84]. Recently, Wei Pan [85] reported that the number of haze days significantly increased in North China (including in the HRB). Therefore, a decrease in sunshine duration may be related to the increase in haze days in the HRB. Regarding the decrease in wind speed, conclusions of various areas also differ; however, the main consensus is that the decrease in wind speed may be related to variations in global circulation [86,87], as well as the increase in surface roughness caused by afforestation and urbanization near the observation sites [88]. Based on the present studies, it is difficult to determine the reasons for the decrease in wind speed, and further studies are required.

5. Conclusions

In this study, the Canopy and *k*-means clustering method was employed to categorize the HRB into six sub-regions. Then, 44 out of the 55 meteorological stations in the surrounding area that had relatively complete data were selected, and the trends and significance of the interannual and seasonal variations of the pan evaporation, temperature, precipitation, relative humidity, sunshine duration, and wind speed for 1961 to 2010 were analyzed using TFPW-MK. Based on this analysis, the sensitivities of the average, maximum, and minimum temperatures, and precipitation, relative humidity, sunshine duration, and average wind speed to E_{pan} were qualitatively analyzed using the Spearman correlation coefficient. In the whole basin, the primary cause of declining E_{pan} was a significant reduction in sunshine duration, followed by a significant reduction in wind speed. In sub-regions, E_{pan} showed a downward trend; however, the influential factors on E_{pan} reduction per sub-region were slightly different from those of the entire region. Except for sub-region II, which was only affected by sunshine duration, reductions in E_{pan} in other sub-regions were due to the joint influence of decreasing sunshine duration and wind speed.

In this paper, only a qualitative analysis was performed on the reduction of sunshine duration and wind speed in the HRB, and an explanation for this reduction is still lacking, and thus further research is needed.

Author Contributions: This paper is a joint effort by several authors. S.W. conceived and designed the paper's structure; B.L. performed the Canopy-Kmeans clustering; Z.Y. compiled the code for the Mann-Kendall Test; Z.Y. and D.M. analyzed the data; H.L. and S.L. drew the figures; and Z.Y. wrote the paper.

Funding: This work was supported by the National Natural Science Foundation of China (Grant No. 71573274; grant No. 51879066), the National Water Pollution Control and Management Science and Technology Major Project of China (2014ZX07203-008), the Science and Technology Project of Hebei Province (Grant No. 15227005D), and the Science and Technology Research and Development Program of Handan (1723209055-4).

Acknowledgments: The authors are very grateful to the editor and the anonymous reviewers whose suggestions significantly contributed to improving this work. The authors are also thankful for support from the Hebei University of Engineering, and Editage [www.editage.cn] for English language editing.

Conflicts of Interest: The authors declare no conflicts of interest.

References

1. IPCC. *Climate Change 2001: The Science Basis, Contribution of Working Group I to the Third Assessment Report of Inter-Government Panel on Climate Change*; The Press Syndicate of University of Cambridge: Cambridge, UK, 2001.
2. IPCC. *Climate Change 2007: Synthesis Report. Contribution of Working Groups I, II and III to the Fourth Assessment Report of the Intergovernmental Panel on Climate Change*; The Intergovernmental Panel on Climate Change: Geneva, Switzerland, 2007.
3. Roderick, M.L.; Farquhar, G.D. The cause of decreased pan evaporation over the past 50 years. *Science* **2002**, *298*, 1410–1411. [PubMed]
4. Kang, S.Z. Towards water and food security in China. *Chin. J. Eco Agric.* **2014**, *22*, 880–885. (In Chinese)
5. Gleick, P.H. Climate change, hydrology, and water resource. *Rev. Geophys.* **1989**, *27*, 329–344. [CrossRef]
6. Limjirakan, B.S.; Limsakul, A. Trends in Thailand pan evaporation from 1970 to 2007. *Atmos. Res.* **2012**, *108*, 122–127. [CrossRef]
7. Liu, X.M.; Zheng, H.X.; Liu, C.M.; Cao, Y.J. Sensitivity of the potential evapotranspiration to key climatic variables in the Haihe River Basin. *Res. Sci.* **2009**, *31*, 1470–1476. (In Chinese)
8. Thornthwaite, C.W. An approach toward a rational classification of climate. *Geogr. Rev.* **1948**, *38*, 55–94. [CrossRef]
9. Jung, M.; Reichstein, M.; Ciais, P.; Seneviratne, S.I.; Sheffield, J.; Goulden, M.L.; Bonan, G.; Cescatti, A.; Chen, J.; De Jou, R.; et al. Recent decline in the global land evapotranspiration trend due to limited moisture supply. *Nature* **2010**, *467*, 951–954. [CrossRef]
10. Eagleman, J.R. Pan evaporation, potential and actual evapotranspiration. *J. Appl. Meteorol.* **1967**, *6*, 482–488. [CrossRef]
11. Wang, S.Q.; Chen, N.X. *Water Resources Evaluation and Management*, 1st ed.; Water Resources and Electric Power Press: Beijing, China, 1996; ISBN 7-120-01998-8. (In Chinese)
12. Jhajharia, D.; Shrivastava, S.K.; Sarkar, D.; Sarkar, S. Temporal characteristics of pan evaporation trends under the humid conditions of northeast India. *Agric. For. Meteorol.* **2009**, *149*, 763–770. [CrossRef]
13. Azorin-Molina, C.; Vicente-Serrano, S.M.; Sanchez-Lorenzo, A.; McVicar, T.R.; Morán-Tejeda, E.; Revuelto, J.; Kenawy, A.E.; Martín-Hernández, N.; Tomas-Burguera, M. Atmospheric evaporative demand observations, estimates and driving factors in Spain (1961–2011). *J. Hydrol.* **2015**, *523*, 262–277. [CrossRef]
14. Tabari, H.; Marofi, S. Changes of pan evaporation in the west of Iran. *Water Resour. Manag.* **2011**, *25*, 97–111. [CrossRef]
15. Cohen, S.; Ianetz, A.; Stanhill, G. Evaporative climate changes at Bet Dagan, Israel, 1964–1998. *Agric. For. Meteorol.* **2002**, *111*, 83–91. [CrossRef]
16. Silva, V.D.P.R.D. On climate variability in Northeast of Brazil. *J. Arid Environ.* **2004**, *58*, 575–596. [CrossRef]
17. Peterson, T.C.; Golubev, V.S.; Groisman, P.Y. Evaporation losing its strength. *Nature* **1995**, *377*, 687–688. [CrossRef]
18. Golubev, V.S.; Lawrimore, J.H.; Groisman, P.Y.; Speranskaya, N.A.; Zhuravin, S.A.; Menne, M.J.; Peterson, T.C.; Malone, R.W. Evaporation changes over the contiguous United States and the former USSR: A reassessment. *Geophys. Res. Lett.* **2001**, *28*, 2665–2668. [CrossRef]
19. Roderick, M.L.; Farquhar, G.D. Changes in New Zealand pan evaporation since the 1970s. *Int. J. Climatol.* **2005**, *25*, 2031–2039. [CrossRef]
20. Zuo, H.C.; Li, D.L.; Hu, Y.Q.; Bao, Y.; Lv, S.H. The relationship between climate change trends and its changes in evaporation observed by evaporating dishes in the past 40 years in China. *Chin. Sci. Bull.* **2005**, *11*, 1125–1130. (In Chinese)
21. Sheng, Q. The Variation and Causes of Small Pan Evaporation over the Past 45 Years in China. Master's Thesis, Nanjing University of Information Science & Technology, Nanjing, China, May 2006. (In Chinese)

22. Liu, M.; Shen, Y.J.; Ceng, Y.; Liu, C.M. Changing trend of pan evaporation and its cause over the past 50 years in China. *Acta Geogr. Sin.* **2009**, *64*, 259–269. (In Chinese)
23. Ren, G.Y.; Guo, J. Change in pan evaporation and the influential factors over China: 1956–2000. *J. Nat. Res.* **2006**, *1*, 31–44. (In Chinese)
24. Tebakari, T.; Yoshitani, J.; Suvanpimol, C. Time-space trend analysis in pan evaporation over Kingdom of Thailand. J. *Hydrol. Eng.* **2005**, *10*, 205–215. [CrossRef]
25. Oguntunde, P.G.; Abiodun, B.J.; Olukunle, O.J.; Olufayoa, A.A. Trends and variability in pan evaporation and other climatic variables at Ibadan, Nigeria, 1973–2008. *Meteorol. Appl.* **2012**, *19*, 464–472. [CrossRef]
26. Roderick, M.L.; Farquhar, G.D. Changes in Australian pan evaporation from 1970–2002. *Int. J. Climatol.* **2004**, *24*, 1077–1090. [CrossRef]
27. Jovanovic, B.; Jones, D.A.; Collins, D. A high-quality monthly pan evaporation data set for Australia. *Clim. Chang.* **2008**, *87*, 517–535. [CrossRef]
28. Brutsaert, W.; Parlange, M.B. Hydrologic cycle explains the evaporation paradox. *Nature* **1998**, *396*, 30. [CrossRef]
29. Lawrimore, J.; Peterson, T.C. Pan evaporation in dry and humid regions of the United States. *J. Hydrometeorol.* **2000**, *1*, 543–546. [CrossRef]
30. Jaswal, A.K.; Rao, G.S.P.; De, U.S. Spatial and temporal characteristics of evaporation trends over India during 1971–2000. *Mausam* **2008**, *59*, 149–158.
31. Burn, D.H.; Hesch, N.M. Trends in evaporation for the Canadian prairies. *J. Hydrol.* **2007**, *336*, 61–73. [CrossRef]
32. Hoffman, M.T.; Cramer, M.D.; Gillson, L.; Wallace, M. Pan evaporation and wind run decline in the Cape Floristic Region of South Africa (1974–2005): Implications for vegetation responses to climate change. *Clim. Chang.* **2011**, *109*, 437–452. [CrossRef]
33. Yang, H.; Yang, D. Climatic factors influencing changing pan evaporation across china from 1961 to 2001. *J. Hydrol.* **2012**, *414*, 184–193. [CrossRef]
34. Cao, Y.Q.; Zhang, T.T.; Xu, D.; Yang, C.X. Analysis of evapotranspiration of temporal-space evolution in the Haihe Basin. *Res. Sci.* **2014**, *36*, 1489–1500. (In Chinese)
35. Bao, Z.X.; Yan, X.L.; Wang, G.Q.; Liu, C.S.; He, R.M. Mechanism of effect of meteorological factors in paradox theory of pan evaporation of Haihe River basin. *J. Water Res. Water Eng.* **2014**, *25*, 1–7. (In Chinese)
36. Li, X.C. Spatio-Temporal Variation of Actual Evapotranspiration in the Pearl, Haihe and Tarim River Basins of China. Ph.D. Thesis, Nanjing University of Information Science & Technology, Nanjing, China, May 2013. (In Chinese)
37. Liu, M.; Shen, Y.J. Change trend of hydrological elements in Haihe River Basin over the last 50 years. *J. China Hydrol.* **2010**, *30*, 74–77. (In Chinese)
38. Rong, Y.S.; Zhang, X.N.; Jiang, H.Y.; Bai, L.Y. Pan evaporation change and its impact on water cycle over the upper reach of the Yangtze River. *Chin. J. Geophys.* **2012**, *55*, 2889–2897. (In Chinese) [CrossRef]
39. Zhao, F.N.; Zhao, M.; Wang, Y.; Zhang, P.F. Variation characteristics of reference evapotranspiration and pan evaporation during 1960–2009 in Shiyang River Basin. *J. Arid Meteorol.* **2014**, *32*, 560–568. (In Chinese)
40. Li, L.P.; Li, Y.Y.; Liu, M.C. Change trend of pan evaporation and its causes in Shiyang River Basin during 1961–2005. *J. Desert Res.* **2012**, *32*, 832–841. (In Chinese)
41. Zhang, T.T. Analysis on Pan Evaporation Trend And Its Impacted Factors in Xiangjiang River Basin. Master's Thesis, Hunan Normal University, Changsha, China, May 2013. (In Chinese)
42. Rong, Y.S.; Zhou, Y.; Wang, W. Analysis of pan evaporation changes in the Huaihe River basin. *Adv. Water Sci.* **2011**, *22*, 15–22. (In Chinese)
43. Xie, P.; Chen, X.H.; Wang, Z.L.; Xie, Y.W. Comparison of actual evapotranspiration and pan evaporation. *Acta Geogr. Sin.* **2009**, *64*, 270–277. (In Chinese)
44. Wang, Z.L.; Qin, J.X.; Chen, X.H. Variation characteristics and impact factors of pan evaporation in Pearl River Basin. *Trans. CSAE* **2010**, *26*, 73–77. (In Chinese)
45. Qiu, X.F.; Liu, C.M.; Zeng, Y. Changes of evaporation in the recent 40 years over the Yellow River Basin. *J. Nat. Res.* **2003**, *18*, 437–441. (In Chinese)
46. Huang, Y.; Wang, Y. Analysis on temporal spatial distribution and inter-annual change of the evaporation capacity in Yunnan Province. *Hydrology* **2003**, *23*, 36–40. (In Chinese)

47. Xu, Z.X.; He, W.L. Analysis on the long-term trend of pan evaporation in the Yellow River Basin over the past 40 years. *Hydrology* **2005**, *06*, 6–11. (In Chinese)

48. Song, M.B.; Chen, J.Q.; Zhang, X.J.; Zhang, W.J. Pan evaporation trend in Yangtze River Basin from 1951 to 2000. *Water Res. Prot.* **2011**, *27*, 24–27. (In Chinese)

49. Zheng, H.; Liu, X.; Liu, C.; Dai, X.; Zhu, R. Assessing contributions to pan evaporation trends in Haihe River Basin, China. *J. Geophys. Res. Atmos.* **2009**, *114*, D24105. [CrossRef]

50. Hao, Z.C.; Yan, L.Z.; Ju, Q.; Dunzhu, J.C. Spatiotemporal characteristics of climate variation in different kinds of landforms of Haihe River Basin. *Res. Soil Water Conserv.* **2014**, *21*, 56–60. (In Chinese)

51. Guo, J.; Ren, G.Y. Recent change of pan evaporation and possible climate factors over the Huang-Huai-Hai watershed, China. *Adv. Water Sci.* **2005**, *16*, 666–672. (In Chinese)

52. Mac Queen, J.B. Some methods for classification and analysis of multivariate observations. In Proceedings of the 5th Berkeley Symposium on Mathematical Statistics and Probability, Berkeley, CA, USA, 21 June–18 July 1965 and 27 December 1965–7 January 1966; Volume 1, pp. 281–297.

53. Guha, S.; Rastogi, R.; Shim, K. Cure: An efficient clustering algorithm for large databases. *Inf. Syst.* **2001**, *26*, 35–58. [CrossRef]

54. Zhang, T.; Ramakrishnan, R.; Livny, M. BIRCH: An efficient data clustering method for very large database. In Proceedings of the 1996 ACM SIGMOD International Conference on Management of data, Montreal, QC, Canada, 4–6 June 1996; Volume 25, pp. 103–114.

55. Mann, H.B. Non-parametric tests against trend. *Econometrica* **1945**, *13*, 245–259. [CrossRef]

56. Kendall, M.G. *Rank Correlation Measures*; Charles Griffin: London, UK, 1975.

57. Kottegoda, N.T. *Stochastic Water Resources Technology*; Wiley: New York, NY, USA, 1980; ISBN 978-1-349-03467-3.

58. Burn, D.H.; Elnur, M.A.H. Detection of hydrologic trends and variability. *J. Hydrol.* **2002**, *55*, 107–122. [CrossRef]

59. Abdul Aziz, O.I.; Burn, D.H. Trends and variability in the hydrological regime of the Mackenzie River Basin. *J. Hydrol.* **2006**, *319*, 282–294. [CrossRef]

60. Mishra, A.K.; Singh, V.P. Changes in extreme precipitation in Texas. *J. Geophys. Res. Atmos.* **2010**, *115*, D14. [CrossRef]

61. Von Storch, H. Misuses of statistical analysis in climate research. In *Analysis of Climate Variability: Applications of Statistical Techniques*; Von Storch, H., Navarra, A., Eds.; Springer: New York, NY, USA, 1995; pp. 11–26.

62. Yue, S.; Pilon, P.; Phinney, B.; Cavadias, G. The influence of autocorrelation on the ability to detect trend in hydrological series. *Hydrol. Process.* **2002**, *16*, 1807–1829. [CrossRef]

63. Theil, H. A rank-invariant method of linear and polynomial regression analysis, I. *Nederlands Akad. Wetensch. Proc.* **1950**, *53*, 386–392.

64. Theil, H. A rank-invariant method of linear and polynomial regression analysis, II. *Nederlands Akad. Wetensch. Proc.* **1950**, *53*, 52–525.

65. Theil, H. A rank-invariant method of linear and polynomial regression analysis, III. *Nederlands Akad. Wetensch. Proc.* **1950**, *53*, 1397–1412.

66. Sen, P.K. Estimates of the regression coefficient based on Kendall's tau. *J. Am. Stat. Assoc.* **1968**, *63*, 1379–1389. [CrossRef]

67. Hirsch, R.M.; Slack, J.R.; Smith, R.A. Techniques of trend analysis for monthly water quality data. *Water Resour. Res.* **1982**, *18*, 107–121. [CrossRef]

68. Demaree, G.R.; Nicolis, C. Onset of Sahelian drought viewed as a fluctuation-induced transition. *Q. J. R. Meteorol. Soc.* **1990**, *116*, 221–238. [CrossRef]

69. Burn, D.H. Hydrologic effects of climatic change in west-central Canada. *J. Hydrol.* **1994**, *160*, 53–70. [CrossRef]

70. Gan, T.Y. Hydroclimatic trends and possible climatic warming in the Canadian Prairies. *Water Resour. Res.* **1998**, *34*, 3009–3015. [CrossRef]

71. Spearman's Rank Correlation Coefficient. Available online: https://en.wikipedia.org/wiki/Spearman%27s_rank_correlation_coefficient (accessed on 20 January 2019).

72. Myers, J.L.; Well, A.D. *Research Design and Statistical Analysis*, 2nd ed.; Lawrence Erlbaum Associates: Mahwah, NJ, USA, 2003; ISBN 0-8058-4037-0.

73. Maritz, J.S. *Distribution-Free Statistical Methods*, 1st ed.; Chapman & Hall: New York, NY, USA, 1981; ISBN 978-0-412-15940-4.
74. Spearman's correlation. Available online: http://www.statstutor.ac.uk/resources/uploaded/spearmans.pdf (accessed on 20 January 2019).
75. Wei, F.Y. *Modern Climate Statistical Diagnosis and Prediction Technology*, 2nd ed.; China Meteorological Press: Beijing, China, 2007; ISBN 9787502942991. (In Chinese)
76. Salinger, M.J.; Griffiths, G.M. Trends in New Zealand daily temperature and rainfall extremes. *Int. J. Climatol.* **2001**, *21*, 1437–1452. [CrossRef]
77. Liu, B.H.; Xu, M.; Henderson, M.; Gong, W.G. A spatial analysis of pan evaporation trends in China, 1955–2000. *J. Geophys. Res. Atmos.* **2004**, *109*, D15102. [CrossRef]
78. Liu, X.M.; Luo, Y.Z.; Zhang, D.; Zhang, M.H.; Liu, C.M. Recent changes in pan-evaporation dynamics in China. *Geophys. Res. Lett.* **2011**, *38*, 142–154. [CrossRef]
79. Liu, M.; Shen, Y.J.; Zeng, Y.; Liu, C.M. Trend in pan evaporation and its attribution over the past 50 years in China. *J. Geogr. Sci.* **2010**, *20*, 557–568. [CrossRef]
80. Kaiser, D.P.; Qian, Y. Decreasing trends in sunshine duration over China for 1954–1998: Indication of increased haze pollution? *Geophys. Res. Lett.* **2002**, *29*, 38-1–38-4. [CrossRef]
81. Yildirim, U.; Yilmaz, I.O.; Akinoglu, B.G. Trend analysis of 41 years of sunshine duration data for Turkey. *Turk. J. Eng. Environ. Sci.* **2013**, *37*, 286–305. [CrossRef]
82. Niroula, N.; Kobayashi, K.; Xu, J. Sunshine duration is declining in Nepal across the period from 1987 to 2010. *J. Agric. Meteorol.* **2015**, *71*, 15–23. [CrossRef]
83. Ren, J.; Lei, X.; Zhang, Y.; Wang, M.; Xiang, L. Sunshine duration variability in Haihe River Basin, China, during 1966–2015. *Water* **2017**, *9*, 770. [CrossRef]
84. Wang, Y.; Wild, M.; Sánchez-Lorenzo, A.; Manara, V. Urbanization effect on trends in sunshine duration in china. *Ann. Geophys.* **2017**, *35*, 839–851. [CrossRef]
85. Pan, W. Interdecadal Variation of Haze Days over China and the Atmospheric Causes in Recent 50 Years. Master's Thesis, Chinese Academy of Meteorological Sciences, Beijing, China, April 2017. (In Chinese)
86. Liu, X.M.; Zheng, H.X.; Zhang, M.H.; Liu, C.M. Identification of dominant climate factor for pan evaporation trend in the Tibetan Plateau. *J. Geogr. Sci.* **2011**, *21*, 594–608. [CrossRef]
87. Wang, Z.Y.; Ding, Y.H.; He, J.H.; Yu, J. An updating analysis of the climate change in China in recent 50 years. *Acta Meteorol. Sin.* **2004**, *62*, 228–236. (In Chinese)
88. Shen, Y.J.; Liu, C.M.; Liu, M.; Zeng, Y.; Tian, C.Y. Change in pan evaporation over the past 50 years in the arid region of China. *Hydrol. Process.* **2010**, *24*, 225–231. [CrossRef]

Article

Effects of the Three Gorges Project on Runoff and Related Benefits of the Key Regions along Main Branches of the Yangtze River

Yanjun Gao [1] and Yongqiang Zhang [2,*]

[1] School of Economics, Henan University of Science and Technology, Kaiyuan Ave. 263, Luoyang 471023, Henan, China; yanjungao@sohu.com

[2] Institute of Geographic Sciences and Natural Resources Research, the Chinese Academy of Science, Beijing 10010, China

* Correspondence: yongqiang.zhang2014@gmail.com

Received: 8 January 2019; Accepted: 31 January 2019; Published: 4 February 2019

Abstract: The Three Gorges Project (TGP) is the largest hydroelectric project in the world. It is crucial to understand the relationship between runoff regime changes and TGP's full operation after 2009 in the Yangtze River Basin (YRB). This paper defines core, extended and buffer areas of YRB, analyzes the effects of TGP on runoff anomaly (RA), runoff variation (RV) and change of coefficient of variation (CCV) between two periods (2003–2008 and 2009–2016), takes percentage of runoff anomaly (PRA) as the evaluation standard, assures alleviation effect on severe dry and wet years of the research area, and finally summarizes related benefits of flood control from TGP. Our results indicate the inter-annual fluctuation of runoff in the core and extended areas expanded, but reduced in the buffer areas, and the frequencies of severe dry and wet years alleviated in the buffer, core and extended areas. Generally, the extended and core areas become less wet, and the buffer areas become less dry. The RV and CCV are both strengthened in the extended and core areas, but are weakened in the buffer areas, and RV is well positively correlated (R^2 = 0.80) to CCV. Furthermore, the main benefits of TGP on flood control are remarkable in the reduction of disaster affected population, the decrease of agricultural disaster-damaged area, and the decline of direct economic loss. However, due to torrentially seasonal and non-seasonal precipitation, the sharp rebounds of three standards for Hubei and Anhui occurred in 2010 and 2016, and the percentage of agricultural damage area of five regions in the core and extended areas did not decline synchronously and performed irregularly. Our results suggest that the five key regions along the main branches of the Yangtze River should establish a flood control system and promote the connectivity of infrastructures at different levels to meet the significant functions of TGP. It is a great challenge for TGP operation to balance the benefits and conflicts among flood control, power generation and water resources supply in the future.

Keywords: Three Gorges Project; dam; runoff changes; flood control; Yangtze River; benefits

---

1. Introduction

Water is one of the most significant resources for a country's social, economic and environmental development, and most nations are facing different degrees of floods and drought threats caused by the imbalance of water distribution in the world [1]. In order to rebalance, redistribute and make full use of potential water resources, many countries made great efforts to construct dams to control floods, generate electricity, improve shipping capacity, and supply water, including Egypt, Japan, U.S., Canada, Australia, China, and so on. By the year 2000, about 45,000 large dams and 800,000 small ones have been built worldwide. Globally, discussions and arguments on large dams have increasingly emerged because of the potential and comprehensive impacts. They basically include enormous environmental changes of habitats inundation and fragmentation, extinction of local species of plants and fishes, ugly landscapes of water level fluctuation zone, sedimentation accumulation and capacity decrease of reseRVoir and regional climate changes, the apparent economic impacts of loss of old industries and enterprises and shortages of new ones, depressed livelihood of involuntary resettlement, high unemployment rates, infertile farmland located far away from economic centers, and the continuous social uncertainties of loss of unique historical and cultural heritage, spiritual sustenance and cultural integration of migration [2–7]. Among the changes, coastal erosion, caused by sediments reduction at river outlet, is a serious environmental problem in many nations. For example, the case study of Nestos River (Greece) indicated the sedimentation effect of construction and operation of two reseRVoirs (Thisavros and Platanovrysi) to coastal erosion, the sharp sediments decrease impacted sediments supply to basin outlet of river delta, the neighboring coast and the coastal morphology, which even inversed the erosion/accretion balance in the deltaic as well as the adjacent shorelines, from accretion predominated erosion to erosion predominates accretion, just within five years after the reseRVoirs' construction [8,9].

China's geographical and climatic location within monsoon zones determines its large difference in precipitation and potential evapotranspiration among the 31 provinces, municipalities and autonomous regions. The main spatial pattern is that more water exists in its southern parts than in its northern parts, and so 'too much water to control in Southern China' is one of the key issues for water resources management. Flooding has resulted in major disasters in both the midstream and downstream parts of YRB. For example, three big recorded floods occurred in YRB in 1934, 1954 and 1998, respectively [10]. The 1998 flood, called the 1998 Great Flood in China, caused significant losses in human lives and properties. Flooding threats became unpredictable under climate change and high intensities of human activities, which have resulted in large changes in hydrological processes, precipitation, runoff and groundwater in YRB.

To solve the flooding issue, China started large-scale dam construction from the 1950s. For instance, Sanmenxia Dam was built in the midstream of the Yellow River in 1954, and was aimed at flood control, irrigation, electric power generation, and shipping improvement; Gezhou Dam built in the downstream of the Yangtze River (YR) in 1971, aiming to provide hydropower generation and shipping improvement; Xiaolangdi Dam built in the downstream of the Yellow River in 1991, intended not only to provide flood and ice control, sedimentation reduction, and electric power generation, but for water supply and irrigation; the Three Gorges Project (TGP) was launched in the midstream of YR in 1994, aiming to provide flood control, shipping improvement, electric power generation, and so on.

Among the dams, TGP is well-known worldwide for its scale, cost, range, migration, ecological and environmental impact, and even for its controversial impacts on sustainability. It achieved primary impoundment in 2003 and the full operation in 2009. Afterwards many ecological and environmental problems emerged. One particular example was more frequent severe droughts that have occurred since its operation [11–16], which have impacted on Sichuan Province and Chongqing Municipality. This resulted in serious hydrological, ecological, and socioeconomic consequences in spatial pattern and temporal process [17–19], and also initiated controversial debates on TGP.

There were three research issues on the hydrological consequences of TGP. The first is the drought frequency of the Three Gorges ReseRVoir (TGR) and the relationship between TGP and the drought trends of YRB [20]. In the last two decades, YRB displayed significant vacillation between droughts and floods of TGR, indicated by the increasing drought trend in the upper reaches of YRB and TGP, with the drought evolution being inseparable from the background of the whole basin level. The second area has been water regulation and storage capacity between the lakes and YR and the response to TGP [21–24]. YR discharges into Dongting Lake in Hubei Province, Poyang Lake in Jiangxi Province, but receives inflow from the lakes from January to March. After the full operation of TGP under different dispatching modes, the weakened water from YR resulted in enhancing of the compensation ability of the lakes into YR in the neighboring provinces. The third research area is the impact of

TGP's impoundment on the flood and low stage adjustment in the midstream of YRB [25–27]. After the impoundment of TGP, the water storage capacities of provinces have experienced no change because the flood stage did not significantly decrease, although the flow discharge compensation of TGR improved the low flow stage, and the wharf change in the water stages was harmful to the improvement of the channel depth and the water storage of TGP. Meanwhile, the flow of YR decreased after flood season, increased in the dry season, and benefited from the peak shaving and flood control of TGP. Flow also decreased from July to August, and the annual runoff allocation also changed, which benefited flood control and water resources use in the midstream and downstream of YRB.

The rest of this paper is organized as follows: Section 2 reviews the basic conditions of TGP; Section 3 introduces the definition of the research area, data sources and methods; Referring to the hierarchy framework of Environmental Impact Assessment of Impacts on TGP [28], Section 4 presents TGP's effects on runoff changes between two periods of 2003–2008 and 2009–2016, including runoff anomaly, percentage of runoff anomaly, runoff variation, change of coefficient of variation, and then sums the related benefits of flood control of TGP from disaster affected populations, direct economic loss, agricultural disaster-damaged areas and the percentage of the agricultural disaster-damaged area; Section 5 proposes the discussions; Section 6 draws the main conclusions.

2. Basic Conditions of TGP

As the largest water generation project in the world, TGP located in the main stream of YR between Chongqing Municipality and Yichang City of Hubei Province (Figure 1), with a general storage capacity of 39.3×10^9 m^3 and a controllable storage capacity of 22.1×10^9 m^3 water, aimed at controlling effectively floods of the upstream of YR, with a general generation electricity of 84.7×10^9 kWh per year, aimed at alleviating the shortage of electricity in East, Central and South China, an amount equal to that produced by burning 50 million tons of coal, with a water depth improvement from 2.9 m for Chaotian Gate of Chongqing Municipality to 3.5–4.5 m for Yichang City, aimed at improving transport capacity from 3000 tons to 10,000 tons, and the transport capacity of the channels of YR with a length of 660 km, from Shanghai Port to Chongqing Port, was promoted from 1000 tons to 5000 tons.

The formal construction of TGP started on 14th December 1994, in Sandouping Town of Yichang City, Hubei Province. Then, TGP was implemented in three stages, 1993–1997 was the first stage, mainly for the construction preparation and damming up of YR, with the water level up to 90 m; 1998–2003 was the second one, mainly for primary impoundment, operation of first generator unit, and perpetual navigation of ship lock, with the water level up to 135 m; 2004–2009 was the third one, mainly for operation of all generator units and the accomplishment of whole project, with the water level up to 156 m (2006) and 175 m (2009).

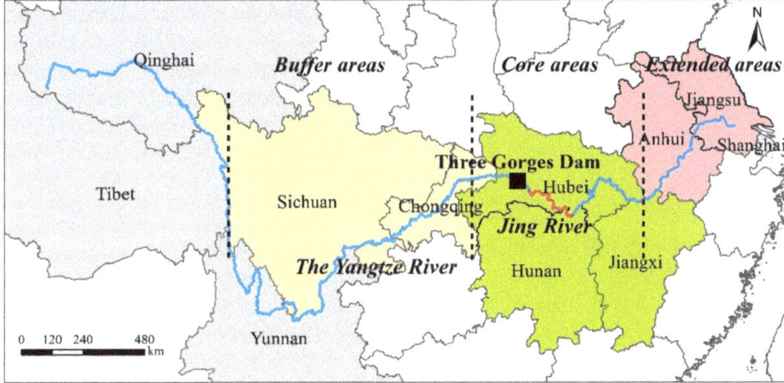

Figure 1. Locations of TGP, the Jing River (red), the Yangtze River (blue), and the definition of the research area, the buffer (yellow), core (green) and extended areas (pink).

3. Research Area, Data Sources and Methods

3.1. Research Area

According to the flow direction of main YRB branches (Qinghai to Shanghai), there are 5 regions in the upstream, which are Qinghai, Tibet, Sichuan, Yunnan and Chongqing, 3 regions in the midstream, which are Hubei, Hunan and Jiangxi, and 3 regions in the downstream, which are Anhui, Jiangsu and Shanghai. Meanwhile, according to China's traditional geographical regionalization of YR, the branches upward of Yichang City belong to the upstream; the branches between Yichang and Hukou County in Jiangxi Province belong to the midstream; and the branches below Hukou belong to the downstream. The major originally scheduled flood control of TGP mainly focused on the Jing River in Hubei Province (Figure 1), which cultivated two Plains of Jiang-Han and Dongting Lake, an important basis for commodity grain and aquatic products for China. In addition to the effect of extending back tail-waters of the Three Gorges Dam wall with 181 m height to the upstream after the full impoundment of 175 m, especially to Chongqing and Sichuan, the 7 key regions impacted by TGP regarding effects of runoff changes of YRB are defined as follows (Figure 1): the core areas including Hubei, Hunan and Jiangxi, the extended areas including Anhui and Jiangsu, and the buffer areas including Sichuan and Chongqing.

3.2. Data Sources

The information on TGP was obtained from the website of TGP (www.3g.gov.cn). The annual runoff data of the 7 regions and YRB were obtained from Changjiang and Southeast Rivers Water Resources Bulletin (2003–2016), the website of Changjiang Water Resources Commission, Ministry of Water Resources of China (www.cjw.gov.cn). The annual disaster data of the 5 regions were obtained from Bulletin of Flood and Drought Disasters in China (2006–2016), the website of Ministry of Water Resources of China (www.mwr.gov.cn).

3.3. Methods

This study compares and evaluates four runoff indices between two periods (2003–2008 and 2009–2016) for the 7 key regions in YRB, including runoff anomaly (RA), percentage of runoff anomaly (PRA), runoff variation (RV), and change of coefficient of variation (CCV).

The classic model on runoff (R) is calculated by Equation (1):

$$R = P - E - \Delta W \tag{1}$$

where R, P, E and ΔW is the annual runoff, precipitation, evaporation, and storage change of groundwater of the typical year of regions, respectively, and here the value of surface water is regarded as runoff [29].

The runoff anomaly (RA) is calculated by Equation (2). Then, to judge the dry or wet state of typical year, the percentage of runoff anomaly (PRA) is defined and categorized according to the standard of Equation (3) [30]:

$$RA = R_y - R_n \tag{2}$$

$$PRA = (R_y - R_n)/R_n \times 100 \tag{3}$$

where R_y and R_n is the runoff of typical year and the mean annual runoff during the period, respectively. The standards and the categories are as follows: the year with PRA < −20% belongs to a dry year; the year with −20% < PRA < −10% belongs to a less dry year; the year with −10% < PRA < 10% belongs to a normal year; the year with 10% < PRA < 20% belongs to a less wet year; the year with 20% < PRA belongs to a wet year.

The runoff variation (RV) is calculated by Equation (4):

$$RV = (R_{n2} - R_{n1})/R_{n1} \times 100 \tag{4}$$

where R_{n1} and R_{n2} is the mean annual runoff of two periods of 2003–2008 and 2009–2016, respectively.

The coefficient of variation (CV) and the change of CV (CCV) are calculated by Equations (5) and (6), respectively:

$$CV = \sigma/\mu \tag{5}$$

$$CCV = CV_{n2} - CV_{n1} \tag{6}$$

here σ and μ is the standard deviation and the mean annual runoff during the period, and CV_{n1} and CV_{n2} is CV of two periods of 2003–2008 and 2009–2016, respectively.

Three indices are used to present the benefits of TGP on flood control, the disaster-affected population (DAP), the direct economic loss (DEL), the agricultural disaster-affected area (ADAA), the agricultural disaster-damaged area (ADDA), and one index, the percentage of agricultural disaster-damaged area (PADDA), is used to evaluate the flood control effect, which is calculated by Equation (7):

$$PADDA = ADDA/ADAA \times 100 \tag{7}$$

where ADDA and ADAA represent the agricultural area with yield reduction including and over 30% and 10% affected by flood and waterlogging, respectively.

4. Results

After the full operation of TGP with 175 m height impoundment in 2009, an enormous flowing reseRVoir of the river-channel type crossing Chongqing and Hubei was formed with a general area of 1084 km^2. Considered the potential influence of water adjustment of TGR between wet and dry season, the runoff changes, including RA, PRA, RV and CCV, are compared between two periods of 2003–2008 and 2009–2016 among the core, extended and buffer areas.

4.1. Impacts of TGP on RA and PRA

4.1.1. Inter-Annual Fluctuation of RA Expanded in the Core and Extended Areas but Reduced in the Buffer Areas

When comparing 2003–2008 to 2009–2016, RA expanded in the core and extended areas but converged in the buffer areas, and the amplification extent was higher in the extended areas than that in the core areas. In the core areas (Figure 2a), RA for Hubei, Hunan and Jiangxi kept a remarkable increase trend, with an amplification of the fluctuation range of 0.25, 1.4 and 1.2 times, respectively. Similarly in extended areas (Figure 2b), RA also amplified, and the increased trend for Jiangsu (2.2 times) was stronger than that for Anhui (1.6 times). In contrast, the buffer areas differed (Figure 2c), Sichuan and Chongqing both converged on 1/3 of the fluctuation range of RA. The results indicate that TGP's effects were diversified among the core, extended and buffer areas [31]. Generally, the inter-annual fluctuation of RA expanded in the extended and core areas but reduced in the buffer areas, compared to pre-2009.

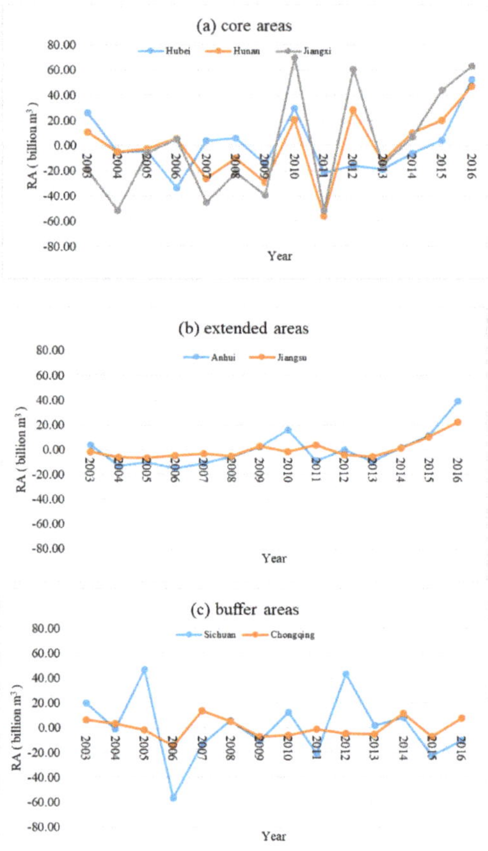

Figure 2. Change trends of the runoff anomaly (RA) among the core (**a**), extended (**b**) and buffer areas (**c**).

4.1.2. Alleviation on the Frequencies of Dry and Wet Years in the Buffer, Core and Extended Areas

YRB went into a drier phase with sharp fluctuations in 2009–2016 (Figure 3a and Table 1), with 62.5% dry and 37.5% wet years. Under the extreme drought, the alleviation effect of TGP at the whole basin scale was remarkable among the core, extended and buffer areas [32]. First, the core areas (Hubei, Hunan and Jiangxi) showed different patterns in PRA (Figure 3b and Table 1). Hubei became drier from 2003–2008 to 2009–2016. In comparison, Hunan and Jiangxi became wetter in 2009–2016. Second, the extended areas (Anhui and Jiangsu) shifted from dry to wet (Figure 3c and Table 1). For Anhui, dry and less dry years both decreased and correspondingly normal and wet years increased by 58%. Then for Jiangsu, the proportions of dry and less dry years decreased and those of normal, less wet and wet years increased by 62%, respectively. Third, the buffer areas (Sichuan and Chongqing) had a drying tendency, but with the majority of years remaining under the normal condition (Figure 3d and Table 1). Chongqing had a drying tendency, with less dry years increasing by 50%. In summary, Hubei, Sichuan and Chongqing went into a less dry or dry period after the full operation of TGP in 2009. Meanwhile, Hunan, Jiangxi, Anhui and Jiangsu went into a less wet or wet period.

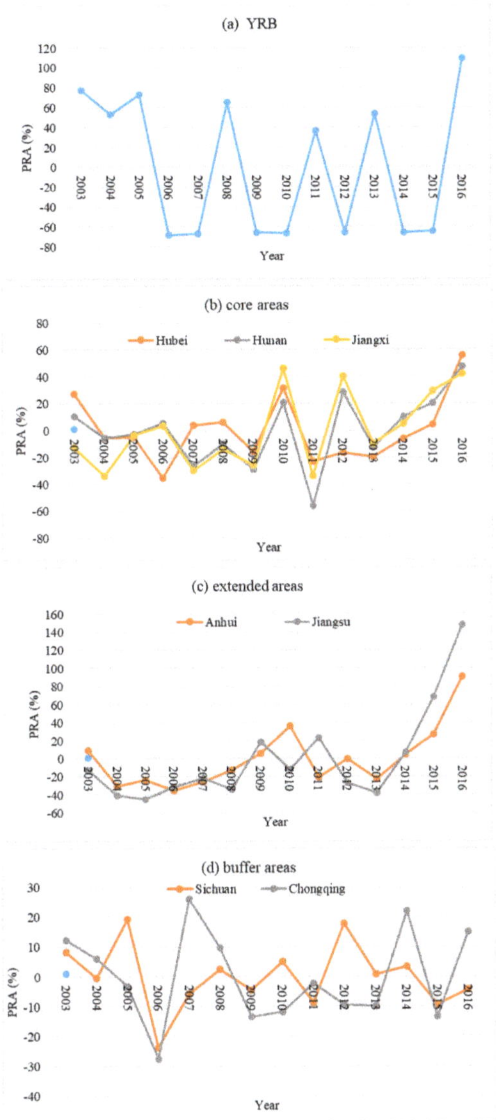

Figure 3. Change trends of the percentage of runoff anomaly (PRA) among YRB (**a**), the core (**b**), extended (**c**) and buffer areas (**d**).

Table 1. Frequencies of dry or wet years of YRB and the 7 regions based on PRA.

	Dry Years Proportion (%)		Less Dry Years Proportion (%)		Normal Years Proportion (%)		Less Wet Years Proportion (%)		Wet Years Proportion (%)	
	2003–2008	2009–2016	2003–2008	2009–2016	2003–2008	2009–2016	2003–2008	2009–2016	2003–2008	2009–2016
YRB	33.33	62.50							66.67	37.50
Sichuan	16.67			12.50	66.67	75.00	16.67	12.50		
Chongqing	16.67			50.00	33.33	25.00	33.33	12.50	16.67	12.50
Hubei	16.67	25.00		25.00	66.67	25.00			16.67	25.00
Hunan	16.67	25.00	16.67	12.50	50.00		16.67	12.50		50.00
Jiangxi	33.33	25.00	33.33	12.50	33.33	12.50				50.00
Anhui	66.67	25.00	16.67			16.67	37.50			37.50
Jiangsu	83.33	25.00	16.67	12.50		12.50		12.50		37.50

The occurrence of these droughts in YRB was related to drought and flood transformation at a large scale and the characteristics of precipitation evolution, and according to statistical data [33], YRB experienced a wet period around the 1980s, and went into a less wet period after 1999. During the last decade or so, the annual precipitation in YRB decreased by 10–12%. YRB's drought occurred just at this background of less wet climate. Our results are consistent with the findings that severe droughts occurred inevitably in the southwestern parts and the midstream and downstream of YRB, including the great drought in Sichuan and Chongqing in 2006, the severe drought in the southwestern China from 2009 to 2010, and the serious drought in midstream and downstream of YRB during 2010–2011. They indicate that there is no direct relationship between the drought disasters in YRB and TGP operation, and the drought in the 7 key regions in the past two decades was mainly driven by climate conditions. The TGP operation alleviated drought severity among the core, extended and buffer areas after 2009.

4.2. RV and CCV Both Strengthened in the Extended and Core Areas but Weakened in the Buffer Areas

The RV and CCV both apparently increased from 2003–2008 to 2009–2016 in all three areas (Figure 4a). The RV kept almost stable in Sichuan, decreased in Chongqing, increased slightly in Hubei and Hunan, but increased sharply in Jiangxi, Anhui and Jiangsu, with RV increasing by 31.4%, 43.1% and 78.5%, respectively. The RV mean in the extended areas is almost 4 and 20 times higher than that in the core and buffer areas, respectively. A similar situation occurred for CCV, which increased by 29.6% in Jiangsu, followed by Anhui, Hunan and Jiangxi, with their increased mean of 11~12%, then followed by Hubei with an increase of 7.0%, but instead decreased by −4.8% and −2.7% in Sichuan and Chongqing, respectively. The results indicate that RV and CCV were strengthened in the extended and core areas, but weakened in the buffer areas. Moreover, RV was strongly correlated ($R^2 = 0.80$) to CCV (Figure 4b), indicating that there existed coherence between the effects of TGP on runoff increase and the period fluctuation among the 7 regions.

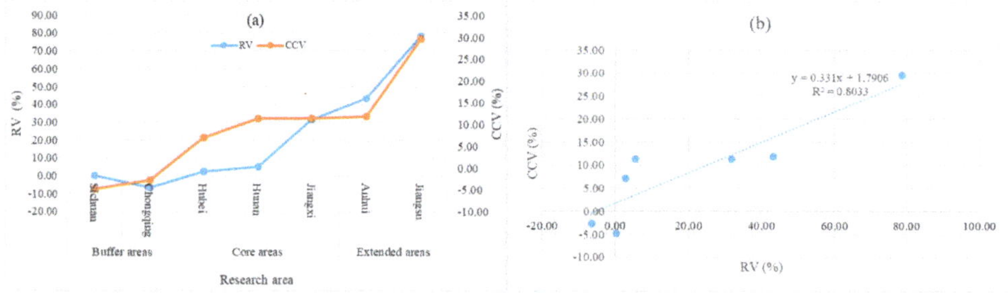

Figure 4. Change trends of RV and CCV (**a**) and their relationships (**b**) among the core, extended and buffer areas.

4.3. Related Benefits of Flood Control of TGP

As mentioned above, the originally scheduled function of flood control of TGP was mainly on the midstream and downstream of YRB, especially the Jing River (Figure 1). The benefits of flood control of TGP started in 2003, and were remarkably exhibited after TGP fully operated with 175m impoundment in 2010. Based on the available disasters data from 2006 to 2016, we analyzed the benefits of flood control in the core and extended areas, including DAP, DEL, ADDA, and PADDA.

4.3.1. Reduction of the Disaster-Affected Population (DAP)

DAP in the core and extended areas decreased sharply in the latter phase, especially after 2010. In the core areas (Figure 5a,b), DAP of the 3 regions remained higher and fluctuated remarkably during 2006–2008. DAP for Hubei declined remarkably from 2010 to 2014, but rebounded to the maximum in 2016; DAP for Hunan declined gradually with fluctuations until 2016, and DAP for Jiangxi declined sharply and became stable until 2016. The mean DAP for Hubei and Hunan decreased by 16% and 34% from 2006–2008 to 2009–2016, respectively. Similarly in the extended areas (Figure 5c), Anhui had a peak in 2007, then kept decreasing from 2010 to 2014, but rebounded during 2015–2016, and Jiangsu went into a steady period during in 2008–2014 after a sharp fall in 2007, then rebounded more than 10 times during 2015–2016, and the corresponding mean for Anhui and Jiangsu decreased by 25% and 67% during 2009–2016, respectively, compared to 2006–2008.

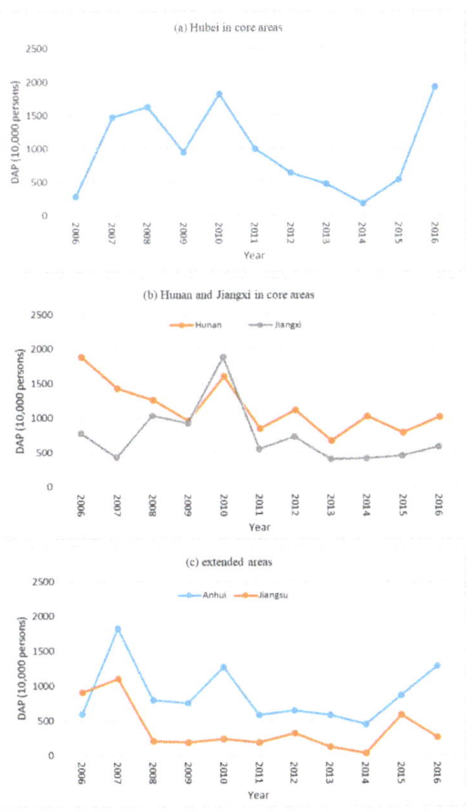

Figure 5. Change trends of DAP for Hubei (**a**), Hunan and Jiangxi (**b**) in the core and extended areas (**c**).

DAP for Hubei had a sharp rebound in 2016, and those of Anhui also emerged during 2015–2016. From the records of Bulletin of Flood and Drought Disasters in China (2015–2016), during the flood season (May–September), the mean precipitation of Hubei in 2016 increased by 19%, compared to the normal year. As a result, three big floods occurred in Yichang and other cities along YR. The mean precipitation of Anhui also increased by 14% and 27% in 2015 and 2016, resulting in three and two floods, respectively, which led to DAP rebounding in Hubei and Anhui.

4.3.2. Decline of the Direct Economic Loss (DEL)

In the core areas (Figure 6a,b), except for 2016, the mean DEL remained stable with little fluctuations, after the 2010 rebounds. The 2011–2015 mean DEL for Hubei, Hunan and Jiangxi accounted for 24%, 40% and 14% of the 2010 DEL, respectively. This is mainly caused by the abnormal conditions of 2010 (Bulletin of Flood and Drought Disasters in China, 2010). There were four heavily intensified precipitations that occurred in YRB in 2010, the first occurred in the southern parts of YRB in 13–28 June, the second and third occurred in the upstream of YRB in 15–25 July and 10–26 August, the fourth occurred in the Han River Basin in 10–26 August. These led to four big floods and the higher DEL in in Hubei, Hunan and Jiangxi, and the second and third precipitations even caused the biggest flooding events in the main branches of YRB since 1987, and the highest floods peak to TGR since TGP was constructed. Moreover, except for Anhui in 2016, the extended areas kept stable (Figure 6c), the DEL mean for Anhui and Jiangsu during 2009–2015 accounted for 56% and 29% of the peak in 2015, respectively. The DAP also illustrated the sharp rebound of DEL for Hubei and Anhui in 2016.

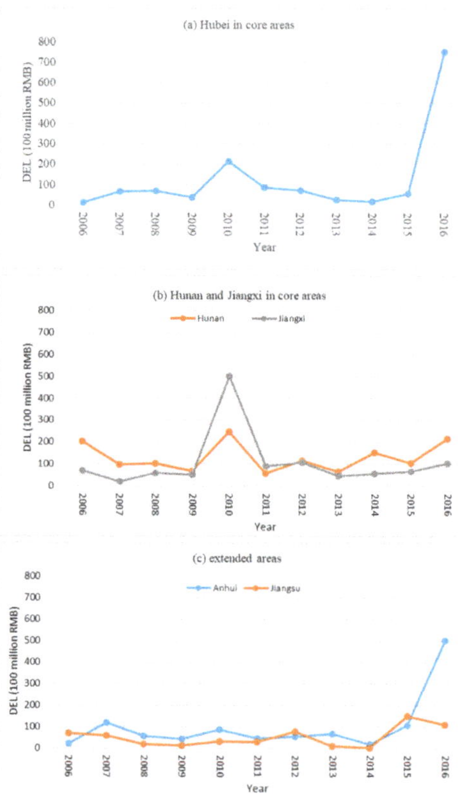

Figure 6. Change trends of DEL for Hubei (**a**), Hunan and Jiangxi (**b**) in the core and extended areas (**c**).

4.3.3. Decrease of the Agricultural Disaster-Damaged Area (ADDA) and Irregularities of the Percentage of Agricultural Disaster-Damaged Area (PADDA)

The ADDA of Hubei, Jiangxi and Hunan fluctuated remarkably in the core areas from 2006 to 2009 (Figure 7a,b). The ADDA for the three regions declined sharply and remained stable from 2011 to 2014 after the high rebound of 2010. The ADDA mean of Hubei decreased by 31% during 2009–2016, compared to 2006–2008; Hunan remained flat; Jiangxi increased by 27%. Nevertheless, in the extended areas (Figure 7c), the ADDA mean of Anhui and Jiangsu both decreased by 42% and 73% in 2009–2016, compared to 2006–2008, respectively. The reason why the ADDA of Hubei (2016) and Anhui (2015 and 2016) rebounded sharply was the same as that for DAP and DEL.

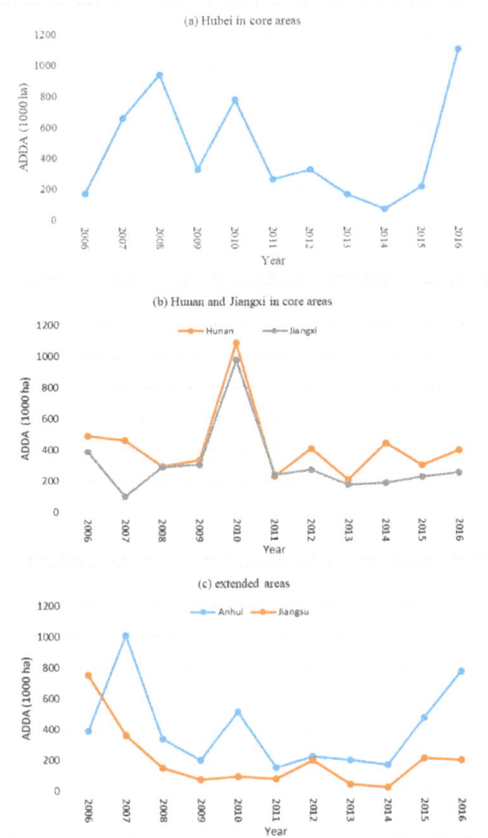

Figure 7. Change trends of ADDA for Hubei (**a**), Hunan and Jiangxi (**b**) in the core and extended areas (**c**).

Except for 2015 and 2016, the ADDA in the core areas after 2009 and in the extended areas after 2007 both showed a descending trend, respectively. However, the PADDA did not decline synchronously. In the core areas (Figure 8a,b), the PADDA of Hubei, Hunan and Jiangxi fluctuated sharply during 2006–2008, and during 2009–2016 Hubei went into a fluctuating period with a difference of 33% between the peak (62%, 2016) and the trough (28%, 2011), and Hunan and Jiangxi both went into a smooth period. The extended areas, however, performed differently (Figure 8c), the PADDA of Anhui and Jiangsu both showed an ascending trend from 2009 to 2016 with an increase of 32% and 18%, respectively. In summary, the TGP operation did not work for the middle and lower basin of YRB, since the PADDA among the five regions showed various trends: one (Hubei) fluctuated sharply, two (Hunan and Jiangxi) remained steady, two (Anhui and Jiangsu) increased.

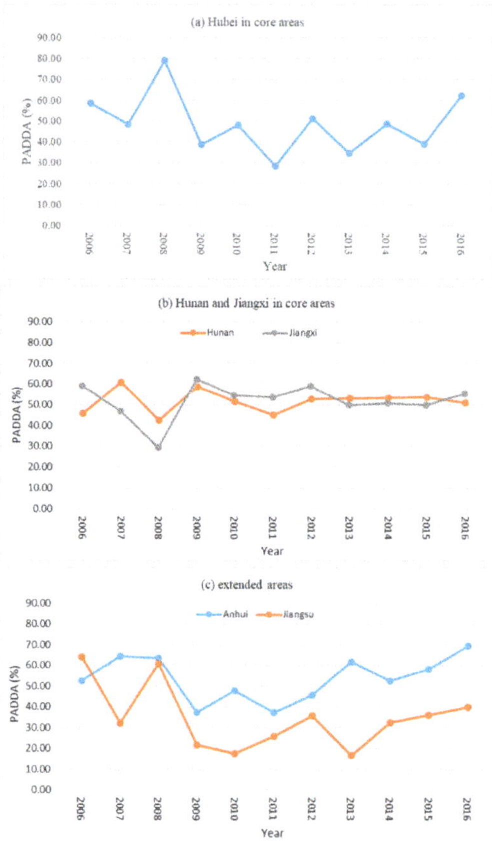

Figure 8. Change trend of PADDA for Hubei (**a**), Hunan and Jiangxi (**b**) in the core and extended areas (**c**).

5. Discussions

Based on the definition of research area and the analysis of four runoff indices, and by comparing the period of 2003–2008 with 2009–2016, we find that: (1) TGP operation after 2009 intensified the inter-annual fluctuation of RA in the extended and core areas, but reduced the fluctuation in the buffer areas. Based on the effective flood control of the upstream and water adjustment between the dry and wet seasons, TGP alleviated the frequencies of severe dry and wet years in the buffer, core and extended areas; (2) TGP strengthened RV and CCV in the extended and core areas but weakened them in the buffer areas, the RV was strongly correlated to CCV in the 7 regions, the runoff of the extended and core areas increased, and the corresponding inter-period of CCV was also amplified, especially in Jiangsu, Anhui and Jiangxi; (3) the benefits of TGP's flood control were mainly reflected by the reduction of DAP, the decline of DEL, and the decrease of ADDA in the core and extended areas.

Our results indicate some preliminary trends regarding TGP's effects on runoff regime changes. The following three aspects should be further considered and strengthened in the future. The first is the operating years of TGP. From 2009 to 2016, TGP was fully operated for only 7 years, while the comprehensive impacts and benefits need a long period of time to demonstrate. The second is the climate condition of YRB. YRB went into a drier period during 2009–2016, with sharp fluctuations. The capacity of water storage and reallocation of TGP between the dry and wet periods results in either drier or wetter influence in the core and extended areas. The third is the influence of seasonal and non-seasonal precipitation. One main function of TGP is flood controlling for the upstream of YRB. Nevertheless, besides the upstream, local precipitation is another flood source for the core and extended areas. Therefore, DAP, DEL, and ADDA for Hubei and Anhui rebounded sharply in 2010, 2015 and 2016, respectively, and the three indices presented a higher consistence when they faced the severe seasonal and non-seasonal precipitations than they would during a normal year, and big floods caused by regional heavy precipitations become a major disaster for Provincial Governments to solve. To cope with the extreme weather conditions under climate change of YRB, the standard for flood control of water conseRVancy infrastructures and projects in Hubei and Anhui need to improve.

Beyond for the climate change and the disasters of YRB from heavy precipitations, runoff variation and water reallocation, two significant problems highly influence the development of TGP. One is the resettlement displacement and economic development of TGR. The regulations developed by the State Council of China to guide the resettlement were the Regulations on Resettlement for the Construction of TGP on YR in 1993 (henceforth the 1993 regulations), which required rural resettlements were to move up the inundation line and draw back to the feasible farmlands in higher altitude mountainous areas, named exploration-oriented migration (*kaifa yimin in Chinese*), which meant rural households lived near their old houses and made a living on the reclaimed farmland, and resettled populations in urban areas were to employed in State-owned Enterprises [34]. However, due to the Great 1998 Flood, the impact of intensive land reclamation, deforestation and environmental degradation on TGR [35], the pressure of a resettlements arrangement increase from 1993 to 1998 of 20% beyond the scheduled 1.13 million registered in 1992, and the ceaseless problems in exploration resettlement, the State Council of China announced the adjustment to the 1993 regulations at a working meeting on TGP resettlement in May 1999 (henceforth the 2001 regulations), and resettled 190,000 rural residents (about 15% of the total) to 11 provinces outside TGR, including Shanghai, Jiangsu, Zhejiang, Anhui, Fujian, Jiangxi, Shandong, Hunan, Guangdong, Hubei and Sichuan. Meanwhile, with China's environmental protection policy being reformed and becoming increasingly strict since the beginning of the 1990s, lots of State-owned Enterprises were closed or restructured [36]. To make matters worse, substandard infrastructure, steep terrain and an unskilled workforce didn't attract investors, so few new enterprises set up in TGR, which led to the economic growth failing to keep pace with non-TGR regions in Chongqing and Hubei. Therefore the State Council of China initiated the Partner-ship Support Program (PSP) in 1992 and 2014, respectively. Another is sedimentation of TGR. In October 2002, in order to solve the problem of sedimentation from the upstream of YR, the State Council of China approved the Three Gorges Corporation building four hydropower stations along the Jinsha River, Wudongde, Baihetan, Xiluodu and Xiangjia Dam, to share the accumulation and decrease the velocity of sedimentation in TGR.

6. Conclusions

This study conducted a comprehensive analysis on TPG operation impacts by comparing the period of 2009–2016 with 2003–2008. Our results indicated the primary effects of TGP operation after 2009 on runoff changes of the 7 regions in the following three aspects. Firstly, the inter-annual fluctuation of the runoff anomaly in the extended (Anhui and Jiangsu) and core areas (Hubei, Hunan and Jiangxi) expanded and the buffer areas (Sichuan and Chongqing) converged, while the decrease trend was remarkable from the extended, core to buffer areas. With a macro-background of climate change, YRB went into a drier period with sharp fluctuations, while TGP alleviated the frequencies of dry or wet years in the research area. The core (except for Hubei) and extended areas both had a tendency of becoming less wet, however, Hubei and the buffer areas went into a less dry period. Secondly, the runoff variation and the change of CV both strengthened in the extended and core areas but weakened in the buffer areas, and the RV presented a highly positive correlation to CCV. The inter-period increase of runoff was sharp in Jiangsu, Anhui and Hunan, but the corresponding

CCV also amplified sharply, which benefited drought alleviation but intensified the flood control risk of the inter-period. Thirdly, the general benefits of flood control of TGP mainly exhibited in the reduction of the DAP, the decline of DEL and the decrease of ADDA in the core and extended areas. Nevertheless, the PADDA of the 5 regions performed in irregularities after 2009, as Hubei went into a sharp fluctuating period, Hunan and Jiangxi both went into a smooth period, Anhui and Jiangsu both showed an ascending trend, respectively, instead of keeping descending trends during 2006–2008.

Moreover, due to big floods caused by heavy precipitations, the sharp rebounds occurred in the DAP, DEL and ADDA for Hubei and Anhui in 2010 and 2016, respectively. So flood control and disaster mitigation capacity within the core and extended areas not only depend on TGP but also rely on the intra system. TGP aims to control the big floods from the upstream of YR, and the intra-system focuses on controlling the seasonal and non-seasonal heavy precipitations. Regional flood control and disaster alleviation is comprehensive and systemic, e.g., agricultural flood control in YRB, one guarantee comes from the regulation and storage of water conseRVancy projects of main branches, tributaries and rivers, while another comes from the storage and reallocation capacity of lakes, reseRVoirs, and ponds connecting with the farmlands.

After experiencing the stages of argument, construction and operation, the Chinese Central Government became cautious towards large dam construction for hydropower exploration [37,38], and the management countermeasures on TGP and YR also became more scientific, including resettlement support [39], sediment sharing in the upstream, securities management, economic supports to TGR, and TGP going into a stage of rehabilitation and improvement. In the future, in order to match and promote the scheduled functions of TGP on flood control, the key regions along the main branches of YR need to build a system and strengthen the connectivity between projects and infrastructure from the county, city, and provincial levels, to the regional level, and even to the national level. Under the drought intensification of YRB in recent years, it is a great challenge for TGP operation to balance the benefits and conflicts among flood control, power generation and water resources distribution in the key regions of the research area.

Author Contributions: Y.Z. conceived this study; Y.G. gathered all the information, data and prepared the first draft of the paper; Y.Z. improved and revised the manuscript.

Acknowledgments: This work was supported by the CAS Pioneer Hundred Talents Program, National Social Science Foundation of China (18BJY068) and National Scholarship Foundation of China (20175087).

Conflicts of Interest: The authors declare no conflicts of interest.

References

1. Hadjigeorgalis, E. A place for water markets: Performance and challenges. *Rev. Agric. Econ.* **2009**, *31*, 50–67. [CrossRef]
2. Baghel, R.; Nüsser, M. Discussing large dams in Asia after the world commission on dams: Is a political ecology approach the way forward? *Water Altern.* **2010**, *3*, 231–248.
3. Dudgeon, D. Requiem for a river: Extinctions, climate change and the last of the Yangtze. *Aquat. ConseRV. Mar. Freshw. Ecosyst.* **2010**, *20*, 127–131. [CrossRef]
4. McDonald-Wilmsen, B.; Webber, M. Dams and displacement: Raising the standards and broadening the research agenda. *Water Altern.* **2010**, *3*, 142–161.
5. Abbink, J. Dam controversies: Contested governance and developmental discourse on the Ethiopian Omo River dam. *Eur. Assoc. Soc. Anthropol.* **2012**, *20*, 125–144. [CrossRef]
6. Yüksel, I. Dams and hydropower for sustainable development. *Energy Sources Part B* **2009**, *4*, 100–110. [CrossRef]
7. Beck, M.W.; Claassen, A.H.; Hundt, P.J. Environmental and livelihood impacts of dams: Common lessons across development gradients that challenge sustainability. *Int. J. River Basin Manag.* **2012**, *10*, 73–92. [CrossRef]
8. Andredaki, M.; Georgoulas, A.; Hrissanthou, V.; Kotsovinos, N. Assessment of reseRVoir sedimentation effect on coastal erosion in the case of Nestos River, Greece. *Int. J. Sediment Res.* **2014**, *29*, 34–48. [CrossRef]

9. Samaras, A.G.; Koutitas, C.G. Modelling the impact on coastal morphology of the water management in transboundary river basins: The case of River Nestos. *Manag. Environ. Qual. Int. J.* **2008**, *19*, 455–466. [CrossRef]

10. Ye, Q.; Glantz, M.H. The 1998 Yangtze Floods: The use of short-term forecasts in the context of seasonal to inter-annual water resource management. *Mitig. Adapt. Strateg. Glob. Chang.* **2005**, *10*, 159–182. [CrossRef]

11. Jiao, L. Scientists line up against dam that would alter protected wetlands. *Science* **2009**, *326*, 508–509. [CrossRef] [PubMed]

12. HeRVe, Y.; Claire, H.; Lai, X.; Stéphane, A.; Li, J.; Sylviane, D.; Muriel, B.; Chen, X.; Huang, S.; Burnham, J.; et al. Nine years of water resources monitoring over the middle reaches of the Yangtze River, with ENVISAT, MODIS, Beijing-1 time series, altimetry data and field measurements. *Lake Reserve Manag.* **2011**, *16*, 231–247. [CrossRef]

13. Huang, S.F.; Li, J.G.; Xu, M. Water surface variations monitoring and flood hazard analysis in Dongting Lake are using long-term Terra/MODIS data time series. *Nat. Hazards* **2012**, *62*, 93–100. [CrossRef]

14. Liu, Y.; Wu, G.; Zhao, X. Recent declines of the China's largest freshwater lake: Trend or regime shift? *Environ. Res. Lett.* **2013**, *8*, 014010. [CrossRef]

15. Wang, J.; Sheng, Y.; Tong, T.S.D. Monitoring decadal lake dynamics across the Yangtze Basin downstream of Three Gorges Dam. *Remote Sens. Environ.* **2014**, *152*, 251–269. [CrossRef]

16. Hu, C.; Fang, C.; Cao, W. Shrinking of Dongting Lake and its weakening connection with the Yangtze River: Analysis of the impact on flooding. *Int. J. Sediment Res.* **2015**, *30*, 256–262. [CrossRef]

17. Yang, G.; Zhu, C.; Jiang, Z. *Yangtze ConseRVation and Development Report 2011*; Yangtze Press: Wuhan, China, 2011.

18. Nicol, S.; Roach, J.K.; Griffith, B. Spatial heterogeneity in statistical power to detect changes in lake area in Alaskan National Wildlife Refuges. *Landsc. Ecol.* **2013**, *28*, 507–517. [CrossRef]

19. Roach, J.; Griffith, B. Climate-induced Lake drying causes heterogeneous reductions in waterfowl species richness. *Landsc. Ecol.* **2015**, *30*, 1005–1022. [CrossRef]

20. Ma, D.; Liu, M.; Ju, Y. Evolution characteristics of droughts and floods in Yangtze River and Three Gorges ReseRVoir area in recent 542 Years. *Meteorol. Sci. Technol.* **2016**, *44*, 622–630.

21. Li, J.; Zhou, Y.; Ou, C.; Cheng, W.; Yang, Y.; Zhao, Z. Evolution of water exchange ability between Dongting Lake and Yangtze River and its response to the operation of the Three Gorges ReseRVoir. *Acta Geogr. Sin.* **2013**, *68*, 108–119.

22. Deng, J.; Fan, S.; Pang, C.; Liu, C. Adjustment of regulation and storage capacity of lakes in the middle Yangtze River Basin during impoundment of Three Gorges ReseRVoir. *J. Yangtze River Sci. Res. Inst.* **2018**, *35*, 147–152.

23. Zhang, W.; Gao, Y.; Xu, Q.; Yuan, J. Changes in dominant discharge and their influential factors in the middle and lower reaches of Yangtze River after the Three Gorges Dam impoundment. *Adv. Water Sci.* **2018**, *29*, 331–338.

24. Huang, W.; Wang, W.D. Effects of three gorges dam project on Dongting lake wetlands. *Acta Ecol. Sin.* **2016**, *36*, 6345–6352.

25. Han, J.; Sun, Z.; Yang, Y. Flood and low stage adjustment in the middle Yangtze River after impoundment of the Three Gorges ReseRVoir (TGR). *J. Lake Sci.* **2017**, *29*, 1217–1226.

26. Zhao, W.; Feng, B.; Chen, Y. Influence of impoundment of upstream reseRVoirs on inflow of Three Gorges ReseRVoir from August to October. *Yangtze River* **2013**, *44*, 1–4.

27. Zhang, D.; He, S.; Hu, G.; Hu, G. Preliminary analysis on runoff variation of middle Yangtze River after impoundment of Three Gorges ReseRVoir. *Yangtze River* **2013**, *1*, 1–3.

28. Tullos, D. Assessing the influence of environmental impact assessments on science and policy: An analysis of the Three Gorges Project. *J. Environ. Manag.* **2009**, *90*, S208–S223. [CrossRef]

29. Rui, X. *The Principles of Hydrology*; Higher Education Press: Beijing, China, 2013.

30. Hu, X. Analysis of time-space distribution regulation and evolution tendency off runoff of main rivers in Gansu Province. *Adv. Earth Sci.* **2000**, *15*, 516–521.

31. Wang, J.; Sheng, Y.; Gleason, C.J.; Wada, Y. Downstream Yangtze River levels impacted by Three Gorges Dam. *Environ. Res. Lett.* **2013**, *8*, 1–9. [CrossRef]

32. Fan, Y.; Zhang, W.; Han, J.; Yu, M. The typical meandering river evolution adjustment and its driving mechanism in the downstream reach of TGR. *Acta Geogr. Sin.* **2017**, *72*, 420–431.

33. Wang, W.; Zheng, G. (Eds.) *Annual Report on Actions to Address Climate Change (2011): Durban Dilemma and China's Strategic Options*; Social Sciences Academic Press: Beijing, China, 2011.

34. McDonald-Wilmsen, B. Development-induced displacement and resettlement: Negotiating fieldwork complexities at the Three Gorges Dam, China. *Asia Pac. J. Anthropol.* **2009**, *10*, 283–300. [CrossRef]

35. Tan, Y.; Bryan, B.; Hugo, G. Development, land-use change and rural resettlement capacity: A case study of the Three Gorges Project, China. *Aust. Geogr.* **2005**, *36*, 201–220. [CrossRef]

36. Duan, Y.; Steil, S. China Three Gorges project resettlement: Policy, planning and implementation. *J. Refugee Stud.* **2003**, *16*, 422–443.

37. Magee, D.; McDonald, K. Beyond Three Gorges: Nu Rivers hydropower and energy decision politics in China. *Asian Geogr.* **2006**, *25*, 39–60.

38. Xie, L.; Van Der Heijden, H.A. Environmental movements and political opportunities: The case of China. *Soc. Mov. Stud.* **2010**, *9*, 51–68. [CrossRef]

39. Wilmsen, B. After the deluge: A longitudinal study of resettlement at the Three Gorges Dam, China. *World Dev.* **2016**, *84*, 41–54. [CrossRef]

 *water*

Article

Analysis of the Recent Trends of Two Climate Parameters over Two Eco-Regions of Ethiopia

Mohammed Gedefaw [1,*] **, Denghua Yan** [2,*]**, Hao Wang** [2]**, Tianling Qin** [2] **and Kun Wang** [2]

[1]   College of Environmental Science & Engineering, Donghua University, Shanghai 200336, China
[2]   State Key Laboratory of Simulation and Regulation of Water Cycle in River Basin,
      China Institute of Water Resource and Hydropower Research, Beijing 100038, China;
      wanghao@iwhr.com (H.W.); tianling406@163.com (T.Q.); pingguo88wangkun@163.com (K.W.)
*   Correspondence: mohammedgedefaw@gmail.com (M.G.); yandh@iwhr.com (D.Y.);
    Tel.: +86-10-68781976 (M.G.); +86-10-68781976 (D.Y.)

Received: 23 December 2018; Accepted: 15 January 2019; Published: 17 January 2019

Abstract: The changes in climatic variables in Ethiopia are not entirely understood. This paper investigated the recent trends of precipitation and temperature on two eco-regions of Ethiopia. This study used the observed historical meteorological data from 1980 to 2016 to analyze the trends. Trend detection was done by using the non-parametric Mann-Kendall (MK), Sen's slope estimator test, and Innovative Trend Analysis Method (ITAM). The results showed that a significant increasing trend was observed in the Gondar, Bahir Dar, Gewane, Dembi-Dolo, and Negele stations. However, a slightly decreasing trend was observed in the Sekoru, Degahabur, and Maichew stations regarding precipitation trends. As far as the trend of temperature was concerned, an increasing trend was detected in the Gondar, Bahir Dar, Gewane, Degahabur, Negele, Dembi-Dolo, and Maichew stations. However, the temperature trend in Sekoru station showed a sharp decreasing trend. The effects of precipitation and temperature changes on water resources are significant after 1998. The consistency in the precipitation and temperature trends over the two eco-regions confirms the robustness of the changes. The findings of this study will serve as a reference for climate researchers, policy and decision makers.

Keywords: trend analysis; precipitation; temperature; eco-region; Ethiopia

---

1. Introduction

Extreme climatic and weather events in recent decades have been a critical global issue due to the severity of the impacts on natural environments, economy, and on human life [1–3]. These extreme events are unpredictable and destructive, especially, on agriculture production. The likelihood of fewer cool days and nights, increasing heavy precipitation events, and droughts has increased since the 1970s [4]. This indicates that the global climate is undergoing a significant change which is manifested by rising temperature, droughts, rainstorms, and flooding. Scientific studies showed that the mean global temperature could rise by 1.4 to 5.8 °C in 2100 with a mean sea level rise of 10 cm over the same period as reported by Intergovernmental Panel for Climate Change in 2008 [5]. However, considerable regional and seasonal changes in the climate are expected, affecting climatic variables differently depending on the regions with great impact on environments and human systems [6]. The recent increasing frequency of heavy rainfall and severe droughts in many parts of the world is an indication of these situations [7]. Any change of mean global and regional temperature will impact the spatial and temporal distribution of rainfall [8]. This, in turn, affects the hydrological cycles and the availability of water resources [9]. The probability of the frequency of extreme events in the near future is very likely to increase and thus understanding the recent trends is crucial in order to predict the future climate changes. Hence, climate change is perceived through extreme events which tend to alter the magnitude of the predicted climate impacts and this may also be supported by severe flood events. The impacts of climate change on different regions are very different. In this regard, different studies have been conducted in many regions of the world such as in China [10–12], Iran [13], Senegal [14], and India [15,16].

Ethiopia is the most vulnerable country with regard to climate change due to its climatic, hydrology, and low economic conditions [17]. Annual rainfall is highly variable, ranging from less than 200 mm in the southeast, east, and northeast borders to 1200 mm in the central and western highlands of the country [18]. Notably, the country mainly depends on rainfed agriculture and available water resources in the highlands, while large parts of its southern and eastern regions are extremely arid and prone to drought and desertification [5]. Hence, the rainfall is determined by seasonal and interannual variability in the country. Changes in precipitation have a direct impact on floods, droughts, and water resources [19].

Climate change threatens to increase temperature and evapotranspiration; and hence, increasing the risks of heat waves associated with drought [20]. Thus, the change in climate is expected to increase vulnerability in all eco-regions through the increased temperature and more erratic rainfall, which will impact food security and economic growth. Some regional analysis was undertaken to understand the extreme climate and trends. However, the trend indices showed significant increases and decreases in seasonal and annual precipitation, for example, Asfaw et al. [21] reported a decreased rainfall in annual, Belg, and Kiremt in the Woleka sub-basin of Ethiopia. On the other hand, Bewket and Conway [22] reported variations in daily rainfall with no consistent trends. Mekasha et al. [23] also reported increasing warm extremes in temperature and increasing precipitation in different stations across Ethiopia.

Thus, extreme climate indices should be tested for future studies on the perception of climate change with a wide coverage within the country. Therefore, it is essential to analyze the recent trends of climatic variables as these show the climate-related adaptation and mitigation strategies employed by different entities to improve the agrarian economy of the country at large. Furthermore, trend analysis of climatic variables is very important to understand the climate system of the country and has become a vital research area for other researchers. The objective of this study was to assess the recent trends of precipitation and temperature between 1980 and 2016. Therefore, the output of this paper will provide insights for concerned body with ecological and sustainable economic development.

2. Materials and Methods

2.1. Study Area Description

Ethiopia lies between 3°–15° N and 33°–48° E. The total area of the country is about 1.13 million km^2 [18], see Figure 1. The country is characterized by a diversified climate due to its equatorial positioning and topography. Its climate is controlled by atmospheric circulations, complex physiography, and the marked contrast in elevation [18]. The country is mainly divided into two eco-regions, namely lowlands and highlands, where the lowest point is at Danakil Depression and the highest point (4543 m) is at Ras Dejen, above sea level [24]. This classification is mainly based on altitudinal classes, precipitation, and temperature variations. We mainly focused on precipitation and temperature variations for this paper. The lowest mean minimum temperature and high precipitation mostly occur in the highland regions of the country. The highest mean temperature and low precipitation occur in lowland parts of the country. The rainfall also showed seasonal and interannual variability [25].

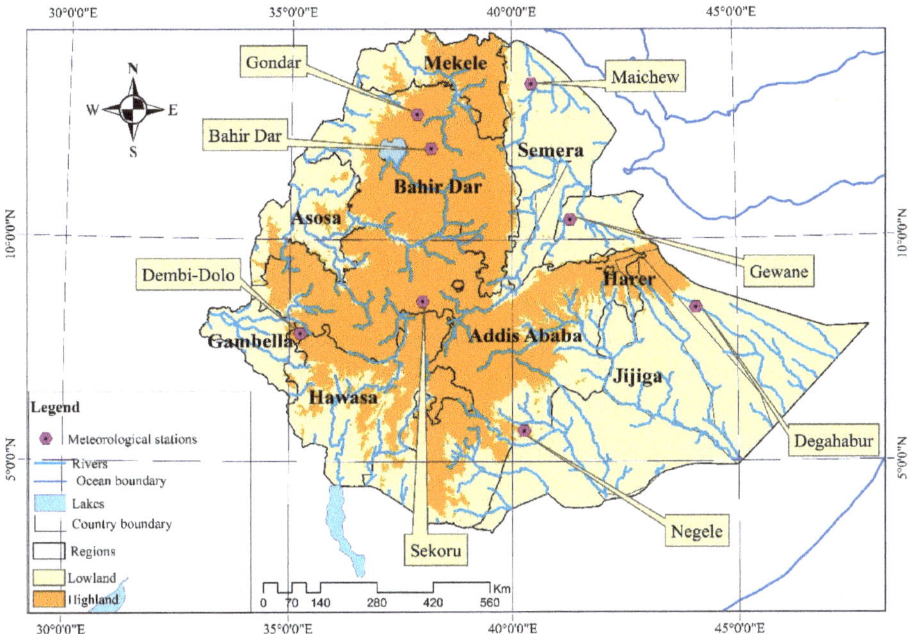

Figure 1. Location map of the study area.

2.2. Data Sources

The raw climatic data were collected from the National Meteorological Services Agency of Ethiopia [26]. As the data series from 1980 to 2016 are complete, the observed precipitation and temperature data were selected as the basic analysis data in this study. All the necessary data for this manuscript were provided after quality control. The stations were also selected based on the completeness of the data during the study periods. We have selected eight stations from two eco-regions (four from highland and four from lowland eco-regions to represent the entire study regions) for this study, see Table 1.

Table 1. Meteorological information's of stations.

Station's Name	Elevation (m)	Latitude (N)	Longitude (E)	Eco-Regions
Dembi-Dolo	1850	34.8°	8.5167°	Highland
Gondar	1973	37.4319°	12.5212°	Highland
Bahir Dar	1827	37.322°	11.6027°	Highland
Sekoru	1928	37.4167°	7.9167°	Highland
Gewane	568	40.633°	10.15°	Lowland
Maichew	2432	39.5337	12.7841°	Lowland
Degahabur	1070	43.55°	8.2167°	Lowland
Negele	1544	39.5667°	5.4167°	Lowland

2.3. Methods

This paper used various methods to detect trends in the precipitation and temperature. The methods are either Parametric or non-parametric which are essential to detect the trends of hydrometeorological observations [27]. Following, are the lists of trend detection non-parametric tests used in this paper, see Figure 2.

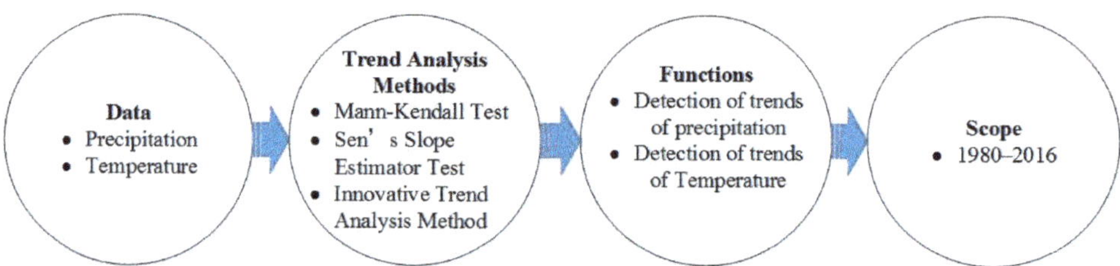

Figure 2. Flow diagram to detect trends of precipitation and temperature.

2.3.1. Mann-Kendall (MK) Test

The Mann-Kendall (MK) test is suited for hydrometeorological observations where the data points are not necessarily uniform [13,28–31]. It is used to detect the presence of either increasing or decreasing monotonic trends in the study area and to see whether the trend is statistically significant or not. Since the test statistics of the MK test are based on plus or minus signs, the determined trends are less affected by the outliers. It is given by:

$$S = \sum_{i=1}^{n-1} \sum_{j=i+1}^{n} sgn\left(X_j - X_i\right) \tag{1}$$

where, X_i ($i = 1, 2, \ldots, n-1$) and X_j ($j = i + 1, 2, \ldots, n$). The observations of each X_i and X_j are calculated as:

$$sgn\left(X_j - X_i\right) = \begin{cases} +1 \text{ if } \left(X_j - X_i\right) > 0 \\ 0 \text{ if } \left(X_j - X_i\right) = 0 \\ -1 \text{ if } \left(X_j - X_i\right) < 0 \end{cases} \tag{2}$$

where X_i and X_j are the data points in i and j years. The variance is calculated with the following equations when the data points ($n \geq 10$) and the mean $E(S) = 0$ [32]:

$$Var(S) = \frac{n \times (n-1) \times (2n+5) - \sum_{q=1}^{p} t_q \times \left(t_q - 1\right) \times \left(2t_q + 5\right)}{18} \tag{3}$$

where p is tied groups in data points, and t_q is the time series in the qth tied groups. The $Z_{(mk)}$ is given as:

$$Z_{(mk)} = \begin{cases} \frac{S-1}{\delta} \text{ if } S > 0 \\ 0 \text{ if } S = 0 \\ \frac{S+1}{\delta} \text{ if } S < 0 \end{cases} \tag{4}$$

When $Z_{(mk)} \geq 10$, it shows an upward trend and when $Z_{(mk)} < 10$, it shows a downward trend. In a time series data sequence, the test statistics are defined separately:

$$UF_k = \frac{d_k - E(d_k)}{\sqrt{var(d_k)}} \ (K = 1, 2, 3,\ldots,n) \tag{5}$$

If $UF_k > UF\alpha/2$, it shows that the trend is significant.

$$UB_k = -UF_k \tag{6}$$

$$K = n + 1 - k \tag{7}$$

Finally, UB_k and UF_k are drawn as UB and UF curves. The intersection is the beginning of mutation between the two curves [33].

2.3.2. Sen's Slope Estimator Test

This test is used to estimate the magnitude of trends in time series data [9]. The slope (Q_i) between two time series data is given as:

$$Q_{i=} = \frac{X_p - X_t}{p - t}, \ for \ i = 1, \ 2, \dots, N \tag{8}$$

where X_p and X_k are time series at period p and t ($p > t$), respectively. If there is single datum in each time, then $N = \frac{n(n-1)}{2}$; n is number of time series. Whereas, if there are many data points, N is computed as $N < \frac{n(n-1)}{2}$; n total number of observations. The N values of the slope estimator are arranged from smallest to biggest.

A positive value of Q_i indicates an upward trend and a negative value of Q_i represents a downward trend in the time series data. The median of these N values of Q_i is represented as Sen's slope estimator. The median of slope (β) is given:

$$\beta = \begin{cases} Q \times [(N+1)/2] & \text{when } N \text{ is odd} \\ Q \times [(N/2) + Q \times (N+2)/(2)/(2)] & \text{when } N \text{ is even} \end{cases} . \tag{9}$$

When β is positive, it indicates the trend is increasing. However, a negative value of β represents a decreasing trend.

2.3.3. Trend Analysis by Innovative Method (ITAM)

Trend Analysis through Innovative Method (ITAM) is also used for trend detection and its reliability was checked with the MK test [9,34]. The observational time series data were classified into two classes and then the data points were arranged independently in ascending order. The mean difference between X_i and X_j would give the magnitude of the trend of the data series. The first observed time series data in this paper were not considered since the total time series data are odd. The test is multiplied by 10 to make the scale similar to MK and Sen's slope estimator tests [9]:

$$\Phi = \frac{1}{n} \sum_{i=1}^{n} \frac{10 \ (X_j - X_i)}{\mu} \tag{10}$$

where, Φ = slope estimator, n = number of time series in the subseries, X_i = observations in the first half subseries, X_j = observations in the second half subseries and μ = mean of data series X_i subseries.

When Φ is positive, it indicates the trend is increasing. However, a negative value of Φ represents a decreasing trend.

3. Results

3.1. Analysis of Mean Annual Precipitation

From 1980 to 2016, the mean annual precipitation of the study area was found to be 834.97 mm, with a CV (coefficient of variation) of 15% and a standard deviation of 122.27 mm. Quantities of 509.93 and 1015.90 mm were the minimum and maximum precipitation per annum, respectively. An increase in the precipitation levels was observed in 2000, 2005, 2007, 2010, and 2013 ($R^2 = 0.01$), with a sharp decreasing trend in 1992. The highest annual precipitation was recorded at the highland eco-region stations (Gondar, Bahir Dar, Sekoru, and Dembi-Dolo), which accounts for approximately 20.3% of lowland eco-regions (Gewane, Degahabur, Negele, and Maichew). The annual precipitation was mainly contributed by the Kiremt months of June–August (47.58%), especially in July and August. These two months contributed 56% of the total annual rainfall.

As far as the seasonal rainfall was concerned, the values varied from 133.82 to 2018.24 mm (Kiremt), from 1176.13 to 1219.32 mm (Meher), from 59.73 to 80.80 mm (Bega), and 551.63 to 1144.75 mm (Belg).

3.2. Trend Analysis of Precipitation

The MK curve annual precipitation (*UF* and *UB* = Changing Parameters) shows the trends of precipitation in highland and lowland eco-environments of the study area. The result showed that the trend in Gondar (Z = 1.69), Dembi-Dolo (Z = 0.28) and Bahir Dar (Z = 0.72) was increasing and the trend in Sekoru (Z = 0.45) was decreasing. On the other hand, in lowland eco-regions, a significant increasing trend was observed in the Gewane (Z = 0.80) and Negele (Z = 0.72) stations, respectively. However, the trend in Degahabur (Z = 0.30) and Maichew (Z = 0.51) was a decreasing one, see Figure 3.

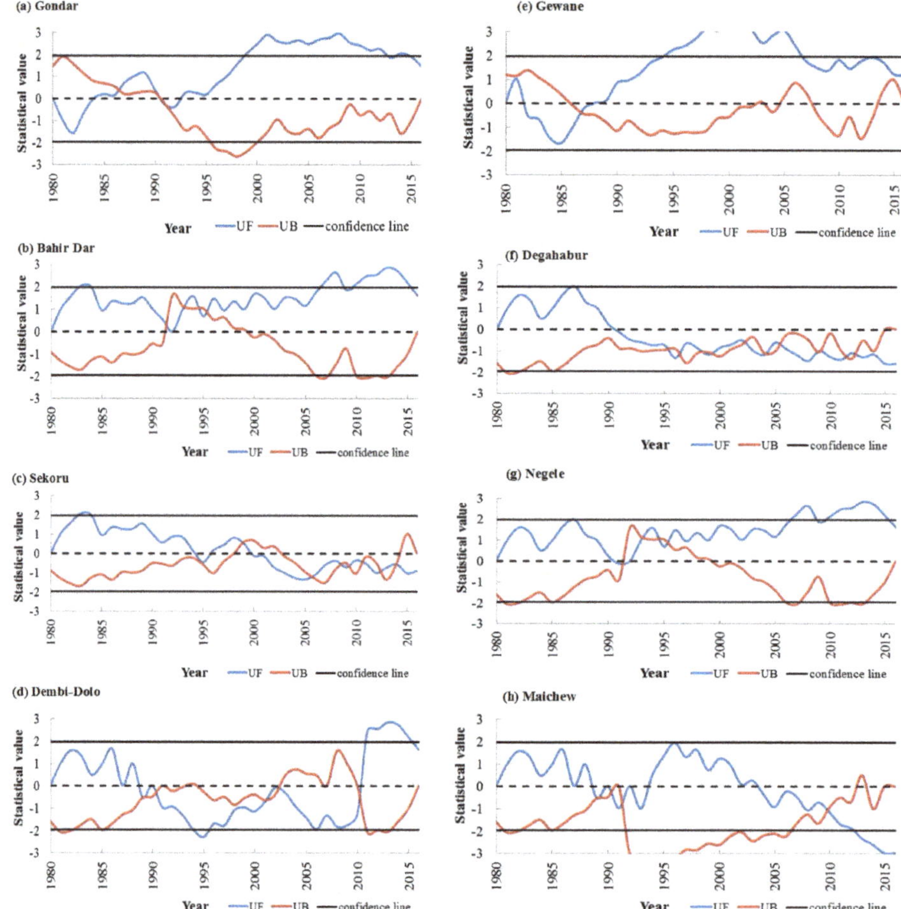

Figure 3. Mean annual precipitation trends of (**a**) Gondar, (**b**) Bahir Dar, (**c**) Sekoru, (**d**) Dembi-Dolo, (**e**) Gewane, (**f**) Degahabur, (**g**) Negele and (**h**) Maichew.

The trend results of precipitation by three trend detection tests are presented in Table 2 with a level of significance α = 5%, α = 10%.

Table 2. Statistical trend results of precipitation.

No.	Stations	Z	Φ	β
1	Gondar	1.69 **	0.54	1.84 **
2	Bahir Dar	−0.07 *	−23.51	1.80 *
3	Sekoru	1.37	0.21	0.01
4	Dembi-Dolo	−0.28	−0.07	−11.55
5	Gewane	5.59 **	0.69	0.10 **
6	Degahabur	0.30	−0.56	4.13
7	Negele	0.72 **	−0.03	23.40 **
8	Maichew	0.51 *	−0.05	18.49 *

Note: * $\alpha = 0.1$; ** $\alpha = 0.05$.

3.3. Analysis of Mean Annual Temperature

The mean annual temperature of the study area was found to be 29.16 °C during the study period. The minimum and maximum recorded temperature were 27.92 and 30.35 °C, respectively. An increasing temperature was recorded in 2010 and 2015 with ($R^2 = 0.67$), and a decreasing trend in the temperature was recorded in 1989. The highest temperature was recorded in the lowland eco-regions (Gewane, Degahabur, Negele, and Maichew). Whereas, a slightly lower temperature was observed in highland eco-regions (Gondar, Bahir Dar, Sekoru, and Dembi-Dolo).

3.4. Trend Analysis of Temperature

The statistical test result of this study showed that the trends of temperature in the Gondar ($Z = 5.68$), Bahir Dar ($Z = 7.59$), Dembi-Dolo ($Z = 3.88$), Maichew ($Z = 6.45$), Gewane ($Z = 5.59$), Degahabur ($Z = 4.78$), and Negele ($Z = 8.01$) stations are significantly increasing. However, a statistically significant decreasing trend was observed in Sekoru ($Z = 1.37$) station, as shown in Figure 4. The trend results of the temperature by three trend detection tests are presented in Table 3.

Table 3. Statistical trend results of temperature.

No.	Stations	Z	Φ	β
1	Gondar	5.68 **	0.35	0.04 **
2	Bahir Dar	7.59 **	0.62	0.08 **
3	Sekoru	1.37 **	0.21	0.01 **
4	Dembi-Dolo	3.88 *	0.22	0.02 *
5	Gewane	5.59 **	0.69	0.10 **
6	Degahabur	4.78 *	0.18	0.03 *
7	Negele	8.01 *	0.48	0.07 *
8	Maichew	6.388 **	0.42	0.06 **

Note: * $\alpha = 0.1$; ** $\alpha = 0.05$.

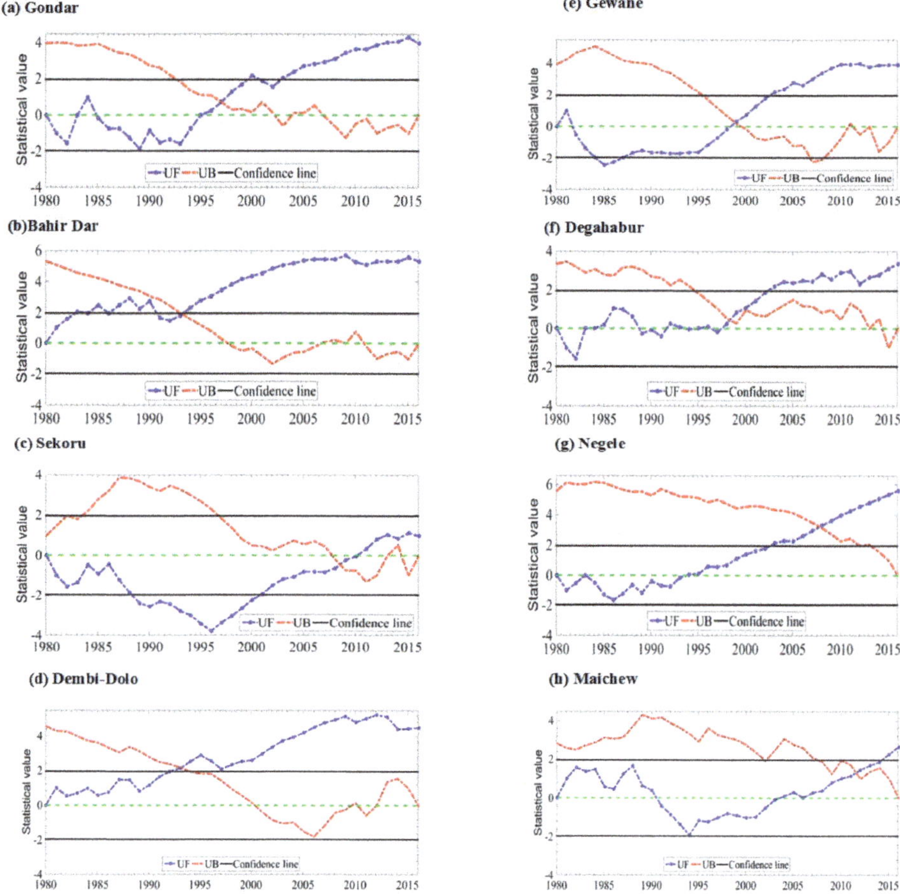

Figure 4. Average annual temperature trends of (**a**) Gondar, (**b**) Bahir Dar, (**c**) Sekoru, (**d**) Dembi-Dolo, (**e**) Gewane, (**f**) Degahabur, (**g**) Negele and (**h**) Maichew.

3.5. Temporal Patterns of Precipitation and Temperature in Individual Stations

The temporal pattern (1980–2016) of precipitation and temperature is illustrated in Figure 5. It is observed that precipitation shows a sharply increasing trend in the Bahir-Dar station, though other stations showed a non-uniform pattern. However, all stations showed an increasing trend in the temperature.

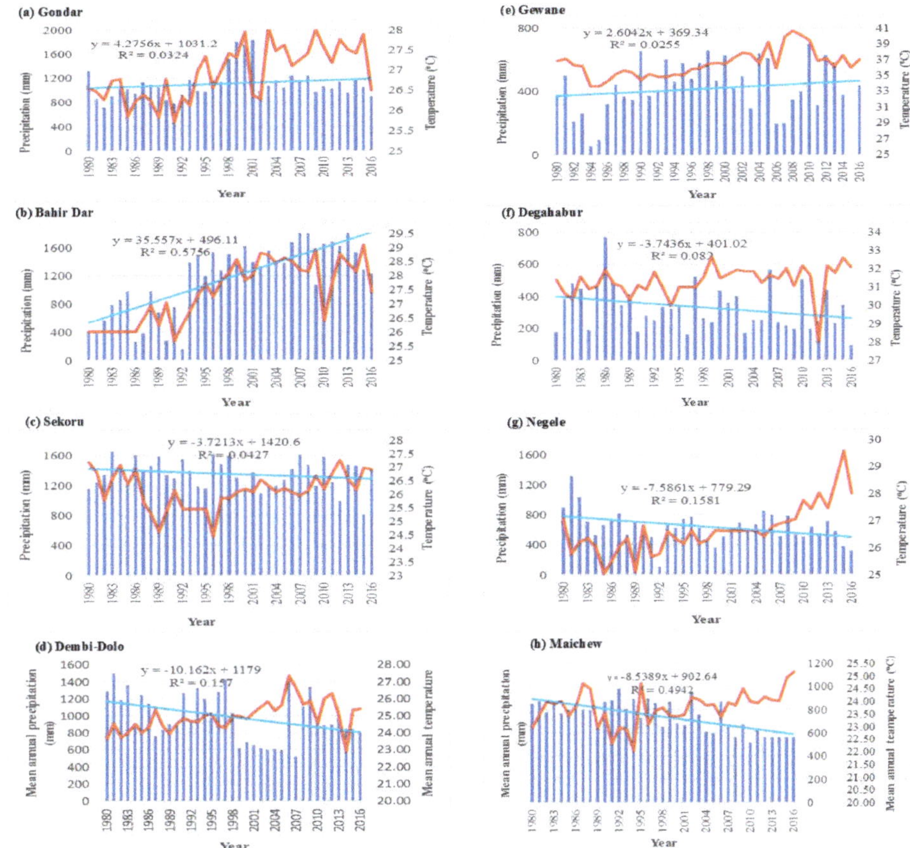

Figure 5. Temporal patterns of precipitation and temperature: (**a**) Gondar, (**b**) Bahir Dar, (**c**) Sekoru, (**d**) Dembi-Dolo, (**e**) Gewane, (**f**) Degahabur, (**g**) Negele and (**h**) Maichew.

4. Discussion

The trends in the precipitation and temperature were analyzed in two eco-regions of Ethiopia. The findings of the study indicated that there is a general tendency towards increasing temperature and a non-uniform pattern of precipitation trends across the stations. Increasing precipitation has been reported in the Gondar, Bahir Dar, Dembi-Dolo, Gewane, and Negele stations. However, slightly decreasing trends were detected in the Sekoru, Maichew, and Degahabur stations. As far as trends of temperature are concerned, almost all stations exhibit a general tendency of increasing temperature. The observed trends have an implication, particularly, on agriculture production of the two eco-regions which are unable to mitigate the impacts of climate change. The observed warming trend may lead to a high energy demand for cooling, high evapotranspiration rate, and weaken the economy at large [35]. Increasing temperature also increases transpiration which increases the chance of rainfall and may interfere with groundwater recharge triggered by reduction in Kiremt season. In the same way, an increasing occurrence of extreme rainfall events impacts the production systems.

The change in trends of precipitation and temperature observed in each station could imply that the variations are more pronounced for certain stations and less for others. It was confirmed that precipitation is mainly caused by a cold summer, and thus correlates to a large extent with temperature in the study area. Therefore, the cause of these variations needs to be studied further to link them with climate variability and change.

Our findings are consistent with previous studies concerning the variations of precipitation and temperature trends [3,7,23,36–41]. However, the causes of such changes of climatic trends across the stations during the study period (1980–2016) will require another detailed investigation.

5. Conclusions

This study analyzed recent changes in precipitation and temperature trends in Ethiopia for the study period from 1980 to 2016. The temporal variability of precipitation and temperature were analyzed. A Mann-Kendall test, Sen's slope estimator test, and Innovative Trend Analysis Methods were used to analyze the trends. Our results showed that five out of eight stations showed increasing trends of precipitation. On the other hand, the Sekoru, Degahabur, and Maichew stations showed decreasing trends of precipitation.

The study eco-regions are characterized by maximum precipitation in Kiremt (June to August) season. The trend is positive in Kiremt season and negative in Bega season which may lead to shifting of the annual cycles of the hydrologic regime. Furthermore, this paper would suggest other studies are conducted to confirm the changing climatic trends over two eco-regions by increasing the sample meteorological stations and, additionally, to investigate the rainfall intensity and frequency of wet and hot days. This finding thus provides insights for policy and decision makers to take proactive measures for climate change mitigation.

Author Contributions: M.G. Wrote the original manuscript. D.Y. Supervision. H.W. Project administrator. T.Q. and K.W. are resource persons. The final manuscript was approved by all the authors.

Funding: This research was funded by National Key Research and Development Project (Grant No. 2016YFA0601503).

Acknowledgments: The authors would like to thank the National Meteorological Service Agency of Ethiopia for providing the raw meteorological data.

Conflicts of Interest: All authors declare no conflict of interest.

References

1. Plisnier, P.; Nshombo, M.; Mgana, H.; Ntakimazi, G. Monitoring climate change and anthropogenic pressure at Lake Tanganyika. *J. Great Lakes Res.* **2018**. [CrossRef]

2. Wang, Y.J.; Qin, D.H. Influence of climate change and human activity on water resources in arid region of Northwest China: An overview. *Adv. Clim. Chang. Res.* **2017**, *8*, 268–278. [CrossRef]

3. Suryabhagavan, K.V. GIS-based climate variability and drought characterization in Ethiopia over three decades. *Weather Clim. Extrem.* **2017**, *15*, 11–23. [CrossRef]

4. Sun, W.; Mu, X.; Song, X.; Wu, D.; Cheng, A.; Qiu, B. Changes in extreme temperature and precipitation events in the Loess Plateau (China) during 1960–2013 under global warming. *Atmos. Res.* **2016**, *168*, 33–48. [CrossRef]

5. Roth, V.; Lemann, T.; Zeleke, G.; Teklay, A. Effects of climate change on water resources in the upper Blue Nile Basin of Ethiopia. *Heliyon* **2018**. [CrossRef] [PubMed]

6. Chen, Z.; Grasby, S.E. Reconstructing river discharge trends from climate variables and prediction of future trends. *J. Hydrol.* **2014**, *511*, 267–278. [CrossRef]

7. Melesse, A.M.; Abtew, W.; Setegn, S.G. *Nile River Basin: Ecohydrological Challenges, Climate Change and Hydropolitics*; Springer: Berlin, Germany, 2013; pp. 1–718.

8. Tesfahunegn, G.B.; Mekonen, K.; Tekle, A. Farmers' perception on causes, indicators and determinants of climate change in northern Ethiopia: Implication for developing adaptation strategies. *Appl. Geogr.* **2016**, *73*, 1–12. [CrossRef]

9. Gedefaw, M.; Wang, H.; Yan, D.; Song, X.; Yan, D.; Dong, G.; Wang, J.; Girma, A.; Ali, B.A.; Batsuren, D.; et al. Trend Analysis of Climatic and Hydrological Variables in the Awash River Basin, Ethiopia. *Water* **2018**, *10*, 1554. [CrossRef]

10. Fischer, T.; Gemmer, M.; Liu, L.; Su, B. Trends in Monthly Temperature and Precipitation Extremes in the Zhujiang River Basin, South China (1961–2007). *Adv. Clim. Chang. Res.* **2010**, *1*, 63–70. [CrossRef]
11. Yang, P.; Xia, J.; Zhang, Y.; Hong, S. Temporal and spatial variations of precipitation in Northwest China during 1960–2013. *Atmos. Res.* **2017**, *183*, 283–295. [CrossRef]
12. Zhao, N.; Yue, T.; Li, H.; Zhang, L.; Yin, X.; Liu, Y. Spatio-temporal changes in precipitation over Beijing-Tianjin-Hebei region. *Atmos. Res.* **2018**, *202*, 156–168. [CrossRef]
13. Feizi, V.; Mollashahi, M.; Frajzadeh, M.; Azizi, G. Spatial and Temporal Trend Analysis of Temperature and Precipitation in Iran. *Ecopersia* **2015**, *2*, 727–742.
14. Diop, L.; Bodian, A. Spatiotemporal Trend Analysis of the Mean Annual Rainfall in Senegal. *Eur. Sci. J.* **2016**, *12*, 231–245. [CrossRef]
15. Kumar, V.; Jain, S.K.; Singh, Y. Analysis of long-term rainfall trends in India. *Hydrol. Sci. J.* **2010**, *55*, 484–496. [CrossRef]
16. Sahu, R.K.; Khare, D. Spatial and temporal analysis of rainfall trend for 30 districts of a coastal State (Odisha) of India. *Int. J. Geol. Earth Environ. Sci.* **2015**, *5*, 40–53.
17. Pingale, S.M.; Khare, D.; Jat, M.K.; Adamowski, J. Spatial and temporal trends of mean and extreme rainfall and temperature for the 33 urban centers of the arid and semi-arid state of Rajasthan, India. *Atmos. Res.* **2014**, *138*, 73–90. [CrossRef]
18. Seleshi, Y.; Zanke, U. Recent changes in rainfall and rainy days in Ethiopia. *Int. J. Climatol.* **2004**, *24*, 973–983. [CrossRef]
19. Wen, X.; Wu, X.; Gao, M. Spatiotemporal variability of temperature and precipitation in Gansu Province (Northwest China) during 1951–2015. *Atmos. Res.* **2017**, *197*, 132–149. [CrossRef]
20. Touré Halimatou, A.; Kalifa, T.; Kyei-Baffour, N. Assessment of changing trends of daily precipitation and temperature extremes in Bamako and S egou in Mali from 1961–2014. *Weather Clim. Extrem.* **2017**, *18*, 8–16. [CrossRef]
21. Asfaw, A.; Simane, B.; Hassen, A.; Bantider, A. Variability and time series trend analysis of rainfall and temperature in northcentral Ethiopia: A case study in Woleka sub-basin. *Weather Clim. Extrem.* **2017**, 1–13. [CrossRef]
22. Bewket, W. Soil and water conservation intervention with conventional technologies in northwestern highlands of Ethiopia: Acceptance and adoption by farmers. *Land Use Policy* **2007**, *24*, 404–416. [CrossRef]
23. Mekasha, A.; Tesfaye, K.; Duncan, A.J. Trends in daily observed temperature and precipitation extremes over three Ethiopian eco-environments. *Int. J. Climatol.* **2014**, *34*, 1990–1999. [CrossRef]
24. Wondie, M.; Schneider, W.; Melesse, A.M.; Teketay, D. Spatial and temporal land cover changes in the simen mountains national park, a world heritage Site in northwestern Ethiopia. *Remote Sens.* **2011**, *3*, 752–766. [CrossRef]
25. Teklu, B.M.; Adriaanse, P.I.; Ter Horst, M.M.S.; Deneer, J.W.; Van den Brink, P.J. Surface water risk assessment of pesticides in Ethiopia. *Sci. Total Environ.* **2015**, *508*, 566–574. [CrossRef] [PubMed]
26. NMA (National Meteorological Agency). *The Federal Democratic Republic of Ethiopia Climate Change National Adaptation Programme of Action (NAPA) of Ethiopia: Climate Change National Adaptation Programme of Action (NAPA) of Ethiopia*; NMA: Addis Ababa, Ethiopia, 2007.
27. Fersi, W.; Lézine, A.-M.; Bassinot, F. Hydro-climate changes over southwestern Arabia and the Horn of Africa during the last glacial–interglacial transition: A pollen record from the Gulf of Aden. *Rev. Palaeobot. Palynol.* **2016**, *233*, 176–185. [CrossRef]
28. Wang, Z.; Luo, Y.; Liu, C.; Xia, J.; Zhang, M. Spatial and temporal variations of precipitation in Haihe River Basin, China: Six decades of measurements. *Hydrol. Process.* **2011**, *25*, 2916–2923. [CrossRef]
29. Fathian, F.; Aliyari, H. Temporal trends in precipitation using spatial techniques in GIS over Urmia Lake Basin, Iran Farshad Fathian * and Hamed Aliyari Ercan Kahya Zohreh Dehghan. *Int. J. Hydrol. Sci. Technol.* **2016**, *6*, 62–81. [CrossRef]
30. Hamisi, J. Study of Rainfall Trends and Variability Over Tanzania—A Research Project Submitted in Partial Fulfilment of the Requirements for the Postgraduate Diploma in Meteorology University of Nairobi. Master's Thesis, University of Nairobi, Nairobi, Kenya, August 2013. Available online: http://erepository.uonbi.ac.ke:8080/xmlui/handle/123456789/55844 (accessed on 23 December 2018).
31. Duhan, D.; Pandey, A. Statistical analysis of long term spatial and temporal trends of precipitation during 1901–2002 at Madhya Pradesh, India. *Atmos. Res.* **2013**, *122*, 136–149. [CrossRef]

32. Ma, X.; He, Y.; Xu, J.; Van Noordwijk, M.; Lu, X. Catena Spatial and temporal variation in rainfall erosivity in a Himalayan watershed. *Catena* **2014**, *121*, 248–259. [CrossRef]

33. Zhang, Q.; Sun, P.; Singh, V.P.; Chen, X. Spatial-temporal precipitation changes (1956–2000) and their implications for agriculture in China. *Glob. Planet. Chang.* **2012**, *82–83*, 86–95. [CrossRef]

34. Gedefaw, M.; Yan, D.; Wang, H.; Qin, T.; Girma, A.; Abiyu, A.; Batsuren, D. Innovative Trend Analysis of Annual and Seasonal Rainfall Variability in Amhara Regional State, Ethiopia. *Atmosphere* **2018**, *9*, 326. [CrossRef]

35. Karmeshu, N. Trend Detection in Annual Temperature; Precipitation using the Mann Kendall Test—A Case Study to Assess Climate Change on Select States in the Northeastern United States. *Mausam* **2015**, *66*, 1–6.

36. Berhe, F.T.; Melesse, A.M.; Hailu, D.; Sileshi, Y. Catena MODSIM-based water allocation modeling of Awash River Basin, Ethiopia. *Catena* **2013**, *109*, 118–128. [CrossRef]

37. Edossa, D.C.; Babel, M.S.; Das Gupta, A. Drought analysis in the Awash River Basin, Ethiopia. *Water Resour. Manag.* **2010**, *24*, 1441–1460. [CrossRef]

38. Hailu, R.; Tolossa, D.; Alemu, G. Water institutions in the Awash basin of Ethiopia: The discrepancies between rhetoric and realities. *Int. J. River Basin Manag.* **2018**, *16*, 107–121. [CrossRef]

39. Belayneh, A.; Adamowski, J.; Khalil, B. Short-term SPI drought forecasting in the Awash River Basin in Ethiopia using wavelet transforms and machine learning methods. *Sustain. Water Resour. Manag.* **2015**. [CrossRef]

40. Mulat, A.G.; Moges, S.A.; Moges, M.A. Journal of Hydrology: Regional Studies Evaluation of multi-storage hydropower development in the upper Blue Nile River (Ethiopia): Regional perspective. *J. Hydrol. Reg. Stud.* **2018**, *16*, 1–14. [CrossRef]

41. Tekleab, S.; Mohamed, Y.; Uhlenbrook, S. Hydro-climatic trends in the Abay / Upper Blue Nile basin, Ethiopia. *Phys. Chem. Earth* **2013**, *61–62*, 32–42. [CrossRef]

 *water*

Article

Evaluating Future Flood Scenarios Using CMIP5 Climate Projections

Narayan Nyaupane [1], Balbhadra Thakur [1], Ajay Kalra [1] and **Sajjad Ahmad [2,*]**

[1] Department of Civil and Environmental Engineering, Southern Illinois University, 1230 Lincoln Drive, Carbondale, IL 62901, USA; narayan.nyaupane@siu.edu (N.N.); balbhadra.thakur@siu.edu (B.T.); kalraa@siu.edu (A.K.)

[2] Department of Civil and Environmental Engineering and Construction, University of Nevada, Las Vegas, NV 89154-4015, USA

* Correspondence: sajjad.ahmad@unlv.edu; Tel.: +1-(702)-895-5456

Received: 30 October 2018; Accepted: 13 December 2018; Published: 17 December 2018

Abstract: Frequent flooding events in recent years have been linked with the changing climate. Comprehending flooding events and their risks is the first step in flood defense and can help to mitigate flood risk. Floodplain mapping is the first step towards flood risk analysis and management. Additionally, understanding the changing pattern of flooding events would help us to develop flood mitigation strategies for the future. This study analyzes the change in streamflow under different future carbon emission scenarios and evaluates the spatial extent of floodplain for future streamflow. The study will help facility managers, design engineers, and stakeholders to mitigate future flood risks. Variable Infiltration Capacity (VIC) forcing-generated Coupled Model Intercomparison Project phase 5 (CMIP5) streamflow data were utilized for the future streamflow analysis. The study was done on the Carson River near Carson City, an agricultural area in the desert of Nevada. Kolmogorov–Smirnov and Pearson Chi-square tests were utilized to obtain the best statistical distribution that represents the routed streamflow of the Carson River near Carson City. Altogether, 97 projections from 31 models with four emission scenarios were used to predict the future flood flow over 100 years using a best fit distribution. A delta change factor was used to predict future flows, and the flow routing was done with the Hydrologic Engineering Center's River Analysis System (HEC-RAS) model to obtain a flood inundation map. A majority of the climate projections indicated an increase in the flood level 100 years into the future. The developed floodplain map for the future streamflow indicated a larger inundation area compared with the current Federal Emergency Management Agency's flood inundation map, highlighting the importance of climate data in floodplain management studies.

Keywords: flood; streamflow; CMIP5; climate change; HEC-RAS

---

1. Introduction

A rise in the mean surface temperature around the globe has been observed in recent climatic records with some warming hole exceptions. The global mean surface temperature in the past three decades has been higher than that of previous decades [1]. This global warming has induced a rise in the evaporation of surface water and evapotranspiration over the land surface, which in turn has increased the average global amount of precipitation. Additionally, the wind and ocean current pattern affects local precipitation trends, which eventually causes fluctuations in the streamflow. Different regions around the globe have already shown signs of adverse effects on water availability due to climate change. The peak streamflow is expected to increase in some parts of the globe [2–7]. At the same time, low flow is also expected to decrease with a greater number of drought days across the globe [8–10]. The occurrence of both high and low flows as a result of climate change is generally placed on the same footing. However, there might be differences in the statistical significance at which low and high flows show variability resulting from climate change [11]. Overall, extreme weather phenomena occur more frequently these days, and this is anticipated to continue in the future.

Flooding is one of the major natural hazards in the U.S., along with tropical cyclones and drought/heatwaves [12]. A reduction in carbon emissions could result in a huge monetary benefit in the long term, as the difference in future flow by the end of the 21st century from a higher emissions pathway to a lower emissions pathway will be billions of dollars per year [13]. Despite these benefits, climate change and its impact on the community have intensified in recent years [14,15]. Flood prevention practice along with a proper understanding of a flooding event can mitigate the risks of this hazard, and floodplain mapping is one of the widely used techniques to quantify the severity of flooding [16].

The Coupled Model Intercomparison Project (CMIP), which is a framework for analyzing and quantifying the results of the Atmosphere–Ocean Coupled General Circulation Model (AOGCM), was first started in 1995. World Climate Research Programme (WCRP) projections through CMIP5 represent future climate projections from new-generation global climate models and advancements in recent climate science [17]. These CMIP5 projections are based on updated global greenhouse gas emission scenarios represented as representative concentration pathways (RCPs). The CMIP5 model, which is large in scale and comprises major climate models from different groups, incorporates a simulation of the 20th century's climate for projecting the climatic scenario of the 21st century [18]. Recently, earth system models have combined conventional Earth system models (ESM) and the AOGCM under an experimental design, where ESM and AOGCM observations were compared. Recent decades were initialized based on the observations and its use for future climate prediction provides the CMIP5 models with enhanced capability [18].

The CMIP5 hydrology projection was released in 2015, and was based on a total of 234 CMIP5 climate projections. These projections were downscaled to the contiguous U.S. utilizing the Bias-Corrected Statistically Downscaled (BCSD) technique [19]. The results of the BCSD projection from phase 3 and phase 5 are known as BCSD3 and BCSD5, respectively. The model results from the BCSD5 hydrology projections were based on a common gridded daily historical meteorology forced simulation [20]. The Constructed Analog (CA) method was applied to spatially downscale a General Circulation Model (GCM) day by matching the same grid-coarsened set of observed days [21]. Projected precipitation changes at spatial and temporal scales show the climate's impact on peak streamflow. The projections from the GCMs need to be translated into similar locally relevant precipitation data before any further use at local scales. This includes, but is not limited to, the selection of an appropriate GCM for a given study area [22], removing biases, and downscaling the GCM to a local resolution [23]. Gangopadhyay et al. [24] translated a downscaled projection into hydrologic projections over a portion of the western U.S., making the projections consistent and making more easy an analysis of climate change's hydrologic impact. Due to the practical limitation on the scope of the hydrologic modeling, only 97 BCSD5 climate projections from 31 CMIP5 climate models with four emission scenarios were available.

Over a long period of time, runoff is equivalent to the tradeoff between precipitation and evapotranspiration. Hence, it is equal to the horizontal water flux that converges at a particular location [25]. For the simulation of hydrology in the future, the Variable Infiltration Capacity (VIC) [26] hydrologic model was utilized. The VIC model is a semi-distributed model in which key aspects of large-scale land surface models are coupled with GCMs [27]. A VIC forcing modeling code and generation process were evaluated before obtaining the VIC forcing results through the modeling code. During the production of the VIC results, we ensured that the forcing generation and VIC simulation process were error-free. For more details on the VIC model, readers are referred to Liang et al. [28] and Nijssen et al. [29]. The correct size and number of output files were produced. After obtaining the output files, BCSD climate monthly data were compared with monthly derived data by aggregating the daily forcing data to check whether any error occurred during the VIC forcing generation process.

There was exact matching in most of the cases [30]. BCSD5 features a larger range compared to BCSD3, as CMIP5 uses a variety of scenarios that mimic the larger range of future greenhouse emissions as compared to CMIP3 [31]. The main difference between BCSD3 and BCSD5 climate projections is in the driving emission scenarios and climate model change, making the projections of temperature and precipitation somewhat different. However, other differences were from model updates on VIC to generate projections with BCSD5 that provide a complete representation of the range of possible future climate and hydrology scenarios. CMIP5 meteorological parameters along with soil parameters, land cover, and vegetation root depth are the main input parameters for the VIC model. Thus, CMIP5 model output from AOGCM was used in a prebuild VIC model to obtain the streamflow data. VIC-generated streamflow is utilized in the current study to predict future streamflow at different recurrence intervals.

The occurrence of extreme events can be estimated from historical flow records by fitting different probability distribution functions [32–36]. Using only the historic flow may not truly reflect the probable future scenario due to climate change. Since, in the stationary approach, the conventional way to predict extreme events in the future is to use historical data only, it is not the best way to deal with the nonstationary climate. To overcome the shortcomings in the design based on the nonstationarity of the climate, climate models and projections are useful. Various climate models based on the Intergovernmental Panel on Climate Change (IPCC)'s fifth assessment report and Special Report on Emission Scenarios (SRES) representing future climate scenarios are available for research and use. Besides the available data, the selection of the distribution method significantly impacts the design value. In most cases, governmental agencies select Generalized Extreme Value (GEV) distribution along with Gumbel and log-Pearson type III distribution. The GEV is mostly used to fit the streamflow distribution and has been shown to be efficient [37]. Further, the streamflow of dry, arid, and semi-arid regions follows the GEV distribution [33,34,38,39]. However, the GEV does not always best fit the annual peak flood. Similarly, studies have utilized other distributions to define a streamflow's behavior [35,36]. Thus, it is worth examining the given sets of yearly flood data and choosing those distributions that produce reliable estimates. Twenty-seven prospective statistical distributions for a semi-arid region are tested in the current study to evaluate which distribution best fits the streamflow. An empirical goodness of fit is one of the criteria for the selection of appropriate distributions. At the same time, the theoretical assumptions associated with all statistical distributions should be taken into account [40]. The selection of the distribution that best fits the peak annual streamflow of Carson River is one of the objectives of this study.

Climate change, which alters the magnitude and frequency of precipitation, ultimately changes the design flood. The comprehension of the changing pattern of the design flood is necessary for flood risk management in the current scenario of a changing climate. This paper uses the VIC-forced CMIP5 streamflow to find the underlying best-fit probability distribution of an area among 27 different statistical distributions. The best-fit distribution was then employed to predict the future streamflow. Finally, a comparison among the existing design parameters was made and the change in hydraulic parameters, such as velocity, top width, and flow area of the river, was estimated. This study will add to the current literature by answering the following research questions regarding variability in streamflow. (1) Is it valid to assume stationarity in streamflow data and to design future structures based on this assumption? (2) What type of statistical distribution does the streamflow follow? (3) By what factor should the occurrence of future flooding events be anticipated to be lesser/greater than that of current flooding events? This paper further presents floodplain mapping for the future design flood that was identified assuming nonstationarity in the climate. The study also evaluates whether, under the changing and nonstationary nature of the climate, the extent of future flooding will be greater than that of the past at the study area. The floodplain map will help us to understand the extent of inundation for the evaluated future flooding events corresponding to different return periods.

1.1. Study Area

The southwestern United States not only experiences extreme heat but is also vulnerable to extreme flooding due to climate change [41]. Carson City, NV has had, since 1852, a historical record of flood, and is currently experiencing some flooding due to extreme storm events. Carson Valley, which lies 4700–5000 feet above the mean sea level, is the rain shadow of the Sierra Nevada. The highest point of the catchment lies on the Sierra Nevada, and is 11,462 feet above the mean sea level. The climate of the area ranges from semi-arid over the valley plain to humid or super humid over the peaks of the catchment. The catchment receives precipitation mostly as rain at the lower altitude and as snow at the higher altitude. Runoff reaches its yearly peak mainly in May. In this study, the downstream end of Carson River at Carson City is examined for future floods. The selected reach was flooded in 2007, and is susceptible to similar events in the future. The spatial location of the Carson River reach at Carson City selected for the current study is shown in Figure 1. Figure 1 also shows the digital elevation map (DEM) that represents the altitude along the river reach that was considered for hydraulic modeling.

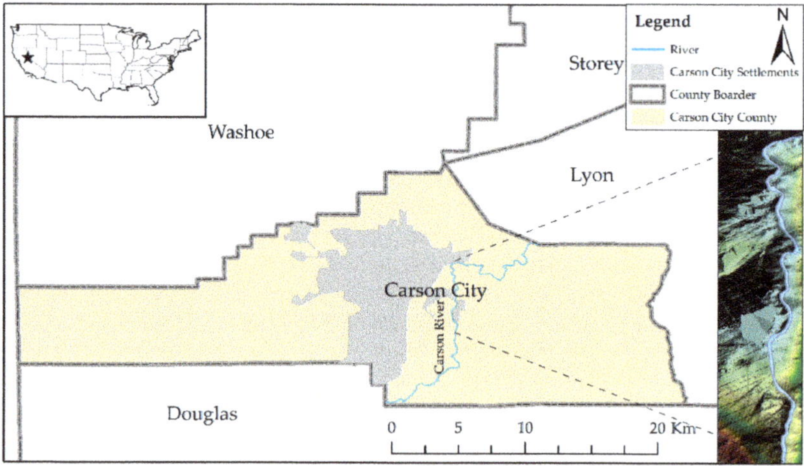

Figure 1. Carson River flowing through Carson City in Nevada.

1.2. Data

The latest daily average runoff from 31 AOGCMs participating in the CMIP5 was used to analyze the change in the extreme runoff for Carson River. These CMIP5-AOGCMs have produced the Bias-Corrected Spatially Downscaled (BCSD) streamflow for different streams in the United States from 1950 to 2099. The data produced by these AOGCMs were routed over a historic period of 1950 to 1999. Thus, in this study, the same period of 1950–1999 is considered to be the historic period. The farthest 50-year period i.e., 2050–2099, is considered to be the future period. The VIC-enforced streamflow for 195 different locations is available, among which 152 locations are co-located with the United States Geological Survey (USGS)'s Hydroclimatic Data Network (HCDN), and 43 locations are co-located with West Wide Climate Risk Assessment spatially downscaled locations. Streamflow data for East Fork Carson River near Gardnerville from total 97 projections that were derived through 31 models and four RCPs were used to estimate the change in streamflow due to climate change. The location of the streamflow is at Latitude 38.844° N and Longitude 119.702° W, which is co-located with the 'East Fork Carson River near Gardnerville' HCDN station (station ID 0071). The projections were used for the future streamflow analysis, while the HCDN station was used for the historic record. The details of the climate model and the relevant institutions are summarized in Table 1.

Table 1. The Coupled Model Intercomparison Project phase 5 (CMIP5) Atmosphere–Ocean Coupled General Circulation Models (AOGCMs) adopted in the study (a total of 31 models with 97 projections).

Modeling Center	Institution	Model	Used Concentration Path (RCP)			
			2.6	4.5	6.0	8.5
CSIRO-BOM	CSIRO (Commonwealth Scientific and Industrial Research Organisation, Australia) and BOM (Bureau of Meteorology, Australia)	ACCESS1.0		√		√
BCC	Beijing Climate Center, China Meteorological Administration	BCC-CSM1.1	√	√	√	√
		BCC-CSM1.1(m)		√		√
CCCma	Canadian Centre for Climate Modelling and Analysis	CanESM2	√	√		√
NCAR	National Center for Atmospheric Research	CCSM4	√	√	√	√
NSF-DOE-NCAR	National Science Foundation, Department of Energy, and National Center for Atmospheric Research	CESM1(BGC)		√		√
		CESM1(CAM5)	√	√	√	√
CMCC	Centro Euro-Mediterraneo per I Cambiamenti Climatici	CMCC-CM		√		√
CNRM-CERFACS	Centre National de Recherches Meteorologiques/Centre Europeen de Recherche et Formation Avancees en Calcul Scientifique	CNRM-CM5		√		√
CSIRO-QCCCE	Commonwealth Scientific and Industrial Research Organisation in collaboration with the Queensland Climate Change Centre of Excellence	CSIRO-Mk3.6.0	√	√	√	√
LASG-CESS	LASG, Institute of Atmospheric Physics, Chinese Academy of Sciences; and CESS, Tsinghua University	FGOALS-g2	√	√		√
FIO	The First Institute of Oceanography, State Oceanic Administration, Beijing, China	FIO-ESM	√	√	√	√
NOAA GFDL	Geophysical Fluid Dynamics Laboratory	GFDL-CM3	√	√	√	√
		GFDL-ESM2G	√	√	√	√
		GFDL-ESM2M	√	√	√	√
NASA GISS	NASA Goddard Institute for Space Studies	GISS-E2-H-CC		√		
		GISS-E2-R	√	√	√	√
		GISS-E2-R-CC		√		
MOHC (additional realizations by INPE)	Met Office Hadley Centre (additional HadGEM2-ES realizations contributed by Instituto Nacional de Pesquisas Espaciais)	HadGEM2-A	√	√	√	√
		HadGEM2-CC		√		√
		HadGEM2-ES	√	√	√	√
INM	Institute for Numerical Mathematics	INM-CM4		√		√
IPSL	Institute Pierre-Simon Laplace	IPSL-CM5A-MR	√	√	√	√
		IPSL-CM5B-LR		√		√
MIROC	Atmosphere and Ocean Research Institute (The University of Tokyo), National Institute for Environmental Studies, and Japan Agency for Marine-Earth Science and Technology	MIROC5	√	√	√	√
MIROC	Japan Agency for Marine-Earth Science and Technology, Atmosphere and Ocean Research Institute (The University of Tokyo), and National Institute for Environmental Studies	MIROC-ESM	√	√	√	√
		MIROC-ESM-CHEM	√	√	√	√
MPI-M	Max Planck Institute for Meteorology (MPI-M)	MPI-ESM-LR	√	√		√
		MPI-ESM-MR	√	√		√
MRI	Meteorological Research Institute	MRI-CGCM3	√	√		√
NCC	Norwegian Climate Centre	NorESM1-M	√	√	√	√

The DEM required for the river terrain was obtained from the USGS National Map viewer. Models using fine-resolution DEM products are more stable and accurate as compared to models that use DEM products of a coarser resolution. Additionally, it is recommended to adopt Light Detecting and Ranging using Remote-Sensing-based products to model a riverine system of higher depths. Due to the limitations of data availability in the study area, the current study utilizes a 1/3 arc-second DEM product for producing the river profile and cross-sections. The river cross-section locations were considered at and in between the Federal Emergency Management Agency (FEMA)'s adopted cross-sections for a comparison purpose. The levee and other existing structures were not adopted in the prepared model as the details of the structures are not readily available. Manning's roughness values were adopted from the Flood Insurance Study (FIS) for the area [42]. Figure 2 represents the Hydrologic Engineering Center's River Analysis System (HEC-RAS) geometric model with the river sections. Eighteen of these cross-sections match with the cross-sections in the FEMA-developed flood map.

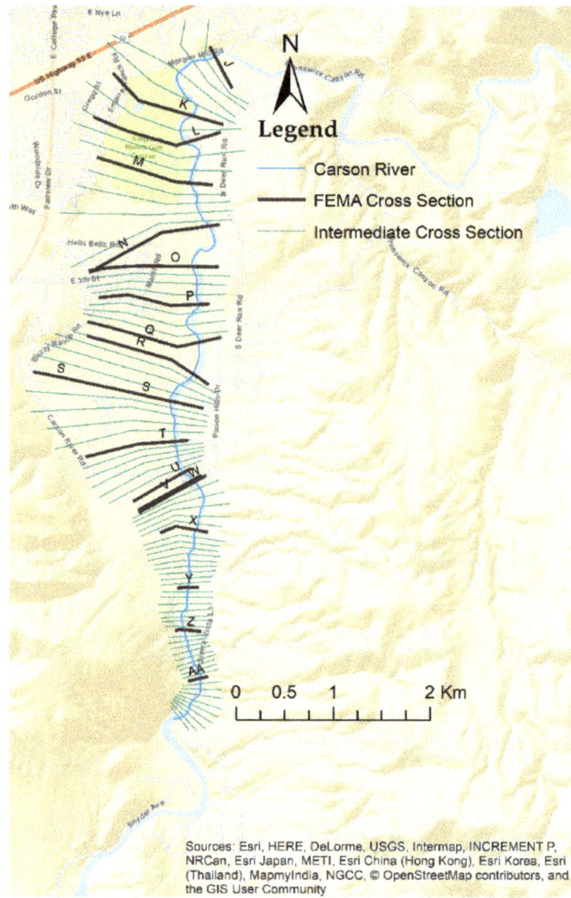

Figure 2. Carson River with the intermediate cross sections and the Federal Emergency Management Agency (FEMA) cross-sections. Sources: Esri, HERE, DeLorme, USGS, Intermap, INCREMENT P, NRCan, Esri Japan, METI, Esri China (Hong Kong), Esri Korea, Esri (Thailand), MapmyIndia, NGCC, [®] OpenStreetMap contributors, and the GIS User Community.

2. Method

The method section is subdivided into three subsections: (i) Frequency analysis and best fit, (ii) Future flow prediction, and (iii) flow routing.

(i) Frequency analysis and best fit: The streamflow projections, along with the nearby real gauge station, were analyzed with a frequency distribution to find the best-fit frequency distribution for the study area. From the 97 streamflow projections for the historic and future periods, a total of 194 projection datasets, each containing 50 years of yearly peak flow, were prepared. These datasets, along with one Carson River gauge dataset, for a total of 195 datasets, were fitted with 27 different distribution methods to obtain the best-fit distribution. The 27 different distributions that were applied to the datasets are listed in Table 2. The data were tested for goodness of fit with Kolmogorov–Smirnov and Pearson Chi-square tests. The tests were implemented over the 195 datasets for the historic and future periods of the model and the historic gauge data. Each best-fit test returns a significance level as an attained value, which is represented as $\alpha_{attained}$ [43]. The significant level for the Pearson Chi-square test and the Kolmogorov–Smirnov test is given, respectively, by

$$\alpha_{attained} = 1 - \chi^2\ (m = k - r - 1,\ q) \tag{1}$$

$$\alpha_{attained} = 1 - \chi^2\ (m,\ q) \tag{2}$$

where, m is the degree of freedom, k is the class interval, r is the number of parameters of the distribution, and q is the computed Pearson parameter, which is given by

$$q = \frac{k}{n} \sum_{j=1}^{k} n_j^2 - n \tag{3}$$

where n is the size of the sample.

These analyses were performed utilizing the statistical Hydrognomon software developed by the National Technical University of Athens [43]. Hydrognomon is a robust tool for performing different time-series analyses: regularization of data, interpolation, regression, fitting a distribution function to a time series, statistical predictions, and homogeneity testing. This tool can handle time-series data at different time scales: daily, monthly, and annual. The current study only uses the capabilities of the Hydrognomon tool to fit a distribution function to a hydrologic time series. Hydrognomon includes 27 different functions that were utilized in the current study to fit the historical records with the help of the Pearson Chi-square and Kolmogorov–Smirnov tests. Hydrognomon utilizes the Monte-Carlo algorithm to determine the confidence interval of any fitted distribution function. The best-fit distribution based on the Hydrognomon tool was selected to generate the future streamflow. All 27 distribution functions included in the Hydrognomon tool are summarized in Table 2.

Table 2. The 27 different distributions used in the best-fit analysis.

Distribution Methods
Normal, Normal(L-Moments), Log Normal, Galton, Exponential, Exponential (L-Moments), Gamma, Pearson III, Log Pearson III, EV1-Max (Gumbel), EV2-Max, EV1-Min (Gumbel), EV3-Min (Weibull), GEV-Max, GEV-Min, Pareto, GEV-Max (L-Moments), GEV-Min (L-Moments), EV2-Max (L-Moments), EV1-Min (Gumbel, L-Moments), EV3-Min (Weibull, L-Moments), Pareto (L-Moments), GEV-Max (Kappa Specified), GEV-Min (Kappa Specified), GEV-Max (Kappa Specified, L-Moments), GEV-Min (Kappa Specified, L-Moments)

(ii) Future flow prediction: Based on the best-fitted distribution method, 100-year flood (the design flood) was calculated for the historic and projected streamflow datasets. The Delta Change Factor (DCF) was used to calculate the future flow at the stream station. The future flow that was estimated using the FEMA and delta change methods depicts a flood under a stationary condition and a flood under climate change conditions in the future, respectively. Among the range of delta change factor future flows, the peak one was selected to represent the maximum increase in the future design flood's condition. In this study, it was assumed that the ratio of the downstream peak flow to the upstream peak flow remains same in the future.

$$\text{Delta change factor} = \frac{\text{Future model daily peak}}{\text{Historic model daily peak}} \tag{3}$$

(iii) Flow routing: A HEC-RAS model was prepared from the available DEM model. Eighteen cross-sections from the Flood Insurance Rate Map (FIRM) and 58 intermediate cross-sections were prepared for the HEC-RAS model using ArcGIS (Version 10.4, Environmental Systems Research Institute (Esri), Redlands, CA, USA). A total of 76 cross-sections were assigned using the FIS-suggested Manning's roughness coefficient. The existing 100-year return period's flow (the design flow) was routed to the prepared model and compared with the existing FIRM map. Finally, the prepared model was routed to the future peak flood, and the hydraulic parameters were compared with the existing design condition of the FEMA map. The aforementioned steps involved in flow routing and flood plain delineation is summarized as the flowchart in Figure 3.

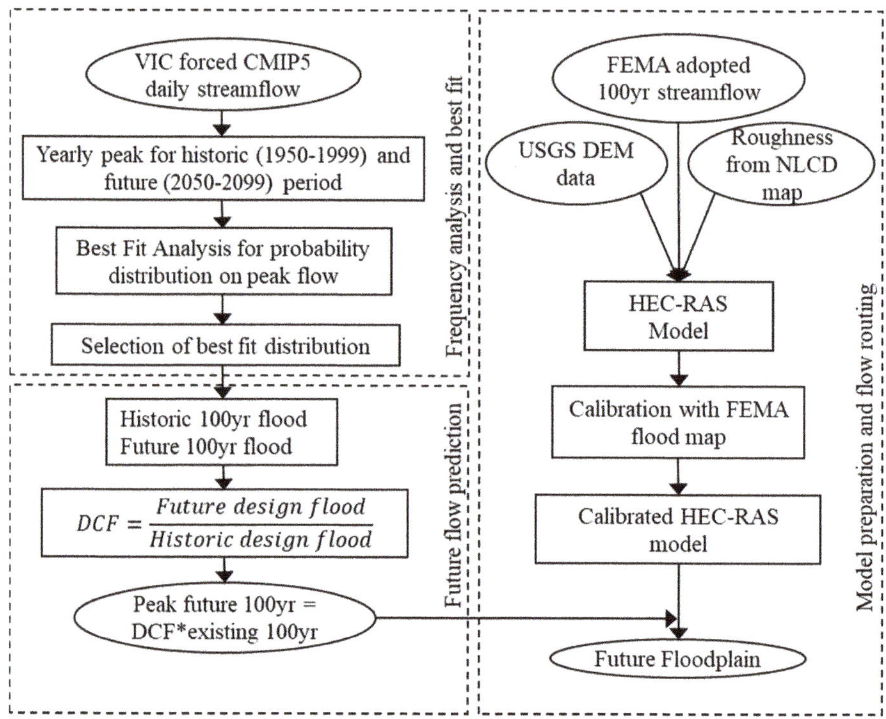

$$DCF = \frac{Future\ design\ flood}{Historic\ design\ flood}$$

Figure 3. The best fit analysis, future design flow prediction, and future design flow routing using HEC-RAS. VIC, Variable Infiltration Capacity; DCF, Delta Change Factor; DEM, digital elevation mode; HEC-RAS, Hydrologic Engineering Center's River Analysis System; NLCD, National Land Cover Data.

3. Results

The daily streamflow series that was derived from the climate model projections shows a clear trend of an increasing future peak streamflow in Carson River and, at the same time, a decreasing yearly peak minimum, as shown in Figure 4. This signifies the occurrence of both more intense high flows and low flows as shown by the spread time series in Figure 4, which increases along the positive abscissa. The maximum and minimum yearly flow plotted in Figure 4 were obtained from 97 streamflow projections of 31 models and four RCPs. In this study, only the probable maximum flow was analyzed, as the study area is subject to a greater flood risk as a result of the high flows as compared to the low flows.

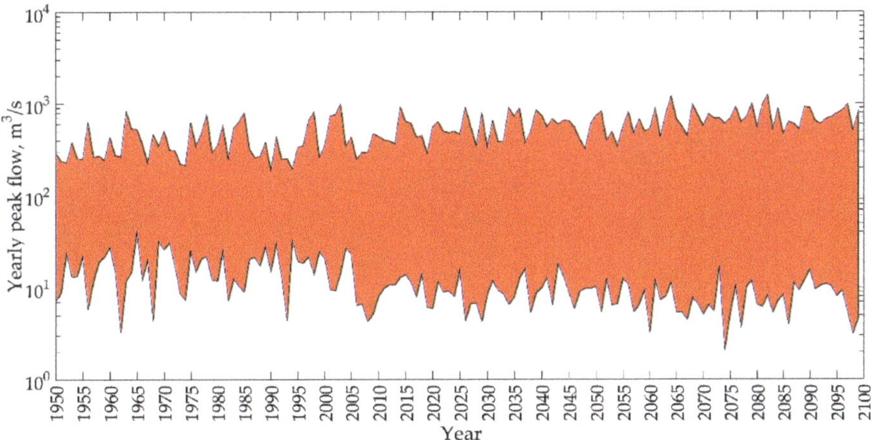

Figure 4. The spread of the band of yearly peak flow from the 97 climate projections, indicating the variability in the future streamflow.

Yearly maximum streamflow data from the 97 climate projections from 1950 to 2099 were selected and analyzed using Pearson Chi-square and Kolmogorov–Smirnov tests. The best fit from both models is presented in Figure 5. The bars in Figure 5 represent the number of projections that was best-fitted with a specific distribution method from the two different tests. From Figure 5, GEV-Max (L-Moments) was selected as the best-fit distribution from the Pearson Chi-square and Kolmogorov–Smirnov tests, with a count of 24 and 53, respectively, out of 97 total projections. Thus, GEV-Max (L-moments) was found to be the best distribution method and selected to analyze future floods in the study area.

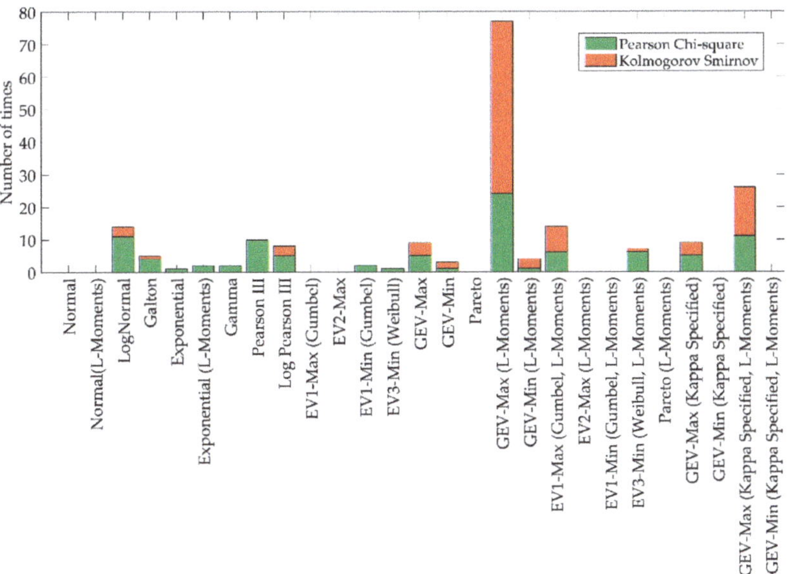

Figure 5. The best fit analysis for the 27 different distributions using Pearson Chi-square and Kolmogorov–Smirnov tests.

The best-fit distribution method, GEV-Max (L-Moments), was used to calculate the peak flow over 100 years (the design flow) for the historic and future period. The selected distribution method was used to calculate the Delta Change Factor. The delta change factor is the ratio of the future to the historic design flow. It was calculated from each climate model, and the values of the delta change factor are summarized in Figure 6a. The inclined lines DCF1, DCF2, and DCF4 represent the delta change factors 1, 2, and 4, respectively, which represent that the future design flood would be the same, two times, and four times the historic period, respectively. In the figure, each emission scenario projection is represented with a different color so that they can be easily distinguished. Figure 6b represents the boxplot of the DCFs corresponding to each RCP. The red-colored horizontal lines in the boxplots represent the median and horizontal edges of the boxes, which represent the 25th and 75th percentiles, respectively, of the DCFs corresponding to each RCP. Similarly, the whiskers in the boxplots represent the 5th and 95th percentiles, and the red data point corresponding to RCP2.6 boxplot is the outlier. From Figure 6a, RCP2.6 has the lowest delta change factor, and RCP8.5 has the highest delta change factor. Figure 6b represents that, with an increase in greenhouse gases, the future extremes of streamflow are expected to increase. Although the two lower RCPs have only a few models with a delta change factor of less than one, the higher RCPs have a DCF that is greater than one. For the flood mapping, the maximum delta change factor of 5.086, which was obtained from the CNRM-CM5 model with RCP8.5, is considered.

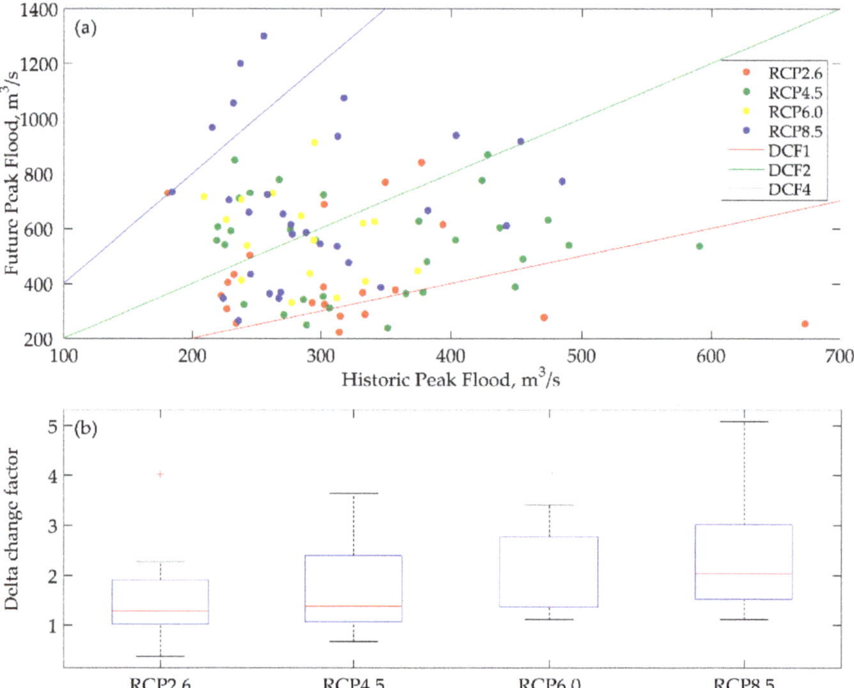

Figure 6. A comparison of the delta change factors from the 97 climate projections. (**a**) Historic versus future design flow (the peak flow over 100 years), (**b**) box plots comparing the delta change factor for different RCPs.

Table 3 shows a hydrological summary of USGS gauge site 1031000 in Carson City, for which a flood analysis has been carried out that developed estimates of flood chance for different return periods. For this study's purpose, only the 1% chance and the 0.2% chance of an annual occurrence, i.e., the 100-year and 500-year return periods, were used. FEMA has developed estimates of the 100-year and 500-year return period flows for the selected study area. A FIRM map with the panel numbers 3200010227E, 3200010112E, and 3200010114F covers the study area.

Table 3. A hydrological summary of flow at USGS gauge site 1031000 (flow in m^3/s) as per the FEMA Flood Insurance Study (FIS).

Flooding Source	Location	Drainage Area (Square Miles)	10% Annual Chance	2% Annual Chance	1% Annual Chance	0.2% Annual Chance
Carson River	5 km Upstream of Lloyds Bridge (USGS 1031000)	2269	238	674	1020	2560

The calibration of the HEC-RAS two-dimensional (2D) model was performed with 100 years of historic streamflow data that was obtained from the gauge site. The calibration was done by comparing the HEC-RAS-generated floodplain map for the past 100 years with FEMA's 100-year flood boundary estimation. A perfect model would result in the same HEC-RAS-generated floodplain map as compared to the FEMA 100-year flood boundary estimation. This result is shown in Figure 7a. Some discrepancy may be attributed to the errors associated with the Manning's roughness coefficient of the reach, which is a function of how a river meanders and channel bed roughness. The robustness of the HEC-RAS model was established with its conformity to the simulated water surface elevation and the observed gauge height at USGS gage station 10311400, which is located downstream of the selected river reach. The Nash–Sutcliffe efficiency coefficient (NSE), the correlation coefficient (R^2), and the percent bias (P-Bias) were computed with the simulated water surface elevation resulting from the observed streamflow that was recorded at USGS gauge station 10311400 corresponding to the observed gauge height. Based on the observed and simulated data, the NSE, R^2, and P-Bias were evaluated as 0.79, 0.98, and −0.008, respectively. These statistical parameters demonstrate the robustness of the calibrated HEC-RAS model, and the water surface elevation predicted by the model can be utilized to generate a future floodplain map with the estimated future streamflow.

The delta change factor that was calculated for the study was used to calculate the future design flood (the flood flow over 100 years). The future design flood flow comes to be 5185 m^3/s, which is more than the current 500-year flood flow. Thus, the climate-generated future design flood flow may be greater than the recent 500-year flood flow. The developed HEC-RAS model routed for the three different flood flows, that is, the existing 100-year, the existing 500-year, and the future 100-year flows, a discharge of 1020 m^3/s, 2560 m^3/s, and 5185 m^3/s, respectively. The flood area developed using HEC-RAS and ArcGIS is presented in Figure 7b. The floodplain for the present design flood flow, the present 500-year flood flow, and the future design flood flow was plotted. The area covered by these three conditions is 3,915,290 m^2, 4,762,168 m^2, and 5,947,893 m^2, respectively, while the FEMA 100-year flood flow covers 4,882,183 m^2. The floodplain for each of these three conditions was compared, and it was observed that the extreme future 100-year floodplain could cover more than 1.5 times the area covered by the present 100-year floodplain.

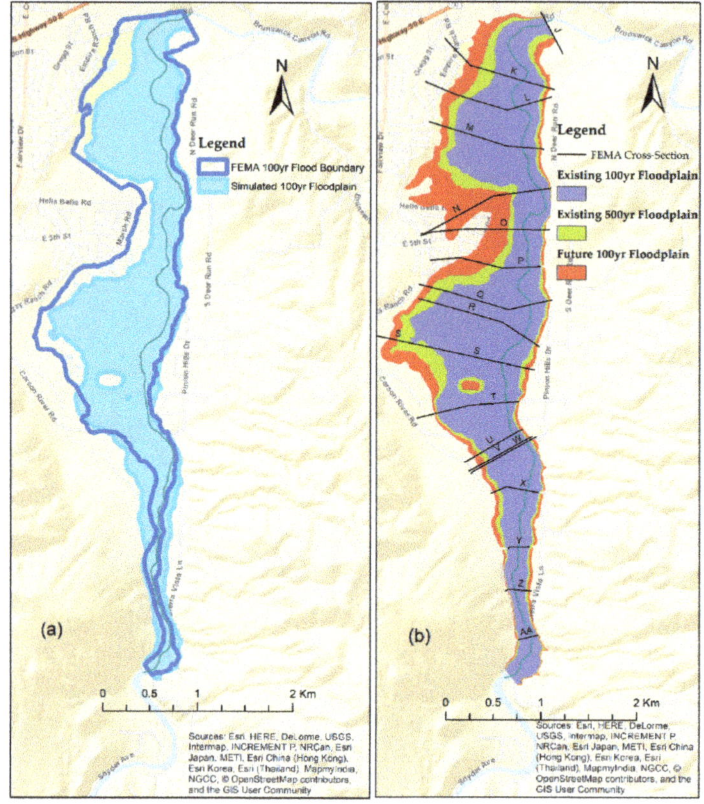

Figure 7. (**a**) A comparison of the Flood Insurance Rate Map (FIRM)'s 100-year flood area versus the baseline scenario (the 100-year flood area obtained from the HEC-RAS model). (**b**) The three layers of flood area for the 100-year historic, 500-year historic, and 100-year future floodplains (from a smaller to a larger area, respectively). Sources: Esri, HERE, DeLorme, USGS, Intermap, INCREMENT P, NRCan, Esri Japan, METI, Esri China (Hong Kong), Esri Korea, Esri (Thailand), MapmyIndia, NGCC, ® OpenStreetMap contributors, and the GIS User Community.

Further, the channel velocity, flow area, and top width were compared between the historic 100-year, the historic 500-year, and the future 100-year floods. The FIRM map has 18 cross-sections within the reach length. Hydraulic parameters, such as channel velocity, flow area, and top width, were compared within this reach length and the FIRM map cross sections, and are presented in Figure 8. The calibration of the HEC-RAS model can also be verified from Figure 7a, which shows a similar floodplain for 100-year event as compared to the 100-year event from the FIRM map. The generated floodplain maps shows that there will be more flooding on the left bank of the river than the right due to its topography. In addition, the city nearby the river might be affected due to this change in future flow. The low-lying agricultural land on the Carson floodplain will be vulnerable to flooding events in the future, as our results suggest that these events will be more intense than in the past.

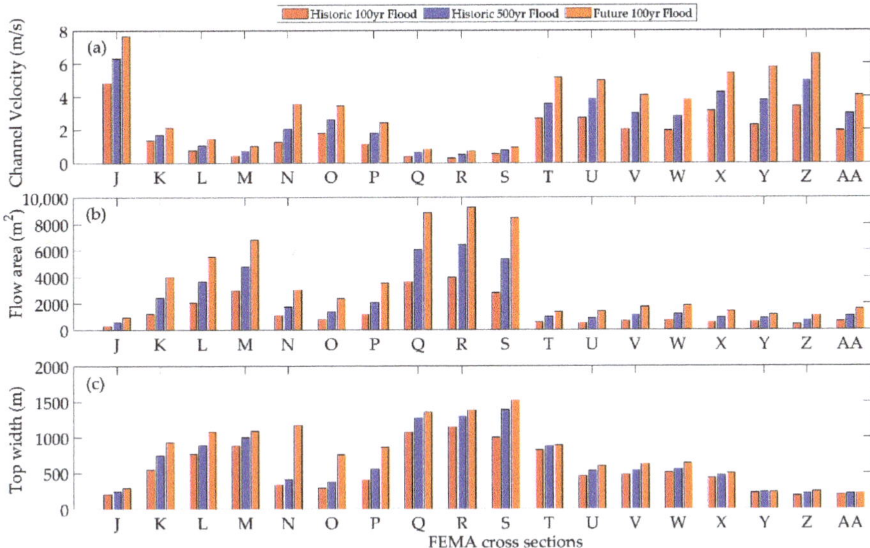

Figure 8. A comparison of the (**a**) channel velocity, (**b**) flow area, and (**c**) top width for different flood scenarios.

Channel velocity, flow area, and top width are the key hydraulic parameters of floods, and were compared under different flow conditions. The future 100-year flood has the highest channel velocity (around 8 m/s). The channel flow area will more than double at most of the cross-sections, and there will be a significant increase in the top width along sections N, O, and P.

4. Discussion

Floodplain management should reduce flood damage in the future with efficient mitigation measures at minimal expense. A flood protection decision can endure for more than a century; thus, it is worth considering long-term changes in environmental conditions. A decision on floodplain management would likely affect the long-term performance of infrastructure [44]. As human beings have evolved, societies have started to inhabit places close to freshwater to ensure water for drinking and agricultural and livestock use. Major civilizations of the past settled very close to rivers with an adequate supply of fresh water. More than half of the total global population resides within 3 km from freshwater bodies, mostly near a river [45]. This population is more vulnerable to changes in streamflow in the future. In this study, such a case is analyzed using CMIP5 hydrologic projections. The CMIP5 data helped to predict the nature of future streamflow without assuming stationarity in the hydro-climatic data. Further, CMIP5 models consider different carbon emission scenarios, allowing us to comprehend a wide range of future streamflow conditions.

As different probability distributions result in different flood frequencies, selecting an appropriate flood distribution method is crucial in flood frequency analysis. To select the best-fit distribution, two different tests—the Pearson Chi-Square and the Kolmogorov–Smirnov—were implemented. The selected distribution was used to predict the design flood for present and future datasets of climate projections. Further, the performance of different distribution methods varies spatially. Thus, one distribution that performs well for a given dataset of an area may not represent another distribution. Thus, it is always recommended to analyze the results that are obtained with different distributions, and the best distribution should be selected with a different approach. It is good to test a variety of statistical distributions; however, at the same time, it is impractical to test all of the possible distributions. In the current study, the selected distributions were based on a previous study that was pertinent to the climatic conditions of the selected study region. Testing other distributions over a wide variety of watersheds is recommended for future studies. Both the Pearson Chi-Square test and the Kolmogorov–Smirnov test showed that GEV-max (L-moment) was the best distribution. Previous studies suggest that GEV is to likely to fit the streamflow across semi-arid regions, which was also established in the current findings [32–34,38,39].

From the 97 total models/selections, the Pearson Chi-square test and the Kolmogorov–Smirnov test selected GEV-max (L-moment) 24 and 53 times, respectively. GEV-max (L-moment) was then used to calculate the design flow for the present and future periods. Among the 97 models, 86 models had a DCF greater than one, suggesting that the design flow for the future period should be more than that for the present period. This suggests that there is a higher possibility that the design flow on the river would be higher in the future. Among the 97 models, 39 (more than 40% of the models) indicated that the design flow in the future will be more than 2 times the present design flow. The DCF results also show that the higher the emission or RCP, the higher the delta change factor will be, which represents an increase in the design flow in the future. The delta change method was adopted to predict the flow of future floods, which was routed on a HEC-RAS one-dimensional (1D) model to compare the floodplain and hydraulic parameters.

The maximum increase in future design flow resulted from the CNRM-CM5 model with the RCP8.5 scenario and the maximum DCF. The CNRM-CM5 model with RCP8.5 scenario was also considered for the flood mapping of the Carson River near Carson City, Nevada as it would represent the most extreme flooding scenario. The model indicates that a 5.086-fold increase in the design flow of the river will occur. This suggests that the future 100-year flood flow will be more than that of the current 500-year flow, while the future 100-year flood flow will be more than 1.5 times the current 100-year flood flow. The floodplain map showed an increase in the extent of the flooding in the future as compared to that of current flooding events. Most of the GCMs suggested an increase in the future design flow when compared to present conditions, which can be attributed to the nonstationary nature of the climate. Further, the floodplain variability in the future may also be affected by other factors, such as a change in land use [46]. An increase in the streamflow in a semi-arid region as a result of climate change has also been documented by previous studies. For example, an increase in streamflow has been projected for the city of Las Vegas [47], and a positive streamflow trend has been documented in the Colorado River basin [48].

The population of Carson City is settled on the left bank of the Carson River. The topography of the river shows that the river has more floodplain on left side than on the right side. The floodplain contains fertile agricultural land that is crucial, as Carson City lies in the desert of Nevada. Due to the predicted increase in the design flood flow in the future, more area than expected might be flooded. Future flooding will not only affect the agricultural supply but also people residing near the river. Thus, a proper analysis of future streamflow will help to minimize the flood risk. The current study's evaluation of the best-fit distribution, and use of a climate distribution, to evaluate the future streamflow frequency suggested that the streamflow data has a nonstationary nature as a result of climate variability and change. Thus, future design streamflow may not be same as the current design streamflow. This should be considered by planners and engineers when planning and building new hydraulic structures to minimize the flood risks associated with the changing climate.

5. Conclusions

The risk of hydrologic extremes as a result of the changing climate is one of the main global challenges of the 21st century. This risk has been increasing in recent years, as limited effort has been made to curb the emission of greenhouse gases. Most governmental agencies apply stationary approaches to flood management. Due to this, the effect of climate change cannot be incorporated into flood flow predictions. Thus, this study suggests the possible approach of applying a nonstationary approach to future streamflow prediction in those regions vulnerable to flooding events. As most flood management structures are constructed for a lifespan of several decades to more than a century, forecasting streamflow while considering the effects of climate change can contribute to the optimization of hydraulic structures. Hence, this study's contribution is to present guidelines to planners, designers, engineers, and policy-makers for incorporating climate change into flood risk management. The current study was focused on the Carson River at Carson City at a regional scale; so, the results may not be similar to those obtained from other watersheds, whilst the proposed algorithm can be utilized elsewhere.

The key findings of the current study are as follows:

1. The best-fit distribution was evaluated utilizing both Pearson Chi-square and Kolmogorov Smirnov tests for the considered study area.
2. A majority of the climate models indicated an increase in the future streamflow in the study region, while 40 percent of the models suggested that the future 100-year streamflow would be more than 2 times the present 100-year streamflow in the selected study area.
3. A higher increase in the future 100-year streamflow was observed for a higher RCP, suggesting that the streamflow in the study region will increase as carbon emissions increase.
4. A majority of climate models depicted DCFs higher than 1, suggesting that the streamflow in the Carson River exhibits nonstationary behavior and that the future streamflow is likely to exceed that of the past.
5. The CNRM-CM5 model with the RCP8.5 scenario showed the maximum increase in future runoff for the Carson River.
6. The future flow, depth of flow, and inundation comparison gave an explicit image of the extent of future flooding due to climate change for the Carson River at Carson City.
7. The extent of flooding of the future 100-year streamflow for the Carson River at Carson City was evaluated to be higher than that of the past 500-year streamflow, highlighting the likelihood of an increase in the extent of future flooding.

To summarize, hydraulic structures are conventionally designed for different return periods assuming the stationarity in the streamflow. The current research highlights the utilization of climate information for evaluating the future streamflow in different recurrence intervals. This future streamflow is then used to evaluate future floodplain maps and design the streamflow. We also recommend testing the application of other statistical distributions to a variety of watersheds with a higher resolution dataset in the future to extend the current study.

Author Contributions: A.K. and S.A. designed the research idea and the results are prepared by N.N. and B.T. All the authors analyzed the results and wrote the paper together.

Funding: This research received no external funding.

Acknowledgments: The authors would like to thank three anonymous reviewers for providing valuable comments. We acknowledge the Program for Climate Model Diagnosis and Intercomparison (PCMDI) and the World Climate Research Programme (WCRP) Working Group on Coupled Modelling (WGCM) for their role in making available the WCRP CMIP5 multi-model dataset. Support for this dataset is provided by the Office of Science, U.S. Department of Energy. The authors would like to acknowledge the Office of the Vice Chancellor for Research at Southern Illinois University, Carbondale for providing the research support. Thanks to Ankit Kumar from the Indian Institute of Technology, Mumbai, India for his valuable suggestions during the preparation of this manuscript.

Abbreviations

AOGCM	Atmosphere-Ocean Coupled General Circulation Model
BCSD	Bias Corrected Statistically Downscaled
CMIP	Coupled Model Intercomparison Project
DCF	Delta Change Factor

DEM	Digital Elevation Model
FEMA	Federal Emergency Management Agency
FIRM	Flood Insurance Rate Map
FIS	Flood Insurance Study
GCM	General Circulation Model
GEV	Generalized Extreme Value
RCP	Representative Concentration Pathways
VIC	Variable Infiltration Capacity
WCRP	World Climate Research Programme

References

1. IPCC. Summary for Policymakers. In *Climate Change 2013: The Physical Science Basis. Contribution of Working Group I to the Fifth Assessment Report of the Intergovernmental Panel on Climate Change*; Stocker, T.F., Qin, D., Plattner, G.-K., Tignor, M., Allen, S.K., Boschung, J., Nauels, A., Xia, Y., Bex, V., Midgley, P.M., Eds.; Cambridge University Press: Cambridge, UK, 2013.

2. Dlamini, N.S.; Kamal, M.R.; Soom, M.A.B.M.; Mohd, M.S.F.; Abdullah, A.F.B.; Hin, L.S. Modeling potential impacts of climate change on streamflow using projections of the 5th assessment report for the Bernam River basin, Malaysia. *Water* **2017**, *9*, 226. [CrossRef]

3. Hirabayashi, Y.; Kanae, S.; Emori, S.; Oki, T.; Kimoto, M. Global projections of changing risks of floods and droughts in a changing climate. *Hydrolog. Sci. J.* **2008**, *53*, 754–772. [CrossRef]

4. Kefi, M.; Mishra, B.K.; Kumar, P.; Masago, Y.; Fukushi, K. Assessment of tangible direct flood damage using a spatial analysis approach under the effects of climate change: Case study in an urban watershed in Hanoi, Vietnam. *ISPRS Int. J Geo-Inf.* **2018**, *7*, 29. [CrossRef]

5. Nohara, D.; Kitoh, A.; Hosaka, M.; Oki, T. Impact of climate change on river discharge projected by multimodel ensemble. *J. Hydrometeorol.* **2006**, *7*, 1076–1089. [CrossRef]

6. Rafiei Emam, A.; Mishra, B.K.; Kumar, P.; Masago, Y.; Fukushi, K. Impact assessment of climate and land-use changes on flooding behavior in the upper Ciliwung River, Jakarta, Indonesia. *Water* **2016**, *8*, 559. [CrossRef]

7. Yerramilli, S. Potential impact of climate changes on the inundation risk levels in a dam break scenario. *ISPRS Int. J. Geo-Inf.* **2013**, *2*, 110–134. [CrossRef]

8. Dankers, R.; Feyen, L. Flood hazard in Europe in an ensemble of regional climate scenarios. *J. Geophys. Res. Atmos.* **2009**, *114*. [CrossRef]

9. Davie, J.C.; Falloon, P.D.; Kahana, R.; Dankers, R.; Betts, R.; Portmann, F.T.; Wisser, D.; Clark, D.B.; Ito, A.; Masaki, Y. Comparing projections of future changes in runoff from hydrological and biome models in ISI–MIP. *Earth Syst. Dyn.* **2013**, *4*, 359–374. [CrossRef]

10. Hirabayashi, Y.; Mahendran, R.; Koirala, S.; Konoshima, L.; Yamazaki, D.; Watanabe, S.; Kim, H.; Kanae, S. Global flood risk under climate change. *Nat. Clim. Chang.* **2013**, *3*, 816–821. [CrossRef]

11. Koirala, S.; Hirabayashi, Y.; Mahendran, R.; Kanae, S. Global assessment of agreement among streamflow projections using CMIP5 model outputs. *Environ. Res. Lett.* **2014**, *9*. [CrossRef]

12. NCEI (National Centers for Environmental Information). U.S. Billion-Dollar Weather and Climate Disasters, 2018. NCEI Web Site. Available online: https://www.ncdc.noaa.gov/billions/ (accessed on 3 October 2018).

13. Wobus, C.; Gutmann, E.; Jones, R.; Rissing, M.; Mizukami, N.; Lorie, M.; Mahoney, H.; Wood, A.W.; Mills, D.; Martinich, J. Modeled changes in 100 year flood risk and asset damages within mapped floodplains of the contiguous United States. *Nat. Hazards Earth Syst. Sci. Discuss.* **2017**. [CrossRef]

14. Papalexiou, S.M.; AghaKouchak, A.; Trenberth, K.E.; Foufoula-Georgiou, E. Global, regional, and megacity trends in the highest temperature of the year: Diagnostics and evidence for accelerating trends. *Earth's Future* **2018**, *6*, 71–79. [CrossRef] [PubMed]

15. Van Aalst, M.K. The impacts of climate change on the risk of natural disasters. *Disasters* **2006**, *30*, 5–18. [CrossRef] [PubMed]

16. Lamichhane, N.; Sharma, S. Development of flood warning system and flood inundation mapping using field survey and LiDAR data for the Grand River near the city of Painesville, Ohio. *Hydrology* **2017**, *4*, 24. [CrossRef]

17. Taylor, K.; Stouffer, R.; Meehl, G. A summary of the CMIP5 experiment design. *January* **2011**, *22*, 33.

18. Taylor, K.E.; Stouffer, R.J.; Meehl, G.A. An overview of CMIP5 and the experiment design. *Bull. Am. Meteorol. Soc.* **2012**, *93*, 485–498. [CrossRef]

19. Wood, A.W.; Leung, L.R.; Sridhar, V.; Lettenmaier, D. Hydrologic implications of dynamical and statistical approaches to downscaling climate model outputs. *Clim. Chang.* **2004**, *62*, 189–216. [CrossRef]

20. Maurer, E.; Wood, A.; Adam, J.; Lettenmaier, D.; Nijssen, B. A long-term hydrologically based dataset of land surface fluxes and states for the conterminous United States. *J. Clim.* **2002**, *15*, 3237–3251. [CrossRef]

21. Hidalgo, H.G.; Dettinger, M.D.; Cayan, D.R. Downscaling with Constructed Analogues: Daily Precipitation and Temperature Fields over the United States. In *California Energy Commission PIER Final Project Report*; CEC-500-2007-123; California Energy Commission: Sacramento, CA, USA, 2008.

22. Mote, P.W.; Salathe, E.P. Future climate in the Pacific Northwest. *Clim. Chang.* **2010**, *102*, 29–50. [CrossRef]

23. Fowler, H.J.; Blenkinsop, S.; Tebaldi, C. Linking climate change modelling to impacts studies: Recent advances in downscaling techniques for hydrological modelling. *Int. J. Climatol.* **2007**, *27*, 1547–1578. [CrossRef]

24. Gangopadhyay, S.; Pruitt, T.; Brekke, L. *West-Wide Climate Risk Assessments: Bias-Corrected and Spatially Downscaled Surface Water Projections*; US Department of the Interior, Bureau of Reclamation, Technical Service Center: Denver, CO, USA, 2011.

25. Milly, P.C.; Dunne, K.A.; Vecchia, A.V. Global pattern of trends in streamflow and water availability in a changing climate. *Nature* **2005**, *438*, 347–350. [CrossRef] [PubMed]

26. Liang, X.; Lettenmaier, D.P.; Wood, E.F.; Burges, S.J. A simple hydrologically based model of land surface water and energy fluxes for general circulation models. *J. Geophys. Res. Atmos.* **1994**, *99*, 14415–14428. [CrossRef]

27. Liang, X. Variable Infiltration Capacity (VIC): Macroscale Hydrologic Model, 2002. VIC Documentation Web Site. Available online: http://www.hydro.washington.edu/Lettenmaier/Models/VIC/VIChome.html (accessed on 20 August 2018).

28. Liang, X.; Wood, E.F.; Lettenmaier, D.P. Surface soil moisture parameterization of the VIC-2L model: Evaluation and modification. *Glob. Planet Chang.* **1996**, *13*, 195–206. [CrossRef]

29. Nijssen, B.; Lettenmaier, D.P.; Liang, X.; Wetzel, S.W.; Wood, E.F. Streamflow simulation for continental-scale river basins. *Water Resour. Res.* **1997**, *33*, 711–724. [CrossRef]

30. Brekke, L.; Wood, A.; Pruitt, T. *Downscaled CMIP3 and CMIP5 Hydrology Projections: Release of Hydrology Projections, Comparison with Preceding Information, and Summary of User Needs*; US Department of the Interior Bureau of Reclamation: Denver, CO, USA, 2014.

31. Brekke, L.; Thrasher, B.; Maurer, E.; Pruitt, T. *Downscaled CMIP3 and CMIP5 Climate Projections: Release of Downscaled CMIP5 Climate Projections, Comparison with Preceding Information, and Summary of User Needs*; US Department of the Interior, Bureau of Reclamation, Technical Service Center: Denver, CO, USA, 2013.

32. Ahmed, S.; Tsanis, I. Hydrologic and hydraulic impact of climate change on Lake Ontario tributary. *Am. J. Water Resour.* **2016**, *4*, 1–15.

33. Thakali, R.; Kalra, A.; Ahmad, S. Understanding the effects of climate change on urban stormwater infrastructures in the Las Vegas Valley. *Hydrology* **2016**, *3*, 34. [CrossRef]

34. Moglen, G.E.; Vidal, G.E.R. Climate change impact and storm water infrastructure in the Mid-Atlantic region: Design mismatch coming. *J. Hydrol. Eng.* **2014**, *19*. [CrossRef]

35. Zhu, J. Impact of climate change on extreme rainfall across the United States. *J. Hydrol. Eng.* **2013**, *18*, 1301–1309. [CrossRef]

36. Zhu, J.; Stone, M.C.; Forsee, W. Analysis of potential impact of climate change on intensity-duration-frequency (IDF) relationships for six regions in the United States. *J. Water Clim. Chang.* **2012**, *3*, 185–196. [CrossRef]

37. Hailegeorgis, T.T.; Alfredsen, K. Regional flood frequency analysis and prediction in ungauged basins including estimation of major uncertainties for mid-Norway. *J. Hydrol. Reg. Stud.* **2017**, *9*, 104–126. [CrossRef]

38. Farquharson, F.A.K.; Meigh, J.R.; Sutcliffe, J.V. Regional flood frequency analysis in arid and semi-arid areas. *J. Hydrol.* **1992**, *138*, 487–501. [CrossRef]

39. Simmers, I. Hydrological Processes and Water Resources Management. In *Understanding Water in a Dry Environment*; CRC Press: Boca Raton, FA, USA, 2005; pp. 17–30.

40. Canfield, R.V.; Olsen, D.; Hawkins, R.; Chen, T. *Use of Extreme Value Theory in Estimating Flood Peaks from Mixed Populations*; Utah State University: Logan, UT, USA, 1980; p. 577.

41. Jardine, A.; Merideth, R.; Black, M.; LeRoy, S. *Assessment of Climate Change in the Southwest United States: A Report Prepared for the National Climate Assessment*; Island press: Washington, DC, USA, 2013.

42. FEMA (Federal Emergency Management Agency). *Flood Insurance Study*; Number 320001V000C, Version Number 2.3.3.0; FEMA: Washington, DC, USA, 2016.

43. Kozanis, S.; Christofides, A.; Mamassis, N.; Efstratiadis, A.; Koutsoyiannis, D. Hydrognomon-Open Source Software for the Analysis of Hydrological Data. *Geophys. Res. Abstr.* **2010**, *12*, 12419.

44. Zhu, T.; Lund, J.R.; Jenkins, M.W.; Marques, G.F.; Ritzema, R.S. Climate change, urbanization, and optimal long-term floodplain protection. *Water Resour. Res.* **2007**, *43*. [CrossRef]

45. Kummu, M.; De Moel, H.; Ward, P.J.; Varis, O. How close do we live to water? A global analysis of population distance to freshwater bodies. *PloS ONE* **2011**, *6*. [CrossRef] [PubMed]

46. Jobe, A.; Kalra, A.; Ibendahl, E. Conservation reserve program effects on floodplain land cover management. *J. Environ. Manag.* **2018**, *214*, 305–314. [CrossRef] [PubMed]

47. Forsee, W.J.; Ahmad, S. Evaluating urban storm-water infrastructure design in response to projected climate change. *J. Hydrol. Eng.* **2011**, *16*, 865–873. [CrossRef]

48. Kalra, A.; Sagarika, S.; Pathak, P.; Ahmad, S. Hydro-climatological changes in the Colorado River basin over a century. *Hydrol. Sci. J.* **2017**, *62*, 2280–2296. [CrossRef]

 *water*

Article

Analyzing the Impacts of Climate Variability and Land Surface Changes on the Annual Water–Energy Balance in the Weihe River Basin of China

Wenjia Deng [1,2]**, Jinxi Song** [1,3,]*****, Hua Bai** [4]**, Yi He** [3]**, Miao Yu** [5]**, Huiyuan Wang** [3] **and Dandong Cheng** [1,2]

[1]   State Key Laboratory of Soil Erosion and Dryland Farming on the Loess Plateau, Institute of Soil and Water Conservation, Chinese Academy of Sciences and Ministry of Water Resources, Yangling, Xianyang 712100, China; wenjiadeng@outlook.com (W.D.); chengdandong@hotmail.com (D.C.)
[2]   University of Chinese Academy of Sciences, Beijing 100049, China
[3]   Shaanxi Key Laboratory of Earth Surface System and Environmental Carrying Capacity, College of Urban and Environmental Sciences, Northwest University, Xi'an 710127, China; yihe@nwu.edu.cn (Y.H.); huiyuanking@126.com (H.W.)
[4]   Jiangxi Key Laboratory of Hydrology-Water Resources and Water Environment, Nanchang Institute of Technology, Nanchang 330099, China; baihua1985@126.com
[5]   School of Urban and Environmental, Liaoning University, Dalian 116029, China; ym8815email@126.com
*****   Correspondence: jinxisong@nwu.edu.cn; Tel./Fax: +86-29-8830-8569

Received: 27 October 2018; Accepted: 3 December 2018; Published: 6 December 2018

Abstract: The serious soil erosion problems and decreased runoff of the Loess Plateau may aggravate the shortage of its local water resources. Understanding the spatiotemporal influences on runoff changes is important for water resource management. Here, we study this in the largest tributary of the Yellow River, the Weihe River Basin. Data from four hydrological stations (Lin Jia Cun (LJC), Xian Yang (XY), Lin Tong (LT), and Hua Xian (HX)) and 10 meteorological stations from 1961–2014 were used to analyze changes in annual runoff. The Mann–Kendall test and Pettitt abrupt change point test diagnosed variations in runoff in the Weihe River basin; the time periods before and after abrupt change points are the base period (period I) and change period (period II), respectively. Within the Budyko framework, the catchment properties (ω in Fu's equation) represent land surface changes; climate variability comprises precipitation (P) and potential evapotranspiration (ET_0). All the stations showed a reduction in annual runoff during the recording period, of which 22.66% to 50.42% was accounted for by land surface change and 1.97% to 53.32% by climate variability. In the Weihe River basin, land surface changes drive runoff variation in LT and climate variability drives it in LJC, XY, and HX. The contribution of land surface changes to runoff reduction in period I was less than that in period II, indicating that changes in human activity further decreased runoff. Therefore, this study offers a scientific basis for understanding runoff trends and driving forces, providing an important reference for social development, ecological construction, and water resource management.

Keywords: climate variability; land surface change; runoff; Budyko framework; elasticity coefficient; Weihe River Basin

---

1. Introduction

In a long-term hydrological system, changes in runoff can be influenced by many factors, such as climate change (including natural and anthropogenic climate change) and land surface changes (including changes in vegetation and agricultural irrigation and changes in water quantity and quality caused by various types of water use), which has triggered a series of questions about the water cycle [1,2]. In recent years, many river runoff values have shown decreasing trends, but it remains unclear whether these changes were caused by climate warming or human activities, and their relative contributions are also poorly understood. In response to this problem, domestic and foreign researchers have begun to attempt to quantitatively distinguish the impacts of human activities and climate change on hydrological processes. Due to the enormous influence of changes in rivers, ecological systems, social development, global climate, and recreation, many people are interested in understanding the trends of river changes and their driving forces. The response of runoff to land surface change in the Loess Plateau of China accounted for more than 50% of the decrease in mean annual runoff [3]. The impacts of climate variability and land surface changes on runoff in the upper reaches of the Yellow River Basin in China indicate that land use changes were responsible for more than 70% of the decrease in runoff in the 1990s [4]. Climate change has also led to changes in global precipitation patterns [5], and human activities have changed the spatial and temporal distribution of water resources [6]. The extent of the total change in runoff in the Loess Plateau and the degrees of influence of various factors on this change are variable. The integrated consequences of climatic variability and human activities are the main drivers impacting runoff change [7]. Climate variability is of vital importance if water resource management systems are to be sustainable [8]. Land use and water resources are closely linked in comprehensive management, in which land use is a key factor in the allocation of water resources in watersheds [9]. A considerable global variability of catchment hydrological processes has been observed in most basins throughout the world, and quantifying the effects of climate variability and human activities is crucial for the management and sustainable development of water resources [10].

Quantitative attribution methods have been developed to isolate the hydrological effects of climate variability and land surface changes [11], including the following: (i) statistical methods [12], (ii) the Budyko-based elasticity coefficient method [13], and (iii) hydrological modelling [14].

It is difficult to explain the physical mechanisms between runoff and climatic factors using statistical methods, thus long-term historical hydrological and meteorological data are usually required [4,15]. Regression relationship analyses [16], time-trend analyses [17,18], and change abrupt point analyses are the main statistical methods used in these studies. Vogel et al. [19] adopted the regional multivariate regression model and suggested that a 1% increase in precipitation would result in a 1.9% increase in annual runoff in the upper reaches of the Colorado River. Xu [12] showed that climate change can explain the observed reductions in natural runoff (72.9%) and annual observed runoff (78.6%) by using linear regression to assess the impact of climate change and human activities on the annual changes in runoff. Zhao et al. [20] used linear regression to analyze sediment load and runoff in the Yangtze River from 1953 to 2010 and showed that 72% of runoff and 14% of sediment load decreased due to the effects of climate change.

Schaake [21] originally proposed utilizing the elasticity coefficient to evaluate the effects of precipitation and potential evapotranspiration on changes in runoff. The most widely utilized methods based on the elasticity coefficient method include the nonparametric method and an analytical method based on the Budyko framework [22,23]. The Budyko hypothesis is a coupled water–energy balance equation developed by Budyko himself when conducting global water and energy balance analysis. The Budyko hypothesis indicates that actual evapotranspiration is controlled by the availability of water and energy by annual precipitation and potential evapotranspiration, respectively, on an annual scale. This approach is based on the principle of water–energy balance in terms of the long-term variability of hydrometeorological variables; it is a simple and practical approach used to study the hydrological response of the basin to environmental changes. The Budyko hypothesis is a useful way to investigate the relationship between hydrological processes, climate change, and land change characteristics [24,25]. The improved Budyko-based elasticity coefficient method [13] was proposed to assess the impact of climate change and human activities on runoff reduction; it was found that human activities accounted for 71–78% and climate change accounted for 22–29% of the changes in runoff reduction. The hydrological response to climate change and human activities in the Jingjiang

River Basin from 1961 to 2009 was determined using the Budyko hypothesis; the results showed that climate change accounted for >63% of the decrease in runoff [26].

The hydrological model was first used to predict changes in runoff caused by changes in land surface [27], which has recently become a major method by which to distinguish the hydrological response between climate variability and land surface change. In hydrological models, which include sound physical mechanisms, time-series continuity, and large amounts of observation data in their model structures and parameters, variations in the uncertainty associated with the model calibration will yield different results; for example, in the SWAT (Soil and Water Assessment Tool), TETIS, and Xinanjiang models, the parameters are often controversial due to the calibration and uncertainty of each model [28–30]. In recent years, a simple water balance model called the Budyko curve has been widely utilized to distinguish the effects of climate variability and land surface changes on runoff response [24,31].

The objectives of this study are as follows: (i) to statistically determine the trend and abrupt change points in the runoff data for the period 1961–2014; (ii) to differentiate between the effects of climate variability and land surface changes in time and space based on the elasticity coefficient; and (iii) to discuss the associations between different factors and their corresponding quantitative attributions. These findings will deepen our understanding of the runoff response to climate variability and land surface changes in the Weihe River Basin, which is essential for improving soil conservation measures and the sustainable development of water resource management.

2. Materials and Methods

2.1. Study Area

The Loess Plateau (35–41° N, 102–114° E) covers a total area of 624,000 km^2; it is located in the middle and upper reaches of the Yellow River, and it comprises an ancient loess deposit. The Wei River Basin (WRB; latitude 33.5° N–37.5° N; longitude 103.5° E–110.5° E), which is the largest tributary of the Yellow River (China), originates north of the Niaoshu Mountains at an altitude of 3485 m in Gansu Province; it then flows for 818 km, with a drainage area of 13.5×10^4 km^2, which accounts for 17.9% of the total area of the Yellow River Basin (Figure 1). The WRB runs from east to west through Gansu, Ningxia, and Shaanxi Provinces, and it plays an important role in the social, ecological, and economic development of these regions [32]. The watershed length of Shaanxi Province is approximately 502.4 km, with a basin area of approximately 6.67×10^4 km^2; this region is where the well-known Guanzhong Plain in northwest China is located. The WRB is characterized by a warm, semihumid continental monsoon climate with high precipitation and temperatures in summer and low precipitation and temperatures in winter [33,34]. The annual mean temperature in the WRB varies from 7.8 °C to 13.5 °C. The precipitation throughout the entire basin ranges from 558–750 mm, with over 60% of the annual precipitation falling in the summer monsoon period between June and September [35].

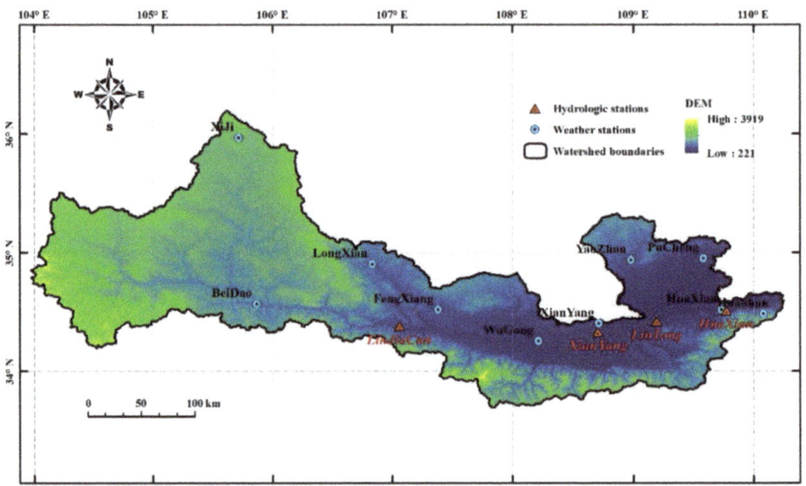

Figure 1. Geographic location of Weihe River basin and its hydrological and meteorological stations.

2.2. Data Collection

The monthly runoff data obtained from the 4 hydrological stations at Lin Jia Cun (LJC), Xian Yang (XY), Lin Tong (LT), and Huan Xian (HX), which were available for the period 1961–2014, were used in this study (WRB) and were collected from the Yellow River Conservancy Commission. LJC is located in the upper reaches of the WRB, XY is located in the middle stream, and LT and HX are located downstream. The control areas of the LJC, XY, LT, and HX stations are 30.661 km^2, 46.827 km^2, 97.299 km^2, and 106.498 km^2, respectively and they account for 22.7%, 34.7%, 72.2%, and 79% of the catchment, respectively. Daily metrological data (1961–2014) were collected from 10 stations in the WRB from the National Meteorological Data Sharing Service Platform (http://data.cma.cn), including daily precipitation and mean, maximum, and minimum temperature data at a height of 2 m; mean relative humidity and wind speed data at a height of 10 m; and daily sunshine duration data. The potential evapotranspiration data were calculated using the Penman–Monteith equation outlined in FAO-56 [36].

2.3. Methodology

2.3.1. Trend Analysis and Breakpoint Test

The nonparametric Mann–Kendall test method [37,38] is an effective tool for identifying trends. In this paper, the Mann–Kendall test method was used to examine the trend of hydrological data variables with the runoff series of the watershed, with a significance level of 0.05, and the magnitude and direction of the trend were also calculated. A positive trend indicates that the sequence has an increasing trend; a negative trend indicates that it has a decreasing trend. The Pettitt abrupt change point test [39] is a nonparametric method that can be used to detect abrupt changes in various variables. A given continuous time series (n1, n2, . . . , nT) is divided into a base period (period I; n1, n2, . . . , nT) and a change period (period II; nt + 1, nt + 2, . . . , nT), which is calculated based on the characteristic hydroclimatic values of the 2 subperiod series, then the degree of hydrologic climate change between the 2 subperiods is estimated.

2.3.2. Budyko Framework

The water–energy balance based on the Budyko hypothesis describes the relationship between precipitation (P), potential evotranspiration (ET_0), and runoff. In this study, the water balance in a given watershed is calculated using the following equation:

$$Q = P - ET - \Delta S \tag{1}$$

where Q is the runoff, P is the precipitation (mm), ET is the actual evapotranspiration, and ΔS is the change in water storage in the basin. The Budyko hypothesis assumes that under stable water balance conditions, the water storage capacity of the watershed can be neglected on a large time scale. According to Budyko [20], ET is a function of P, ET_0, and the controlling parameter ω (which represents land surface conditions). For simplicity, Fu's equation is expressed in the following form [40]:

$$\frac{ET}{P} = 1 + \frac{ET_0}{P} - \left[1 + \left(\frac{ET_0}{P}\right)^{\omega}\right]^{1/\omega} \tag{2}$$

or

$$\frac{E}{ET_0} = 1 + \frac{P}{ET_0} - \left[1 + \left(\frac{P}{ET_0}\right)^{\omega}\right]^{1/\omega} \tag{3}$$

where ω is a model parameter that is based on land cover, vegetation, soil infiltration, topography, and hydraulic properties. The water–energy coupled balance equation can be expressed in the following form:

$$Q = P \times \left[1 + \left(\frac{ET_0}{P}\right)^{\omega}\right]^{\frac{1}{\omega}} - ET_0 \tag{4}$$

or

$$Q = ET_0 \times \left[1 + \left(\frac{P}{ET_0}\right)^{\omega}\right]^{\frac{1}{\omega}} - ET_0 \tag{5}$$

Thus, the parameter ω in Fu's equation can be calculated based on ET_0, Q, and P, according to Equations (4) and (5).

2.3.3. Sensitivity Analysis

Based on Equations (4) and (5), the following differential forms can be used to assess the change in runoff:

$$dQ = \frac{\partial Q}{\partial p}dp + \frac{\partial Q}{\partial E_0}dET_0 + \frac{\partial Q}{\partial \omega}d\omega \tag{6}$$

Schaake [19] first proposed a method using climate elasticity to predict the impact of climate change on runoff and expressed the p, ET_0, and land surface change elasticity coefficients of runoff as $\varepsilon_P = \frac{dQ/Q}{dP/P}$, $\varepsilon_{ET_0} = \frac{dQ/Q}{dET_0/ET_0}$, and $\varepsilon_\omega = \frac{dQ/Q}{d\omega/\omega}$, respectively.

In this method, $\varnothing = \frac{ET_0}{P}$, ε_P, ε_{ET_0}, and ε_ω can be obtained as follows:

$$\varepsilon_P = \frac{(1+\varnothing^\omega)^{1/\omega+1} - \varnothing^{\omega+1}}{(1+\varnothing^\omega)\left[(1+\varnothing^\omega)^{1/\omega} - \varnothing\right]} \tag{7}$$

$$\varepsilon_{ET_0} = \frac{1}{(1+\varnothing^\omega)\left[1 - (1+\varnothing^{-\omega})^{1/\omega}\right]} \tag{8}$$

$$\varepsilon_\omega = \frac{\ln(1+\varnothing^\omega) + \varnothing^\omega \ln(1+\varnothing^{-\omega})}{\omega(1+\varnothing^\omega)[1 - (1+\varnothing^{-\omega})^{1/\omega}} \tag{9}$$

According to Equations (7)–(9), the ε_P, ε_{ET_0}, and ε_ω elasticity coefficients of annual runoff can be deduced based on P, ET_0, and land surface data, which reflect the average hydrological climate and underlying surface characteristics of the basin over many years.

2.3.4. Contribution Analysis

To differentiate between the effects of climate variability and human activities on runoff, the study period was divided into 2 subperiods: the base period (period I) and the change period (period II). According to Equation (6), the changes in mean annual runoff are caused by changes in climate and human activities, and total changes can be expressed as:

$$Q\,(\text{all}) = \text{Con}(P) + \text{Con}(ET_0) + \text{Con}(\omega) \tag{10}$$

where Q (all) is the total change in annual mean runoff due to climate variability and human activities and Con(P), Con(ET_0), and Con(ω) are the contributions of changes in P, ET_0, and land surface, respectively, to changes in Q. The contribution of each variable to the change in runoff can be expressed as follows:

$$E_{\_\text{Con}(P)} = \frac{\text{Con}(P)}{Q\,(\text{all})} \times 100\% \tag{11}$$

$$E_{\_\text{Con}(ET_0)} = \frac{\text{Con}(ET_0)}{Q\,(\text{all})} \times 100 \tag{12}$$

$$E_{\_\text{Con}(\omega)} = \frac{\text{Con}(\omega)}{Q\,(\text{all})} \times 100\% \tag{13}$$

where $E_{\_\text{Con}(P)}$, $E_{\_\text{Con}(ET_0)}$, and $E_{\_\text{Con}(\omega)}$ are the percentages representing the contribution of each variable to the total decrease in runoff. These contributions can be explored and used to provide information for water resource management.

3. Results

3.1. Changes in Long-Term Hydrometeorological Variables

The Mann–Kendall trend was applied to detect the trend and significance of variables during the period 1961–2014. As indicated in Table 1, at all four hydrological stations in the WRB, the annual runoff showed a significant downward trend (at confidence levels of 0.05 (*) and 0.01 (**)). The annual runoff at LJC, XY, LT, and HX decreased by 0.45, 0.71, 0.75, and 0.83 mm yr^{-1}, respectively, representing 2.4%, 1.9%, 1.2%, and 3.0%, respectively, of the corresponding annual runoff. In addition, the four stations showed minor downward trends in P, which decreased in LJC, XY, LT, and HX by 0.66, 0.77, 0.50, and 1.08 mm yr^{-1}, respectively. ET_0 showed an insignificant upward trend in LJC and significant decreases in XY and HX. To summarize, runoff and precipitation exhibit the same downward trend, thus indicating that climate variability may have an important impact on the reduction of runoff.

The Pettitt abrupt change point test was applied to detect the change points in the annual runoff series during the period 1961–2014 at these four hydrological stations. The results show that the abrupt changes in annual runoff at LJC, XY, and HX occurred in 1993 and at LT in 1990. These four abrupt change points are thus basically consistent, as they all occurred in the early 1990s. The entire research period was then divided into two subperiods to quantify the effects of changes in climate variability and human activities on runoff. The runoff data at the four hydrological stations before the abrupt change points exhibited a downward trend, with LJC and XY exhibiting a significant downward trend. After the abrupt change points, the runoff data exhibited an upward trend, with significant increases at XY and HX.

Table 1. Averages and slope of runoff (R), precipitation (P), and potential evotranspiration (ET_0) in the Weihe River basin during 1961–2014. LJC, Lin Jia Cun; XY, Xian Yang; LT, Lin Tong; HX, Hua Xian.

		LJC			XY			LT			HX		
		Period I 1961–1992	Period II 1993–2014	Whole	Period I 1961–1992	Period II 1993–2014	Whole	Period I 1961–1989	Period II 1990–2014	Whole	Period I 1961–1992	Period II 1993–2014	Whole
R (mm yr^{-1})	Average	25.2128	10.4614	19.203	47.4119	23.6098	37.7147	77.511	49.6476	64.6113	77.5528	45.9295	64.6693
	Slope	−0.40782 *	0.14971	−0.45334 **	−0.85665 *	0.94667 **	−0.708 **	−0.99952	0.68827	−0.75 **	−0.96486	1.3153 *	−0.83407 **
P (mm yr^{-1})	Average	543.6715	508.4231	529.311	576.6696	531.1751	558.1348	574.4519	535.9655	556.6342	606.7371	551.992	584.4335
	Slope	0.26493	4.375	−0.66121	−0.23257	6.8705 *	−0.76634	−0.30765	3.8812	−0.49851	−0.53298	6.001 *	−1.0773
ET_0 (mm yr^{-1})	Average	952.904	976.4785	962.5085	976.0936	968.8985	973.1623	988.01	995.2711	991.3716	1020.1251	1013.678	1017.4987
	Slope	−2.6822 *	1.2068	0.23939	−4.4722 **	−0.40235	−1.2491 *	−3.2497 *	1.9376	−0.054167	−4.3096 **	−1.6731	−1.175 *

$* p < 0.05; ** p < 0.01.$

At all four hydrological stations, P (except at LJC) and ET_0 showed a downward trend in period I, and ET_0 showed a significant downward trend; in period II, P and ET_0 both exhibited an increasing trend at all stations except for HX.

3.2. Elasticity Coefficients on Runoff

The catchment property (ω) characterizes the water balance parameters based on the long-term mean annual values of ET_0, runoff depth, P, and the actual evapotranspiration data ($E = P - Q$) covering the entire period of the study (1961–2014). In Fu's equation, the optimization problem of parameter ω is solved by using the programming solution tool in Excel to more accurately express the elasticity coefficient. As shown in Table 2, the optimal values of ω for LJC, XY, LT, and HX are 2.94, 3.10, 2.47, and 2.92, respectively.

Table 2. Elasticity coefficients of runoff to climate variability and land surface change; the best calibration value of ω.

Station	ω	Period	ε_{ET0}	ε_P	ε_ω
LJC	2.94	Whole	−2.62	3.62	−2.60
		1961–1992	−2.58	3.58	−2.49
		1993–2014	−2.66	3.66	−2.77
XY	3.10	Whole	−2.74	3.74	−2.59
		1961–1992	−2.71	3.71	−2.49
		1993–2014	−2.78	3.78	−2.74
LT	2.47	Whole	−2.11	3.11	−2.25
		1961–1989	−2.08	3.08	−2.16
		1990–2014	−2.14	3.14	−2.35
HX	2.92	Whole	−2.56	3.56	−2.47
		1961–1992	−2.53	3.53	−2.36
		1993–2014	−2.61	3.61	−2.64

Table 2 shows that runoff is positively correlated with P and negatively correlated with ET_0 and ω; these trends are exactly the same in each sub-basin and subperiod. The ranges of the elasticity coefficients of ET_0, P, and ω are −2.08 to −2.78, 3.08 to 3.78, and −2.16 to −2.77, respectively. From these results, we can conclude that when P, ET_0, and ω all decrease by 1%, the annual runoff of each typical watershed will decrease by 3.08–3.78, increase by 2.08–2.74, and increase by 2.16–2.77, respectively. The changes in the elasticity coefficient due to climate change decreased from big to small, followed by XY, LJC, HX, and LT; in contrast, the changes in the elasticity coefficient due to human activities decreased from LJC to XY, HX, and LT. The elasticity coefficients of the four hydrological stations, ET_0, P, and ω were greater after than before the abrupt change. It can be seen from Figure 2 that the elasticity coefficients (P, ET_0, and ω) are higher in the upstream region of ZJS and the vicinity of the XY hydrological station than in the XY and HX stations, which is consistent with the data shown in Table 2.

Figure 2 shows the annual elasticity coefficients of the four hydrological stations from 1961 to 2014; this not only shows the annual changes in the elasticity coefficients due to climate variability and land surface changes, but also reflects the spatial variability of the sensitivity of the WRB to these variables. In terms of the absolute values of the elasticity coefficients, those of climate change (ET_0 and P) and land surface change (ω) in LJC, XY, and HX are higher than those in LT. The absolute value observed at the LT hydrological station is smaller than those observed at the other three stations, which may indicate that the effects of human activities at the LT station are relatively small.

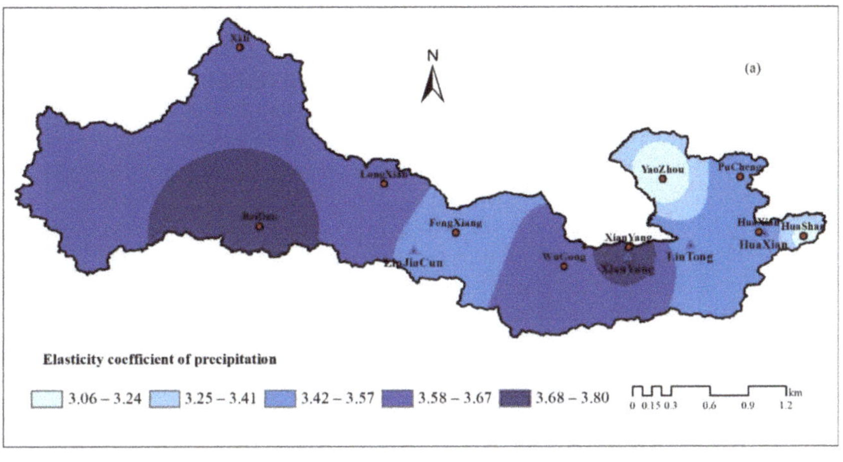

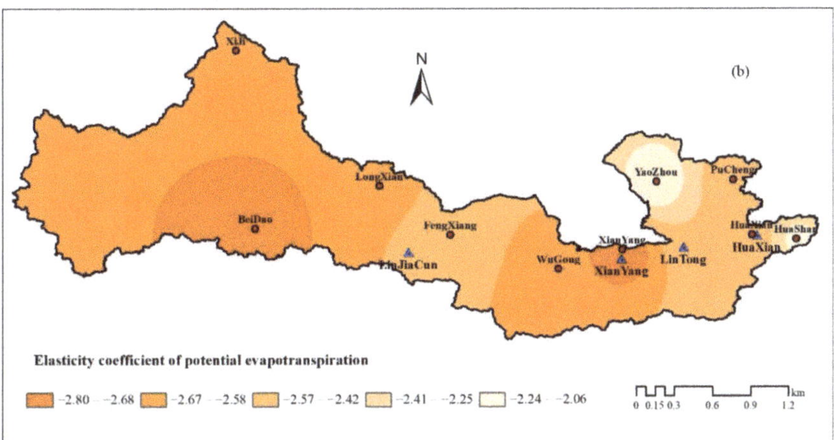

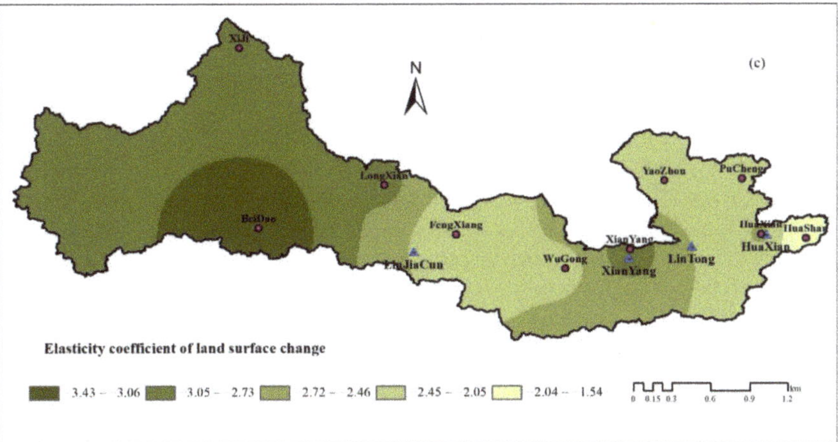

Figure 2. Spatial variation characteristics of the elasticity coefficient of runoff in the Weihe River Basin: elasticity coefficient of (**a**) precipitation; (**b**) potential evapotranspiration; (**c**) land surface change.

3.3. Attribution Analysis of Runoff Reduction

Changes in either of these factors, i.e., climate change (P and ET_0) or land surface change (ω), may lead to changes in runoff. The influence of changes in all variables on runoff was estimated using hydrological sensitivity analysis and Equations (11)–(13). Table 3 shows that from 1961 to 2014 in the WRB, the decrease in P had a positive effect on runoff, with an average contribution rate of 47.6%; ET_0 and ω mainly had negative effects on runoff, with average contribution rates of 14.8% and 37.6%, respectively. Table 3 also shows that at the LJC, XY, and HX hydrological stations, the average proportions of P, ET_0, and ω decreased in the order p > ω > ET_0, in which the proportion of P at LJC and HX always exceeded 50%. It can be seen that in terms of the three indicators (P, ET_0, ω), the change in P has the greatest effect on the change in runoff, the change in ω has an intermediate effect, and the change in ET_0 has the smallest effect.

Table 3. Contributions of P, ET_0, and ω to changes in runoff.

Station	Period		RC_(P) (%)	RC_(ET0) (%)	RC_(ω) (%)
	Period I	1961–1992	9.38	39.05	51.58
LJC	Period II	1993–2014	50.51	5.27	44.22
	Whole	1961–2014	51.76	7.46	40.78
	Period I	1961–1992	5.6	46.46	47.94
XY	Period II	1993–2014	47.72	1.13	51.15
	Whole	1961–2014	37.58	25.74	36.68
	Period I	1961–1989	12.07	50.07	37.87
LT	Period II	1990–2014	52.31	9.59	38.1
	Whole	1961–2014	47.61	1.97	50.42
	Period I	1961–1989	14.22	49.02	36.76
HX	Period II	1993–2014	43.51	4.77	51.71
	Whole	1961–2014	53.32	24.02	22.66

The main factors affecting the changes in runoff during periods I and II at different hydrological stations are different. At all stations except LJC, the proportions of P and ω in period II increased relative to those in period I; however, the proportions of ET_0 in period I were significantly lower than those in period II. The decrease in runoff at LJC and LT for period II (the change period, i.e., 1993–2014 for LJC and 1990–2014 for LT) was controlled by changes in P (which contributed 50.51% and 52.31%, respectively). However, land surface change was the main controlling factor at XY and HX for period II (the change period, i.e., 1993–2014 for XY and HX). Based on these results, we conclude that precipitation and land surface changes were the main driving forces of the changes in runoff observed at these hydrological stations.

4. Discussion

Assessing the continuity and integrity of data is becoming a fundamental problem that cannot be ignored in data-based research. During the preparation, analysis, and presentation of data, confirming its continuity and integrity guarantees the completeness, validity, and accuracy of the results. Data from four hydrological stations and 10 meteorological stations were used in the study area; however, these data may not be sufficient to cover the Weihe River Basin, which has a drainage area of 13.5×10^4 km^2. This may thus limit the calculation of the PET data and the accuracy of the estimated runoff. The quantitative estimation was based on the assumption that climate variability and land surface changes are independent. However, land surfaces interact with climate systems, especially at the catchment scale [4]. The data used in the hydrological sensitivity analysis method were derived from periodic runoff data, which are not affected by long-term human activities and are thus used for calibration of the model. In the Weihe River Basin during the runoff observation period, disturbances were caused by human activities, such as the building of reservoirs and dams [41,42], which may restrict the accuracy of the observation data and model parameters. The expansion of the Taylor series based on the Budyko equation may be another cause of data uncertainty. The first-order Taylor expansion has been widely studied and used in many studies; however, with the decrease in P and increases in ET_0 and ω, uncertainty in the data will increase. Therefore, eliminating this uncertainty to improve the accuracy of prediction is a key issue that needs to be resolved in future research.

Based on the Budyko equation of Fu and taking P, ET_0, and Q as variables, the parameter ω was determined using the programming solution function in Excel. The precision of the hydrological model parameters can be improved by using the ω value determined by the programming solution function to better explain the accuracy of runoff reduction in the WRB. The catchment characteristic parameter ω in the Budyko equation is related to topography [43], soil [44], vegetation [45,46], and climate seasonality [47]. The land surface changes in the WRB were mainly due to the implementation of water and soil conservation and ecological restoration activities. In the short term, the soil properties and topographic changes remain unchanged, while changes in the vegetation cover and climate are considered to influence the characteristic parameter ω. The normalized difference vegetation index (NDVI) is the most widely used index to characterize vegetation coverage (V) status and can adequately reflect vegetation cover, growth vitality, and biomass. All other conditions being similar, the parameter ω of a drainage basin with a larger vegetation cover is generally larger than that of a drainage basin with smaller vegetation coverage. Generally, the average NDVI has fluctuated upward in recent decades in the WRB. The relationship between V and ω showed a similar interannual change from 1961 to 2014, which indicated that the change in vegetation had a positive effect on ω.

In addition to changes in vegetation, changes in soil and water conservation measures have significantly altered the surface, including biological measures (e.g., afforestation and pasture improvement) and engineering measures (e.g., dams, terraces, and reservoirs). For example, by the end of 2000, more than 3150 reservoirs had been built in the Yellow River Basin, including 171 large- and medium-sized reservoirs with a total capacity of 22.6 km^3. A study conducted by Liang et al. [48] indicated that key dams (e.g., the LJC dam) have altered land surface changes and resulted in noticeable changes in the hydrological regime [49], which could intercept stormwater runoff for a short period during flood seasons and allow more time for infiltration [50]. Land surface changes caused by the impact of soil and water conservation measures, such as the construction of reservoirs and dams, could significantly change the availability of natural water resources in the region. Moreover, many studies have shown that climate change cannot be regarded as a unilateral phenomenon, which means that the climate drive and feedback of local infrastructure such as dams cannot be ignored. Soil and water conservation, especially the implementation of the "returning farmland to forests and grassland" policy in 1999, has thus affected the hydrological processes in this area. From 1959 to 2006, the terraced field area increased from 5.30 to 285.40 104 km^2, and the growth rate has only increased since then. Zhang et al. [51] previously showed that the conversion of sloping farmland to terraces can result in a significant decrease in runoff. It can be seen that the implementation of soil and water conservation measures has achieved significant ecological benefits and has had temporal and spatial effects on hydrological processes in recent years; however, these measures are also one of the most important reasons for the decrease in runoff.

Due to the high-scale dependence of runoff on rainfall, the impact of land surface change has been limited; however, climate change has been the main driving force of large-scale changes [12]. Table 3 shows that the relationship between runoff, average annual precipitation, and ω was relatively close and that this change has exhibited good consistency in the WRB. The runoff values of almost all hydrological stations in the WRB obviously decreased in the 1990s; in terms of climate change, runoff was more sensitive to changes in precipitation than it was to changes in potential evapotranspiration. Zuo et al. [52] concluded that the impacts of climate variability and land surface changes on runoff were 31%–51% and 33%–65%, respectively, which is consistent with the results obtained in this study. The WRB contains rainfall-recharge rivers, thus changes in its precipitation-runoff have continually been a focus of scholars. However, as can be seen in Table 3, whether in the total period or in the subperiod, the proportion of ω is always relatively small (i.e., below 50%). There is no significant relationship between ω and V. At the four hydrological stations, the proportion of ω was greater than that before the abrupt change (i.e., the abrupt change for LJC, XY, and HX occurred in 1993 and at LT occurred in 1990). Previous studies have proposed that the best vegetation in the WRB occurred in 1990, indicating that the ecological environment there generally improved from the 1980s to the 1990s. In 2000, V (vegetation coverage) was significantly worse than it was in other years, but it improved from 2000 to 2005; nevertheless, there was a small decline in 2007. Generally speaking, the average V value has fluctuated upward in recent decades in the WRB.

When comparing period I and period II (in Table 3), P and ω increased almost at all stations in period II, which may be due to the fact that the ω part represents the V value, and with the increase of V, the precipitation significantly increased. Wang et al. [11] concluded that the contribution of climate variability and land surface change to runoff evolution temporally varies, which is in line with the results of periods I and II. This has considerable significance for water resources utilization management. The traditional water management strategies do not adequately take into account hydrological regime changes over time, especially changes in climatic conditions. Climate variability also greatly impacts stakeholders, e.g., farming households and government decision makers. Stakeholders, in return, deal with climate variability in a host of ways, e.g., with water use rights and regulations. Incorporating climate variability puts forward new thinking on the development of water resources [9,53]. Changes in climate conditions alter the processes of the hydrological cycle and affect the structure of the water resources system, bringing new challenges to the development and utilization of water resources worldwide. Consequently, we recommend that adaptive management of water resources should be implemented to match environmental changes. Adaptive management enables regional river and reservoir systems to sustain and even strengthen the interests of all stakeholders in the context of climate change [54]. Finally, through the implementation of a strict water resources management system, adaptive countermeasures including extensive development of water resources could be achieved.

5. Conclusions

The annual changes in climate variability and land surfaces in the WRB are based on data collected from meteorological and hydrologic stations from 1961 to 2014, and quantitative analysis of the contributions of runoff change can be an effective part of the water management of the Loess Plateau. Taking stations LJC, XY, LT, and HX as examples, the runoff data were analyzed by trend analysis and the mutation point test, and the entire period was divided into two subperiods (base period and change period). The elasticity coefficient method, which is based on the Budyko framework, was used in this study. The runoff data of all hydrological stations showed a significant downward trend. The abrupt change points of the four stations were basically consistent and occurred in the early 1990s. Runoff was positively correlated with precipitation and negatively correlated with potential evapotranspiration and ω, which were exactly the same in each sub-basin and subperiod. Three indicators (P, ET_0, ω) contribute the most to changes in runoff; the change in P has the largest contribution, the change in ω has an intermediate contribution, and the change in ET_0 has the smallest contribution. The proportion of land surface change is relatively small, owing partly to the construction of soil and water conservation facilities, such as check dams and terraces, and partly to the measures of returning farmland to forests and grassland.

Two avenues should be explored in future research. For one thing, the empirical formula of parameter ω should be calculated through a stepwise regression analysis, and subsequently, mean annual and interannual changes should be predicted. Such measures would ultimately propose a universal equation of to apply to different watersheds, providing a powerful tool for assessing water–energy balance using the Budyko hypothesis. For another, during the snowy season, a warming climate may lead to less precipitation in the form of snow, which not only alters the temporal distribution of annual runoff, but also leads to a reduction in the total annual runoff. Therefore, the effect of snow on runoff as well as its related process on the catchment scale will be discussed. How snow couples with the Budyko hypothesis needs to be answered in future research.

Author Contributions: J.S. contributed to conceiving and designing the research; W.D. and M.Y. analyzed the data; H.B. and H.W. contributed analysis tools; Y.H. and D.C. provided some guiding suggestions; W.D. wrote the paper.

Funding: This study was jointly supported by the National Natural Science Foundation of China (Grant Nos. 51679200 and 51379175), the Hundred Talents Project of the Chinese Academy of Sciences (Grant No. A315021406), the National Key R&D Program on Monitoring, Early Warning, and Prevention of Major Natural Disasters (Grant No. 2017YFC1502506), and the Program for Key Science and Technology Innovation Team in Shaanxi Province (Grant No. 2014KCT-27). Part of the data were provided by the Science Data Centre of Loess Plateau from the National Data Sharing Infrastructure of Earth System Science.

Acknowledgments: We are grateful to the editor and anonymous reviewers who provided numerous comments and suggestions, resulting in an improved manuscript.

Conflicts of Interest: The authors declare no conflict of interest.

References

1. Hao, X.; Chen, Y.; Xu, C.; Li, W. Impacts of climate change and human activities on the surface runoff in the Tarim River Basin over the last fifty years. *Water Resour. Manag.* **2007**, *22*, 1159–1171. [CrossRef]

2. Murray, S.J.; Foster, P.N.; Prentice, I.C. Evaluation of global continental hydrology as simulated by the land-surface processes and exchanges Dynamic Global Vegetation Model. *Hydrol. Earth Syst. Sci.* **2011**, *15*, 91–105. [CrossRef]

3. Zhang, X.; Zhang, L.; Zhao, J.; Rustomji, P.; Hairsine, P. Responses of streamflow to changes in climate and land use/cover in the Loess Plateau, China. *Water Resour. Res.* **2008**, *44*. [CrossRef]

4. Zheng, H.; Zhang, L.; Zhu, R.; Liu, C.; Sato, Y.; Fukushima, Y. Responses of streamflow to climate and land surface change in the headwaters of the Yellow River Basin. *Water Resour. Res.* **2009**, *45*. [CrossRef]

5. Dore, M.H. Climate change and changes in global precipitation patterns: What do we know? *Environ. Int.* **2005**, *31*, 1167–1181. [CrossRef] [PubMed]

6. Yang, D.; Sun, F.; Liu, Z.; Cong, Z.; Ni, G.; Lei, Z. Analyzing spatial and temporal variability of annual water-energy balance in nonhumid regions of China using the Budyko hypothesis. *Water Resour. Res.* **2007**, *43*, 436–451. [CrossRef]

7. Li, L.; Zhang, L.; Wang, H.; Wang, J.; Yang, J.; Jiang, D.; Li, J.; Qin, D. Assessing the impact of climate variability and human activities on streamflow from the Wuding River Basin in China. *Hydrol. Process.* **2007**, *21*, 3485–3491. [CrossRef]

8. Zhao, G.; Tian, P.; Mu, X.; Jiao, J.; Wang, F.; Gao, P. Quantifying the impact of climate variability and human activities on streamflow in the middle reaches of the Yellow River basin, China. *J. Hydrol.* **2014**, *519*, 387–398. [CrossRef]

9. Akiyama, T.; Kharrazi, A.; Li, J.; Avtar, R. Agricultural water policy reforms in China: A representative look at Zhangye City, Gansu Province, China. *Environ Monit Assess.* **2018**, *190*, 9. [CrossRef] [PubMed]

10. Zhang, L.; Podlasly, C.; Ren, Y.; Feger, K.; Wang, Y.; Schwärzel, K. Separating the effects of changes in land management and climatic conditions on long-term streamflow trends analyzed for a small catchment in the Loess Plateau region, NW China. *Hydrol. Process.* **2014**, *28*, 1284–1293. [CrossRef]

11. Wang, S.; Zhang, Z.; McVicar, T.R.; Guo, J.; Tang, Y.; Yao, A. Isolating the impacts of climate change and land use change on decadal streamflow variation: Assessing three complementary approaches. *J. Hydrol.* **2013**, *507*, 63–74. [CrossRef]

12. Xu, J. Variation in annual runoff of the Wudinghe River as influenced by climate change and human activity. *Q. Int.* **2011**, *244*, 230–237. [CrossRef]

13. Zhan, C.; Jiang, S.; Sun, F.; Jia, Y.; Niu, C.; Yue, W. Quantitative contribution of climate change and human activities to runoff changes in the Wei River basin, China. *Hydrol. Earth Syst. Sci.* **2014**, *18*, 3069–3077. [CrossRef]

14. Bronstert, A.; Niehoff, D.; Bürger, G. Effects of climate and land-use change on storm runoff generation: Present knowledge and modelling capabilities. *Hydrol. Process.* **2002**, *16*, 509–529. [CrossRef]

15. Sankarasubramanian, A.; Vogel, R.M.; Limbrunner, J.F. Climate elasticity of streamflow in the United States. *Water Resour. Res.* **2001**, *37*, 1771–1781. [CrossRef]

16. Ma, Z.; Kang, S.; Zhang, L.; Tong, L.; Su, X. Analysis of impacts of climate variability and human activity on streamflow for a river basin in arid region of northwest China. *J. Hydrol.* **2008**, *352*, 239–249. [CrossRef]

17. Zhang, L.; Zhao, F.; Chen, Y.; Dixon, R.N.M. Estimating effects of plantation expansion and climate variability on streamflow for catchment in Australia. *Water Resour. Res.* **2011**, *47*. [CrossRef]

18. Zhang, Y.; Guan, D.; Jin, C.; Wang, A.; Wu, J.; Yuan, F. Analysis of impacts of climate variability and human activity on streamflow for a river basin in northeast China. *J. Hydrol.* **2011**, *410*, 239–247. [CrossRef]

19. Vogel, R.M.; Wilson, I.; Daly, C. Regional regression models of annual streamflow for the United States. *J. Irrig. Drain Eng.* **1999**, *125*, 148–157. [CrossRef]

20. Zhao, Y.; Zou, X.; Gao, J.; Xu, X.; Wang, C.; Tang, D.; Wang, T.; Wu, X. Quantifying the anthropogenic and climate contributions to changes in water discharge and sediment load into the sea: A case study of the Yangtze River, China. *Sci. Total Environ.* **2015**, *536*, 803–812. [CrossRef]

21. Schaake, J.C. From climate to flow. In *Climate Change and U.S. Water Resources*; Waggoner, P.E., Ed.; John Wiley: New York, NY, USA, 1990; pp. 177–206.

22. Budyko, M.I. *Climate and Life*; Academic Press: New York, NY, USA, 1974.

23. Dooge, J.C. Sensitivity of runoff to climate change: A Hortonian approach. *Bull. Am. Meteorol. Soc.* **1992**, *73*, 2013–2024. [CrossRef]

24. Milly, P.C.D.; Dunne, K.A. Macroscale water fluxes 2. Water and energy supply control of their interannual variability. *Water Resour. Res.* **2002**, *38*, 1287–1295. [CrossRef]

25. Yokoo, Y.; Sivapalan, M.; Oki, T. Investigating the roles of climate seasonality and landscape characteristics on mean annual and monthly water balances. *J. Hydrol.* **2008**, *357*, 255–269. [CrossRef]

26. Gao, G.; Fu, B.; Wang, S.; Liang, W.; Jiang, X. Determing the hydrological responses to climate variability and land use/cover change in the Loess Plateau with the Budyko framework. *Sci. Total Environ.* **2016**, *557–558*, 331–342. [CrossRef] [PubMed]

27. Onstad, C.A.; Jamieson, D.G. Modeling the effect of land use modifications on runoff. *Water Resour. Res.* **1970**, *6*, 1287–1295. [CrossRef]

28. Serpa, D.; Nunes, J.P.; Santos, J.; Sampaio, E.; Jacinto, R.; Veiga, S.; Lima, J.C.; Mpreira, M.; Corte-Real, J.; Keizer, J.J.; et al. Impacts of climate and land use changes on the hydrological and erosion processes of two contrasting Mediterranean catchments. *Sci. Total Environ.* **2015**, *538*, 64–77. [CrossRef] [PubMed]

29. Buendia, C.; Bussi, G.; Tuset, J.; Vericat, D.; Sabater, S.; Palau, A.; Batalla, R.J. Effects of afforestation on runoff and sediment load in an upland Mediterranean catchment. *Sci. Total Environ.* **2015**, *540*, 144–157. [CrossRef]

30. Lü, H.; Hou, T.; Horton, R.; Zhu, Y.; Chen, X.; Jia, Y.; Wang, W.; Fu, X. The streamflow estimation using the Xinan jiang rainfall runoff model and dual state-parameter estimation method. *J. Hydrol.* **2013**, *480*, 102–114. [CrossRef]

31. Wang, W.; Shao, Q.; Yang, T.; Peng, S.; Xing, W.; Sun, F.; Luo, Y. Quantitative assessment of the impact of climate variability and human activities on runoff changes: A case study in four catchments of the Haihe River Basin, China. *Hydrol. Process.* **2013**, *27*, 1158–1174. [CrossRef]

32. Song, J.; Xu, Z.; Hui, Y.; Li, H.; Li, Q. Instream flow requirements for sediment transport in the lower Weihe River. *Hydrol. Process.* **2010**, *24*, 3547–3557. [CrossRef]

33. Huang, S.; Hou, B.; Chang, J.; Huang, Q.; Chen, Y. Spatial-temporal change in precipitation patterns based on the cloud model across the Wei river basin, China. *Theor. Appl. Climatol.* **2014**, *120*, 391–401. [CrossRef]

34. Zhu, Y.; Huang, S.; Chang, J.; Leng, G. Spatial–temporal changes in potential evaporation patterns based on the Cloud model and their possible causes. *Stoch. Environ. Res. Risk Assess* **2017**, *31*, 2147–2158. [CrossRef]

35. Liu, S.; Huang, S.; Huang, Q.; Xie, Y.; Leng, G.; Luan, J.; Song, X.; Wei, X.; Li, X. Identification of the non-stationarity of extreme precipitation events and correlations with large-scale ocean-atmospheric circulation patterns: A case study in the Wei River Basin, China. *J. Hydrol.* **2017**, *548*, 184–195. [CrossRef]

36. Allen, R.G.; Pereira, L.S.; Raes, D.; Smith, M. *Crop Evapotranspiration: Guidelines for Computing Crop Water Requirements*; Food and Agriculture Organization: Rome, Italy, 1998.

37. Mann, H.B. Non-parametric test against trend. *Econometrica* **1945**, *13*, 245–259. [CrossRef]

38. Kendall, M.G. *Rank Correlation Measures*; Charles Griffin: London, UK, 1975.

39. Pettitt, A. A nonparametric approach to the change-point problem. *Appl. Stat.* **1979**, *28*, 126–135. [CrossRef]

40. Fu, B.P. The calculation of the evaporation from land surface. *Sci. Atmos. Sin.* **1981**, *5*, 23–31.

41. Chen, J.; Wang, F.; Meybeck, M.; He, D.; Xia, X.; Zhang, L. Spatial and temporal analysis of water chemistry records (1958–2000) in the Huanghe (Yellow River) basin. *Glob. Biogeochem. Cycles* **2005**, *19*. [CrossRef]

42. Wang, H.; Yang, S.; Saito, Y.; Liu, P.; Sun, X. Interannual and seasonal variation of the Huanghe (Yellow River) water discharge over the past 50 years: Connections to impacts from ENSO events and dams. *Glob. Planet. Chang.* **2006**, *50*, 212–225. [CrossRef]

43. Yang, H.; Qi, J.; Xu, X.; Yang, D.; Lv, H. The regional variation in climate elasticity and climate contribution to runoff across China. *J. Hydrol.* **2014**, *517*, 607–616. [CrossRef]

44. Price, K. Effects of watershed topography, soils, land use and climate on baseflow in humid regions: A review. *Prog. Phys. Geogr.* **2011**, *35*, 465–492. [CrossRef]

45. Donohue, R.J.; Roderick, M.L.; McVicar, T.R. On the importance of including vegetation dynamics in Budyko's hydrological model. *Hydrol. Earth Syst. Sci. Discuss.* **2006**, *3*, 1517–1551. [CrossRef]

46. Li, D.; Pan, M.; Cong, Z.; Zhang, L.; Wood, E. Vegetation control on water and energy balance within the Budyko framework. *Water Resour. Res.* **2013**, *49*, 969–976. [CrossRef]

47. Ning, T.; Li, Z.; Liu, W. Vegetation dynamics and climate seasonality jointly control the interannual catchment water balance in the Loess plateau under the Budyko framework. *Hydrol. Earth Syst. Sci.* **2017**, *21*, 1515–1526. [CrossRef]

48. Liang, W.; Bai, D.; Wang, F.; Fu, B.; Yan, J.; Wang, S.; Yang, Y.; Long, D.; Feng, M. Quantifying the impacts of climate change and ecological restoration on streamflow changes based on a Budyko hydrological model in China's Loess Plateau. *Water Resour. Res.* **2015**, *51*, 6500–6519. [CrossRef]

49. Mu, X.; Zhang, L.; McVicar, T.; Chille, B.; Gau, P. Analysis of the impact of conservation measures on stream flow regime in catchments of the Loess Plateau, China. *Hydrol. Process.* **2007**, *21*, 2124–2134. [CrossRef]

50. Polyakov, V.; Nichols, M.; McClaran, M.; Nearing, M. Effect of check dams on runoff, sediment yield, and retention on small semiarid watersheds. *J. Soil Water Conserv.* **2014**, *69*, 414–421. [CrossRef]

51. Zhang, L.; Podlasly, C.; Feger, K.; Wang, Y.; Schwaerzel, K. Different land management measures and climate change impacts on the runoff—A simple empirical method derived in a mesoscale catchment on the Loess Plateau. *J. Arid Environ.* **2015**, *120*, 42–50. [CrossRef]

52. Zuo, D.; Xu, Z.; Wu, W.; Zhao, J.; Zhao, F. Identification of streamflow response to climate change and human activities in the Wei river basin, China. *Water Resour. Manag.* **2014**, *28*, 833–851. [CrossRef]

53. Urcola, H.; Elverdin, J.; Mosciaro, M.; Albaladejo, C.; Manchado, J.; Giussepucci, J. Climate change impacts on rural societies: Stakeholders perceptions and adaptation strategies in Buenos Aires, Argentina. In *Innovation & Sustainable Development in Agriculture & Food*; FRA: Montpellier, France, 2010; pp. 489–492.

54. Georgakakos, K.; Graham, N.; Cheng, F.; Spencer, C.; Shamir, E.; Georgakakos, A.; Yao, H.; Kistenmacher, M. Value of adaptive water resources management in northern California under climatic variability and change: Dynamic hydroclimatology. *J. Hydrol.* **2012**, *412–413*, 47–65. [CrossRef]

Article

Quantifying the Impact of Climate Change and Human Activities on Streamflow in a Semi-Arid Watershed with the Budyko Equation Incorporating Dynamic Vegetation Information

Lei Tian [1,3], Jiming Jin [2,3,*], Pute Wu [1,4,*] and Guo-yue Niu [5,6]

1   Institute of Soil and Water Conservation, Northwest A&F University, Yangling 712100, China; lei.tian@aggiemail.usu.edu
2   College of Water Resources and Architectural Engineering, Northwest A&F University, Yangling 712100, China
3   Department of Watershed Sciences, Utah State University, Logan, UT 84322, USA
4   National Engineering Research Center of Water Saving and Irrigation Technology at Yangling, Yangling 712100, China
5   Biosphere 2, University of Arizona, Tucson, AZ 85623, USA; niug@email.arizona.edu
6   Department of Hydrology & Atmospheric Sciences, University of Arizona, Tucson, AZ 85721, USA
*   Correspondence: Jiming.Jin@usu.edu (J.J.); gjzwpt@vip.sina.com (P.W.);
    Tel.: +1-435-213-5183 (J.J.); +86-29-8701-1354 (P.W.)

Received: 13 November 2018; Accepted: 1 December 2018; Published: 4 December 2018

Abstract: Understanding hydrological responses to climate change and land use and land cover change (LULCC) is important for water resource planning and management, especially for water-limited areas. The annual streamflow of the Wuding River Watershed (WRW), the largest sediment source of the Yellow River in China, has decreased significantly over the past 50 years at a rate of 5.2 mm/decade. Using the Budyko equation, this study investigated this decrease with the contributions from climate change and LULCC caused by human activities, which have intensified since 1999 due to China's Grain for Green Project (GFGP). The Budyko parameter that represents watershed characteristics was more reasonably configured and derived to improve the performance of the Budyko equation. Vegetation changes were included in the Budyko equation to further improve its simulations, and these changes showed a significant upward trend due to the GFGP based on satellite data. An improved decomposition method based on the Budyko equation was used to quantitatively separate the impact of climate change from that of LULCC on the streamflow in the WRW. Our results show that climate change generated a dominant effect on the streamflow and decreased it by 72.4% in the WRW. This climatic effect can be further explained with the drying trend of the Palmer Severity Drought Index, which was calculated based only on climate change information for the WRW. In the meantime, although human activities in this watershed have been very intense, especially since 1999, vegetation cover increase contributed a 27.6% decline to the streamflow, which played a secondary role in affecting hydrological processes in the WRW.

Keywords: climate change; LULCC; Budyko equation; streamflow; drought

---

1. Introduction

Climate change and land use and land cover change (LULCC) have had profound influences on global and regional hydrological processes [1–3]. Understanding the hydrological responses in watersheds to climate change and LULCC is important for water resource planning and management throughout the world, especially in arid and semi-arid areas where water is the primary limiting factor for environmental services and social development [4–6]. Climate change causes changes in different components of hydrological processes [2,7]. These components include evapotranspiration, infiltration, streamflow, soil moisture, etc. Global evapotranspiration has shown a significant upward trend over the past three decades, caused partly by the increasing atmospheric moisture demand [8]. In particular, hydrological processes are very sensitive to climate change in arid and semi-arid areas. In the Middle East, acceleration of hydrological processes induced by climate change has caused more severe droughts and flooding events, affecting the region's well-being [9].

In addition to climate change, LULCC also alters hydrological processes. Reforestation/afforestation or deforestation changes surface evapotranspiration, canopy water interception, and soil water infiltration capacity, changing the hydrological processes within watersheds. Many previous studies have shown that reforestation results in a decrease in streamflow due to greater infiltration into the soil and higher precipitation interception by vegetation [10,11]. Deforestation can reduce root density and depth, and lower leaf mass, resulting in decreased vegetation water consumption, weaker evapotranspiration, and higher streamflow [12,13]. These changes within a watershed lead to a redistribution among the components of hydrological processes [14].

As mentioned, climate change and LULCC are two important factors that significantly affect hydrological processes at different temporospatial scales. Streamflow observations around the world have indicated varying levels of climate change and LULCC impact, particularly in basins located in arid and semi-arid climate zones [15,16]. Modeling techniques have been adopted to evaluate the contributions of climate change and LULCC to streamflow changes. The Budyko equation is a commonly used and effective tool to address such contributions due to its simplicity and physical background [17,18]. The Budyko equation, based on the water and energy balance at a watershed scale, demonstrates the physical distribution among precipitation, evapotranspiration, and streamflow at a long-term temporal scale [19]. Since it was established, the Budyko equation has been widely used to answer water and energy balance questions throughout the world [20–22].

However, limitations still exist for applications of the Budyko equation, which assumes non-changing water storage in a watershed over an application period. This assumption is often very difficult to satisfy due to the lack of sufficient observations. Yang et al. [23] used the Budyko equation to derive the elasticity of streamflow in relation to climatic variables in China at an annual timescale. Jiang et al. [24] used a time length of 11 years to satisfy the non-changing water storage assumption without observed evidence. Donohue et al. [25] asserted that 30 years of data were required to meet the criterion of the Budyko non-changing water storage for their study watersheds. In addition, many studies assume that the physical properties of a watershed do not exhibit significant changes by setting the Budyko parameter that represents such properties to a constant [26,27]. Nevertheless, vegetation as a key component in the watershed often changes significantly under climate change and/or through human activities. In this study, variable vegetation was introduced to the Budyko equation to improve understanding of the influence of LULCC on hydrological processes. Thus, we applied the Budyko equation to a watershed in the Loess Plateau, China, where vegetation cover has been significantly altered by climate and human activities. In Section 2, the study methods are described, Section 3 introduces the study area and data, Section 4 describes the results, and conclusions are given in Section 5.

2. Methods

2.1. Budyko Parameter Estimation

With the Budyko equation's assumption that changes in water storage in a watershed are negligible over a sufficiently long time, precipitation (P) is partitioned into evapotranspiration (E) and streamflow (R) for a watershed [19]. The ratio of actual evapotranspiration to precipitation ($\theta = E/P$, the evapotranspiration ratio) is controlled principally by the ratio of potential evapotranspiration to precipitation ($\varepsilon = E_p/P$, the climatic dryness index) on a long-term timescale. For humid watersheds

$(P > E_p)$, the actual evapotranspiration is controlled predominantly by the energy supply (E_p), while for non-humid watersheds $(P < E_p)$, it is controlled mainly by the water supply (P), as shown in Figure 1. Different functional forms of the Budyko equation have been developed [28]; one of the most widely used forms, the Choudhury-Yang (CY hereafter) equation, was selected for this study [29]. E_p was estimated using the Penman-Monteith method [30], and P, E_p, and R were used as inputs for the CY equation:

$$E = P * E_p / \left(P^\eta + E_p^\eta \right)^{\frac{1}{\eta}} \tag{1}$$

where η is the Budyko parameter that represents the average state of watershed characteristics such as vegetation cover, soil properties, topography, etc.

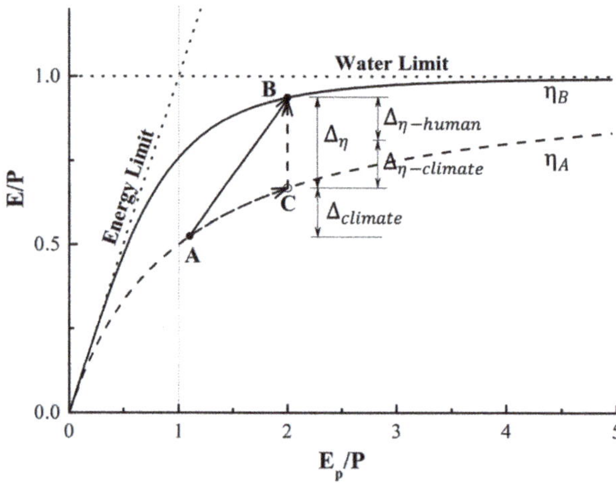

Figure 1. Schematic of water-energy balance changes in a watershed as indicated by the Choudhury-Yang (CY) Budyko-type equation, and the decomposition method.

Traditionally, η can be derived from climate and streamflow data [27], but η cannot be calculated for ungauged watersheds using such a method (e.g., lack of streamflow data). Thus, determining η for ungauged watersheds is a challenge. For this study, we propose a polynomial equation to calculate η using climate, soil, topography, vegetation, and other available data (e.g., remote sensing data) for ungauged watersheds (without streamflow measurements) as follows:

$$\eta = \beta_0 + \beta_h H + \beta_c C \tag{2}$$

where H represents explanatory variables defining LULCC caused by human activities, C represents explanatory variables defining climate change; β_0 is a constant term, and β_h and β_c are the corresponding regression coefficients. Through the maximum likelihood estimation method, β_h and β_c are estimated, and η is then estimated.

2.2. Quantifying the Contributions of Different Factors to Streamflow Changes

The Budyko parameter η might change for a watershed, implying a change in the watershed's characteristics. To quantify the contribution of each factor to a change in a watershed's water-energy balance, we adopted the decomposition method [14,24], described in Figure 1. There are two assumed paths to change a watershed from Point A to Point B: (1) a move from A to C along the dashed line, and (2) a vertical move from C to B. The first (A to C) shows that the η value for the watershed does not change, implying that the watershed ecosystem automatically adapts itself to climate change. The second (C to B) indicates a change in η, implying that external forcing alters the watershed's physical features such as vegetation. Such external forcing could stem from human influences and/or climate change, but in the original decomposition method this external forcing is wholly attributed to human activities, assuming that all the factors that influence η originate from human activities. In our study, the contribution represented by the second path is decomposed based on the polynomial equation (Equation (2)) in Section 2.1.

2.3. Calculation of Vegetation Fraction and Relative Infiltration Capacity

The accuracy of the Budyko equation can be improved if vegetation changes are included [31–33]. To study the effect of vegetation on the hydrological processes, the green vegetation fraction (F_g) was introduced in the Budyko equation. F_g can be derived from the normalized difference vegetation index (*NDVI*) based on satellite data. In this study, a quadratic equation was adopted to calculate F_g using *NDVI* [34]:

$$F_g = (NDVI_i - NDVI_s / NDVI_\infty - NDVI_s)^2 \tag{3}$$

where $NDVI_i$ is the *NDVI* value on a remote sensing map, $NDVI_s$ is for bare soil, and $NDVI_\infty$ is for dense green vegetation. For this study, $NDVI_\infty$ and $NDVI_s$ were set to 0.05 and 0.68, respectively, based on remotely sensed data and land use types [35,36].

Besides vegetation, water infiltration into the soil also affects the production of streamflow. The infiltration rate is controlled by rainfall intensity and soil infiltration capacity. In this study, the relative infiltration capacity was used to describe the relationship between the soil and the parameter η. The relative infiltration capacity is defined as the ratio of the saturation hydraulic conductivity, K_s, to the average rainfall intensity, i_r, within a period of 24 h [37]; i_r is the average value for rainy days, and K_s is obtained from the soil type database for the Wuding River Watershed (WRW) [38].

3. Study Area and Data Sources

3.1. Study Area

To control soil erosion and improve environmental conditions, many soil conservation measures have been applied in the Loess Plateau (Figure 2) since the 1960s, one of which is the Grain for Green Project (GFGP) [39]. This project has remarkably increased the vegetation cover in the Loess Plateau through afforestation/reforestation [40]. Furthermore, this water-limited, environmentally fragile area is vulnerable to climate change at different temporospatial scales [41]. For this study, we selected a typical watershed in the Loess Plateau, the WRW (Figure 2), to explore how afforestation/reforestation due to the GFGP affects hydrological processes under climate change.

Covering an area of approximately 30,261 km^2, the WRW, located at 37.04°–39.03° N and 108.04°–110.57° E, is in the center of the Loess Plateau. The Wuding River is a first-order tributary of the Yellow River. Streamflow data for this study were obtained from the Baijiachuan gauging station, which is located 100 km from the outlet of the WRW and has a drainage area accounting for 98% of the WRW. The WRW is in a semi-arid temperate continental climate zone, with average annual precipitation of 405 mm, a mean annual temperature of 8.0 °C, and potential evapotranspiration of 1007 mm, based on observational data over 1960–2011 (http://data.cma.cn/). Affected by the East Asian monsoon, approximately 75% of the annual rainfall occurs between June and September and is characterized by a significant number of heavy rain events. The topography is a typical loess hilly/gullied landscape with elevation ranging from 579 m to 1824 m.

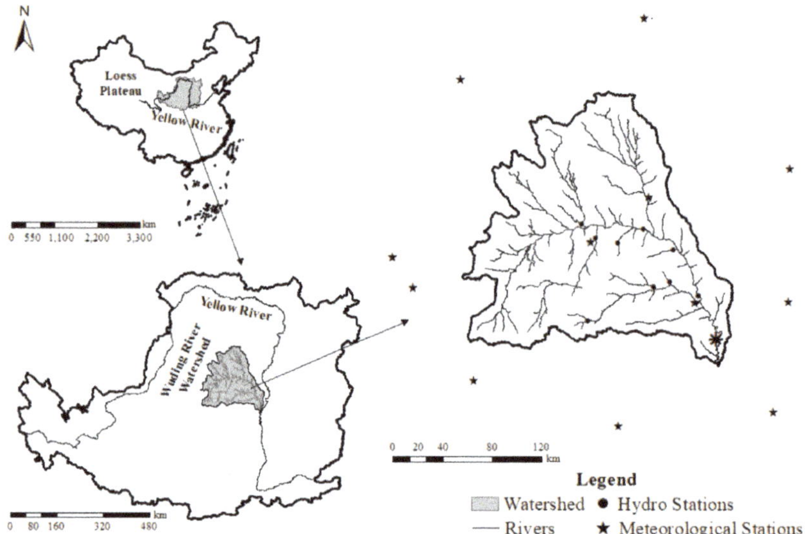

Figure 2. Locations of the study area and hydrometeorological stations (asterisk shows the river outlet).

3.2. Data Sources

3.2.1. Hydrometeorological Data

Monthly streamflow data from gauge stations located in the main stream and first-order tributaries in the WRW were obtained from the Yellow River Hydrological Bureau. Only data covering at least 50 years were used in this study; data from eight stations met this criterion. Thus, all streamflow data used in this study covered the period from 1960 to 2011. Daily meteorological data from 12 stations in and around the WRW were obtained from the National Meteorological Information Center, China Meteorological Administration (http://data.cma.cn/), for the study period. These meteorological data include precipitation, maximum and minimum temperature, relative humidity, wind speed, sunshine duration, and solar radiation. We used the nonparametric Mann-Kendall (MK) test to detect the significance of temporal trends with a 95% confidence interval [42].

3.2.2. Digital Elevation Model (DEM) and Soil Data

A DEM dataset at 30-m resolution was provided by the Geospatial Data Cloud site, Computer Network Information Center, Chinese Academy of Sciences (http://www.gscloud.cn). A soil dataset at 1-km resolution, containing soil property data and the spatial distribution of each soil type in the WRW, was provided by the Ecological Environment Database of the Loess Plateau (http://www.loess.csdb. cn/pdmp/index.action). The saturation hydraulic conductivity was verified with site observations from the WRW.

3.2.3. Satellite Remote Sensing Data

As one of the most useful indices for vegetation monitoring in terrestrial ecosystems, *NDVI* derived from remote sensing data was used. This study selected the Global Inventory Modeling and Mapping Studies Normalized Difference Vegetation Index 3rd generation dataset (NDVI3g) for the WRW [43]. The NDVI3g covers the period from 1982 to 2011 at a 0.083° spatial resolution and a semi-monthly time step. NDVI3g data have been examined and compared with other *NDVI* products [44] and were found to be consistent with these data. The maximum value composite method was used to obtain the monthly and annual *NDVI* values [45]. Therefore, this dataset was used to analyze the long-term vegetation trends and the relationship between vegetation and climate variability.

4. Results

4.1. Hydrometeorological Trends Analysis

4.1.1. Temporal Trends of Streamflow

Figure 3a shows the changes in annual streamflow in the WRW from 1960 to 2011. The annual streamflow in the WRW experienced a significant decrease over this 52-year period. The observed downward trend of 5.2 mm/decade passes the 95% significance level using the MK test. Moreover, the annual streamflow shows two significant abrupt points in 1972 and 1998, which were detected using the nonparametric multiple change-points detection method [46]. These abrupt points divide the study period into three stages, i.e., 1960–1972, 1973–1998, and 1999–2011, defined as Stages 1, 2, and 3, respectively. Figure 3a also shows that the amplitude of streamflow variation over the study period becomes weaker with time.

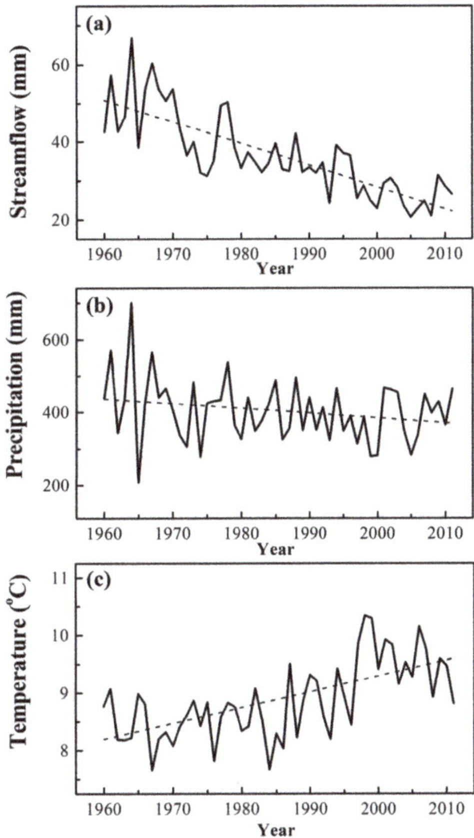

Figure 3. Annual changes in hydrometeorological variables in the Wuding River Watershed (WRW) from 1960 to 2011: (**a**) streamflow; (**b**) precipitation; and (**c**) temperature.

These three stages are consistent with water and soil conservation activities in the WRW, according to a survey of the WRW [47] (p. 385), which shows that soil and water conservation activities over the

WRW involve approximately three stages. The first stage spans from the 1950s to early 1970s, during which small-scale experimental field tests were performed to explore suitable ways of controlling soil erosion. The second stage lasts from the mid-1970s to the end of the 1990s, when the WRW was used as a national water and soil erosion management area. The last stage begins in 1999, when the WRW was one of the first GFGP pilot areas and more intensive conservation was performed. Watershed management records prove the validity of the abrupt statistical tests employed; thus, streamflow changes are closely related to human activities in the watershed.

4.1.2. Temporal Trends of Precipitation and Temperature

Climate change is one of the main factors affecting hydrological processes in the WRW. In Figure 3b, the annual precipitation shows a downward trend of 10 mm/decade, but this trend does not reach the 95% statistical significance level. A comparison of Figure 3a,b demonstrates that streamflow variability is controlled partly by precipitation changes. The correlation coefficients between precipitation and streamflow over the three stages are 0.8, 0.5, and 0.4, respectively, implying that the response by streamflow to precipitation becomes weaker. There must be other factors causing the decline in streamflow.

Figure 3c shows the time series of area-averaged annual temperature for the WRW. An upward trend of 0.27 °C/decade at the 95% significance level was detected by the MK test. This remarkable trend is five times the global average temperature change [48]. A rising temperature could contribute to the reduced streamflow in this area by increasing evapotranspiration [49], as will be discussed again in Section 4.5.

4.2. Determination of Timescale in the Budyko Equation

In the Budyko equation, water storage change in a watershed is assumed to be zero or very close to zero over a long-term period. However, the length of this period is watershed dependent, and it is impossible to accurately measure water storage change in almost any watershed. In some studies, researchers have arbitrarily set water storage change to be zero over a period ranging from 1 to 30 years with no support from observed evidence [23–25]. For this study, we made a series of sensitivity tests to determine the timescale at which the water storage change is reasonably close to zero in the WRW. In these tests, we calculated the Budyko parameter η on timescales of 1 to 52 years with increments of one year. For each of the 52 tests, the water storage change was set to zero. We found that the Budyko parameter η stabilized on timescales longer than 13 years, although there was a slight upward trend between timescales of 13 and 52 years (Figure 4). Therefore, we derived the value of η on a timescale of 13 years, at which water storage change can be reasonably assumed to be zero.

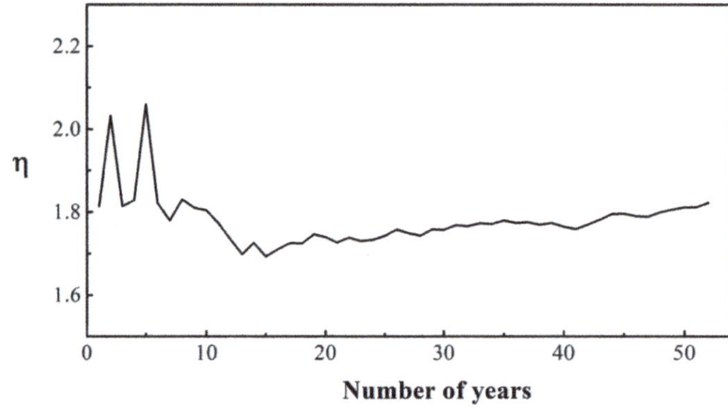

Figure 4. Sensitivity tests determining the timescale used to calculate the Budyko parameter η.

4.3. Temporospatial Changes in Vegetation

NDVI is an effective parameter representing vegetation cover in the WRW. Figure 5 shows that the area-averaged annual *NDVI* for the WRW increased from 1982 to 2011, indicating a growth in vegetation over this period. There was a pronounced change around 2000, which divided the period into two stages. These two stages fall within Stages 2 and 3, characterized by significant water conservation activities in the watershed. The *NDVI* trends for these two stages are 3.6×10^{-3}/yr and 11.8×10^{-3}/yr, respectively. The significant increase in the latter stage indicates remarkable vegetation growth in the WRW associated with the GFGP since 1999 [41].

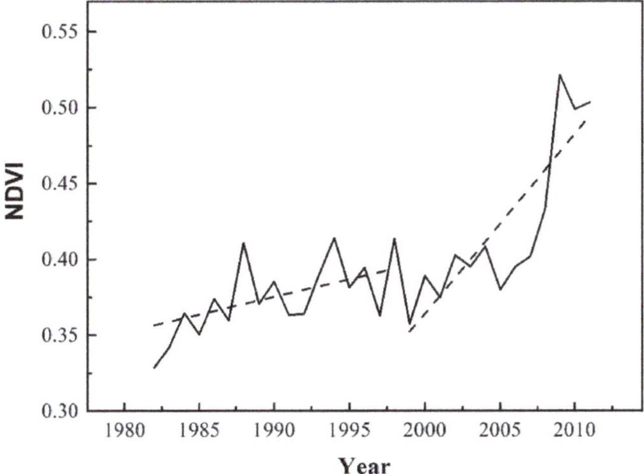

Figure 5. Variation in area-averaged annual normalized difference vegetation index (*NDVI*) from 1982 to 2011 in the WRW (dashed lines are the trends of periods before and after the Grain for Green Project (GFGP)).

The average *NDVI* spatial distributions over the two stages are shown in Figure 6a,b. Generally, the *NDVI* increases from southeast to northwest in the watershed during both stages, consistent with a change from a humid climate to a semi-arid one. The vegetation cover increases quite dramatically from the 1982–1998 to 1999–2011 periods. Figure 6c,d show the trends of *NDVI* in the WRW (pixels) and their 95% significance levels (black dots) for the same two stages, where 29% of the WRW in 1982–1998 and 83% of the WRW in 1999–2011 pass the 95% significance level. Particularly in the second stage, the middle and lower reaches of the WRW have the most significant *NDVI* increases, where the most severe soil erosion often occurs, and thus where reforestation/afforestation has been focused. In addition, pixels that did not pass the 95% significance level are predominantly urban areas.

Based on the above analysis, the WRW has experienced remarkable vegetation growth, particularly from 1999 to 2011, due to reforestation/afforestation. Such a substantial landscape change goes against the rules of the Budyko equation application, which assumes minimal landscape changes in a watershed. In this study, we made a significant effort to include landscape changes in the Budyko equation, with a focus on vegetation.

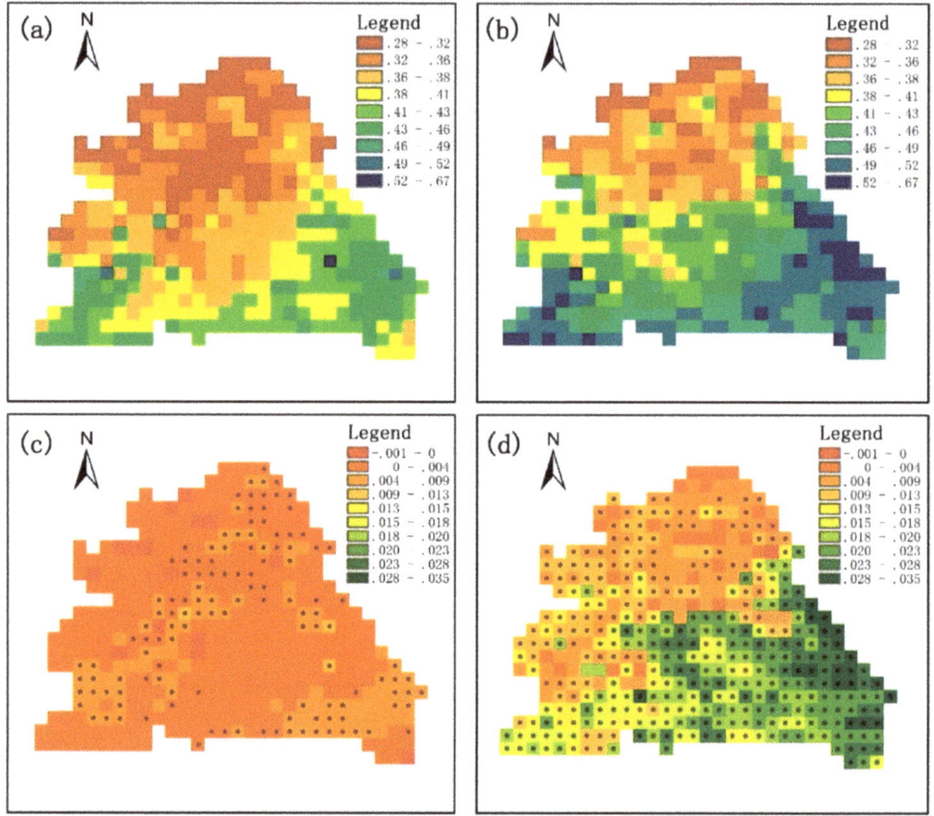

Figure 6. Spatial distribution of average annual *NDVI* in the WRW: (**a**) average annual *NDVI* from 1982 to 1998; (**b**) average annual *NDVI* from 1999 to 2011; (**c**) linear regression slope of *NDVI* from 1982 to 1998; and (**d**) linear regression slope of *NDVI* from 1999 to 2011 (dots denote the slope at the 95% significance level).

4.4. Estimation of Parameter η in the Budyko Equation

By considering the vegetation changes in the WRW, we used covariate analysis with the Akaike information criterion [50] to develop an empirical scheme to estimate η. In this scheme, we parameterized η as a function of explanatory variables including vegetation cover, F_g, the relative infiltration capacity, irrigation area, and terrace area. The Budyko parameter η was optimized using the above method to quantify the relationship between η and the explanatory variables. Finally, η was estimated as follows:

$$\eta = 2.21 + 0.19 \times log_{10}F_g - 1.29 \times 10^{-5} \times exp(K_s/i_r) \tag{4}$$

where F_g reflects the vegetation conditions as one of the most important landscape factors in a watershed and is derived from *NDVI* through the conversion model discussed above. The relative infiltration capacity denotes the infiltration property that influences streamflow generation.

The multiple R-squared of the regression equation is 0.86 and passes the 95% significance level, indicating that these factors can realistically explain η. These factors represent vegetation, soil, and climate conditions, in which vegetation changes are induced mainly by human activities. The result reveals a significant positive correlation (0.76) between F_g and η in the WRW. Figure 7 illustrates the η estimated with Equation (4) versus the η calculated based on the Budyko equation with the observed input variables. For the WRW, the η value generated with the above regressed polynomial equation agrees very well with that derived from the Budyko equation.

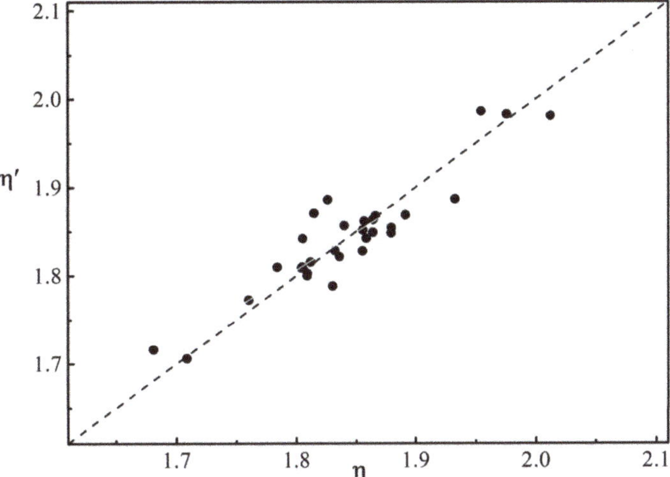

Figure 7. Comparison of η calculated by observed input variables with η' estimated by the regression equation (dashed line is the 1:1 line).

By inputting this estimated η into the Budyko equation, we calculated the streamflow for the WRW. As shown in Figure 8, the root mean square error and Nash Sutcliffe efficiency coefficient are 1.22 mm and 0.91, respectively. The streamflow results calculated by a constant η are also displayed in Figure 8, and the root mean square error and Nash Sutcliffe efficiency coefficient are 2.95 mm and 0.49, respectively. The constant η was derived on a timescale of the entire period, indicating no watershed landscape change over the study period. A comparison between these two calculated results indicates that the Budyko equation is more accurate when changes in landscape factors, especially vegetation, are included. The streamflow calculated by the η that considers vegetation changes reflects not only the streamflow trend but also the magnitude. Conversely, the streamflow calculated by a constant η greatly underestimates streamflow during the first several years and overestimates streamflow in the last several years of the period. This implies that a constant η cannot reflect dynamic changes in watershed landscape characteristics. However, it is worth noting that the constant η case also substantially demonstrates the streamflow trend. This case is useful for situations where vegetation data are insufficient, especially on the large timescale of future climate scenarios.

4.5. Contributions of Climate Change and Vegetation to Streamflow

To quantify the contributions of different factors to streamflow changes, the improved decomposition method mentioned in Section 2 was applied. In view of the good performance of explanatory variables at interpreting the Budyko parameter η, Equation (4) was used to calculate the change in mean annual streamflow in each 13-year period, together with the Budyko equation (Equation (1)). Therefore, the streamflow changes in each period are compared with the baseline period 1970–1982, which is the first 13-year period containing vegetation information. The baseline period is denoted as the pre-stage, and other lengths are denoted as the post-stage. The calculated result of the decomposition method is shown in Figure 9, indicating that a combination of climate and human activities (mainly from vegetation changes) led to the streamflow decline in recent years. From the average contributions of climate and vegetation during different periods (Figure 9a), the conclusion can be made that climate change is the dominant factor affecting streamflow, accounting for nearly 76% of the total streamflow reduction. Vegetation changes are also important factors, accounting for about 24% of the streamflow decrease. Further, the streamflow reduction induced by climate increased substantially after 1999 (Figure 9a), which is attributed to the increasingly dry climate. The relationship between drought and streamflow change is discussed in the following.

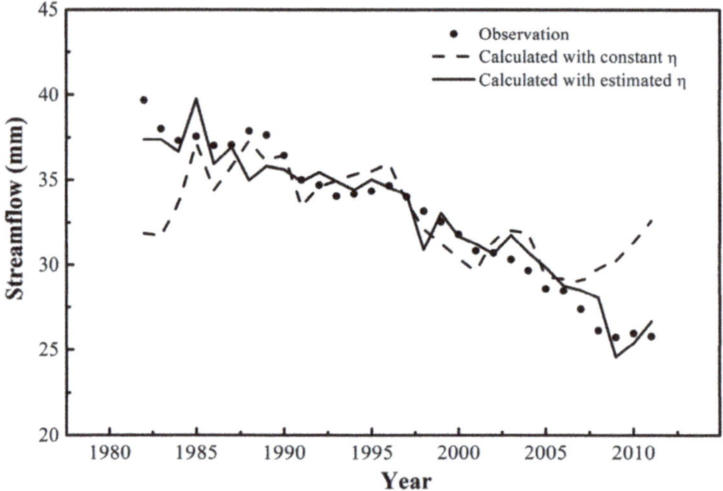

Figure 8. Comparison of two types of calculated streamflow with the observed streamflow.

Climate and landscape changes in a watershed have an important effect on hydrological processes; this effect can be reflected in the Budyko equation by altering parameter η. The contribution from climate can be divided into two parts: the first part is caused by change in the meteorological input of precipitation and potential evapotranspiration to the WRW (direct climate change); the other part is caused by climate change through modification of the watershed landscape characteristics (indirect climate change). In this study, the impact of climate change on η (indirect climate change) originates from the change in average rainfall intensity, which influences the relative infiltration capacity. This is a crucial factor to consider in landscape characteristics, because infiltration excess overland flow is the main mechanism for streamflow generation in a typical loess soil watershed [51]. In order to distinguish the impact of climate and vegetation changes on η, the streamflow reduction induced by these two factors via altering η is compared in Figure 9b. In Figure 9b, streamflow reduction caused by climate change remains steady with little variation and is smaller than that caused by vegetation changes. This indicates that changes in η induced by climate change are not negligible, which has not previously been considered [52]. Figure 9b also indicates that vegetation is the primary factor affecting η; streamflow reduction induced by vegetation changes represents the majority of streamflow reduction caused by altering parameter η. This implies that vegetation is vital to the hydrology in this semi-arid watershed, and growth in vegetation cover increases the evapotranspiration ratio and reduces the streamflow ratio to precipitation. It also demonstrates the significance of introducing a vegetation factor into streamflow estimates in the Budyko equation.

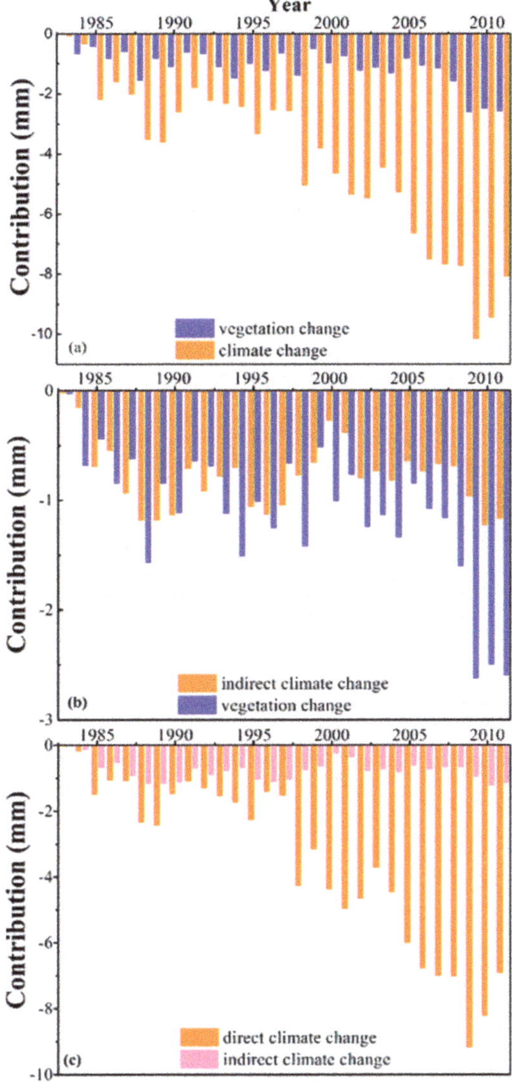

Figure 9. Contributions separation results of the decomposition method: (**a**) comparison of the contributions of climate change and vegetation; (**b**) comparison of the contributions of climate change and vegetation by altering the Budyko parameter; and (**c**) comparison of the contributions of two types of climate change.

The two different components of climate change contribution to the streamflow are shown in Figure 9c. One represents the contribution directly induced by climate change (direct climate change), and the other is the contribution induced by altering η by climate change (indirect climate change). The direct climate change contribution accounts for the majority (88%) of the total climate change contribution, and the indirect climate change contribution accounts for 12%. In order to test the validity and rationality of the improved decomposition method, the contribution was also quantified using another mainstream method called the elasticity method [27,53]. The elasticity method results are not shown here, but our results with this method are similar to those with the improved decomposition method.

There are 74 dams with a storage capacity greater than one million cubic meters in the WRW with the purpose of flooding control [54] (p. 705). Nevertheless, these dams mainly affect the seasonal variations of streamflow in the WRW, and they do not have a significant influence on the volume of annual streamflow. In addition, most of these dams lost normal function in the end of 1980s due to the sediment deposition caused by severe soil erosion [55] (pp. 428–429). Thus, our study did not include the influence of dam regulation on the streamflow, and focused on the change in annual streamflow in WRW over the period of 1982 to 2011.

Precipitation is the only source of water input to a closed watershed and is partitioned into different parts, such as soil water storage and evapotranspiration. The results with our improved Budyko equation application indicate that hydrological processes are the result of the long-term co-evolution of a watershed's vegetation and climate [14]. The contribution analysis results of the WRW demonstrate the dominant role of climate in this complex evolved system. These findings were further confirmed with the Palmer Drought Severity Index (PDSI) [56,57], a physically based hydrometeorological index. The calculation of PDSI does not consider interference from human activities in this watershed, and thus this index explains hydrological drought patterns regardless of human influences [58]. The annual changes in streamflow and PDSI in the WRW from 1960 to 2011 are shown in Figure 10. These two variables derived from independent datasets exhibit similar trends and variations. The downward trend of streamflow is −0.048 mm/yr and that of PDSI is −0.047. The MK test results indicate that both show a significant downward trend at the 95% significance level. The decreasing PDSI indicates that the WRW has experienced increasingly serious droughts since the early 1980s. Moreover, this similarity shows that PDSI captures the trend of streamflow change and the dominant role of climate in streamflow reduction. However, the performance of PDSI deteriorates in Stage 2 and Stage 3 compared to Stage 1, which demonstrates that human activities play a non-negligible role in streamflow reduction in the WRW.

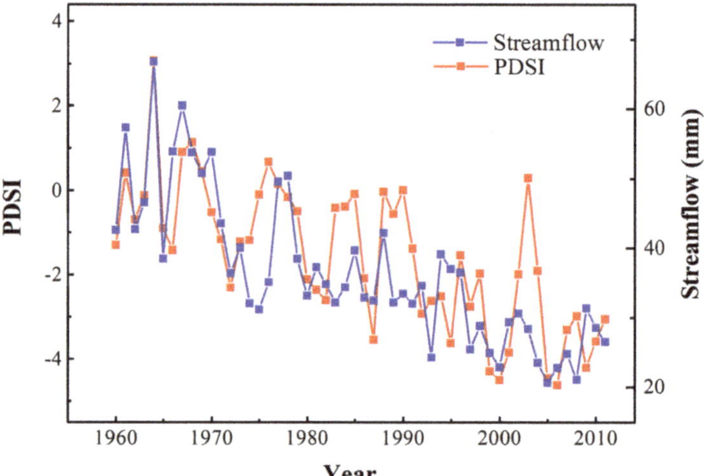

Figure 10. Annual streamflow and Palmer Drought Severity Index (PDSI) changes in the WRW from 1960 to 2011.

5. Conclusions

In this study, we diagnosed hydrometeorological changes in the WRW, with a focus on vegetation cover changes. Over recent years, streamflow has dramatically declined, regional climate change has become evident, and the watershed has experienced more severe drought. Vegetation cover changes are the main reason for underlying surface changes in the WRW. The timing of abrupt changes indicates that *NDVI* changes are closely tied to water and soil conservation activities in the WRW and streamflow changes. Intense variations of *NDVI* in such a short time reveal that human activities are the main driving force of vegetation cover changes.

Using the moving average method with a timescale of 13 years, an optimized model was established, incorporating the Budyko parameter, vegetation cover, and relative infiltration capacity. The main factors that influence watershed landscape characteristics were then determined, i.e., climate change and vegetation changes. The good performance of the estimated streamflow implies that the Budyko parameter can be explained by these variables. Based on this optimized model, an improved decomposition method was used to separate the impact of climate change and vegetation cover changes on streamflow. It should be noted that we considered the climatic impact on the Budyko parameter η, which has previously been ignored. Furthermore, introducing the main factors that affect the Budyko parameter improved the performance of the Budyko equation by incorporating physical mechanisms.

Author Contributions: L.T. conducted the modeling, performed the analysis, and drafted the manuscript; J.J. and P.W. designed the study, interpreted the results, and supervised the research; G.-y.N. contributed ideas during analysis and interpretation, and edited the paper.

Funding: This research was funded by the National Natural Science Foundation of China (No. 41571030, No. 91637209, No. 91737306), and it was also partially funded by Utah Agricultural Experiment Station.

Conflicts of Interest: The authors declare no conflict of interest.

References

1. Neupane, R.P.; Kumar, S. Estimating the effects of potential climate and land use changes on hydrologic processes of a large agriculture dominated watershed. *J. Hydrol.* **2015**, *529*, 418–429. [CrossRef]
2. Serpa, D.; Nunes, J.P.; Santos, J.; Sampaio, E.; Jacinto, R.; Veiga, S.; Lima, J.C.; Moreira, M.; Corte-Real, J.; Keizer, J.J.; et al. Impacts of climate and land changes on the hydrological and erosion processes of two contrasting Mediterranean catchments. *Sci. Total Environ.* **2015**, *538*, 64–77. [CrossRef] [PubMed]
3. Tan, M.L.; Ibrahim, A.L.; Yusop, Z.; Duan, Z.; Ling, L. Impacts of land-use and climate variability on hydrological components in the Johor River basin, Malaysia. *Hydrol. Sci. J.* **2015**, *60*, 873–889. [CrossRef]
4. Kahil, M.T.; Dinar, A.; Albiac, J. Modeling water scarcity and droughts for policy adaptation to climate change in arid and semiarid regions. *J. Hydrol.* **2015**, *522*, 95–109. [CrossRef]
5. Salazar, A.; Baldi, G.; Hirota, M.; Syktus, J.; McAlpine, C. Land use and land cover change impacts on the regional climate of non-Amazonian South America: A review. *Glob. Planet. Chang.* **2015**, *128*, 103–119. [CrossRef]
6. Yin, J.; He, F.; Xiong, Y.J.; Qiu, G.Y. Effects of land use/land cover and climate changes on surface runoff in a semi-humid and semi-arid transition zone in northwest China. *Hydrol. Earth Syst. Sci.* **2017**, *21*, 183–196. [CrossRef]
7. Hattermann, F.F.; Krysanova, V.; Gosling, S.N.; Dankers, R.; Daggupati, P.; Donnelly, C.; Flörke, M.; Huang, S.; Motovilov, Y.; Buda, S.; et al. Cross-scale intercomparison of climate change impacts simulated by regional and global hydrological models in eleven large river basins. *Clim. Chang.* **2017**, *141*, 561–576. [CrossRef]
8. Zhang, K.; Kimball, J.S.; Nemani, R.R.; Running, S.W.; Hong, Y.; Gourley, J.J.; Yu, Z. Vegetation greening and climate change promote multidecadal rises of global land evapotranspiration. *Sci. Rep.* **2015**, *5*, 15956. [CrossRef]
9. Sowers, J.; Vengosh, A.; Weinthal, E. Climate change, water resources, and the politics of adaptation in the Middle East and North Africa. *Clim. Chang.* **2011**, *104*, 599–627. [CrossRef]
10. Greenwood, W.J.; Buttle, J.M. Effects of reforestation on near-surface saturated hydraulic conductivity in a managed forest landscape, southern Ontario, Canada. *Ecohydrology* **2014**, *7*, 45–55. [CrossRef]
11. Locatelli, B.; Catterall, C.P.; Imbach, P.; Kumar, C.; Lasco, R.; Marín-Spiotta, E.; Mercer, B.; Powers, J.S.; Schwartz, N.; Uriarte, M. Tropical reforestation and climate change: Beyond carbon. *Restor. Ecol.* **2015**, *23*, 337–343. [CrossRef]

12. Coe, M.T.; Marthews, T.R.; Costa, M.H.; Galbraith, D.R.; Greenglass, N.L.; Imbuzeiro, H.M.A.; Levine, N.M.; Malhi, Y.; Moorcroft, P.R.; Muza, M.N.; et al. Deforestation and climate feedbacks threaten the ecological integrity of south-southeastern Amazonia. *Philos. Trans. R. Soc. B* **2013**, *368*, 20120155. [CrossRef] [PubMed]

13. Panday, P.K.; Coe, M.T.; Macedo, M.N.; Lefebvre, P.; de Almeida Castanho, A.D. Deforestation offsets water balance changes due to climate variability in the Xingu River in eastern Amazonia. *J. Hydrol.* **2015**, *523*, 822–829. [CrossRef]

14. Wang, D.; Tang, Y. A one-parameter Budyko model for water balance captures emergent behavior in darwinian hydrologic models. *Geophys. Res. Lett.* **2014**, *41*, 4569–4577. [CrossRef]

15. Guardiola-Claramonte, M.; Troch, P.A.; Breshears, D.D.; Huxman, T.E.; Switanek, M.B.; Durcik, M.; Cobb, N.S. Decreased streamflow in semi-arid basins following drought-induced tree die-off: A counter-intuitive and indirect climate impact on hydrology. *J. Hydrol.* **2011**, *406*, 225–233. [CrossRef]

16. Hughes, J.D.; Petrone, K.C.; Silberstein, R.P. Drought, groundwater storage and stream flow decline in southwestern Australia. *Geophys. Res. Lett.* **2012**, *39*. [CrossRef]

17. Cong, Z.; Zhang, X.; Li, D.; Yang, H.; Yang, D. Understanding hydrological trends by combining the Budyko hypothesis and a stochastic soil moisture model. *Hydrol. Sci. J.* **2015**, *60*, 145–155. [CrossRef]

18. Yang, H.; Yang, D.; Hu, Q. An error analysis of the Budyko hypothesis for assessing the contribution of climate change to runoff. *Water Resour. Res.* **2014**, *50*, 9620–9629. [CrossRef]

19. Budyko, M. *Climate and Life*; Academic Press: New York, NY, USA, 1974; ISBN 9780080954530.

20. Abatzoglou, J.T.; Ficklin, D.L. Climatic and physiographic controls of spatial variability in surface water balance over the contiguous United States using the Budyko relationship. *Water Resour. Res.* **2017**, *53*, 7630–7643. [CrossRef]

21. Koppa, A.; Gebremichael, M. A Framework for validation of remotely sensed precipitation and evapotranspiration based on the Budyko hypothesis. *Water Resour. Res.* **2017**, *53*, 8487–8499. [CrossRef]

22. Wang, C.; Wang, S.; Fu, B.; Zhang, L. Advances in hydrological modelling with the Budyko framework: A review. *Prog. Phys. Geogr.* **2016**, *40*, 409–430. [CrossRef]

23. Yang, H.; Yang, D. Derivation of climate elasticity of runoff to assess the effects of climate change on annual runoff. *Water Resour. Res.* **2011**, *47*. [CrossRef]

24. Jiang, C.; Xiong, L.; Wang, D.; Liu, P.; Guo, S.; Xu, C.-Y. Separating the impacts of climate change and human activities on runoff using the Budyko-type equations with time-varying parameters. *J. Hydrol.* **2015**, *522*, 326–338. [CrossRef]

25. Donohue, R.J.; Roderick, M.L.; McVicar, T.R. Assessing the differences in sensitivities of runoff to changes in climatic conditions across a large basin. *J. Hydrol.* **2011**, *406*, 234–244. [CrossRef]

26. Han, S.; Hu, H.; Yang, D.; Liu, Q. Irrigation impact on annual water balance of the oases in Tarim Basin, Northwest China. *Hydrol. Process.* **2011**, *25*, 167–174. [CrossRef]

27. Xu, X.; Yang, D.; Yang, H.; Lei, H. Attribution analysis based on the Budyko hypothesis for detecting the dominant cause of runoff decline in Haihe basin. *J. Hydrol.* **2014**, *510*, 530–540. [CrossRef]

28. Du, C.; Sun, F.; Yu, J.; Liu, X.; Chen, Y. New interpretation of the role of water balance in an extended Budyko hypothesis in arid regions. *Hydrol. Earth Syst. Sci.* **2016**, *20*, 393–409. [CrossRef]

29. Zhang, S.; Yang, H.; Yang, D.; Jayawardena, A.W. Quantifying the effect of vegetation change on the regional water balance within the Budyko framework. *Geophys. Res. Lett.* **2016**, *43*, 1140–1148. [CrossRef]

30. Allen, R.G.; Pereira, L.S.; Raes, D.; Smith, M.J.F. *Crop Evapotranspiration-Guidelines for Computing Crop Water Requirements-FAO Irrigation and Drainage Paper 56*; FAO: Rome, Italy, 1998; Volume 300, p. D05109.

31. Gentine, P.; D'Odorico, P.; Lintner, B.R.; Sivandran, G.; Salvucci, G. Interdependence of climate, soil, and vegetation as constrained by the Budyko curve. *Geophys. Res. Lett.* **2012**, *39*. [CrossRef]

32. Liu, Q.; McVicar, T.R.; Yang, Z.; Donohue, R.J.; Liang, L.; Yang, Y. The hydrological effects of varying vegetation characteristics in a temperate water-limited basin: Development of the dynamic Budyko-Choudhury-Porporato (dBCP) model. *J. Hydrol.* **2016**, *543*, 595–611. [CrossRef]

33. Troch, P.A.; Carrillo, G.; Sivapalan, M.; Wagener, T.; Sawicz, K. Climate-vegetation-soil interactions and long-term hydrologic partitioning: Signatures of catchment co-evolution. *Hydrol. Earth Syst. Sci.* **2013**, *17*, 2209–2217. [CrossRef]

34. Carlson, T.N.; Ripley, D.A. On the relation between NDVI, fractional vegetation cover, and leaf area index. *Remote Sens. Environ.* **1997**, *62*, 241–252. [CrossRef]

35. Montandon, L.M.; Small, E.E. The impact of soil reflectance on the quantification of the green vegetation fraction from NDVI. *Remote Sens. Environ.* **2008**, *112*, 1835–1845. [CrossRef]

36. Yang, H.; Yang, Z. A modified land surface temperature split window retrieval algorithm and its applications over China. *Glob. Planet. Chang.* **2006**, *52*, 207–215. [CrossRef]

37. Yang, D.; Shao, W.; Yeh, P.J.-F.; Yang, H.; Kanae, S.; Oki, T. Impact of vegetation coverage on regional water balance in the nonhumid regions of China. *Water Resour. Res.* **2009**, *45*. [CrossRef]

38. Gao, X.; Wu, P.; Zhao, X.; Zhou, X.; Zhang, B.; Shi, Y.; Wang, J. Estimating soil moisture in gullies from adjacent upland measurements through different observation operators. *J. Hydrol.* **2013**, *486*, 420–429. [CrossRef]

39. Qiu, L.; Wu, Y.; Wang, L.; Lei, X.; Liao, W.; Hui, Y.; Meng, X. Spatiotemporal response of the water cycle to land use conversions in a typical hilly-gully basin on the Loess Plateau, China. *Hydrol. Earth Syst. Sci.* **2017**, *21*, 6485–6499. [CrossRef]

40. Zhang, B.; Long, B.; Wu, Z.; Wang, Z. An evaluation of the performance and the contribution of different modified water demand estimates in drought modeling over water-stressed regions. *Land Degrad. Dev.* **2017**, *28*, 1134–1151. [CrossRef]

41. Fan, X.; Ma, Z.; Yang, Q.; Han, Y.; Mahmood, R.; Zheng, Z. Land use/land cover changes and regional climate over the Loess Plateau during 2001–2009. Part I: Observational evidence. *Clim. Chang.* **2015**, *129*, 427–440. [CrossRef]

42. Hamed, K.H. Trend detection in hydrologic data: The Mann-Kendall trend test under the scaling hypothesis. *J. Hydrol.* **2008**, *349*, 350–363. [CrossRef]

43. Ibrahim, Y.; Balzter, H.; Kaduk, J.; Tucker, C. Land degradation assessment using residual trend analysis of GIMMS NDVI3g, soil moisture and rainfall in Sub-Saharan West Africa from 1982 to 2012. *Remote Sens.* **2015**, *7*, 5471–5494. [CrossRef]

44. Pinzon, J.; Tucker, C. A non-stationary 1981–2012 AVHRR NDVI3g time series. *Remote Sens.* **2014**, *6*, 6929–6960. [CrossRef]

45. Zhang, B.; Wu, P.; Zhao, X.; Wang, Y.; Gao, X. Changes in vegetation condition in areas with different gradients (1980–2010) on the Loess Plateau, China. *Environ. Earth Sci.* **2013**, *68*, 2427–2438. [CrossRef]

46. Zou, C.; Yin, G.; Feng, L.; Wang, Z. Nonparametric maximum likelihood approach to multiple change-point problems. *Ann. Stat.* **2014**, *42*, 970–1002. [CrossRef]

47. Wang, G.; Fan, Z. *Study on Changes of Water and Sediment of the Yellow River*; The Yellow River Water Conservancy Press: Zheng Zhou, China, 2002; Volume 3, p. 385, ISBN 9787806215708.

48. Root, T.L.; Price, J.T.; Hall, K.R.; Schneider, S.H.; Rosenzweig, C.; Pounds, J.A. Fingerprints of global warming on wild animals and plants. *Nature* **2003**, *421*, 57–60. [CrossRef] [PubMed]

49. Zhang, B.; He, C.; Burnham, M.; Zhang, L. Evaluating the coupling effects of climate aridity and vegetation restoration on soil erosion over the Loess Plateau in China. *Sci. Total Environ.* **2016**, *539*, 436–449. [CrossRef]

50. Akaike, H. Akaike's information criterion. In *International Encyclopedia of Statistical Science*; Springer: Berlin, Germany, 2011; p. 25, ISBN 9783642048975.

51. Kang, S.; Zhang, L.; Song, X.; Zhang, S.; Liu, X.; Liang, Y.; Zheng, S. Runoff and sediment loss responses to rainfall and land use in two agricultural catchments on the Loess Plateau of China. *Hydrol. Process.* **2001**, *15*, 977–988. [CrossRef]

52. Wang, D.; Hejazi, M. Quantifying the relative contribution of the climate and direct human impacts on mean annual streamflow in the contiguous United States. *Water Resour. Res.* **2011**, *47*. [CrossRef]

53. Sankarasubramanian, A.; Vogel, R.M.; Limbrunner, J.F. Climate elasticity of streamflow in the United States. *Water Resour. Res.* **2001**, *37*, 1771–1781. [CrossRef]

54. Wang, G.; Fan, Z. *Study on Changes of Water and Sediment of the Yellow River*; The Yellow River Water Conservancy Press: Zheng Zhou, China, 2002; Volume 2, p. 705, ISBN 9787806215692.

55. Wang, G.; Fan, Z. *Study on Changes of Water and Sediment of the Yellow River*; The Yellow River Water Conservancy Press: Zheng Zhou, China, 2002; Volume 3, pp. 428–429, ISBN 9787806215708.

56. Dai, A. Characteristics and trends in various forms of the Palmer Drought Severity Index during 1900–2008. *J. Geophys. Res. Atmos.* **2011**, *116*. [CrossRef]

57. Dai, A. Drought under global warming: A review. *Wiley Interdiscip. Rev. Clim. Chang.* **2011**, *2*, 45–65. [CrossRef]

58. Wells, N.; Goddard, S.; Hayes, M.J. A self-calibrating Palmer Drought Severity Index. *J. Clim.* **2004**, *17*, 2335–2351. [CrossRef]

 *water*

Article

Spatiotemporal Variation of Snowfall to Precipitation Ratio and Its Implication on Water Resources by a Regional Climate Model over Xinjiang, China

Qian Li [1,2] , Tao Yang [1,2], Zhiming Qi [1] and Lanhai Li [1,3,4,5,*]

1   State Key Laboratory of Desert and Oasis Ecology, Xinjiang Institute of Ecology and Geography, CAS, Urumqi 830011, China; liqian0109@mails.ucas.ac.cn (Q.L.); yangtao515@mails.ucas.ac.cn (T.Y.); qzhiming@ms.xjb.ac.cn (Z.Q.)
2   University of Chinese Academy of Sciences, Beijing 100049, China
3   Ili Station for Watershed Ecosystem Research, Urumqi 830011, China
4   Xinjiang Regional Center of Resources and Environmental Science Instrument, Chinese Academy of Sciences, Urumqi 830011, China
5   CAS Research Center for Ecology and Environment in Central Asia, 818 South Beijing Road, Urumqi 830011, China
*   Correspondence: lilh@ms.xjb.ac.cn; Tel.: +86-991-782-3125

Received: 13 September 2018; Accepted: 15 October 2018; Published: 17 October 2018

Abstract: Snow contributes one of the main water sources to runoff in the arid region of China. A clear understanding of the spatiotemporal variation of snowfall is not only required for climate change assessment, but also plays a critical role in water resources management. However, in-situ observations or gridded datasets hardly meet the requirement and cannot provide precise spatiotemporal details on snowfall across the region. This study attempted to apply the Weather Research and Forecasting (WRF) model to clarify the spatiotemporal variation of snowfall and the ratio of snowfall to total precipitation over Xinjiang in China during the 1979–2015 period. The results showed that the snowfall increased in the southern edge of the Tarim Basin, the Ili Valley, and the Altay Mountains, but decreased in the Tianshan Mountains and the Kunlun Mountains. The snowfall/precipitation (S/P) ratio revealed the opposite trends in low-elevation regions and mountains in the study area. The S/P ratio rose in the Tarim Basin and the Junggar Basin, but declined in the Altay Mountains, the Tianshan Mountains, and the west edge of the Junggar Basin. The study area comprises two major rivers in the middle of the Tianshan Mountains. Both the runoff magnitude increase and earlier occurrence of snowmelt recharge in runoff identified for the 1980s were compared with the 2000s level in decreasing S/P ratio regions.

Keywords: snowfall to precipitation ratio; WRF model; arid region; Xinjiang; water resources management

---

1. Introduction

Snow plays an important role in balancing radiation and generating streamflow in arid and semi-arid regions. It regulates the energy balance and hydrological cycle, which exerts a large influence on atmospheric circulation and the climatic system [1]. An increase in temperature will reduce the fraction of precipitation that falls as snowfall, shorten snow cover duration in the cold season, bring early timing of snowmelt in spring, and increase snowmelt intensity [2,3]. Recent model simulations, satellite-derived records, and in-situ observations demonstrated that the snow cover extent experienced a strong negative trend in North America and Eurasia, particularly in spring time [4–7], which is consistent with an increase in the mean winter temperature in the Northern Hemisphere. In addition, the changes in the fraction of snowfall to precipitation alter snow dynamics and runoff amount [8,9], leading to more rain-on-snow floods [10,11]. Thus, a combined indicator, defined as the ratio of snowfall to total precipitation (S/P), has been developed to represent the simultaneous change of snowfall and precipitation for the sake of clarifying dynamics of snow with distinct characteristics on a regional scale [9,12,13].

Many efforts have focused on the variability of the S/P ratio along with the rising temperature. For instance, Huntington et al. [14] found a significant reduced trend of the S/P ratio in New England from 1949 to 2000. Similar trends of the winter total snowfall-water-equivalent (SWE) and S/P ratio in the United States were also confirmed during the period from 1949 to 2004 [15], but there was no significant change in the S/P ratio for the Canadian Arctic, except for the summer [16]. Serquet et al. [17] analyzed the S/P ratio at 76 meteorological stations in Switzerland for up to 100 years and discovered a clear decreasing trend, especially at lower elevations. The same trend was found in the middle altitude of 1500 to 2500 m in the Tianshan Mountains [13]. Yang et al. [18] and Littell et al. [19] found that the S/P ratio significantly decreased at the end of 21st century in the Tianshan Mountains and Alaska. Long term change in the S/P ratio over time is of importance to the extent that it influences the magnitude and timing of spring runoff and recession to summer baseflow. As a consequence, it is important to expound the distinct regional variation of snowfall and S/P ratio in pursuit of realizing the change of snowfall and precipitation under climate change. However, studies on long-term and large-scale variations of snowfall and S/P ratio are insufficient, such as in the arid region of China, where snowfall is an important water source and observations are scarce.

Situated far from oceans, the arid region in western China has a typical continental climate marked by generally low precipitation, high evaporation potential, wide temperature fluctuations, and strong winds [20,21]. Mountainous precipitation, water from snowmelt, and glacier-melt are the main water sources in this region, which are highly sensitive to climate change [22]. During the last five decades, the rise in annual temperature in this region was greater than that of China's national average, and has been in a state of high variability since 1997 [23]. A higher warming rate was also observed in the Tianshan Mountains [24,25]. Such variation would greatly influence the snowfall regimes and change the supplying patterns of runoff in both spring and summer in the arid region [26]. As a good indicator of climate change, the S/P ratio also has a significant influence on annual runoff [8]. Thus, effectively monitoring the snowfall and S/P ratio would make clear sense for local climate change and water resources management.

The distribution of precipitation has been mainly statistically assessed on the basis of observed data. However, due to the scarcity of meteorological stations in spatial distribution and the complex topography in Xinjiang, China, the dataset from observations (including the gridded dataset from the extrapolation by in-situ observation) does not represent the local climate feature well, nor provide spatiotemporal details of precipitation over the region, especially in the high altitude where stations are not installed. Furthermore, quality problems in time series data resulting from observation metric and systematic errors of the equipment during conventional meteorological observation are serious, particularly in the cold season. The under-catch errors in precipitation gauge records can be as large as 50–100% at a high latitude [27–29]. Dynamical downscaling is therefore an inevitable alternative in order to characterize the spatiotemporal variation of snowfall and S/P ratio, especially in the region with a large vertical gradient from basin to mountain and a lack of observations. Some studies have reported that precipitation patterns simulated by the Weather Research and Forecasting (WRF) model were highly accurate at different spatial scales, including high-elevation and complex topography regions [30–35]. Consequently, the WRF model could provide the most accurate available estimation of the mesoscale precipitation distribution [33], enabling research on long-term variation of snowfall and the S/P ratio over complex terrain. This study attempted to employ the outputs of the WRF model to: (1) investigate the ability of the WRF model to reproduce the temporal and spatial distribution of the S/P ratio over the region, (2) characterize the regional variation pattern of both snowfall and the S/P ratio over complex topography and a large vertical gradient region, and (3) reveal the changes in hydrological processes in the study area and discuss the implications on water resources based on (1) and (2). The results could help to assess the impact of regional climate change on snowfall and provide information for water resources management in arid regions.

2. Data and Methodology

2.1. Study Area

This study focuses on Xinjiang, a typical arid region in northwestern China. The region is located within the range of 34°25′–48°10′ N and 73°40′–96°18′ E, with an area of about 1.66×10^6 km², and is characterized by its highly vulnerable water resources and fragile environment. Complex topographic and geomorphologic features build mountains, plains, and basins in this region. The elevation of this region ranges from below sea level with 161 m a.s.l. to as high as 7906 m a.s.l. (Figure 1). The annual mean precipitation is less than 200 mm. Due to diverse and extreme terrains in the region, the spatial and temporal distributions of precipitation are rather heterogeneous. The river runoff in Xinjiang is generated in the mountainous region and mainly depends on glacier-melt, snowmelt, and precipitation. The study area comprises two major rivers in the middle of the Tianshan Mountains: the Manas River in the north slope and the Kaidu River in the south slope. The Kaidu River Basin is gauged at Bayanbulak (2458 m a.s.l.) for monitoring runoff, where the area above Bayanbulak is 18,725 km² [36]. The Manas River Basin is gauged at Kensiwate (940 m a.s.l.) for monitoring runoff, where the area above Kensiwate is 4637 km², which consists of 608 km² with glacier [37]. The area above both gauges is little influenced by human activities. The summer precipitation accounts for about 60–80% of total annual precipitation in both rivers [22]. Annual runoffs over the period 1958–2007 were 9.18×10^8 m³ and 12.37×10^8 m³ for the Kaidu River Basin and Manas River Basin, respectively [22,38]. Snowmelt water is the main recharge source in both rivers in spring. Maximum monthly discharge occurs in July at Bayanbulak gauge and in August at Kensiwate gauge, when runoff is generated by snowmelt, glacier-melt, and rainfall.

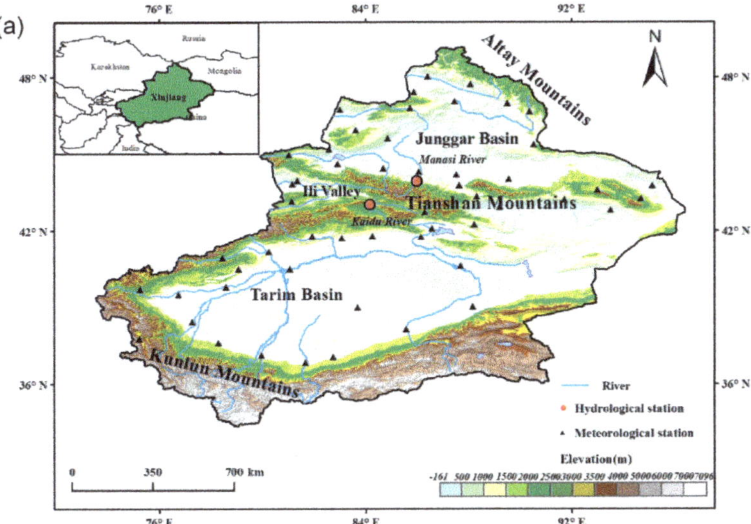

Figure 1. *Cont.*

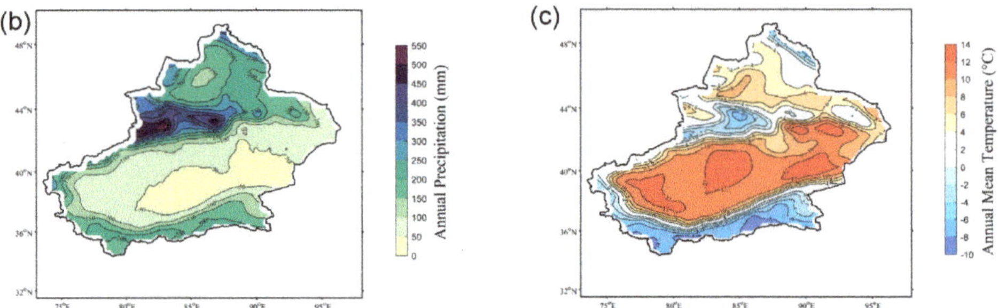

Figure 1. Study area. (**a**) Distribution of meteorological stations, (**b**) annual precipitation (mm), (**c**) annual temperature (°C).

2.2. WRF Model Set Up

The WRF model used in this study runs at a resolution of 12 km across western China. The simulation domain covers 73.447°–127.017° E, 26.855°–53.585° N. The model contains a large number of physical processes, and their parameterizations are as follows: the top of atmosphere has been set to 50 hpa, with 31 layers along the vertical direction. The base period of the climatic simulation started from 06:00:00 BJT 1 January 1979 (8 h earlier than the UTC) and ended at 23:00:00 BJT 31 December 2015, with a 3-h interval for outputs. The initial lateral boundary condition was forced using the NCEP/DOE dataset, and the sea surface temperature was obtained from the ERA-Interim dataset. Other parameterization included the Kain-Fritsch Cumulus Scheme [39], the Rapid Radiation Transfer Model (RRTM) for longwave radiation [40], the Dudhia shortwave radiation model [41], the WRF Single-Moment 3-Class microphysics model, the Noah Land Surface Model (Noah LSM) [42], and the Yonsei University model for the planetary boundary layer (YSU) [43].

2.3. Meteorological and Streamflow Datasets

Before further analysis was carried out for the S/P ratio in this study, bias analysis between simulated results and in-situ observations, as well as the distribution of CN05.1, a dataset extrapolated with 2480 meteorological stations across the whole of China in terms of the Thin Plate Spline (TPS) method, was implemented. Fifty-one stations of in-situ observations in Xinjiang, China, selected from a total of 793 stations in the China Meteorological dataset version 3.0 (http://www.escience.gov.cn), were collected to evaluate the performance of the WRF model. In addition, the CN05.1 dataset was applied for spatial validation of the model outputs. The dataset contains daily precipitation and daily temperature at the resolution of 0.25 degree from 1961 to 2015. The accuracy analysis on WRF outputs showed small differences over eastern China with dense observation stations, but larger differences over western China, where there were less stations [44,45].

Daily discharge data for the Bayanbulak and Kensiwate gauges are available for the period of 1979 to 2011. The daily data were checked for homogeneity and continuity, and then monthly sums of daily data were used for analyzing the changes in water resources in the study area.

2.4. Snowfall Calculation

In this study, snowfall was calculated from precipitation. Based on the study from Dai [46], the conditional frequencies of snowfall (F) can be calculated using a hyperbolic tangent function. Snowfall occurs when the temperature-dependence exceeds 50%. Therefore, a sigmoidal hyperbolic tangent curve was used to fit the observations of snow conditional frequency per 0.3 °C Ta bin from -10 °C to 10 °C in this study.

$$F = a[tanh(b(T_a - c)) - d]$$

(1)

where T_a is the temperature, and parameters *a*, *b*, *c*, and d are calculated by least squares fitting observations. This method was estimated in the North Hemisphere for rain-snow partition [47]. The parameters were fitted best as −49.7648, 0.3146, 1.6540, and 0.9786 in the study area during the cold season (from October to April, respectively).

2.5. Assessment of Performance of the WRF Model

Bias (BIAS) and root mean square error (RMSE) were used to evaluate the performance of the WRF model before further analysis. BIAS measures the average tendency of the simulated data with observations [48], while RMSE measures the deviation between the simulated data and observations. For the BIAS, positive values indicate model overestimation bias, and negative values indicate model underestimation bias.

$$BIAS = \sum_{i=1}^{n} (Y_i^{obs} - Y_i^{sim}) \tag{2}$$

$$RMSE = \sqrt{\frac{1}{n}\sum_{i=1}^{n} (Y_i^{obs} - Y_i^{sim})^2} \tag{3}$$

where Y^{obs} and Y^{sim} are the in-situ observations and simulated data, respectively. The optimal value of BIAS and RMSE is 0.0, with low values indicating accurate model simulation.

2.6. Trend Analysis

The Mann-Kendall (MK) test was employed to detect the trends of snowfall and its ratio to total precipitation in this study. This test has been widely used in hydro-meteorological time series analysis as a non-parametric statistical test [49,50]. Compared with parametric statistical tests, non-parametric tests are more suitable to analyze the monotonic trends for non-normally distributed data [49]. To reduce the effect of autocorrelation, all data went through pre-whitening before the MK test [49].

3. Results

3.1. Performance of the WRF Model

Table 1 reveals the mean BIAS and RMSE of monthly total precipitation and mean temperature between WRF outputs and observations of the selected 51 stations in Xinjiang. The simulated temperature agreed well with observations. The model displayed a cold bias in summer, but warm bias in spring, winter, and annual mean. Overestimation in precipitation was more frequently noted in all seasons, and especially occurred in high elevation areas such as the Altay Mountains and the Tianshan Mountains. The distribution of RMSE was similar to that of bias, with the largest value in high altitude regions.

Table 1. Bias and RMSE over Xinjiang for annual and seasonal mean temperature and precipitation between WRF simulation and observation. A stands for annual, MAM for March-April-May, SON for September-October-November, and DJF for December-January-February.

	Temperature Bias (°C month^{-1})	RMSE (°C month^{-1})	Precipitation Bias (mm month^{-1})	RMSE (mm month^{-1})
A	0.33	2.44	7.54	15.56
MAM	−2.02	3.51	11.12	16.85
SON	0.81	2.51	5.65	13.01
DJF	1.16	4.00	11.45	17.22

Figure 2 shows the spatial distribution patterns of two datasets for climatology snowfall during the cold season. Both the outputs of WRF and the gridded dataset (CN05.1) shared similar distributions. High snowfall mainly occurred in the mountainous areas, such as the Tianshan Mountains, the Kunlun

Mountains, and the Altay Mountains, while low values took place in the Tarim Basin and the Junggar Basin. Although the snowfall in both datasets had very similar distributions in the context of amount, the WRF output was higher than that of CN05.1. The result of WRF was overestimated in the alpine area, especially in the Tianshan Mountains, the Kunlun Mountains, and the Altay Mountains.

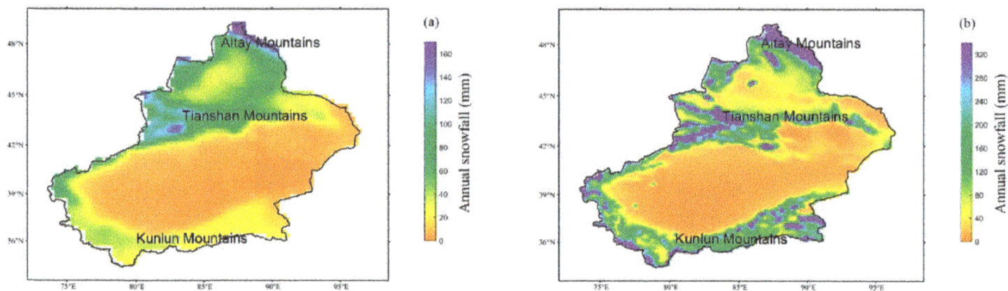

Figure 2. Spatial distributions of climatology snowfall during the cold season (mm). (**a**) is CN05.1, (**b**) is WRF.

3.2. Climatology and Changes of Snowfall over Xinjiang

Figure 2b shows the spatial distribution of climatology snowfall during the cold season over Xinjiang from 1979 to 2015. High snowfall mainly occurred in the mountainous areas, such as the Tianshan Mountains, the Kunlun Mountains, and the Altay Mountains, while low values were exhibited in the Tarim Basin and the Junggar Basin. The spatial distribution of changes in snowfall during 1979–2015 is shown in Figure 3. Changes were defined as the linear regression slope. The snowfall varied in different regions, although the snowfall revealed no significant decreasing trend in Xinjiang during last decades. Significantly decreasing trends were found in the high-elevation regions of the Kunlun Mountains, and the middle and south slope of the Tianshan Mountains. However, dramatic increases occurred in the Altay Mountains, as well as in the Ili Valley and low-elevation regions of the Kunlun Mountains. There also existed a slightly increasing trend in the Junggar Basin.

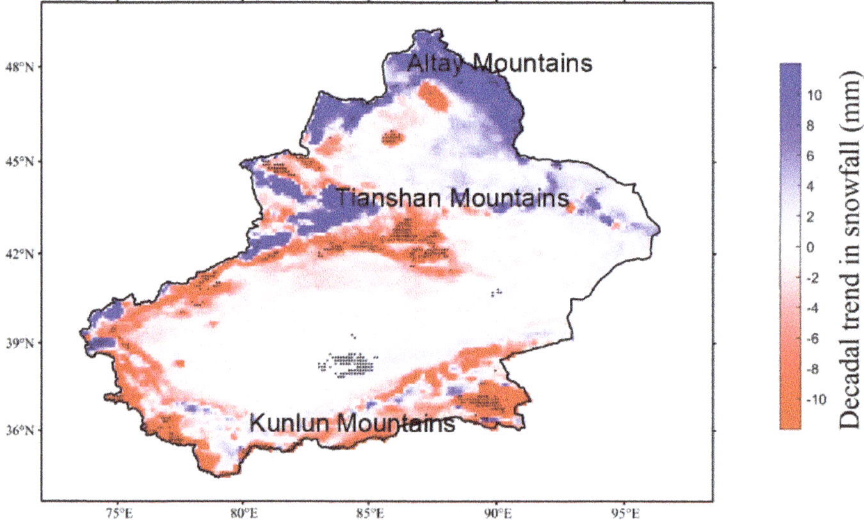

Figure 3. Spatial distribution variation of snowfall time series of WRF during the cold season from 1979 to 2015. Black dots represent significance at the 0.05 level.

3.3. Climatology and Changes of S/P Ratio over Xinjiang

Figure 4 displays the spatial distribution of climatology and changes in the S/P ratio during the cold season from 1979 to 2015 over Xinjiang. The distribution of S/P ratio is similar to that of snowfall, with high values occurring in the Mountains and high-elevation regions, but low values observed in basins and valleys. The value of the S/P ratio was near 1 at an elevation above 3500 m, where the temperature is often below 0 °C. The S/P ratio slightly increased during the past decades in Xinjiang. Based on the results of the Mann-Kendall test, the S/P ratios estimated by the WRF model did not show a significant change at a level of 0.10 in the study area. As illustrated in Figure 4b, the changes of S/P ratio also varied in different regions, which was similar to the distribution of the changes of snowfall. The ratio had a rising trend in low-elevation regions, such as the Tarim Basin and the Junggar Basin, but a decreasing trend in relative high-elevation regions, such as the Tianshan Mountains, the Altay Mountains, and the western edge of the Junggar Basin.

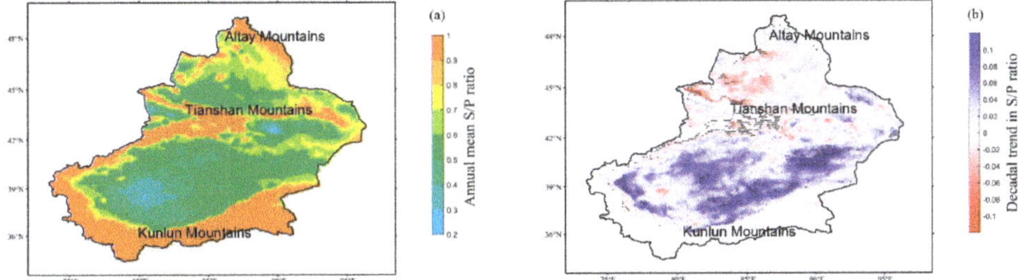

Figure 4. Spatial distribution of climatology (**a**) and variation (**b**) of the S/P ratio in Xinjiang revealed by WRF from 1979 to 2015. Black dots represent significance at the 0.05 level.

3.4. Changes in Hydrological Processes in the Tianshan Mountains

The runoff time series between the 1980s and 2000s from two hydrological stations at the two river basins in the Tianshan Mountains are shown in Figure 5. The annual runoff significantly increased from 1979 to 2011 by 17.4% and 20.4% in Manas River Basin and Kaidu River Basin, respectively. The runoffs in the two river basins all exhibited decreasing trends in spring, but increasing trends in summer and autumn. For both river basins, insignificant trends were found in the runoff during winter. In addition, an earlier melt was found in the Bayanbulak station at Kaidu River Basin. Compared with the 1980s, about five days in advance were observed in Bayanbulak stations in the 2000s.

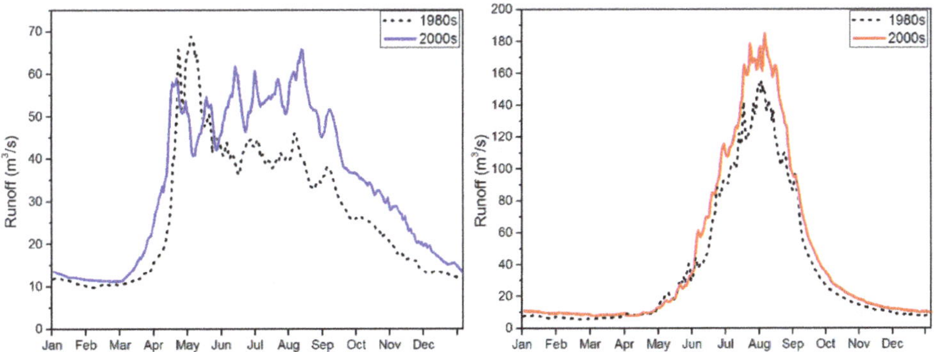

Figure 5. Runoff time series between the 1980s and 2000s. Left: the Kaidu River Basin, right: the Manas River Basin.

4. Discussion

4.1. Performance of WRF

The WRF outputs were evaluated with in-situ observation and the CN05.1 dataset on point-scale and spatial distribution, respectively. The bias values of temperature ranged from -2.02 °C to 1.16 °C in the WRF simulations, which was more consistent with the observations from -0.5 °C to 4 °C of Tang et al. [32]. It is noted that WRF generally overestimated precipitation compared with in-situ observations, especially in the alpine region and lake area. It indicated that the simulated precipitation in this study was very reliable in the basins and low-elevation regions, while the overestimation leading to more uncertainty occurred in high-elevation regions. This uncertainty may result from various sources, not only by overestimation of the model, but also possibly from the underestimation of in-situ observations that were caused by the snowdrift and sublimation in the alpine area. The underestimation of rain gauges caused by measurement errors for solid precipitation has frequently exceeded 50% in alpine areas [51,52]. On the other hand, some studies reported that the WRF model produced strong wet biases in precipitation over China, with overestimation exceeding 150% against in-situ observations in some regions [30,32,53,54]. However, the simulation from WRF was close to the snowfall records which were required for the maintenance of glaciers in the Karakoram Mountains (above 4 km a.s.l.) [33].

The distribution of in-situ observations is uneven over the study area, especially in alpine regions. Therefore, the observation datasets might increase the uncertainty and representativeness of data in the study area. The same problem existed in observation-based gridded rainfall data (CN05.1). Although more than 2400 stations were used to extrapolate the data across China in the CN05.1 dataset, stations were still not enough in the western China for efficient extrapolation. The gridded rainfall data is also too coarse to capture the orographic precipitation patterns, due to the complex topography of Xinjiang. Previous studies documented that the WRF model performed reasonably at rainfall predicting of a single precipitation event [30], as well as the climatological precipitation pattern and interannual precipitation variability with a fine resolution [33,55]. In this study, the results of WRF and CN05.1 have similar distributions in snowfall amount, and the WRF model could achieve higher-resolution grids data (12 km in this study) than the CN05.1 dataset, and capture more details of spatial snowfall features. Additionally, based on dynamic processes, the WRF model can exhibit precipitation events in areas that lack in-situ observations. As a result, WRF output prevailed over in-situ observations and CN05.1 in characterizing the long-term spatiotemporal distribution of snowfall and the S/P ratio, and offered a convincing basis for analysis of the long-term and large-scale variation of snowfall, particularly over large vertical gradient and complex topography regions.

4.2. Spatiotemporal Variations of Snowfall and S/P Ratio

A decreased S/P ratio could be explained by snowfall decreases that were proportionally larger than decreases in rainfall, constant snowfall, and increasing rainfall, or increases in both, but larger increases in rainfall than snowfall. However, an increasing S/P ratio may be caused by increasing snowfall or decreasing rainfall. Although changes in snowfall more closely paralleled the pattern of S/P ratio trends, the total precipitation had a weak correlation with the S/P ratio [15]. In addition, the relative changes in snowfall and precipitation contributed together to the variation of the S/P ratio [13]. According to Figures 3b and 6a, the variations of snowfall and precipitation were similar, but had different magnitudes, in the study area. The precipitation mainly occurred as snowfall at an elevation above 3500 m.

In this study, the trends of the S/P ratio were the opposite for the low-elevation regions and mountains. The decreasing variability of the S/P ratio was because of the decline in precipitation more than that of snowfall in the relative high-elevation Tianshan Mountains. Additionally, although both snowfall and precipitation increased at the western edge of the Junggar Basin, the S/P ratio exhibited a downward trend. This could be because increasing snowfall would offset a portion of decreased snowfall that was caused by the warming temperature in this region (see Figure 6b). The fall of the S/P ratio in the Tianshan Mountains was also reported in a previous study by Guo and Li [13]. The rise in S/P ratio was caused by the increase in both the snowfall and precipitation, but was larger for snowfall than precipitation in the Tarim Basin. However, the S/P ratios have also changed a little at elevations above 3500 m, where the temperature is often far below 0 °C, such as the Kunlun Mountains.

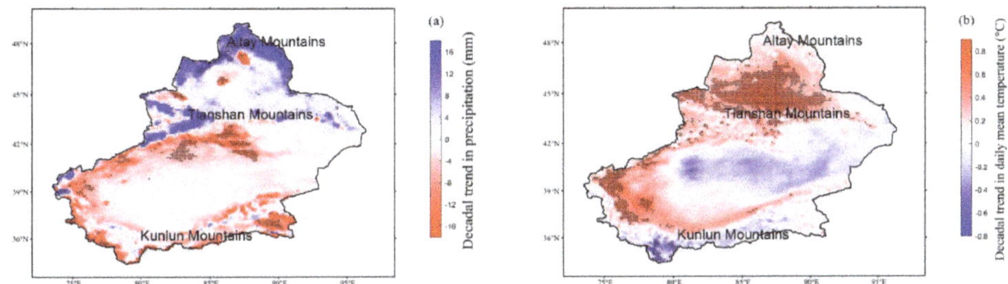

Figure 6. Decadal trends of precipitation (**a**) and temperature (**b**) time series of WRF during the cold season from 1979 to 2015. Black dots represent significance at the 0.05 level.

4.3. Water Resources Management in Different Regions Based on Current Findings

The Manas River Basin and Kaidu River Basin are located in the decreasing S/P ratio regions. The correlation coefficients between the S/P ratio and runoff were -0.52 ($p < 0.05$) and -0.65 ($p < 0.05$) in the Manas River Basin and Kaidu River Basin, respectively. This negative correlation indicated that the larger runoff was associated with the lower S/P ratio during the cold season, with a larger impact on the runoff of the Kaidu River Basin. Under climate change, the change in snowfall and S/P ratio has implications for the water resources, which poses a serious challenge to the water resources management authorities.

In the decreasing S/P ratio regions, by the 2000s, both the changes in runoff magnitude and shift of intra-annual patterns in runoff were found in the river basins in the Tianshan Mountains with respect to the 1980s level. More rainfall than snowfall in late winter and spring implies an increase in temperature, which may influence the peak river runoff and promote the risks of early snowmelt in spring in the decreasing S/P ratio regions [10,11,56]. For the snowmelt-recharged river, the runoff of the Kaidu River showed an earlier occurrence of maximum snow melt and glacial recharge. However, for the glacial melt-recharged river, the runoff of the Manas River showed an increasing trend when comparing the 1980s with the 2000s level. Both the changes in runoff magnitude and shift of intra-annual patterns in runoff were in line with the decreasing S/P ratio. On the other hand, in the increasing S/P ratio regions, the runoffs of headwaters of the Tarim River significantly increased over the past decades in winter, spring, and summer [57]. In addition, strong fluctuation in river runoffs from the three main water systems of the Tarim River were observed during the past decades [58]. However, the runoff from Tarim River is expected to increase in the period 2010–2039, but reduce in 2070–2099 with the shrinking of glaciers [59].

As a result, it is important to allocate the annual water resources in the decreasing S/P ratio regions, including the Tianshan Mountains, which can reduce the influence on farm land and grass land. By contrary, more snowfall in winter would trigger more floods in spring and summer in the increasing S/P ratio regions, which occurred in the Tarim Basin and the region to its east. Although flooding could cause damage to agricultural development and the rural residential area, it would also provide water resources for these regions where precipitation hardly contributed throughout the year, and play an important role in maintaining ecosystem stability in these regions.

5. Conclusions

In this study, the WRF model outputs were used to dynamically downscale and evaluate the long-term variation of snowfall and S/P ratio in Xinjiang, China. The results showed that the snowfall displayed complex spatial patterns concerning its long-term variation, with an increase in the southern edge of the Tarim Basin, the Ili Valley, and the Altay Mountains, but a decrease in the Tianshan Mountains and the Kunlun Mountains. The S/P ratio experienced an increasing trend in low-elevation regions, such as the Tarim Basin and the Junggar Basin, but the opposite trends were identified in relatively high-elevation regions, such as the Altay Mountains, the Tianshan Mountains, and the western edge of the Junggar Basin. However, the S/P ratios have also changed a little at elevations above 3500 m, where the temperature is often far below 0 °C, such as the Kunlun Mountains. The increasing runoff was in line with the decreasing S/P ratio in Kaidu River Basin and Manas River Basin, due to the negative correlation between the S/P ratio and the runoff during the cold season. It is important to allocate the annual water resources in the decreasing S/P ratio regions, because both the runoff magnitude and intra-annual patterns in runoff were changed in this region. Validation of WRF outputs proved that the outputs agreed fairly well with in-situ observations and the CN05.1 dataset. The model captured most details of the precipitation distribution and the results can be used for assessing the impacts of climate change on snowfall and water resources. However, since the uncertainty exists in the evaluation of simulating precipitation at high altitude and a complex topography area, more parameterization schemes and boundary conditions are also needed for investigating the regional climate change in complex topography areas. A series of sensitivity studies are also needed to improve the accuracy of the simulations. In addition, it is necessary to couple a physically-based hydrological model with the WRF model for the estimation of runoff changes.

Author Contributions: L.L. and Q.L. conceived the original design of the work. Q.L. and L.L. wrote the paper. Q.L. and T.Y. analyzed all of the data and created all the tables and figures. Z.Q. provided comments on the paper.

Funding: This research was funded by the projects jointly-funded by Xinjiang and National Natural Science Foundation of China (Grant No: U1703241), the National Natural Science Foundation of China (Grant No. 41501050), and the National Project of Investigation of Basic Resources for Science and Technology (Grant No. 2017FY100501). We are grateful to the support from the Tianshan Station for Snow cover and Avalanche Research, Chinese Academy of Sciences, for data collection and analysis.

Conflicts of Interest: The authors declare no conflict of interest.

References

1. Groisman, P.Y.; Karl, T.R.; Knight, R.W.; Stenchikov, G.L. Changes of snow cover, temperature, and radiative heat balance over the Northern Hemisphere. *J. Clim.* **1994**, *7*, 1633–1656. [CrossRef]
2. Simonovic, S.P.; Li, L.H. Sensitivity of the Red River basin flood protection system to climate variability and change. *Water Resour. Manag.* **2004**, *18*, 89–110. [CrossRef]
3. Brown, R.D.; Mote, P.W. The response of Northern Hemisphere snow cover to a changing climate. *J. Clim.* **2009**, *22*, 2124–2145. [CrossRef]
4. Brown, R.D. Northern Hemisphere Snow Cover Variability and Change, 1915–1997. *J. Clim.* **2000**, *13*, 2339–2355. [CrossRef]
5. Derksen, C.; Brown, R. Spring snow cover extent reductions in the 2008–2012 period exceeding climate model projections. *Geophys. Res. Lett.* **2012**, *39*, 1–6. [CrossRef]
6. Mudryk, L.R.; Kushner, P.J.; Derksen, C. Interpreting observed northern hemisphere snow trends with large ensembles of climate simulations. *Clim. Dyn.* **2014**, *43*, 345–359. [CrossRef]
7. McCabe, G.J.; Wolock, D.M. Long-term variability in Northern Hemisphere snow cover and associations with warmer winters. *Clim. Chang.* **2010**, *99*, 141–153. [CrossRef]
8. Berghuijs, W.R.; Woods, R.A.; Hrachowitz, M. A precipitation shift from snow towards rain leads to a decrease in streamflow-supplement. *Nat. Clim. Chang.* **2014**, *4*, 583–586. [CrossRef]
9. Hatchett, B.J.; Daudert, B.; Garner, C.B.; Oakley, N.S.; Putnam, A.E.; White, A.B. Winter snow level rise in the Northern Sierra Nevada from 2008 to 2017. *Water (Switzerland)* **2017**, *9*, 899. [CrossRef]

10. Beniston, M.; Stoffel, M. Rain-on-snow events, floods and climate change in the Alps: Events may increase with warming up to 4 °C and decrease thereafter. *Sci. Total Environ.* **2016**, *571*, 228–236. [CrossRef] [PubMed]

11. McCabe, G.J.; Clark, M.P.; Hay, L.E. Rain-on-Snow Events in The Western United States. *Bull. Am. Meteorol. Soc.* **2007**, *88*, 319–328. [CrossRef]

12. Harpold, A.A.; Kaplan, M.L.; Zion Klos, P.; Link, T.; McNamara, J.P.; Rajagopal, S.; Schumer, R.; Steele, C.M. Rain or snow: Hydrologic processes, observations, prediction, and research needs. *Hydrol. Earth Syst. Sci.* **2017**, *21*, 1–22. [CrossRef]

13. Guo, L.; Li, L. Variation of the proportion of precipitation occurring as snow in the Tian Shan Mountains, China. *Int. J. Climatol.* **2015**, *35*, 1379–1393. [CrossRef]

14. Huntington, T.G.; Hodgkins, G.A.; Keim, B.D.; Dudley, R.W. Changes in the proportion of precipitation occurring as snow in New England (1949–2000). *J. Clim.* **2004**, *17*, 2626–2636. [CrossRef]

15. Knowles, N.; Dettinger, M.D.; Cayan, D.R. Trends in snowfall versus rainfall in the western United States. *J. Clim.* **2006**, *19*, 4545–4559. [CrossRef]

16. Przybylak, R. Variability of total and solid precipitation in the Canadian Arctic from 1950 to 1995. *Int. J. Climatol.* **2002**, *22*, 395–420. [CrossRef]

17. Serquet, G.; Marty, C.; Dulex, J.-P.; Rebetez, M. Seasonal trends and temperature dependence of the snowfall/precipitation-day ratio in Switzerland. *Geophys. Res. Lett.* **2011**, *38*, 14–18. [CrossRef]

18. Yang, J.; Fang, G.; Chen, Y.; De-Maeyer, P. Climate change in the Tianshan and northern Kunlun Mountains based on GCM simulation ensemble with Bayesian model averaging. *J. Arid Land* **2017**, *9*, 622–634. [CrossRef]

19. Littell, J.S.; McAfee, S.A.; Hayward, G.D. Alaska snowpack response to climate change: Statewide snowfall equivalent and snowpack water scenarios. *Water (Switzerland)* **2018**, *10*. [CrossRef]

20. Li, Z.; Chen, Y.; Li, W.; Deng, H.; Fang, G. Potential impacts of climate change on vegetation dynamics in Central Asia. *J. Geophys. Res. Atmos.* **2015**, *120*, 345–356. [CrossRef]

21. Li, X.; Jiang, F.; Li, L.; Wang, G. Spatial and temporal variability of precipitation concentration index, concentration degree and concentration period Xinjiang, China. *Int. J. Climatol.* **2011**, *31*, 1679–1693. [CrossRef]

22. Zhang, F.; Bai, L.; Li, L.; Wang, Q. Sensitivity of runoff to climatic variability in the northern and southern slopes of the Middle Tianshan Mountains, China. *J. Arid Land* **2016**, *8*, 681–693. [CrossRef]

23. Chen, Y.; Li, Z.; Fan, Y.; Wang, H.; Deng, H. Progress and prospects of climate change impacts on hydrology in the arid region of northwest China. *Environ. Res.* **2015**, *139*, 11–19. [CrossRef] [PubMed]

24. Xu, M.; Kang, S.; Wu, H.; Yuan, X. Detection of spatio-temporal variability of air temperature and precipitation based on long-term meteorological station observations over Tianshan Mountains, Central Asia. *Atmos. Res.* **2018**, *203*, 141–163. [CrossRef]

25. Chen, Y.; Li, W.; Deng, H.; Fang, G.; Li, Z. Changes in Central Asia's Water Tower: Past, Present and Future. *Sci. Rep.* **2016**, *6*, 1–12. [CrossRef] [PubMed]

26. Chen, Y.; Li, Z.; Fang, G.; Li, W. Large Hydrological Processes Changes in the Transboundary Rivers of Central Asia. *J. Geophys. Res. Atmos.* **2018**, 5059–5069. [CrossRef]

27. Groisman, P.Y.; Koknaeva, V.V.; Belokrylova, T.A.; Karl, T.R. Overcoming Biases of Precipitation Measurement: A History of the USSR Experience. *Bull. Am. Meteorol. Soc.* **1991**, *72*, 1725–1733. [CrossRef]

28. Adam, J.C. Adjustment of global gridded precipitation for systematic bias. *J. Geophys. Res.* **2003**, *108*, 1–15. [CrossRef]

29. Adam, J.C.; Clark, E.A.; Lettenmaier, D.P.; Wood, E.F. Correction of global precipitation products for orographic effects. *J. Clim.* **2006**, *19*, 15–38. [CrossRef]

30. Maussion, F.; Scherer, D.; Finkelnburg, R.; Richters, J.; Yang, W.; Yao, T. Sciences WRF simulation of a precipitation event over the Tibetan Plateau, China—An assessment using remote sensing and ground observations. *Hydrol. Earth Syst. Sci.* **2011**, *15*, 1795–1817. [CrossRef]

31. Gao, Y.; Xu, J.; Chen, D. Evaluation of WRF mesoscale climate simulations over the Tibetan Plateau during 1979–2011. *J. Clim.* **2015**, *28*, 2823–2841. [CrossRef]

32. Tang, J.; Niu, X.; Wang, S.; Gao, H.; Wang, X.; Wu, J. Statistical downscaling and dynamical downscaling of regional climate in China: Present climate evaluations and future climate projections. *J. Geophys. Res. Atmos.* **2016**, *121*, 2110–2129. [CrossRef]

33. Norris, J.; Carvalho, L.M.V.; Jones, C.; Cannon, F.; Bookhagen, B.; Palazzi, E.; Tahir, A.A. The spatiotemporal variability of precipitation over the Himalaya: Evaluation of one-year WRF model simulation. *Clim. Dyn.* **2016**, 1–26. [CrossRef]

34. Lee, J.; Choi, J.; Lee, O.; Yoon, J.; Kim, S. Estimation of probable maximum precipitation in Korea using a regional climate model. *Water (Switzerland)* **2017**, *9*, 240. [CrossRef]

35. Rasmussen, R.; Liu, C.; Ikeda, K.; Gochis, D.; Yates, D.; Chen, F.; Tewari, M.; Barlage, M.; Dudhia, J.; Yu, W.; et al. High-resolution coupled climate runoff simulations of seasonal snowfall over Colorado: A process study of current and warmer climate. *J. Clim.* **2011**, *24*, 3015–3048. [CrossRef]

36. Zhang, F.; Ahmad, S.; Zhang, H.; Zhao, X.; Feng, X.; Li, L. Simulating low and high streamflow driven by snowmelt in an insufficiently gauged alpine basin. *Stoch. Environ. Res. Risk Assess.* **2016**, *30*, 59–75. [CrossRef]

37. Yu, M.; Chen, X.; Li, L.; Bao, A.; De, M.J. Incorporating accumulated temperature and algorithm of snow cover calculation into the snowmelt runoff model. *Hydrol. Process.* **2013**, *3595*, 3589–3595. [CrossRef]

38. Li, X.; Li, L.; Guo, L.; Zhang, F.; Adsavakulchai, S.; Shang, M. Impact of climate factors on runoff in the Kaidu River watershed: Path analysis of 50-year data. *J. Arid Land* **2011**, *3*, 132–140. [CrossRef]

39. Kain, J.S. The Kain–Fritsch Convective Parameterization: An Update. *J. Appl. Meteorol.* **2004**, *43*, 170–181. [CrossRef]

40. Mlawer, E.J.; Taubman, S.J.; Brown, P.D.; Iacono, M.J.; Clough, S.A. Radiative transfer for inhomogeneous atmospheres: RRTM, a validated correlated-k model for the longwave. *J. Geophys. Res.* **1997**, *102*, 16663. [CrossRef]

41. Dudhia, J. Numerical Study of Convection Observed during the Winter Monsoon Experiment Using a Mesoscale Two-Dimensional Model. *J. Atmos. Sci.* **1989**, *46*, 3077–3107. [CrossRef]

42. Chen, F.; Dudhia, J. Coupling an Advanced Land Surface–Hydrology Model with the Penn State–NCAR MM5 Modeling System. Part I: Model Implementation and Sensitivity. *Mon. Weather Rev.* **2001**, *129*, 569–585. [CrossRef]

43. Hong, S.; Lim, J. The WRF single-moment 6-class microphysics scheme (WSM6). *J. Korean Meteorol. Soc.* **2006**, *42*, 129–151.

44. Xie, P.; Chen, M.; Fukushima, Y.; Yang, S.; Liu, C.; Yatagai, A.; Hayasaka, T. A Gauge-Based Analysis of Daily Precipitation over East Asia. *J. Hydrometeorol.* **2007**, *8*, 607–626. [CrossRef]

45. Xu, Y.; Gao, J.; Shen, Y.; Xu, C.; Shi, Y.; Giorgi, F. A Daily Temperature Dataset over China and Its Application in Validating a RCM Simulation. *Adv. Atmos. Sci.* **2009**, *26*, 763–772. [CrossRef]

46. Dai, A. Temperature and pressure dependence of the rain-snow phase transition over land and ocean. *Geophys. Res. Lett.* **2008**, *35*, 1–7. [CrossRef]

47. Jennings, K.S.; Winchell, T.S.; Livneh, B.; Molotch, N.P. Spatial variation of the rain-snow temperature threshold across the Northern Hemisphere. *Nat. Commun.* **2018**, *9*, 1–9. [CrossRef] [PubMed]

48. Gupta, H.V. Status of automatic calibration for hydrologic models: Comparison with multilevel expert calibration. *J. Hydrol. Eng.* **1999**, 135–143. [CrossRef]

49. Yue, S.; Pilon, P.; Cavadias, G. Power of the Mann-Kendall and Spearman's rho tests for detecting monotonic trends in hydrological series. *J. Hydrol.* **2002**, *259*, 254–271. [CrossRef]

50. Hamed, K.H. Trend detection in hydrologic data: The Mann-Kendall trend test under the scaling hypothesis. *J. Hydrol.* **2008**, *349*, 350–363. [CrossRef]

51. Rasmussen, R.; Baker, B.; Kochendorfer, J.; Meyers, T.; Landolt, S.; Fischer, A.P.; Black, J.; Thériault, J.M.; Kucera, P.; Gochis, D.; et al. How well are we measuring snow: The NOAA/FAA/NCAR winter precipitation test bed. *Bull. Am. Meteorol. Soc.* **2012**, *93*, 811–829. [CrossRef]

52. Grossi, G.; Lendvai, A.; Peretti, G.; Ranzi, R. Snow precipitation measured by gauges: Systematic error estimation and data series correction in the central Italian Alps. *Water (Switzerland)* **2017**, *9*, 461. [CrossRef]

53. Yu, E.; Sun, J.; Chen, H.; Xiang, W. Evaluation of a high-resolution historical simulation over China: Climatology and extremes. *Clim. Dyn.* **2015**, *45*, 2013–2031. [CrossRef]

54. Ma, J.; Wang, H.; Fan, K. Dynamic downscaling of summer precipitation prediction over China in 1998 using WRF and CCSM4. *Adv. Atmos. Sci.* **2015**, *32*, 577–584. [CrossRef]

55. Marteau, R.; Richard, Y.; Pohl, B.; Smith, C.C.; Castel, T. High-resolution rainfall variability simulated by the WRF RCM: Application to eastern France. *Clim. Dyn.* **2014**, *44*, 1093–1107. [CrossRef]

56. Barnett, T.P.; Adam, J.C.; Lettenmaier, D.P. Potential impacts of a warming climate on water availability in snow-dominated regions. *Nature* **2005**, *438*, 303–309. [CrossRef] [PubMed]

57. Duethmann, D.; Bolch, T.; Farinotti, D.; Kriegel, D.; Vorogushyn, S.; Merz, B.; Pieczonka, T.; Jiang, T.; Su, B.; Güntner, A. Attribution of streamflow trends in snow-and glacier melt dominated catchments of the Tarim River, Central Asia. *Water Resour. Res.* **2015**, *51*, 4727–4750. [CrossRef]

58. Wang, H.; Chen, Y.; Li, W. Characteristics in streamflow and extremes in the Tarim River, China: Trends, distribution and climate linkage. *Int. J. Climatol.* **2015**, *35*, 761–776. [CrossRef]

59. Duethmann, D.; Menz, C.; Jiang, T.; Vorogushyn, S. Projections for headwater catchments of the Tarim River reveal glacier retreat and decreasing surface water availability but uncertainties are large. *Environ. Res. Lett.* **2016**, *11*. [CrossRef]

Article

Observed Trends of Climate and River Discharge in Mongolia's Selenga Sub-Basin of the Lake Baikal Basin

Batsuren Dorjsuren [1,2], Denghua Yan [1,3,4,*], Hao Wang [1,3,4], Sonomdagva Chonokhuu [2], Altanbold Enkhbold [5], Xu Yiran [6], Abel Girma [1,7], Mohammed Gedefaw [1,7] and Asaminew Abiyu [1]

[1] College of Environmental Science and Engineering, Donghua University, Shanghai 201620, China; Batsuren@seas.num.edu.mn (B.D.); Wanghao@iwhr.com (H.W.); Abelethiopia@yahoo.com (A.G.); Mohammedgedefaw@gmail.com (M.G.); Asaminewab@yahoo.com (A.A.)

[2] Department of Environment and Forest Engineering, School of Engineering and Applied Sciences, National University of Mongolia, Ulaanbaatar 210646, Mongolia; ch_sonomdagva@num.edu.mn

[3] State Key Laboratory of Simulation and Regulation of Water Cycle in River Basin, China Institute of Water Resources and Hydropower Research (IWHR), Beijing 100038, China

[4] Water Resources Department, China Institute of Water Resources and Hydropower Research (IWHR), Beijing 100038, China

[5] Department of Geography, School of Art & Sciences, National University of Mongolia, Ulaanbaatar 210646, Mongolia; altanbold@num.edu.mn

[6] School of Resources and Earth Science, China University of Mining and Technology, Xuzhou 221116, China; hsuijan@163.com

[7] Department of Natural Resource Management, University of Gondar, Gondar 196, Ethiopia

[*] Correspondence: Yandh@iwhr.com; Tel.: +86-10-68781976

Received: 11 September 2018; Accepted: 9 October 2018; Published: 12 October 2018

Abstract: Mongolia's Selenga sub-basin of the Lake Baikal basin is spatially extensive, with pronounced environmental gradients driven primarily by precipitation and air temperature on broad scales. Therefore, it is an ideal region to examine the dynamics of the climate and the hydrological system. This study investigated the annual precipitation, air temperature, and river discharge variability at five selected stations of the sub-basin by using Mann-Kendall (MK), Innovative trend analysis method (ITAM), and Sen's slope estimator test. The result showed that the trend of annual precipitation was slightly increasing in Ulaanbaatar ($Z = 0.71$), Erdenet ($Z = 0.13$), and Tsetserleg ($Z = 0.26$) stations. Whereas Murun ($Z = 2.45$) and Sukhbaatar ($Z = 1.06$) stations showed a significant increasing trend. And also, the trend of annual air temperature in Ulaanbaatar ($Z = 5.88$), Erdenet ($Z = 3.87$), Tsetserleg ($Z = 4.38$), Murun ($Z = 4.77$), and Sukhbaatar ($Z = 2.85$) was sharply increased. The average air temperature has significantly increased by 1.4 °C in the past 38 years. This is very high in the semi-arid zone of central Asia. The river discharge showed a significantly decreasing trend during the study period years. It has been apparent since 1995. The findings of this paper could help researchers to understand the annual variability of precipitation, air temperature, and river discharge over the study region and, therefore, become a foundation for further studies.

Keywords: precipitation; air temperature; river discharge; Mann-Kendall test; Selenga river basin; Lake Baikal basin; Mongolia

---

1. Introduction

The Lake Baikal basin (LBB) is a suitable area to study climate change impacts. The climate is a long-term prevailing weather condition. Weather parameters include air temperature, precipitation, humidity, sunshine hours, cloudiness, atmospheric pressure, the number of rainy days, wind velocity, etc. These parameters interact directly or indirectly, greatly affecting the environment and the living organisms [1]. The semi-arid environment is highly vulnerable to climate change [2,3]. Land surface temperature is an important ecological factor and its warming trend will influence the topsoil [4,5]. Evapotranspiration and precipitation rates may change due to changes in soil temperature and air temperature.

Precipitation change may greatly affect the hydrological system of the basin. Due to climate change and intensive human activities in recent decades, the runoff of many rivers in the world has been changing. About 22% of the world's rivers were shown to have a significant decrease in the annual runoff because of increasing water consumptions and diversion [6,7].

The most sensitive areas for climate change are arid and semi-arid regions of central Asia [2,8,9]. LBB is the largest representative of these regions [3,10]. Lake Baikal, as the world's largest natural freshwater lake and its corresponding catchments, is already affected by climate change and the water quantity becomes erratic [10]. In recent decades, changes in hydrological and water quantity are primarily attributed to climate change, land use change, contaminant influx from mining areas and urban settlements as well [11].

As the largest sub-basin of LBB, Selenga River basin is located in the Mongolian and Russian Federation. The Mongolian plateau is spatially extensive, with pronounced environmental gradients driven primarily by precipitation and air temperature on large scales. Therefore, it is an ideal region to examine the dynamics of the landscape structures and hydrology parameters [9]. This area, hydro-climatic changes can also lead to a shift in hydrology parameters, ecosystem and lake conditions in these areas. In addition, human activity effects on water pollution, water resources, sedimentary, and river discharge. In particular, in the Mongolian's Selenga River basin high socio-economic activities are taking place. In the study region, there is high population density particularly, around Tuul and Kharaa sub basin. The sub basin is located in the biggest cities of Mongolia (Ulaanbaatar, Darkhan, and Erdenet), thus the river basin is highly consumed by the city residence. They consumed the water for agriculture, recreation and domestic use. In addition, the Selenga river basin water resource is used by the mining industry. Thus, it has a great role in reducing the quantity of the basin water flow [3,11,12].

This paper aims to investigate spatial and temporal changes in climate and river discharge changes in Mongolia's Selenga sub-basin of the LBB understanding from 1979 to 2016. The overall objectives of the present study are (i) to identify historical climate trends, (ii) to identify trends of spatial and temporal changes in the river discharge, (iii) to assess the relationship and to which extent can climate change trends affect the river discharge.

2. Materials and Methods

2.1. Study Area

Lake Baikal is the oldest (about 25 million years old), deepest (1637 m) and largest freshwater lake (23,000 km^3) and is located in the southern part of East Siberia [13]. The transboundary basin of Lake Baikal is located on the boundary of North and Central Asia (96°52′–113°50′ N, 46°28′–56°42′ W) [12]. The longest stretch of the basin from southwest to north-east is 1470 km, and from west to east is 962 km. The minimal length from west to east is 193 km (Figure 1). The total area of the basin is 573,478 km^2, and 52% of which belong to Mongolia and the remaining belongs to the Russian Federation [14]. In recognition of its biodiversity and endemism, United Nations Educational, Scientific and Cultural Organization (UNESCO) declared Lake Baikal as a World Heritage Site in 1996. The lake contains an outstanding variety of endemic flora and fauna, which is an exceptional value to evolutionary science. It is also surrounded by a system of protected areas with good scenery and natural values [10,15,16].

As the largest sub-basin of LBB, Selenga River basin is located in the Mongolian and Russian Federation. The Selenga is a Trans-boundary river, the largest tributary of Lake Baikal. On the average, it discharges into Lake Baikal ~30 km^3 of water, i.e., half of the total inflow into the lake.

Forty-six percent of Selenga annual runoff forms in Mongolian territory. The Mongolian plateau is spatially extensive, with pronounced environmental gradients driven primarily by precipitation and air temperature on large scales. The length of the river is 1024 km, its drainage area is 447.06 thousand km^2, of which 148.06 thousand km^2 are in the territory of Russia.

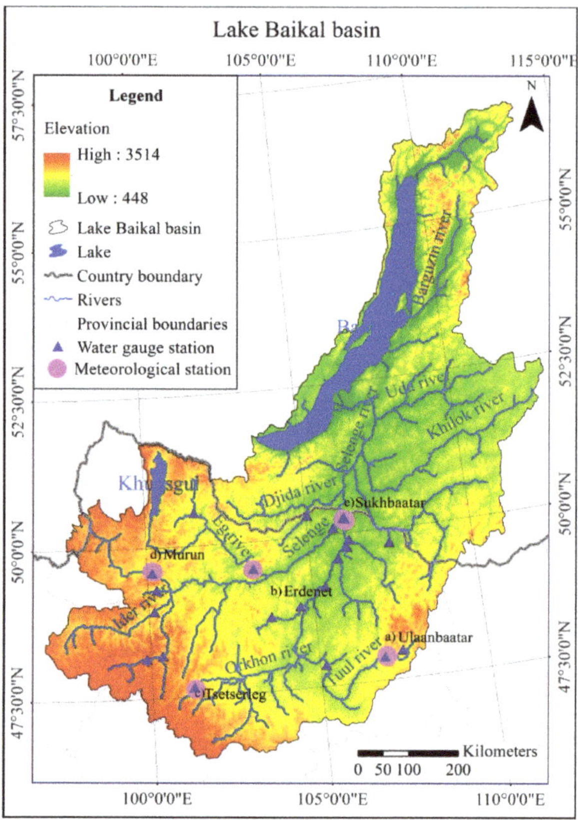

Figure 1. Location of Lake Baikal basin (LBB), Water gauge station and meteorological stations of the Selenga River basin located in the Mongolian.

2.2. Data Sources

The data of air temperature, precipitation and hydrologic data of in Mongolia's Selenga river basin were taken from Information and Research Institute of Meteorology, Hydrology, and Environment (IRIMHE) hosts (http://irimhe.namem.gov.mn/), and National Centers for Environmental Information NOAA's National Centers for Environmental Information (NCEI) hosts (https://ngdc.noaa.gov/). The location of the 5 water gauge stations used in this study is shown in Figure 1. For the selection of climate and water gauge stations in Mongolia's Selenga sub-basin of the LBB, the following factors were taken into consideration: (1) spatial distribution; (2) capacity of stations and (3) whether it is near to the water system (Figure 1) [8,17].

2.3. Methods

Analyses of long-term trends in both the observed and adjusted data were done using the Mann-Kendall test, with linear changes in the data represented by Kendall-Theil Robust Lines. This non-parametric approach is well suited for evaluating changes in hydrologic regimes.

The appendix of [18] provides a concise explanation of the statistics as applied to river discharge data [19], to remove the influence of serial correlations on the trend analyses. Trend analysis is used to investigate whether the trend is upward, downward, or no trend in data value points. The non-parametric Mann-Kendall (MK) test has been applied in studies to detect the trends in hydro-meteorological observations that do not need the normal distribution of data points. This paper used the Mann-Kendall (MK) test method to detect the trends in climate and river discharge time series data. To evaluate the reliability of Mann-Kendall (MK), the results were compared with ITAM and Sen's slope estimator test. In addition, annual and seasonal precipitation variability time series data were investigated by statistical analysis. The study region has four distinct seasons: summer (June–August), autumn (September–November), winter (December–February) and spring (March–May). Significance levels at 10%, 5%, and 1% were taken to assess the climate and river discharge time's series data by MK, ITAM, and Sen's slope estimator method (Figure 2).

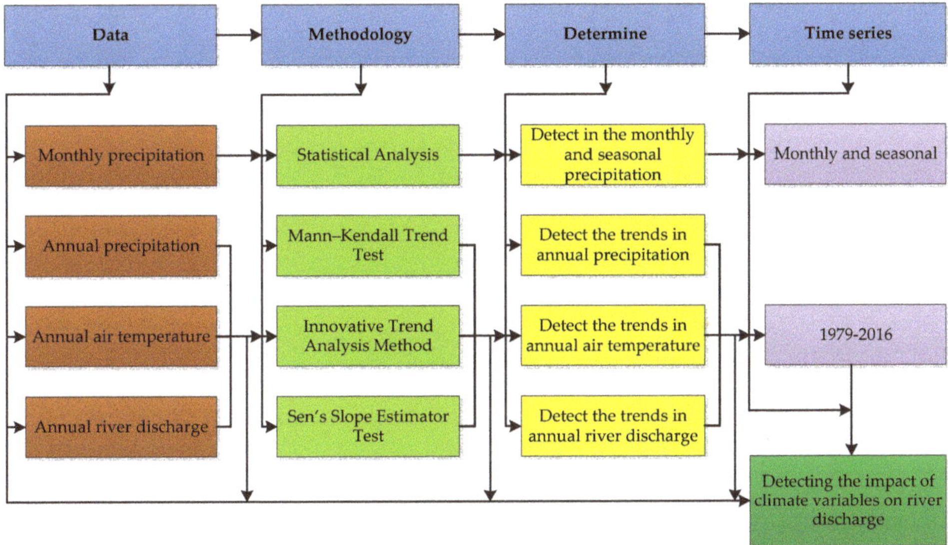

Figure 2. Workflow diagram to detect the changing trends in hydro-meteorological.

2.3.1. Mann-Kendall Trend Test

The Mann-Kendall (MK) test method also shows upward and downward trends with statistical significance. The strength of the trend depends on the magnitude, sample size, and variations of data series. The trends in the MK test is not significantly affected by the outliers occurred in the data series since the MK test statistic depends on positive or negative signs [20–22].

Annual and seasonal data series are used for trend analysis in this study. The trends of annual precipitation, air temperature, and river discharge have been also analyzed separately.

Individual time series data of climate and discharge are compared with all corresponding time series data of the year. When the data point of later year is larger than the data point of the previous year, the MK statistics is increased by one otherwise the MK statistics decreased by one. Thus, the MK statistics is the cumulative result of all the data values. The Mann-Kendall test statistics "S" is then equated as:

$$S = \sum_{i=1}^{n-1} \sum_{j=i+1}^{n} sgn\left(x_j - x_i\right) \tag{1}$$

The trend test is applied to x_i data values ($i = 1, 2, \ldots, n-1$) and x_j ($j = i+1, 2, \ldots, n$). The data value of each x_i is used as a reference point to compare with the data value of x_j which is given as:

$$sgn(x_j - x_i) = \begin{cases} +1 & \text{if } (x_j - x_i) > 0 \\ 0 & \text{if } (x_j - x_i) = 0 \\ -1 & \text{if } (x_j - x_i) < 0 \end{cases} \tag{2}$$

where x_j and x_i are the values in period j and i. When the number of data series greater than or equal to ten ($n \geq 10$), MK test is then characterized by a normal distribution with the mean $E(S) = 0$ and variance $Var(S)$ is equated as [23]:

$$E(S) = 0 \tag{3}$$

$$Var(S) = \frac{n(n-1)(2n+5) - \sum_{k=1}^{m} t_k(t_k-1)(2t_k+5)}{18} \tag{4}$$

where m is the number of the tied groups in the time series, and t_k is the number of ties in the kth tied group.

The test statistics Z is as follows:

$$Z = \begin{cases} \frac{S-1}{\delta} & \text{if } S > 0 \\ 0, & \text{if } S = 0 \\ \frac{S+1}{\delta} & \text{if } S < 0 \end{cases} \tag{5}$$

when Z is greater than zero, it indicates an increasing trend and when Z is less than zero, it is a decreasing trend.

In time sequence, the statistics are defined independently:

$$UF_k = \frac{d_k - E(d_k)}{\sqrt{var(d_k)}} (k = 1, 2, \ldots, n) \tag{6}$$

Firstly, given the confidence level α, if the $UF_k > UF_\alpha/2$, indicates that the sequence has the significant trend. Then, the time sequence is arranged in reverse order. According to the equation calculation, while making

$$UB_k = -UF_k \tag{7}$$

$$K = n+1-k \tag{8}$$

Finally, UB_k and UF_k are drawn as UB and UF curve. If there is an intersection between the two curves, the intersection is the beginning of the mutation [24].

2.3.2. Innovative Trend Analysis Method (ITAM)

Innovative trend analysis method (ITAM) has been used in many studies to detect the hydrometeorological observations and its accuracy was compared with the results of MK method [25,26]. The ITAM divides a time series into two equal parts, and it sorts both sub-series in ascending order. Then after, the two halves placed on a coordinate system ($x_i : i = 1, 2, 3, \ldots, n/2$) on X-axis and ($x_j : j = n/2 + 1, n/2 + 2, \ldots, n$) on Y-axis. If the time series data on a scattered plot are collected on the 1:1 (45°) straight line, it indicates no trend. However, the trend is increasing when data points accumulate above the 1:1 straight line and decreasing trend when data points accumulate below the 1:1 straight line.

The mean value difference between x_i and x_j could give the trend magnitude of data series. The first observed data point was not considered in this study when classifying the time series data into x_i and x_j data plots since the total number of observed data points are 38 from 1979–2016.

The direction of the trend is also affected by x_i data series. The trend indicator of ITAM is multiplied by 10 to make the scale similar to the other two tests. The trend indicator is given as:

$$\Phi = \frac{1}{n} \sum_{i=1}^{n} \frac{10\,(x_j - x_i)}{\mu} \tag{9}$$

where Φ = trend indicator, n = number of observation on the subseries, x_i = data series in the first half subseries class, x_j = data series in the second half subseries part and μ = mean of data series in the first half subseries part.

A positive value of Φ indicates an increasing trend. However, a negative value of Φ indicates a decreasing trend. However, when the scatter points closest around the 1:1 straight line, it implies the non-existence of a significant trend.

2.3.3. Sen's Slope Estimator Test

The trend magnitude is calculated by [27–30] slope estimator methods. The slope Q_i between two data points is given by the equation:

$$Q_i = \frac{x_j - x_k}{j - k}, \text{ for } i = 1, 2, \ldots, N \tag{10}$$

where x_j and x_k are data points at time j and $(j > k)$, respectively. When there is only single datum in each time, then $N = \frac{n(n-1)}{2}$; n is number of time periods. However, if the number of data in each year is many, then $N < \frac{n(n-1)}{2}$; n total number of observations. The N values of slope estimator are arranged from smallest to biggest. Then, the median of slope (β) is computed as:

$$\beta = \begin{cases} Q[(N+1)/2] & \textit{when N is odd} \\ Q[(N/2) + Q(N+2)/(2)/(2)] & \textit{when N is even} \end{cases} \tag{11}$$

The sign of β shown whether the trend is increasing or decreasing.

3. Results

3.1. Analysis of Precipitation

Annual mean precipitation of the study region from 1979 to 2016 was found to be 295.2 mm. The minimum and maximum recorded annual average precipitations were 175.0 and 380.0 mm respectively. The seasons of the study region are divided into four categories: Spring, summer, autumn, and winter seasons. The summer season has the largest proportion of precipitation. The seasonal precipitation varied from spring 39.27 mm (13.3%) to Summer 204.11 mm (69.15%), autumn 43.51mm (14.74%) to Winter 8.29 mm (2.81%) (Table 1).

The MK curve annual precipitation (changing parameters) shows a sharp decreasing trend in Ulaanbaatar 1994 to 2010 (Z = 0.71), a sharp decreasing trend in Erdenet from 1994 to 2005 (Z = 0.13), also, a sharp decreasing trend in Tsetserleg from 1994 to 2005 (Z = 0.26), a statistically significant increasing trend in Murun from 1984 to 1995 (Z = 2.45), in Sukhbaatar a significant increasing trend was observed with (Z = 1.06) from 1981 to 2016 and finally a statistically significant increasing trend was observed in Average (five stations) from 1984 to 1987 (Z = 0.68) (Figure 3).

The annual trend analysis of precipitation in all station using the Mann Kendall test, ITAM, Sen's slope estimator test result are presented in (Table 2). The trend in ITAM test shows an increasing trend in Murun and decreasing trend in other stations. Hence, the increase and decrease in innovative trend analysis Φ test value predict that the magnitude becomes strong and weak, respectively.

Table 1. The monthly and seasonal precipitation of stations.

Months, Season	Ulaanbaatar (mm)	Erdenet (mm)	Tsetserleg (mm)	Murun (mm)	Sukhbaatar (mm)	Average Precipitation (mm)	Z-Score
January	2.38	2.54	2.52	1.47	3.09	2.40	(−0.83)
February	2.45	2.56	2.90	1.10	2.18	2.24	(−0.84)
March	4.60	6.93	7.37	1.40	2.66	4.59	(−0.75)
April	8.39	14.03	13.47	7.71	9.74	10.67	(−0.52)
May	20.53	24.13	32.52	17.89	24.98	24.01	(−0.02)
June	48.23	69.73	58.95	48.19	48.78	54.78	1.13
July	69.37	99.39	86.13	69.86	64.01	77.75	1.99
August	66.80	86.26	75.71	57.60	71.55	71.58	1.76
September	25.52	34.71	25.98	19.78	32.45	27.69	0.12
October	8.94	12.34	13.07	5.62	10.02	10.00	(−0.55)
November	5.48	7.61	6.27	3.06	6.73	5.83	(−0.70)
December	3.52	4.28	3.02	2.98	4.44	3.65	(−0.79)
Spring	33.52	45.09	53.36	26.99	37.38	39.27 (13.3%)	0.55
Summer	184.41	255.38	220.79	175.65	184.34	204.11 (69.15%)	6.73
Autumn	39.94	54.66	45.31	28.46	49.20	43.51 (14.74%)	0.71
Winter	8.35	9.39	8.43	5.56	9.71	8.29 (2.81%)	(−0.61)
Annual precipitation	266.22	364.52	327.89	236.66	280.62	295.18 (100%)	10.14

Note: The number in the brackets indicates low precipitation rates.

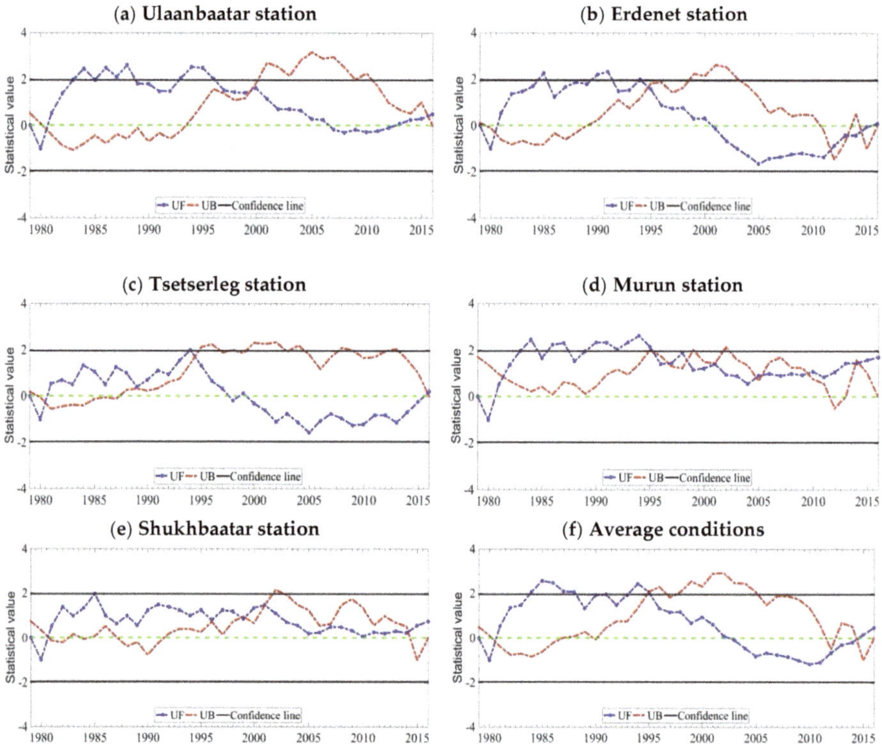

Figure 3. Trends of annual precipitation across stations (note: UF and UB are changing parameters where UB = −UF).

Table 2. The result of Z-statistic of Mann-Kendall (MK), Innovative Trend Analysis Method (ITAM) (ɸ), and Sen's slope estimator test (β).

S/No.	Name of Stations	Z (MK)	ɸ	β
1	Ulaanbaatar	0.71	−0.53	0.63
2	Erdenet	0.13	−0.41	0.28
3	Tsetserleg	0.26	−0.49	0.13
4	Murun	2.45 **	0.25	1.21 *
5	Sukhbaatar	1.06 *	−0.03	0.62
6	Average	0.68	−0.28	0.31

* Trends at 0.1 significance level; ** Trends at 0.05 significance level.

3.2. Analysis of Air Temperature

The MK curve annual air temperature (changing parameters) shows a statistically sharply increasing trend in Ulaanbaatar from 1994 to 2016 (Z = 5.88), a statistically sharp increasing trend in Erdenet from 1988 to 2016 (Z = 3.87), a statistically sharply increasing trend in Tsetserleg from 1993 to 2016 (Z = 4.38), a statistically sharp increasing trend in Murun from 1992 to 2016 (Z = 4.77), in Sukhbaatar a statistically significant increasing trend was observed with (Z = 2.85) from 1986 to 2013 and finally a statistically significant increasing trend was observed in Average (five stations) (Z = 4.71) (Figure 4).

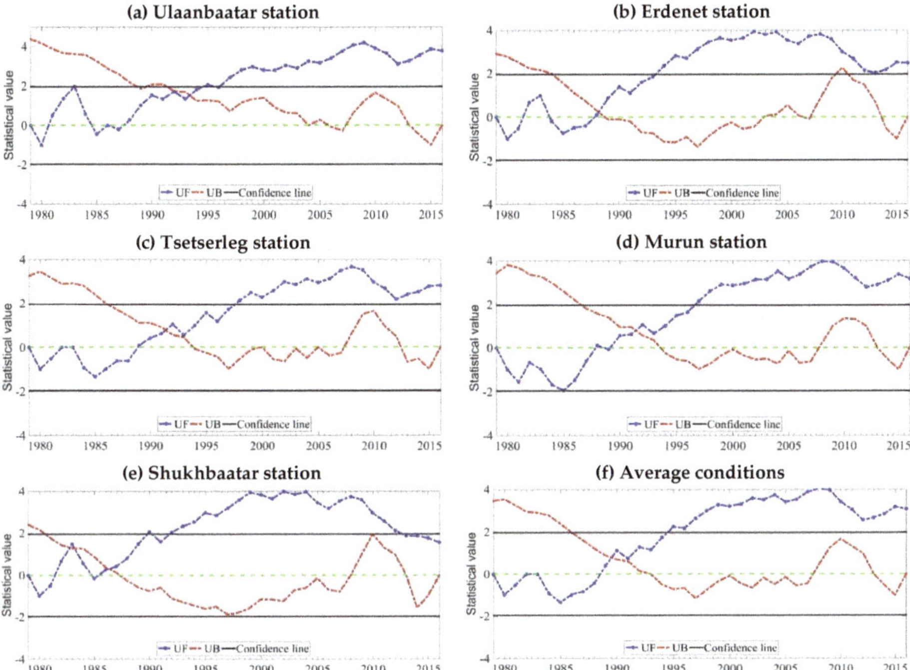

Figure 4. Trends of annual air temperature across stations (note: UF and UB are changing parameters where UB = −UF).

The annual trend analysis of air temperature in all station using the Mann Kendall test, ITAM, Sen's slope estimator test result is presented in (Table 3). The trend in ITAM test shows an increasing trend in all stations. Hence, the increase and decrease in innovative trend analysis ɸ test value predict that the magnitude becomes strong.

Table 3. The result of Z-statistic of MK, ITAM (ϕ), and Sen's slope estimator test (β).

S/No.	Name of Stations	Z (MK)	ϕ	β
1	Ulaanbaatar	5.88 ***	−54.55 ***	0.05
2	Erdenet	3.87 ***	8.39 **	0.03
3	Tsetserleg	4.38 ***	7.26	0.04
4	Murun	4.77 ***	−47.56	0.06
5	Sukhbaatar	2.85 **	7.13 ***	0.02
6	Average	4.71 ***	18.66 ***	0.04

** Trends at 0.05 significance level; *** Trends at 0.01 significance level.

3.3. Analysis of River Discharge

The MK curve annual river discharge (changing parameters) shows a sharply decreasing trend in Ulaanbaatar 1994 to 2016 (Z = −3.32), a statistically sharp decreasing trend in Tsetserleg from 1982 to 2016 (Z = −3.84), a significant decreasing trend in Murun from 1986 to 2016 (Z = −1.28), in Sukhbaatar a significant decreasing trend was observed with (Z = −2.05) from 1993 to 2016 and finally a significant decreasing trend was observed in Average (five stations) (Z = −2.05) (Figure 5). The annual trend analysis of river discharge in all station using the Mann Kendall test, ITAM, Sen's slope estimator test result are presented in (Table 4). The trend in ITAM test shows a decreasing trend in all stations. Hence, the increase and decrease in innovative trend analysis ϕ test value predict that the magnitude becomes strong.

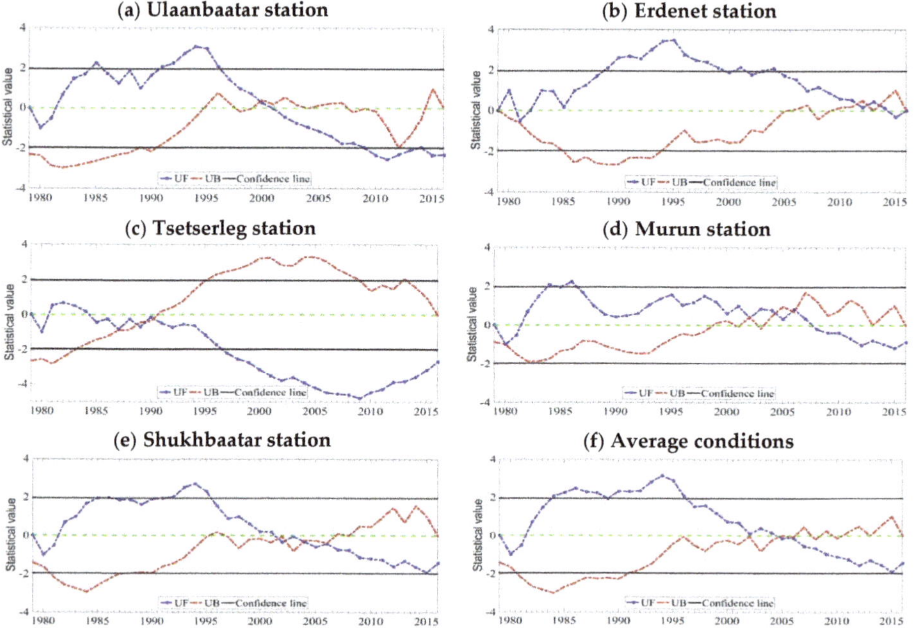

Figure 5. Trends of annual discharge across stations (note: UF and UB are changing parameters where UB = −UF).

Table 4. The result of Z-statistic of MK, ITAM (ф), and Sen's slope estimator test (β).

S/No.	Name of Stations	Z (MK)	Φ	β
1	Ulaanbaatar	−3.32 ***	−5.63	−144.12
2	Erdenet	0.00	−0.96	4.46
3	Tsetserleg	−3.84 ***	−4.65	−56.31
4	Murun	−1.28 *	−1.01	−64.15
5	Sukhbaatar	−2.05 **	−2.00	−550.33
6	Average	−2.05 **	−2.00	−169.80

* Trends at 0.1 significance level; ** Trends at 0.05 significance level; *** Trends at 0.01 significance level.

River discharge trend is generally exhibited a downward trend from 1979 to 2016. Especially, river discharges show a sharp decreasing trend in all stations since 1995.

3.4. Relationship of Climate and River Discharge

The annual average air temperature of the study region from 1979 to 2016 was found to be 0.83 °C. The minimum and the maximum recorded air temperature were −0.9 °C and 2.9 °C per year, respectively. A dramatic increase in air temperature was observed from 1984 to 2007. In the study region, the observed air temperature was increased from 1979 to 2016 (R^2 = 0.2632) (Figure 6f). The warmest year was in 2007 (2.9 °C). The air temperature most increasing area is Ulaanbaatar city, it was increasing 1.9 °C (R^2 = 0.4226) (Figure 6a). The annual average air temperature increased significantly by 1.4 °C. The mean annual air temperature is 16 °C to 18 °C in July and −16 °C to −22 °C in January. In the Mongolian's Selenga river basin, precipitation varies both in time and space scale. The average precipitation is 295.2 mm/year. About 85% of the total precipitation falls from April to September, of which about 69.15% falls during June, July and August. The air temperature and precipitation changes within the five stations show a different value (Figure 6). Overall precipitation showed a slightly increasing trend during the period from 1979 to 2016 (Figure 6f).

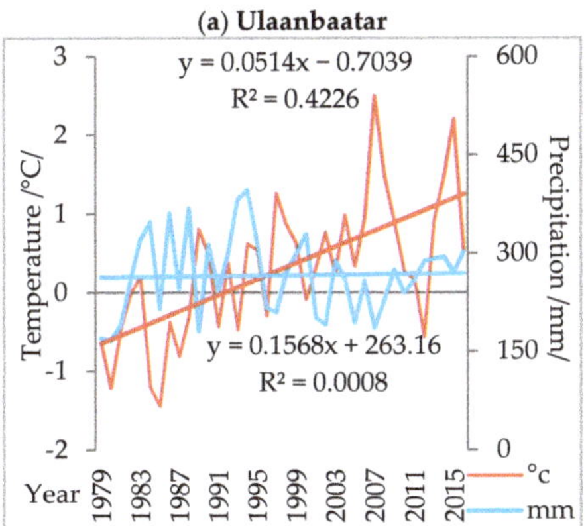

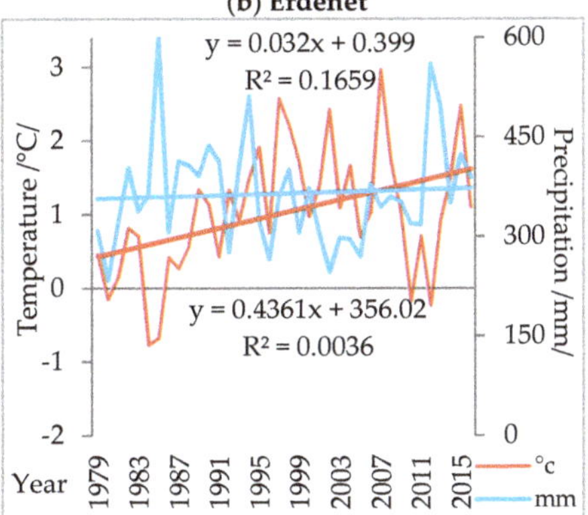

Figure 6. *Cont.*

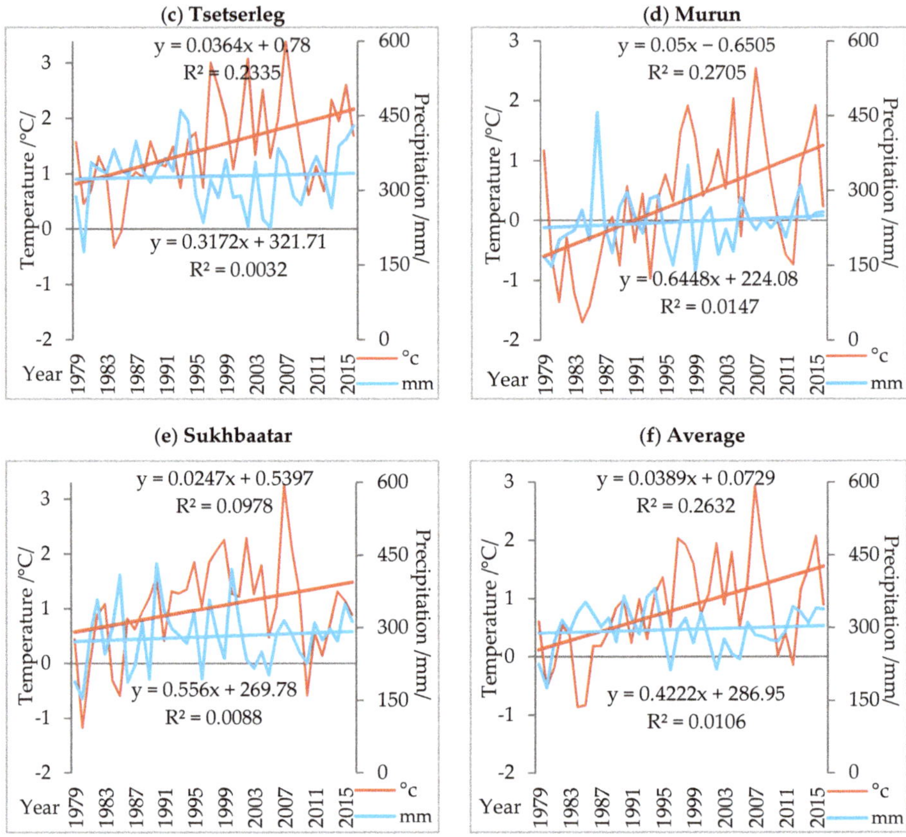

Figure 6. The air temperature and precipitation trend for the period 1979–2016. The vertical column is air temperature and precipitation change, and fluctuations line indicates annual values and solid lines indicate period running averages.

The ratio of precipitation and river discharge to this basin is calculated by the location of the five meteorological stations and water gauge stations (Figure 7). However, precipitation has been relatively stable ranging from 1979 to 2016 (Figure 7f).

The trend of air temperature change and the trend of river discharge were estimated. A statistically significant increase in average air temperature (five stations average) was from 1994 to 2016 (Z = 4.71). Also, the air temperature increased on all meteorological stations. River discharge exhibited a decreasing pattern (Figure 8).

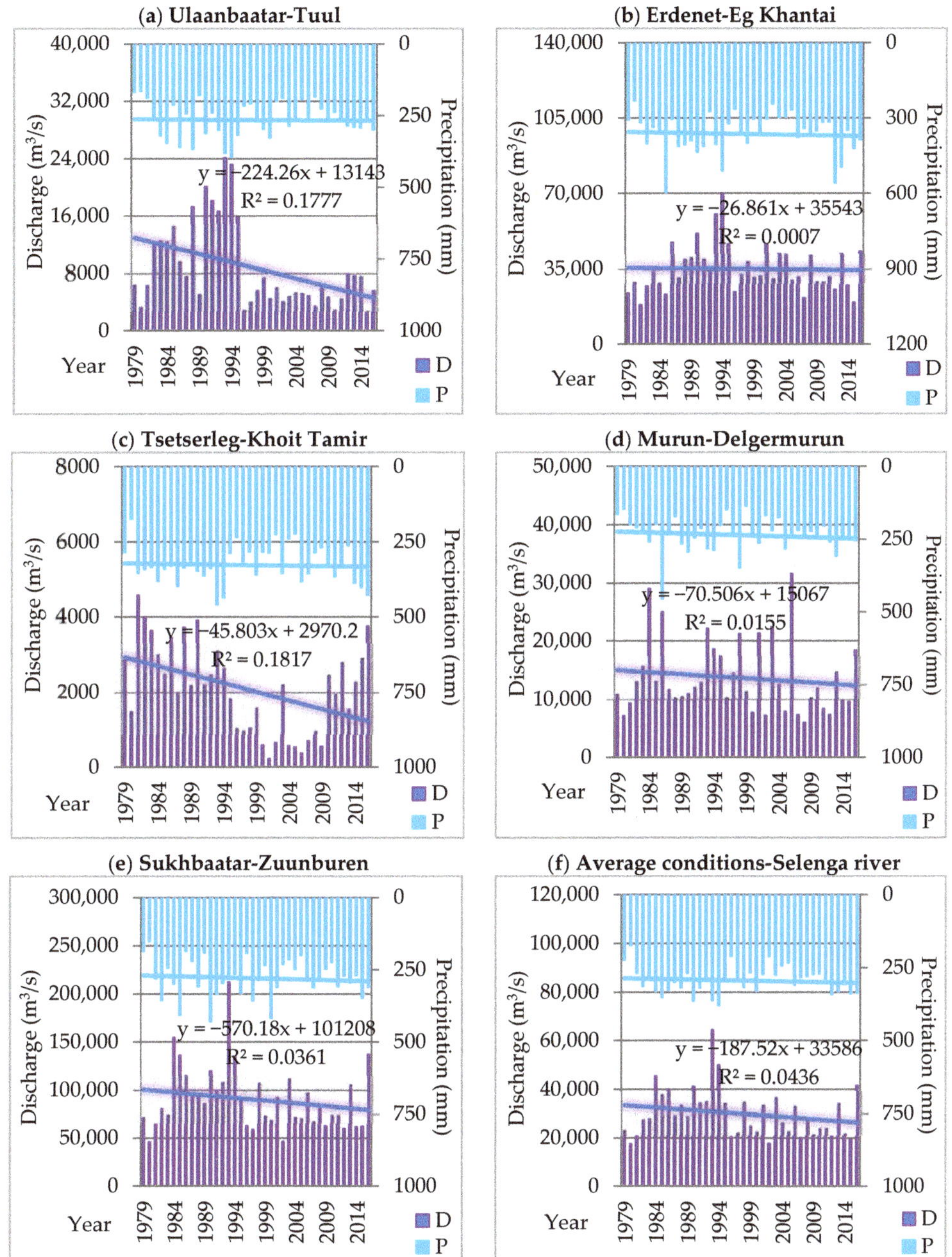

Figure 7. Long-term in precipitation and discharge change, in during 1979–2016.

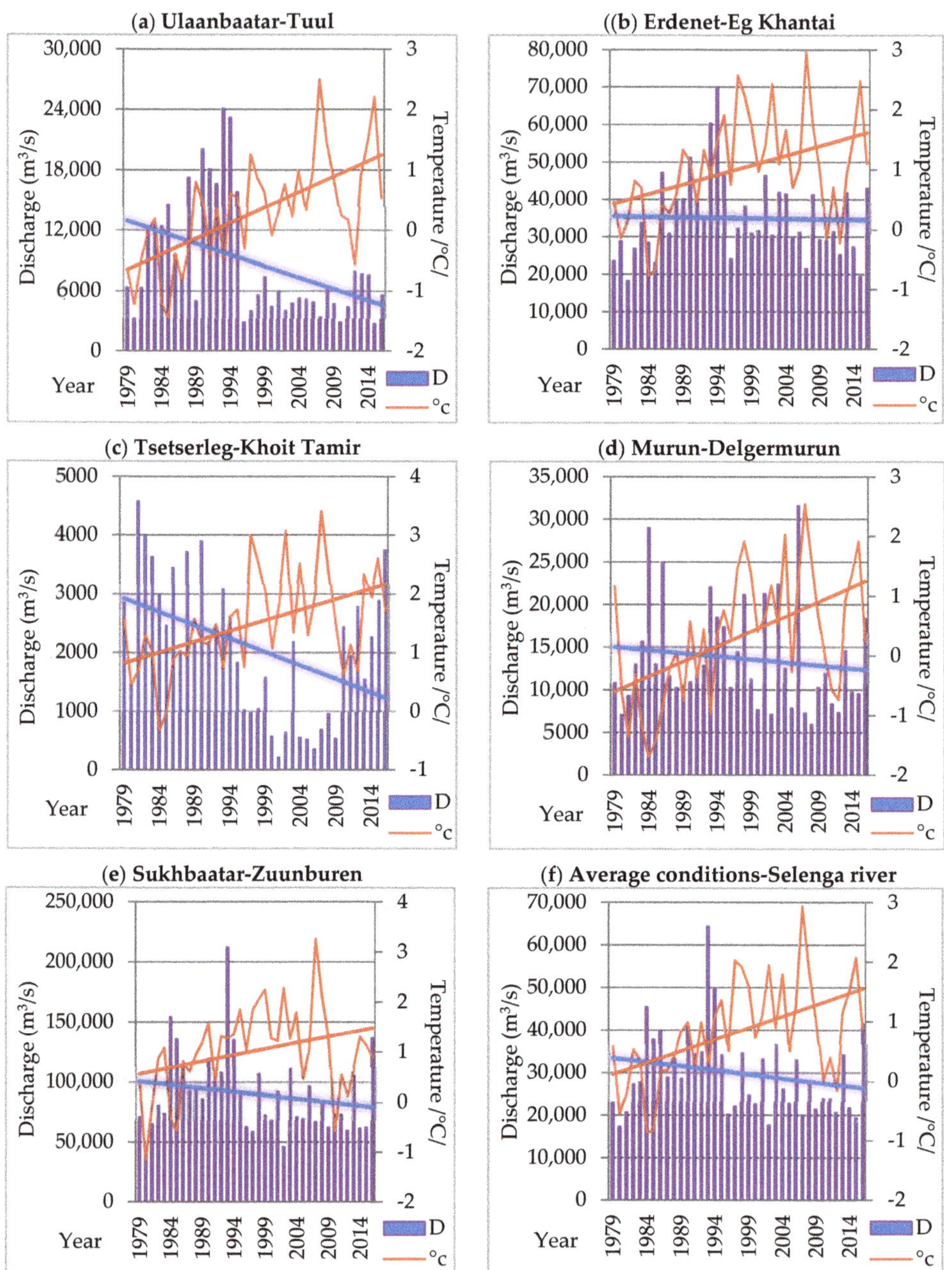

Figure 8. Long-term in air temperature and discharge change, in during 1979–2016.

Potential linkages between climate variables and the observed changes in river discharge are the subject of ongoing debate. To determine this, it was the estimation of the relationship between climate parameters and river discharge (Figure 9).

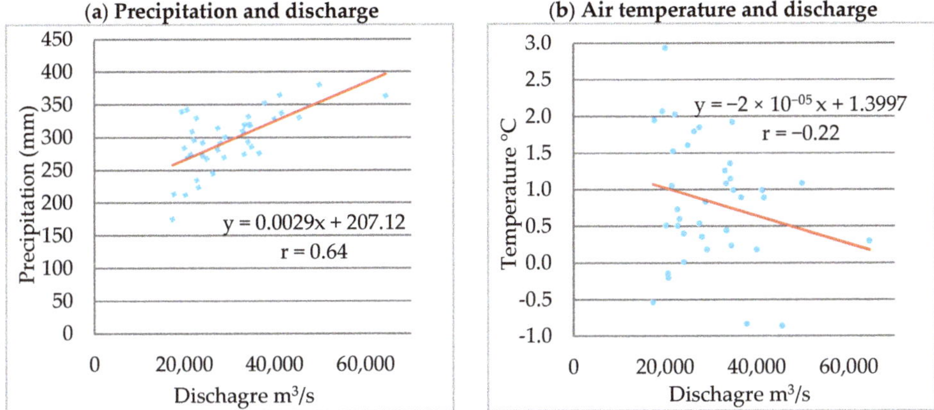

Figure 9. Correlation coefficient: climate and river discharge.

The correlation coefficient between precipitation and river discharge has a strong positive correlation (r = 0.64) from 1979 to 2016. In this case, the volume of the river discharge will increase when the number of precipitation increases. During this period, precipitation has increased. The correlation coefficient between air temperature and river discharge has a weak negative relationship (r = −0.22) from 1979 to 2016. In this case, the volume of the river discharge will decrease when the air temperature increases. However, Figures 7 and 8 shows that the river discharge has a sharp decreasing trend significantly since 1995, it may be related to the impact of other factors. During this period, quantities of the river discharge passing through bigger cities are dramatically decreasing. Climate change and river discharges are interdependent [31]. Especially in the rivers fed by precipitation, precipitation can directly affect the hydrological changes in the basin. Changes in river discharge are different at the five stations. In particular, River discharge decreased at the Tuul river water gauge station in Ulaanbaatar city. It has been decreased apparently since 1995. The water shortage was (y = −224.26x + 13,143). Also, The River discharge has been decreased in Zuunburen station near Sukhbaatar city. The water shortage was (y = −570.18x + 101,208). This may be due to the high consumption of the river water (Figure 7).

The average change in climate and river discharge was categorized by 10 years period. These include: from 1979 to 1988 (I), from 1989 to 1998 (II), from 1999 to 2008 (III), from 2009 to 2016 (IV) (Figure 10). During the III period, precipitation decreases, the temperature increases and also the river discharge decrease this is in illustrated in Figure 9.

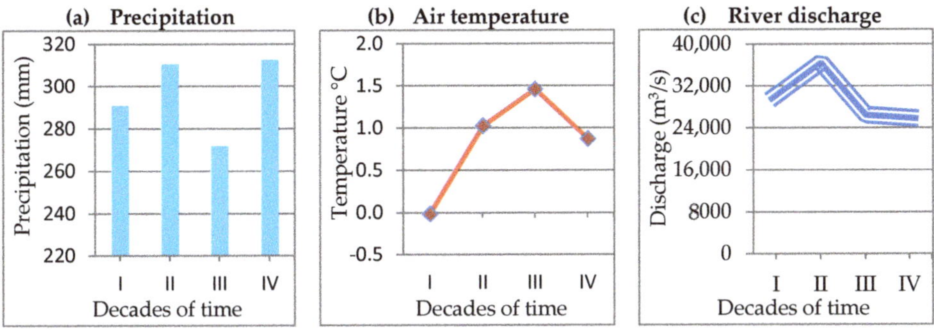

Figure 10. Decades changing of climate and river discharge.

However, during in period IV, the amount of precipitation increased, the air temperature decreased, and the river discharge also decreased. This could be related to other factors rather than climatic factors.

4. Discussion

An increase in air temperature is among the manifestations of global climate change. The global average air temperature has increased by 0.85 °C from 1880 to 2012, and this may even accelerate in the near future. The air temperature of worldwide large inland water bodies has been rapidly warming since 1985 with an average rate of 0.045 ± 0.011 °C/year and with the highest rate of 0.10 ± 0.01 °C/year [32]. There has been an observed increasing trend mean annual air temperature in the Selenga River basin by 1.4 °C or 0.036 °C/year during the considered historical period from 1979 to 2016 ($p < 0.05$). This is almost twice as much as the global average warming rate of 0.012 °C/year (0.72 °C increase during the period from 1951 to 2012). The climate of the Mongolian's Selenga river basin is characterized by long and cold winters, dry and hot summers, less precipitation, and high-temperature fluctuations [10,33]. The annual mean precipitation is 300–400 mm/year in the Khangai, Khentein, and Huvsgul mountainous regions 150–250 mm/year in the steppe and river valleys. The results of this study are generally consistent with other research results which reported increased air temperature and changes precipitation [33,34].

The Selenga River basin lies in the zone of extremely continental climate and a considerable portion of the basin is occupied by permafrost [10]. Runoff formation conditions in the Selenga River basin are very diverse. The southern part of the Selenga River basin shows low soil moisture content and steppe vegetation, while its northern part is covered by dense taiga vegetation and permafrost—an important source of soil water in summer. The high elevation difference (from 418 to 3514 m) also has its effect on runoff formation conditions. Rains are the main source of Selenga River basin nourishment. Snow cover in its drainage basin is not rich, hence the low share of snow in river nourishment. About half of Selenga annual runoff is the runoff occurred in summer season (June–August), the role of groundwater in river nourishment is also small [35]. Selenga river basin runoff varies mostly because of variations of summer precipitation. The rivers of the Selenga River basin show pronounced winter low-water period from November to March (3–10% of the annual runoff volume), a relatively low spring snow-melt flood and a series of rain floods in summer and autumn. Many rivers freeze in the winter season [36]. Especially hydrological processes are very sensitive. The MK, ITAM, and Sen's slope estimator test analysis showed that decreasing trend of river discharge was observed across the stations. The river discharge has a sharp decreasing trend significantly since 1995. It is maybe related to the impact of other factors. Especially, this may be due to the socioeconomic activities including mining, industry, agriculture, and urbanization in the basin [8,37–40].

5. Conclusions

In this study, the Mann–Kendall trend test, ITAM, and Sen's slope estimator test methods were used to analyze the variability of precipitation, air temperature, and river discharge on an annual basis in the study basin.

Seasonal variability of precipitation was investigated in all stations. The small significant increasing trend was observed in Ulaanbaatar, Erdenet, and Tsetserleg stations, whereas other Murun and Sukhbaatar stations demonstrated a significant increasing trend. The average annual air temperature for Mongolian's Selenga river basin is 0.83 °C. The average air temperature has significantly increased by 1.4 °C in the past 38 years. This is very high in the semi-arid zone of central Asia during the past 40 years. This is almost twice the global average warming rate. The river discharge trend has significantly decreased in the determined study periods. It has been particularly apparent since 1995. There was conformity in the results obtained from the Mann-Kendall, ITAM test, Sen's slope estimator test and the trend line for all stations during the specified study period.

In the near future, it's vital to conduct scientific studies on the causes of river discharge change and its potential influences on the Ecohydrological systems in the basin area.

Author Contributions: B.D. made substantial contributions to the design, idea generating, analysis, interpretation, and drafting of the manuscript. Y.D. assisted and commented on the draft manuscript and supervised the whole work. W.H. advising and operated the MK test for data analysis and is a resource person. S.C. and A.E. interpreted the results. X.Y., M.G., A.A. and A.G. are participated in the design of the study and supervised all methodologies utilized. The final manuscript before submission was checked and approved by all the authors.

Funding: This research was funded by The China, National Key Research and Development Project (grant No. 2016YFA0601503).

Acknowledgments: The authors would like to thank the Information and Research Institute of Meteorology, Hydrology, and Environment of Mongolian for providing the raw meteorological data. We also thank the China Institute of Water Resources and Hydropower Research for financing this research.

Conflicts of Interest: The authors declare no conflict of interest.

References

1. Palmate, S.S.; Pandey, A.; Kumar, D.; Pandey, R.P.; Mishra, S.K. Climate change impact on forest cover and vegetation in Betwa Basin, India. *Appl. Water Sci.* **2017**, *7*, 103–114. [CrossRef]
2. Malsy, M.; Aus der Beek, T.; Eisner, S.; Flörke, M. Climate change impacts on Central Asian water resources. *Adv. Geosci.* **2012**, *32*, 77–83. [CrossRef]
3. Malsy, M.; Flörke, M.; Borchardt, D. What drives the water quality changes in the Selenga Basin: Climate change or socio-economic development? *Reg. Environ. Chang.* **2017**, *17*, 1977–1989. [CrossRef]
4. Kayet, N.; Pathak, K.; Chakrabarty, A.; Sahoo, S. Spatial impact of land use/land cover change on surface temperature distribution in Saranda Forest, Jharkhand. *Model. Earth Syst. Environ.* **2016**, *2*, 127. [CrossRef]
5. Zhuo, L.; Han, D.; Dai, Q. Exploration of empirical relationship between surface soil temperature and surface soil moisture over two catchments of contrasting climates and land covers. *Arabian J. Geosci.* **2017**, *10*, 410. [CrossRef]
6. Wang, G.; Zhang, J.; Li, X.; Bao, Z.; Liu, Y.; Liu, C.; He, R.; Luo, J. Investigating causes of changes in runoff using hydrological simulation approach. *Appl. Water Sci.* **2017**, *7*, 2245–2253. [CrossRef]
7. Walling, D.E.; Fang, D. Recent trends in the suspended sediment loads of the world's rivers. *Glob. Planet. Chang.* **2003**, *39*, 111–126. [CrossRef]
8. Ma, X.; Yasunari, T.; Ohata, T.; Natsagdorj, L.; Davaa, G.; Oyunbaatar, D. Hydrological regime analysis of the Selenge River basin, Mongolia. *Hydrol. Process.* **2003**, *17*, 2929–2945. [CrossRef]
9. Fang, J.; Bai, Y.; Wu, J. Towards a better understanding of landscape patterns and ecosystem processes of the Mongolian Plateau. *Landsc. Ecol.* **2015**, *30*, 1573–1578. [CrossRef]
10. Törnqvist, R.; Jarsjö, J.; Pietroń, J.; Bring, A.; Rogberg, P.; Asokan, S.M.; Destouni, G. Evolution of the hydro-climate system in the Lake Baikal basin. *J. Hydrol.* **2014**, *519*, 1953–1962. [CrossRef]
11. Karthe, D.; Chalov, S.; Moreido, V.; Pashkina, M.; Romanchenko, A.; Batbayar, G.; Kalugin, A.; Westphal, K.; Malsy, M.; Flörke, M. Assessment of runoff, water and sediment quality in the Selenga River basin aided by a web-based geoservice. *Water Resour.* **2017**, *44*, 399–416. [CrossRef]
12. Kasimov, N.; Karthe, D.; Chalov, S. Environmental change in the Selenga River—Lake Baikal Basin. *Reg. Environ. Chang.* **2017**, *17*, 1945–1949. [CrossRef]
13. Troitskaya, E.; Blinov, V.; Ivanov, V.; Zhdanov, A.; Gnatovsky, R.; Sutyrina, E.; Shimaraev, M. Cyclonic circulation and upwelling in Lake Baikal. *Aquat. Sci.* **2015**, *77*, 171–182. [CrossRef]
14. United Nations Development Programme (UNDP). Chapter I. General Characteristics of Lake Baikal Basin. In *State of the Environment Report. The Lake Baikal Basin 2012–2013*; United Nations Development Programme: New York, NY, USA, 2013.
15. Moore, M.V.; Hampton, S.E.; Izmest'eva, L.R.; Silow, E.A.; Peshkova, E.V.; Pavlov, B.K. Climate Change and the World's "Sacred Sea"—Lake Baikal, Siberia. *BioScience* **2009**, *59*, 405–417. [CrossRef]
16. United Nations Educational, Scientific and Cultural Organization (UNESCO). *Convention Concerning the Protection of the World Cultural and Natural Heritage*; Report; United Nations Educational, Scientific and Cultural Organization: Paris, France, 1996. Available online: http://whc.unesco.org/archive/repcom96.htm#754 (accessed on 10 March 1997).

17. Sharma, K.P.; Moore, B.; Vorosmarty, C.J. Anthropogenic, Climatic, and Hydrologic Trends in the Kosi Basin, Himalaya. *Clim. Chang.* **2000**, *47*, 141–165. [CrossRef]
18. Déry, S.J.; Wood, E.F. Decreasing river discharge in northern Canada. *Geophys. Res. Lett.* **2005**, *32*. [CrossRef]
19. McClelland, J.W.; Déry, S.J.; Peterson, B.J.; Holmes, R.M.; Wood, E.F. A pan-arctic evaluation of changes in river discharge during the latter half of the 20th century. *Geophys. Res. Lett.* **2006**, *33*. [CrossRef]
20. Asfaw, A.; Simane, B.; Hassen, A.; Bantider, A. Variability and time series trend analysis of rainfall and temperature in northcentral Ethiopia: A case study in Woleka sub-basin. *Weather Clim. Extremes* **2018**, *19*, 29–41. [CrossRef]
21. Gedefaw, M.; Yan, D.; Wang, H.; Qin, T.; Girma, A.; Abiyu, A.; Batsuren, D. Innovative Trend Analysis of Annual and Seasonal Rainfall Variability in Amhara Regional State, Ethiopia. *Atmosphere* **2018**, *9*, 326. [CrossRef]
22. Wu, L.; Wang, S.; Bai, X.; Luo, W.; Tian, Y.; Zeng, C.; Luo, G.; He, S. Quantitative assessment of the impacts of climate change and human activities on runoff change in a typical karst watershed, SW China. *Sci. Total Environ.* **2017**, *601–602*, 1449–1465. [CrossRef] [PubMed]
23. Ma, X.; He, Y.; Xu, J.; van Noordwijk, M.; Lu, X. Spatial and temporal variation in rainfall erosivity in a Himalayan watershed. *CATENA* **2014**, *121*, 248–259. [CrossRef]
24. Wu, H.S.; Liu, D.F.; Chang, J.X.; Zhang, H.X.; Huang, Q. Impacts of climate change and human activities on runoff in Weihe Basin based on Budyko hypothesis. *IOP Conf. Ser. Earth Environ. Sci.* **2017**, *82*, 012063. [CrossRef]
25. Wu, H.; Qian, H. Innovative trend analysis of annual and seasonal rainfall and extreme values in Shaanxi, China, since the 1950s. *Int. J. Climatol.* **2017**, *37*, 2582–2592. [CrossRef]
26. Cui, L.; Wang, L.; Lai, Z.; Tian, Q.; Liu, W.; Li, J. Innovative trend analysis of annual and seasonal air temperature and rainfall in the Yangtze River Basin, China during 1960–2015. *J. Atmos. Sol. Terr. Phys.* **2017**, *164*, 48–59. [CrossRef]
27. Gocic, M.; Trajkovic, S. Analysis of changes in meteorological variables using Mann-Kendall and Sen's slope estimator statistical tests in Serbia. *Glob. Planet. Chang.* **2013**, *100*, 172–182. [CrossRef]
28. Gu, X.; Zhang, Q.; Singh, V.P.; Shi, P. Changes in magnitude and frequency of heavy precipitation across China and its potential links to summer temperature. *J. Hydrol.* **2017**, *547*, 718–731. [CrossRef]
29. St Laurent, J.; Mazumder, A. Influence of seasonal and inter-annual hydro-meteorological variability on surface water fecal coliform concentration under varying land-use composition. *Water Res.* **2014**, *48*, 170–178. [CrossRef] [PubMed]
30. Wang, H.; Zhang, M.; Zhu, H.; Dang, X.; Yang, Z.; Yin, L. Hydro-climatic trends in the last 50years in the lower reach of the Shiyang River Basin, NW China. *CATENA* **2012**, *95*, 33–41. [CrossRef]
31. Sorg, A.; Bolch, T.; Stoffel, M.; Solomina, O.; Beniston, M. Climate change impacts on glaciers and runoff in Tien Shan (Central Asia). *Nat. Clim. Change* **2012**, *2*, 725. [CrossRef]
32. Li, X.; Zhang, L.; Yang, G.; Li, H.; He, B.; Chen, Y.; Tang, X. Impacts of human activities and climate change on the water environment of Lake Poyang Basin, China. *Geoenviron. Disasters* **2015**, *2*, 22. [CrossRef]
33. Batima, P.; Natsagdorj, L.; Gombluudev, P.; Erdenetsetseg, B. Observed Climate Change in Mongolia. *AIACC Work. Pap.* **2005**, *12*, 1–26.
34. Nandintsetseg, B.; Greene, J.S.; Goulden, C.E. Trends in extreme daily precipitation and temperature near lake Hövsgöl, Mongolia. *Int. J. Climatol.* **2007**, *27*, 341–347. [CrossRef]
35. Berezhnykh, T.V.; Marchenko, O.Y.; Abasov, N.V.; Mordvinov, V.I. Changes in the summertime atmospheric circulation over East Asia and formation of long-lasting low-water periods within the Selenga river basin. *Geogr. Nat. Resour.* **2012**, *33*, 223–229. [CrossRef]
36. Frolova, N.L.; Belyakova, P.A.; Grigor'ev, V.Y.; Sazonov, A.A.; Zotov, L.V. Many-year variations of river runoff in the Selenga basin. *Water Resour.* **2017**, *44*, 359–371. [CrossRef]
37. Chalov, S.R.; Jarsjö, J.; Kasimov, N.S.; Romanchenko, A.O.; Pietroń, J.; Thorslund, J.; Promakhova, E.V. Spatio-temporal variation of sediment transport in the Selenga River Basin, Mongolia and Russia. *Environ. Earth Sci.* **2015**, *73*, 663–680. [CrossRef]
38. Stubblefield, A.; Chandra, S.; Eagan, S.; Tuvshinjargal, D.; Davaadorzh, G.; Gilroy, D.; Sampson, J.; Thorne, J.; Allen, B.; Hogan, Z. Impacts of gold mining and land use alterations on the water quality of central Mongolian rivers. *Integr. Environ. Assess. Manag.* **2005**, *1*, 365–373. [CrossRef] [PubMed]

39. Sorokovikova, L.M.; Popovskaya, G.I.; Tomberg, I.V.; Sinyukovich, V.N.; Kravchenko, O.S.; Marinaite, I.I.; Bashenkhaeva, N.V.; Khodzher, T.V. The Selenga River water quality on the border with Mongolia at the beginning of the 21st century. *Russ. Meteorol. Hydrol.* **2013**, *38*, 126–133. [CrossRef]

40. Karthe, D.; Heldt, S.; Houdret, A.; Borchardt, D. IWRM in a country under rapid transition: Lessons learnt from the Kharaa River Basin, Mongolia. *Environ. Earth Sci.* **2015**, *73*, 681–695. [CrossRef]

Article

Multiple Linear Regression Models for Predicting Nonpoint-Source Pollutant Discharge from a Highland Agricultural Region

Jae Heon Cho [1],* and Jong Ho Lee [2]

[1] Department of Biosystems and Convergence Engineering, Catholic Kwandong University, 24 Beomil-ro, 579 beon-gil, Gangneung-si, Gangwon-do 25601, Korea
[2] Department of Urban Planning and Real Estate, Cheongju University, 298 Daeseongro, Cheongwon-gu, Cheongju, Chungbuk 28503, Korea; jhlee1013@cju.ac.kr
* Correspondence: jhcho@cku.ac.kr; Tel.: +82-33-6497534

Received: 25 July 2018; Accepted: 27 August 2018; Published: 29 August 2018

Abstract: Sediment runoff from dense highland field areas greatly affects the quality of downstream lakes and drinking water sources. In this study, multiple linear regression (MLR) models were built to predict diffuse pollutant discharge using the environmental parameters of a basin. Explanatory variables that influence the sediment and pollutant discharge can be identified with the model, and such research could play an important role in limiting sediment erosion in the dense highland field area. Pollutant load per event, event mean concentration (EMC), and pollutant load per area were estimated from stormwater survey data from the Lake Soyang basin. During the wet season, heavy rains cause large amounts of suspended sediment and the occurrence of such rains is increasing due to climate change. The explanatory variables used in the MLR models are the percentage of fields, subbasin area, and mean slope of subbasin as topographic parameters, and the number of preceding dry days, rainfall intensity, rainfall depth, and rainfall duration as rainfall parameters. In the MLR modeling process, four types of regression equations with and without log transformation of the explanatory and response variables were examined to identify the best performing regression model. The performance of the MLR models was evaluated using the coefficient of determination (R^2), root mean square error (RMSE), coefficient of variation of the root mean square error (CV(RMSE)), the ratio of the RMSE to the standard deviation of the observed data (RSR) and the Nash–Sutcliffe model efficiency (NSE). The performance of the MLR models of pollutant load except total nitrogen (TN) was good under the condition of RSR, and satisfactory for the NSE and R^2. In the EMC and load/area models, the performance for suspended solids (SS) and total phosphorus (TP) was good for the RSR, and satisfactory for the NSE and R^2. The standardized coefficients for the models were analyzed to identify the influential explanatory variables in the models. In the final performance evaluation, the results of jackknife validation indicate that the MLR models are robust.

Keywords: highland agricultural field area; diffuse pollutant discharge; multiple regression model; climate change; jackknife validation

---

1. Introduction

In Lake Soyang basin of South Korea, large amounts of sediment are discharged from highland agricultural field regions in the wet season. To develop environmental preservation measures that protect water resources from the turbid water problem and diffuse pollution, prediction models are necessary to estimate the amount of pollutants discharged from subbasins. In this study, a multiple linear regression (MLR) model is established to predict the pollutant runoff discharge using environmental parameters, such as land use, rainfall, and topography.

In South Korea, rainfall events of 200 mm or more occurred only once annually, on average, until the end of the 1970s, but increased to a frequency of two per year in the 1980s and thereafter occurred five times in both 1984 and 1998. And, annual precipitation increased by 19% in the past decade, compared to the first half of the 20th century [1]. Conditions in Lake Soyang, located in the upper reaches of the Han River, greatly affect the water quality of the water supply of the capital region of South Korea. Discharged sediments from the highland field area flow into Lake Soyang in the wet season. Consequently, the turbidity of the lake increases to high levels and persists for a long time. In July 2006, a heavy rain event occurred in the Lake Soyang watershed. Overall, 670 mm fell over 8 days, with a maximum intensity of 66 mm per hour. The suspended sediment stayed in Lake Soyang for an extended period of time because of stratification; thus, the turbidity of the lake remained high and was measured at over 20 nephelometric turbidity units (NTU) for 168 days [2,3].

Regression models have been developed to estimate the sediment discharge using the subbasin environmental parameters in many areas. Valtanen et al. [4] applied stepwise multiple linear regression (SMLR) analysis to identify the variables that best explained the variation in event mass loads (EMLs) in each study catchment during cold and warm periods. Runoff duration, peak flow, antecedent dry period, mean runoff intensity, total suspended solids (TSS), TN, TP and total organic carbon (TOC) were used as explanatory variables. Another SMLR analysis was also carried out to assess whether catchment variables explain the EML and EMC values during the cold and warm periods [4]. The catchment variables included total impervious area and land use type. All data were log10-transformed to obtain approximately normal distributions. Bian et al. [5] proposed a procedure combining different statistical methods and a hydrological model to quantify the annual runoff response to spatial and temporal variations in impervious surface areas in an urbanized basin. A hydrological model relating annual runoff depth to precipitation, potential evapotranspiration and spatial metrics of the impervious area for baseline periods and periods of change was built using stepwise multiple regression analysis. Roman et al. [6] developed multivariate regression models to enable the prediction of mean annual suspended sediment discharge on the basis of basin characteristics, which is useful for many ungaged river locations in the eastern United States. The models are based on long-term mean sediment discharge estimates and explanatory variables, such as drainage area, mean elevation, and urban area, obtained from a combined dataset of 1201 US Geological Survey (USGS) stations. Tuset et al. [7] analyzed rainfall, runoff and sediment transport relationships in a meso-scale Mediterranean mountain catchment. The relationships among rainfall, runoff and suspended sediment transport were analyzed with Pearson correlations and multivariate regression analysis. The multivariate regression method was used to analyze the relationship between the independent variables (pre-event conditions, rainfall and runoff) and suspended sediment transport for all flood events. Seasonal relationships between total surface runoff and total sediment transport indicate that the sediment transport magnitude shows a clear seasonality influenced by rainfall intensity and sediment availability.

Buendia et al. [8] attempted to use empirical relationships to assess the relationship between sediment yield and basin scale and to provide an update on the main drivers controlling sediment yield in these particular river systems. Quantile regression analysis was used to assess the correlation between basin area and sediment yield, while additional basin-scale descriptors were related to sediment yield by means of multiple regression analysis. The performance of the model was tested through the jackknife validation method [8–14]. Paule-Mercado et al. [15] used MLRs to identify the significant parameters affecting fecal-indicator bacteria concentrations and to predict the response of bacteria concentrations to changes in land use and land cover. Stormwater temperature, 5-day biochemical oxygen demand (BOD$_5$), turbidity, TSS, and antecedent dry days were the most influential independent variables for the bacteria concentrations at the monitoring sites. Several studies have utilized linear regression techniques to predict bacteria concentrations in rivers [16–19]. Furthermore, regression models have been widely used to predict and characterize rainfall and runoff characteristics and to determine the relationship between these two variables [20–26]. Process-based erosion prediction models have also been established to predict the intensity of soil erosion in a particular area [27–30].

In this study area, two types of environmental parameters affect the stormwater sediment runoff: meteorological factors, such as rainfall depth, rainfall intensity, rainfall duration; and number of preceding dry days and topographic factors, such as percentage of upland field area, subbasin area, and subbasin slope. In this study, the Pearson correlation test was employed to identify the linear relationship between the explanatory environmental parameters and the observed stormwater discharge. SMLR analysis was applied to identify the best performing regression model. Four types of regression equations were examined to determine the best MLR model. Explanatory and dependent variables with and without log e-transformation were tested. Then, the MLR models were validated via a jackknife validation procedure.

2. Materials and Methods

2.1. Study Area and Field Data

Lake Soyang formed following the construction of the Soyang River Dam. The dam was built to provide irrigation water, flood control and hydroelectric power. The dam has a height of 123 m, a length of 530 m, and a total storage capacity of 2900 million m^3. The basin area is 2969.3 km^2; the forest occupies 86.4%; and the dry field, paddy field and residential areas occupy 4.4%, 1.58%, and 1.60% of the basin, respectively.

In Lake Soyang basin (Figure 1), sediment discharge occurred mainly from the upper part of the basin. Land use in the tributary watersheds in the dense highland upland field area is shown in Table 1. In the case of Mandaecheon, the percentage of agricultural area is 27.7%, and upland fields represent 75% of the agricultural land. The Jungjohangcheon, Johangcheon and Jauncheon subbasins contain very small paddy field areas. In the highland area, to decrease the damage caused by the continuous cultivation of economic crops, manage pests, maximize crop productivity and improve soil fertility, 30–50 cm of soil dressing has been applied to the top layer of soil. This soil dressing is a significant contributor to the sediment discharge from the highland fields.

During rainfall, water sampling and flow measurements were performed at the same time. Field surveys were conducted to perform flow measurements at most of the survey points. For the remaining points, such as Inbukcheon, Bukcheon, and Soyang River, real-time water level and flow measurements were obtained from the Ministry of Land Infrastructure and Transport and the Korea Water Resources Corporation. The sampled water from the measurement sites was delivered to the laboratory as quickly as possible, and BOD, chemical oxygen demand (COD), SS, TP, TN, and TOC were analyzed using standard methods [2].

Table 1. Land use of the tributary watersheds in the dense highland fields area.

Stream	Subbasin Area (ha)	Land Use				
		Forest (ha)	Upland Field (ha)	Paddy Field (ha)	Others (ha)	Proportion of Agricultural Land (%)
Jungjohangcheon	1022	860	150	0	12	14.6
Johangcheon	4161	3556	489	0.3	117	12.0
Jauncheon	13,641	11,703	1445	0.4	493	11.0
Mandaecheon	6079	1261	1261	420	3137	27.7
Gaahcheon	4732	1104	578	298	2752	18.5

In the statistical analysis of this study, the results of stormwater runoff surveys from 2013 to 2016 at nine points in the Jaun, Mandae and Gaha area [31] were used. Of the 79 rainfall events, nine data were too high or too low for rainfall amount due to runoff load measurements and calculation errors; those data were excluded. The discharge load survey was performed from the beginning of the rainfall to the point where it returned to the normal water level after the end of the rainfall. Runoff discharge data of 70 storm events were used to build the MLR models to predict the pollutant load, EMC and pollutant load per area. The range of the rainfall depth used in the MLR model construction was from 10 mm to 215 mm.

Figure 1. Soyang River basin and stormwater survey sites.

2.2. Data Analysis

Using water quality and runoff flow data from 70 rainfall events in the Lake Soyang basin, pollutant load per event, EMC, and pollutant load per area were estimated for each rainfall event. The total pollutant load during a rainfall event was calculated using Equation (1). The EMC was defined as the pollutant mass contained in the runoff event divided by the total flow volume of the event. The total pollutant load was divided by the subbasin area to estimate the pollutant load per area.

$$Total\ pollution\ load/rainfall\ event = \sum_{i=1}^{n} C_i Q_i \Delta t_i \tag{1}$$

$$EMC = \frac{\sum Q_i C_i \Delta t}{\sum Q_i \Delta t} \tag{2}$$

where n represents the number of total measurements, Q_i is the runoff flow at n number of time steps (Δt) and C_i is the concentration of a water quality measurement.

The distribution of the nonpoint pollutant discharge for the 70 rainfall events from 2013 to 2016 is presented in Table 2. Figure 2 shows box plots of pollutant load per event, EMC, and pollutant load per area at the survey points in the Lake Soyang basin. The maximum, minimum and median values of the suspended sediment (SS) load/event were 46,125,100 kg, 613 kg and 263,083 kg, respectively. The maximum, minimum and median values of the TP load/event were 32,406 kg, 1.83 kg and 480 kg, respectively. As shown in Figure 2, all the mean values of the pollutant loads are larger than the values of the third quartile, and the distributions are biased toward the high values. The maximum, minimum and median values of the SS EMC were 1437 mg/L, 3.8 mg/L and 157 mg/L, respectively. The maximum, minimum and median values of the TP EMC were 1.96 mg/L, 0.011 mg/L and

0.27 mg/L, respectively. All the mean values of the EMCs lie between the 50th percentile and the 75th percentile, and the distributions are relatively uniform.

The explanatory variables that are considered to explain the nonpoint pollutant discharge in the MLR models are the percentage of fields (% field), subbasin area (SA), and mean slope of subbasin (slope) as topographic parameters, and the number of preceding dry days (Ndry), rainfall intensity (Rint), rainfall depth (Rain), and rainfall duration (Dur) as rainfall parameters (Table 3).

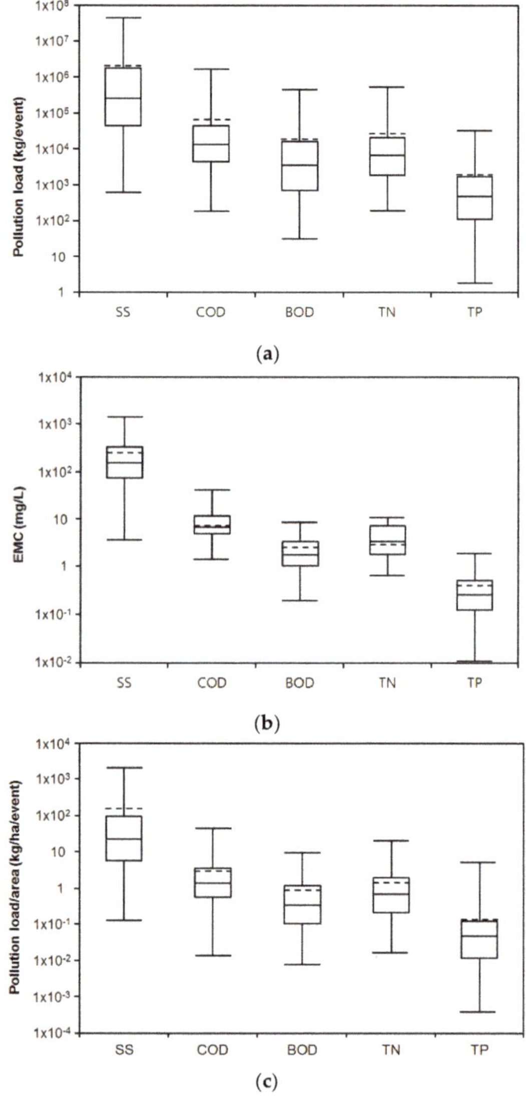

Figure 2. Box plots of nonpoint pollutant discharge in the Lake Soyang basin. The top (**a**) and bottom (**c**) of each box represent the third and first quartiles, the solid line inside the box is the second quartile (**b**), and the dotted line inside the box is the mean. One whisker stretches from the third quartile to the maximum, and the other whisker stretches from the first quartile to the minimum.

Table 2. Distribution of the nonpoint pollutant discharge for the 70 rainfall events in the Lake Soyang basin.

Pollutant	Min.	25th Percentile	50th Percentile	75th Percentile	Max.	Mean
SS load (kg)	613	44,839	263,083	1,802,409	46,125,100	1,843,969
COD load (kg)	186	4410	13,192	44,608	1,686,594	83,565
BOD load (kg)	31	689	3477	15,982	456,773	20,233
TN load (kg)	187	1852	6563	20,745	541,563	28,277
TP load (kg)	1.8	111	480	1685	32,406	2010
SS (EMC) (mg/L)	3.8	75.6	157	338	1437	266
COD (EMC) (mg/L)	1.5	5.13	7.21	12.2	43.6	9.16
BOD (EMC) (mg/L)	0.20	1.1	1.85	3.47	9.0	2.49
TN (EMC) (mg/L)	0.67	1.89	3.48	7.63	11.4	4.75
TP (EMC) (mg/L)	0.011	0.13	0.27	0.54	1.96	0.37
SS (load/area) (kg/ha)	0.129	5.81	22.6	95.37	2118	130
COD (load/area) (kg/ha)	0.0139	0.57	1.38	3.54	44.0	3.12
BOD (load/area) (kg/ha)	0.0078	0.10	0.34	1.18	9.5	0.87
TN (load/area) (kg/ha)	0.0164	0.21	0.69	1.93	20.4	1.54
TP (load/area) (kg/ha)	0.00038	0.012	0.047	0.12	5.19	0.16

Table 3. Explanatory variables considered in the regression models to predict pollutant discharge.

Variables	Description	Units
% field	Percentage of fields	%
SA	Subbasin area	km^2
Ndry	Number of preceding dry days	day
Rint	Rainfall intensity	mm/h
Slope	Mean slope of the subbasin	°
Rain	Rainfall depth	mm
Dur	Rainfall duration	h

A Pearson correlation coefficient matrix was used to identify the correlations among the surveyed pollutant discharge estimates and the explanatory variables. The correlations among natural log-transformed variables were also tested using Pearson correlation.

2.3. MLR Model Building

MLR modeling was performed to predict the pollutant discharge from the subbasins in the Soyang River. The models were built to explain the pollutant discharge using the subbasin topographic and rainfall data. In the MLR modeling, four types of regression equations are examined:

$$\text{Type1} : Y = a_0 + \sum_{i=1}^{n} a_i X_i \tag{3}$$

$$\text{Type2} : Ln(Y) = a_0 + \sum_{i=1}^{n} a_i X_i \tag{4}$$

$$\text{Type3} : Y = e^{a_0} X_1^{a_1} X_2^{a_2} \cdots X_n^{a_n} \text{ or } Ln(Y) = a_0 + \sum_{i=1}^{n} (a_i Ln(X_i)) \tag{5}$$

$$\text{Type4} : Y = e^{a_0} X_1^{a_1} X_2^{a_2} \cdots X_m^{a_m} e^{a_{m+1} X_{m+1}} \cdots e^{a_n X_n} \text{ or } Ln(Y) = a_0 + \sum_{i=1}^{m} (a_i Ln(X_i)) + \sum_{i=m+1}^{n} (a_i X_i) \tag{6}$$

where a_0 is the regression constant and a_i is the regression coefficient of the explanatory variable X_i. In type 1, the original variables are used to build the MLR model. In type 2, dependent variables, such as pollutant load, EMC and load/area, are log e-transformed to reduce skewness. In type 3, all the explanatory and dependent variables are log e-transformed. In type 4, the dependent variables and some of the explanatory variables are log e-transformed. The fitness of the four regression equations was evaluated by the coefficients of determination of the MLR models. The MLR models were examined in terms of their ability to predict the runoff pollutant discharge for each water quality variable (SS, COD, BOD, TN, and TP).

Collinearity may introduce serious stability problems, such as high mean square errors, in a regression model. Therefore, the collinearity of the predictor variables in the created MLR model were tested by calculating the variance inflation factor (VIF) [32]. Collinearity is present when the largest VIF is greater than 10 or the average VIF value is substantially greater than 1 [32,33]. VIFs were calculated to analyze the multicollinearity in this MLR model.

The MLR model performance was evaluated using the R^2, RMSE, CV(RMSE), RSR and the NSE.

$$R^2 = \frac{\left[\sum_{i=1}^n \left(P_i - \overline{P}\right)\left(O_i - \overline{O}\right)\right]^2}{\sum_{i=1}^n \left(P_i - \overline{P}\right)^2 \sum_{i=1}^n \left(O_i - \overline{O}\right)^2} \tag{7}$$

$$RMSE = \left[\frac{1}{N} \sum_{i=1}^n \left(P_i - O_i\right)^2\right]^{1/2} \tag{8}$$

$$CV(RMSE) = \frac{\left[\sum_{j=1}^n \left(P_{ij} - O_{ij}\right)^2 / n\right]^{1/2}}{\left(\sum_{i=1}^n O_{ij}/n\right)} \tag{9}$$

$$RSR = \frac{\left[\sqrt{\sum_{i=1}^n \left(O_i - P_i\right)^2}\right]}{\left[\sqrt{\sum_{i=1}^n \left(O_i - \overline{O}\right)^2}\right]} \tag{10}$$

$$NSE = 1 - \frac{\sum_{i=1}^n \left(O_i - P_i\right)^2}{\sum_{i=1}^n \left(O_i - \overline{O}\right)^2} \tag{11}$$

where O_i is the observed daily load, $\overline{O}$ is the mean of the observed daily load, p_i is the calculated daily load, and n is the number of data values. The R^2 index describes the ability of the model to explain variability among the data. RSR incorporates the benefits of error index statistics and includes a scaling/normalization factor; the lower the RSR is, the better the model simulation performance. The performance ratings for stream flow proposed by Moriasi et al. [33] were 'very good' ($0.00 \leq$ RSR ≤ 0.50), 'good' ($0.50 <$ RSR ≤ 0.60), or 'satisfactory' ($0.60 <$ RSR ≤ 0.70). NSE is a normalized statistic that reflects the relative magnitude of the residual variance compared with the variance in the observed data (good (NSE > 0.7), satisfactory ($0.4 <$ NSE ≤ 0.7) and unsatisfactory (NSE ≤ 0.4)) [30,34].

Finally, the performance of the MLR model was tested using the jackknife validation method [8,11,14]. This method consists of deleting one site and carrying out the multiple regression analysis with the same dependent variables and the remaining sites. The pollutant discharge of the deleted site is calculated with the equation resulting from the multiple regression associated with the remaining sites. This process is repeated, deleting one site each time.

3. Results and Discussion

3.1. Correlation Analysis between Nonpoint Pollutant Discharge and Explanatory Variables

Table 4 shows the Pearson correlation between runoff discharge and subbasin characteristics, without log transformation of the variables. In Table 5, the Pearson correlation matrix between log-transformed variables is introduced. As the values of the correlation coefficients between log-transformed variables were slightly higher (r < 0.69; Table 5) than those of the non-log-transformed variables (r < 0.65; Table 4), we used log-transformed variables as explanatory variables in the MLR models. Compared to other environmental parameters, the rainfall depth and subbasin area showed a relatively significant correlation with most response discharge variables. The rainfall intensity had a relatively significant positive correlation with the response variables of EMC and load/area because the rainfall intensity directly affects the EMC and load/area of each storm event. On the other hand, the subbasin slope had a negative correlation with the response variables of EMC and load/area.

Table 4. Pearson correlation matrix between stormwater runoff discharge and subbasin characteristics.

Variables	% field	SA	Rain	Dur	Ndry	Rint	Slope
SS (load)	−0.187	0.102	**0.524**	0.236	−0.135	0.260	0.088
COD (load)	−0.211	**0.563**	**0.317**	0.297	0.015	0.022	0.239
BOD (load)	−0.217	**0.458**	**0.352**	0.353	−0.080	0.011	0.213
TN (load)	−0.214	**0.488**	**0.374**	0.316	−0.053	0.057	0.220
TP (load)	−0.205	**0.417**	**0.514**	0.356	−0.150	0.160	0.178
SS (EMC)	**0.498**	−0.293	**0.387**	0.181	−0.115	0.251	**−0.585**
COD (EMC)	0.125	−0.199	0.166	−0.098	−0.099	**0.397**	−0.150
BOD (EMC)	0.120	**−0.317**	0.187	0.095	−0.049	0.147	−0.227
TN (EMC)	0.196	−0.065	0.157	−0.022	0.227	0.195	−0.207
TP (EMC)	**0.355**	**−0.366**	0.223	−0.115	−0.216	**0.404**	**−0.391**
SS (load/area)	0.198	−0.166	**0.632**	0.313	−0.185	0.283	−0.240
COD (load/area)	0.102	−0.095	**0.599**	0.367	−0.175	0.224	−0.108
BOD (load/area)	0.077	−0.147	**0.652**	0.476	−0.212	0.172	−0.124
TN (load/area)	0.132	−0.180	**0.583**	0.348	−0.172	0.210	−0.137
TP (load/area)	0.108	−0.114	**0.441**	0.190	−0.148	0.212	−0.106

Note: Bold marked correlations are significant at $p < 0.01$.

Table 5. Pearson correlation matrix between natural log-transformed stormwater runoff discharge and subbasin characteristics.

Variables	ln(% field)	ln(SA)	ln(Rain)	ln(Dur)	ln(Ndry)	ln(Rint)	Slope
ln(SS(load))	−0.12	**0.42**	**0.58**	**0.44**	−0.12	0.29	−0.16
ln(COD(load))	**−0.38**	**0.69**	**0.43**	**0.49**	−0.12	0.08	0.25
ln(BOD(load))	**−0.34**	**0.60**	**0.51**	**0.49**	−0.11	0.18	0.14
ln(TN(load))	**−0.32**	**0.61**	**0.47**	**0.48**	−0.12	0.14	0.17
ln(TP(load))	−0.16	**0.48**	**0.59**	**0.43**	−0.16	0.30	−0.06
ln(SS(EMC))	**0.40**	**−0.37**	**0.47**	0.03	−0.11	**0.48**	**−0.63**
ln(COD(EMC))	0.19	**−0.40**	0.23	−0.12	−0.06	**0.33**	−0.14
ln(BOD(EMC))	0.20	**−0.44**	0.37	−0.06	−0.02	**0.44**	**−0.36**
ln(TN(EMC))	**0.33**	**−0.49**	0.23	−0.13	0.09	**0.35**	**−0.38**
ln(TP(EMC))	**0.45**	**−0.56**	0.36	−0.17	−0.08	**0.52**	**−0.59**
ln(SS(load/area))	**0.34**	−0.30	**0.64**	0.28	−0.16	**0.47**	**−0.54**
ln(COD (load/area))	0.17	−0.18	**0.62**	0.39	−0.22	**0.37**	−0.23
ln(BOD (load/area))	0.18	−0.24	**0.65**	0.35	−0.18	**0.43**	**−0.33**
ln(TN(load/area))	0.30	**−0.37**	**0.59**	0.29	−0.19	**0.41**	**−0.36**
ln(TP(load/area))	**0.37**	**−0.37**	**0.65**	0.24	−0.21	**0.52**	**−0.52**

Note: Bold marked correlations are significant at $p < 0.01$.

3.2. MLR Analysis

Four types of MLR models corresponding to Equations (3)–(6) were tested to identify the most suitable models (Table 6). The R^2 values for SS, COD, BOD, TN, and TP in the type 1 MLR of pollutant load ranged from 0.275 to 0.447. The R^2 values for SS, COD, BOD, TN, and TP in the type 1 MLR of EMC and load/area were also low, indicating poor performance of the regression models. The R^2 values for SS, COD, BOD, TN, and TP in the type 2 MLR of pollutant load were 0.76, 0.67, 0.64, 0.65, and 0.80, respectively. The R^2 values of the type 2 MLR were quite high, but most of the VIF values were larger than 5, with a few values greater than 10. Thus, the VIF showed that multicollinearity was observed in the established models and that the type 2 MLR was not adapted. Although the R^2 values of the type 2 MLR for load/area were acceptable, the VIF values were high, indicating multicolinearity. VIF values and other statistics of MLRs were presented only for the selected model. The results of the MLR model employing the type 4 equation are listed in Tables 7–9.

Table 6. Coefficients of determination from four types of MLR analysis.

Runoff Discharge Type	MLR Type	SS	COD	BOD	TN	TP
Load	Type 1	0.275	0.425	0.340	0.386	0.447
	Type 2	0.764	0.672	0.641	0.654	0.801
	Type 3	0.720	0.687	0.688	0.614	0.689
	Type 4	0.736	0.687	0.694	0.614	0.741
EMC	Type 1	0.477	0.157	0.100	0.254	0.33
	Type 2	0.646	0.123	0.273	0.321	0.584
	Type 3	0.536	0.226	0.324	0.539	0.592
	Type 4	0.655	0.226	0.324	0.539	0.662
Load/Area	Type 1	0.448	0.359	0.460	0.340	0.195
	Type 2	0.734	0.503	0.526	0.497	0.686
	Type 3	0.640	0.424	0.496	0.471	0.651
	Type 4	0.695	0.427	0.509	0.471	0.675

The R^2 values for SS, COD, BOD, TN, and TP in the type 3 MLR of the pollutant load were also fairly high, but all VIF values were less than 5. Among the type 3 MLR models, the SS, TN, and TP in the MLR of EMC and the SS and TP in the MLR of load/area showed acceptable R^2 values. The values of R^2 for SS, COD, BOD, TN, and TP in the type 4 MLR of pollutant load were 0.74, 0.69, 0.69, 0.61, and 0.74 respectively. The R^2 values of the type 4 MLR were a little better than those of the type 3 MLR, and all VIF values were less than 5. Thus, we selected the type 4 equation as the MLR model to predict the runoff pollutant discharge in the study area. However, the COD and BOD in the MLR of EMC and COD and TN in the MLR of load/area could not explain the variance in the pollutant discharge properly.

Using the stepwise variable selection method, two to five variables were retained in the pollutant load model, as shown in Table 7. In the case of the SS model, given the R^2 value, 73.6% of the variability of the dependent variable ln(SS load) is explained by the four explanatory variables. The MLR models indicated in Tables 7–9 are statistically significant at $p < 0.0001$ except for the ln(COD EMC) model ($p = 0.00019$). The R^2 values for SS, COD, BOD, TN, and TP in the type 4 MLR of pollutant load were fairly high ($0.614 < R^2 < 0.741$), as indicated in Table 7. The performance evaluation by CV(RMSE) [35] shows that the SS model was the best and that the other models of the water quality variables were also acceptable. The range of RSR for SS, COD, BOD, and TP in the MLR models of pollutant load (Table 7) was from 0.509 to 0.559, and the performance of the MLR for these variables was good [34]. The RSR for TN was 0.622, and the performance of the TN model was satisfactory. The NSE values for the MLR models of pollutant load ranged from 0.61 to 0.74, and the MLR models of the pollutant load had good performance. As a special case, in linear regression forecasting models like this study, NSE is equal to the coefficient of determination, R^2 [36]. Overall, all the MLR models of the pollutant load had good prediction performance.

All VIF values in Table 7 are lower than 5, and the mean VIF values are not large. These results suggest that the coefficient of regression for the explanatory variables could be statistically acceptable and that multicollinearity was not present in the established models.

Standardized coefficients refer to how many standard deviations a dependent variable will change in response to an increase of one standard deviation in the predictor variable. This statistic allows us to compare the relative contribution of each independent variable in the prediction of the dependent variable. The higher the absolute value of a coefficient is, the more important the weight of the corresponding variable. Standardized coefficients are useful for comparing effects across different measures. The standardized regression coefficients of Table 7 indicate that subbasin area ($0.576 < \beta_i < 0.709$) and rainfall depth ($0.453 < \beta_i < 0.563$) are important influential parameters for all the load predictions. In addition, % field has relatively small effects on the SS, BOD and TP models.

Table 7. MLR models for pollutant load discharge during stormwater events.

Response Variable	Explanatory Variables	a_0	a_i	β_i	VIF	DW	R^2	RMSE	CV(RMSE)	RSR	NSE
ln(SS load)	Intercept	12.50				1.773	0.736	1.230	0.099	0.514	0.736
	ln(% field)		−2.31	−0.40	4.017						
	ln(SA)		0.81	0.58	1.610						
	ln(Rain)		1.78	0.56	1.008						
	Slope		−0.25	−0.75	3.377						
ln(COD load)	Intercept	0.44				1.746	0.687	1.135	0.119	0.559	0.687
	ln(SA)		0.86	0.71	1.001						
	ln(Rain)		1.24	0.45	1.001						
ln(BOD load)	Intercept	4.92				1.938	0.694	1.172	0.145	0.553	0.694
	ln(% field)		−1.53	−0.30	4.017						
	ln(SA)		0.80	0.64	1.610						
	ln(Rain)		1.44	0.51	1.008						
	Slope		−0.12	−0.41	3.377						
ln(TN load)	Intercept	0.72				1.873	0.614	1.121	0.127	0.622	0.614
	ln(SA)		0.67	0.63	1.001						
	ln(Rain)		1.19	0.49	1.001						
ln(TP load)	Intercept	3.44				1.539	0.741	1.047	0.174	0.509	0.741
	ln(% field)		−1.36	−0.28	4.025						
	ln(SA)		0.77	0.64	1.620						
	ln(Rain)		1.50	0.56	1.039						
	ln(Ndry)		−0.24	−0.14	1.049						
	Slope (°)		−0.17	−0.60	3.423						

Notes: a_0 is the regression constant; a_i is the regression coefficient of the explanatory variable X_i; β_i is the standardized regression coefficient.

Table 8. MLR models for EMCs of stormwater events.

Response Variables	Explanatory Variables	a_0	a_i	β_i	VIF	DW	R^2	RMSE	CV(RMSE)	RSR	NSE
ln(SSEMC)	Intercept	10.94				1.654	0.655	0.875	0.180	0.587	0.655
	ln(% field)		−1.79	−0.50	4.017						
	ln(SA)		−0.17	−0.19	1.610						
	ln(Rain)		0.83	0.42	1.008						
	Slope		−0.19	−0.93	3.377						
ln(CODEMC)	Intercept	2.46				1.338	0.226	0.515	0.252	0.880	0.226
	ln(SA)		−0.12	−0.35	1.054						
	ln(Rint)		0.22	0.26	1.054						
ln(BODEMC)	Intercept	−0.07				1.129	0.324	0.723	1.223	0.822	0.324
	ln(SA)		−0.23	−0.43	1.001						
	ln(Rain)		0.42	0.36	1.001						
ln(TNEMC)	Intercept	2.47				0.843	0.539	0.504	0.384	0.679	0.539
	ln(SA)		−0.29	−0.65	1.054						
	ln(Rint)		0.25	0.23	1.054						
ln(TPEMC)	Intercept	4.62				0.919	0.662	0.723	−0.481	0.581	0.662
	ln(% field)		−1.14	−0.38	4.017						
	ln(SA)		−0.25	−0.34	1.693						
	ln(Ndry)		−0.19	−0.19	1.057						
	ln(Rint)		0.75	0.43	1.100						
	Slope		−0.12	−0.69	3.396						

Table 9. MLR models for pollutant load per area during stormwater events.

Response Variables	Explanatory Variables	a_0	a_i	β_i	VIF	DW	R^2	RMSE	CV(RMSE)	RSR	NSE
ln(SS load/area)	Intercept	5.70				1.728	0.695	1.246	0.408	0.552	0.695
	ln(% field)		−1.81	−0.33	3.365						
	ln(Rain)		1.79	0.60	1.007						
	Slope		−0.25	−0.79	3.376						
ln(COD load/area)	Intercept	−3.88				1.779	0.427	1.123	4.710	0.757	0.427
	ln(Rain)		1.22	0.61	1.003						
	Slope		−0.04	−0.19	1.003						
ln(BOD load/area)	Intercept	−5.65				1.927	0.509	1.206	−0.995	0.701	0.509
	ln(Rain)		1.47	0.63	1.003						
	Slope		−0.07	−0.29	1.003						
ln(TN load/area)	Intercept	−3.88				1.873	0.471	1.121	−2.108	0.727	0.471
	ln(SA)		−0.33	−0.36	1.001						
	ln(Rain)		1.19	0.58	1.001						
ln(TP load/area)	Intercept	−6.57				1.519	0.675	1.090	−0.328	0.570	0.675
	ln(Rain)		1.53	0.60	1.033						
	ln(Ndry)		−0.25	−0.15	1.036						
	Slope		−0.13	−0.49	1.011						

The area with the high density of highland fields in Lake Soyang basin has steeper slopes than the other areas. However, Lake Soyang basin also contains highly mountainous terrain; thus, the mean slopes of the dense highland field subbasins are lower than the average slope of the entire Lake Soyang basin. Therefore, the standardized regression coefficients of mean slope for the SS and TP load models have ($-$) signs, and the mean slope has a negative influence on the SS and TP loads.

The explanatory variables for the SS and TP models explained 65.5% and 66.2% of variation in the response variables of EMC. The R^2 values were fairly high, as indicated in Table 8. The R^2 value for the TN model of EMC was 0.539, and the TN model was acceptable [34]. The CV(RMSE) value of the BOD model was quite high, and the model was not acceptable. The RSR values for SS and TP in the MLR models of EMC (Table 8) were 0.587 and 0.581, respectively, and the performances of these models were good. The RSR for the TN model was 0.679, and the performance of the TN model was satisfactory. However, the RSR values for the COD and BOD models were high, and these models were unsatisfactory. The NSE values for the MLR models of the EMC show that the SS, TP, and TN models were satisfactory but that the COD and BOD models were not satisfactory. The VIF values for the EMC models were lower than 5, and the mean VIF values were not large. Overall, the MLR models for SS and TP have good prediction performance, and the TN model has acceptable performance.

The standardized regression coefficients in Table 8 indicate that rainfall intensity and rainfall depth are influential explanatory variables for the EMC response variables. Rainfall intensity (0.234 < β_i < 0.426) is an important factor for the TP, TN, and COD models, and rainfall depth is important for the SS and BOD models. In the pollutant load model, rainfall depth is a very important parameter, whereas rainfall intensity is not an important explanatory variable. However, rainfall intensity is an influential parameter for the EMC of a storm event. From the Pearson correlation matrix between natural log-transformed stormwater runoff discharge and subbasin characteristics in Table 5, we also can see that EMCs are better correlated to rainfall intensity than rainfall depth, and pollutant loads are much better correlated to rainfall depth than rainfall intensity. In agricultural areas such as the study area, the larger the rainfall intensity, the more nutrients are released from fertilizer and vegetation roots. The standardized regression coefficients of the mean slope for the SS and TP load models have ($-$) sign, and the mean slope has a large negative influence on the SS and TP EMC. Additionally, % field also has a negative impact on the SS and TP EMC.

The explanatory variables for the SS and TP models explained 69.5% and 67.5% of the variation in the load/area response variables, and the R^2 values were fairly high, as indicated in Table 9. The R^2 value for the BOD model of load/area was 0.51; thus, the BOD model was acceptable. The RSR values for SS and TP in the MLR models of load/area (Table 9) were 0.55 and 0.57, respectively, and the performances of these models were good. The RSR for the BOD model was 0.70, and the performance of the TN model was satisfactory. The NSE values in the MLR models of the load/area show that the SS, TP, and BOD models were satisfactory. The VIF values for the load/area models were less than 5, and the mean VIF values were not large. Overall, the MLR models of load/area for SS and TP have good performance, and the BOD model has acceptable performance.

The standardized regression coefficients in Table 9 indicate that rainfall depth (0.576 < β_i < 0.634) is a highly influential parameter for all response variables in the load/area prediction. The β coefficients of the mean slope for the SS and TP load/area models are -0.79 and -0.49, respectively, and the absolute values of the coefficients are comparable to the coefficients of rainfall depth, indicating that the mean slope is a remarkable negative parameter on the SS and TP load/area results.

3.3. Jackknife Validation of the MLR Model

The performance of the jackknife validation was evaluated using R^2, RSR and NSE (Table 10). The R^2 values were calculated by the linear regression between observed and jackknife validation values, and RSR and NSE were also calculated. The R^2 (Figure 3) and NSE values associated with the jackknife procedure were slightly lower than the results of the MLR models, whereas the RSR values were slightly higher than the MLR models. Therefore, the performance of the jackknife validation was slightly worse than that of the MLR models. The results of jackknife validation indicate that the MLR models are robust.

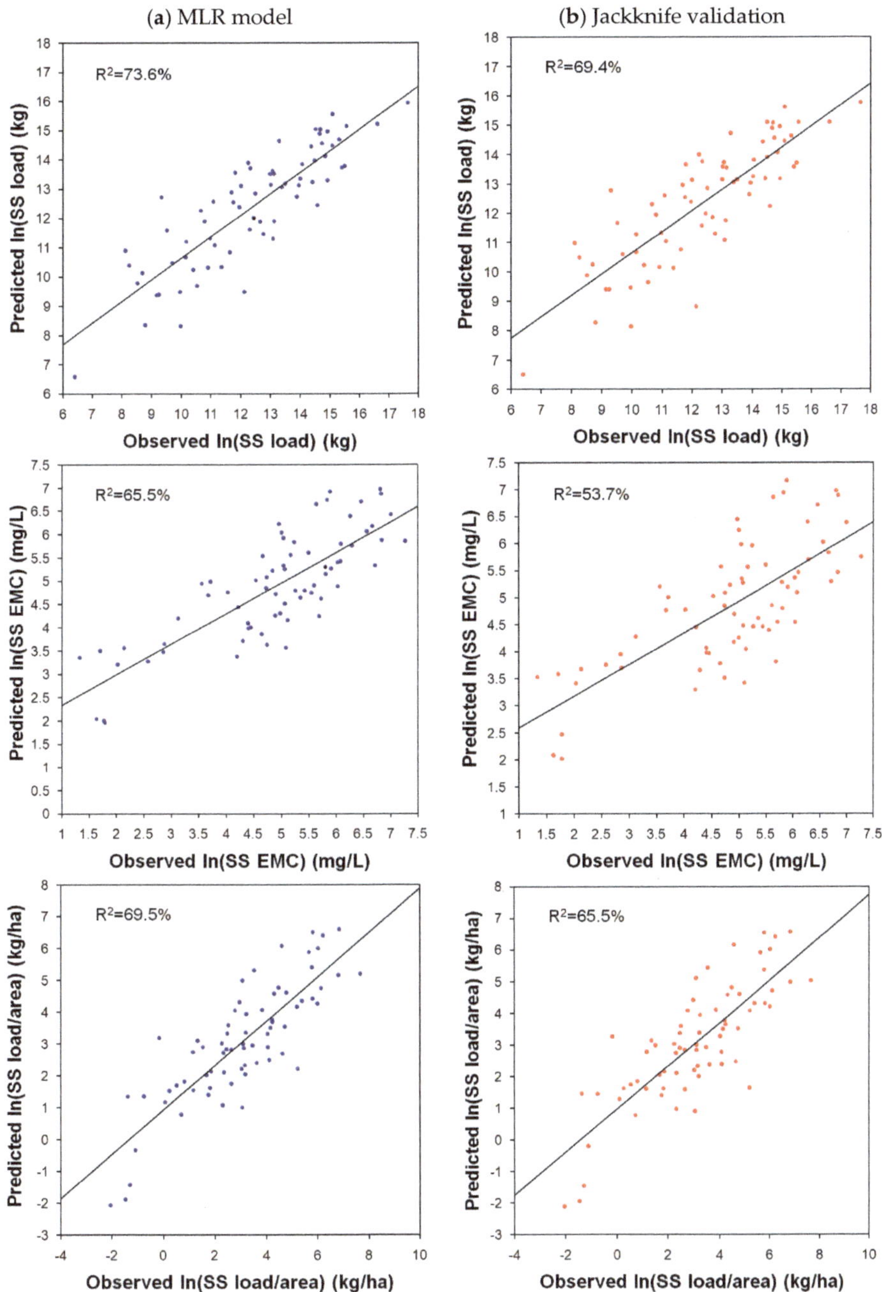

Figure 3. Comparison between the observed and predicted values during storm events based on (**a**) MLR models and (**b**) the results of jackknife validation.

Table 10. Three performance indicators for the stormwater runoff discharge values based on jackknife validation.

Response Variable (Jackknife)	R^2	RSR	NSE
ln(SS load)	0.694	0.554	0.693
ln(COD load)	0.630	0.611	0.627
ln(BOD load)	0.607	0.630	0.603
ln(TN load)	0.550	0.674	0.545
ln(TP load)	0.609	0.629	0.605
ln(SS EMC)	0.537	0.684	0.533
ln(COD EMC)	0.155	0.924	0.147
ln(BOD EMC)	0.211	0.894	0.202
ln(TN EMC)	0.503	0.730	0.468
ln(TP EMC)	0.601	0.633	0.599
ln(SS load/area)	0.655	0.588	0.654
ln(COD load/area)	0.305	0.845	0.287
ln(BOD load/area)	0.478	0.723	0.477
ln(TN load/area)	0.413	0.768	0.410
ln(TP load/area)	0.602	0.632	0.600

4. Conclusions

MLR models were built to predict the nonpoint-source pollutant discharge in the highland field area in the wet season using environmental parameters as explanatory variables. Runoff discharge data from 70 storm events were used to build the MLR models to predict the pollutant load, EMC and pollutant load per area. Pearson correlation tests were employed to identify the linear relationships between subbasin environmental parameters and the observed stormwater discharge. As the values of correlation coefficients between log-transformed variables were slightly higher than those of variables that had not been log transformed, the log-transformed variables were selected as explanatory variables in the MLR models.

The R^2 values for SS, COD, BOD, TN, and TP in the type 4 MLR of pollutant load were quite high (the best among the four examined MLR types), and all VIF values were less than 5. Thus, the type 4 equation was chosen as the MLR model to predict the runoff pollutant discharge.

The R^2 values for the five water quality variables in the MLR of pollutant load were fairly high ($0.614 < R^2 < 0.741$), and the RSR values for SS, COD, BOD, and TP in the MLR models of pollutant load ranged from 0.509 to 0.559. Hence, the performance of the MLR for these variables was good [34]. The RSR for TN was 0.622, and the performance of the TN model was satisfactory. The NSE values for the MLR models of the pollutant load indicated good performance. Hence, most of the MLR models of the pollutant load have good prediction performance.

The MLR models of EMC for SS and TP also have good prediction performance, and TN model has acceptable performance. The MLR models of load/area for SS and TP have relatively good performance, and the BOD model has acceptable performance. Based on the R^2, RSR and NSE values, the performance of the jackknife validation was slightly worse than that of the MLR models. Thus, the results of jackknife validation indicate that the MLR models are robust.

The results of the standardized coefficients for the MLR models indicate that subbasin area and rainfall depth are important influential parameters for all the load predictions. The mean slope exerts a negative influence on the SS and TP loads on account of topographic characteristics, as previously explained. For the pollutant load model, rainfall depth is a very important parameter, whereas rainfall intensity was not chosen as an explanatory variable. However, rainfall intensity has an influence on the EMC of the storm event. The mean slope has a large negative influence on the SS and TP EMC, and % field has a negative impact on the SS and TP EMC. Additionally, the rainfall depth is a highly influential parameter for all response variables of the load/area predictions, similar to the pollutant load models. The mean slope has a large negative influence on the SS and TP load/area.

The average slope of fields, rather than the average slope of the whole sub-basin, can be an important explanatory variable for the pollutant discharge load of each subbasin. Similar or even better MLR results for EMC could have been obtained using peak rainfall intensity as explanatory variables. Therefore, future studies on MLR need to consider this.

Author Contributions: J.H.C. designed research, conducted data analysis and wrote the paper. J.H.L. contributed to the discussion and writing of the paper.

Funding: This research was supported by the Wonju Regional Environmental Office and the Basic Science Research Program through the National Research Foundation of Korea (NRF) funded by the Ministry of Education (grant number: NRF-2017R1D1A1B03032816).

Conflicts of Interest: The authors declare no conflicts of interest.

References

1. Kim, H.J.; Lee, K.K. A comparison of the water environment policy of Europe and South Korea in response to climate change. *Sustainability* **2018**, *10*, 384. [CrossRef]
2. Cho, J.H.; Lee, J.H. Stormwater runoff characteristics and effective management of nonpoint source pollutants from a highland agricultural region in the Lake Soyang watershed. *Water* **2017**, *9*, 784. [CrossRef]
3. Kim, B.; Jung, S. Turbid storm runoffs in Lake Soyang and their environmental effect. *J. Korean Soc. Environ. Eng.* **2007**, *29*, 1185–1190.
4. Valtanen, M.; Sillanpää, N.; Setälä, H. Key factors affecting urban runoff pollution under cold climatic conditions. *J. Hydrol.* **2015**, *529*, 1578–1589. [CrossRef]
5. Bian, G.D.; Du, J.K.; Song, M.M.; Xu, Y.P.; Xie, S.P.; Zheng, W.L.; Xu, C.Y. A procedure for quantifying runoff response to spatial and temporal changes of impervious surface in Qinhuai River basin of southeastern China. *Catena* **2017**, *157*, 268–278. [CrossRef]
6. Roman, D.C.; Vogel, R.M.; Schwarz, G.E. Regional regression models of watershed suspended-sediment discharge for the eastern United States. *J. Hydrol.* **2012**, *472–473*, 53–62. [CrossRef]
7. Tuset, J.; Vericat, D.; Batalla, R.J. Rainfall, runoff and sediment transport in a Mediterranean mountainous catchment. *Sci. Total Environ.* **2016**, *540*, 114–132. [CrossRef] [PubMed]
8. Buendia, C.; Herrero, A.; Sabater, S.; Batalla, R.J. An appraisal of the sediment yield in western Mediterranean river basins. *Sci. Total Environ.* **2016**, *572*, 538–553. [CrossRef] [PubMed]
9. Castiglioni, S.; Lombardi, L.; Toth, E.; Castellarin, A.; Montanari, A. Calibration of rainfall-runoff models in ungauged basins: A regional maximum likelihood approach. *Adv. Water Resour.* **2010**, *33*, 1235–1242. [CrossRef]
10. Tramblay, Y.; Saint-Hilaire, A.; Ouarda, T.B.M.J.; Moatar, F.; Hecht, B. Estimation of local extreme suspended sediment concentrations in California Rivers. *Sci. Total Environ.* **2010**, *408*, 4221–4229. [CrossRef] [PubMed]
11. Lombardi, L.; Toth, E.; Castellarin, A.; Montanari, A.; Brath, A. Calibration of a rainfall-runoff model at regional scale by optimising river discharge statistics: Performance analysis for the average/low flow regime. *Phys. Chem. Earth* **2012**, *42–44*, 77–84. [CrossRef]
12. Ali, M.; Seeger, M.; Sterk, G.; Moore, D. A unit stream power based sediment transport function for overland flow. *Catena* **2013**, *101*, 197–204. [CrossRef]
13. Heng, S.; Suetsugi, T. Comparison of regionalization approaches in parameterizing sediment rating curve in ungauged catchments for subsequent instantaneous sediment yield prediction. *J. Hydrol.* **2014**, *512*, 240–253. [CrossRef]
14. Zhao, J.; Vanmaercke, M.; Chen, L.; Govers, G. Vegetation cover and topography rather than human disturbance control gully density and sediment production on the Chinese Loess Plateau. *Geomorphology* **2016**, *274*, 92–105. [CrossRef]
15. Paule-Mercado, M.A.; Ventura, J.S.; Memon, S.A.; Jahng, D.; Kang, J.H.; Lee, C.H. Monitoring and predicting the fecal indicator bacteria concentrations from agricultural, mixed land use and urban stormwater runoff. *Sci. Total Environ.* **2016**, *550*, 1171–1181. [CrossRef] [PubMed]
16. Eleria, A.; Vogel, R.M. Predicting fecal coliform bacteria levels in the Charles River, Massachusetts, USA. *J. Am. Water Resour. Assoc.* **2005**, *41*, 1195–1209. [CrossRef]

17. David, M.M.; Haggard, B.E. Development of regression-based models to predict fecal bacteria numbers at select sites within the Illinois River Watershed, Arkansas and Oklahoma, USA. *Water Air Soil Pollut.* **2011**, *215*, 525–547. [CrossRef]

18. Motamarri, S.; Boccelli, D.L. Development of a neural-based forecasting tool to classify recreational water quality using fecal indicator organisms. *Water Res.* **2012**, *46*, 4508–4520. [CrossRef] [PubMed]

19. Herrig, I.M.; Böer, S.I.; Brennholt, N.; Manz, W. Development of multiple linear regression models as predictive tools for fecal indicator concentrations in a stretch of the lower Lahn River, Germany. *Water Res.* **2015**, *85*, 148–157. [CrossRef] [PubMed]

20. Khan, S.; Lau, S.-L.; Kayhanian, M.; Stenstrom, M.K. Oil and grease measurement in highway runoff—Sampling time and event mean concentrations. *J. Environ. Eng.* **2006**, *132*, 415–422. [CrossRef]

21. Kayhanian, M.; Suverkropp, C.; Ruby, A.; Tsay, K. Characterization and prediction of highway runoff constituent event mean concentration. *J. Environ. Manag.* **2007**, *85*, 279–295. [CrossRef] [PubMed]

22. Ha, S.J.; Stenstrom, M.K. Predictive modeling of storm-water runoff quantity and quality for a large urban watershed. *J. Environ. Eng.* **2008**, *134*, 703–711. [CrossRef]

23. Maniquiz, M.C.; Lee, S.; Kim, L.H. Multiple linear regression models of urban runoff pollutant load and event mean concentration considering rainfall variables. *J. Environ. Sci.* **2010**, *22*, 946–952. [CrossRef]

24. Madarang, K.J.; Kang, J.-H. Evaluation of accuracy of linear regression models in predicting urban stormwater discharge characteristics. *J. Environ. Sci.* **2014**, *26*, 1313–1320. [CrossRef]

25. Feng, X.; Cheng, W.; Fu, B.; Lü, Y. The role of climatic and anthropogenic stresses on long-term runoff reduction from the Loess Plateau, China. *Sci. Total Environ.* **2016**, *571*, 688–698. [CrossRef] [PubMed]

26. Hou, X.; Zhou, F.; Leip, A.; Fu, B.; Yang, H.; Chen, Y.; Gao, S.; Shang, Z.; Ma, L. Spatial patterns of nitrogen runoff from Chinese paddy fields. *Agric. Ecosyst. Environ.* **2016**, *231*, 246–254. [CrossRef]

27. Smith, R.E.; Goodrich, D.C.; Quinton, J.N. Dynamic, distributed simulation of watershed erosion—The KINEROS2 and EUROSEM models. *Trans. ASAE* **1995**, *50*, 517–520.

28. De Roo, A.P.J.; Offermans, R.J.E.; Cremers, N.H.D.T. LISEM: A single-event, physically based hydrological and soil erosion model for drainage basins. II: Sensitivity analysis, validation and application. *Hydrol. Process.* **1996**, *10*, 1119–1126. [CrossRef]

29. Morgan, R.P.C.; Quinton, J.N.; Smith, R.E.; Govers, G.; Poesen, J.W.A.; Auerswald, K.; Chisci, G.; Torri, D.; Styczen, M.E.; Folly, A.J. The European soil erosion model (EUROSEM): Documentation and user guide. *Earth Surf. Process. Landf.* **1998**, *23*, 527–544. [CrossRef]

30. Wu, B.; Wang, Z.; Shen, N.; Wang, S. Modelling sediment transport capacity of rill flow for loess sediments on steep slopes. *Catena* **2016**, *147*, 453–462. [CrossRef]

31. Wonju Regional Environmental Office. *Monitoring and Assessment for the Nonpoint Source Pollution Management Area of Mandae, Gaah and Jaun Region*; Ministry of Environment: Wonju, Korea, 2016.

32. Cho, K.H.; Kang, J.H.; Ki, S.J.; Park, Y.; Cha, S.M.; Kim, J.H. Determination of the optimal parameters in regression models for the prediction of chlorophyll-a: A case study of the Yeongsan Reservoir, Korea. *Sci. Total Environ.* **2009**, *407*, 2536–2545. [CrossRef] [PubMed]

33. Gonzalez, R.A.; Noble, R.T. Comparisons of statistical models to predict fecal indicator bacteria concentrations enumerated by qPCR- and culture-based methods. *Water Res.* **2014**, *48*, 296–305. [CrossRef] [PubMed]

34. Moriasi, D.N.; Arnold, J.G.; Van Liew, M.W.; Bingner, R.L.; Harmel, R.D.; Veith, T.L. Model evaluation guidelines for systematic quantification of accuracy in watershed simulations. *Trans. ASABE* **2007**, *50*, 885–900. [CrossRef]

35. Chong, A.; Lam, K.P.; Pozzi, M.; Yang, J. Bayesian calibration of building energy models with large datasets. *Energy Build.* **2017**, *154*, 343–355. [CrossRef]

36. Hwang, S.H.; Ham, D.H.; Kim, J.H. A new measure for assessing the efficiency of hydrological data-driven forecasting models. *Hydrol. Sci. J.* **2012**, *57*, 1257–1274. [CrossRef]

 *water*

Article

Analysis of Natural Streamflow Variation and Its Influential Factors on the Yellow River from 1957 to 2010

Jie Wu [1,2], Zhihui Wang [2,3,]*, Zengchuan Dong [1], Qiuhong Tang [3] , Xizhi Lv [2] and Guotao Dong [2]

[1] Hydrology and Water Resources Institute, Hohai University, Nanjing 210000, China; 18251825285@126.com (J.W.); zcdong@hhu.edu.cn (Z.D.)
[2] Yellow River Institute of Hydraulic Research, Yellow River Conservancy Commission, Zhengzhou 450003, China; lvxizhi@hky.yrcc.gov.cn (X.L.); dongguotao@hky.yrcc.gov.cn (G.D.)
[3] Institute of Geographic Sciences and Natural Resources Research, Chinese Academy of Sciences, Beijing 100101, China; tangqh@igsnrr.ac.cn
* Correspondence: wangzhihui@hky.yrcc.gov.cn; Tel.: +86-371-66025361

Received: 10 July 2018; Accepted: 24 August 2018; Published: 29 August 2018

Abstract: In this study, variation characteristics of hydrometeorological factors were explored based on observed time-series data between 1957 and 2010 in four subregions of the Yellow River Basin. For each region, precipitation–streamflow models at annual and flood-season scales were developed to quantify the impact of annual precipitation, temperature, percentage of flood-season precipitation, and anthropogenic interference. The sensitivities of annual streamflow to these three climatic factors were then calculated using a modified elasticity coefficient model. The results presented the following: (1) Annual streamflow exhibited a negative trend in all regions; (2) The reduction of annual streamflow was mainly caused by a precipitation decrease and temperature increase for all regions before 2000, whereas the contribution of anthropogenic interference increased significantly—more than 45%, except for Tang-Tou region after 2000. The percentage of flood-season precipitation variation can also be responsible for annual streamflow reduction with a range of 7.36% (Tang-Tou) to 21.88% (Source); (3) Annual streamflow was more sensitive to annual precipitation than temperature in the humid region, and the opposite situation was observed in the arid region. The sensitivities to intra-annual climate variation increased after 2000 for all regions, and the increase was more significant in Tou-Long and Long-Hua regions.

Keywords: intra-annual climate change; variation in percentage of flood-season precipitation; natural streamflow variation; contribution and sensitivity analysis; Yellow River

---

1. Introduction

A number of studies have reported streamflow reduction in several rivers throughout the world [1–5], putting enormous stress on ecological and socioeconomic systems. This is especially stressful for semiarid and semihumid regions, where the hydrological cycle and water yield will be more vulnerable to climate change and anthropogenic interference [6]. Climatic changes include temperature changes and the redistribution of precipitation, which together affect streamflow discharge [7]. Anthropogenic interference mainly consists of land use/cover change (LUCC), urbanized and industrialized extension, and hydropower development and irrigation intensification, which greatly alter the underlying surface and water resource reapportionment [8]. Quantification of streamflow changes and identification of the various contributing factors are of considerable importance for a better understanding of the hydrologic mechanisms, which is beneficial for planning suitable adaptation strategies and water management.

There are various methods to separate the impacts of climate change and anthropogenic interference on streamflow, mainly including catchment experiments, hydrological models, and statistical methods [9]. Catchment experiments are the most rigorous empirical research design for estimating the effects of land use on aquatic systems [10], but they can be influenced by the variation in experimental conditions and the presentation of results [11]. Most relevant studies indicate that catchment streamflow decreased significantly after afforestation and increased after deforestation [10,12,13]. Hydrological models, both distributed and lumped, have been widely used [7,14–16]. Hu et al. applied the water and energy budget-based distributed hydrological model (WEB-DHM) to diagnose and quantify climate and human impacts on streamflow change [17]. Hundecha et al. applied a conceptual rainfall–runoff model to 95 catchments in the Rhine basin to model the effect of land use change on runoff [18]. Statistical methods such as streamflow elasticity have also been used in regions specifically with available long-term climate and hydrologic data [9,19,20]. Tian et al. used regression analysis to illustrate runoff decline via comparison of precipitation–runoff correlation for the period prior to and after sharp runoff decline [21].

The semiarid and arid Yellow River Basin (YRB) is the main source of surface water in the northwest and northern part of China. The annual streamflow is about 58 billion m^3, and the water resource per capita is 905 m^3—only a third of the national average, which poses a threat to the YRB's water resources availability. The climbing industry, agriculture, and household demand for water induced by rapid economic development and expanding urbanization is also a challenge [22]. In addition, some ecological programs launched by the Chinese government since 1999 have greatly altered the regional water cycle, including the Natural Forest Conservation Program (NFCP) and Grain for Green Project (GFGP) (http://tghl.forestry.gov.cn/) [6,23], and therefore, the basin is very sensitive to climate change and anthropogenic interference. Attempts have been made to understand the long-term streamflow variation and the sensitivity of streamflow to climate change in the Yellow River Basin. Tang et al. used a distributed biosphere hydrological (DHB) model system to simulate hydroclimate connections in the Yellow River Basin and found that climate change dominated the predicted changes in the upper and middle reaches, but anthropogenic interference dominated the lower reaches [24]. Liu et al. found that streamflow was more sensitive to precipitation in humid regions or wet years than in arid regions or dry years by means of streamflow elasticity [25]. Li et al. investigated the changing properties and underlying causes for decreased streamflow by both the Budyko framework and hydrological modeling techniques [26]. However, most of these previous studies focused on the entire basin or a local scale of catchments instead of comparing different subregions, let alone the comparison before and after the implementation of Natural Forest Conservation Program and Grain for Green Project. Moreover, few studies have paid attention to the contribution made by variations in the intra-annual distribution of precipitation, with only the annual precipitation considered.

The objectives of this paper are as follows: (1) to explore the spatial–temporal variation of annual precipitation, average temperature, the percentage of flood-season precipitation and natural streamflow in different subregions of YRB; (2) to quantitatively analyze the spatial–temporal characteristics of the contribution made by different meteorological factors and anthropogenic interference to streamflow changes in different subregions; (3) to analyze the spatial–temporal characteristics of the sensitivity of annual streamflow to various meteorological factors.

2. Study Area and Data

2.1. Study Area

The Yellow River (Figure 1) originates in the Qinghai Province of China and flows into Bohai Bay, forming the Yellow River Basin, which covers a total watershed area of 795,000 km^2 (including endoreic inter flow area). The main stream is 5464 km long with a slope of 4480 m. It can be divided into three parts. The upper reach travels 3472 km and drains 428,000 km^2 of land. The middle reach flows for 1206 km, with a drainage area of 344,000 km^2. When the middle reach flows through the Loess Plateau, the tributaries transport vast amounts of sediment, proclaiming the Yellow River as having the highest sediment content in the world. The remaining down reach has a length of 786 km and a drainage area of 23,000 km^2. The climatic and hydrologic conditions of the YRB are complex because of the large geographical extent and elevation difference. The precipitation exhibits high spatial and temporal variabilities: the ratio of rainfall between the North and South is greater than 5, 70% of precipitation falls between June and September, and the variation coefficient (C_v) is between 0.15 and 0.4. Temperature disparity is one of the major climate features in the YRB, with an annual mean temperature fluctuating from -4 °C to 14 °C. Considering the critical role played by the Yellow River in regional water supply and the tremendous challenges posed by water shortages, an analysis of the variation and sensitivity of annual streamflow is both important and imperative.

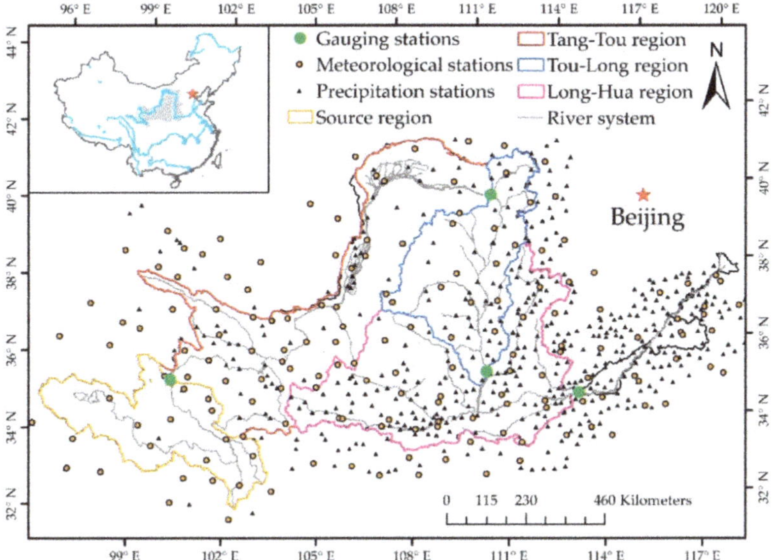

Figure 1. Location and national network of the meteorological and hydrological stations in the YRB.

2.2. Data Collection and Preprocessing

The datasets used in this study include climate, streamflow, leaf area index (LAI), and Digital Elevation Model (DEM) data.

Climate data were obtained from the China Meteorological Administration (CMA), including daily precipitation from 582 rainfall gauges and the daily mean, maximum, and minimum temperatures from 97 meteorological stations inside and near the Yellow River basin from 1957 to 2010.

The monthly naturalized streamflow time-series for four hydrological stations (Tangnaihai, Toudaoguai, Longmen, and Huayuankou) between 1957 and 2010 were obtained from the Yellow River Hydrographic Bureau (YRHB). These four-gauge stations were selected with the intent of determining streamflow changes in four different subregions. Specifically, streamflow at Tangnaihai was considered the source region; the streamflow from the upper reach was the difference between Toudaoguai and Tangnaihai; the difference between Huayuankou and Toudaoguai was declared as the middle reach streamflow. Due to the complex hydrogeological conditions in middle reach (Loess Plateau), Longmen station was added to separate the middle reach into two detailed parts. Four regions were thus formed:

the source region, Tang-Tou region, Tou-Long region, and Long-Hua region. Particularly, this dataset was the naturalized streamflow, having removed the variation caused by artificial water intake and reservoir storage and streamflow. That is to say, different from the broad sense, the anthropogenic interference defined in this study mainly included soil and water conservation measures.

GLASS LAI, one of the five typical global LAI products, was chosen for this study because it includes the longest duration (1982–2013) LAI product. Additionally, compared with those of the current MODIS and CYCLOPES LAI products, it provides temporally continuous LAI profiles with much better quality and accuracy [27].

Furthermore, a 30 × 30 m digital elevation model (DEM) was used for the interpolation of climatic variables. ANUSPLIN, a well-performed spatial interpolation package based on thin-plate smooth-spline interpolation, was selected to interpolate climatic variables. Developed by Australian National University, it is a tool mainly used for the transparent analysis and interpolation of noisy multi-variants data [28]. Using the longitude, latitude, and elevation of the meteorological stations as variables, daily precipitation and mean temperature datasets were aggregated to obtain mean monthly and annual values in four different subregions.

3. Methodology

3.1. Time-Series Analysis Method

3.1.1. Change-Point Detection and Trend Analysis of Hydrological and Climate Data

Both the change-point detection of annual streamflow data and trend analysis of hydrometeorological data were conducted by a Mann–Kendall (MK) test, which is widely used for its simplicity, robustness and the ability to deal with non-normal and missing data distributions [29,30]. After estimating the test statistics UF_i and UB_i, the curve of these two test statistics are plotted. If a match point of the two curves exists and the trend is statistically significant, the match point can be regarded as a change-point of the time series [17]. In terms of trend analysis, the MK test statistic Z was calculated. A positive and negative Z value represent increasing and declining trends, respectively. The null hypothesis, H_0, states that there is no statistically significant trend in the series for a given significance level α. In this paper, α was set to be 0.05 and the $1-\alpha/2$ quantile of the standard normal distribution for α ($Z_{(1-\alpha/2)}$) was 1.96. If $|Z| > Z_{(1-\alpha/2)}$, the null hypothesis is rejected, indicating the trend is significant. Otherwise, the H_0 hypothesis is accepted.

In addition, the precipitation–runoff double cumulative curve (DCC) was also used as an auxiliary confirmation of the change-points by providing a visual representation of the consistency of the precipitation and streamflow data [31].

3.1.2. Trend Analysis of LAI

The temporal and spatial variation of the mean LAI were analyzed using a linear regression analysis method in this study. Using overall LAI trend computations to identify spatial patterns of directions and rates of change, a least squares regression was fit through the time series of each pixel and the slope coefficient that represent trends was calculated [27]. The slope of the trend coefficient was defined as follows:

$$slope = \frac{n \times \sum_{i=1}^{n} i \times LAI_i - \sum_{i=1}^{n} i \sum_{i=1}^{n} LAI_i}{n \times \sum_{i=1}^{n} i^2 - (\sum_{i=1}^{n} i)^2} \tag{1}$$

where n is the cumulative number of years in the study periods, i is the order of year, and LAI_i is the value of LAI in the ith year. In general, if slope > 0, LAI will increase, suggesting better vegetation in this pixel.

3.2. Multitemproal-Scale Precipitation–Runoff Model

Since access to very limited information and data for basin geometry can hardly satisfy the minimal requirements of basin-scale models, statistical methods were employed to determine the relationship of streamflow and other climatic factors for the baseline period. Both an annual scale model and flood and nonflood season model were built.

At annual scale, the precipitation–runoff model was built using multiple linear regression analysis. At flood and nonflood season scale, given the fact that linear regression analysis method may not satisfy the requirement of model accuracy, a statistical model based on the Random Forest (RF) regression, which is one of the most effective machine learning models for predictive analytical approaches [32,33], was trained to reconstruct streamflow data in the human-affected period. RFs were developed as a method of improving the predictions of classification and regression trees by alleviating the overfitting concern of regression trees [34]. It has proved to be more robust and accurate than traditional linear (e.g., multiple linear regression) or more complex methods [35]. Two parameters need to be set in order to produce the forest trees: the number of decision trees to be generated (N_{tree}) and the number of variables to be selected and tested for the best split when growing the trees (M_{try}) [36]. In this paper, N_{tree} was set as 200, and M_{try} was set as the default value in the R package for random forests.

In this study, the correlation of different climatic variables (annual precipitation, mean temperature, precipitation in the former years, flood-season precipitation and mean temperature, nonflood season precipitation and mean temperature, precipitation of the last month of the flood season) between annual streamflow, flood-season streamflow and nonflood season streamflow were analyzed using Pearson correlation coefficient analysis respectively, and those with a high correction coefficient between streamflow were chosen as the independent variable [36] for developing annual, flood and nonflood seasonal precipitation–runoff models. Furthermore, variance analysis and an F-test were conducted to test the accountability of the statistical models.

3.3. Contribution Calculation of Climatic and Anthropogenic Factors on Annual Streamflow

Model simulation, along with the hypothesis that climate fluctuations and anthropogenic interference are independent, was employed to separate the impacts on streamflow variation. Several scenarios were designed to reconstruct natural streamflow and then separate the impact of climatic fluctuations and anthropogenic interference on natural streamflow:

S1: Conducting the control simulation based on the annual precipitation–runoff model with observed changes in precipitation and temperature over the human-affected period;

S2: Using the same forcing data as the control simulation S1, except the mean value of the temperature was fixed to the mean of the baseline period;

S3: Conducting the control simulation based on the flood and nonflood season precipitation–runoff model with observed changes in precipitation and temperature over the human-affected period;

S4: Using the same forcing data as the control simulation S3, except the mean of the percentage of flood-season precipitation was fixed at the level of the baseline period and the annual precipitation remained as the S3 observations.

The total streamflow change (ΔR_{total}) can be obtained by the difference between the observed streamflow in baseline period (R_{ob}) and that in human-affected period (R_{oh}), which can be expressed as:

$$\Delta R_{total} = R_{oh} - R_{ob} = \Delta R_C + \Delta R_H = \Delta R_P + \Delta R_T + \Delta R_H \tag{2}$$

where ΔR_{total} includes two main parts, the streamflow change caused by climate fluctuations ΔR_C and anthropogenic interference ΔR_H, and the former ΔR_C is made up of precipitation-induced change ΔR_P and temperature-induced variation ΔR_T.

The difference between S1 and S2 (S1–S2) was used to estimate the change magnitude of the simulated annual streamflow caused by the temperature variation ΔR_T:

$$\Delta R_T = R_{S1} - R_{S2} \tag{3}$$

where R_{S1} is the mean of simulated annual natural streamflow in the scenario S1, and R_{S2} is the mean of simulated annual natural streamflow for the scenario S2.

ΔR_C can be calculated by the following equation:

$$\Delta R_C = R_{S1} - R_{ob} \tag{4}$$

The streamflow change magnitudes caused by anthropogenic factor (ΔR_H) and annual precipitation variation (ΔR_P) are calculated using Equation (2).

The contribution rate of each factor, which is defined as η_k, is quantitatively estimated by:

$$\eta_k = \frac{\Delta R_k}{\Delta R_{total}} \times 100\% \tag{5}$$

where k can be referred to as precipitation (P), temperature (T), and anthropogenic interference (H).

The difference between S3 and S4 (S3–S4) was used to estimate the change magnitude of the simulated annual streamflow caused by the variation of the percentage of flood-season precipitation:

$$\Delta R_{P\_dis} = R_{S3} - R_{S4} \tag{6}$$

$$\eta_{P\_dis} = \frac{\Delta R_{P\_dis}}{\Delta R_{total}} \times 100\% \tag{7}$$

where R_{S3} is the mean of simulated annual natural streamflow in S3, and R_{S4} is the mean of simulated annual natural streamflow for S4. ΔR_{P_dis} is the change magnitude of natural streamflow caused by the variation of the percentage of flood-season precipitation.

3.4. Sensitivity Calculation of Annual Streamflow to Climatic Factors

Contribution assessment alone cannot fully explain the response of streamflow to different variables. For example, certain variable contributions may be greater because of the larger change magnitude of this variable. Therefore, to better understand the streamflow response to climatic factor changes in different regions and periods, a modified sensitivity coefficient was defined that reflects sensitivity of streamflow to various climatic variables: The formula uses simulated streamflow data in different scenarios and observed meteorological factor data to calculate the sensitivity of streamflow to different meteorological factors. The specific calculation formulas are as follows:

$$\frac{R_{S1i} - R_{ob}}{R_{ob}} = f'_P \times \frac{P_i - P_{ob}}{P_{ob}} + f'_T \times \frac{T_i - T_{ob}}{T_{ob}} \tag{8}$$

$$\frac{R_{S3i} - R_{S4}}{R_{S4}} = f'_{P\_dis} \times \frac{\gamma_i - \gamma_{ob}}{\gamma_{ob}} \tag{9}$$

where R_{ob}, P_{ob}, T_{ob}, γ_{ob} are the means of annual natural streamflow, precipitation, temperature, and percentage of flood-season precipitation over the baseline period, respectively. R_{S1_i}, R_{S3_i}, R_{S4_i} are the simulated natural streamflow in the ith year in S1, S3, and S4, respectively. R_{S4} is the mean of simulated annual natural streamflow for S4. P_i, T_i, γ_i are the observed annual precipitation, temperature, and percentage of flood-season precipitation over the human-affected period, respectively.

4. Results

4.1. Spatial–Temporal Variation Characters for Hydrometeorological Variables

4.1.1. Change-Point Detection

The change-point detection of annual streamflow was mainly conducted using MK mutation analysis, combined with the auxiliary annual precipitation–streamflow double cumulative curve. Figure 2a,b demonstrate a change in the relationship between annual precipitation and streamflow in the Long-Hua region in 1989. Consequently, the study period in Long-Hua region was separated into two parts: the baseline period (1957–1989) and human-affected period (1990–2010). The change-points of the source region, Tang–Tou region, and Tou-Long region were 1989, 1991, and 1982, respectively.

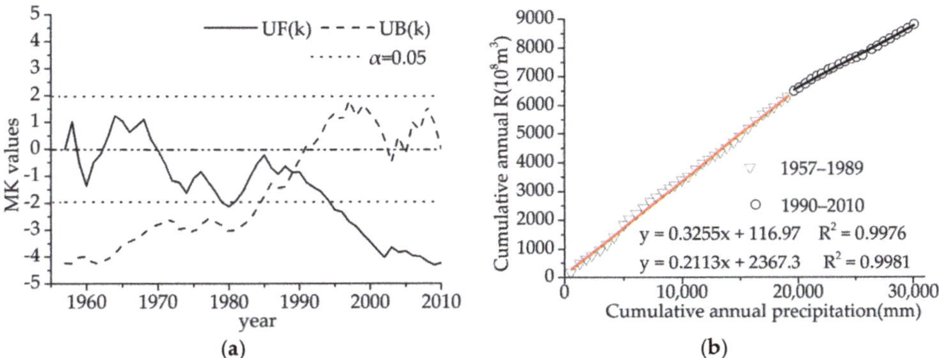

Figure 2. Mann–Kendall test values (**a**) and precipitation–streamflow double cumulative curve (**b**) for change-point detection in Long-Hua region of YRB (1957–2010).

Particularly, the Chinese government launched the Grain for Green Project (GFGP) in 1999. Since then, the land cover and vegetation in the middle reaches of the Yellow River have undergone drastic changes, which may also affect streamflow. In addition, it is found that there is a significant change point in 2000 in Tou-Long region by DCC in Figure 3, indicating that the impacts of human activities became more prominent in the Yellow River Basin, especially in the Loess Plateau after 2000. Therefore, the year 2000 was added to further divide the human-affected period into two parts: from the change-point year to 2000 (Period I) and from 2001 to 2010 (Period II), with the intent of analyzing how much non-meteorological factors have affected the streamflow after 2000 when land cover and vegetation change became more intense.

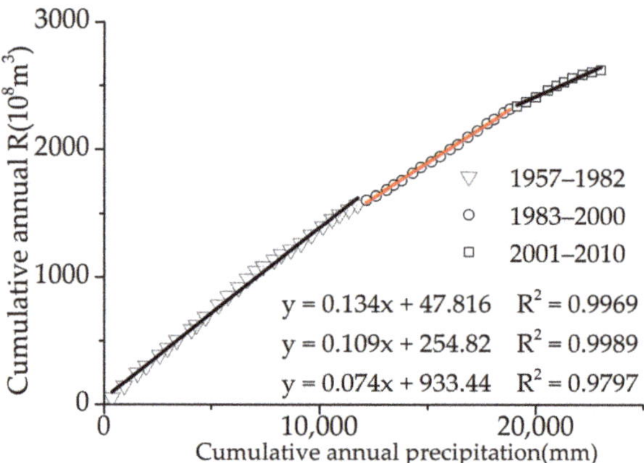

Figure 3. Precipitation–streamflow double cumulative curve for change-point detection in Tou-long region of YRB (1957–2010).

4.1.2. Trend Analysis of Annual Precipitation, Mean Temperature, and Naturalized Streamflow

Overall, a drying and warming trend was apparent in the YRB throughout the past 54 years. Figure 4 plots the annual time series of precipitation, mean temperature, naturalized streamflow, and their mean value before and after the change-point across the Long-Hua region. Both the annual precipitation and streamflow decreased, whereas mean temperature increased.

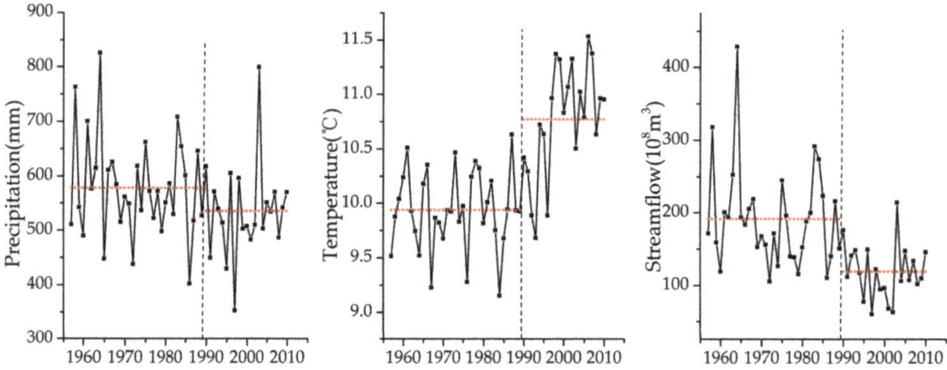

Figure 4. Long-term variations in annual precipitation (P), temperature (T), and streamflow (R) in the Long-Hua region in YRB.

The other three regions in the YRB also exhibited a similar trend. According to the MK analysis results in Figure 5, a decreasing trend of precipitation was detected in the YRB, excluding the source region, which had a positive MK value. As for the reduction rate, Table 1 shows that the precipitation of Tou-Long and Long-Hua in the middle reach reduced faster than that of Tang–Tou in the upper reach. However, none of the decreasing trends of precipitation were statistically significant. Conversely, the obvious warmer trend was statistically significant in the whole study area, with an average increasing rate of about 0.04 °C/a. A decreasing streamflow occurred in the whole basin, with the reduction rate ranging from 0.57×10^8 m^3/a to 2.21×10^8 m^3/a.

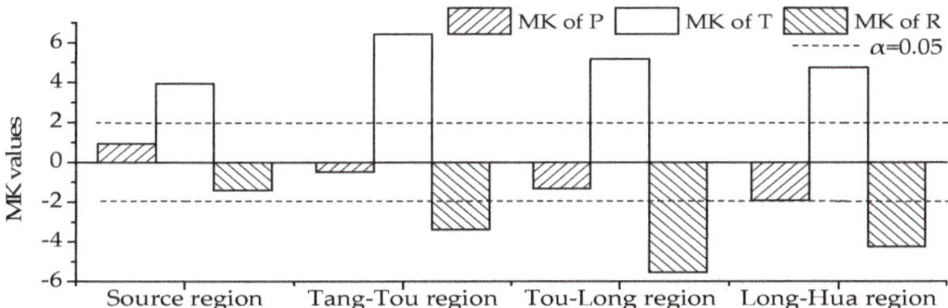

Figure 5. MK results of the trend analysis for annual precipitation, mean temperature, and streamflow in four regions of YRB.

Table 1. Average of annual precipitation, mean temperature, and streamflow and their change rate in four regions of YRB.

Sub Basin	Precipitation		Mean Temperature		Streamflow	
	$\overline{P}$ (mm)	Δ (mm/a)	$\overline{T}$ (°C)	Δ (°C/a)	$\overline{R}$ (10^8 m^3)	Δ (10^8 m^3/a)
Source	528.93	0.42	0.39	0.04	201.78	−0.57
Tang–Tou	342.27	−0.43	6.23	0.05	126.50	−1.03
Tou-Long	433.85	−1.39	7.93	0.04	49.02	−0.86
Long-Hua	560.92	−1.38	10.26	0.03	162.91	−2.21

4.1.3. Trend Analysis of Percentage of Flood-Season Precipitation

Table 2 presents the MK test results and the average change in the percentage of flood-season precipitation (γ). The overall declining trend of γ indicates that the intra-annual distribution of precipitation had changed. Spatially, the absolute value of MK decreased from the upper reach to the middle reach, with a significant trend in the source region (significance level = 0.05). For the entire human-affected period, γ dropped 4.63%, 1.18%, 6.89%, and 3.21%. It should be noted that γ increased by 2.70% during period I in the Tang-Tou region.

Table 2. MK results and average change of the percentage of flood-season precipitation in four regions of YRB.

Sub Basin	M-K Test from 1957 to 2010		Average Change (%) Compared with Baseline Period		
	Z	H_0	Period I	Period II	Human-Affected Period
Source region	−2.69	R	−5.54	−3.52	−4.63
Tang-Tou region	−1.37	A	2.70	−5.06	−1.18
Tou-Long region	−1.27	A	−6.75	−7.18	−6.89
Long-Hua region	−0.81	A	−4.52	−1.61	−3.21

R: reject H_0; A: accept H_0.

4.2. Precipitation–Runoff Model Calibration and Validation

4.2.1. Annual Model

According to the correlation coefficients in Table 3, annual precipitation was positively related to annual streamflow, the largest coefficient among the influential factors, suggesting that annual precipitation dominated the annual streamflow change in the Yellow River Basin. To better reflect the condition of soil moisture content, precipitation from the former year was introduced as a factor [37] and was also positively related to annual streamflow, with a varying correlation of 0.03 to 0.27. In contrast, temperature was negatively related to annual streamflow. Among the four regions, the correlation coefficient between temperature and annual streamflow in the source region was far less than that of the other three regions, revealing a spatial difference. Thus, annual precipitation (P), precipitation of the former year (P_{-1}), and mean temperature (T) were considered the main factors in the construction of the annual-scale model.

Table 3. The correlation coefficient of each element to annual streamflow in four regions of the YRB.

Sub Basin	Basin Scope	P	P_{-1}	T
Source region	Source region	0.84	0.27	−0.13
Tang-Tou region	Upper	0.84	0.10	−0.53
Tou-Long region	Midstream	0.84	0.03	−0.35
Long-Hua region	Midstream	0.87	0.11	−0.45

Based on the correlation coefficients analysis results, a three-parameter linear regression model was built for each region. All models were calibrated in the period of 1957–1977 with climatic data and then validated in the period from 1978 to the change-point year. The observed annual streamflow and simulated streamflow during calibration and validation period in each region are plotted in Figure 6. The relative bias (BIAS), relative root-mean-square error (RRMSE), and Nash–Sutcliffe efficiency coefficient (NSE) are given in Table 4. All four models performed reasonably well: their NSE values were in the range of 0.75–0.89, and BIAS and RRMSE were within the range of 2.90% and 0.34, respectively. Moreover, nearly all the trends were captured, and all models passed the F-test. Overall, the model performance was acceptable within the study domain.

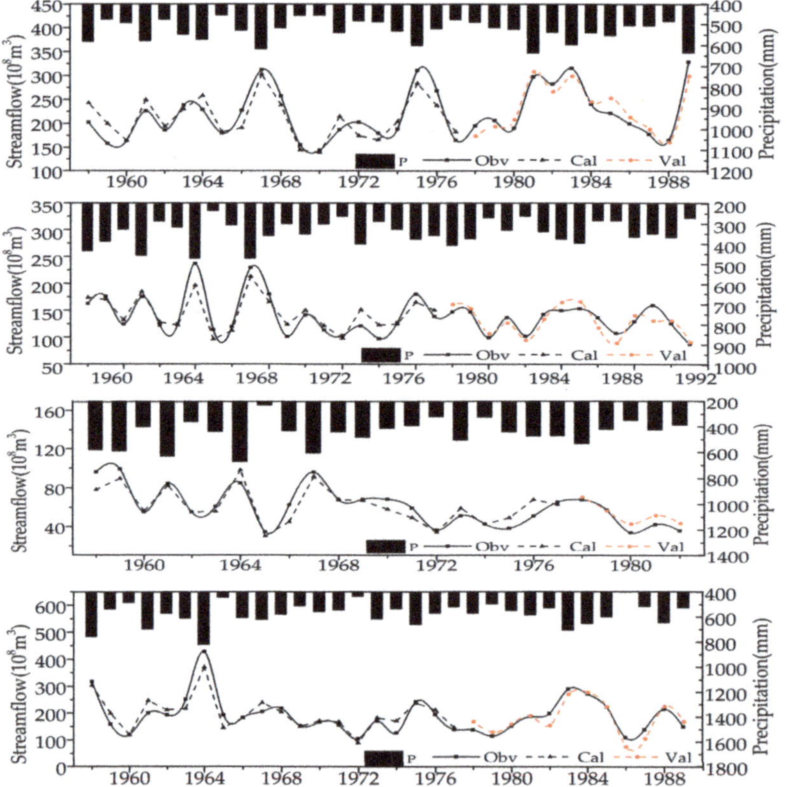

Figure 6. Observed and simulated annual streamflow in four regions during calibration and validation period.

Table 4. Calibration and validation of the annual, flood and nonflood season model in four subregions of the YRB.

Sub Basin	Period	Annual			Flood Season			Non-Flood Season		
		BIAS (%)	RRMSE	NSE	BIAS (%)	RRMSE	NSE	BIAS (%)	RRMSE	NSE
Source	Calibration	0.40	0.22	0.76	1.19	0.29	0.93	0.47	0.21	0.91
	Validation	0.41	0.22	0.89	2.60	0.25	0.92	0.38	0.22	0.90
Tang-Tou	Calibration	0.57	0.25	0.83	0.28	0.32	0.93	1.71	0.21	0.92
	Validation	1.05	0.18	0.81	0.23	0.28	0.94	2.68	0.15	0.83
Tou-Long	Calibration	2.90	0.30	0.78	1.53	0.49	0.94	2.09	0.17	0.87
	Validation	0.11	0.24	0.75	8.55	0.40	0.89	9.70	0.21	0.76
Long-Hua	Calibration	1.14	0.34	0.85	1.30	0.45	0.92	2.65	0.34	0.90
	Validation	2.22	0.32	0.83	5.24	0.40	0.93	2.49	0.24	0.91

4.2.2. Flood and Nonflood Seasonal Model

To further identify the contribution made by the variation of the flood-season precipitation percentage, a flood season model and nonflood season model were built, calibrated and validated after identifying the contributing factor by correlation coefficients analysis. Specifically, flood season precipitation and mean temperature were introduced to build the flood season model. In the construction of nonflood season model, in addition to nonflood season precipitation and average temperature, precipitation in October (the last month of the flood season in YRB) was also introduced as a factor [38]. The model performance in Table 4 and Figure 7, as well as the successful F-test, suggest that these models were reliable when reconstructing the natural streamflow, despite some peak differences.

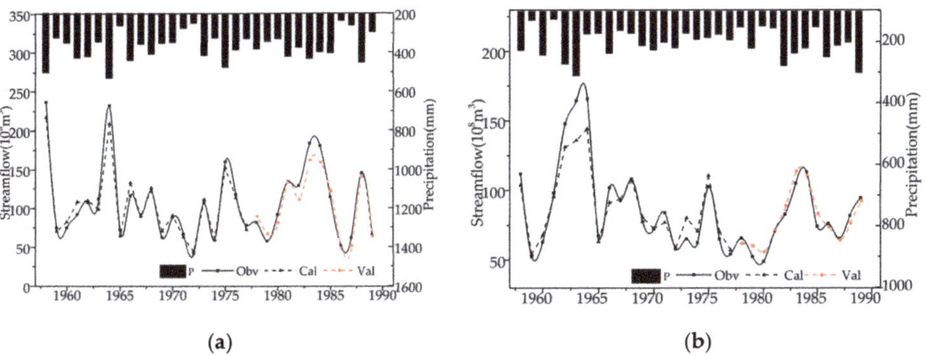

(a) (b)

Figure 7. Observed and simulated streamflow in Long-Hua region during the calibration and validation period in the flood season model (**a**) and nonflood season model (**b**).

4.3. Contribution Assessment

The quantitative assessment of the effects on streamflow due to climate fluctuations and anthropogenic interference was carried out in four regions of the YRB. It can be seen from Table 5 that a tremendous reduction had taken place in annual streamflow, up to 41.64×10^8 m^3/a, 31.31×10^8 m^3/a, 21.96×10^8 m^3/a, and 72.24×10^8 m^3/a, during the entire human-affected period. The reduction accounted for 19.00%, 22.77%, 36.35%, and 37.82%, from the source region to the middle reach. Both climate change and anthropogenic interference had a negative effect on annual streamflow in the YRB, except for the annual precipitation in the source region from 2001 to 2010, which contributed to a rise in annual streamflow of 20.72×10^8 m^3/a. The climate fluctuations and anthropogenic interference effects varied spatially and temporally. From the Tang-Tou to the Long-Hua regions, the negative effect of precipitation increased from 6.68% to 53.07%, while the temperature impact decreased from

78.10% to 17.55%. Anthropogenic interference had a greater contribution in the middle-reach Tou-Long and Long-Hua regions than in the upper Tang-Tou region. Temporally, the dominant contributor shifted from precipitation before 2000, to anthropogenic interference after 2000, in the Tou-Long and Long-Hua regions.

Table 5. Contribution of precipitation, temperature, and anthropogenic interference on annual streamflow variation in four regions of the YRB.

Sub Basin	Period	Total Change		Precipitation		Temperature		Anthropogenic Interference	
		ΔR_{total}	%	ΔR_P	%	ΔR_T	%	ΔR_H	%
Source region	1990–2000	−43.78	−19.98	−21.95	50.14	−3.73	8.52	−18.10	41.34
	2001–2010	−39.28	−17.92	20.72	−52.74	−18.89	48.08	−41.11	104.66
	1990–2010	−41.64	−19.00	−1.63	3.92	−10.95	26.29	−29.06	69.78
Tang-Tou region	1992–2000	−31.04	−22.57	−4.22	13.60	−19.46	62.67	−7.36	23.72
	2001–2010	−31.54	−22.94	−0.18	0.56	−28.94	91.77	−2.42	7.68
	1992–2010	−31.31	−22.77	−2.09	6.68	−24.45	78.10	−4.76	15.21
Tou-Long region	1983–2000	−17.74	−29.36	−9.28	52.34	−4.50	25.37	−3.95	22.29
	2001–2010	−29.55	−48.93	−5.36	18.14	−10.52	35.60	−13.67	46.26
	1983–2010	−21.96	−36.35	−7.88	35.90	−6.65	30.29	−7.42	33.82
Long-Hua region	1990–2000	−73.21	−38.33	−53.23	72.71	−9.22	12.59	−10.76	14.70
	2001–2010	−71.17	−37.26	−21.96	30.86	−16.48	23.16	−32.73	45.98
	1990–2010	−72.24	−37.82	−38.34	53.07	−12.68	17.55	−21.22	29.38

It is obvious that the combination of decreased precipitation and increased mean temperature caused the annual streamflow reduction in the YRB. Furthermore, the change in the percentage of flood-season precipitation also affected streamflow variation. Table 6 shows that the flood-season precipitation percentage variation mainly caused the streamflow reduction, up to 9.45×10^8 m^3/a, 2.32×10^8 m^3/a, 2.01×10^8 m^3/a, and 14.81×10^8 m^3/a—comprising 21.88%, 7.36%, 10.28%, and 18.24%, respectively, of the total streamflow variation. The effects in the Tang-Tou region in Period I were an exception, causing a 0.70×10^8 m^3/a rise in streamflow. With the variation of the flood-season precipitation percentage (Table 2) taken into consideration, a decline in the percentage of flood-season precipitation led to a corresponding drop in streamflow. In contrast, a rise of γ resulted in an increase in streamflow in the Tang-Tou region during Period I. This indicated that the greater the percentage of precipitation in flood season, the greater the simulated streamflow in YRB, assuming the same annual precipitation.

Table 6. Contribution of the variation of flood-season precipitation percentage in four regions of the YRB.

Sub Basin	Period I		Period II		Human-Affected Period	
	ΔR	%	ΔR	%	ΔR	%
Source region	−10.04	22.08	−8.72	21.60	−9.45	21.88
Tang-Tou region	0.70	−2.11	−4.00	12.44	−2.32	7.36
Tou-Long region	−1.87	12.61	−2.50	8.56	−2.01	10.28
Long-Hua region	−13.40	16.86	−14.92	20.04	−14.81	18.24

Specifically, the contribution made by the variation in the percentage of flood-season precipitation was relatively higher in the source and Long-Hua regions. Furthermore, the contribution lowered after 2000 in source and Tou-Long region. To conclude, the streamflow was affected not only by the amount of annual precipitation but also the intra-annual distribution of precipitation.

4.4. Sensitivity Assessment

In this study, Equations (8) and (9) were used to calculate the sensitivity of streamflow to different climatic factors in four regions and the results were presented in Table 7. In general, the absolute value of the sensitivity coefficient to precipitation was larger than that to mean temperature in the humid regions including Source and Long-Hua regions, whereas an opposite situation was observed in the relatively arid regions including Tang-Tou and Tou-Long regions, indicating that the sensitivity of streamflow to various climatic factors are different for regions with different hydrothermal conditions in YRB.

Table 7. Sensitivity of streamflow to various elements in four regions.

Sub-District	P		T		γ	
	Period I	Period II	Period I	Period II	Period I	Period II
Source region	1.73	2.09	−0.01	−0.02	0.63	0.78
Tang-Tou region	0.99	1.06	−0.99	−1.07	0.34	0.36
Tou-Long region	1.09	1.23	−1.16	−1.54	0.19	0.42
Long-Hua region	1.64	1.88	−1.12	−1.15	0.37	0.80

In Period I, regarding the four regions of YRB, a 1% increase in annual precipitation would generate a 0.99–1.73% (1.36% on average) annual streamflow increase, whereas a 1% increase in temperature would produce a 0.01–1.16% (0.79% on average) decrease in annual streamflow. In addition, the sensitivity to the percentage of flood-season precipitation ranged from 0.19 to 0.63 in Period I. It should also be noted that the sensitivity of streamflow to various factors was not a constant. The sensitivity to precipitation and temperature increased considerably in Period II compared to that of Period I, suggesting that streamflow would be more sensitive to climate change. The sensitivities to intra-annual climate variation increased after 2000 as well, and the increase in sensitivity to percentage of flood-season precipitation was more significant in Tou-Long and Long-Hua regions.

5. Discussion

5.1. Analysis of the Impact of Anthropogenic Interference on Natural Streamflow

Overall, the findings of this study agreed with the results of other research in the contribution assessment [8,29,32,39]. Table 5 shows that anthropogenic interference had a greater contribution in Tou-Long and Long-Hua regions after 2000, which was mainly due to the ecological program launched by the Chinese government in these two regions. Table 8 displays the implementation of different soil and water conservation measures in the Tou-Long region (including 25 tributaries) in 1997, 2000, 2003, and 2006 [40]. According to the table, the amount of all kinds of soil and water conservation had risen gradually. Among them, forestland and grassland increased more sharply. By 2006, 28,540 km^2 of terraced area, 1310.26 km^2 of dammed land, 58,613.53 km^2 of forest land, 14,072.64 km^2 of grassed land, and 8380.18 km^2 of closed hillside area had been constructed in the Tou-Long region. Moreover, according to the analysis of LAI data before and after 2000 in Figure 8, the change intensity in forest and grass vegetation was largest in the Tou-Long region, followed by Long-Hua.

Table 8. Implementation of different soil and water conservation in the Tou-Long region in 1997, 2000, 2003, and 2006 (unit: km^2).

Year	Terraced Area	Dammed Land	Forest Land	Grassland	Hillsides Closed for Erosion Control	Total
1997	19,977.25	968.77	27,176.62	6939.60	2601.75	55,062.24
2000	23,291.18	1085.55	36,920.30	8929.06	3346.84	70,226.09
2003	26,166.33	1213.42	49,011.68	11,738.26	5528.36	88,129.69
2006	28,540.36	1310.26	58,613.53	14,072.64	8380.18	102,536.79

Studies have demonstrated that such large-scale land use and land-cover change, driven by soil and water conservation measures, were closely related to streamflow reduction [41,42]. For example, the terraced area can reduce the hillside slope and prolong streamflow retention, reducing surface streamflow [17]. The increasing forests and grasslands play an important role in intercepting rainfall and alleviating streamflow [43]. To conclude, all of these can explain why anthropogenic interference played a more critical role in streamflow reduction in the Tou-Long and Long-Hua regions, especially during Period II.

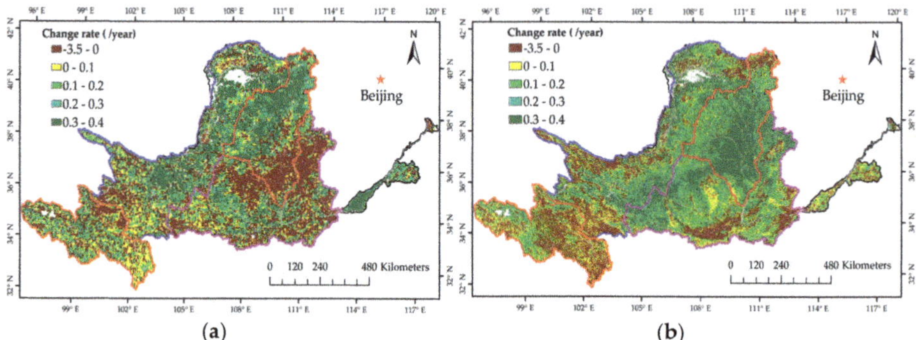

Figure 8. Overall change trend in annual LAI in 1982–2000 (**a**) and in 2001–2013 (**b**) in the YRB.

5.2. Analysis of the Sensitivity of Streamflow to Annual Precipitation and the Percentage of Flood-Season Precipitation

This study modified the traditional formula of the elasticity coefficient and used the calculated sensitivity coefficient to analyze the sensitivity of streamflow to meteorological factors. It should be pointed out that the advantage of this modified sensitivity coefficient calculation formula is that it can be combined with the runoff derived from scenario simulation. That is, the dynamic streamflow time series under the influence of the target factor (such as precipitation, mean temperature, the percentage of flood season precipitation, etc.) can be constructed by the scenario simulation method, and then the sensitivity of streamflow to the certain target factor can be directly calculated based on the sensitivity formula, and the underlying mechanics for calculating the sensitivity of streamflow to the specific factor is relatively easy to understand. In contrast, the traditional elastic coefficient model can only use the original observed streamflow, and the simulation accuracy of the elastic coefficient is affected by the type and number of factors selected in the construction of the elastic model, leading to the greater uncertainty in the sensitivity analysis of specific factors.

According to the results shown in Table 7, the sensitivity to annual precipitation exhibited both temporal and spatial differences. The annual streamflow after 2000 became more sensitive to annual precipitation in the whole basin. Chiew [44] reported a strong negative correlation between the elasticity coefficient to precipitation and streamflow coefficient (R_C) for 219 catchments in Australia. Inspired by Zheng et al. [39], the relationship between the sensitivity to annual precipitation (f'_p) and R_C, which were estimated within a moving window of 10 years, was analyzed in this study. The relationship in Figure 9a shows that the sensitivity to annual precipitation (f'_p) was positively related to the inverse of the runoff coefficient ($1/R_c$), indicating that streamflow was more sensitive to precipitation in catchments or periods with low streamflow coefficients. The declining trend of streamflow coefficients of four regions, illustrated in Figure 8b, successfully explained the rise in sensitivity to annual precipitation from period I to period II. This also shows that the sensitivity coefficient calculated in this study has similar properties to the traditional elastic coefficient and can appropriately reflect the sensitivity of streamflow to meteorological elements. Furthermore, part of the reason for a streamflow coefficient decline may be that these regions experienced a significant rise in forest and grass vegetation, as demonstrated in previous studies [42].

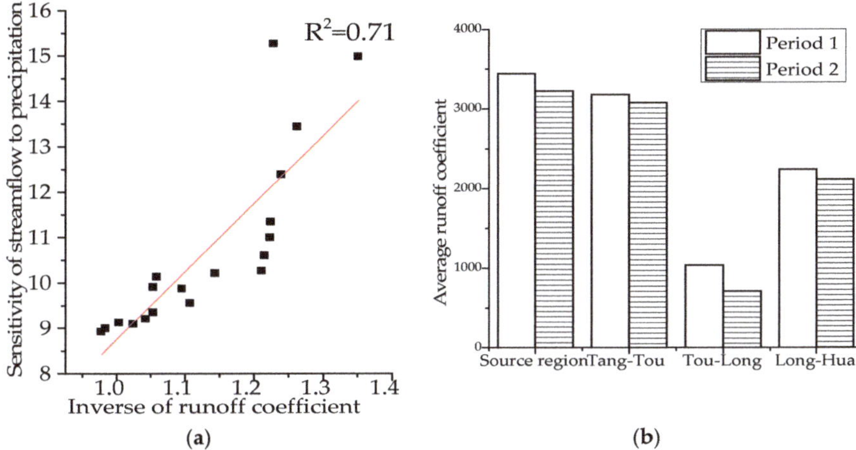

Figure 9. Correlation between sensitivity to precipitation and streamflow coefficient in the Tou-Long region (**a**) and the streamflow coefficients in different periods in four regions (**b**).

Table 7 also shows that streamflow in the source, Tou-Long, and Long-Hua regions was more sensitive to precipitation than that in the Tang-Tou region. This spatial difference partly accorded with their streamflow coefficient except for the source area, which was expected to have a similar sensitivity to that of Tang-Tou. In fact, the value of the sensitivity of streamflow depends on many factors, such as the stochastic nature of climate, vegetation conditions, field capacity of soils, soil moisture levels, length of soil water depletion, and saturated hydraulic conductivity [24]. Compared with other regions, the source region has a relatively saturated soil condition ascribed to good vegetation cover (Figure 8) and humid climate conditions (Table 1), making it easy to form streamflow. Thus, streamflow in the source region is more sensitive to precipitation change than that in other dry regions, such as the arid Tang-Tou region.

As for the sensitivity to the percentage of flood-season precipitation, it increased more in the Tou-Long and Long-Hua regions than in the source and Tang-Tou regions. Given the same annual precipitation, streamflow yield would not be affected by the precipitation temporal pattern in the regions with less forest and grass vegetation. However, as illustrated in Figure 8, forest and grass vegetation increased in the Tou-Long and Long-Hua regions, playing a crucial role in streamflow yield: streamflow would become more sensitive to flood-season precipitation, leading to greater sensitivity to the percentage of flood-season precipitation.

5.3. Uncertainties Analysis

There are some uncertainties associated with the contribution assessment and sensitivity analysis. First, combining two different-scaled models—namely, the annual model and flood and nonflood season model—generates some uncertainties. Another uncertainty in the results exists in the assumption that climate change is independent of anthropogenic interference. In fact, these two factors are interrelated. For example, land cover change and vegetation increase caused by afforestation would also lead to climatic changes.

Despite the uncertainties and limitations, this study provides a relatively easy way to analyze the contribution of climate variables and the sensitivity of streamflow to these factors, including the intra-annual distribution of precipitation. More detailed work, such as distributed models, should be introduced to improve the understanding of the monthly streamflow response mechanism.

6. Conclusions

With the intent of distinguishing the effect of climate fluctuations and anthropogenic interference on streamflow reduction and analyzing the sensitivity of streamflow, an improved three-parameter annual precipitation–streamflow model and flood and nonflood season models were built to simulate natural streamflow. The major findings from this study are summarized below.

The MK results demonstrated a decreasing trend in annual precipitation, significant increasing trend in mean temperature, and decreasing trend in annual streamflow across the Yellow River, excluding the increasing annual precipitation in the source region during 1957–2010. Abrupt change did not take place simultaneously, with the earliest change-point detected in the Tou-Long region. On average, the percentage of flood-season precipitation exhibited a decreasing trend, with each of the four regions experiencing a different level of decline.

The intensity of streamflow reduction improved spatially from 19.00% in the source region to 37.82% in Long-Hua. The contribution made by the climatic fluctuations and anthropogenic interference varied spatially and temporally. From Tang-Tou to Long-Hua, the impacts of annual precipitation and anthropogenic interference increased, while the temperature effect decreased. Temporally, the dominant factor in the Tou-Long and Long-hua regions had shifted from precipitation to anthropogenic interference after 2000. Further, the variation in the percentage of flood-season precipitation was responsible for streamflow variation. The greater the percentage of flood-season precipitation, the greater the simulated streamflow will be.

The sensitivity of streamflow to various climatic factors are different for regions with different hydrothermal conditions in YRB: annual streamflow was more sensitive to annual precipitation than temperature in the humid regions, whereas an opposite situation was observed in the relatively arid regions. Sensitivity to precipitation and temperature both increased in the whole basin after 2000, indicating that substantial challenges and uncertainties might be introduced to regional water availability. The sensitivity of streamflow to the percentage of flood-season precipitation increased most significantly in the Tou-Long and Long-Hua regions, where the highest change intensity of forest and grass vegetation occurred after 2000. These research conclusions can provide a scientific reference for future Yellow River water resource management and ecological construction planning.

Author Contributions: All authors contributed to the design and development of this manuscript. J.W. performed the data processing and wrote the first draft of the manuscript. Z.W. and Z.D. are the graduate supervisors of J.W. and gave constructive suggestions on the design and modification of the manuscript. Z.W. also helped process the LAI data. Q.T., X.L., and G.D. helped edit the manuscript prior to submission.

Funding: This research was funded by the National Key R&D Program of China (2016YFC0402402), China Postdoctoral Science Foundation (2017M610458), Young Elite Scientists Sponsorship by CAST (2017QNRC023), Foundation of development on science and technology by YRIHR (HKF201709) and National Natural Science Foundation of China (41701509, 51509102, 51779099).

Acknowledgments: We gratefully thank the anonymous reviewers for their critical comments and constructive suggestions on the manuscript.

Conflicts of Interest: The authors declare no conflicts of interest.

References

1. Razmara, P.; Bavani, A.R.M.; Motiee, H. Investigating uncertainty of climate change effect on entering runoff to Urmia Lake Iran. *Hydrol. Earth Syst. Sci. Discuss.* **2013**, *10*, 2183–2214. [CrossRef]
2. Stahl, K.; Hisdal, H.; Tallaksen, L. Trends in low flows and streamflow droughts across Europe. In Proceedings of the UNESCO, Paris, France, 14 March 2008.
3. Rood, S.B.; Samuelson, G.M.; Weber, J.K. Twentieth-century decline in streamflows from the hydrographic apex of North America. *J. Hydrol.* **2005**, *306*, 215–233. [CrossRef]
4. Xu, X.Y.; Yang, D.W.; Yang, H.B. Attribution analysis based on the Budyko hypothesis for detecting the dominant cause of runoff decline in Haihe basin. *J. Hydrol.* **2014**, *510*, 530–540. [CrossRef]

5. Hughes, J.D.; Petrone, K.C.; Silberstein, R.P. Drought, groundwater storage and stream flow decline in southwestern Australia. *Geophys. Res. Lett.* **2012**, *39*, L03408. [CrossRef]

6. Li, Y.Z.; Liu, C.M.; Zhang, D. Reduced Runoff Due to Anthropogenic Intervention in the Loess Plateau, China. *Water* **2016**, *8*, 458. [CrossRef]

7. Wang, D.; Hejazi, M. Quantifying the relative contribution of the climate and direct human impacts on mean annual streamflow in the contiguous United States. *Water Resour. Res.* **2011**, *47*, 411. [CrossRef]

8. Feng, A.Q.; Li, Y.Z.; Gao, J.B. The determinants of streamflow variability and variation in Three-River Source of China: Climate change or ecological restoration? *Environ. Earth Sci.* **2017**, *76*, 696. [CrossRef]

9. Chang, J.X.; Zhang, H.X.; Wang, Y.M.; Zhu, Y.L. Assessing the impact of climate variability and human activity to streamflow variation. *Hydrol. Earth Syst. Sci.* **2016**, *20*, 1547–1560. [CrossRef]

10. Tamai, K. A paired-catchment experiment in the Tatsunokuchiyama experimental forest, Japan: the influence of forest disturbance on water discharge. In *River Basin Management III*; WIT Press: Southampton, UK, 2005; Volume 2, pp. 173–182.

11. Bosch, J.M.; Hewlett, J.D. A review of catchment experiments to determine the effect of vegetation changes on water yield and evapotranspiration. *J. Hydrol.* **1982**, *55*, 3–23. [CrossRef]

12. Tuteja, N.K.; Vaze, J.; Teng, J.; Mutendeudzi, M. Partitioning the effects of pine plantations and climate variability on streamflow from a large catchment in southeastern Australia. *Water Resour. Res.* **2007**, *43*, W08415. [CrossRef]

13. Buendia, C.; Batalla, R.J.; Sabater, S. Runoff Trends Driven by Climate and Afforestation in a Pyrenean Basin. *Land Degrad. Dev.* **2016**, *27*, 823–838. [CrossRef]

14. Adam, J.C.; Haddeland, I.; Su, F. Simulation of reservoir influences on annual and seasonal streamflow changes for the Lena, Yenisei, and Ob' rivers. *J. Geophys. Res.-Atmos.* **2007**, *112*, D24114. [CrossRef]

15. Tesfa, T.K.; Li, H.Y.; Leung, L.R. A subbasin-based framework to represent land surface processes in an Earth system mode. *Geosci. Model Dev. Dis.* **2013**, *6*, 2699–2730.

16. Petchprayoon, P.; Blanken, P.D.; Ekkawatpanit, C. Hydrological impacts of land use/land cover change in a large river basin in central-northern Thailand. *Int. J. Clim.* **2010**, *30*, 1917–1930. [CrossRef]

17. Hu, Z.D.; Wang, L.; Wang, Z.J.; Yang, H.; Hang, Z. Quantitative assessment of climate and human impacts on surface water resources in a typical semiarid watershed in the middle reaches of the Yellow River from 1985 to 2006. *Int. J. Clim.* **2015**, *35*, 97–113. [CrossRef]

18. Hundecha, Y.; Bardossy, A. Modeling of the Effect of Land Use Changes on the Runoff Generation of a River Basin Through Parameter Regionalization of a Watershed Model. *J. Hydrol.* **2004**, *292*, 281–295. [CrossRef]

19. Ye, B.S.; Yang, D.Q.; Kane, D.L. Changes in Lena River streamflow hydrology: Human impacts versus natural variations. *Water Resour. Res.* **2003**, *39*, SWC14. [CrossRef]

20. Liu, X.M.; Liu, C.M.; Luo, Y.Z. Dramatic decrease in streamflow from the headwater source in the central route of China's water diversion project: Climatic variation or human influence? *Geophys. Res.-Atmos.* **2012**, *117*, D06113–D06122. [CrossRef]

21. Tian, F.; Yang, Y.H.; Han, S.M. Using runoff slope-break to determine dominate factors of runoff decline in Hutuo River Basin, North China. *Water Sci. Technol.* **2009**, *60*, 2135–2144. [CrossRef] [PubMed]

22. Zhang, L.; Yang, X. Applying a Multi-Model Ensemble Method for Long-Term streamflow Prediction under Climate Change Scenarios for the Yellow River Basin, China. *Water* **2018**, *10*, 301. [CrossRef]

23. Zhang, P.C.; Shao, G.F.; Zhao, G.; Master, D.C.L.; Parker, G.R.; Dunning, J.B.J.; Li, Q. China's forest policy for the 21st century. *Science* **2000**, *288*, 2135–2136. [CrossRef] [PubMed]

24. Tang, Q.H.; Oki, T.; Kanae, S. Hydrological Cycles Change in the Yellow River Basin during the Last Half of the Twentieth Century. *J. Clim.* **2008**, *21*, 1790–1806. [CrossRef]

25. Liu, Q.H.; Cui, B. Impacts of climate change/variability on the streamflow in the Yellow River Basin, China. *Ecol. Model.* **2011**, *222*, 268–274. [CrossRef]

26. Li, B.; Li, C.Y.; Liu, J.Y.; Zhang, Q.; Duan, L.M. Decreased Streamflow in the Yellow River Basin, China: Climate Change or Human-Induced? *Water* **2017**, *9*, 116. [CrossRef]

27. Guli, J.; Liang, S.L.; Yi, Q.X. Vegetation dynamics and responses to recent climate change in Xinjiang using leaf area index as an indicator. *Ecol. Indic.* **2015**, *58*, 64–76.

28. Shu, S.J.; Liu, C.S.; Shi, R.H.; Gao, W. Research on spatial interpolation of meteorological elements in Anhui Province based on ANUSPLIN. In Proceedings of the SPIE 8156, Remote Sensing and Modeling of Ecosystems for Sustainability VIII, San Diego, CA, USA, 22–23 August 2011.

29. Mann, H.B. Non-parametric tests against trend. *Econ. J. Econ. Soc.* **1945**, *13*, 245–259.
30. Kendall, M.G. *Rank Correlation Methods*; Griffin: London, UK, 1975.
31. Matouškov, M.; Kliment, Z. Runoff changes in the Šumava Mountains (black forest) and the foothill regions: Extent of influence by human impact and climate change. *Water Resour. Manag.* **2009**, *23*, 1813–1834.
32. Breiman, L. Random forests. *Mach. Learn.* **2001**, *45*, 5–32. [CrossRef]
33. Liu, X.H.; Guanter, L.; Liu, L.Y. Downscaling of solar-induced chlorophyll fluorescence from canopy level to photosystem level using a random forest model. *Remote Sens. Environ.* **2018**. [CrossRef]
34. Carlisle, D.M.; Falcone, J.; Wolock, D.M. Predicting the natural flow regime: Models for assessing hydrological alteration in streams. *River Res. Appl.* **2010**, *26*, 118–136. [CrossRef]
35. Belgiu, M.; Dragut, L. Random forest in remote sensing: A review of applications and future directions. *ISPRS J. Photogramm.* **2016**, *114*, 24–31. [CrossRef]
36. He, Y.H.; Lin, K.R.; Chen, X.H. Effect of Land Use and Climate Change on Runoff in the Dongjiang Basin of South China. *Math. Probl. Eng.* **2013**, *2013*, 14–26. [CrossRef]
37. Xu, X.Y. Hydrological Response of Typical Watersheds under Climate Change. Ph.D. Thesis, Tsinghua University, Beijing, China, 2012.
38. Bao, W.M. *Hydrological Forecast*, 4st ed.; China Water Resources and Hydropower Press: Beijing, China, 2009; pp. 250–257, ISBN 9787508461922. (In Chinese)
39. Zheng, H.X.; Lu, Z.; Zhu, R.R. Responses of streamflow to climate and land surface change in the headwaters of the Yellow River Basin. *Water Resour. Res.* **2009**, *45*, 641–648. [CrossRef]
40. Yao, W.Y.; Xu, J.H.; Ran, D.C. *Analysis and Evaluation of Water and Sediment Changes in the Yellow River Basin*, 1st ed.; Yellow River Conservancy Press: Zhengzhou, China, 2011; pp. 37–39, ISBN 9787550901414. (In Chinese)
41. Mu, X.; Gao, P.; Basangchilie. Application of flow duration curves to analyze the effect of water and soil conservation measures on river runoff in the Loess Plateau. *Adv. Earth Sci.* **2008**, *23*, 382–389. (In Chinese)
42. Liu, X.; Liu, C.; Yang, S. Effect of remote sensing on changes of forest and grass vegetation in the Loess Plateau to river runoff. *Acta Geogr. Sin.* **2014**, *69*, 1595–1603. (In Chinese)
43. Chen, L.D.; Wei, W.; Fu, B.J. Soil and Water Conservation on the Loess Plateau in China: Review and Perspective. *Prog. Phys Geogr.* **2007**, *31*, 3547–3554. [CrossRef]
44. Chiew, F.S. Estimation of rainfall elasticity of streamflow in Australia. *Int. Assoc. Sci. Hydrol. Bull.* **2006**, *51*, 613–625. [CrossRef]

Article

Four Decades of Estuarine Wetland Changes in the Yellow River Delta Based on Landsat Observations Between 1973 and 2013

Changming Zhu [1,2], Xin Zhang [2,*] and Qiaohua Huang [3]

[1] Department of Geography and Geomatics, Jiangsu Normal University, Xuzhou 221116, China;
    zhuchangming@jsnu.edu.cn
[2] State Key Laboratory of Remote Sensing Science, Institute of Remote Sensing and Digital Earth,
    Chinese Academy of Sciences, Beijing 100101, China
[3] Department of Geography and Geomatics, Jiangsu Normal University, Xuzhou 221116, China;
    huangqh@jsnu.edu.cn
* Correspondence: zhangxin@radi.ac.cn; Tel. +86-131-6137-3356

Received: 30 April 2018; Accepted: 3 July 2018; Published: 13 July 2018

Abstract: Yellow River Delta wetlands are essential for the migration of endangered birds and breeding. The wetlands, however, have been severely damaged during recent decades, partly due to the lack of wetland ecosystem protection by authorities. To have a better historical understanding of the spatio-temporal dynamics of the wetlands, this study aims to map and characterize patterns of the loss and degradation of wetlands in the Yellow River Delta using a time series of remotely sensed images (at nine points in time) based on object-based image analysis and knowledge transfer learning technology. Spatio-temporal analysis was conducted to document the long-term changes taking place in different wetlands over the four decades. The results showed that the Yellow River Delta wetlands have experienced significant changes between 1973 and 2013. The total area of wetlands has been reduced by 683.12 km^2 during the overall period and the trend of loss continues. However, the rates and trends of change for the different types of wetlands were not the same. The natural wetlands showed a statistically significant decrease in area during the overall period (36.04 km$^2 \cdot$year^{-1}). Meanwhile, the artificial wetlands had the opposite trend and showed a statistically significant increase in area during the past four decades (18.96 km$^2 \cdot$year^{-1}). According to the change characteristics revealed by the time series wetland classification maps, the evolution process of the Yellow River Delta wetlands could be divided into three stages: (1) From 1973–1984, basically stable, but with little increase; (2) from 1984–1995, rapid loss; and (3) from 1995–2013, slow loss. The area of the wetlands reached a low point around 1995, and then with a little improvement, the regional wetlands entered a slow loss stage. It is believed that interference by human activities (e.g., urban construction, cropland creation, and oil exploitation) was the main reason for wetland degradation in the Yellow River Delta over the past four decades. Climate change also has long-term impacts on regional wetlands. In addition, due to the special geographical environment, the hydrological and sediment conditions and the location of the Yellow River mouth also have a significant influence on the evolution process of the wetlands.

Keywords: Yellow River Delta; estuarine wetlands; spatiotemporal change analysis; remote sensing

---

1. Introduction

The Yellow River Delta wetlands are a critical ecological system and are a transitional buffer zone between the sea and inland areas [1–3]. They play a very important role because they affect nitrogen absorption, geochemical cycles, climate regulation, and act as a carbon sink; they also provide a breeding ground and habitat for birds and land for bioconservation [4,5]. This area is also a very important migratory route for nationally rare and endangered birds [6]. In 1992, the Chinese government created the largest national wetland nature reserve (Yellow River Delta National Nature Reserve, YRDNNR) as a stopover and breeding ground for globally endangered birds. In addition, the Yellow River Delta is also one of the biggest and fastest growing deltas in the world due to continuous sedimentation [7,8]. The average rate of newly created land is about 21.3 km^2 per year [8]. It is characterized by extensive coverage of saline and wet soils due to a high groundwater table and inundation by sea water [7]. Within the delta, there are a large number of shallow water areas with an abundance of wetland vegetation and aquatic biological resources. The area of the wetlands is 4167 km^2, and it occupies over 50% of the total area of the delta [9].

Globally, nearly half of all wetlands have been lost over the past century because of human disturbance and climate change [7,8,10–13]. China's wetlands have also suffered tremendous loss in recent decades [14,15]. These trends are still generally continuing and result in serious environmental problems [4,16–21]. Coastal wetland systems have, in particular, become one of the most vulnerable ecosystems around the world [4,16–21]. The Yellow River delta is not only a highly dynamic area, but it is also very important for ecological functions. Therefore, the conservation of wetland resources has attracted the attention of scientists and local authorities. Existing research has found that the degradation of the Yellow River Delta wetlands has triggered numerous environmental problems and seriously threatened sustainable economic development in the region [7,8,22,23]. In a study from the perspective of landscape ecology, using remote-sensing data from three different years (2000, 2005, 2010), Liu confirmed that the regional wetland landscape patches became more complex and decentralized due to rapid economic development [22]. However, studies examining long-term time series of change are lacking. The distribution of wetlands and their historical changes in this region are not very clear. Timely surveying and mapping of wetlands is a fundamental task for the research, management, and conversation of wetlands. So, it is imperative to examine the current status of wetlands, how it has changed, and trends in the delta.

However, because of the complexity and poor accessibility of wetland areas, traditional field surveys are limited by either spatial or temporal coverage. As a result, there is a lack of long-term time series of wetland maps and no systematic temporal analyses of wetlands. Until now, we have not had adequate knowledge of the characteristics of the majority of wetlands over the past few decades. Therefore, it is imperative to quantify the spatio-temporal characteristics and status-trends of regional wetlands and to determine the best ways to develop scientific strategies for wetland management, conservation, and restoration. With the development of remote-sensing technology, there are abundant multi-source satellite data (e.g., Landsat1~8, H-J1/2, MODIS) available online at no cost. These data may provide accurate, reliable, and economical means for a wetland inventory. Currently, more than 90% of wetland maps in China were obtained by means of remote-sensing surveys [24], and remote sensing has become one of the most efficient means for wetland monitoring [25].

Numerous methods have been developed to automatically identify and classify wetlands from multisource remotely sensed imagery; they include unsupervised/supervised classification, spectral angle mapping (SAM), artificial neural networks (ANN), Support vector machine (SVM), and "TUPU" coupled hybrid classification [26–37]. The automatic extraction technique has been pursued by researchers to extract the geographic area of wetlands. Due to the complexity, wetland remote sensing has undergone manual interpretation, semi-automatic, and intelligent extraction periods [14,29,36,38–44]. Wetland remote sensing has experienced changes in terms of data sources; it has gone from a single data source to multi-phase, multi-angle, and multi-data sources for wetland detection [36,37,45,46]. In terms of the scale of information extraction, wetland classifications have been performed from the individual pixel level to the level of objects, which are equivalent to wetland patches [2,28]. Although wetland extraction has been moderately successful in some applications, automatic mapping remains a challenge, largely due to the ambiguous spectral characteristics and complex shapes, sizes, and forms of wetlands [10,47,48].

In order to assess the change characteristics and dynamics of wetlands in the Yellow River Delta over the past four decades, we proposed a per-parcel level classification approach based on the knowledge transfer classification and machine learning for automatic classification of wetlands. This study produced time-series maps that document how these reginal wetlands have changed. We examined the spatial distribution, change process, and drive mechanism of wetlands in the Yellow River Delta over the past four decades. The results provide data and a scientific reference for wetland management, conservation, and restoration.

2. Study Area

The Yellow River Delta is the fastest growing delta in the world (Figure 1); it is located in the north-east of Shandong province, China, in the estuary of the Yellow River. The wetlands of the Yellow River Delta are not only important for protection of the eco-environmental system but also for internationally endangered bird species [22,49,50]. Generally, the modern Yellow River Delta is located from Wu Hao Zhuang (north) to Song Chun Rong Gou (south) (longitude 118.30~119.30° E, latitude 37.05~38.20° N) [51]. This area has a temperate monsoon climate; the annual mean temperature is 11.9 °C and the average precipitation is about 640 mm with 196 frostless days [22,52,53]. Because of the sedimentation in the Yellow River, the sediment builds and rebuilds the delta continuously. There is an abundance of coastal wetland resources, and it has been named one of the most beautiful six wetlands in China [2].

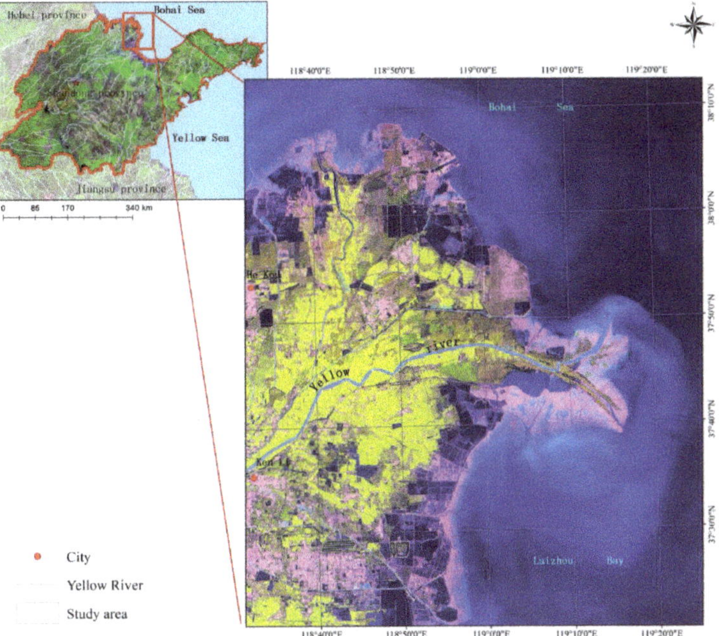

Figure 1. Location of study area.

Historically, wetlands have covered 80% of the total land area in the Yellow River Delta [2]. Natural wetlands (including lakes, rivers, marshes, swamps, lagoons, and intertidal zones) are mainly distributed in the east of the delta from the Xiaodao River to the Tuhai River. Artificial wetlands (including cultivated areas and salt pans) are primarily distributed in the mid-west of the delta. The Yellow River Delta is an important breeding ground for birds and it is a migratory stopover location.

In 1992, the Yellow River Delta National Nature Wetland Reserve (YRDNNW) was established to protect the native wetland ecosystem and rare and endangered birds in the estuary. Hundreds of kinds of rare and endangered birds are found in this area every year, including nationally protected birds [22,50]. However, according to the second wetland survey report of Shandong province, the wetland has suffered tremendous degradation, but how this occurred and the causes are not clearly understood. This study addressed this challenge by using four decades of Landsat observations ranging from 1973 to 2013.

3. Materials and Methods

3.1. Data Used for Wetlands Mapping

In this research, Landsat satellite series data (MSS: Multi spectral scanner, TM: Thematic mapper, and OLI: Operational Land Imager) from the period of 1973–2013 were used to map regional wetlands. Because they are affected by phenology and precipitation, wetlands are dynamic and have seasonal changes throughout the year. Because wetlands have seasonal changes throughout the year, annual seasonal changes impact time-series change analysis. In order to ensure comparability among the results, the image acquisition time is very important. In order to ensure the comparability among the results, Landsat images from nine different points in time were selected as the main data source. These images' acquisition times are mainly distributed in September and October (Table 1). During this period, the regional hydrological characteristics are relatively stable. These images were used to assess changes in the Yellow River Delta wetlands over the past four decades. These images, downloaded from the United States Geological Survey (USGS) website (http://glovis.usgs.gov/), have gone through radiometric calibration and geometric correction and were saved as 8-bit digital numbers (DNs) with a coordinate system of Universal Transverse Mercator (UTM).

Table 1. List of remotely sensed images of the Yellow River Delta wetlands used for mapping.

No.	Acquisition Time	Orbits Path/Row	Satellite No.	Sensor Type	Spatial Resolution	Cloud Cover	Image Quality
1	1973/12/06	130/034	Landsat 1	MSS	60 m	0.0%	9
2	1977/10/05	130/034	Landsat 3	MSS	60 m	0.0%	9
3	1984/10/05	121/034	Landsat 4	MSS	60 m	0.0%	5
4	1991/09/03	121/034	Landsat 5	TM	30 m	0.0%	7
5	1995/09/18	121/034	Landsat 5	TM	30 m	0.0%	9
6	2000/09/15	121/034	Landsat 5	TM	30 m	3.0%	9
7	2006/10/02	121/034	Landsat 5	TM	30 m	0.0%	9
8	2010/09/11	121/034	Landsat 5	TM	30 m	0.3%	9
9	2013/10/05	121/034	Landsat 8	OLI	30 m	2.5%	9

In the preprocessing stage, all images were converted to radiance (or reflectance). The image DN-to-L_λ transform calculation was carried out using Formula (1) or (2) [54–56]. All images were re-projected to Lambert azimuthal equal area projection for wetlands area qualitative statistics at different times.

$$L_\lambda = Gains * DN + Bias \tag{1}$$

which is also expressed as,

$$L_\lambda = (L_{\lambda max} - L_{\lambda min})/(QCAL_{\lambda max} - QCAL_{\lambda min}) * (DN - QCAL_{\lambda min}) + L_{\lambda min} \tag{2}$$

where, L_λ is the spectral radiance at the sensor's aperture in $mW/(cm^2 \cdot sr \cdot \mu m)$. DN is the digital number of the quantized calibrated pixel value. *Gains* is the band-specific rescaling gain factor in $(mW/(cm^2 \cdot sr \cdot \mu m))/DN$. *Bias* is the band-specific rescaling bias factor in $mW/(cm^2 \cdot sr \cdot \mu m)$. $L_{\lambda max}$ is the maximum spectral radiance that is scaled to $QCAL_{\lambda max}$. $L_{\lambda min}$ is the minimum spectral radiance that is scaled to $QCAL_{\lambda min}$. $QCAL_{\lambda max}$ is the maximum quantized calibrated pixel value in DN (corresponding to $L_{\lambda max}$). $QCAL_{\lambda min}$ is the minimum quantized calibrated pixel value in DN (corresponding to $L_{\lambda min}$).

3.2. Classification System for the Yellow River Delta Wetlands

Several classification systems have been developed to describe coastal wetlands (e.g., Ramsar Convention (Convention on Wetlands of Importance Especially as Waterfowl Habitat) wetlands classification system, Technical Specification of Survey National Wetlands Resources 2008, Chinese Wetland Remote-Sensing mapping Classification System) [14]. Each classification system has its specific purpose. In this study, the aim is to examine the status and change process of the regional wetlands. Therefore, our main concerns are the number, distribution, and composition of wetlands. In accordance with the regional characteristics and wetland types [57] and the international wetland classification system (i.e., the Ramsar), and taking into account the spectral separability on the medium resolution remotely sensed image, we classified the Yellow River Delta wetlands as natural wetlands or artificial wetlands with six subclasses (Table 2). The minimum mapping unit of the wetlands was 0.01 km^2.

Table 2. Classification system for regional wetland remote-sensing maps.

Wetland	Type	Code	Subtype
The Yellow River Delta wetlands	01 Natural wetlands	11	River wetland
		12	Marsh and swamp
		13	Intertidal zone
	02 Artificial wetlands	14	Reservoir and pond
		15	Salt pan (salt evaporation pond)
		16	Aquaculture

3.3. Wetland Automatic Classification and Updating via Transfer Learning

The object-oriented classification technology has many advantages, including that there are more features that can be used, it avoids salt-and-pepper noise, it is convenient for spatial relations and reasoning, and it reduces post-processing work [28]. However, the structure and spectral signature of the different wetlands are heterogeneous across space and time [27,29,58]. It is hard to select representative samples at the per-parcel level. This process requires a lot of manual intervention, which greatly affects the classification accuracy and comparability of the classification results. Therefore, the way in which to automatically select the classification of samples is the current technical obstacle [59]. This study proposed a hybrid method for automatic classification of wetland thematic maps and updating these by integrating the use of knowledge transfer and machine learning (Figure 2). To improve the automation of wetland information extraction, knowledge transfer was used to create classification training samples to aid in automatic selection.

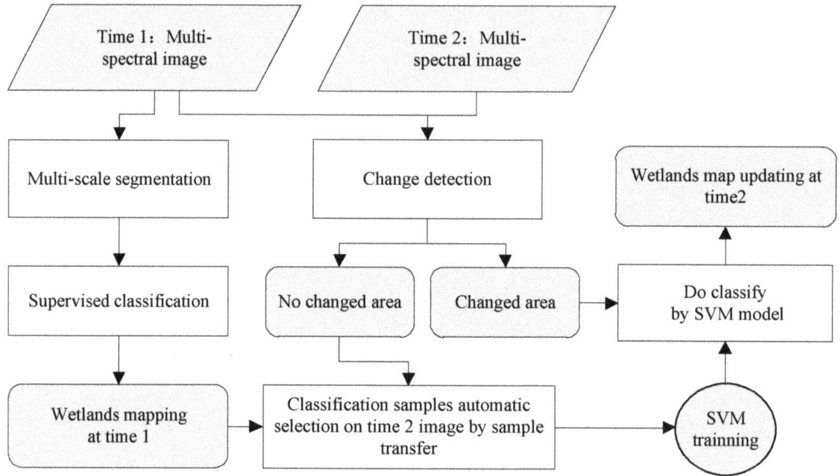

Figure 2. Diagram of wetland classification and updating process based on sample transfer learning.

The whole procedure of wetland classification and updating through sample transfer learning can be summarized by the following six steps. In the first step, the first phase (Time 1) image is segmented and classified by a supervised method based on object-oriented classification technology. Next, change detection is calculated between the first (Time 1) and second phase (Time 2) data. The objective of this step is to detect the changed region and the unaltered area on the second phase (Time 2) image. For the third step, if there is no change area between the Time 1 and Time 2 images, then the classification of the Time 1 image is used to label the samples on the Time 2 image as unchanged areas and the classification sample automatic transfer is complete. The fourth step is the machine (SVM) learning and training. Cross-validation (CV) is mainly used to evaluate the generalization ability of machine-learning algorithms. The most commonly used cross-validation method is K-fold cross-validation. The initial sample is divided into K sub-samples. A single sub-sample is retained as the data for the validation model. The other K-1 samples are used for training. So, the new labeled samples on the Time 2 image are classified into K parts (datasets); one part is used for the SVM training and the other is used for model validation. Each time the results are verified once, and k = 10 is the optimal parameter through the trial-and-error method (Figure 3). K is greater than 10, the improvement of the classification accuracy is limited, and the training time of the model is too long. So, through the 10-fold cross-validation, the SVM model initialization training was completed. The fifth step is wetlands identification. The Time 2 image changed areas were reclassified by the trained SVM model. At this point, the wetland updating is performed automatically. The last step is wetland remote-sensing mapping on the Time 2 imagery followed by accuracy checking and validation. The results are checked by overlaying the vector of wetland thematic map on the remotely sensed image. We performed detailed manual editing of the classification results. Misclassified and missing points are corrected one by one. This is done to test the accuracy of the classification of the wetlands and to revised the intertidal zone manually according to optical remote elevation data (DEM: Digital elevation model) and the average tidal range (0.9 m). The accuracy of the wetland thematic map was assessed by comparing the map with reference data (obtained from visual interpretation and field surveys).

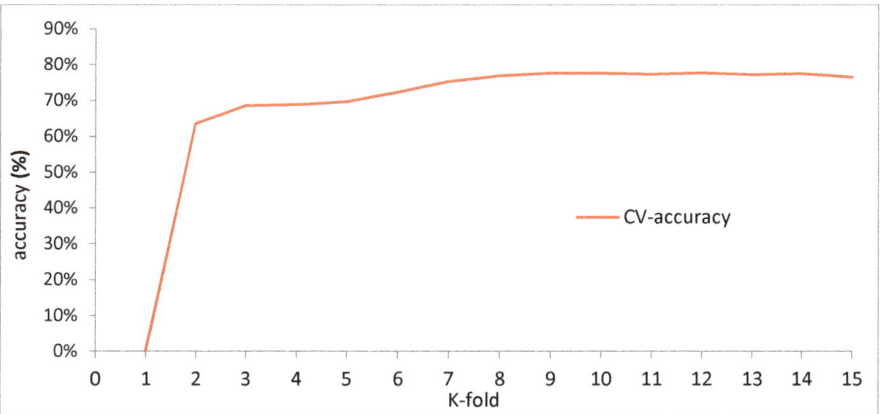

Figure 3. Accuracy of K-fold cross-validation.

3.4. Wetland Change Analysis

3.4.1. Dynamic Change

To quantify the degradation of wetlands in the Yellow River Delta, the dynamic change (DC) model was implemented [60,61]. Dynamic change, also known as the change degree, is an important model for the study of wetland changes. The DC is defined as:

$$DC = (U_a - U_b)/U_b/T * 100\%$$ (3)

where DC is the dynamic change, U_b is the initial wetland area, U_a is the wetland area at a later time, and T is the time. The DC model can describe not only the degree of wetland area variation but also the temporal features of the wetland changes.

3.4.2. Land-Use Transition Matrix

In order to better grasp the regional wetland change characteristics and reasons for change, change intensity needs to be understood, but it is also essential to know the spatial change and transformation. In this study, the land-use/cover transition matrix [62,63] was employed to analyze the theoretical frequency of inter-class conversions. Through the superposition of wetland maps from two points in time, we can find the number of mutual conversions between different types of land cover in a period of time and then reveal the spatial variation between the different types. Generally, a land-use transition matrix is expressed as a table (Table 3). In Table 3, the rows are the land types in time T_1 and the columns are the land types in time T_2. P_{ij} was the area or percentage of the type A_i transformed into the type A_j between T_1 and T_2. P_{ij} indicates that there was no change in area for the type A_i during the period of the time from T_1 to T_2.

Table 3. Land-use/cover transition matrix.

Time	Types	T_2						
		A_1	A_2	A_3	...	A_j	...	A_n
	A_1	P_{11}	P_{12}	P_{13}	...	P_{1j}	...	P_{1n}
	A_2	P_{21}	P_{22}	P_{23}	...	P_{2j}	...	P_{2n}
T_1	A_3	P_{31}	P_{32}	P_{33}	...	P_{3j}	...	P_{3n}
	...	...	...	...	...	...	...	...
	A_i	P_{i1}	P_{i2}	P_{i3}	...	P_{ij}	...	P_{in}
	A_n	P_{n1}	P_{n2}	P_{n3}	...	P_{nj}	...	P_{nn}

4. Results: Long-Term Wetland Changes in the Yellow River Delta

4.1. 1973–2013 Wetlands Remote-Sensing Mapping

The pattern of wetlands in the Yellow River Delta has undergone a significant change over the past four decades (Figure 4). Initially (1973–1984), there were abundant natural wetlands and the landscape was relatively simple. The main wetland types were marsh and intertidal zone. There were almost no artificial wetlands. Later (1984–2013), wetland coverage was reduced markedly with a large amount of marsh and swamp degradation. Large areas of artificial wetlands appeared and added to the regional wetland landscape. The natural wetland patches were shrinking. Also, due to the sedimentation in the Yellow River and the movements of the channel, the delta area has been increasing constantly. The newly formed land contained natural wetlands at the beginning and was later dominated by human activities. This caused the shape of the delta to change continuously and the local shoreline to erode into the sea.

To ensure classification accuracy is reliable and classification results are credible, accuracy verification of results is an essential task. All results are checked by overlaying the vector of the wetland thematic map on the remotely sensed image. On that basis, through the accuracy analysis of classification maps and manual samples (manual interpretation), each phase of the classification results has a detailed accuracy assessment report (Table 4). Overall accuracy reaches 85% or more. Average producer accuracy is above 86%. Average user accuracy is above 85%. The Kappa coefficient is more than 0.80.

The Yellow River Delta wetlands have experienced considerable degradation from 1973 to 2013 (Figure 5 and Table 5). The total area of wetlands in this region has decreased 683.12 km^2, but not all kinds of wetlands have decreased in the same way. Different types of wetlands have their own individual change characteristics. The natural wetlands decreased by 1441.5 km^2 (36.04 km^2·year^{-1} on average). The largest reduction was marsh and swamp, its total area declined by 1036.22 km^2. The second largest decrease was intertidal zone (beach) and the total area decrease by 425.16 km^2. Although the river wetlands experienced fluctuations from 1991–2006, the area remained relatively stable. Meanwhile, the extent of artificial wetlands has increased by 758.47 km^2 (18.96 km^2·year^{-1} on average). Specifically, aquaculture pond, salt pan, and reservoir and pond area have increased by 526.7 km^2, 128.2 km^2 and 103.57 km^2, respectively.

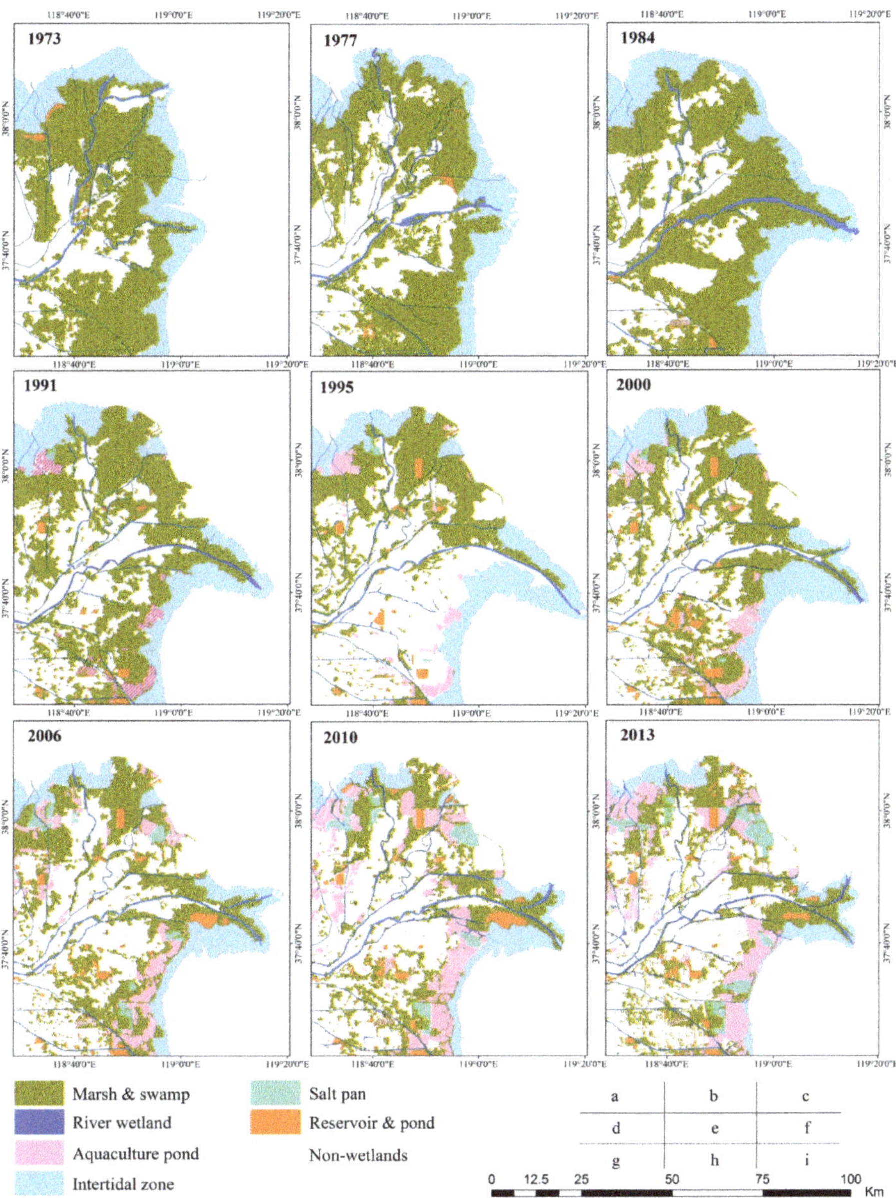

Figure 4. A time series of wetland distribution in the Yellow River Delta from 1973 to 2013. Maps were created using remote sensing.

Table 4. Wetlands classification accuracy evaluation in 2013.

Types	Manual Interpretation						Total	User Accuracy/%
	Aquaculture Pond	River Wetland	Marsh and Swamp	Salt Pan	Reservoir and Pond	Intertidal Zone		
Aquaculture pond	185	6	10	5	0	0	206	89.81
River wetland	5	158	7	0	2	2	174	90.80
Marsh and swamp	7	20	162	9	2	5	205	79.02
Salt pan	0	4	7	168	8	1	188	89.36
Reservoir and pond	3	10	11	16	180	8	228	78.95
Intertidal zone	0	2	3	2	8	184	199	92.46
Total	200	200	200	200	200	200	1200	
Producer accuracy/%	92.50	79.00	81.00	84.00	90.00	92.00		
Overall accuracy = 86.0%					Kappa coefficient = 0.837			

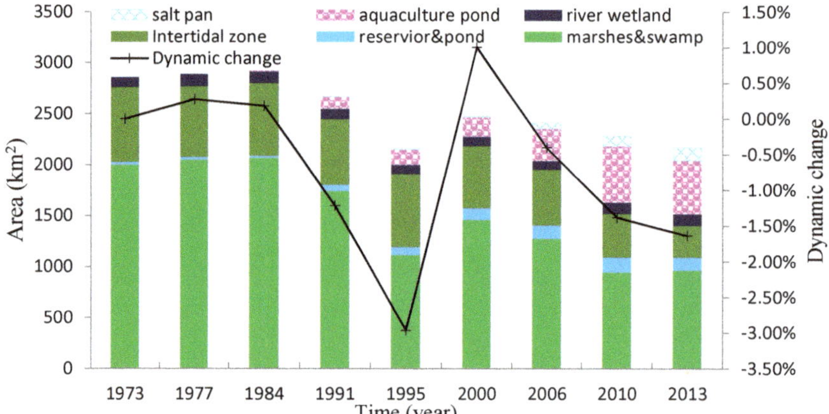

Figure 5. Accumulation histogram of area and dynamic change for all types of wetlands through time.

Table 5. Wetland area by type in the Yellow River Delta at different periods from 1973 to 2013 (units: km²).

Time	Marsh and Swamp	River Wetland	Reservoir and Pond	Aquaculture Pond	Salt Pan	Intertidal Zone	Total Area
1973	1996.44	92.18	30.23	0.00	0.00	732.90	2851.75
1977	2046.74	113.28	31.29	0.00	0.00	691.86	2883.16
1984	2064.86	113.22	23.08	8.13	0.20	710.64	2920.23
1991	1738.98	100.03	63.13	117.52	6.49	645.76	2671.90
1995	1114.18	87.80	79.78	149.09	7.60	716.60	2355.05
2000	1453.79	91.12	119.86	188.20	9.68	610.64	2473.29
2006	1271.25	83.78	136.43	313.22	60.36	549.09	2414.13
2010	942.42	112.68	150.18	552.67	98.77	423.88	2280.60
2013	960.22	111.97	133.8	526.70	128.20	307.74	2168.63

4.2. Wetlands Spatio-Temporal Changes Analysis

The total area of wetland was reduced by 683.12 km²; the natural wetland area decreased by 1441.59 km², while the artificial wetland area increased by 758.47 km² (Figure 6). The total area of wetlands in this region had a statistically significant decreasing trend (18.29 km²·year⁻¹, $p < 0.05$). At the same time, the natural wetlands also had a statistically significant decreasing trend (38.52 km²·year⁻¹, $p < 0.05$). Meanwhile, the area of artificial wetlands had a statistically significant increasing trend during the entire period (20.2 km^2·year^{-1}, $p < 0.05$). Additionally, natural wetlands had a fluctuating downward trend, but the artificial wetlands showed a steadily rising trend.

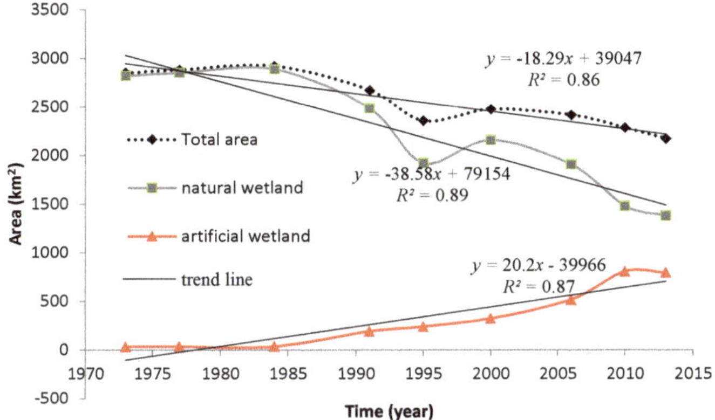

Figure 6. Area of natural and artificial wetlands in theYellow River Delta from 1973–2013.

The change process had three stages: steady increase (1973–1984), sharp decline (1984–1995), and slow decrease (2000–2013). The average dynamic change rate was −1.61% (Figure 7). The marsh and swamp had the most degradation, the intertidal zone area had a steady and slow decline over the time period, and the river wetland area was relatively stable. For the artificial wetlands, the average dynamic change rate was 12.09% (Figure 8). The area of the three types of artificial wetlands increased consistently and aquaculture ponds had the biggest increase.

Figure 9 shows the wetland gain, loss and unchanged area. Wetland reduction areas were mainly distributed in the center of the Yellow River Delta. The area where wetlands were expanding predominantly occurred in the coastal zone and the mouth of the river. Also, natural wetland landscapes were severely destroyed. The new wetlands formed from the deposition of sediment and the creation of aquaculture. The newly formed land became colonized by natural wetlands (e.g., swamp and intertidal zone), but it was later occupied by humans and converted into other land uses.

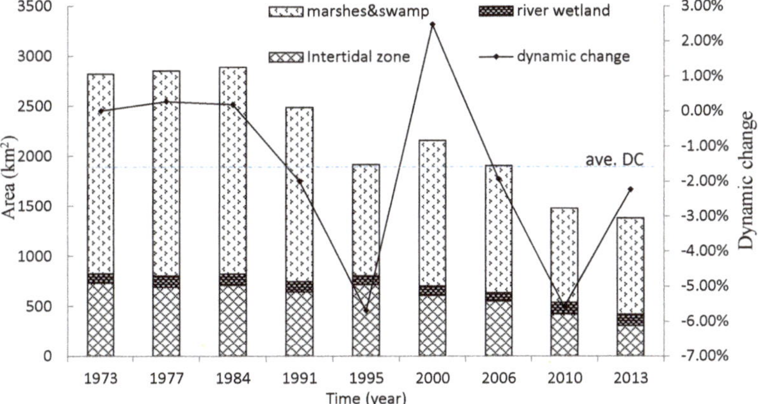

Figure 7. Accumulation histogram of natural wetland area and dynamic change. The blue line is the average dynamic change (ave. DC).

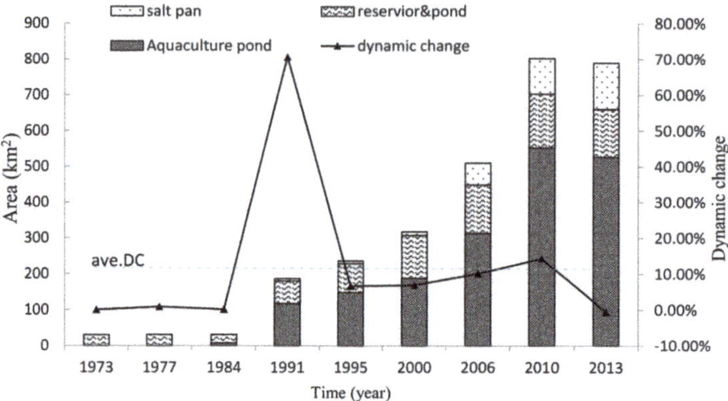

Figure 8. Accumulation histogram of artificial wetland area and dynamic change. The blue line is the average dynamic change (ave. DC).

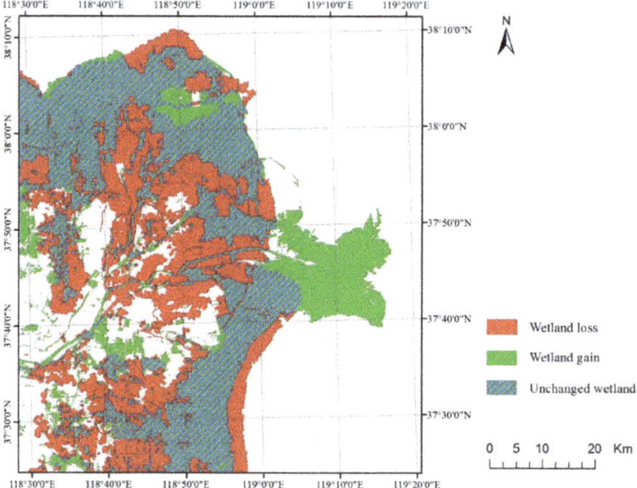

Figure 9. Map showing changes in area of wetlands in the Yellow River Delta over the past four decades. Three scenarios of wetland change are shown: wetland loss, wetland expansion (wetland gain), and unchanged wetland areas.

Various natural wetlands were converted into artificial wetlands, farmland, city, and other land uses (Table 6). Firstly, a large area of river wetland became marsh and swamp (33.85%). The original ponds were primarily transformed into marsh and swamp (44.67%), aquaculture pond (16.39%), and cropland (27.46%). A large area of marsh and swamp were converted into cropland (30.08%), city (15.16%), and artificial wetland (23.18%). The intertidal zone was mainly transformed into marsh and swamp, salt pan, aquaculture pond, and oil fields. In addition, there was a large area of intertidal zone (18.84%) that converted to sea because of sediment erosion. This was another factor in the reduction of regional natural wetlands. In total, more than half of the natural wetlands have been damaged. Meanwhile, some of the shallow sea (26.99%) in the study area transitioned into marsh and swamp, intertidal zone, aquaculture pond, and other land types. This is an incremental increase for regional wetland area and effectively slows down the rate of wetland loss.

Table 6. The land cover transition matrix for the Yellow River Delta 1973–2013 (units: km^2).

Types	Sea	River	Reservoir and Pond	Marsh and Swamp	Intertidal Zone	City	Cropland	Aquaculture	Salt Pan	Oil Field	Others
Sea	1300.91	14.38	26.65	133.04	134.77	3.26	37.38	19.94	5.94	99.49	0.19
River	1.25	2.88	2.20	24.83	0.00	5.27	29.33	5.37	0.00	0.00	2.20
Reservoir and pond	0.00	0.58	0.00	10.45	0.00	2.01	6.42	3.83	0.00	0.00	0.10
Marsh and swamp	5.18	41.31	73.42	506.87	33.55	285.06	565.53	295.03	67.29	3.74	2.97
Intertidal zone	136.36	17.16	8.63	130.17	120.87	14.76	57.61	149.05	60.96	26.74	0.00
City	0.00	1.44	1.44	6.71	0.00	41.31	7.28	1.25	0.00	0.00	0.00
Cropland	0.00	9.68	5.66	39.20	0.00	63.26	216.05	11.50	0.00	0.00	1.73
Others	1.15	0.67	9.20	47.83	0.19	3.26	16.87	18.88	0.29	0.00	0.10

5. Discussion

5.1. Change Mode and Trend

From a comprehensive analysis of the change process of the total area of wetland in the Yellow River Delta from 1973 to 2013, there are two significant turning points on the change curve (Figure 10). One is in 1984 and the other in 1995. From 1973 to 1984, the regional wetland area was steady with a small increase. Then, the total area of the regional wetlands rapidly dropped to the lowest point in 1995, followed by a little rebound in 2000. Since then, the area of regional wetlands has slowly decreased. According to these change characteristics, the entire change process can be divided into three stages: (1) Stable with a little increase (1973–1984). In this stage, the regional wetlands area steadily increased. This was possibly linked to the change in channel of the Yellow River which occurred two times during this period (1976 and 1979). (2) Rapid decline (1984–1995). In this stage, the regional natural wetlands showed a sharp decline in area. At the same time, artificial wetlands were gaining significant area. This was possibly due to the high intensity of human activity (e.g., urban construction, cropland creation, and oil exploitation). From reviewing historical literature, the lowest point of wetland area was likely around 1997, because in 1997 the amount of precipitation was very low, which resulted in the longest duration of no flow in the lower Yellow River and the river went dry in this area. Also, the area suffered a major storm surge, which resulted in the destruction of a large number of natural wetlands [57,64–67]. (3) Slow decline (1995–2013). In this stage, the wetland experienced a little improvement and then slowly decreased in area. This may be associated with the creation of regional wetland reserves and climate change. The regional wetlands had a small increase in area around 2000, which could be related to the unified water resources scheduling that began in 1999 within the watershed.

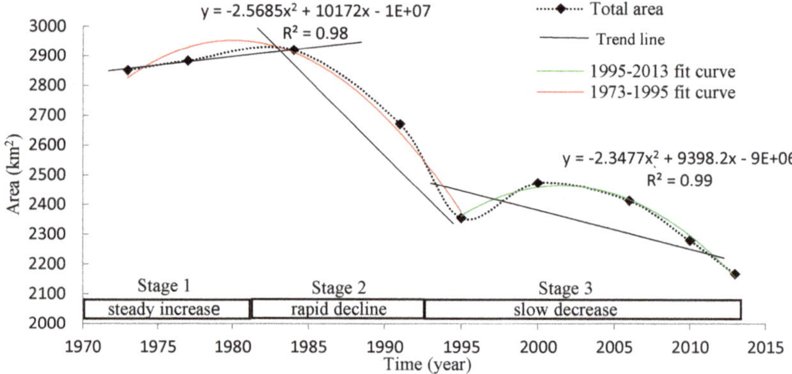

Figure 10. Change trends and mode of total wetlands area in the Yellow River Delta from 1973–2013.

5.2. Attribution Analysis

There are many reasons for wetland degradation in the Yellow River Delta over the past decades (e.g., farmland reclamation, beach development, coastal erosion, and climate change). The original natural wetlands in the Yellow River Delta were mainly converted to cropland, aquaculture ponds, salt pans, oil fields, city, residential areas, and infrastructure (Figure 11). The reasons for this can be summarized into four major categories: (1) Human activities (e.g., farmland reclamation, beach development, hydraulic engineering, oil exploitation, infrastructure construction); (2) climate change (e.g., temperature increase, precipitation decrease, evaporation enhancement, sea-level rise); (3) the Yellow River diversion (e.g., river mouth migration, coastal erosion); (4) policies (e.g., conservation policy and management regulations, restoration projects).

5.2.1. Human Activities

Land reclamation for farming, the census data showed that the Yellow River Delta has been a key area for land reclamation. This directly resulted in a large number of low-lying swamps being drained and converted into farmland. In the late 1990s, in particular, there was an increase in cotton production, so a lot of swamp land was transformed into cotton fields [68]. Beach resource development, aquaculture, and salt production have been the major development activities in the Yellow River Delta tidal flat for a long time. Beach aquaculture was developing rapidly (mainly shrimp farming, tilapia, and artemia) in the mid-to-late 1980s, and the area of the aquaculture ponds increased exponentially. A census of the aquaculture area showed that the area increased 10-fold in the estuary region between 1995 and 2002 [67]. Additionally, Dongying city has a long history of salt production. The city had 34 raw salt production fields, a total salt area of 10.82 km^2 in 1983, and by 2000, the area reached 70.72 km^2 [67]. To create these salt fields, a large area of natural wetlands was changed to artificial wetland. This is why the natural wetland loss rate was 50.55% from 1973 to 2013, but the total area of wetland only decreased by 23.95%.

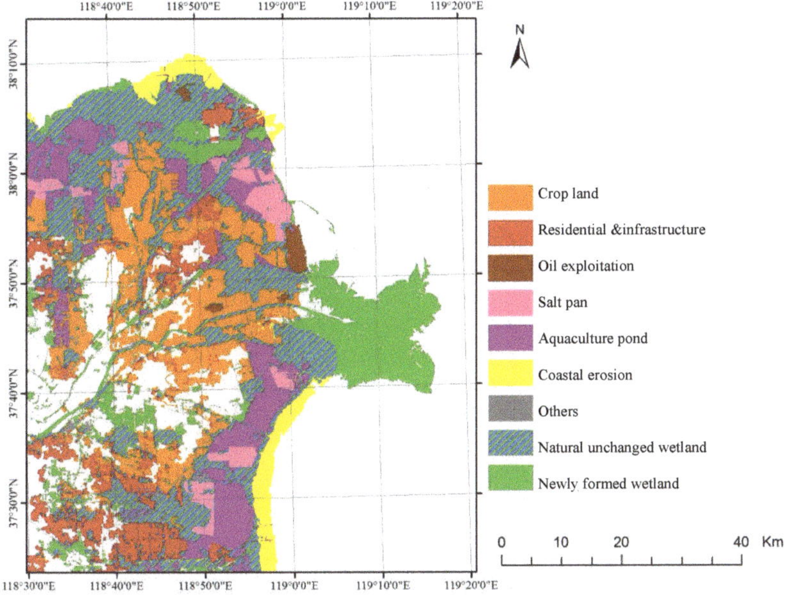

Figure 11. Changes in land use for natural wetland areas from 1973 to 2013.

Oil exploitation, the Yellow River Delta was a critical oil production location in China. Shngli oilfield has been exploring and developing since 1961 and has become China's second largest crude oil production location. With the discovery and exploitation of new oilfields, the invasion and destruction of wetlands were inevitable [22,57]. The area of the Shengli oilfield increased by 2.28 times from 1984 to 2001, with a total area of 853 km^2 in Dongying City [69]. The competition between oil production and wetland protection cannot be reconciled in the short-term. Water conservancy construction, as a result of an increase in water demand by industry and agriculture, a large number of water storage and water diversion projects have been built in the Yellow River Basin. At present, the Yellow River Basin has more than 10,000 water storage projects and over 3 million water pumping projects [67]. The volume of water diverted reached 4.7×10^{10} m^3 in the 1990s [67,70]. The various water conservation projects allowed a rapid increase in water consumption and increased the competition between water supply and demand in the basin. In 1972, part of the Yellow River channel dried up. Thereafter, periods when no water reached the lower part of the Yellow River became a common phenomenon; from 1972–1999, this occurred 22 times. The most extreme occurrence of the river water drying up was in 1997, lasting 226 days and with no water in the channel up to 704 km up stream [65–67,71]. The absence of water in the Yellow River fundamentally changed the water environment of the delta and directly resulted in loss in the regional ecosystem, deterioration, and irreversible damage [72]. Because there is less water, it is bringing less sediment to the area [67,73]. This not only slowed down the sedimentation, but also reduced the area of newly formed coastal wetlands. It directly threatened the coastal wetland ecosystem [57,72,74].

Additionally, economic development, population growth, and urban expansion, especially the development of the Shengli oilfield, caused a number of wetlands to be destroyed by the building of roads, dams, ports, and other infrastructure [7,8,22,75,76]. The integrity of the wetlands has been undermined due to the infrastructure construction and the original wetland landscape has become increasingly degraded and fragmented. Overall, this research has shown that human activities were the main driver of wetland landscape changes in the Yellow River Delta [22,77,78].

5.2.2. Climate Change

Global warming has been accelerating sea-level rise, which has led to the submersion of low-lying areas of coastal wetlands [79,80]. According to the China Sea Level Bulletin 2016, China's sea-level rise is generally higher than the global average over the period from 1980 to 2016 (3.2 mm/a) [81]. In the Yellow River estuary coastal district, the relative increase in sea level over the past 30 years was 4.5–5.5 mm/a, including regional crustal subsidence [82]. Climate warming has also led to the redistribution of water resources and exacerbated regional water resource imbalances. During the period from 1961 to 2010, the annual average temperature in the coastal wetlands of the Yellow River Delta increased by 1.85 °C, while the annual precipitation decreased by 121.42 mm [83]. A large number of studies have shown that over the past four decades precipitation decreased in the Yellow River Basin and the area showed a warm and dry trend [83–86]. The precipitation in the upper and middle reaches of the river has decreased due to the warm and dry regional climate. Meanwhile, the water demand for industry and resident life increased, which accelerated the crisis of the downstream ecological water resources. The annual flow of the Yellow River has decreased dramatically in the past decade, and the flow has a significant negative correlation with the temperature and positive correlation with precipitation [49,86–90]. Climate change was a long-term critical impact factor on the reduction of flow. However, it is understood that water diversion was the main cause of the flow changes in the lower reaches of the Yellow River [91].

5.2.3. Yellow River Diversion

From 1973 to 2013, the Yellow River Delta had a great increase in area (495.77 km^2) due to sedimentation (Figure 12b), with an annual expansion rate of 21.3 km^2/a. These newly formed lands are the source of increased wetland area. Also, the Yellow River channel location has changed periodically. The site of the river mouth not only affected the speed of siltation but it influenced the direction of the delta's expansion. Once the river mouth changed location to the other sites, the source of sediment in the old Yellow River channel was cut off and the coastal erosion increased. Over the period from 1973–2013, the mouth of the Yellow River has moved four times (Figure 12).

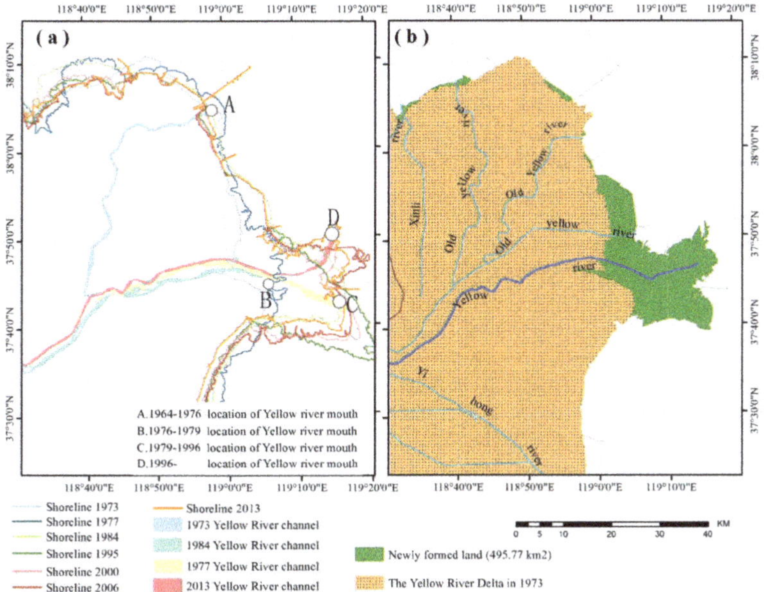

Figure 12. (**a**) Map showing different locations of the Yellow River channel and mouth and the coastal shoreline at different points in time. (**b**) Map showing past locations of the Yellow River and new areas of land that formed in the Yellow River Delta since 1973.

From 1964–1976, the Yellow River flowed north into the sea from the Diao Mouth (Figure 12a site A). The Yellow River was rechanneled in 1976 at the west river mouth. Between 1976 and 1979, it flowed into the sea from the Qingshuigou (Figure 12a, site B). In the period from 1979 to 1996, the location of the Yellow River channel was in an unstable condition and with little change at flood season (Figure 12a, site C). In 1996, the Yellow River was diverted by human intervention at the location of the Qing No.8 Brouch, which blocked the main channel and forced the river to veer north-east. Since then, the Yellow River has flowed into the sea north-east of the Qing No.8 Brouch (Figure 12a, site D). With all of these change in the estuary, the coastline also varied constantly over the past four decades (Figure 12a). Wherever the mouth of the Yellow River was located, the delta expanded. Since the mouth of the Yellow River moved from the Diao Mouth, the coastline there eroded severely.

Additionally, due to the special terrain and geographical environment, the Yellow River estuary was prone to storm surge. There was a serious storm surge every 3 to 4 years on average [67]. Storm surges led to coastal erosion and coastal wetlands being destroyed. According to the weather record data, the storm surge in 1992 contributed to part of the coastline retreating up to 30 m, the direct loss of land was 4.66 km^2 [92]. The storm surge in 1997 directly destroyed the region's largest area of *Robinia pseudoacacia* [67]. From the discussion above, the mouth migration and coastal erosion obviously had major impacts on the regional wetland changes.

5.2.4. Conservation Policies and Projects

Over past decades, natural wetlands in China have suffered a great loss of area and degradation. This is inextricably linked to the development of agriculture, the lack of awareness of the importance of wetlands, and government policies and laws [7,14,15,93,94]. The Yellow River Delta wetlands were no exception; in the middle stage of wetland change (stage 2: 1984–1995), because of the lack of ecological protection awareness and the excessive pursuit of economic interests, a large area of wetland was converted into farmland, aquaculture ponds, salt pans and other uses (Table 2). After the regional wetland ecosystem was severely degraded, relevant government departments and authorities realized the importance of wetlands and developed a series of protection policies and laws. For example, in 1990, Dongying Municipal People's Government approved the establishment of the Yellow River Delta Nature Reserve (YRDNR). In 1991, the Yellow River Delta Nature Reserve was upgraded to a provincial protected area. In 1992, the State Council of China approved the establishment of the Yellow River Delta National Nature Reserve (YRDNNR). In 1999, the implementation of the Yellow River basin unified water resources scheduling rectified the lack of flow (the river drying up) in the lower reaches of the river and this has been fundamentally reversed. In 2004, the Yellow River Delta Nature Reserve Phase II project was completed and passed the national acceptance. In the same year, the Yellow River Delta National Nature Reserve wetlands monitoring project began, and it used remote sensing, global position system (GPS), geographic information system (GIS), and other technology to determine the timing and locations of fixed-point monitoring for the study area. In 2006, the Yellow River Delta National Geo-park was opened.

In stage 3 (1996–2013) of wetland change, the degradation rate of the regional wetlands has slowed down, which was potentially linked to these protection projects and laws. It should be noted that there may have been some time lag between the implementation and effectiveness of the protection policy. From the results of this study, in the late 1990s the effects of the conservation efforts began to be visible. There was probably a five- to 10-year time lag. The unified water resources scheduling which was begun in 1999 seems to be very effective, because the area of wetland rebounded in about 2000 (Figure 3). There were likely links between these two events. However, the effective protection of wetland resources was not a simple matter, it involved trade-offs among the interests and behaviors of the various stakeholders (e.g., the petroleum industry, government departments at all levels, urban developers). There is still a lot to be done to protect and restore these regional wetlands. However, from the results of this study, great success in slowing down the rate of regional wetland degradation has been achieved so far. The above discussion showed some possible, preliminary reasons for the Yellow River Delta wetland degradation over the past four decades. To eventually confirm the cause–effect relationship in this system, there needs to be more integrated and systematic analysis using comprehensive technologies, such as multisource remote sensing and metrological and hydrological models.

5.3. Credibility and Error Sources

5.3.1. Validity

Creating time-series maps over an extended period of time for the Yellow River Delta wetlands is essential for wetland research and change analysis. It is also necessary to determine if it is possible that the time-series changes came from misclassification. To insure that the classification results are credible, the wetland information inferred from remotely sensed images should be validated. All of the multispectral images used in the study have been carefully selected and strictly preprocessed. The radiometric calibration and atmospheric correction was applied to polish the path radiance and improve the clarity based on the ENVI 5.3 FLAASH model. The classification method used in this study addressed the changed area map-updating by samples transfer learning rather than reclassification of the whole image. The changed area machine identification improved the efficiency of classification, and it also ensured classification accuracy. In addition, each section of the data change area was manually checked carefully. Logical judgment was essential (e.g., wetlands can be transformed into impervious surfaces, but the city's impervious surfaces do not change into wetlands) to ensure that errors of commission and omission were as small as possible. Therefore, we have confidence in the accuracy of the wetlands classification maps.

The wetland area changes were possibly caused by inter-annual variation or linked to intra-annual fluctuation. Sometimes, these two variables tend to interact. To discriminate the two different changes in order to ensure that inter-annual variability was comparable in this study, the images used for wetland detection were from the same season (autumn—September to October), except for the two earliest images. The river flow was relatively stable during this period each year [83,95]. So, the uncertainties of the inter-annual observed changes caused by intra-annual fluctuation was very little and acceptable.

5.3.2. Error Source

Regardless of the methodology, to perform wetland classification and updating successfully, the limitations (e.g., map scales, data acquisition time, spatial and spectral resolution, characteristics of wetlands, algorithms) of remote sensing inevitably affected the accuracy of the mapping [96]. From analyzing potential error sources in this study, the cartographic errors were mainly from scale and technical error.

Scale Errors

The scale factors were the inherent errors, which included both time scale and space scale errors [97]. Time scale errors were due to the variation in precipitation and wetness from year to year (i.e., drought or flood years). Accordingly, the ideal time for observations was at a time with average weather conditions [96,98,99], but that is only an ideal condition. The time and weather conditions of the images were not controllable. It cannot be known whether images captured during weather extremes would affect the mapping of the wetlands, but using nine observations to characterize and infer the change process and mode over the past four decades of the wetlands. This was a sample survey to some extent because the images were spread out over so many years. It was not possible to determine the exact time of a change. Therefore, the time scale error was not only related to the image acquisition time but also linked to the time density of the observations.

The space scale factors included the spatial resolution of the images and the map scale. The spatial resolution of the image reflects the ability to distinguish different category information, but it also has a negative impact on the classification accuracy [97,100–102]. The richness of wetland information derived from an image positively relates to its spatial resolution [100]. However, the specifications necessary for a "good" wetland map is heavily dependent on what the map was to be used for [96,98,99]. Therefore, neither a high nor low spatial resolution is better [102,103]. In this study, the image resolution of the earlier multiple spectral scanner (MSS) images was 60 m and the later (TM/OLI) images was 30 m. The difference in spatial resolution between MSS and TM is the size of one TM pixel (30 m). The lower spatial resolution increased the mixed pixels problem and decreased the separability of wetlands spectra. Also, the mis-registration between the different spatial resolution images is 15 m, which has a direct impact on the transitions between wetland type. Furthermore, the minimum mapping unit was 0.01 km^2, which means an area of wetland less than that size was ignored. The map scale also affects the map accuracy.

Technical Errors

No matter how good the algorithm, the classification accuracy cannot be 100% and errors are unavoidable. The technical errors were mainly caused by the characteristics of the wetlands, such as changes in different season, they have complex spectra, they are heterogeneous, and the same land cover types have multiple spectrums. To improve the classification accuracy in the future, more research can be done on the following aspects: (1) To discover more effective features, not just in spectrum, new technologic methods maybe good alternative choices (e.g., synthetic aperture radar, lidar, and geospatial modeling); (2) to enhance the intensity of machine learning; taking into account the all possible situations via the new learning structures (e.g., deep convolutional artificial neural network (ANN) and deep learning) [104–112]. The deep convolutional neural network algorithm, in particular, has better learning and generalization performance for multiple variables and large datasets.

Nevertheless, this is the first time that the Yellow River Delta wetlands have been mapped and had a long time-series change analysis performed on them. The results provided important information and scientific support to help the local government agencies develop robust strategies for wetland management, conservation, and restoration in the future. The study attempted to label classification samples automatically by the transfer learning method. This methodology emphasized the importance of existing thematic map knowledge. The entire technological procedure has greatly improved the efficiency of the work and the level of automatic wetland classification. The hybrid method used in this research could be extended to other regions for wetland surveys and mapping.

6. Conclusions

This study developed a new methodology to identify wetlands automatically and documented the time-series changes from 1973 to 2013 in the Yellow River Delta using remotely sensed data. Spatio-temporal change analysis was conducted to examine the long-term change processes and modes of different wetlands over the last four decades using nine images for classification and mapping. The results quantitatively assessed the temporal and spatial changes within the wetlands as well as between wetland and non-wetland areas to determine the main reason for degradation. Several findings were revealed and communicated in this study.

(1) The Yellow River Delta wetlands have been severely damaged over the past four decades (683.12 km^2 were lost). Over half (50.55%) of the original natural wetland area was lost between 1973 and 2013. Meanwhile, the regional artificial wetland area had a significant increasing trend (20.2 $km^2 \cdot year^{-1}$, $p < 0.05$).

(2) The whole change process can be divided into 3 stages: relatively stable (1973–1984), rapid reduction (1984–1995), and slow degradation (1995–2013). Moreover, the total area of the regional wetlands dropped to the lowest point in the second stage and then began to rebound around 1995.

(3) The designed hybrid method for wetland map automatic updating based on sample transfer and machine learning has greatly improved the work efficiency. This approach provided a new way to make full use of existing thematic maps and could be extended to other areas.

(4) Regional human activities (e.g., farmland creation, salt development, oilfield exploration, aquaculture, industrialization) were the main cause of the regional wetland degradation in the short-term. Climate change was a long-term factor that has been affecting the evolution of regional wetlands. The hydrological factors and the channel diversion of the Yellow River directly affected the formation and development of the regional wetlands. Regional wetland protection policies and engineering have played an important role in slowing down the process of regional wetland degradation to a certain extent. In order to fully understand the cause–effect relationship of the wetland change, more integrated and systematic analysis by comprehensive technologies is needed.

Author Contributions: Changming Zhu conducted data collection and processing, result analysis and preparation of the manuscript. Xin Zhang designed the research procedure and conducted result analysis and preparation of the manuscript. Qiaohua Huang provided the editing and discussion of this manuscript.

Funding: This research was funded by the Natural Science Foundation of China (Grant No. 61473286 and 401201460) and the National R&D Program of China (Granted No. 2017YFB0504201).

Acknowledgments: The authors would like to thank the anonymous reviewers for their constructive comments and suggestions.

Conflicts of Interest: The authors declare no conflict of interest.

References

1. Zhao, H.; Wang, L. Classification of the coastal wetland in China. *Mar. Sci. Bull.* **2000**, *19*, 72–82.
2. Zhang, X.; Wang, L.; Jiang, X.; Zhu, C. *Modeling with Digital Ocean and Digital Coast*; Springer International Publishing: Cham, Switzerland, 2017.
3. Bao, K. Coastal wetlands: Coupling between ocean and land systems. *J. Coast. Res.* **2015**, *313*, iv. [CrossRef]
4. Boesch, D.F.; Josselyn, M.N.; Mehta, A.J.; Morris, J.T.; Nuttle, W.K.; Simenstad, C.A.; Swift, D.J. Scientific assessment of coastal wetland loss, restoration and management in Louisiana. *J. Coast. Res.* **1994**, *20*, 1–103.

5.   Sklar, F.H.; Costanza, R.; Day, J.W. Dynamic spatial simulation modeling of coastal wetland habitat succession. *Ecol. Model.* **1985**, *29*, 261–281. [CrossRef]

6.   Sun, Z.; Mou, X.; Chen, X.; Wang, L.; Song, H.; Jiang, H. Actualities, problems and suggestions of wetland protection and restoration in the Yellow River delta. *Wetl. Sci.* **2011**, *9*, 107–116.

7.   Wang, M.; Qi, S.; Zhang, X. Wetland loss and degradation in the Yellow River delta, ShanDong Province of China. Environ. *Earth Sci.* **2011**, *67*, 185–188. [CrossRef]

8.   Zhang, J.; Sun, Q. Causes of wetland degradation and ecological restoration in the Yellow River delta region. *For. Stud. China* **2005**, *7*, 15–18. [CrossRef]

9.   Xu, X.; Lin, H.; Fu, Z. Probe into the method of regional ecological risk assessment a case study of wetland in the Yellow River delta in China. *J. Environ. Manag.* **2004**, *70*, 253–262. [CrossRef] [PubMed]

10.   Zhu, P.; Gong, P. Suitability mapping of global wetland areas and validation with remotely sensed data. *Sci. China Earth Sci.* **2014**, *57*, 2283–2292. [CrossRef]

11.   Development Assistance Committee Organisation for Economic Co-operation Development (OECD). *Guidelines for Aid Agencies for Improved Conservation and Sustainable Use of Tropical and Sub-Tropical Wetlands*; Development Assistance Committee Organisation for Economic Co-operation Development (OECD): Paris, France, 1996.

12.   O'Connell, M.J. Detecting, measuring and reversing changes to wetlands. *Wetl. Ecol. Manag.* **2003**, *11*, 397–401. [CrossRef]

13.   Hu, S.; Niu, Z.; Zhang, H.; Chen, Y.; Gong, N. Simulation of spatial distribution of China potential wetland. *Chin. Sci. Bull.* **2015**, *60*, 3251.

14.   Gong, P.; Niu, Z.; Cheng, X.; Zhao, K.; Zhou, D.; Guo, J.; Liang, L.; Wang, X.; Li, D.; Huang, H.; et al. China's wetland change (1990–2000) determined by remote sensing. *Sci. China Earth Sci.* **2010**, *53*, 1036–1042. [CrossRef]

15.   Sun, Z.; Sun, W.; Tong, C.; Zeng, C.; Yu, X.; Mou, X. China's coastal wetlands: Conservation history, implementation efforts, existing issues and strategies for future improvement. *Environ. Int.* **2015**, *79*, 25–41. [CrossRef] [PubMed]

16.   Nicholls, R.J. Coastal flooding and wetland loss in the 21st century: Changes under the sres climate and socio-economic scenarios. *Glob. Environ. Chang.* **2004**, *14*, 69–86. [CrossRef]

17.   Gedan, K.B.; Kirwan, M.L.; Wolanski, E.; Barbier, E.B.; Silliman, B.R. The present and future role of coastal wetland vegetation in protecting shorelines: Answering recent challenges to the paradigm. *Clim. Chang.* **2011**, *106*, 7–29. [CrossRef]

18.   Li, Q.; Wu, Z.; Chu, B.; Zhang, N.; Cai, S.; Fang, J. Heavy metals in coastal wetland sediments of the Pearl River estuary, China. *Environ. Pollut.* **2007**, *149*, 158–164. [CrossRef] [PubMed]

19.   Simenstad, C.; Reed, D.; Ford, M. When is restoration not?: Incorporating landscape-scale processes to restore self-sustaining ecosystems in coastal wetland restoration. *Ecol. Eng.* **2006**, *26*, 27–39. [CrossRef]

20.   Cahoon, D.R.; Hensel, P.F.; Spencer, T.; Reed, D.J.; McKee, K.L.; Saintilan, N. Coastal wetland vulnerability to relative sea-level rise: Wetland elevation trends and process controls. In *Wetlands and Natural Resource Management*; Springer: Berlin, Germany, 2006; pp. 271–292.

21.   Lee, S.Y.; Dunn, R.J.K.; Young, R.A.; Connolly, R.M.; Dale, P.; Dehayr, R.; Lemckert, C.J.; McKinnon, S.; Powell, B.; Teasdale, P. Impact of urbanization on coastal wetland structure and function. *Aust. Ecol.* **2006**, *31*, 149–163. [CrossRef]

22.   Liu, G.; Zhang, L.; Zhang, Q.; Musyimi, Z.; Jiang, Q. Spatio–temporal dynamics of wetland landscape patterns based on remote sensing in Yellow River delta, China. *Wetlands* **2014**, *34*, 787–801. [CrossRef]

23.   Ye, Q.; Tian, G.L.; Liu, G.H.; Ye, J.; Lou, W. Tupu analysis on the land cover evolving patterns in the new-born wetland of the Yellow River delta. *Geogr. Res.* **2004**, *23*, 257–264.

24.   Liu, H.; Li, Y.; Cao, X.; Hao, J.; Junna, H.; Zheng, N. The current problems and perspectives of landscape research of wetlands in China. *Acta Geogr. Sin.* **2009**, *64*, 1394–1401.

25.   Adam, E.; Mutanga, O.; Rugege, D. Multispectral and hyperspectral remote sensing for identification and mapping of wetland vegetation: A review. *Wetl. Ecol. Manag.* **2009**, *18*, 281–296. [CrossRef]

26.   Frohn, R.C.; Autrey, B.C.; Lane, C.R.; Reif, M. Segmentation and object-oriented classification of wetlands in a karst florida landscape using multi-season landsat-7 ETM+ imagery. *Int. J. Remote Sens.* **2011**, *32*, 1471–1489. [CrossRef]

27. Dronova, I.; Gong, P.; Wang, L. Object-based analysis and change detection of major wetland cover types and their classification uncertainty during the low water period at Poyang lake, China. *Remote Sens. Environ.* **2011**, *115*, 3220–3236. [CrossRef]

28. Zhu, C.; Li, J.; Zhang, X.; Luo, J. Wetlands information automatic extraction from high resolution remote sensing imagery based on object-orient technology (in Chinese with English Abstract). *Bull. Surv. Mapp.* **2014**, *10*, 23–28.

29. Han, X.; Chen, X.; Feng, L. Four decades of winter wetland changes in poyang lake based on landsat observations between 1973 and 2013. *Remote Sens. Environ.* **2015**, *156*, 426–437. [CrossRef]

30. Augusteijn, M.F.; Warrender, C.E. Wetland classification using optical and radar data and neural network classification. *Int. J. Remote Sens.* **1998**, *19*, 1545–1560. [CrossRef]

31. Islam, M.A.; Thenkabail, P.S.; Kulawardhana, R.W.; Alankara, R.; Gunasinghe, S.; Edussriya, C.; Gunawardana, A. Semi-automated methods for mapping wetlands using landsat ETM+ and SRTM data. *Int. J. Remote Sens.* **2008**, *29*, 7077–7106. [CrossRef]

32. Bwangoy, J.-R.B.; Hansen, M.C.; Roy, D.P.; Grandi, G.D.; Justice, C.O. Wetland mapping in the Congo Basin using optical and radar remotely sensed data and derived topographical indices. *Remote Sens. Environ.* **2010**, *114*, 73–86. [CrossRef]

33. Davranche, A.; Lefebvre, G.; Poulin, B. Wetland monitoring using classification trees and spot-5 seasonal time series. *Remote Sens. Environ.* **2010**, *114*, 552–562. [CrossRef]

34. Zhang, Y.; Wang, G.; Wang, Y. Changes in alpine wetland ecosystems of the Qinghai-Tibetan plateau from 1967 to 2004. *Environ. Monit. Assess.* **2011**, *180*, 189–199. [CrossRef] [PubMed]

35. Zhang, Y.; Lu, D.; Yang, B.; Sun, C.; Sun, M. Coastal wetland vegetation classification with a landsat thematic mapper image. *Int. J. Remote Sens.* **2011**, *32*, 545–561. [CrossRef]

36. Margono, B.A.; Bwangoy, J.R.B.; Potapov, P.V.; Hansen, M.C. Mapping wetlands in indonesia using landsat and palsar data-sets and derived topographical indices. *Geo-Spat. Inf. Sci.* **2014**, *17*, 60–71. [CrossRef]

37. Lunetta, R.S.; Balogh, M.E. Application of multi-temporal LANDSAT 5 Tm imagery for wetland identification. *Photogr. Eng. Remote Sens.* **1999**, *65*, 1303–1310.

38. Klemas, V. Remote sensing of wetlands: Case studies comparing practical techniques. *J. Coast. Res.* **2011**, *27*, 418–427. [CrossRef]

39. Bai, Q.; Chen, J.; Chen, Z.; Dong, G.; Dong, J.; Dong, W.; Fu, V.W.K.; Han, Y.; Lu, G.; Li, J.; et al. Identification of coastal wetlands of international importance for waterbirds: A review of China coastal waterbird surveys 2005–2013. *Avian Res.* **2015**, *6*, 12. [CrossRef]

40. Best, R.G.; Moore, D.G. Landsat interpretation of prairie lakes and wetlands of eastern South Dakota. *Can. J. Math.* **1979**, *59*, 421–426.

41. Szantoi, Z.; Escobedo, F.; Abd-Elrahman, A.; Smith, S.; Pearlstine, L. Analyzing fine-scale wetland composition using high resolution imagery and texture features. *Int. J. Appl. Earth Obs. Geoinf.* **2013**, *23*, 204–212. [CrossRef]

42. O'Hara, C.G. Remote sensing and geospatial application for wetland mapping, assessment and mitigation. In Proceedings of the Integrating Remote Sensing at the Global, Regional and Local Scale, Pecora 15/Land Satellite Information IV Conference, Denver, CO, USA, 10–15 November 2002.

43. Gluck, M.J.; Rempel, R.S.; Uhlig, P. *An Evaluation of Remote Sensing for Regional Wetland Mapping Applications*; Ontario Forest Research Institute: Sault Ste. Marie, ON, Canada, 1996.

44. Hinson, J.M.; German, C.D.; Pulich, W., Jr. Accuracy assessment and validation of classified satellite imagery of Texas coastal wetlands. *Mar. Technol. Soc. J.* **1994**, *28*, 4–9.

45. Niu, Z.; Zhang, H.; Wang, X.; Yao, W.; Zhou, D.; Zhao, K.; Zhao, H.; Li, N.; Huang, H.; Li, C.; et al. Mapping wetland changes in China between 1978 and 2008. *Chin. Sci. Bull.* **2012**, *57*, 2813–2823. [CrossRef]

46. Töyrä, J.; Pietroniro, A. Towards operational monitoring of a northern wetland using geomatics-based techniques. *Remote Sens. Environ.* **2005**, *97*, 174–191. [CrossRef]

47. Schmidt, K.; Skidmore, A. Spectral discrimination of vegetation types in a coastal wetland. *Remote Sens. Environ.* **2003**, *85*, 92–108. [CrossRef]

48. Mwita, E.; Menz, G.; Misana, S.; Nienkemper, P. Detection of small wetlands with multi sensor data in East Africa. *Adv. Remote Sens.* **2012**, *1*, 64–73. [CrossRef]

49. Li, S.N.; Wang, G.X.; Deng, W.; Hu, Y.M.; Hu, W.W. Influence of hydrology process on wetland landscape pattern: A case study in the Yellow River delta. *Ecol. Eng.* **2009**, *35*, 1719–1726. [CrossRef]

50. Jiang, Q.O.; Deng, X.; Zhan, J.; Yan, H. Impacts of economic development on ecosystem risk in the Yellow River delta. *Procedia Environ. Sci.* **2011**, *5*, 208–218. [CrossRef]

51. Yang, H.; Guo, H.; Wang, C. Coast line dynamic inspect and land cover classification at Yellow River mouth using TM-SAR data fusion method. *Geogr. Terrance Res.* **2001**, *17*, 185–191.

52. Cui, B.; He, Q.; Zhao, X. Ecological thresholds of suaeda salsa to the environmental gradients of water table depth and soil salinity. *Acta Ecol. Sin.* **2008**, *28*, 1408–1418.

53. Cui, B.; Tang, N.; Zhao, X.; Bai, J. A management-oriented valuation method to determine ecological water requirement for wetlands in the Yellow River delta of China. *J. Nat. Conserv.* **2009**, *17*, 129–141. [CrossRef]

54. Chander, G.; Markham, B.L.; Helder, D.L. Summary of current radiometric calibration coefficients for landsat mss, tm, ETM+, and EO-1 ali sensors. *Remote Sens. Environ.* **2009**, *113*, 893–903. [CrossRef]

55. McFeeters, S.K. The use of normalized difference water index (NDWI) in the delineation of open water features. *Int. J. Remote Sens.* **1996**, *17*, 1425–1432. [CrossRef]

56. Zhu, C.; Zhang, X.; Qi, J. Detecting and assessing spartina invasion in coastal region of China: A case study in the Xiangshan Bay. *Acta Oceanol. Sin.* **2016**, *35*, 35–43. [CrossRef]

57. Zhang, G.; Li, K.; Zhan, L. Dynamics of wetland and protection measures for the modern Yellow River delta. *Ecol. Environ. Sci.* **2009**, *18*, 394–398.

58. Houlahan, J.E.; Keddy, P.A.; Makkay, K.; Findlay, C.S. The effects of adjacent land use on wetland species richness and community composition. *Wetlands* **2006**, *26*, 79–96. [CrossRef]

59. Wu, T.; Luo, J.; Xia, L.; Yang, h.; Shen, Z.; Hu, X. An automatic sample collection method for object-oriented classification of remotely sensed imageries based on transfer learning. *Acta Geod. Cartogr. Sin.* **2014**, *43*, 908–916.

60. Jiang, Y.; Cui, H.; Li, Y. Study on dynamic change of wetland in Sanjiang Plain, northeast area. *J. Jilin Univ. (Earth Sci. Ed.)* **2009**, *39*, 1127–1136.

61. Xing, Y.; Jiang, Q.G. Landscape spatial patterns changes of the wetland in Qinghai-Tibet plateau. *Ecol. Environ. Sci.* **2009**, *18*, 1010–1015.

62. Qiao, W.; Fang, B.; Wang, Y. Land use change information mining in highly urbanized area based on transfer matrix: A case study of Suzhou, Jiangsu Province. *Geogr. Res.* **2013**, *32*, 1497–1507.

63. Liu, R.; Zhu, D. Methods for detecting land use changes based on the land use transition matrix. *Resour. Sci.* **2010**, *32*, 1544–1550.

64. Wang, J.; Wang, L.; Shi, G.; Liu, X.; Yao, J.; Han, X. The situation and strategy of Huanghe runoff drying up. *China Water Resour.* **1999**, *04*, 10–13.

65. Liu, C.; Cheng, L. Anlysis on runoff series with special reference to drying up courses of Huanghe River. *Acta Geogr. Sin.* **2000**, *55*, 257–265.

66. Dai, T.; Chen, S. Causes, effects and countermeasures of Huanghe drying up. Shandong Water Conserv. *Sci. Technol.* **1997**, *2*, 2–5.

67. Zhang, X.; Li, P.; Liu, L.; Li, P. *China's Coastal Wetland Degradation*; Maritime Press: Beijing, China, 2010.

68. Li, X.; Wang, S.; Li, P.; Zhang, Z. Problems of wetland agricultural exploitation and sustainable developing countermeasures for Yellow River delta. *Mod. Agric. Sci. Technol.* **2013**, *15*, 272–275.

69. Mu, C.; Yang, L.; Hu, Y. Yellow River delta swamp protection and the concord progresswith oil field development. Environ. *Prot. Oil Gas Fields* **1998**, *8*, 34–37.

70. Xu, J.; Li, X.; Yang, H.; Luo, J. Water and sediment reduction effects through hydraulic and water and soil conservation measures in the middle Yellow River and its influence on the lower Yellow River. *Yellow River* **1997**, *7*, 45–47.

71. Wang, J.; Meng, Y.; Zhang, L. Remote sensing monitoring and change analysis of Yellow River estuary coastline in the past 42 years. *Remote Sens. Land Resour.* **2016**, *28*, 188–193.

72. Ye, Q. Flow interruptions and their environmentalimpact on the Yellow River delta. *Acta Geogr. Sin.* **1998**, *53*, 385–392.

73. Shi, F.; Zhang, R. Cause analysis and recognitions on the recent sharp decreasing ofthe Yellow River water and sediment amount. *Yellow River* **2013**, *35*, 1–3.

74. Cui, S. Influence of water discharge cutoff ofhuanghe on environment of its delta. *Mar. Sci.* **2002**, *26*, 42–46.

75. Chen, J.; Wang, S.; Mao, Z. Monitoring wetland changes in Yellow River delta byremote sensing during 1976–2008. *Prog. Geogr.* **2011**, *30*, 585–592.

76. Zhang, Q.; Zhu, C.; Liu, C.L.; Jiang, T. Environmental change and its impacts on human settlement in the Yangtze delta, China. *Catena* **2005**, *60*, 267–277. [CrossRef]

77. Li, Y.; Yu, J.; Han, G.; Wang, X.; Wang, Y.; Guan, B. Dynamic evolution of natural wetlands and its driving factors in the Yellow River delta. *Chin. J. Ecol.* **2011**, *30*, 1535–1541.

78. Wang, Y.; Yu, J.; Dong, H.; Li, Y.; Zhou, D.; Fu, Y.; Han, G.; Mao, P. Spatial evolution of landscape pattern of coastal wetlands in Yellow River delta. *Sci. Geogr. Sin.* **2012**, *32*, 717–724.

79. Stocker, T. *Climate Change 2013: The Physical Science Basis: Working Group I Contribution to the Fifth Assessment Report of the Intergovernmental Panel on Climate Change*; Cambridge University Press: Cambridge, UK, 2014.

80. Solomon, S. *Climate Change 2007—The Physical Science Basis: Working Group I Contribution to the Fourth Assessment Report of the IPCC*; Cambridge University Press: Cambridge, UK, 2007; Volume 4.

81. Wang, K. China rate of sea level rise than the global average. *Sci. Technol. Commun.* **2017**, *6*, 2. [CrossRef]

82. Wu, Q.; Zheng, X.; Ying, Y.; Hou, Y.; Xie, X. Relative sea level rise in coastal areas of China in the 21st century and its control strategies. *Sci. China* **2002**, *32*, 760–766.

83. Song, D.; Yun, J.; Wang, G.; Han, G.; Guan, B.; Li, Y. Change characteristics of average annual temperature and annual precipitationin costal wetland region of the Yellow River delta from 1961 to 2010. *Wetl. Sci.* **2016**, *14*, 248–253.

84. Jiang, D.; Fu, X.; Wang, K. Vegetation dynamics and their response to freshwater inflow and climate variables in the Yellow River delta, China. *Quat. Int.* **2013**, *304*, 75–84. [CrossRef]

85. Wang, D.; Ma, F.; Hou, D. Multi-time scales analysis of precipitation in Yellow River delta reserve. *J. Southwest For. Univ.* **2010**, *30*, 33–37.

86. Li, G.; Han, M.; Zhang, D. Climatic change tendency in Yellow River delta during 1961–2013. *Yellow River* **2017**, *39*, 30–37.

87. Li, X.; Xue, Z.; Gao, J. Dynamic changes of plateau wetlands in madou county, the Yellow River source zone of China: 1990–2013. *Wetlands* **2016**, *36*, 299–310. [CrossRef]

88. Saito, Y.; Yang, Z.; Hori, K. The Huanghe (Yellow River) and changjiang (Yangtze River) deltas: A review on their characteristics, evolution and sediment discharge during the holocene. *Geomorphology* **2001**, *41*, 219–231. [CrossRef]

89. Lan, Y.; Zhao, G.; Zhang, Y.; Wen, J.; Hu, X.; Liu, J.; Gu, M.; Chang, J.; Ma, J. Response of runoff in the headwater region of the Yellow River to climate change and its sensitivity analysis. *J. Geogr. Sci.* **2010**, *20*, 848–860. [CrossRef]

90. Li, L.; Shen, H.; Dai, S.; Xiao, J.; Shi, X. Response to climate change and prediction of runoffin the source region of Yellow River. *Acta Geogr. Sin.* **2011**, *66*, 1261–1269.

91. Ding, Y.; Pan, S. Evolutionary characteristics of runoff into the sea ofthe Huanghe River and their causes in recent 50 years. *Quat. Sci.* **2007**, *27*, 709–717.

92. Wu, S.; Wang, W.; Wu, G. The calamity caused by the storm tide of the No.9216 typhoon in Shandong Province. *J. Catastrophol.* **1994**, *9*, 44–47.

93. Liu, J. China's road to sustainability. *Science* **2010**, *328*, 50. [CrossRef] [PubMed]

94. Wang, Z.; Wu, J.; Madden, M.; Mao, D. China's wetlands: Conservation plans and policy impacts. *Ambio* **2012**, *41*, 782–786. [CrossRef] [PubMed]

95. Zhang, C.; Han, M.; Shi, L. Runoff into the sea of the Yellow River: Characteristics and its response to climate change. *Yellow River* **2015**, *37*, 10–14.

96. Tiner, R.W. *Wetland Indicators: A Guide to Wetland Identification, Delieation, Classification and Mapping*; Taylor and Francis: New York, NY, USA, 1999. [CrossRef]

97. Ming, D.; Wang, Q.; Yang, J. Spatial scale of remote sensing image and selection of optimal spatial resolution. *J. Remote Sens.* **2008**, *12*, 529–537.

98. Klemas, V.V. Remote sensing of landscape-level coastal environmental indicators. *Environ. Manag.* **2001**, *27*, 47–57. [CrossRef] [PubMed]

99. Koeln, G.; Bissonnette, J. Cross-correlation analysis: Mapping landcover change with a historic landcover database and a recent, single-date multispectral image. In Proceedings of the 2000 ASPRS Annual Convention, Washington, DC, USA, 22–26 May 2000.

100. Gong, H.; Jiao, C.; Zhou, D.; Li, N. Scale issues of wetland classification and mapping using remote sensing images: A case of honghe national nature reserve in Sanjiang Plain, northeast China. *Chin. Geogr. Sci.* **2011**, *21*, 230–240. [CrossRef]

101. Woodcock, C.E.; Strahler, A.H. The factor of scale in remote sensing. *Remote Sens. Environ.* **1987**, *21*, 311–332. [CrossRef]
102. Bo, Y.; Wang, J. Exploring the scale effecting thematic classification of remotely sensed data: The statistical separability based method. Remote Sens. *Technol. Appl.* **2004**, *19*, 443–450.
103. Pan, J.; Zhu, D.; Yan, T.; Sun, L. Application of the relationship between spatial resolution of rs image and mapping scale. *Trans. CASE* **2005**, *21*, 124–128.
104. Arvor, D.; Jonathan, M.; Meirelles, M.S.P.; Dubreuil, V.; Durieux, L. Classification of modis EVI time series for crop mapping in the state of Mato Grosso, Brazil. *Int. J. Remote Sens.* **2011**, *32*, 7847–7871. [CrossRef]
105. Brown, J.C.; Kastens, J.H.; Coutinho, A.C.; Victoria, D.D.C.; Bishop, C.R. Classifying multiyear agricultural land use data from Mato Grosso using time-series modis vegetation index data. *Remote Sens. Environ.* **2013**, *130*, 39–50. [CrossRef]
106. Kehl, T.N.; Todt, V.; Veronez, M.R.; Cazella, S.C. Amazon rainforest deforestation daily detection tool using artificial neural networks and satellite images. *Sustainability* **2012**, *4*, 2566–2573. [CrossRef]
107. Zheng, B.; Myint, S.W.; Thenkabail, P.S.; Aggarwal, R.M. A support vector machine to identify irrigated crop types using time-series landsat NDVI data. *Int. J. Appl. Earth Obs. Geoinf.* **2015**, *34*, 103–112. [CrossRef]
108. Chen, Y.; Lin, Z.; Zhao, X.; Wang, G.; Gu, Y. Deep learning-based classification of hyperspectral data. *IEEE J. Sel. Top. Appl. Earth Obs. Remote Sens.* **2014**, *7*, 2094–2107. [CrossRef]
109. Yue, J.; Zhao, W.; Mao, S.; Liu, H. Spectral–spatial classification of hyperspectral images using deep convolutional neural networks. *Remote Sens. Lett.* **2015**, *6*, 468–477. [CrossRef]
110. Castelluccio, M.; Poggi, G.; Sansone, C.; Verdoliva, L. Land use classification in remote sensing images by convolutional neural networks. *J. Mol. Struct.* **2015**, *537*, 163–172.
111. Hu, X.; Weng, Q. Estimating impervious surfaces from medium spatial resolution imagery using the self-organizing map and multi-layer perceptron neural networks. *Remote Sens. Environ.* **2009**, *113*, 2089–2102. [CrossRef]
112. Hu, F.; Xia, G.S.; Hu, J.; Zhang, L. Transferring deep convolutional neural networks for the scene classification of high-resolution remote sensing imagery. *Remote Sens.* **2015**, *7*, 14680–14707. [CrossRef]

 *water*

Article

The Effects of Litter Layer and Topsoil on Surface Runoff during Simulated Rainfall in Guizhou Province, China: A Plot Scale Case Study

Qiuwen Zhou [1,2,*] , **Xu Zhou** [1,2], **Ya Luo** [1,2] **and Mingyong Cai** [3]

1 School of Geographic and Environmental Science, Guizhou Normal University, Guiyang 550001, China; zxzy8178@163.com (X.Z.); luoya2002@163.com (Y.L.)
2 State Engineering and Technology Institute for Karst Desertification Control, Guizhou Normal University, Guiyang 550001, China
3 Satellite Environment Center of MEP, Beijing 100094, China; caimingyong@126.com
* Correspondence: zhouqiuwen@gznu.edu.cn; Tel.: +86-851-8322-7361

Received: 25 May 2018; Accepted: 9 July 2018; Published: 11 July 2018

Abstract: Litter layers and topsoil have important effects on surface runoff. To investigate these effects at the plot scale, artificial rainfall experiments were conducted on micro-runoff plots in Guizhou Province, China. Three types of plots were selected, the thin litter layer with low soil bulk density type (T-L type), the thick litter layer with high soil bulk density type (T-H type), and the moderate litter depth and soil bulk density type (M type), and three artificial rainfall intensities (30 mm/h, 70 mm/h, 120 mm/h) were used. The runoff volume was largest in the T-H type plot at different rainfall intensities and durations. Runoff in the M type plot had characteristics of both the T-L and T-H type plots. The runoff yielding speed was significantly higher and the runoff yielding time was significantly lower in the T-H type plot. In general, the runoff coefficient was the smallest in the T-L type plot and largest in the T-H type plot. The variations in the runoff coefficient were 15.6%, 19.3%, and 5.8% for the T-L, T-H, and M type plots respectively. The results of this study can improve the understanding of surface runoff processes at the plot scale under different litter and surface soil conditions.

Keywords: runoff; simulated rainfall; plot scale; litter layer; topsoil; karst

---

1. Introduction

A litter layer is typically composed of dead leaves, twigs, small branches, and other fragmented organic material, and influences the hydrological processes that operate in forested watersheds [1]. The regulation of the litter layer includes the interception, throughfall, and stemflow, which regulate soil evaporation, increase permeability, reduce overland flow, and create a rapid-flow component within the litter layer [2–4]. The simultaneous operation of these processes causes the litter layer to affect both short-term runoff and long-term water balance within a hydrological cycle. Not only the litter layer but also the topsoil has a notable impact on hydrological processes. The topsoil state regarding water movement into the soil mass may affect evaporation, infiltration, and distribution of topsoil [5]. Further, various runoff generating processes (saturation excess overland flow, infiltration excess overland flow, and return flow) are highly regulated by the topsoil state [6,7]. In short, as one of the hydrological elements, the generation and dynamics of surface runoff are significantly affected by the state of the litter layer and the topsoil.

Several studies examined how surface runoff is regulated by the extent of litter coverage. To investigate the influence of the litter layer and undergrowth intercrops on surface runoff and soil erosion, experimental field plots were monitored by Liu et al. [8] over one rainy season in a rubber monoculture and a rubber/tea agroforestry system. In a simulated rainfall experiment in runoff plots, Li et al. [4] investigated the effect of litter cover on surface runoff in northern China. Miyata et al. [9] examined the effects of forest floor coverage on overland flow generation and soil erosion in mature Japanese cypress plantations with different coverage conditions. Prosdocimi et al. [10] evaluated the immediate effectiveness of the litter layer in reducing surface runoff generation in Mediterranean vineyards. These works have greatly enriched the understanding of the relationship between the litter layer and surface runoff. In addition to litter layers, the runoff effect of surface soils has attracted the attention of many researchers in the hydrology field. In terms of runoff generating processes, overland flow generation mechanisms affected by topsoil treatment and Hortonian overland flow is responsible for significant amounts of soil loss in Mediterranean geomorphological systems [11]. Sorbotten et al. [12] observed that in an Acrisol on a forested hillslope with a monsoonal subtropical climate, the topsoil responded quickly to rainfall events, causing frequent cycles of saturation and aeration of soil pores. Compaction and destruction of the topsoil structure by machinery, especially at harvest, is important in initiating runoff [13]. In fact, the runoff effect of topsoil also has a scale effect. At sites with intensive grazing small-plot devices deliver significantly higher runoff coefficients than large-plot devices, due to topsoil compaction and the shortened flow path [14]. In terms of the impact of revegetation on runoff, Zhang et al. [15] showed that revegetation with artificial plants improved topsoil hydrological properties but intensified deep-soil drying in northern Loess Plateau, China.

Guizhou Province has the largest area of karst landforms in China, with the karst area accounting for 64.2% of the total area of the province [16]. The karst hydrological system consists of two systems, surface and underground, and surface water flow is connected with an extensive subsurface drainage or karst conduit network by fissures, sinkholes, and swallets [17,18]. In a karst environment, due to the slow rate of formation of soil, rainfall and subsequent surface runoff may scour the topsoil, leading to soil erosion and consequent rocky desertification [19,20]. Precipitation in a karst area quickly infiltrates the ground and enters the underground system. Thus, it cannot support plant growth. Areas with a low density of surface streams often have high soil erosion and a more severe problem of rocky desertification [21]. Thus, surface runoff is one of the key factors affecting the ecological problems associated with karst regions, such as rocky desertification and sparse vegetation that may require restoration. Furthermore, due to the shallowness of the soil in the karst area, the litter layer and topsoil affect the formation of surface runoff. Therefore, it is necessary to conduct in-depth research on surface runoff in Guizhou Province.

Since the 1990s, there has been a series of studies on karst surface runoff [22–27]. These studies focused on the runoff generation on limestone slopes [28,29], the effects of the proportion of bare bedrock and degree of underground pore fissures on surface runoff [30], and the research methods for surface runoff in karst areas [31,32]. However, to the best of our knowledge, there have not been many studies on the combined effects of litter layers and topsoil on surface runoff processes, especially in karst areas.

In this study, three plots with similar climatic conditions and vegetation, but with different litter and topsoil states, were selected in Guizhou Province, China. Simulated rainfall experiments were conducted and the runoff characteristics of the three plots were compared. Our goal was to determine the differences in surface runoff characteristics under different litter and surface soil combinations at the plot scale in a case study at the three sites in Guizhou Province.

2. Materials and Methods

2.1. Study Area

The experimental sites were located in Huaxi District in Guiyang City, Guizhou Province in southwestern China. The region has a subtropical moist monsoon climate and an annual precipitation of 1129 mm. The mean annual temperature is 15.3 °C and the monthly averages range from −7.3 °C to

35.1 °C. The elevation varies from 1002 to 1627 m above sea level and the relief is dominated by hills. The landforms of the study area can be divided into three types: karst landforms, non-karst landforms, and semi-karst landforms. In the karst area, the soils are calcareous soils developed from limestone and the soil layer is thin and discontinuous. In the non-karst area, the soil type is yellow soil developed from sandstone and the soil layer is thick. In the semi-karst area, the soil type is yellow soil developed from dolostone and the soil layer is thinner than that in the non-karst area. The locations of the three plots are shown in Figure 1.

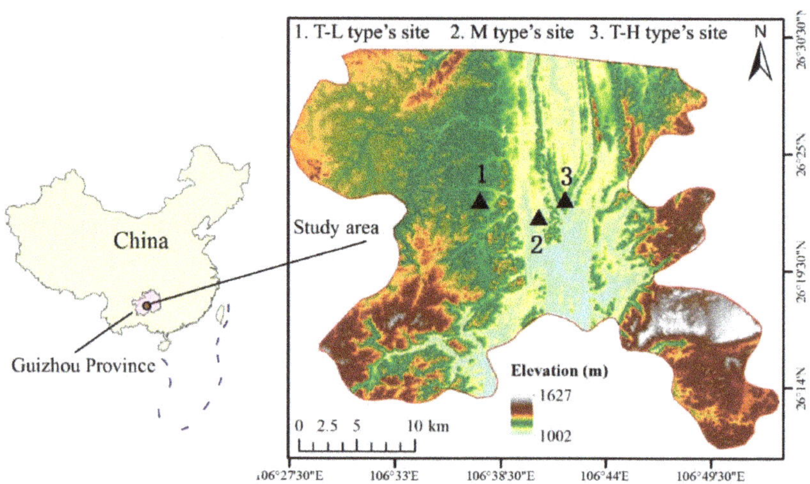

Figure 1. The location and elevation of the study area.

2.2. Description of Plots

In this work, three experimental sites corresponding to the three litter and surface soil combination types were selected. The first type is the thin litter layer with low soil bulk density (T-L type), the second type is the thick litter layer with high soil bulk density (T-H type), and the third type is the moderate litter depth and soil bulk density (M type). To ensure the consistency of climatic conditions, the three sites were selected such that the distance between any sites was not more than 10 km. The main tree species at all sites is *Pinus massoniana*. The T-L, T-H, and M types have an average tree age of 20, 12, and 17 y, an average tree height of 15, 13, and 16 m, and a canopy density of 0.85, 0.95, and 0.95, respectively. All three sites have a small amount of natural understory vegetation on the forest floor and the soil is mostly covered by a litter layer of needles. We set a group of nine runoff plots at each site for a grand total of 27 runoff plots. The runoff plots at each site were adjacent to each other. The average characteristic values of each runoff plot group at the sites are listed in Table 1. The experiments were conducted in October 2016, which is in the early dry season. The soil moisture was generally low and changed slightly because of the seasonal drought. Hence, the antecedent soil moisture differences did not affect the simulation experiments.

Table 1. Characteristics of liter layer and topsoil.

Site	Litter Depth (mm)	Litter Mass (t/hm^2)	Soil Bulk Density (g/cm^3)	Slope Gradient
T-L type	54.16	30.52	1.515	40°
T-H type	133.617	40.60	2.098	39°
M type	68.85	45.26	1.742	37°

The experiments were carried out using a portable rainfall simulator. The nozzle was set on a steel frame at the height of 2 m. At this height, the nozzle generated a constant intensity of rainfall within a 1 m radius around the rainfall simulator. The circular area was marked on the soil surface, and then a 60 cm × 100 cm rectangular runoff plot was set in the center of the circle. The plot boundaries were defined using 30 cm high iron sheets that were inserted into the ground to a depth of 10 cm. A V-shaped groove was placed at the lower end of the plot to collect runoff (Figure 2). Finally, clear water was supplied through a 2.5 cm diameter high-pressure hose by a pump. The pump was powered by a 2.0 Kw/220 V gasoline generator. The rainfall intensity was controlled by a flow meter attached to the hose. The correspondence between rainfall intensity and flow has been determined by our previous experiments.

2.3. Data Acquisition and Processing

Each simulated rainfall event lasted 30 min and runoff volume data were collected every 3 min. When continuous water flow began to appear in the V-shaped groove, the runoff generation time was recorded. Based on the rainfall intensity and rainfall frequency in this area, three rainfall intensity values were selected: 30 mm/h, 75 mm/h, and 125 mm/h. The corresponding three rainfall intensity experiments were conducted on each plot and three replications were made for each rainfall intensity. Thus, a total of 27 rainfall experiments were conducted (3 litter layer and topsoil combination types × 3 intensities × 3 replicates). Finally, to reduce the experimental error and obtain valid data for the analysis, we averaged the data of each replicate.

Figure 2. Micro-runoff plot setup in the forest.

3. Results

3.1. Surface Runoff Changes under Different Litter Layer and Topsoil Combination Conditions

Figure 3 shows that runoff is typically greater in the T-H type than in the T-L type or M type at various rainfall intensities. At the three sites, runoff first increased notably and then reached a steady state. When the rainfall intensity was 30 mm/h, the runoff at an early stage at three different litter layer and topsoil combination conditions was roughly the same. The runoff peaked at 9 min in the T-L type, but it reached its maximum at 18 min in the T-H type and at 24 min in the M type. The runoff in the T-H type fluctuated the most during the entire rainfall event and this runoff value was always the largest one. The maximum value of the T-H type was 1.62 times that of the M type and 1.78 times that of the T-L type. When the rainfall intensity was 75 mm/h, the increasing trend of the runoff in three litter layer and topsoil combination conditions was consistent and stable. The T-H type had the largest runoff, followed by the M type and the T-L type, respectively. When the rainfall intensity was 120 mm/h, the runoff at the T-H type increased rapidly at the beginning of the rainfall event and reached its maximum at 18 min. This maximum was 1.55 times that at the M type and 1.99 times that at the T-L type. The runoff at the T-L and M type increased steadily with a small gap. When the continuous runoff reached a peak, the runoff at the T-L and M type maintained the steady state. The runoff at the T-H type was always the highest under different rainfall intensities and durations. The runoff at the T-L type was typically the smallest. The increase in runoff became smaller with the increase in the rainfall duration. The runoff at the M type was typically between the runoff values at the T-L type and T-H type.

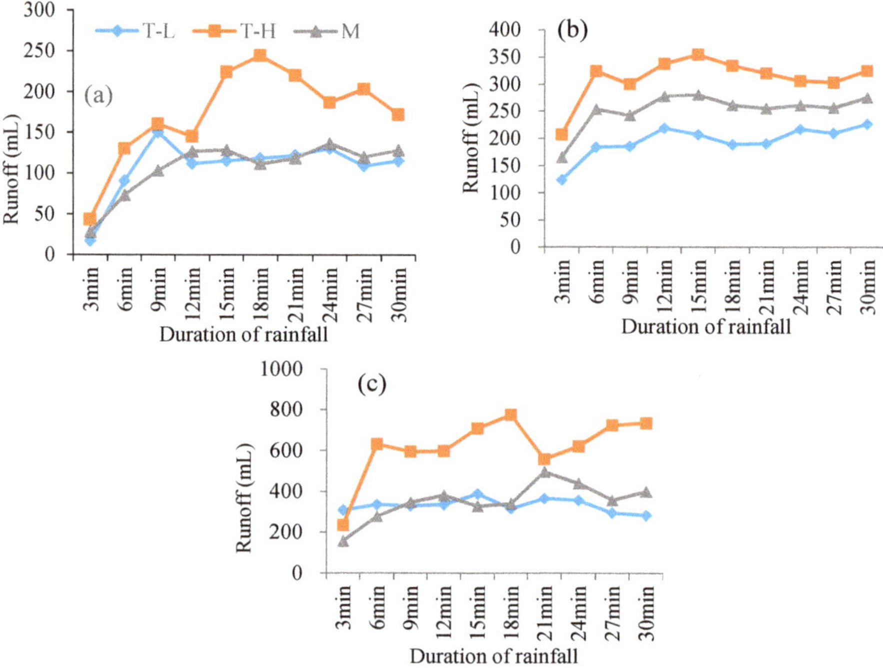

Figure 3. Runoff changes under different surface landscape conditions. (**a**) 30 mm/h rainfall intensity; (**b**) 75 mm/h rainfall intensity; and (**c**) 120 mm/h rainfall intensity. T-L stands for thin litter layer with low soil bulk density, T-H stands for thick litter layer with high soil bulk density, and M stands for moderate litter depth and soil bulk density.

3.2. Surface Runoff Changes under Different Rainfall Intensity

To understand the effect of rainfall intensity on surface runoff, this section analyzes the variations in surface runoff for the T-L, M, and T-H types for the rainfall intensities of 30 mm/h, 75 mm/h, and 120 mm/h, respectively (Figure 4). For the T-L type (Figure 4a), under the condition of heavy rainfall (120 mm/h), the runoff reached a large value at the early stage of rainfall. In contrast, under the conditions of small and medium rainfall intensity (30 mm/h and 75 mm/h, respectively), the runoff took a long time to reach a large value. For the T-H type (Figure 4b), the runoff was relatively stable when the rainfall intensity was 30 mm/h and 75 mm/h. However, the runoff fluctuated greatly when the rainfall intensity was 120 mm/h and the maximum runoff at this intensity was 2.18 times the maximum runoff at 75 mm/h and 3.1 times the maximum runoff at 30 mm/h. In the early stage of rainfall (3 min), the runoff for the T-H type under the heavy rain intensity was not significantly different from the runoff at the medium and small rainfall intensities. However, after a certain period (6 min), the runoff under the highest rainfall intensity reached a much larger value and increased rapidly with rainfall duration (Figure 4b). For the M type (Figure 4c), the trend of runoff was similar for three rainfall intensities. When the rainfall intensity was 120 mm/h, the maximum runoff was 1.6 times the maximum runoff at 75 mm/h and 3.6 times the maximum runoff at 30 mm/h. Figure 4c shows that the runoff for the M type increased with the increase in rainfall intensity and reached a steady state after attaining its maximum value. The effect of rainfall intensity on surface runoff for the M type was limited to a change of magnitude. There was no significant difference in the surface runoff pattern between the rainfall intensities for the M type (Figure 4c).

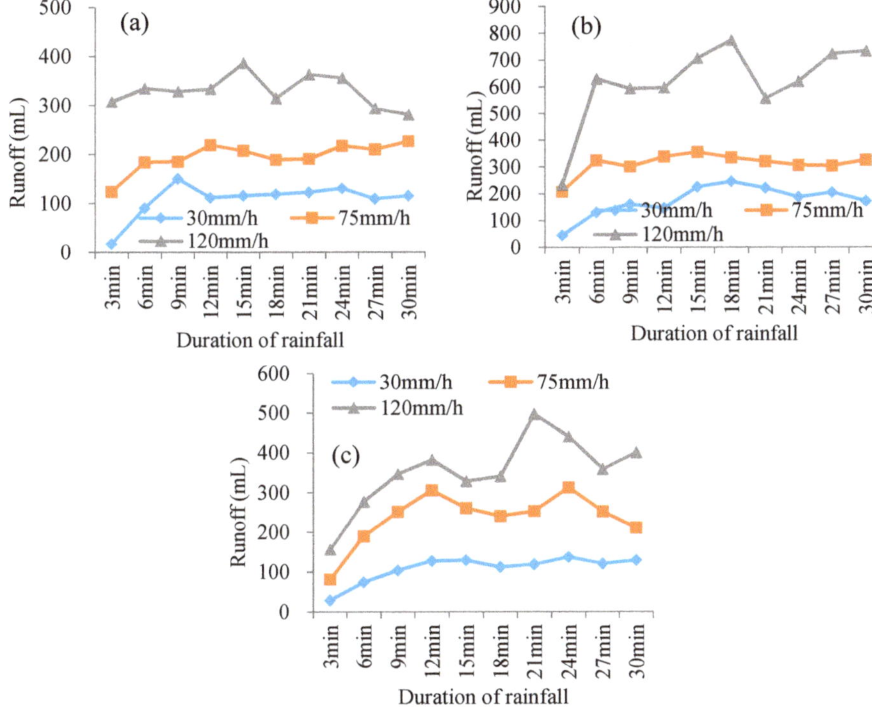

Figure 4. Runoff variations under different rainfall intensities. (**a**) T-L type; (**b**) T-H type; and (**c**) M type.

3.3. Runoff Yield Characteristics under Different Litter Layer and Topsoil Combination Conditions

Figure 5a shows that the runoff yielding times for different litter layer and topsoil combination conditions are 1–4 min. Under the same rainfall intensity, the T-H type had the shortest runoff yield time. When the rainfall intensity was 120 mm/h, this site generated runoff in 0.9 min. The runoff yielding time for the M type was longer than that for the T-H type, but runoff was generated within 1.51 min at the rainfall intensity of 120 mm/h. The T-L type had the longest runoff yielding time at the rainfall intensity of 30 mm/h. Under this intensity, it took 3.27 min to generate runoff for the T-L type, 2.37 min for the M type, and 1.7 min for the T-H type. With increasing rainfall intensity, the runoff yielding time of the three sites gradually shortened and the runoff yielding speed (Figure 5b) gradually increased. The T-H type had the highest runoff yielding speed and the T-L type had the smallest runoff yielding speed. In summary, the runoff yielding time for the T-H type was significantly shorter than the time for the T-L type or the M type and the runoff yielding speed for the T-H type was significantly higher than the rate for the T-L type or the M type.

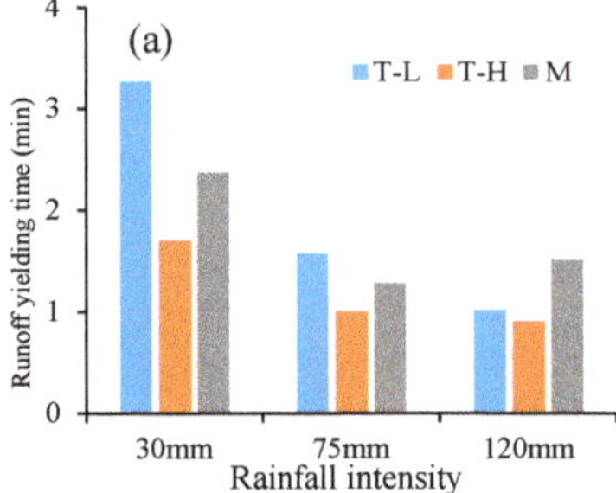
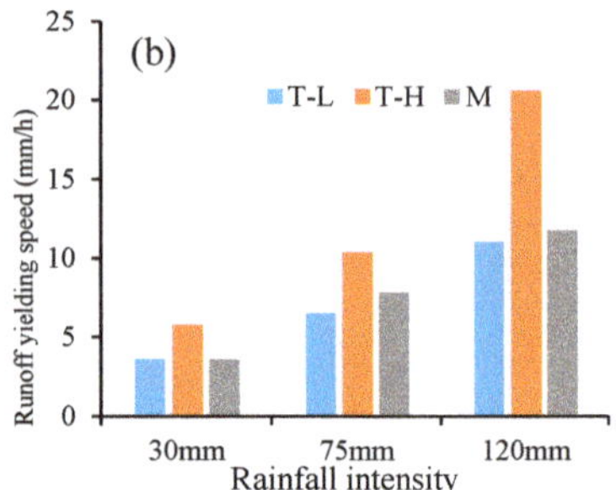

Figure 5. Runoff yielding time (**a**) and rate (**b**) at different sites. T-L stands for thin litter layer with low soil bulk density, T-H stands for thick litter layer with high soil bulk density, and M stands for moderate litter depth and soil bulk density.

3.4. Surface Runoff Coefficient under Different Litter Layer and Topsoil Combination Conditions

Figure 6a shows that the runoff coefficient of the T-L type was the smallest and the runoff coefficient of the T-H type was the largest. However, at the rainfall intensity of 30 mm/h, the runoff coefficient of the M type was slightly lower than the coefficient of the T-L type. Under the rainfall intensity of 120 mm/h, the runoff coefficient of the M type was not significantly different from that of the T-L type. At the T-L type and the T-H type, the runoff coefficient at the rainfall intensity of 75 mm/h was significantly lower than the rainfall coefficients of the other two rainfall intensities. This difference was not obvious for the M type. The effect of rainfall intensity on the runoff coefficient was insignificant. Under three rainfall intensities, the variation in the runoff coefficient was 15.6% for the T-L type, 19.3% for the T-H type, and 5.8% for the M type.

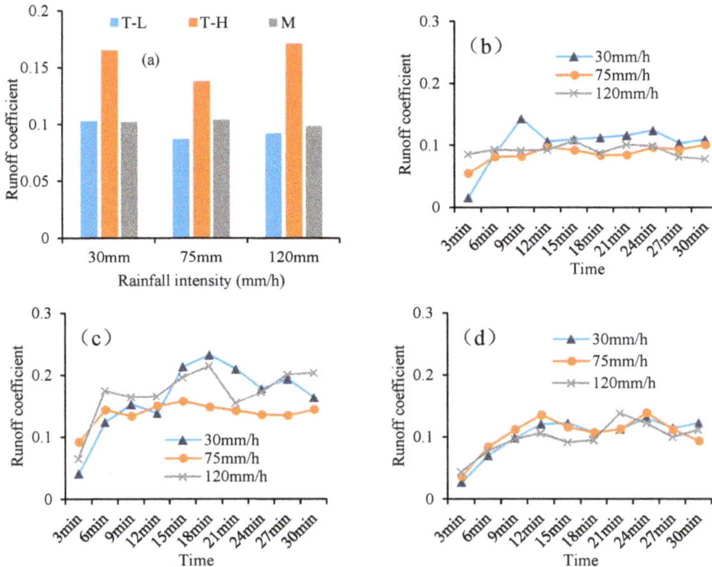

Figure 6. The general characteristics of the runoff coefficient (**a**) and the variations in the runoff coefficient with time for the T-L type (**b**); T-H type (**c**); and M type (**d**) plots. T-L stands for thin litter layer with low soil bulk density, T-H stands for thick litter layer with high soil bulk density, and M stands for moderate litter depth and soil bulk density.

In general, there were some differences in the variation of the runoff coefficient at different sites. After 6 min of rainfall, in general, the runoff coefficient was close to its maximum value for the T-L type (Figure 6b). Thereafter, the runoff coefficient basically remained stable. For the T-H type (Figure 6c), the runoff coefficients peaked at ~18 min after the rainfall began. For the M type (Figure 6d), the runoff coefficient reached its maximum at 12 min and remained basically stable thereafter. The runoff coefficient of the T-H type fluctuated substantially with different rainfall intensity values. The runoff coefficient of the T-L type fluctuated less and the runoff coefficient produced by different rainfall intensities at different points in time showed no obvious differences. In summary, the runoff coefficients of the T-H type were significantly larger than those of the T-L and M types.

4. Discussion

4.1. Influence of Litter Layer Conditions on Surface Runoff

Under normal circumstances, as rainfall intensity increases, the runoff coefficient continually increases, decreases, or remains stable, and the rainfall intensity and runoff coefficient reflect a stable relationship. In this study, the runoff coefficients for the T-H type and the T-L type were smaller at the rainfall intensity of 75 mm/h than at 30 mm/h or 120 mm/h and the relationship between the rainfall intensity and the runoff coefficient was not stable. However, the runoff coefficient for the M type was basically stable under different rainfall intensities. The analysis of the differences in the individual plots showed that there were some differences in litter thickness and the volume between the three types. The litter thickness was higher at the T-H and T-L plots than at the M plots and the litter thickness was the largest at the T-H plots.

The analysis of the characteristics of the litter showed that litter coverage plays an important role in runoff generation in a small-area runoff plot. When the rainfall intensity is low, most of the rainfall remains on the surface of the undecomposed litter and flows down the slope as it accumulates on the surface. In this case, the litter plays the role of a conduit and the continuous leaf litter on the slope surface forms a flow channel for surface runoff. As a result, more surface runoff is generated and the runoff coefficient becomes larger.

The sample sites selected in this study were all in the Pinus massoniana forest. Because of the elongated undecomposed litter of this species (pine needles), the guiding effect on water flow was more obvious. However, the ability of the litter to guide the flow of water was limited. Under the medium rainfall condition (75 mm/h), the amount of rainwater on the litter reaches its upper limit and water moves downward under the force of gravity, escaping the surface water flow channel formed by the litter. The water provided by the subsequent rainfall continues to drip beneath the water flow, thereby cutting off the water flow channel on the slope surface, increasing the volume of infiltration and reducing surface runoff and the runoff coefficient. When the rainfall intensity increases further (120 mm/h), the larger rainfall intensity makes it easier for the moisture content of the mineral soil layer to reach saturation or the rainfall intensity exceeds the infiltration capacity of the soil layer. In this case, the infiltration capacity may decrease and a large amount of rainwater may become surface runoff. Under these circumstances, the runoff coefficient may be relatively large. However, this discussion is based only on a small number of existing research results, combined with experimental observations and an analysis of possible causes. Thus, experimental errors, sample selection, and other accidental uncertainties cannot be completely ruled out. In follow-up studies the role and significance of the litter layer, especially the undecomposed layer, for surface runoff at the plot scale needs to be further explored, and the nonlinear relationships of runoff with rainfall intensity and slope gradient need to be analyzed. This will further reveal the formation and evolution mechanisms of slope runoff under different landscape conditions.

4.2. Influence of Topsoil Conditions on Surface Runoff

The results show that the surface runoff volume and runoff yielding speed of T-H are greater than those of the other types (Figures 3 and 5) and the runoff yielding time is less than that of the other two types (Figure 5). The soil bulk density for the T-H type is relatively large, meaning that the soil porosity is small and saturation is easily reached or the permeability coefficient is small and the rainfall intensity easily exceeds the infiltration rate, resulting in conversion of most of the precipitation to surface runoff. In contrast, the small bulk density of the other two types leads to the easy infiltration of the precipitation, resulting in a relatively small surface runoff value. Due to the large runoff volume of the T-H type, within the same time frame, the runoff yielding speed of the T-H type is also greater than the other two types. Similarly, due to the large soil bulk density of the T-H type and the low infiltration rate, surface runoff is relatively easily generated. Thus, the runoff yielding time is shorter and the runoff coefficient is larger in the T-H type than in the other types.

Although soil bulk density affects surface runoff, surface runoff is mainly determined by geological conditions. The study area is in a karst region, but karst is an extremely heterogeneous medium with varying lithology and sporadic non-carbonate areas. Although it is a carbonate rock, the landscape and soil conditions for limestone, dolomite, and argillaceous limestone areas are different. The selected three types (sites) have different lithological features. The T-H type lithology is sandstone, which can usually develop relatively thick soils, and the texture in the humid subtropical region of China is more viscous, resulting in a relatively large soil bulk density. The lithology for type T-L is limestone, which is easily dissolved by water, resulting in a thin layer of soil and small bulk density. The lithology for the M type is dolomite; dolomite and argillaceous limestone usually form a semi-karst landscape that differs from the karst [33,34]. Dolomite belongs to carbonate rocks, which can retain a part of the weathering products, forming a soil thickness and bulk density between those of the former two types. Therefore, under the same climatic conditions, lithology largely determines the soil conditions and thus affects the surface runoff process.

4.3. Possible Impact of Spatial Scale on Surface Runoff

In this study, the micro-runoff plot was taken as the unit of study. If the spatial scale is expanded to the slope or the small-watershed scale, the surface runoff characteristics of each landscape type can change substantially. Karst areas contain both surface and subsurface hydrological systems. As the scale of the study is expanded to the slope scale, it is found that surface runoff can penetrate the ground along the rocky fractures at low-lying sites. When the study scale is extended to the small catchment scale, it is found that surface runoff can become groundwater by being diverted into underground rivers along the loopholes and sinkholes. This expansion of the research scale leads to significant changes in surface runoff coefficients and other surface runoff characteristics. Therefore, although there is abundant research on the characteristics of surface runoff, the variations in surface runoff characteristics at different spatial scales in different litter layers and topsoil combination conditions remain the important problems to be solved.

5. Conclusions

A case study on three types of litter layer and topsoil combination (T-L, T-H, and M) was conducted to evaluate the characteristics of surface runoff at the plot scale in Guizhou Province, China. The results showed that the runoff volume was the largest for the T-H type under different rainfall intensities and durations. Runoff was the smallest for the T-L type and the increase in runoff with the increase in rainfall duration was also small. Runoff for the M type had similar characteristics to the runoff for the T-L and T-H types. In the early stages of heavy rainfall, the response of surface runoff to rainfall intensity for the M type was not significant. The runoff yielding speed was significantly higher in the T-H type than in the T-L and M types, and the runoff yielding time was significantly lower. In general, the runoff coefficient was smallest in the T-L type and largest in the T-H type. Under the three rainfall intensities, the variations in the runoff coefficient were 15.6% for the T-L type, 19.3% for the T-H type, and 5.8% for the M type. In the T-L type, the runoff coefficient fluctuated slightly and the runoff coefficients produced by different rainfall intensities at different points in time showed no obvious differences. The results showed that the M type was between the T-L and T-H types in terms of surface runoff, runoff yielding time, runoff yielding speed, and runoff coefficient. In addition, although soil conditions notably affected surface runoff, it was eventually determined by geological conditions. Further, despite a small number of field experiments, this work enriches knowledge of the effects of different litter and topsoil conditions on runoff processes at the plot scale.

Author Contributions: Q.Z. conceived and designed the experiments; X.Z. and Y.L. conducted the experiments; Q.Z. and M.C. analyzed the data; Q.Z. wrote the paper.

Funding: This work was supported by the Key Project of Science and Technology Program of Guizhou Province: Key Technology and Demonstration for the State Engineering Technology Institute for Karst Desertification Control (Qiankehe Zhongdazhuanxiang Zi [2014] 6007), the National Science Foundation of China (grant Nos. 41761003 and 41601471), the Science and Technology Support Program of Guizhou Province, China (grant No. [2017]2855), the Project for National Top Discipline Construction of Guizhou Province: Geography in Guizhou Normal University (85-01 2017 Qianjiao Keyan Fa), and the Scientific Research Starting Funds for Doctoral Scholars at Guizhou Normal University.

Conflicts of Interest: The authors declare no conflict of interest.

References

1. Gomyo, M.; Kuraji, K. Effect of the litter layer on runoff and evapotranspiration using the paired watershed method. *J. For. Res.* **2016**, *21*, 306–313. [CrossRef]
2. Walsh, R.P.D.; Voigt, P.J. Vegetation litter: An underestimated variable in hydrology and geomorphology. *J. Biogeogr.* **1977**, *4*, 253–274. [CrossRef]
3. Gerrits, A.M.J.; Pfister, L.; Savenije, H.H.G. Spatial and temporal variability of canopy and forest floor interception in a beech forest. *Hydrol. Processes* **2010**, *24*, 3011–3025. [CrossRef]
4. Li, X.; Niu, J.; Xie, B. Study on hydrological functions of litter layers in north china. *PLoS ONE* **2013**, *8*, e70328. [CrossRef] [PubMed]
5. Udom, B.E.; Omovbude, S.; Abam, P.O. Topsoil removal and cultivation effects on structural and hydraulic properties. *Catena* **2018**, *165*, 100–105. [CrossRef]
6. Castillo, V.M.; Gomez-Plaza, A.; Martınez-Mena, M. The role of antecedent soil water content in the runoff response of semiarid catchments: A simulation approach. *J. Hydrol.* **2003**, *284*, 114–130. [CrossRef]
7. Braud, I.; Desprats, J.F.; Ayral, P.A.; Bouvier, C.; Vandervaere, J.P. Mapping topsoil field-saturated hydraulic conductivity from point measurements using different methods. *J. Hydrol. Hydromech.* **2017**, *65*, 264–275. [CrossRef]
8. Liu, W.; Luo, Q.; Lu, H.; Wu, J.; Duan, W. The effect of litter layer on controlling surface runoff and erosion in rubber plantations on tropical mountain slopes, SW China. *Catena* **2017**, *149*, 167–175. [CrossRef]
9. Miyata, S.; Kosugi, K.; Gomi, T.; Mizuyama, T. Effects of forest floor coverage on overland flow and soil erosion on hillslopes in Japanese cypress plantation forests. *Water Resour. Res.* **2009**, *45*, 192–200. [CrossRef]
10. Prosdocimi, M.; Jordán, A.; Tarolli, P.; Keesstra, S.; Novara, A.; Cerdà, A. The immediate effectiveness of barley straw mulch in reducing soil erodibility and surface runoff generation in mediterranean vineyards. *Sci. Total Environ.* **2016**, *547*, 323–330. [CrossRef] [PubMed]
11. Hueso-González, P.; Ruiz-Sinoga, J.D.; Martínez-Murillo, J.F.; Lavee, H. Overland flow generation mechanisms affected by topsoil treatment: Application to soil conservation. *Geomorphology* **2015**, *228*, 796–804. [CrossRef]
12. Sorbotten, L.E.; Stolte, J.; Wang, Y.; Mulder, J. Hydrological Responses and Flow Pathways in an Acrisol on a Forested Hillslope with a Monsoonal Subtropical Climate. *Pedosphere* **2017**, *27*, 1037–1048. [CrossRef]
13. Evans, R. Factors controlling soil erosion and runoff and their impacts in the upper Wissey catchment, Norfolk, England: A ten year monitoring programme. *Earth Surf. Processes Landf.* **2017**, *42*, 2266–2279. [CrossRef]
14. Mayerhofer, C.; Meißl, G.; Klebinder, K.; Kohl, B.; Markart, G. Comparison of the results of a small-plot and a large-plot rainfall simulator–Effects of land use and land cover on surface runoff in Alpine catchments. *Catena* **2017**, *156*, 184–196. [CrossRef]

15. Zhang, Q.; Jia, X.; Zhao, C.; Shao, M.A. Revegetation with artificial plants improves topsoil hydrological properties but intensifies deep-soil drying in northern Loess Plateau, China. *J. Arid Land* **2018**, *10*, 335–346. [CrossRef]

16. Jiang, Z.; Lian, Y.; Qin, X. Rocky desertification in Southwest China: Impacts, causes, and restoration. *Earth-Sci. Rev.* **2014**, *132*, 1–12. [CrossRef]

17. Bonacci, O.; Roje-Bonacci, T. Sea water intrusion in coastal karst springs: Example of the Blaž Spring (Croatia). *Hydrol. Sci. J.* **1997**, *42*, 89–100. [CrossRef]

18. Ford, D.; Williams, P. Introduction to Karst. In *Karst Hydrogeology and Geomorphology*; John Wiley & Sons Ltd.: Hoboken, NJ, USA, 2007; pp. 1–8.

19. Bailey, G.N.; Ioakim, C.; King, G.P.C.; Turner, C.; Sanchez-Goni, F.; Sturdy, D.; Winder, N.P. Norwest epirus in the Palaeolithic. In *Understanding the Natural and Anthropogenic Causes of Degradation and Desertification in the Mediterranean Basin*; van der Leeuw, S.E., Ed.; EUR 18181 EN; European Communities: Luxembourg, 1998; pp. 329–359.

20. Zhang, D.F.; Wang, S.J.; Zhou, D.Q. Internal driving mechanism of karst rocky desertification in Guizhou Province. *Soil Water Conserv.* **2001**, *21*, 1–5. (In Chinese)

21. Chen, S.Z.; Zhou, Z.; Yan, L.; Li, B. Quantitative evaluation of ecosystem health in a karst area of south China. *Sustainability* **2016**, *8*, 975. [CrossRef]

22. Cerdà, A. The effect of patchy distribution of *Stipa tenacissima* L. on runoff and erosion. *J. Arid Environ.* **1997**, *36*, 37–51. [CrossRef]

23. Imeson, A.C.; Lavee, H.; Calvo, A.; Cerdà, A. The erosional response of calcareous soils along a climatological gradient in southeast Spain. *Geomorphology* **1998**, *24*, 3–16. [CrossRef]

24. Labat, D.; Ababou, R.; Mangin, A. Rainfall–runoff relations for karstic springs. Part ii: Continuous wavelet and discrete orthogonal multiresolution analyses. *J. Hydrol.* **2000**, *238*, 149–178. [CrossRef]

25. Li, Z.; Xu, X.; Xu, C.; Liu, M.; Wang, K.; Yu, B. Annual runoff is highly linked to precipitation extremes in karst catchments of southwest China. *J. Hydrometeorol.* **2017**, *18*, 2745–2759. [CrossRef]

26. Rooij, R.D.; Perrochet, P.; Graham, W. From rainfall to spring discharge: Coupling conduit flow, subsurface matrix flow and surface flow in karst systems using a discrete–continuum model. *Adv. Water Resour.* **2013**, *61*, 29–41. [CrossRef]

27. Wu, L.; Wang, S.; Bai, X.; Luo, W.; Tian, Y.; Zeng, C. Quantitative assessment of the impacts of climate change and human activities on runoff change in a karst watershed, SW China. *Sci. Total Environ.* **2017**, *601–602*, 1449–1465. [CrossRef] [PubMed]

28. Calvo-Cases, A.; Boix-Fayos, C.; Imeson, A.C. Runoff generation, sediment movement and soil water behaviour on calcareous (limestone) slopes of some Mediterranean environments in southeast Spain. *Geomorphology* **2003**, *50*, 269–291. [CrossRef]

29. Peng, T.; Wang, S.J. Effects of land use, land cover and rainfall regimes on the surface runoff and soil loss on karst slopes in southwest China. *Catena* **2012**, *90*, 53–62. [CrossRef]

30. Dai, Q.; Peng, X.; Yang, Z.; Zhao, L. Runoff and erosion processes on bare slopes in the karst rocky desertification area. *Catena* **2017**, *152*, 218–226. [CrossRef]

31. Majone, B.; Bellin, A.; Borsato, A. Runoff generation in karst catchments: Multifractal analysis. *J. Hydrol.* **2004**, *294*, 176–195. [CrossRef]

32. Li, X.Y.; Contreras, S.; Solé-Benet, A.; Cantón, Y.; Domingo, F.; Lázaro, R. Controls of infiltration–runoff processes in Mediterranean karst rangelands in SE Spain. *Catena* **2011**, *86*, 98–109. [CrossRef]

33. Gulley, J.D.; Martin, J.B.; Spellman, P.; Moore, P.J.; Screaton, E.J. Influence of partial confinement and Holocene river formation on groundwater flow and dissolution in the Florida carbonate platform. *Hydrol. Processes* **2014**, *28*, 705–717. [CrossRef]

34. Gunn, J.D.; Folan, W.J.; Robichaux, H.R. A landscape analysis of the Candelaria watershed in Mexico: Insights into paleoclimates affecting upland horticulture in the southern Yucatan Peninsula semi-karst. *Geoarchaeology* **1995**, *10*, 3–42. [CrossRef]

Article

Integrating Field Experiments with Modeling to Evaluate the Freshwater Availability at Ungauged Sites: A Case Study of Pingtan Island (China)

Xiaocong Liu [1,2] , Zhonggen Wang [1,]*, Yin Tang [1], Zehua Wu [3], Yuhan Guo [1,2] and Yashan Cheng [1,2]

[1] Key Laboratory of Water Cycle and Related Land Surface Processes, Institute of Geographic Sciences and Natural Resources Research, Chinese Academy of Sciences, Beijing 100101, China; liuxc.14b@igsnrr.ac.cn (X.L.); tangyin@igsnrr.ac.cn (Y.T.); guoyh.16b@igsnrr.ac.cn (Y.G.); chengyashan77@163.com (Y.C.)

[2] University of Chinese Academy of Sciences, Beijing 100049, China

[3] Fujian Research Institute of Water Conservancy and Hydropower, Fuzhou 350001, China; wzh0299@126.com

* Correspondence to: wangzg@igsnrr.ac.cn

Received: 7 May 2018; Accepted: 30 May 2018; Published: 6 June 2018

Abstract: Predictions in ungauged basins (PUB) has been always a focus of hydrological research. The problem presented by ungauged basins is how to reasonably estimate water resource availability. To solve the issues of data scale, this study combines field experiments and hydrological models to estimate freshwater availability in a typical ungauged sea island located in southeastern China. The free parameters in the hydrological model were derived from the point-scale rainfall-runoff experiments rather than calibration using river discharge observations. The rainfall-runoff experiments were performed on six sites covering 11 land cover types. Model validation at a sub-catchment showed that the combined method could successfully reproduce monthly streamflow, with a Nash–Sutcliffe Efficiency of 0.82, correlation coefficient of 0.85, and flow volume error of 6.5%. The simulation results indicate high heterogeneity and distinct seasonal dynamics in freshwater availability across the entire island. This pioneering PUB study for Chinese islands could provide reference for planning and management of freshwater in a water shortage area.

Keywords: PUB; rainfall-runoff experiments; distributed hydrological model; Hydro-Informatic Modelling System (HIMS); freshwater availability

---

1. Introduction

China has been implementing a national strategy called the Belt and Road Initiative, where the "Belt" stands for the Silk Road Economic Belt and the "Road" stands for the 21st Century Maritime Silk Road. The development of the costal islands is an important part of the 21st Century Maritime Silk Road. There are numerous islands off the continent along the coastline, and most of them suffer from fresh water resource shortages and a lack of hydrologic monitoring. How to make a scientific assessment of available water resources in these ungauged islands is not only a forefront of hydrology research, but also a basic issue of reasonable exploitation and the utilization of water resources.

Predictions in ungauged basins (PUB) has always been a frontier and focus [1]. In the beginning of the 21st century, the International Association of Hydrological Sciences (IAHS) began an initiative of PUB focused on reducing the uncertainty in hydrological models and hydrological forecasts [2]. Many studies have been conducted since then, including hydrological simulation and parameter optimization for data acquisition methods [3–7]. Among this research, distributed hydrological models have been widely used in hydrological process simulations in ungauged basins due to their multi-resource information processing.

The problem presented by the ungauged basins in the utilization of water resources is how to evaluate the regional water resource scientifically and reasonably. Conventional methods for calculating water resources generally consider the total amount based on data from stations, which is difficult to implement in areas where there are no data, and the lumped total estimate cannot clarify the spatial distribution of water resources. Additionally, conventional water resource assessments only include liquid water such as surface water and groundwater, known as blue water [8]. However, the blue water only represents one-third of the freshwater resources if rainfall is taken as the original source of freshwater, since rainfall is converted eventually into three parts—surface water, soil water and groundwater—through hydrological processes such as vegetation interception, infiltration, runoff generation, and confluence [9,10].

The importance of soil water as a part of the water resource was recognized during the 19th century [11], and it has been an important starting point for alleviating the global water crisis in recent years [12], especially in an agricultural country like China. Therefore, a scientific evaluation of water resources needs to consider unconventional water resources such as soil water, based on the water cycle process. However, the definition of soil water resources has not been fully unified [13]. The authors refer to the renewable soil water in the unsaturated zone recharged by natural sources (including precipitation, condensed water, and submersion), which can be used by vegetation and has a certain effect on maintaining a stable circle of the natural ecological environment [14,15].

Many studies have been conducted on the estimation and evaluation of soil water resources, and corresponding calculation methods attained from farmland [9,10] and regional scales [13]. The main methods are based on the principle of water balance [16], although there are no widely used evaluation models. Regional-scale assessment of soil water resources can use hydrological models, especially distributed models, to simulate the spatial and temporal distribution, which can be combined with the estimation of surface water resources and groundwater resources to evaluate the total water resource [17].

However, not all of the water can be extracted to serve the water supply. Research needs to consider the water requirements of ecosystems that rely on freshwater [18]. It is necessary to ensure the sustainable state of the ecosystems when estimating freshwater availability. Additionally, the use of water resources in a region is also limited by the conditions of contemporary engineering measures, especially flood control during the flood season. Therefore, for the foreseeable period, abandoned floodwaters that cannot be controlled by engineering measures are difficult for localities to use, and should be deducted from the total water resource availability.

This study selected Pingtan Island in southeast China as the study area, which is the fifth-largest island in China, and is one of the starting points of the 21st Century Maritime Silk Road [19]. The ungauged island is suffering from a freshwater shortage, and the freshwater availability in space and time is not clear due to a lack of hydrological observations. The existing studies of the freshwater on the entire island are lacking fine spatial and temporal distribution due to a dependence on measured streamflow data, and the evaluation object is only surface and ground water. This work intends to reveal the freshwater availability, including soil water availability, by integrating field experiments with modeling, and provide a basis for future freshwater resource management on this ungauged island.

Against this background, the main objective of this study is to provide a basis for future freshwater resource management in Pingtan and a pattern to assess available freshwater in ungauged areas. The temporal and spatial distribution of freshwater availability in Pingtan Island will be estimated by integrating field experiments with modelling.

To satisfy the objectives of this study, the authors managed to get the spatial distribution of rainfall-runoff relation by ArcGIS based on field investigation and field rainfall-runoff experiments. Then, a distributed hydrological model of the island based on HIMS (Hydro-Informatic Modeling System) [20] was constructed and the model parameters were determined according to the spatial distribution of rainfall-runoff relation. Applying this distributed model, this study quantitatively determined the volumes of components of available water resources, temporally and spatially, with consideration of the environmental flow requirements and abandoned floodwater.

2. Study Area and Data

Pingtan Island is located in southeast China precisely between longitudes 119°40′ E and 119°53′ E and latitudes 25°23′ N and 25°40′ N (Figure 1a). The total area is 275.2 km². There are undulating hills and low mountains in the south and north, and plains comprise the middle part. The average annual temperature is 19.8 °C and the annual precipitation is 1236.1 mm. Due to the impact of typhoons, the distribution of precipitation during the year is extremely uneven. The precipitation is generally the highest from March to June, while from July to September most of the precipitation is caused by typhoons. The surface layer of the island is mainly an aeolian layer, dominated by aeolian sand, and has strong infiltration capacity. The change of topography in the island is relatively small. The highest elevation is only 440 m, and the overall surface water storage capacity is weak [21]. Owing to the restrictions of the topography and soil conditions, the water system of Pingtan Island is extremely underdeveloped. The streams are seasonal, and the surface water is very limited. Pingtan Island lacks hydrological monitoring stations.

This study collected the daily rainfall data, maximum and minimum temperature data of the only national meteorological station on the island (data are monitored by the Pingtan Meteorological Service, and are available at http://data.cma.cn/), and the observed daily rainfall data of a rainfall station from 1986–2012 (data are monitored by the Hydrology and Water Resources Survey Bureau of Fujian Province). The authors also collected a set of observed monthly inflow data of the largest lake (36-Foot Lake, with a catchment area of 11.6 km²) on the island from the Hydrology and Water Resources Survey Bureau of Fujian Province, which were only used for model validation.

Based on the interpretation of a scene of remote sensing images from Landsta8 (https://earthexplorer. usgs.gov/), the land use has five categories: cultivated land (41.6%), wood land (14.1%), grass land (18.0%), water (5.5%), and construction land (20.8%) (Figure 1b). According to the national soil census data (1:50,000, 1983), the soil is classified into three types: amorphic soil (54%), ferralsol (37.8%), and alkali-saline soil and others (8.2%) (Figure 1c). According to the Second National Land Survey Technical Regulations of China [22], the slope derived from the DEM data (spatial resolution 2 m × 2 m, came from the Fujian Research Institute of Water Conservancy and Hydropower) is classified into four grades: (0°–6°), (6°–15°), (15°–25°), (25°–88°) (the slope in the island ranges from 0° to 88°) (Figure 1d).

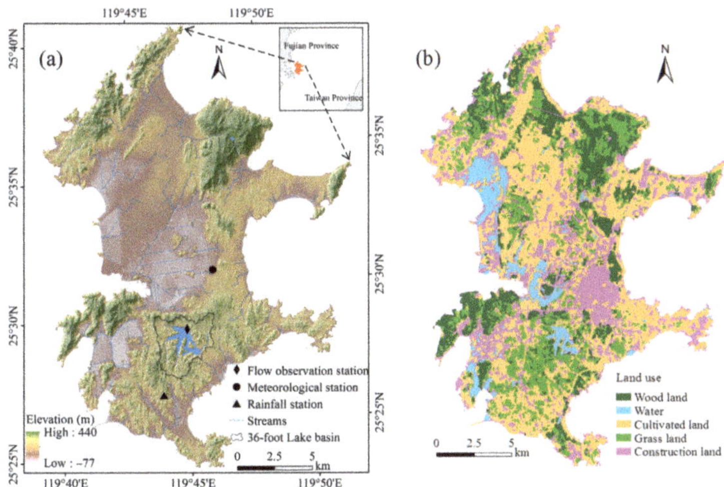

Figure 1. *Cont.*

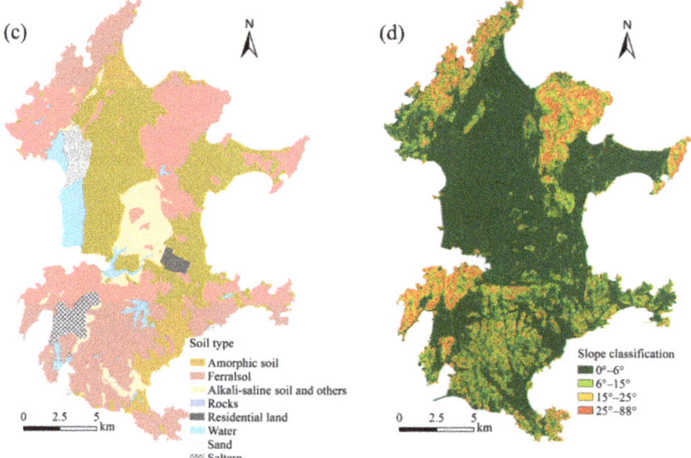

Figure 1. (**a**) The study area and location of meteorological and rainfall stations; (**b**) Spatial distribution of land use; (**c**) Spatial distribution of soil types; and (**d**) Map of slope classification.

3. Methods

3.1. Field Rainfall-Runoff Experiments

Due to the lack of hydrological monitoring data in Pingtan Island, this study carried out rainfall-runoff field experiments in order to understand the hydrological characteristics and provide a basis for the construction and parameter calibration of a HIMS model. Considering the main underlying surface factors affecting the runoff, this study first identified the typical types of underlying surface in the island, and then combined this information with a field survey to determine the experimental sites that could cover the typical underlying surfaces.

Among the land use types, the cultivated land runoff coefficient was nearly zero due to the sandy soil. The construction land runoff coefficient was high due to the high imperviousness degree, so its runoff coefficient was determined according to China's Code for the Design of Outdoor Wastewater Engineering [23] as 0.8. Thus, the grassland and woodland were left for field experiments. Subsequent to combining the woodland and grassland, soil, and the slope of the island, the underlying surface was classified into 24 types. Then, the poorly represented types (less than 1.5% of the area) were merged into those types which were similar, and the underlying surface was eventually classified into 11 types (Table 1). According to the field research, this study identified six representative experimental sites (Figure 2a). A 1 m × 1 m field was set under different underlying surfaces in the experimental area, and a portable rainfall simulator device was applied [24], developed by the Institute of Geographic Sciences and Natural Resources Research (IGSNRR) in the Chinese Academy of Sciences (Figure 2b,c). The runoff volumes and duration were observed under different rainfall intensities (150 mm/h, 200 mm/h, 250 mm/h, 300 mm/h, 350 mm/h).

The artificial rainfall stopped when the runoff came to a stable state for a couple of minutes, and the runoff monitoring stopped when there was no runoff generated for each experiment. The runoff coefficient of each experiment was the ratio of total runoff to total rainfall, and the average runoff coefficient of each underlying surface was the mean of runoff coefficients under five rainfall intensities. The observed runoff coefficient was used for model calibration, and was used for generating the spatial distribution map of runoff coefficients in the study area, together with an ArcGIS spatial assignment and area-weighted statistics. Using ArcGIS, the authors joined the polygon-layer attribute table of the underlying surfaces with the runoff coefficients table, and then intersected this layer with the sub-basins layer. Underlying surfaces polygons inside each sub-basin could be identified based on the intersect result, and the comprehensive runoff coefficient of each sub-basin was determined from the area-weighted statistic of these polygons' runoff coefficients.

Table 1. Typical underlying surface types and their areas.

No.	Typical Underlying Surface Type	Area (km^2)
1	Grassland + Ferralsol + (6°–15°)	13.86
2	Woodland + Ferralsol + (15°–25°)	11.25
3	Grassland + Amorphic soil + (0°–6°)	9.73
4	Woodland + Amorphic soil + (0°–6°)	10.81
5	Grassland + Ferralsol + (15°–25°)	10.04
6	Grassland + Ferralsol + (0°–6°)	8.65
7	Woodland + Ferralsol + (6°–15°)	8.15
8	Woodland + Ferralsol + (25°–88°)	5.96
9	Grassland + Ferralsol + (25°–88°)	3.90
10	Grassland + Alkali-saline soil and others + (0°–6°)	2.50
11	Woodland + Ferralsol + (0°–6°)	2.96

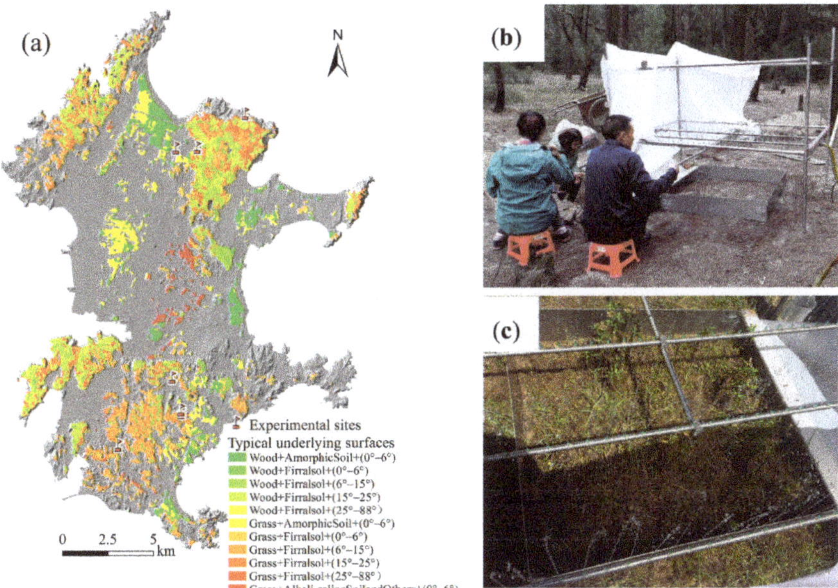

Figure 2. (**a**) Spatial distribution of underlying surface types and selected sites for the field rainfall-runoff experiments; (**b**) Conducting field experiments; and (**c**) A field site for experiments (1 m × 1 m).

3.2. Hydrological Modeling

3.2.1. HIMS Model and Input Data

This study built a Pingtan Island distributed hydrological model based on HIMS [20,24]. This model is described briefly here, and further details can be found in Liu et al. (2008) [20]. The HIMS model is a distributed hydrological model for simulating the hydrological processes with appropriate hydrological modules for the study area. Its hydrological processes included interception, infiltration, evapotranspiration, snowmelt, surface runoff, sub-surface flow, and baseflow generation and concentration. It contains various choices of equations for each hydrological process and combines two runoff generation mechanisms of saturation excess and intensity excess.

The input data included time series data (meteorological and hydrological data) and spatial data. The spatial data input for modelling included land use data, soil data, and slope data. The rainfall data from the national meteorological station and the local rainfall station were interpolated into surface data with the Kriging method and then extracted to a sub-basin scale.

3.2.2. Sub-Basin Division

The plains in the middle of the island occupy 32% of the area of Pingtan Island, and the stream channels influenced by human alterations cannot characterize the catchments properly. Therefore, this study applied grids (1 km × 1 km) rather than sub-basins in the plain area to make discretization, and the flow direction of each grid was determined according to actual river flow. The sub-basins' boundaries were derived from the DEM in the mountainous area and were set as the baseline when joining the grids of the plain area (Figure 3). Thus, the study area was divided into 286 sub-basins and grids in total, with 119 estuaries as shown in Figure 3.

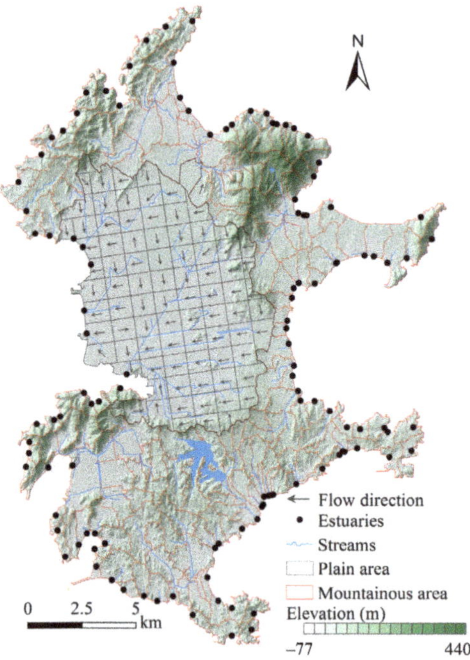

Figure 3. Discretization in the plain area with grids, the mountainous area with sub-basins, and the distribution of estuaries.

3.2.3. Calibration and Validation

The HIMS hydrological model was calibrated using local rainfall-runoff relations, and validated against monthly streamflow observation. Due to the lack of hydrological monitoring, the authors calibrated model parameters (Table A1 in Appendix A1) according to the comparison of rainfall-runoff relations between experimental results and simulations. According to the geological survey of local authorities, the groundwater is mostly shallow groundwater on Pingtan Island and is discharged into channels in the same basin, so we referred to this condition when calibrating the parameters. The model was validated at the 36-Foot Lake sub-basin by using the observed monthly streamflow data (1986–2012). The efficiency was evaluated by the Nash–Sutcliffe Efficiency (*NSE*), the flow volume error (*WE*) and the correlation coefficient (*r*).

$$NSE = 1 - \frac{\sum_{i=1}^{n}(Q_{obs,i} - Q_{sim,i})^2}{\sum_{i=1}^{n}(Q_{obs,i} - \overline{Q_{obs}})^2},\tag{1}$$

$$WE = \frac{\sum_{i=1}^{n}(Q_{obs,i} - Q_{sim,i})}{\sum_{i=1}^{n}Q_{obs,i}} \cdot 100\%,\tag{2}$$

$$r = \frac{\sum_{i=1}^{n}(Q_{obs,i} - \overline{Q_{obs}})(Q_{sim,i} - \overline{Q_{sim}})}{\sqrt{\sum_{i=1}^{n}(Q_{obs,i} - \overline{Q_{obs}})^2 \cdot \sum_{i=1}^{n}(Q_{sim,i} - \overline{Q_{sim}})^2}},\tag{3}$$

where $Q_{obs,i}$ is the *i*th observation, $Q_{sim,i}$ is the *i*th simulated value, $\overline{Q_{obs}}$ and $\overline{Q_{sim}}$ are the means of observed data and simulated values, respectively, and n is the number of datasets. *NSE* describes the fitting extent between simulation and observation datasets of streamflow and *r* describes the correlation between simulated and observed streamflow. The simulating accuracy was high if *NSE* and *r* were very close to 1. The simulation performance was not good if *NSE* was close to 0. The simulation accuracy was high if the absolute value of *WE* was close to 0. Generally, if *WE* was within the range of ±10%, the simulation results were regarded as satisfactory.

3.3. Quantification of Freshwater Availability

3.3.1. Surface Water and Groundwater Availability

Monthly surface and ground water were estimated using the summation of water yield and deep aquifer recharge in the model. The availability of this part of the water resource was calculated by subtracting the environmental flow requirement and flood discharge from the surface water.

This study assumed that precipitation in humid areas could meet the needs of the terrestrial ecosystem function outside the river, and thus only the environmental flow requirement in the river was considered. Moreover, the streams in the study area were seasonal and flowed into the sea independently. Therefore, to maintain the normal environment of estuaries, this study referred to the Tennant method [25] and took 20% of the average annual flows in flood season (April–September) and non-flood season (October–March) as the environmental flows.

The flood discharge was constrained by engineering measures such as ponds and reservoirs, which were useful to collect water during the flood season. This study assumed that these ponds and reservoirs could effectively collect floods, although many of them were not used due to a lack of management, according to our field survey. Thus, the flood discharge came from basins without water collection facilities. The annual monthly amount of flood discharge was determined based on the ratio of the annual monthly maximum flow to the multi-year average monthly flow [26].

3.3.2. Soil Water Availability

The soil water resource was calculated based on the methods of Feng [27] and Shen [15]. The amount of soil water resources within a period was:

$$W_{s,total} = \sum_{i=1}^{n} W_{s,i},\tag{4}$$

where

$$W_{s,i} = E_i + T_i + \Delta W_{SR,i}.\tag{5}$$

$W_{s,total}$ is the total amount of soil water resource within a period; i is the sub period from the start of the ith rainfall to the start of the $(i + 1)$th rainfall within the period; $W_{s,i}$ is the soil water resource of the ith sub period, E_i is the soil evaporation, T_i is the transpiration; and $\Delta W_{SR,i}$ is the soil water storage change.

Soil water availability refers to the renewable soil water that can be used by crops. All the soil water resources of agricultural land (cultivated land) were considered as available water resources because the evaporation and transpiration could be used by applying mulch or using greenhouses [28], while the woodland and grassland were regarded as part of the eco-environment with the main function of maintaining a good state of the environment and were unable to be arbitrarily changed. Therefore, their soil water resources were not included in the available water resource.

3.3.3. Total Freshwater Availability

The total freshwater availability for each basin in this study was the sum of the surface water availability, groundwater availability, and the soil water availability.

4. Results and Discussion

4.1. Spatial Distribution of Runoff Coefficients Based on Field Experiments

The runoff coefficients of typical underlying surfaces and the corresponding distribution in the island were obtained based on the rainfall-runoff experimental data (Table 2, and Figure 4a). The results showed the spatial heterogeneity of runoff coefficients. The grassland was relatively poor in surface water permeability, in terms of land use type, and its runoff coefficient was greater than that of forest. Regarding underlying surfaces, with the same land use type and the same slope classification in terms of soil type, the ferralsol types had the smallest runoff coefficients, while the Alkali-saline soil and others types had the largest. Concerning the slope classification, the larger the slope, the larger the runoff coefficient. The runoff coefficients based on field experiments in the study area ranged from 0.01 to 0.85, and the overall runoff coefficient was 0.44.

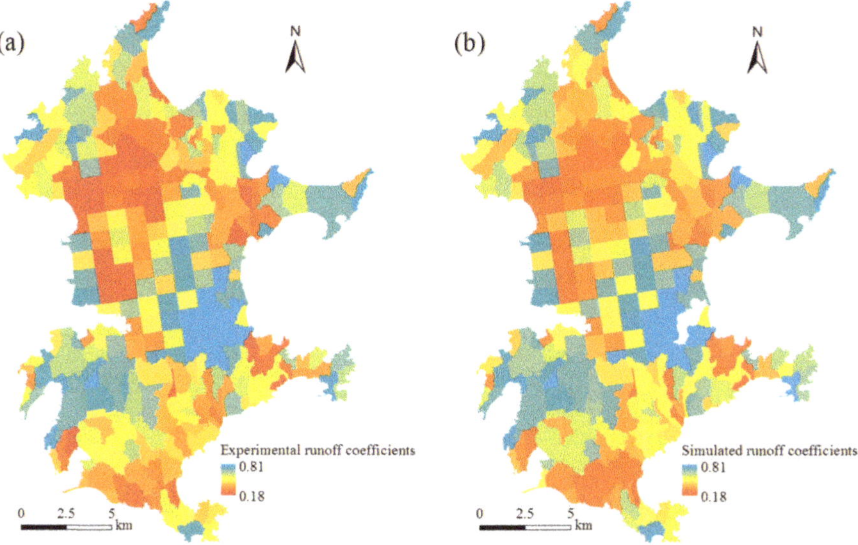

Figure 4. Spatial distribution of runoff coefficients derived from (**a**) field experiments and (**b**) HIMS (Hydro-Informatic Modeling System) model simulation.

Table 2. Runoff coefficients of typical underlying surfaces based on field experiments.

No.	Typical Underlying Surface Type	Runoff Coefficient
1	Grassland + Ferralsol + (6°–15°)	0.72
2	Woodland + Ferralsol + (15°–25°)	0.13
3	Grassland + Amorphic soil + (0°–6°)	0.29
4	Woodland + Amorphic soil + (0°–6°)	0.25
5	Grassland + Ferralsol + (15°–25°)	0.80
6	Grassland + Ferralsol + (0°–6°)	0.59
7	Woodland + Ferralsol + (6°–15°)	0.08
8	Woodland + Ferralsol + (25°–88°)	0.14
9	Grassland + Ferralsol + (25°–88°)	0.85
10	Grassland + Alkali-saline soil and others + (0°–6°)	0.44
11	Woodland + Ferralsol + (0°–6°)	0.01

4.2. Calibration and Validation of HIMS Model

This study completed the HIMS model calibration based on the comparison of the spatial distribution of runoff coefficients derived from the model simulation and field rainfall-runoff experiments. The simulated overall average runoff coefficient of the island was 0.41, with only a 3% difference from the experimental results. Moreover, the runoff coefficients' spatial distribution of each sub-basin between the simulated and experimental results showed a good consistency, with an average error of 13% (Figure 5). The HIMS model not only simulated the average runoff coefficient, but also accurately captured the spatial distribution of the runoff characteristics of the island.

A time series comparison between simulated and observed streamflows of 36-Foot Lake is shown in Figure 6. The r was 0.85, *NSE* was 0.82, and *WE* was 6.5%, which meant the HIMS model simulation was satisfactory in this sub-basin. The results showed that the parameters of the distributed HIMS model established by this method were reasonable and the HIMS model was suitable for the island and could be used for water resource estimation.

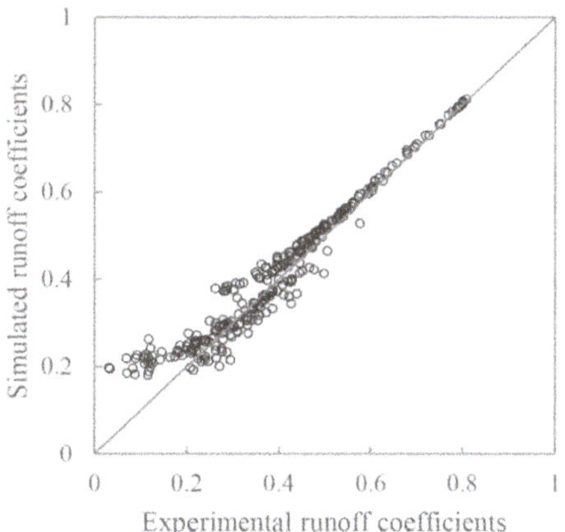

Figure 5. Comparison of sub-basin runoff coefficients between experimental and simulated values.

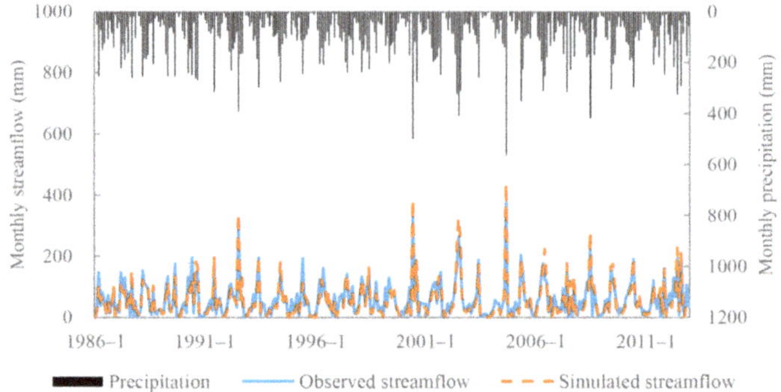

Figure 6. Comparison of monthly streamflows between measured and simulated values at 36-Foot Lake.

4.3. Freshwater Availability

4.3.1. Surface Water and Groundwater Availability

The mean annual (from 1986 to 2012) surface water and groundwater was 1.40×10^8 m^3 (or 509 mm) based on the HIMS model results, which was less than the mean annual 1.79×10^8 m^3 (or 652 mm) from the Fujian Province Water Resources Bulletin (from 2012 to 2016). The surface water and groundwater availability were 0.96×10^8 m^3 (or 348 mm) after cutting off the environmental flow and flood discharge.

Due to the strong weathering on the island, the rocks were loose and the surface soil sandy with strong infiltration, making runoff generation difficult when the rain intensity was low and the duration was short. According to the distribution within the year, the total surface and groundwater availability was concentrated from May to September (Figure 7a). Specifically, summer (June–August) was the highest of the four seasons, accounting for 45.22% of the whole year, followed by spring and autumn, while winter was the least, only 7.34% (Figure 7b).

The spatial distribution of annual surface and groundwater availability showed a spatial differentiation (Figure 8a). There was more water in the central and northern parts of the island than in the small watersheds scattered in the east, south, and on the edge of the island.

4.3.2. Soil Water Availability

The soil water availability in the study area is shown in Figures 7 and 8b. The mean annual availability was 0.74×10^8 m^3 (or 269 mm), which showed a considerable potential for agricultural water use, especially in water shortage areas. Unlike the surface and groundwater, the soil water availability was highest in spring, followed by summer, while autumn and winter were still the lowest. The proportions of soil water availability were higher than those of surface and groundwater availability during November to May, while the opposite was true during June to October. The soil water availability was higher in the central and northern plains and small parts in the southern area.

4.3.3. Total Freshwater Availability

The mean annual total freshwater availability was 1.70×10^8 m^3 (or 617 mm) on this island, and summer accounted for the highest proportion (39.71%), followed by spring (31.18%). Spatially, the central plain accounted for 0.60×10^8 m^3, and the mountainous area accounted for 1.10×10^8 m^3. The available freshwater resources were concentrated in the central plain and the intermountain basins on the inland of the island. The sub-basins around the island with short and dispersive flow channels had small streamflow, which are difficult to collect and use. Moreover, the plains and intermountain basins had good formation conditions for groundwater due to the distribution of loose rocks of aeolian deposits and marine deposits, and cultivated lands were concentrated in these areas. Thus, the total available freshwater resources in the plain area and the intermountain basins were larger than in the mountains.

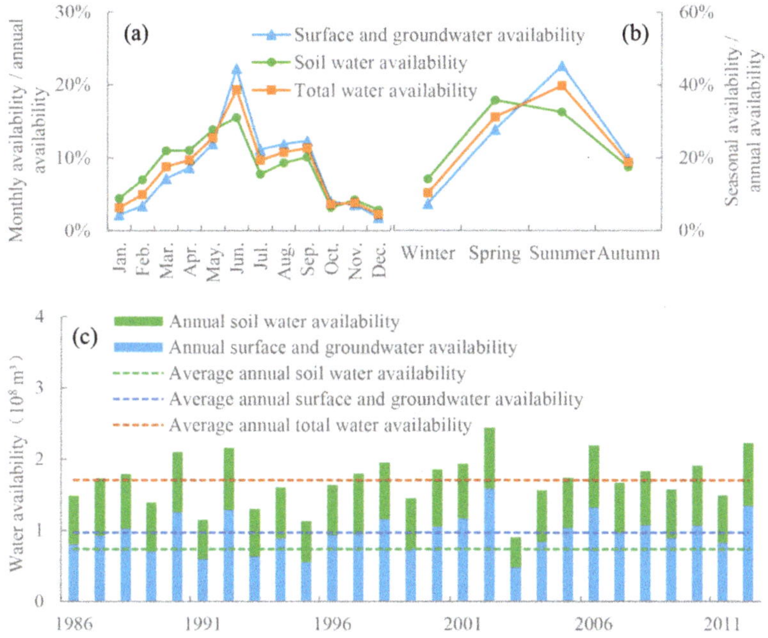

Figure 7. Temporal distribution of (**a**) monthly proportions of freshwater availability; (**b**) seasonal proportions of freshwater availability; and (**c**) yearly freshwater availability.

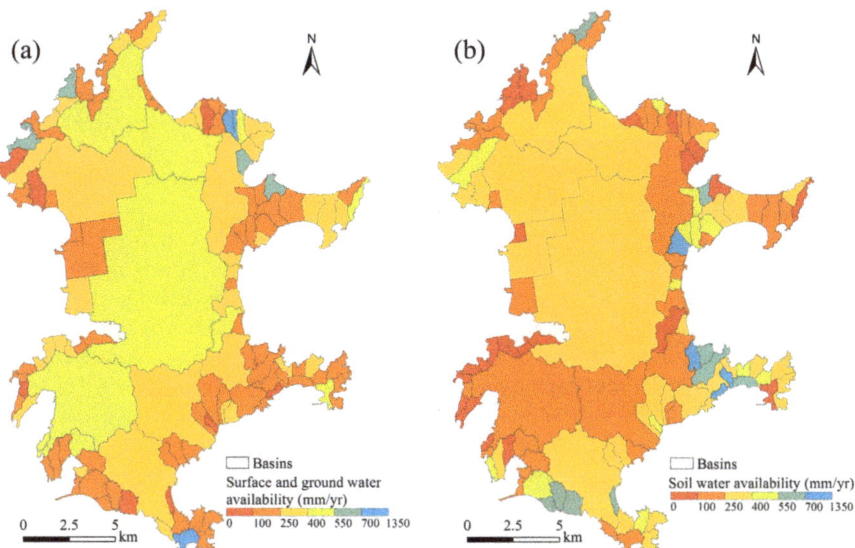

Figure 8. *Cont.*

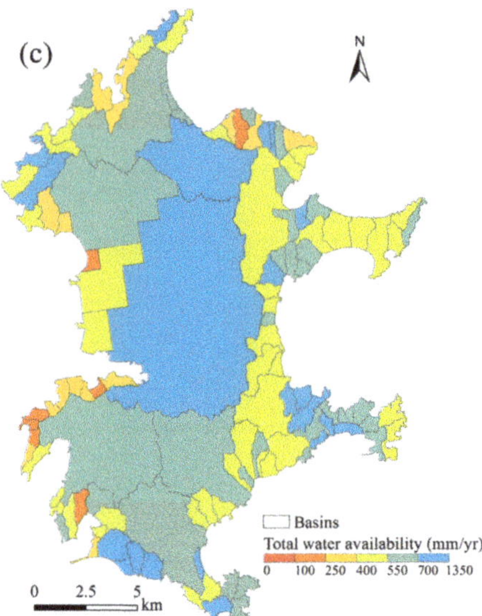

Figure 8. Spatial distribution of (**a**) surface and groundwater availability; (**b**) soil water availability; and (**c**) total freshwater availability.

5. Conclusions

This study provides a basis for freshwater management in Pingtan Island while demonstrating the spatial and temporal distribution of freshwater availability. Additionally, this work provides a method of estimating freshwater resources in an ungauged area. It showed good performance by using the rainfall-runoff relations derived from field experiments to calibrate the distributed hydrological model, and non-specific basin flow data to validate the model. Thus, freshwater resource distribution could be estimated based on the model output. Noteworthy are model parameter adjustments that referred to the local underlying surface and geological characteristics; for example, the groundwater was mostly shallow groundwater on Pingtan Island and discharged into channels in the same basin, according to the geological survey of local authorities.

The field experiment results revealed the spatial distribution of rainfall-runoff relations on ungauged Pingtan Island, with an overall runoff coefficient of 0.44. The HIMS-based hydrological model constructed and calibrated based on experimental results showed good performance in the validation of sub-basin monthly data with an *NSE* of 0.82. Results revealed the spatial heterogeneity and temporal characteristics of freshwater resources. Freshwater resources were concentrated in plains and intermountain basins rather than mountains and small basins around the island. The freshwater availability was concentrated from May to September within the year, and the highest availability proportions were in summer, spring, and summer for surface and groundwater, soil water, and the total freshwater resources, respectively. The temporal and spatial distribution of freshwater could provide references for local flood control and water supply management.

Author Contributions: X.L., Z.W., Z.W., Y.G. performed the experiments. X.L. analyzed the data, performed the model simulation and wrote the manuscript. Y.T. and Y.C. revised the paper.

Funding: This research was funded by the Special Fund for Water Resources Scientific Research in the Public Welfare Profession of China (No. 201401024), the Young Scientists Fund of the National Natural Science Foundation of China (No. 41501046), the Water Resource Science and Technology Innovation Program of Guangdong Province (No. 2016-14).

Conflicts of Interest: The authors declare no conflict of interest.

Appendix A

HIMS (Hydro-Informatic Modelling System)

HIMS was developed to facilitate water resources management and water environment protection. It includes a hydrologic information system (HIS) and a hydrologic model library (HML). The HIS provides functions to deal data with different sources and obtain geographical characters from DEM by integrating with GIS and remote sensing. The HML incorporated most of the commonly used methods and some new models developed by Liu's team in modelling runoff generation and flow routing. These include hydraulic methods and hydrologic methods whether physical, statistical, or conceptual one. In this study, the surface runoff model was LCM model, which is a conceptual model developed by Liu et al. (1965) [29] and has been the core module for runoff calculation in HIMS. Other models dealing with sediment, water quality, ecology, and agriculture can also be incorporated into the framework. Based on spatial topological relationships among the channel network, HIMS divides a catchment into different sub-basins with different soil, vegetation and land use properties. Each sub-basin can include a channel, and sub-basins are linked through stream network. The main parameters of HIMS model are shown in Table A1.

Table A1. Model parameters and range.

Parameters	Meaning	Range
W_{sm}	maximum soil storage capacity	50~1200
R *	parameter related to land use and soil moisture	0~2
r *	parameter related to land use and soil moisture	0~1
L_a	coefficient of sub-surface runoff	0~1
R_c	coefficient of recharge into the groundwater	0~1
ε	actual evapotranspiration index	0~10
K_b	coefficient of base flow	0~1
C_1, C_2	Muskingum model parameters	0~1

* More details about LCM model can be found in Li et al. (2015) [30].

HIMS has been widely tested for catchments under different natural conditions in both northern and southern China, Australia, and some parts of the United States. The modeling results were satisfactory. The HIMS system have been applied to simulate hourly streamflow in small and medium watersheds and daily runoff in medium and large watersheds in case studies [20,31]. The system is still developing and is available by contacting the Key Laboratory of Water Cycle and Related Land Surface Processes.

References

1. Hrachowitz, M.; Savenije, H.H.G.; Blöschl, G.; McDonnell, J.J.; Sivapalan, M.; Pomeroy, J.W.; Arheimer, B.; Blume, T.; Clark, M.P.; Ehret, U. A decade of Predictions in Ungauged Basins (PUB)—A review. *Hydrol. Sci. J.* **2013**, *58*, 1198–1255. [CrossRef]
2. Sivapalan, M.; Takeuchi, K.; Franks, S.W.; Gupta, V.K.; Karambiri, H.; Lakshmi, V.; Liang, X.; McDonnell, J.J.; Mendiondo, E.M.; O'Connell, P.E.; et al. IAHS decade on predictions in ungauged basins (PUB), 2003–2012: Shaping an exciting future for the hydrological sciences. *Hydrol. Sci. J.* **2003**, *48*, 857–880. [CrossRef]
3. Pumo, D.; Conti, F.; Viola, F.; Noto, L. An automatic tool for reconstructing monthly time-series of hydro-climatic variables at ungauged basins. *Environ. Model. Softw.* **2017**, *95*, 381–400. [CrossRef]
4. Qamar, M.U.; Azmat, M.; Cheema, M.J.M.; Shahid, M.A.; Khushnood, R.A.; Ahmad, S. Model swapping: A comparative performance signature for the prediction of flow duration curves in ungauged basins. *J. Hydrol.* **2016**, *541*, 1030–1041. [CrossRef]
5. Gibbs, M.S.; Maier, H.R.; Dandy, G.C. A generic framework for regression regionalization in ungauged catchments. *Environ. Model. Softw.* **2011**, *27*, 1–14. [CrossRef]

6. El-Hames, A.S. An empirical method for peak discharge prediction in ungauged arid and semi-arid region catchments based on morphological parameters and SCS curve number. *J. Hydrol.* **2012**, *456*, 94–100. [CrossRef]

7. Pumo, D.; Viola, F.; Noto, L. Generation of natural runoff monthly series at ungauged sites using a regional regressive model. *Water* **2016**, *8*, 209. [CrossRef]

8. Falkenmark, M.; Rockström, J. The new blue and green water paradigm: Breaking new ground for water resources planning and management. *J. Water Resour. Plan. Manag.* **2006**, *132*, 129–132. [CrossRef]

9. Liu, C.; Wei, Z. *Agricultural Hydrology and Water Resources in the North China Plain*; Science Press: Beijing, China, 1989.

10. You, M.; Wang, H. *Assessment of Field Soil Water Reources*; China Meteorological Press: Beijing, China, 1996.

11. Xia, J.; Pan, X.; Chen, X.; Liu, Y. A review of soil water resource research and application in China. *IAHS Publ. Ser. Proc. Rep.* **2007**, *315*, 215–220.

12. Rockström, J.; Falkenmark, M.; Karlberg, L.; Hoff, H.; Rost, S.; Gerten, D. Future water availability for global food production: The potential of green water for increasing resilience to global change. *Water Resour. Res.* **2009**, *45*. [CrossRef]

13. Wang, H.; Yang, G.; Jia, Y.; Wang, J. Connotation and assessment index system of soil water resources. *J. Hydraul. Eng.* **2006**, *37*, 389–394.

14. Zhang, W.; Zhang, P.; Li, J.; Ma, Y. Evaluation of soil water resource and study of ecological recovery in Loess Plateau. *Yellow River* **2012**, *34*, 100–102.

15. Shen, R. Investigation on conception and evaluation method of soil water resources. *J. Hydraul. Eng.* **2008**, *39*, 1395–1400.

16. Liu, L.; Xu, X.; Wang, H.; Yin, J. Advances in soil resource assessment. *J. Beijing Norm. Univ.* **2009**, *Z1*, 621–625.

17. Anoh, K.A.; Koua, T.J.J.; Eblin, S.G.; Kouamé, K.J.; Jourda, J.P. Modelling freshwater availability using SWAT model at a catchment-scale in ivory coast. *J. Geosci. Environ. Protect.* **2017**, *5*, 70–83. [CrossRef]

18. Smakhtin, V.; Revenga, C.; Döll, P. A pilot global assessment of environmental water requirements and scarcity. *Water Int.* **2004**, *29*, 307–317. [CrossRef]

19. National Development and Reform Commission of the People's Repeblic of China; Ministry of Foreign Affairs of the People's Repeblic of China; Ministry of Commerce of the People's Repeblic of China. Vision and proposed actions outlined on jointly building Silk Road Economic Belt and 21st-Century Maritime Silk Road. Available online: http://www.ndrc.gov.cn/gzdt/201503/t20150328_669091.html (accessed on 17 January 2018).

20. Liu, C.; Wang, Z.; Zheng, H.; Zhang, L.; Wu, X. Development of hydro-informatic modelling system and its application. *Sci. China Ser. E Technol. Sci.* **2008**, *51*, 456–466. [CrossRef]

21. Fujian Province Pingtan County Local Records Committee. *Pingtan County Local Records*; China Local Records Publishing: Beijing, China, 2000.

22. Ministry of Land and Resources, People's Republic of China. *The Second National Land Survey Technical Regulations*; TD/T 1014–2007; Ministry of Land and Resources, People's Republic of China: Beijing, China, 2007.

23. Committee of Construction and Traffic in Shanghai City. *Code for Design of Outdoor Wastewater Engineering (GB50014-2006)*; China Planning Press: Beijing, China, 2006.

24. Wang, Z.; Zheng, H.; Liu, C. A modular framework of distributed hydrological modeling system: HydroInformatic modeling system. *HIMS. Prog. Geogr.* **2005**, *24*, 109–115.

25. Orth, D.J.; Leonard, P.M. Comparison of discharge methods and habitat optimization for recommending instream flows to protect fish habitat. *River Res. Appl.* **1990**, *5*, 129–138. [CrossRef]

26. Yao, S.; Guo, Z.; Ren, J.; Wang, S.; Liu, H. On calculation of amount of available surface water. *J. Zhejiang Univ.* **2005**, *31*, 479–482.

27. Feng, Q.; Wang, H. Discussion on evaluation method of soil water resource. *J. China Hydrol.* **1990**, *4*, 28–32.

28. Ji, S.; Unger, P.W. Soil water accumulation under different precipitation, potential evaporation, and straw mulch conditions. *Soil Sci. Soc. Am. J.* **2001**, *65*, 442–448. [CrossRef]

29. Liu, C.; Hong, B.; Zeng, M.; Cheng, Y. An experimental study on relation of rainfall forecasting in the Loess Plateau. *Sci. Bull.* **1965**, *2*, 158–161.

30. Li, J.; Liu, C.; Wang, Z.; Liang, K. Two universal runoff yield models: SCS vs. LCM. *J. Geogr. Sci.* **2015**, *3*, 311–318. [CrossRef]

31. Liu, C.; Wang, Z.; Yang, S.; Sang, Y.; Liu, X.; Li, J. Hydro-Informatic Modeling System: Aiming at water cycle in land surface material and energy exchange processes. *Acta Geogr. Sin.* **2014**, *5*, 579–587.

 *water*

Article

Assessing the Influence of the Three Gorges Dam on Hydrological Drought Using GRACE Data

Fupeng Li [1], Zhengtao Wang [1], Nengfang Chao [2,*] and Qingyi Song [3]

[1] School of Geodesy and Geomatics, Wuhan University, Wuhan 430079, China;
     fpli@whu.edu.cn (F.L.); ztwang@whu.edu.cn (Z.W.)
[2] College of Marine Science and Technology, China University of Geosciences, Wuhan 430074, China
[3] Faculty of Science, University of Alberta, Edmonton, AB T6G 2R3, Canada; qingyi@ualberta.ca
*   Correspondence: chaonf@cug.edu.cn; Tel.: +86-180-6210-1797

Received: 17 April 2018; Accepted: 19 May 2018; Published: 22 May 2018

Abstract: With worldwide economic and social development, more dams are being constructed to meet the increasing demand for hydropower, which may considerably influence hydrological drought. Here, an index named the "Dam Influence Index" (DII) is proposed to assess the influence of the Three Gorges Dam (TGD) on hydrological drought in the Yangtze River Basin (YRB) in China. First, the total terrestrial water storage (TTWS) is derived from Gravity Recovery and Climate Experiment data. Then, the natural-driven terrestrial water storage (NTWS) is predicted from the soil moisture, precipitation, and temperature data based on an artificial neural network model. Finally, the DII is derived using the empirical (Kaplan-Meier) cumulative distribution function of the differences between the TTWS and the NTWS. The DIIs of the three sub-basins in the YRB were 1.38, −4.66, and −7.32 between 2003 and 2008, which indicated an increase in TTWS in the upper sub-basin and a reduction in the middle and lower sub-basins. According to the results, we concluded that impoundments of the TGD between 2003 and 2008 slightly alleviated the hydrological drought in the upper sub-basin and significantly aggravated the hydrological drought in the middle and lower sub-basins, which is consistent with the Palmer Drought Severity Index. This study provides a new perspective for estimating the effects of large-scale human activities on hydrological drought and a scientific decision-making basis for the managing water resources over the operation of the TGD.

Keywords: hydrological drought; Three Gorges Dam; GRACE

1. Introduction

By implementing an operational definition of drought, three main physical drought types have been established: Meteorological, agricultural, and hydrological [1–3]. Hydrological drought is caused by the shortage of surface water and groundwater [4,5], affecting human society and economic development. The severe hydrological droughts in the Yangtze River Basin (YRB) during the summer of 2006 and the spring of 2011 [6,7] caused significant economic losses to the local people.

In 1993, the Chinese government decided to construct the Three Gorges Dam (TGD) in order to mitigate the effects of global climate change including floods and to make full use of the water resources for hydroelectric power generation [8–11]. The TGD began operation in 2003, and after three major impoundments, construction was completed in 2009 [12]. With the construction and operation of the TGD, YRB hydrological drought has been intensively affected by human activities [6], and the reservoir operations may seriously affect the ecological balance of the downstream environments [13].

Dai et al. [6], Zhang et al. [7], Li et al. [14], Lai et al. [15], and Liu et al. [16] analyzed the effects of the TGD on water levels in the middle and lower reaches of the Yangtze River using observational data from hydrological stations and the TGD impounding data based on many mathematical models [17]. The results of their studies indicated that the construction and operation of the TGD had a non-negligible impact on the water level changes of the Yangtze River. The influence of the TGD on the hydrological drought in the YRB has been investigated using surface water level data from in-situ hydrological stations, and it was found to not only be affected by the shortage of surface water but also by the groundwater changes [4,5]. Moreover, the short-term variation in the total terrestrial water storage (TTWS) was found to be mainly caused by changes in groundwater and surface water [18].

Since the Gravity Recovery and Climate Experiment (GRACE) mission was successfully launched in March 2002 [19], GRACE has been widely used to estimate hydrological drought. Yaraw et al. [20], Chen et al. [21], and Frappart et al. [22] estimated hydrological drought events based on GRACE data. Thomas et al. [23] proposed a method to quantify hydrological drought events using GRACE satellite gravity data. This method detects the beginning, end, duration, recovery, and severity of hydrological drought events. Chao et al. [24] proposed a new GRACE-based index called the "non-seasonal storage deficit" to quantify the hydrological drought characteristics in Southwestern China. Therefore, the GRACE-based TTWS can accurately estimate the influence of the TGD on hydrological drought in the YRB.

Here, the TTWS in the upper, middle, and lower sub-basins of the YRB were calculated using GRACE data. Then, the naturally driven terrestrial water storage (NTWS) in the three sub-basins (upper, middle, and lower sub-basins of the YRB) were estimated from the soil moisture, precipitation, and temperature data based on an artificial neural network (ANN) model. Additionally, an index named the "Dam Influence Index" (DII), which reflects the effects of large-scale dam impoundments on the TTWS, was created by using the NTWS and the TTWS between 2003 and 2008. Finally, the effects of TGD impoundments on the hydrological drought between 2003 and 2008 in the YRB were estimated.

2. Study Area

The world's third largest river, the Yangtze River, with a total length of 6300 km, flows across 11 provinces in China and finally flows into the East China Sea. The YRB [25], with an area of 1.8 million km^2, which accounts for approximately 18.8% of China's territory, is divided into upper, middle, and lower sub-basins by the Yichang and Hukou hydrological stations [26–28]. The region above the Yichang hydrological station is the upper sub-basin of the YRB, and the world's largest dam, the TGD, is located in Yichang, approximately 100 km away from the Yichang hydrological station. The region below the Hukou hydrological station is the lower sub-basin and the region between Yichang and Hukou hydrological station is the middle sub-basin of the YRB. The upper, middle, and lower sub-basins areas are 98, 51, and 29 km^2, respectively. The major surface bodies of the middle and lower sub-basins are Dongting Lake in the central YRB and Poyang Lake near the Hukou hydrological station (Figure 1).

The Yangtze River, between the TGD and Chongqing, with a length of 663 km and an area of 1084 km^2, composes the Three Gorges reservoirs (TGR) (Figure 1), which has a total water storage capacity of 39.3 km^3 [16]. In this study, the effects of the TGD on each sub-basin of the YRB will be estimated based on GRACE data from April 2002 to July 2015.

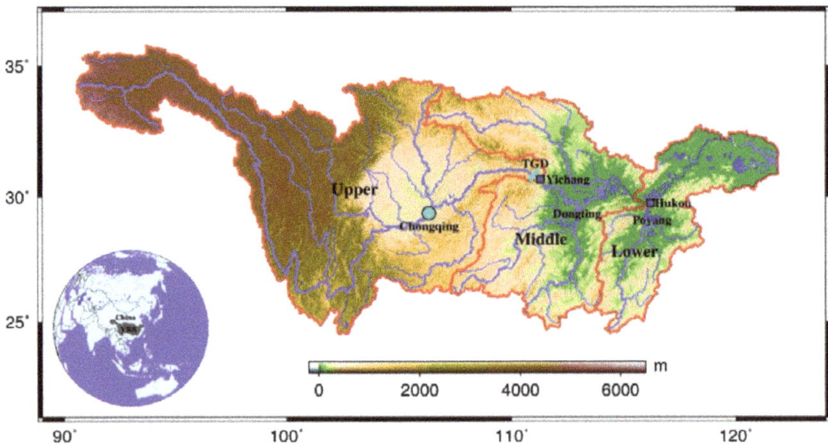

Figure 1. Map of the upper, middle, and lower sub-basins and the mainstream and tributaries of Yangtze River Basin (YRB).

3. Data and Methods

3.1. GRACE

The GRACE monthly time-variable gravity field models from the Center for Space Research (CSR) between April 2002 and July 2015 were used to infer the terrestrial water storage in the YRB after reducing the effects of the atmosphere, tide, and solid tide. Moreover, the monthly time-variable gravity field model with the maximum degree and order of the spherical harmonic coefficients was 60. To eliminate the influence of the geocentric motions on the time-variable gravity field model, the first-order coefficient was estimated from Swenson et al. [29]. To improve the accuracy of the second order of the spherical harmonic coefficient, the C_{20} term of the GRACE time-variable gravity field models was replaced by the satellite laser ranging (SLR) observation data [30]. The glacial isostatic adjustment (GIA) was removed by using the ICE-5G (VM2) model (Peltier, W.R, Toronto, ON, Canada) [31,32]. The north-south strips and high-degree noises [33–35] in the TTWS estimated by the GRACE monthly time-variable gravity field models were removed by de-striping [35] and 300-km Gaussian filtering [33]. Although noise was removed by two-step filtering, the true signal was also attenuated [36]. However, the GRACE signal attenuation due to filtering can be recovered by a scale factor k [37–40]. Here, we used the Global Land Data Assimilation System (GLDAS) and the same method used by Landerer and Swenson [40] to restore the signal, which included attenuation and leakage.

3.2. GLDAS and TRMM

The Global Land Data Assimilation System (GLDAS) [41] model is a hydrological model that contains soil moisture, surface temperature, accumulated snow, water/energy flux, and other hydrological components on land between 60° S and 90° N. The uncertainty of GLDAS model vary in different regions. According to Wang et al. [42], GLDAS precipitation and air temperature data match the ground observations well in most areas of China, and the GLDAS TWS changes and GRACE TWS changes are well correlated in wet eastern China (including the YRB), which demonstrating that the precipitation, temperature, and TWS changes in the YRB from the GLDAS-NOAH model (L4 Monthly 1.0 × 1.0 degree, NASA, Washington, DC, USA) are reliable. In this study, the GLDAS-NOAH 10 M series model from National Aeronautical and Spatial Administration (NASA) from April 2002 to July 2015 with a spatial resolution of 1 × 1° was selected to calculate the average soil moisture and temperature of the YRB.

Furthermore, the scale factor k was calculated using the GLDAS hydrological model following the same method used by Landerer and Swenson. [40], which is summarized as follows: (1) the total water storage of the GLDAS model, named GLDAS TWS (GTWS), is extracted; (2) the GTWS extracted in the first step is converted into a spherical harmonic coefficient with a degree of 60; (3) similar to the GRACE data, de-striping (P5M8) and 300-km Gaussian filtering are applied to the GLDAS spherical harmonic coefficients; and (4) the filtered GTWS time series of a basin is calculated using the filtered GLDAS spherical harmonic coefficients, and the scale factor k is estimated using the following equation [40]:

$$M = \sum (\Delta S_T - k \Delta S_F) \tag{1}$$

where ΔS_T is the true GTWS time series and ΔS_F is the filtered GTWS time series. The parameter k that minimizes M is the scale factor needed in this study, and the TTWS is recovered by multiplying the filtered TTWS time series by the estimated scale factor k [40].

The Tropical Rainfall Measuring Mission (TRMM) satellite [43] from NASA and the National Space Development Agency (NSDA) provide global precipitation data between 50° S and 50° N. In this study, the third-grade monthly precipitation data (3B43), which merged the Global Precipitation Climatology Centre (GPCC) rain gauge data and other satellite precipitation data from April 2002 to July 2015 with a spatial resolution of $0.25 \times 0.25°$, and a temporal resolution of one month was used to estimate the precipitation of the YRB.

3.3. ANN Approach

The ANN approach was proposed to extend the GRACE TTWS time series beyond the GRACE observation period using a set of training samples based on the ANN model [44]. The reliability of the ANN model depends on the correlation between the predictors and the target variable of the training samples: The stronger the correlation, the higher the accuracy of the ANN model. Huang et al. [45] found that a strong correlation between soil moisture and TTWS, and the GRACE TTWS time series has been extended beyond the GRACE observation period based on the soil moisture, precipitation, temperature data and the ANN model [46–48].

The mathematical ANN model is as follows [44]:

$$y = f(x) + \varepsilon \tag{2}$$

where x is the input data (predictors), y is the output data (target variable), ε is the process noise, and f is the function mapping of the input and output data. A more detailed description of the ANN model can be found in the study of Long et al. [44]. After obtaining the function mappings from the training samples, Equation (2) was used to extrapolate and predict the target variable (NTWS) using the predictors (soil moisture, precipitation, and temperature). The accuracy of the ANN model was evaluated based on three criteria: Nash–Sutcliff efficiency (NSE), the coefficient of determination (R^2), and mean absolute error (MAE) [44]:

$$NSE = 1 - \frac{\sum\limits_{i=1}^{n} (y_i - o_i)^2}{\sum\limits_{i=1}^{n} (o_i - \bar{o})^2} \tag{3}$$

$$R^2 = \left[\frac{\sum\limits_{i=1}^{n} (y_i - \bar{y})(o_i - \bar{o})}{\sqrt{\sum\limits_{i=1}^{n} (y_i - \bar{y})^2 \sum\limits_{i=1}^{n} (o_i - \bar{o})}} \right] \tag{4}$$

$$MAE = \frac{1}{n}\sum_{i=1}^{n}|y_i - o_i|$$

(5)

where y is the predictions of the target data based on ANN model, o is the true value of the target data, and n is the number of training samples. As the values of NSE and R^2 increase and the value of MAE decreases, the accuracy of the ANN model increases.

3.4. DII Based on TTWS and NTWS

After obtaining the basin-mean NTWS time series in the upper, middle, and lower sub-basins of the YRB based on the ANN model, the differences $x(t)$ of the basin-mean TTWS and the basin-mean NTWS time series of the three sub-basins was calculated with:

$$x(t) = TTWS(t) - NTWS(t)$$

(6)

where $t = 1, 2, 3, ..., n$ and n are the number of TWS observations. The NTWS is defined so that it cannot be affected by the TGD, and the inconsistency of the TTWS and the NTWS is mainly due to the TGD impoundments. The DII for a specific region (YRB) was calculated by the following equation:

$$DII = \int_{0}^{x_{\max}} [f(x) - f(0)]dx - \int_{x_{\min}}^{0} [f(x) - f(x_{\min})]dx$$

(7)

where $f(x)$ is the empirical (Kaplan-Meier) cumulative distribution function (ECDF) of $x(t)$ from June 2003 to December 2008. The first formula in Equation (7) is the integration of $f(x)$ minus $f(0)$ in the positive interval, and the second formula is the integration of $f(x)$ minus $f(x_{\min})$ in the negative interval. Taking Equation (6) into account, a smaller value of the first formula in Equation (7) and a larger value of the second formula indicate a more severe hydrological drought caused by the TGD impoundments.

DII is an index developed in this study to evaluate anomalies of the TTWS due to the TGD impoundments. High and low DII values represent high and low drought conditions, respectively. For instance, a positive DII value indicates an alleviation of hydrological drought, whereas a negative value represents an aggravation.

4. Results

4.1. Recovered TTWS in the YRB

The scale factors, calculated by using the basin-mean GLDAS TWS time series based on Equation (1), of the upper, middle, and lower sub-basins of the YRB were 1.32, 1.15, and 1.54, respectively [40]. Then, the TTWS of the three sub-basins were recovered by multiplying the filtered TTWS time series by the scale factors (Figure 2). Moreover, the GRACE measurement error and the leakage error of the recovered TTWS were estimated based on the method used by Wahr et al. [49] and Landerer and Swenson [40], respectively (Table 1). Figure 2 shows the filtered and recovered TTWS time series estimated using the GRACE data. For verification, the GLDAS TWS time series in the three sub-basins are also shown in this figure. The correlation coefficients (R) between the GLDAS TWS and the recovered TTWS time series during the TGD impoundment period from June 2003 to December 2008 were 0.89, 0.61, and 0.65 in the upper, middle, and lower sub-basins, respectively, indicating that the recovered TTWS is more consistent with the GLDAS TWS in the upper sub-basin than in the middle and lower sub-basins.

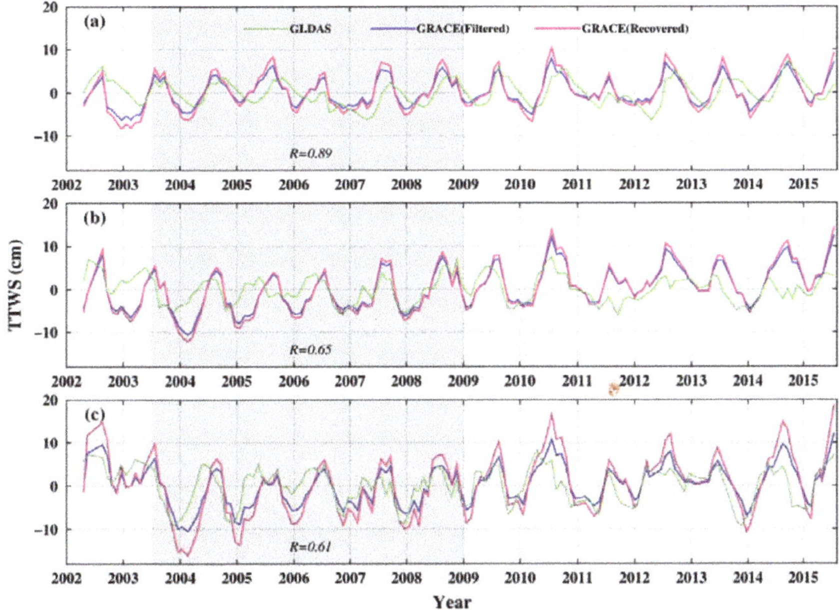

Figure 2. The Global Land Data Assimilation System (GLDAS) terrestrial water storage (TWS), filtered total terrestrial water storage (TTWS), and recovered TTWS time series in the (**a**) upper, (**b**) middle, and (**c**) lower sub-basins of the Yangtze River Basin (YRB). The green, blue, and red lines represent the GLDAS TWS, filtered TTWS, and recovered TTWS time series, respectively. The correlation coefficient (R) between the GLDAS TWS and recovered GRACE TTWS time series during the Three Gorges Dam (TGD) impoundment period are also shown.

Table 1. Measurement and leakage errors of the total terrestrial water storage (TTWS) in the upper, middle and lower sub-basins of the YRB.

Sub-Basin	Upper (cm)	Middle (cm)	Lower (cm)
Measurement error	1.66	2.28	2.94
Leakage error	2.03	2.77	2.69

4.2. NTWS in the YRB

Zhang et al. [47] extended the TTWS of the YRB (2002–2012) to a longer time series (1979–2012) based on the ANN model. According to their study, the TTWS time series in the middle and lower sub-basins of the YRB between 2003 and 2008 was significantly lower than in other years, and the TTWS returned to a normal level in 2009. There are two interpretations for this phenomenon: (1) A natural drought due to precipitation and evaporation anomalies occurred between 2003 and 2008, and (2) during this period, significant human activities, such as the TGD impoundment, aggravated the hydrological drought in the middle and lower sub-basins of the YRB.

Figure 3 shows the time series of precipitation and evaporation obtained from TRMM and GLDAS, respectively, in the upper, middle, and lower sub-basins of the YRB from April 2002 to July 2015.

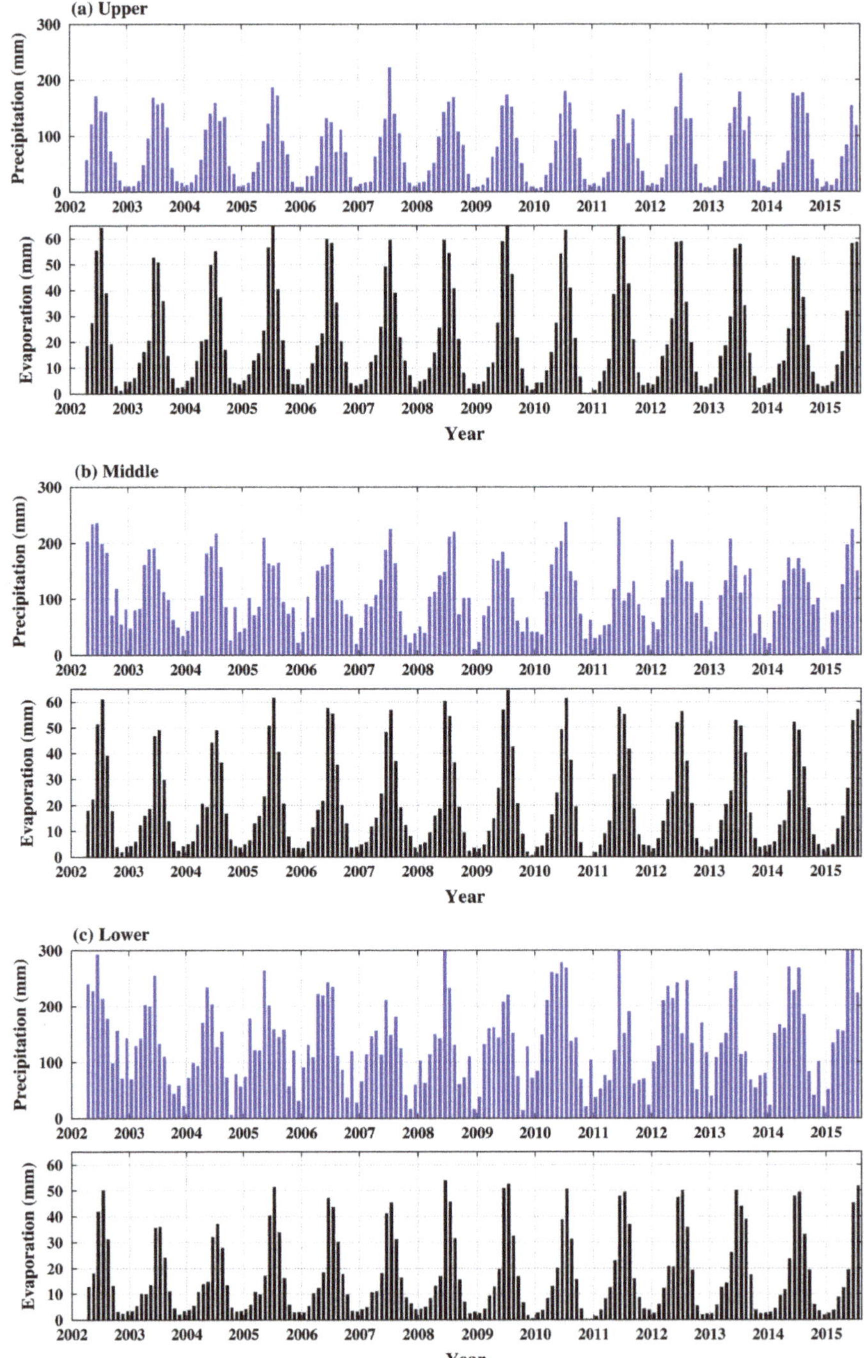

Figure 3. Time series of precipitation and evaporation in the (**a**) upper; (**b**) middle; and (**c**) lower sub-basins of the YRB from April 2002 to July 2015.

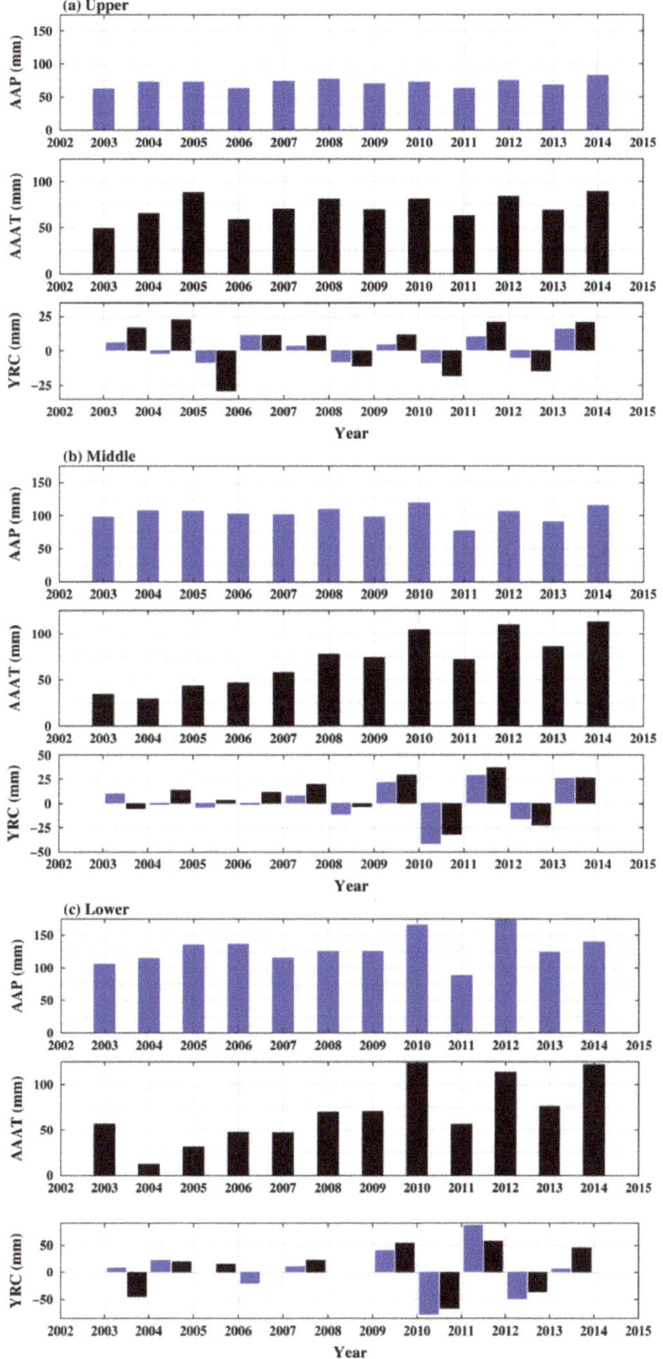

Figure 4. Annual Average Precipitation (AAP), Adjusted Annual Average TTWS (AAAT), and yearly relative change (YRC) of the AAP and AAAT in the (**a**) upper; (**b**) middle; and (**c**) lower sub-basins of the YRB from 2003 to 2014.

The precipitation and evaporation in the YBR were regular from 2003 to 2008. Hence, the possibility of natural drought was excluded. Because changes in the TTWS and droughts are closely related to that of precipitation without significant human activity [50,51], the Annual Average Precipitation (AAP) and Annual Average TTWS (AAT) in the YRB were calculated to determine whether TTWS in the YRB were influenced by non-natural factors. For a more intuitive comparison, the AAT values of the three sub-basins were adjusted by adding a constant, to ensure all Adjusted AAT (AAAT) values were positive (Figure 4), and the yearly relative change (YRC) of the AAP and AAAT were calculated as described in Ahmadalipour et al. [52]. The gap in the YRC between AAP and AAAT in the upper sub-basin from 2003 to 2005 is significant, and the gap gradually narrowed between 2006 and 2008 (Figure 4a). The inconsistency between AAP and AAAT YRCs in the middle and lower sub-basins from 2003 to 2008 was obvious (Figure 4b,c), and the AAP and AAAT in the three regions were highly consistent after 2009. The results demonstrate that the TTWS in the YRB from 2003 to 2008 was significantly affected by human activities and returned to normal in 2009.

The TGD began to operate in 2003 and achieved full capacity in 2009 after three major impoundments. The major impoundments of the TGD between 2003 and 2008 are listed in Table 2 [12,14], and the water level of the TGR changed from 66 to 172.3 m after the three major impoundments, which were the main human activities that may have significantly affected the TTWS and hydrological drought in the YRB. After construction was completed in 2009, the TGD impounded water from September to October and then discharged it from April to May every year. In this period, the operation of the TGD was geared toward adjusting floods and droughts in the middle and lower sub-basins of the YRB, and there were no significant water level changes in the TGR [12]. Thus, the operation of the TGD had little influence on the change in the TTWS in the YRB after 2009, as shown in Figure 4.

Table 2. Three major impoundments of the TGD between 2003 and 2008.

Date	Start Level (m)	End Level (m)	Change (m)
June 2003	66.0	135.0	69.0
October 2006	135.0	156.0	21.0
November 2008	145.0	172.3	27.3

To estimate the effects of the three major TGD impoundments on the TTWS and hydrological drought in the YRB between 2003 and 2008, the TTWS time series from 2009 to 2015 unaffected by the TGD was extended to a longer time series (2002–2015) using natural data (soil moisture, precipitation, and surface temperature) based on the ANN approach [43]. The predicted TTWS time series is a NTWS time series that is unaffected by the TGD.

Figure 5 shows the TTWS time series and the NTWS time series predicted by the ANN approach. The predicted NTWS time series in the upper sub-basin had the highest accuracy among the three basins (Table 3). Compared with the area of the three regions and the accuracy of the ANN model, the accuracy of the ANN model is related to the area of the basin. The larger the area of the basin, the higher the accuracy of the ANN model. This is consistent with the result of Zhang et al. [47]. The NSE and R^2 of the three sub-basins were greater than found by Zhang et al. [47]. They used the GRACE data (2003–2015) affected by the TGD in training the ANN model, which reduced the correlation between inputs and target data, whereas we used the "normal" GRACE data (2009–2015), which revealed results that are more accurate.

In this study, the GRACE TWS time series from 2009 to 2015 was defined not affected by the TGD construction and was used as a true value to text the accuracy of the NTWS based on Equations (3)–(5) (Table 3). However, the GRACE TWS contains measurement and the leakage error (Table 1). Therefore, the uncertainty (Table 3) of the NTWS was calculated by:

$$\sigma = \sqrt{\sigma^2{}_m + \sigma^2{}_l + MAE^2} \tag{8}$$

where σ is the uncertainty of the NTWS; σ_m and σ_l represent the measurement and leakage error of the GRACE TWS respectively.

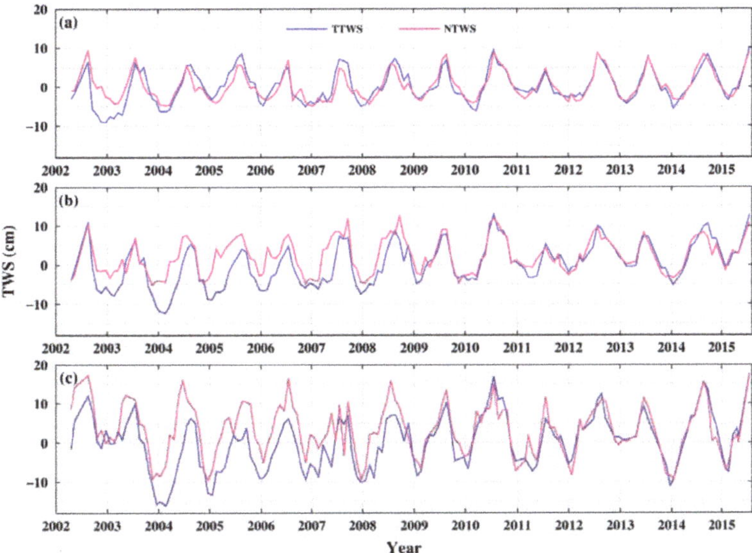

Figure 5. The natural-driven terrestrial water storage (NTWS) and TTWS time series in the (**a**) upper; (**b**) middle; and (**c**) lower sub-basins of the YRB. The red line and the blue lines represent the NTWS and TTWS time series, respectively. Notably, NTWS is predicted by natural data including precipitation, temperature, and soil moisture based on an artificial neural network (ANN) model and TTWS was calculated using GRACE data.

Table 3. The area and the accuracy of the predicted NTWS in the three sub-basins of the YRB.

Sub-Basin	Area (km^2)	NSE	R^2	MAE (cm)	Uncertainty (cm)
Upper	983,118	0.92	0.96	0.93	2.78
Middle	512,733	0.87	0.93	1.29	3.76
Lower	288,205	0.85	0.92	1.89	4.41

The change in the TTWS is mainly caused by changes in soil moisture, groundwater, and surface water (Swenson and Wahr [34]); however, only groundwater may change rapidly with surface water in the short term [18]. Therefore, the inconsistency between the two (NTWS and TTWS) time series in the YRB is mainly due to the changes in groundwater and surface water.

4.3. DII of the YRB

After obtaining the TTWS and the NTWS, the differences $x(t)$ of the two time series from April 2002 to July 2015 were derived using Equation (6). The DIIs of the YRB between June 2003 and December 2008 were calculated based on Equation (7). To provide a decent validation for the DII, the Palmer Drought Severity Index (PDSI; Dai et al. [53]) of the YRB from April 2002 to December 2014 [54] was obtained (Figure 6). For comparison, the PDSI anomaly (PDSIA) in the YRB between June 2003 and December 2008 were estimated using the following equation:

$$PDSIA = ave_{in} - ave_{out} \tag{9}$$

where ave_{in} is the average PDSI during the TGD impoundment period between June 2003 and December 2008, and ave_{out} is the average PDSI of this period, calculated respectively by:

$$\begin{cases} ave_{in} = \dfrac{\sum\limits_{t=t_1}^{t_2} PDSI(t)}{n_1} \\ ave_{out} = \dfrac{\sum\limits_{t=t_0}^{t_1} PDSI(t) + \sum\limits_{t=t_2}^{t_3} PDSI(t)}{n_2} \end{cases} \tag{10}$$

where t_0 is April 2002, t_1 is June 2003, t_2 is December 2008, t_3 is December 2014, and n_1 and n_2 are sum of the PDSIs in the two periods.

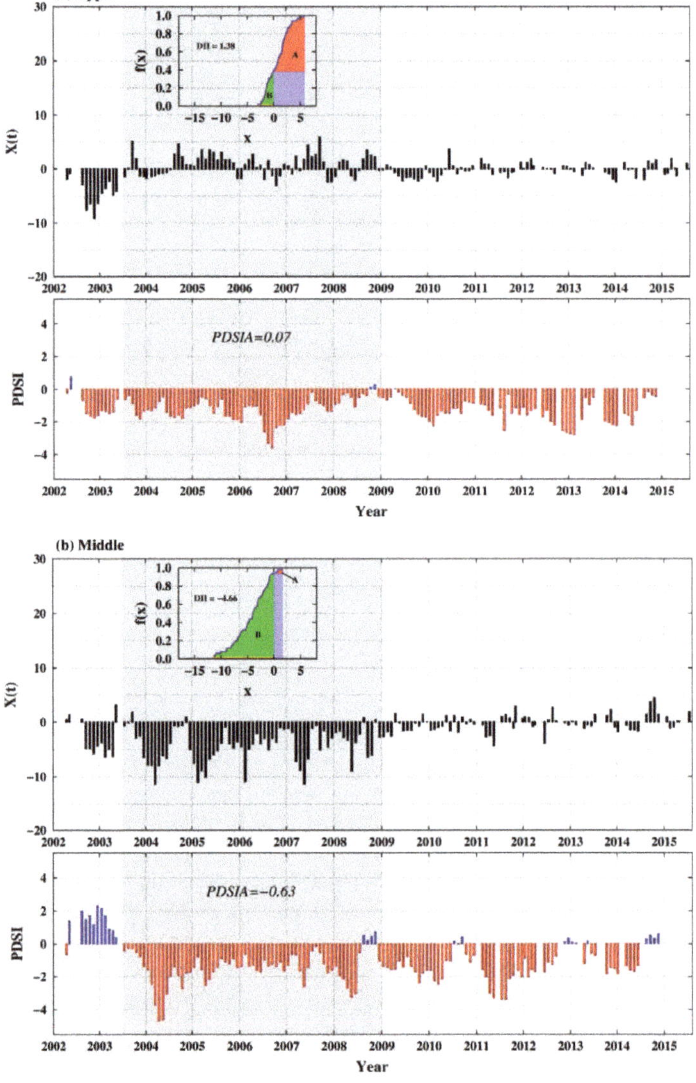

Figure 6. *Cont.*

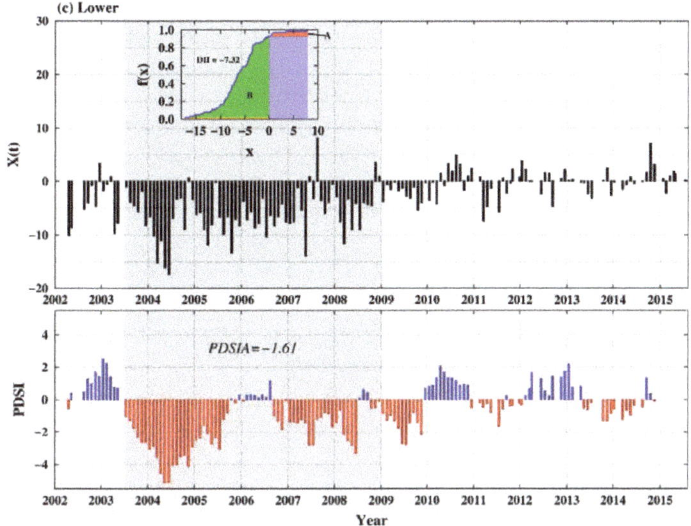

Figure 6. Differences $x(t)$ between TTWS and NTWS (black cylinders) from April 2002 to July 2015 and the Palmer Drought Severity Index (PDSI) (blue and red cylinders) between April 2002 and December 2014 in the (**a**) upper, (**b**) middle, and (**c**) lower sub-basins of the YRB. The blue line in the top window represents the empirical cumulative distribution function (ECDF) of $x(t)$ between June 2003 and December 2008. The Dam Influence Indexes (DIIs) relative to the PDSIAs in the three regions between June 2003 and December 2008 are also shown. Note: DIIs were calculated by subtracting the area of the green block from the red block (i.e., DII = A(area) − B(area)), and PDSIAs were calculated using Equation (9).

The DIIs of the upper, middle and lower sub-basins of the YRB between June 2003 and December 2008 were 1.38, −4.66, and −7.32, respectively (Figure 6). During this period, the DIIs in the middle and lower sub-basins of the YRB were far below zero, demonstrating that the impoundment of the TGD significantly reduced the TTWS in the middle and lower sub-basins. In the upper sub-basin, the DII was 1.38, which is an increase in TTWS.

The PDSIAs of the upper, middle and lower sub-basins of the YRB between June 2003 and December 2008 were 0.07, −0.63, and −1.61 (Figure 6), indicating that the PDSI slightly increased in the upper sub-basin and was greatly reduced in the middle and lower sub-basins during the TGD impoundment period, which is consistent to the DIIs of the three regions.

4.4. Characterization of the Hydrological Drought Events

According to Thomas et al. [23], the hydrological drought signals are calculated by removing the annual and seasonal cycles from the TTWS and the NTWS time series. Any instance in which the negative residuals (hydrological drought signals) last three or more consecutive months is designated a hydrological drought "event". The severity $S(t)$ of a drought event is calculated using the following equation [23]:

$$S(t) = \overline{M}(t) \times D(t) \tag{11}$$

where $M(t)$ and $D(t)$ are the average water storage deficit and the duration of the hydrological drought event, respectively [23].

Figure 7 shows the hydrological drought signals from the NTWS and TTWS time series in the three sub-basins obtained using the method described in Thomas et al. [23]. Table 4 lists the frequency, duration, and severity of the hydrological drought events shown by the dark area in Figure 7 in the three sub-basins of the YRB calculated by the NTWS and the TTWS time series from 2003 to 2008. The total duration of hydrological drought in the upper sub-basin was reduced by 12 months, whereas the total duration in the middle and lower sub-basins increased by 25 and 27 months, respectively. The total severity of the hydrological drought in the upper sub-basin decreased by 3.49 km^3/month, and the total severity increased by 473.16 and 381.24 km^3/month in the middle and lower sub-basins, respectively. The results indicated that the impoundments of the TGD between 2003 and 2008 slightly alleviated the hydrological drought in the upper sub-basin of the YRB but significantly aggravated hydrological drought in the middle and lower sub-basins, which coincides with the DIIs and the PDSIAs.

Table 4. The frequency, duration, and severity of hydrological drought in the upper, middle, and lower sub-basins of the YRB from 2003 to 2008 based on the NTWS and the TTWS.

Sub-Basin	Data	Time Span of Each Event	Duration (Months)	Severity (km^3 Months)	No. of Total Months	Total Severity (km^3 Months)
Upper Area: 983,118 km^2	NTWS	Jan. 2003 to Apr. 2003	4	−134.39	38	−616.21
		Aug. 2003 to Jun. 2004	11	−62.14		
		Aug. 2004 to Oct. 2004	3	−17.61		
		Dec. 2004 to Feb. 2005	3	−207.58		
		Aug. 2006 to Jan. 2007	6	−165.79		
		Mar. 2007 to Oct. 2007	8	−14.07		
		Dec. 2007 to Feb. 2008	3	−134.39		
	TTWS	Jan. 2003 to May 2003	5	−203.39	26	−612.88
		Oct. 2003 to Jul. 2004	10	−142.92		
		Jul. 2006 to Feb. 2007	8	−222.61		
		Oct. 2007 to Jan. 2008	4	−43.80		
Middle Area: 512,733 km^2	NTWS	Apr. 2003 to Feb. 2004	11	−185.43	32	−440.97
		Jul. 2004 to Jan. 2005	7	−69.22		
		May 2006 to Feb. 2007	10	−126.87		
		Dec. 2007 to Mar. 2008	4	−59.45		
	TTWS	Jan. 2003 to Apr. 2003	4	−58.92	57	−914.13
		Jun. 2003 to Aug. 2004	15	−335.13		
		Oct. 2004 to Feb. 2006	17	−242.72		
		Jun. 2006 to Feb. 2007	9	−145.17		
		Apr. 2007 to Jun. 2007	3	−49.41		
		Oct. 2007 to Mar. 2008	6	−60.99		
		May 2008 to Jul. 2008	3	−21.78		
Lower Area: 288,205 km^2	NTWS	Aug. 2003 to Feb. 2004	7	−88.60	24	−270.40
		Nov. 2004 to Jan. 2005	3	−33.66		
		May 2005 to Jul. 2005	3	−31.30		
		Aug. 2006 to Nov. 2006	4	−21.41		
		Jun. 2007 to Aug. 2007	3	−67.86		
		Oct. 2007 to Jan. 2008	4	−27.58		
	TTWS	Aug. 2003 to Aug. 2004	13	−263.31	51	−651.64
		Oct. 2004 to Apr. 2006	19	−193.32		
		Jun. 2006 to Nov. 2006	6	−68.57		
		Apr. 2007 to Aug. 2007	5	−52.27		
		Oct. 2007 to May 2008	8	−74.17		

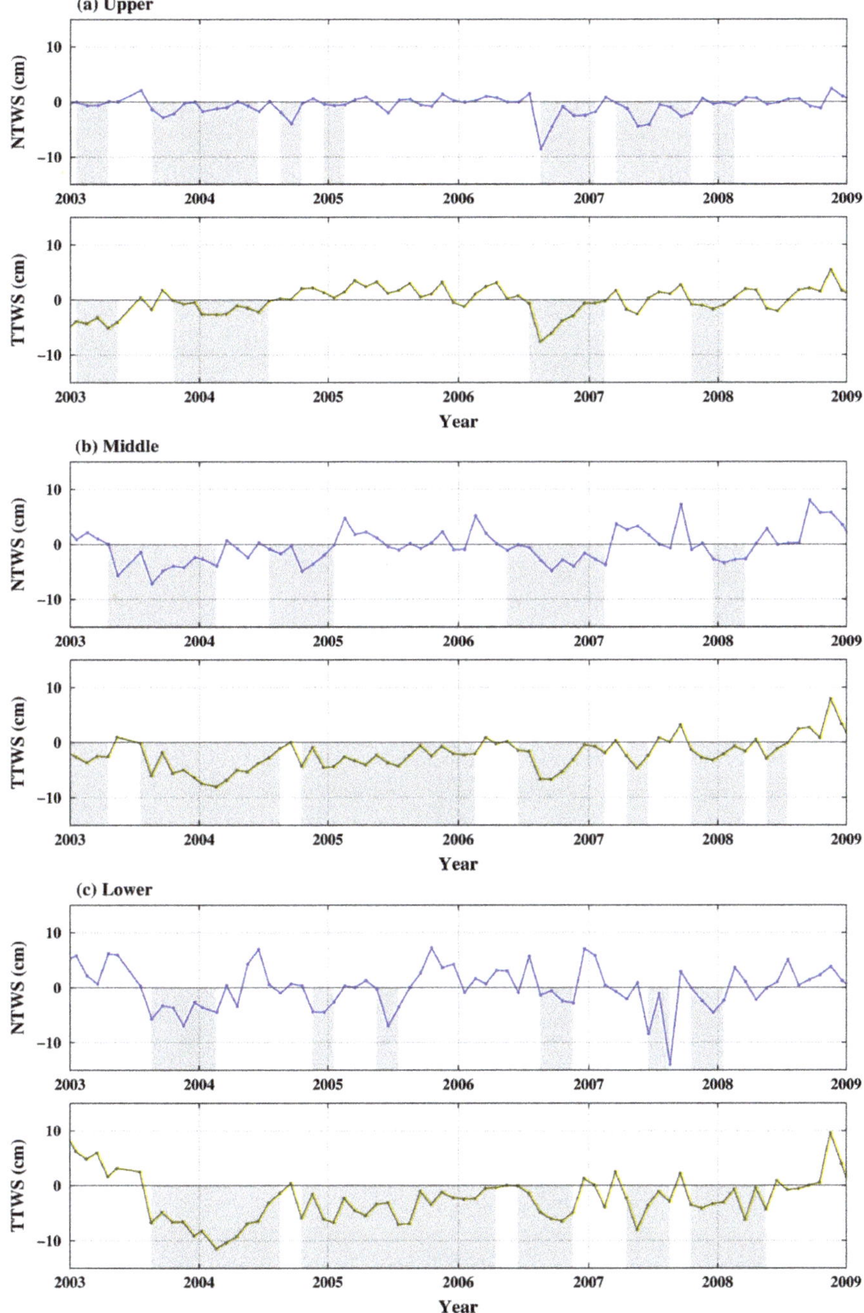

Figure 7. Hydrological drought signals from 2003 to 2008 in the (**a**) upper; (**b**) middle; and (**c**) lower sub-basins. The hydrological drought signals were obtained by removing the annual and seasonal cycles from the TTWS and the NTWS time series. The blue and green lines represent hydrological drought signals from the NTWS and TTWS time series, respectively, and the dark areas indicate hydrological drought events.

5. Conclusions

The effects of the TGD impoundments on the YRB hydrological drought from 2003 to 2008 were evaluated based on the GRACE time-variable gravity field data. An ANN model was used to predict the NTWS of the YRB based on soil moisture, precipitation, and temperature data. From 2003 to 2008, the NTWS time series in the upper sub-basin was in good agreement with the TTWS time series, whereas the time series did not relate well to the middle and lower sub-basins, demonstrating the considerable effect of the TGD impoundments on the TTWS in the middle and lower sub-basins.

The DIIs that reflect the influence of the TGD on the TTWS in the three sub-basins were calculated based on the NTWS and the TTWS. The DIIs between June 2003 and December 2008 in the upper, middle, and lower sub-basins of the YRB were 1.38, −4.66, and −7.32, respectively. These results indicated that the three major TGD impoundments increased the TTWS in the upper sub-basin and reduced the TTWS in the middle and lower sub-basins. For verification, the PDSIA of the YRB during the TGD impoundments period between June 2003 and December 2008 were calculated. The PDSIAs in the upper, middle, and lower sub-basins of the YRB were 0.07, −0.63, and −1.61, respectively, which was consistent to the DIIs in the same regions.

The influence of the TGD impoundments on hydrological drought in the YRB from 2003 to 2008 was estimated based on the method used by Thomas et al. [23]. The total duration of hydrological drought events in the upper sub-basin decreased by 12 months, whereas in the middle and lower sub-basins, the drought duration increased by 25 and 27 months, respectively. The total drought severity was reduced by 3.49 km^3/month in the upper sub-basin and increased by 473.16 and 381.24 km^3/month in the middle and lower sub-basins during 2003–2008, respectively, thereby indicating that the TGD impoundments between 2003 and 2008 had little influence on the upper sub-basin but significantly aggravated hydrological drought in the middle and lower sub-basins. These results coincide with the results estimated by the DIIs and PDSIAs.

Author Contributions: N.C. initiated this study. F.L., Z.W., and N.C. carried out the data analyses and wrote the first draft; F.L. finalized the data interpretations and the draft; Q.S. helped with some of the plots and data analyses.

Funding: This study is supported by the NSFC (China), under Grants 41704011, 41674015, 41274032, 41474018, 41774019, and 41429401, By the China Postdoctoral Science Foundation funded project (2017M622451), by the National 973 Project of China, under Grants 2013CB733301 and 2013CB733302, by the Basic Research Foundation 16-01-01 of the Key Laboratory of Geospace Environment and Geodesy of Ministry of Education, Wuhan University, and by the Open Research Fund Program of the State Key Laboratory of Geodesy and Earth's Dynamics (Grant no. SKLGED2017-2-2-E).

Acknowledgments: The authors thank the following data providers for making the data available: GRACE-CSR, GRGS; GLDAS and TRMM.

Conflicts of Interest: The authors declare no conflict of interest.

References

1. Wilhite, D.A.; Glantz, M.H. Understanding the drought phenomenon: The role of definitions. *Water Int.* **1985**, *10*, 111–120. [CrossRef]
2. Keyantash, J.; Dracup, J.A. The Quantification of Drought: An Evaluation of Drought Indices. *Bull. Am. Meteorol. Soc.* **2002**, *83*, 1167–1180. [CrossRef]
3. Leblanc, M.J.; Tregoning, P.; Ramillien, G.; Tweed, S.O.; Fakes, A. Basin-scale, integrated observations of the early 21st century multiyear drought in Southeast Australia. *Water Resour. Res.* **2009**, *45*, 546–550. [CrossRef]
4. Hisdal, H.; Tallaksen, L.M. Estimation of regional meteorological and hydrological drought characteristics: A case study for Denmark. *J. Hydrol.* **2003**, *281*, 230–247. [CrossRef]
5. Akyuz, D.E.; Bayazit, M.; Onoz, B. Markov Chain Models for Hydrological Drought Characteristics. *J. Hydrometeorol.* **2012**, *13*, 298–309. [CrossRef]
6. Dai, Z.; Du, J.; Li, J.; Li, W.; Chen, J. Runoff characteristics of the Changjiang River during 2006: Effect of extreme drought and the impounding of the Three Gorges Dam. *Geophys. Res. Lett.* **2008**, *35*, 521–539. [CrossRef]

7. Zhang, Q.; Li, L.; Wang, Y.G.; Werner, A.D.; Xin, P.; Jiang, T.; Barry, D.A. Has the Three-Gorges Dam made the Poyang Lake wetlands wetter and drier? *Geophys. Res. Lett.* **2012**, *39*, L20402.1–L20402.7. [CrossRef]

8. Graf, W.L. Dam Nation: A Geographic Census of American Dams and Their Large-Scale Hydrologic Impacts. *Water Resour. Res.* **1999**, *35*, 1305–1311. [CrossRef]

9. Kittinger, J.N.; Coontz, K.M.; Yuan, Z.; Han, D.; Zhao, X.; Wilcox, B.A. Toward holistic evaluation and assessment: Linking ecosystems and human well-being for the Three Gorges Dam. *EcoHealth* **2009**, *6*, 601–613. [CrossRef] [PubMed]

10. Mccartney, M. Living with dams: Managing the environmental impacts. *Water Policy* **2009**, *11*, 121–139. [CrossRef]

11. Bai, Y.; Xie, J.; Wang, X.; Li, C. Model fusion approach for monthly reservoir inflow forecasting. *J. Hydroinform.* **2016**, *18*, 634–650. [CrossRef]

12. Lai, X.; Liang, Q.; Huang, Q.; Jiang, J.; Lu, X.X. Numerical evaluation of flow regime changes induced by the Three Gorges Dam in the Middle Yangtze. *Hydrol. Res.* **2016**, *47*, 149–160. [CrossRef]

13. Li, R.; Chen, Q.; Ye, F. Modelling the impacts of reservoir operations on the downstream riparian vegetation and fish habitats in the Lijiang River. *J. Hydroinform.* **2011**, *13*, 229–244. [CrossRef]

14. Li, S.; Xiong, L.; Dong, L.; Zhang, J. Effects of the Three Gorges Reservoir on the hydrological droughts at the downstream Yichang station during 2003–2011. *Hydrol. Process.* **2013**, *27*, 3981–3993. [CrossRef]

15. Lai, X.; Jiang, J.; Yang, G.; Lu, X.X. Should the Three Gorges Dam be blamed for the extremely low water levels in the middle–lower Yangtze River? *Hydrol. Process.* **2014**, *28*, 150–160. [CrossRef]

16. Liu, Y.; Wu, G.; Guo, R.; Wan, R. Changing landscapes by damming: The Three Gorges Dam causes downstream lake shrinkage and severe droughts. *Landsc. Ecol.* **2016**, *31*, 1883–1890. [CrossRef]

17. Lai, X.; Jiang, J.; Liang, Q.; Huang, Q. Large-scale hydrodynamic modeling of the middle Yangtze River Basin with complex river–lake interactions. *J. Hydrol.* **2013**, *492*, 228–243. [CrossRef]

18. Cai, X.; Feng, L.; Wang, Y.; Chen, X. Influence of the Three Gorges Project on the Water Resource Components of Poyang Lake Watershed: Observations from TRMM and GRACE. *Adv. Meteorol.* **2015**, 1–7. [CrossRef]

19. Tapley, B.D.; Bettadpur, S.; Watkins, M.; Reigber, C. The gravity recovery and climate experiment mission overview and early results. *Geophys. Res. Lett.* **2004**, *31*, L09607. [CrossRef]

20. Yirdaw, S.Z.; Snelgrove, K.R.; Agboma, C.O. GRACE satellite observations of terrestrial moisture changes for drought characterization in the Canadian Prairie. *J. Hydrol.* **2008**, *356*, 84–92. [CrossRef]

21. Chen, J.L.; Wilson, C.R.; Tapley, B.D.; Yang, Z.L.; Niu, G.Y. 2005 drought event in the Amazon River basin as measured by GRACE and estimated by climate models. *J. Geophys. Res. Atmos.* **2009**, *114*, B05404. [CrossRef]

22. Frappart, F.; Papa, F.; da Silva, J.S.; Ramillien, G.; Prigent, C.; Seyler, F.; Calmant, S. Surface freshwater storage and dynamics in the Amazon basin during the 2005 exceptional drought. *Environ. Res. Lett.* **2012**, *7*, 44010–44017. [CrossRef]

23. Thomas, A.C.; Reager, J.T.; Famiglietti, J.S.; Rodell, M. A GRACE-based water storage deficit approach for hydrological drought characterization. *Geophys. Res. Lett.* **2014**, *41*, 1537–1545. [CrossRef]

24. Chao, N.; Wang, Z.; Jiang, W.; Chao, D. A quantitative approach for hydrological drought characterization in Southwestern China using GRACE. *Hydrogeol. J.* **2016**, *24*, 1–11. [CrossRef]

25. Gong, L.; Xu, C.Y.; Chen, D.; Halldin, S.; Chen, Y.D. Sensitivity of the Penman–Monteith reference evapotranspiration to key climatic variables in the Changjiang (Yangtze River) Basin. *J. Hydrol.* **2006**, *329*, 620–629. [CrossRef]

26. Hu, X.; Chen, J.; Zhou, Y.; Huang, C.; Liao, X. Seasonal water storage change of the Yangtze River Basin detected by GRACE. *Sci. China Earth Sci.* **2006**, *49*, 483–491. [CrossRef]

27. Long, D.; Yang, Y.; Wada, Y.; Hong, Y.; Liang, W.; Chen, Y.; Yong, B.; Hou, Z.; Wei, J.; Chen, L. Deriving scaling factors using a global hydrological model to restore GRACE total water storage changes for China's Yangtze River Basin. *Remote Sens. Environ.* **2015**, *168*, 177–193. [CrossRef]

28. Zhang, Z.; Chao, B.F.; Chen, J.; Wilson, C.R. Terrestrial water storage anomalies of Yangtze River Basin droughts observed by GRACE and connections with ENSO. *Glob. Planet. Chang.* **2015**, *126*, 35–45. [CrossRef]

29. Swenson, S.; Chambers, D.; Wahr, J. Estimating geocenter variations from a combination of GRACE and ocean model output. *J. Geophys. Res.* **2008**, *113*, B08410. [CrossRef]

30. Cheng, M.; Ries, J.C.; Tapley, B.D. Variations of the Earth's figure axis from satellite laser ranging and GRACE. *J. Geophys. Res. Solid Earth* **2011**, *116*. [CrossRef]

31. Peltier, W.R. global glacial isostasy and the surface of the ice-age earth: The ICE-5G (VM2) model and GRACE. *Annu. Rev. Earth Planet. Sci.* **2004**, *20*, 111–149. [CrossRef]

32. Wahr, A.; Zhong, S. Computations of the viscoelastic response of a 3-D compressible Earth to surface loading: An application to Glacial Isostatic Adjustment in Antarctica and Canada. *Geophys. J. Int.* **2013**, *192*, 557–572. [CrossRef]

33. Wahr, J.; Molenaar, M.; Bryan, F. Time variability of the Earth's gravity field: Hydrological and oceanic effects and their possible detection using GRACE. *J. Geophys. Res. Solid Earth* **1998**, *103*, 30205–30230. [CrossRef]

34. Swenson, S.; Wahr, J.; Milly, P.C.D. Estimated accuracies of regional water storage variations inferred from the Gravity Recovery and Climate Experiment (GRACE). *Water Resour. Res.* **2003**, *39*, 375–384. [CrossRef]

35. Swenson, S.; Wahr, J. Post-processing removal of correlated errors in GRACE data. *Geophys. Res. Lett.* **2006**, *33*, 2006. [CrossRef]

36. Swenson, S.C.; Wahr, J.M. Estimating signal loss in regularized GRACE gravity field solutions. *Geophys. J. Int.* **2011**, *185*, 693–702. [CrossRef]

37. Swenson, S.; Wahr, J. Multi-sensor analysis of water storage variations in the Caspian Sea. *Geophys. Res. Lett.* **2007**, *34*, L16401. [CrossRef]

38. Rodell, M.; Famiglietti, J.S.; Chen, J.; Seneviratne, S.I.; Viterbo, P.; Holl, S.; Wilson, C.R. Basin scale estimates of evapotranspiration using GRACE and other observations. *Geophys. Res. Lett.* **2004**, *31*, 183–213. [CrossRef]

39. Klees, R.; Zapreeva, E.A.; Winsemius, H.C.; Savenije, H.H.G. The bias in GRACE estimates of continental water storage variations. *Hydrol. Earth Syst. Sci.* **2007**, *11*, 1227–1241. [CrossRef]

40. Landerer, F.W.; Swenson, S.C. Accuracy of scaled GRACE terrestrial water storage estimates. *Water Resour. Res.* **2012**, *48*. [CrossRef]

41. Rodell, M. The global land data assimilation system. *Bull. Am. Meteorol. Soc.* **2004**, *85*, 381–394. [CrossRef]

42. Wang, W.; Cui, W.; Wang, X.; Chen, X. Evaluation of GLDAS-1 and GLDAS-2 forcing data and Noah model simulations over China at monthly scale. *J. Hydrometeorol.* **2016**, *17*, 2815–2833. [CrossRef]

43. Huffman, G.J.; Adler, R.F.; Bolvin, D.T.; Gu, G. Improving the global precipitation record: GPCP Version 2.1. *Geophys. Res. Lett.* **2009**, *36*, 153–159. [CrossRef]

44. Long, D.; Shen, Y.; Sun, A.; Hong, Y.; Longuevergne, L.; Yang, Y.; Longuevergne, L.; Yang, Y.; Li, B.; Chen, L. Drought and flood monitoring for a large karst plateau in Southwest China using extended GRACE data. *Remote Sens. Environ.* **2014**, *155*, 145–160. [CrossRef]

45. Huang, Y.; Salama, M.S.; Krol, M.S.; van der Velde, R.; Hoekstra, A.Y.; Zhou, Y.; Su, Z. Analysis of long-term terrestrial water storage variations in the Yangtze River basin. *Hydrol. Earth Syst. Sci. Discuss.* **2012**, *9*, 11487–11520. [CrossRef]

46. Sun, Y. Predicting groundwater level changes using GRACE data. *Water Resour. Res.* **2013**, *49*, 5900–5912. [CrossRef]

47. Zhang, D.; Zhang, Q.; Werner, A.D.; Liu, X. GRACE-Based Hydrological Drought Evaluation of the Yangtze River Basin, China. *J. Hydrometeorol.* **2016**, *17*, 811–828. [CrossRef]

48. Huang, Y.; Salama, M.S.; Krol, M.S.; Su, Z.; Hoekstra, A.Y.; Zeng, Y.; Zhou, Y. Estimation of human-induced changes in terrestrial water storage through integration of GRACE satellite detection and hydrological modeling: A case study of the Yangtze River Basin. *Water Resour. Res.* **2015**, *51*, 8494–8516. [CrossRef]

49. Wahr, J.; Swenson, S.; Velicogna, I. Accuracy of GRACE mass estimates. *Geophys. Res. Lett.* **2006**, *33*, 178–196. [CrossRef]

50. Hossein, S.Z.; Han, M.S.; Gyewoon, C. Evaluation of Regional Droughts Using Monthly Gridded Precipitation for Korea. *J. Hydroinform.* **2012**, *14*, 1036–1050. [CrossRef]

51. Zhang, N.; Xia, Z.; Zhang, S.; Jiang, H. Temporal and spatial characteristics of precipitation and droughts in the upper reaches of the Yangtze River basin (China) in recent five decades. *J. Hydroinform.* **2012**, *14*, 221–235. [CrossRef]

52. Ahmadalipour, A.; Rana, A.; Moradkhani, H.; Sharma, A. Multi-criteria evaluation of CMIP5 GCMs for climate change impact analysis. *Theor. Appl. Climatol.* **2015**, 1–17. [CrossRef]

53. Dai, A.; Trenberth, K.E.; Qian, T. A Global Dataset of Palmer Drought Severity Index for 1870–2002: Relationship with soil moisture and effects of surface warming. *J. Hydrometeorol.* **2004**, *5*, 1117–1130. [CrossRef]

54. Dai, A. Drought under global warming: A review. *Wiley Interdiscip. Rev. Clim. Chang.* **2011**, *2*, 45–65. [CrossRef]

 *water*

Review

Explaining Water Pricing through a Water Security Lens

Paula Cecilia Soto Rios [1,2,*] **, Tariq A. Deen** [1]**, Nidhi Nagabhatla** [1,3] **and Gustavo Ayala** [4]

[1]   Institute for Water, Health and Environment, United Nations University, 204-175 Longwood Road South, Hamilton, ON L8P 0A1, Canada; tariq.deen92@gmail.com (T.A.D.); Nidhi.Nagabhatla@unu.edu (N.N.)
[2]   Graduate School of Life Sciences, Tohoku University, Aoba-ku, Sendai 980-8578, Japan
[3]   School of Geography and Earth Science, McMaster University, 1280 Main Street West, Hamilton, ON L8S 4L8, Canada
[4]   Ministry of Environment and Water, 20 de Octubre St. 1628 San Pedro, La Paz, Bolivia; gustavo_ayala_t@hotmail.com
*   Correspondence: paulaceciliasoto@hotmail.com

Received: 8 July 2018; Accepted: 27 August 2018; Published: 1 September 2018

Abstract: Can water security serve as a platform for developing a long-term solution to ongoing water crises? Many regions around the world are experiencing severe water problems, including water scarcity, water-borne diseases, water-related natural hazards, and water conflicts. These issues are expected to increase and intensify in the future. Both developed and developing economies face a water supply and demand imbalance that will potentially influence their water pricing structures. Institutions and policies that govern the pricing of this natural capital remain crucial for driving food production and providing services. The complex and multifaceted issues of sustainable water management call for a standard set of tools that can capture and create desired water security scenarios. Water pricing is an important contributing factor for achieving these scenarios. In this paper, we analyze how water pricing can be used as a tool to enact the water security agenda. This paper addresses these issues from three facets: (1) Economic aspects—the multiple processes through which water is conceptualized and priced; (2) analysis of water pricing considering its effect in water consumption; and (3) arguments for assessing the potential of water pricing as a tool to appraise water security.

Keywords: water security; water pricing; sustainable water management; trends and patterns; economics

---

1. Introduction

Water is a crucial and valuable global resource, and its sustainable use is one of the most important challenges of our time. Unfortunately, water will soon be considered a scarce commodity as demand, driven by pressures of economic and population growth and the impacts of climate change, exceeds availability [1]. Among several definitions of water security available in the literature [2], we require a specific framework in which we can analyze water as a commodity, linking water to its price (the price of its access and distribution to consumers) and social good (the necessity for universal access). The definition provided by The Hague Ministerial Declaration described water security as: Ensuring that freshwater, coastal, and related ecosystems are protected and improved; that sustainable development and political stability are promoted, that every person has access to enough safe water at an affordable cost to lead a healthy and productive life, and that the vulnerable are protected from the risks of water-related hazards [3]. Under this definition, discussion regarding water pricing as a tool to achieve the objectives of sustainable development (affordable cost) can be created, considering not only the water resources sector but also different socioeconomic aspects. However, utilizing water pricing as a tool to specifically manage water demand can dangerous when considering the unstable political situations some developing countries have suffered by insulated pricing policies. Water pricing as a tool must therefore be analyzed so that not only the tool itself, but also a set of other strategies, are placed into action to acknowledge the complexity of the socio-hydrological relationship. The scale of this research considers national policies that can be applied at the local level, such as for cities and specific industrial and productive sectors within a country.

The United Nations defines water security as: The capacity of a population to safeguard sustainable access to adequate quantities of acceptable quality water for sustaining livelihoods, human well-being, and socio-economic development, for ensuring protection against water-borne pollution and water-related disasters, and for preserving ecosystems in a climate of peace and political stability [4]. Along with this definition, UN-Water also developed a conceptual framework (Figure 1) outlining eight key aspects that form the larger nexus of water security with respect to good governance, drinking water and human well-being, transboundary cooperation, ecosystems, financing, water-related hazards and climate change, peace and political stability, and economic activities and development. These aspects represent two dimensions: a core (key) with the elements under the control of local and regional authorities within a country, and a boundary (enabling) that reflects the external links governments need to carry out in order to advance national plans in the field of water management. This is a cyclical feedback loop, reflecting inter-dependence and interconnectivity among aspects such as governance, financing, peace, and political stability, and issues that extend beyond national borders with issues such as shared water systems and hydro-diplomacy [5]. The usefulness of this framework, for our purposes, is that the UN definition tackles water security with different levels of interventions.

The water security framework highlights the critical importance of an interdisciplinary approach if truly sustainable water management is to be achieved. The financial aspects of water management have gained significant attention in recent years as a method to tackle economic and societal interdependence [6,7]. Whereas water pricing has been long viewed from a supply standpoint, there has been a shift to use demand management as a tool for potentially achieving the water security agenda, as it acknowledges the unsustainable use of water and promotes the necessity of water conservation [8,9]. Given the growing threat of water insecurity [10], the narratives of water necessity versus desirability also need re-examination.

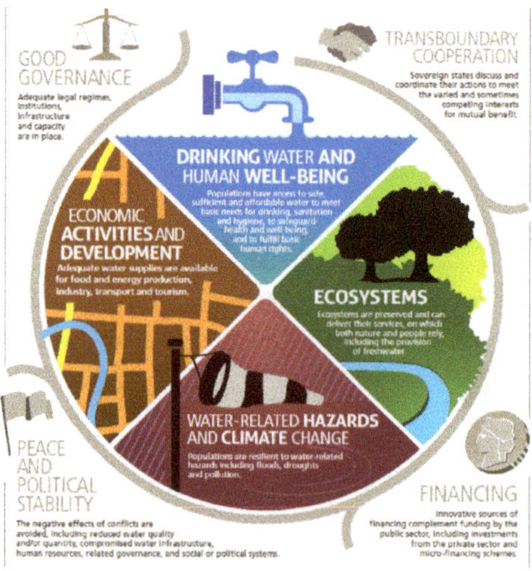

Figure 1. United Nation (UN)-Water's conceptual framework [11].

Achieving water security remains critical to meeting the water aspect of the Sustainable Development Goals (SDGs). Numerous alternatives have been studied to reduce water use, and several utilities service providers across the world are making strong efforts to advance water conservation [12–14] not only in the urban water sector but also the agriculture sector. Research has focused on the agricultural sector to improve water use efficiency (another water security tool). It is anticipated that conceptual, empirical, and case study approaches on water pricing could serve as a smart strategy to ensure long-term water security. The discussions presented in this paper explain how a utility service provider (private or public) can employ water pricing policies to encourage customers (users) to manage consumption, maintain financial viability, and promote the mindset of water conservation. In this sense, allocations should account for actual water use and could be adaptable to changes caused by other water security strategies—like water use efficiency—to prevent and manage as many aspects as possible of the increase in water demand [15].

This paper is structured into the following sections: (1) how water is conceptualized and priced, in which we describe the importance of the characteristics of water as a good, and the implication of its human rights aspect (i.e., universal access), which, in many cases, generates misunderstandings between society and policy makers when discussing the price to access water; (2) We then analyze water pricing considering its effect on water demand, in which we provide a summary of sectoral and temporal dynamics regarding the application of water pricing (e.g., price structure variation for agricultural sector and drinking water sector), and spatial aspects (e.g., the availability of resources). In order to evaluate the performance of water pricing policies already applied, we summarize five case studies, two from the agricultural sector, and three from the urban drinking water sector (two from a water scarce perspective, and one from an economic and water plentiful perspective); (3) Finally, we critically analyze the potential of water pricing as a tool for water security, in which we analyze the aforementioned study cases and establish tradeoffs in water pricing as a tool to regulate water demand, while also evaluating the consideration of water a good (special economic good) in outcomes from the applied case studies' policies. The paper concludes with a summary of all previously discussed points and describes how water pricing could be used as a tool to ensure water security objectives given the framework (UN-Water approach) we outline, considering the scale and level of application stablished at the beginning of the paper.

2. Materials and Methods

The complexity of applied water pricing policy reflects the efforts of governments and utility service providers put into effect actions that will address water demand. Thus, we completed a systematic review of academic and grey literature on the subject. During our review, we accessed 112 studies, with 40 being excluded due to their irrelevance to objective of this paper.

For the case studies, agriculture and urban water consumption were considered for analysis because of their complexity and diversity with respect to water pricing. Both national and municipal/district level examples were chosen. The paper focuses on the agriculture and urban drinking water sectors because 86% of the total global water consumption is due to agricultural activities [16] and given the importance urban water access to societal development.

3. Managing Water Resources: Conceptualization and Pricing Structures

3.1. Water as a "Good"

In 1992, the Fourth Dublin Principle established under principle (1) and (4) that water is a finite and vulnerable resource that has an economic value and should be recognized as an economic good [17]. In addition to this, the Rio Principle expressed that water is a social good and that humans are entitled to, at least, a minimal quantity and quality of safe water. However, the emerging pluralism in the valuation and interpretation of water could lead to scenarios of competing and conflicting conceptualizations. The emerging concepts influence the existing debate by experts from various disciplines over water pricing. When water is priced as an economic good, its economic value can vary depending on buyer and seller willingness to pay [18]. However, if the concept of social good is applied to the pricing structure, then water should be affordable to the poor, benefiting the largest number of people in the best possible way. To add further complexity with respect to the economic principles, the true "economic" value of a good and its "financial" value seldomly correspond; hence, the competitive market prices reflect only the financial and not necessarily the economic values of water [19].

Briscoe [20] tried to connect the concept of water as an economic good by using water pricing. Figure 2 shows the optimal consumption for water as X*, on the left. The graph on the right shows the water consumption when the marginal cost of supply differs from the charged price, then the consumer will not consume X* (optimal consumption) but X^1 (new consumption), but still pay price P^1 (charged price). The increase in cost (the area under the cost curve) exceeds the increase in benefit (the area under the benefit curve) and there is a corresponding loss of net benefit, called the deadweight loss. However, this concept is only applicable to people who can actually afford the price of water, like any other good. In this context, the deadweight loss is seen as a human gesture of tax-payers who are willing to subsidize water to avoid water poverty. Thus, the supply curve will shift down to the point where it intersects the demand curve at P^1, X^1.

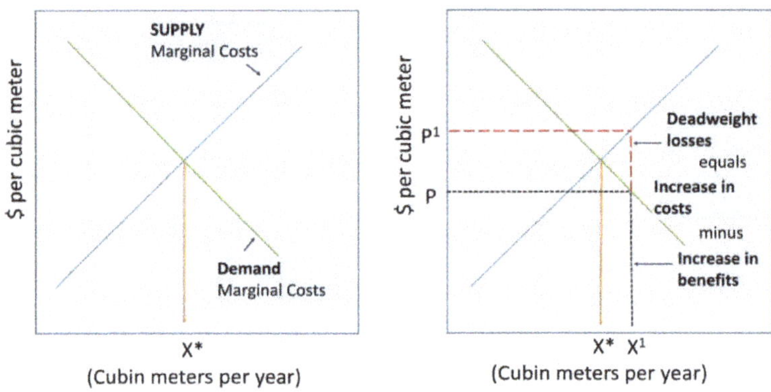

Figure 2. Optimal consumption (left) and "deadweight losses" if water is underpriced (right) [19].

From the previous explanation and in order to continue this analysis, we must consider water as a special economic good (discussed below). However, water utilities are usually a natural monopoly and the marginal costs are usually lower than average costs. Such pricing would lead to a unit price that is less than the average cost and the utility will not generate enough funds to cover all costs (operational, management, quality, maintenance, or future events). From this, the question of how to use water pricing as a strategy by using deadweight losses efficiently is answered using two main theories in economics. One theory involves using subsidies and taxation as a form of lump sum transfer to make up the loss as explained before. The second is to use price discrimination to recover costs through Ramsey pricing and Pareto Superior Non-Linear Outlay Schedule [21]. The answer will vary according to different conditions of time and place, especially as conflicts over resources increase values and people's rationale changes.

3.2. Commodifying the "Good"—Water Pricing Structures

The application of water pricing declined in the early 2000s due to a social resistance in many countries and the lack of ability of governments to properly implement water pricing structures. Each sector faces challenges with respect to pricing methods, not only between countries but within regions of the same country. Water pricing is considered a crucial issue for decision makers, water services providers, and consumers. For this reason, assessing a socially fair average water price that would be acceptable according to the reality of each area is recommended [22,23]. Water security could be severely affected if innovations in water pricing are not developed and implemented. Some efforts, such as the European Union Water Framework Directive (WFD), have been introduced to create structures in which water users' taxation reflects the complexity of valuing the water [24,25].

Water pricing can vary among and within different economic sectors. For example, water pricing within the agricultural context can vary based on agricultural area (i.e., area-based pricing), agricultural output, the value of the land (i.e., betterment levy pricing), or the amount of material used for agricultural production (i.e., input pricing), or pricing can be uniform for all users [26,27]. However, the most basic types of water pricing are: (1) flat rate, which implies no requirement for controlling the demand with the price; (2) uniform (volumetric) rate, which is the basic structure used to put into consideration the value of the quantity of delivered water to the final user; (3) block or tire rate, which is used in order to focus the water consumption in specific values; and (4) complex rate structures, based on behavioral analysis of water users to optimize not only revenues but also water consumption itself [25]. These water pricing structures are illustrated in Figure 3.

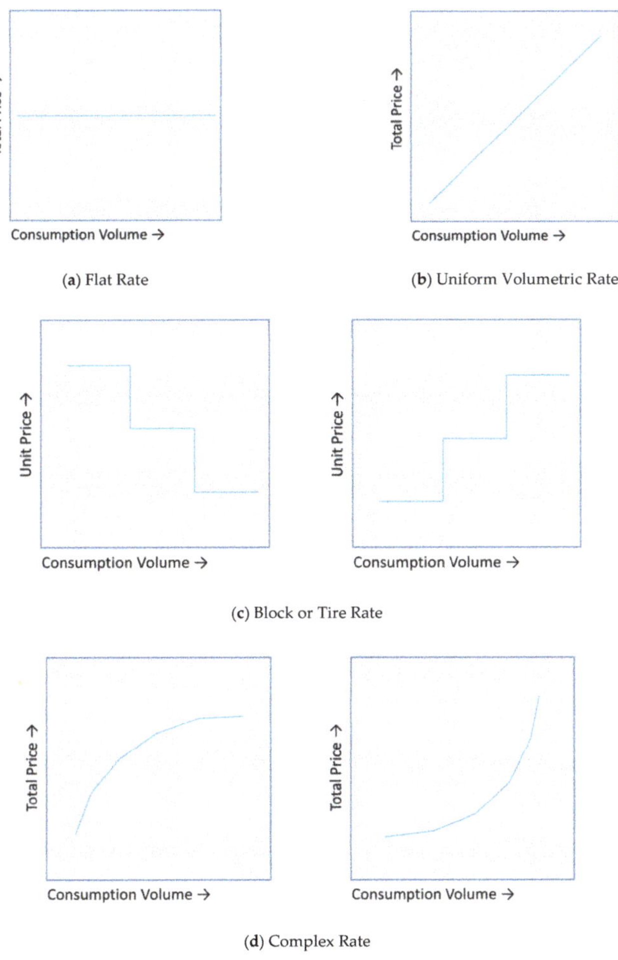

Figure 3. *Cont.*

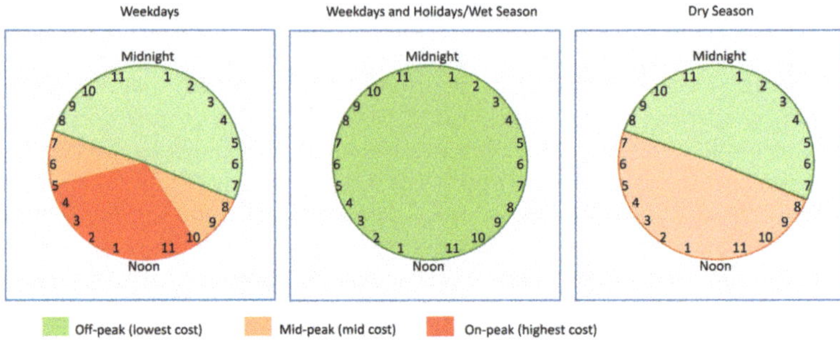

(**e**) Alternative Structures: Time of Day or Seasonal rates.

Figure 3. Overview of water pricing design and structures: (**a**) All customers pay the same price regardless of the amount they consume; (**b**) the total price of water is dependent on the amount of water a customer uses. The unit price of water does not change but the total monthly price will change. (**c**) Customers are charged for water based on increasing (or decreasing) flat rates; i.e., the price of water changes once a limit has been reached. (**d**) Water price changes once a certain limit has been reached and constantly increases. The total water price will always be different depending on water consumption. (**e**) Water pricing is adjusted based on the time of day. The unit price of water is higher during peak hours to encourage water conservation; water pricing is adjusted based on climatic conditions, for example, wet and dry season.

3.3. Models for Water Provisioning

At the national scale, water resources are managed by different entities: normally public utilities provide urban water services, while irrigation water is managed by irrigation districts, and water pricing in both sectors is under the regulation of different government agencies [28].

For service delivery, there are three basic models: (1) Common property management; (2) municipal companies; and (3) forms of privatization.

Common property management is controlled by the user's group that arranges the delivery and supply of the resource. However, water resources are mostly managed by the public sector (i.e., natural monopoly) to maintain the balance between water conservation and revenue stability in water structures. In other cases, privatization is used as a model to enhance access to water, but for some economists, privatization is no more than a means to create competition, since a monopoly is expected to be inefficient whether it is private or public [29]. In Latin America, some cases of privatization have been a polemical topic, leading to a series of protests and riots when the population was not informed properly [30].

Given these realities of rivalries, externalities, and monopolies, as well as equity considerations, some governments require a higher authority that can establish rules regarding the rights of individual access and management of water. These standards basically refer to the amount (i.e., volume) of water extracted from a primary source and the quality of the water both supplied to the end users and released as a secondary source after its use. Overall, water management (although mostly focused on provision) focuses on determining how much demand for water would inevitably increase in the future and how increased demand will be addressed in different sectors (e.g., agriculture, urban water, commercial, etc.). One of the challenges some governments are facing with respect to the conceptualization of water as a special economic good is to identify who should be responsible for its management considering that almost any intervention in managing water affects the environment, society, and the economy, either intentionally or incidentally [29]. This discussion raises the question of whether water is a special economic good that can be reasonably left to free market forces or if some amount of extra-market management is required to effectively and efficiently serve social objectives [24]. Researchers have considered the complex relationship of promoting water conservation and ensuring a stable revenue stream to cover the predominantly fixed charges of running a water utility. They state that, to achieve maximum social efficiency and minimize deadweight losses, the pricing should be at the level of long-run marginal cost [24]. The other challenge is fitting water provision within the overall mandate of sustainable water management.

Van der Zaag and Savenije [31] stated that there is no other economic good that has the complicated characteristics of water. For this reason, water should be called a special economic good that is an essential, non-substitutable resource that needs reasonable pricing structures that aim at cost recovery, and simultaneously ensure access to safe water for the poor, while considering ecological requirements. The future predictions of global water scarcity demands highlight that a reasonable water pricing structure should be assigned that encourages cost recovery and resilience by sending a clear signal to water users that the resource should be used wisely. However, setting appropriate prices depends on water use, making the selection process difficult since the nature of water's cost varies in different economic sectors and geographic locations. For this reason, if water pricing is to be defined as an economic policy intervention tool that can be used to achieve the environmental, social, and economically efficient management of water, then it is essential to understand its implications in real world situations to ensure its applicability as a water security tool.

4. Analysis of Water Pricing Considering its Effect on Water Consumption

Water pricing strategies vary between economic sectors and states. For example, pricing of agriculture or urban water could be based on completely different underlying principles. To further understand the use of water pricing as a tool for water security, it is an imperative to explore how the concept is applied.

4.1. Sectoral and Temporal Dynamics

Water pricing in the agricultural sector has attracted special interest from researchers and policy makers due to the complexity of water consumption concerning the food security nexus [8,9]. Irrigation water supply is often subsidized given that food security and rural livelihood issues take precedence over the cost recovery model [32]. Studies report negative impacts of water pricing on income and agro-production trends due to farmers reducing water consumption by changing crop plants or introducing less profitable crops. Berbel and Gómez-Limón, Elnaboulsi, Kanakoudis et al., and Aidam [33–36] proposed a base-price design through mathematical models (e.g., Linear Programming or Multi-Analysis Tool for the Agricultural, MATA) to establish a tariff that would not negatively affect farmers' income, and in turn encourage saving water by reducing waste and improving efficiency. The MATA model promises a strategy to address potential conflicts to achieve positive results on water rights.

In the urban context, planning water supply methods while employing a based-price design agreement and embedding scenarios with respect to droughts, water stress, or the "incremental cost" of operation creates a challenge for proposing price reforms without polemic consequences or physical conflicts [37]. With this in mind, we selected some case studies to support our argument. Case studies 1 and 2 were selected to analyze the agricultural sector, specifically within the national scale of policy intervention. Case studies 3 and 4 focus on urban water consumption, and the fifth case study exemplifies the importance of water pricing in societies which has both water and resources availability surplus.

4.2. Case Studies

4.2.1. Irrigation Sector

Case Study 1: Smart Water Pricing in the African Continent—Ghana

Although the agricultural sector in Ghana has declined over the years [38], but it remains an important contributor to Ghana's export earnings and a major source of inputs to the manufacturing sector and the most important sector for jobs and livelihoods in the rural areas. [39]. Major Ghanaian crops include cocoa, coffee, oil palm, cashew, and rubber [38,40]. During the dry season, Ghanaian farmers rely heavily on irrigation and are the dominant source of water resource consumption in the country [41,42]. Reports suggest that the country is experiencing significant water scarcity, especially in the agricultural sector, and poor management of agricultural irrigation pricing has resulted in poor economic returns [43–45]. There are also reports of urban farmers using wastewater for irrigation as an attempt to meet growing demand for vegetables due to inadequate health education [46,47]. Therefore, the need for different systems or an appropriate model to price water in the agricultural sector remains crucial.

Aidam suggested a Multi-Analysis Tool for the Agricultural Sector (MATA) approach to understand the relationship between the farmer's activities and water pricing policy [36]. The model is comprised of different modules and takes note of the behavior of the process for creating agricultural products and the behavior of consumers with a price expectation and risk attitudes, while incorporating the environment in which farmers, processors, and consumers make decisions. Finally, water supply variables are also incorporated to add the impact of water demand on the agricultural sector in Ghana [36,47]. This study selected areas according to their resource characteristics (e.g., land, labor, capital, and management) and farmer management decisions were simulated to provide a guideline on water pricing for a homogeneous agricultural sector with identical socio-economic and agro-climatic environments. The principal assumption was income maximization as priority goal for each farmer activities; incomes for each farm were GH¢ 449,867.00 (USD \$101,550.63) for large farms, GH¢ 454,081.00 (USD \$102,501.88) for medium farms, and GH¢ 359,666.00 (USD \$81,189.13) for small farms. The crops used were rice, maize, and vegetables. If water prices increased significantly, then water pricing policy would have a negative impact on the demand for water resources, resulting in a subsequent reduction in both farmers' income as well as labor employment, thus negatively affecting agricultural and social business. However, after using the MATA tool, it was established that adopting 2 cedi/m^3 (0.43 USD/m^3) in that year as the uniform volumetric rate was a measure that both engaged farmers to consciously use less water resources and motivated them to implement water saving technologies.

This analysis provides a different approach to manage water consumption in the sector and focuses on irrigation efficiency. Linstead [15] examined the improvements in irrigation efficiency in environmental services and concluded that allocation of water use should account for actual consumptive use (i.e., irrigation consumption) and should be adaptable to changes over time in irrigation efficiency, i.e., withdrawal or use allocation should decline as the irrigation efficiency increases to prevent overall increases in consumptive use. In this sense, management and planification strategy have a positive effect when water pricing is used as a tool by reducing and regulating consumption growth by MATA modelling, and at the same time ensuring the incentive in the sector remains focused on irrigation efficiency. This allows governments to implement strategies in other sectors, such as the environment, by using the remaining water to improve conditions as a result of the application of these two tools. This non-explicit behavior in the outcomes from the strategies acknowledges the complexity of non-linear water—social response, a characteristic of a water security tool [48].

Case Study 2: National Strategies in Water Pricing—Spain

Agriculture production in Spain relies on the 60% of irrigated areas and only 19% of cultivated areas; therefore, 80% of the water supply is consumed by agriculture. The old system for irrigation was flat rate systems imposed by the Comunidades de Regantes (CR). These communities are assigned by the government agencies to manage water; however, farmers are charged only part of the total distribution cost. This fixed cost is calculated by hectare and used to irrigate subsidized extensive crops. Common Agriculture Policy (CAP) is considered detrimental as it causes losses because of the high consumption of water. Berbel and Gómez-Limón [33] applied Linear Programming mathematical modulation by variables and interrelationships into a matrix algebra to better understand the implication on decisions of policymakers and farmer's objectives in annual herbaceous crops (e.g., alfalfa, wheat, etc.) [33]. This study found that, for almost all crops (under CAP), adding a price to water will cause unemployment and negatively affect farmer income, reducing it to 25–40% before achieving the reduction in water consumption and making the region vulnerable, as farmers would reduce the range of crop production. Berbel and Gómez-Limón simulated that a fixed tariff of 2 Ptas/m^3 (former Spanish currency) could help reduce water consumption and persuade farmers to use water-saving technologies that do not affect the crop [33]. The authors stated that this fixed tariff should be analyzed continuously considering seasonal changes in the consumption and annual income, so the water pricing policy could be implementing properly. In this example, we see the problematic aspects and challenges of applying narrow perspectives (e.g., subsidized policy and flat rate price) given a new national strategy of preserving water resources.

4.2.2. Urban Water Consumption Sector

Case Study 3: Incentive-Based Approach—São Paulo, Brazil

Companhia de Saneamento Básico do Estado de São Paulo S.A. (SABESP) is a Brazilian water and waste management company owned by São Paulo State, considered the largest waste management company in the world by market capitalization, providing water and sewage services. The 2014–2015 drought created a water crisis in Brazil and forced SABESP to implement a subsidy program based on the combination of two forms of incentives to promote sustainable water use. A subsidy was offered to members of the community who reduced their water consumption. By implementing the program, environmental conditions improved and consumers received a government reward (i.e., discount on the water price and sewage tariff). In addition, a contingency fee was imposed on consumers as a form of punishment where an increase in demand was noted. This was implemented in the through taxes, fees, and high charges per unit (Figure 4).

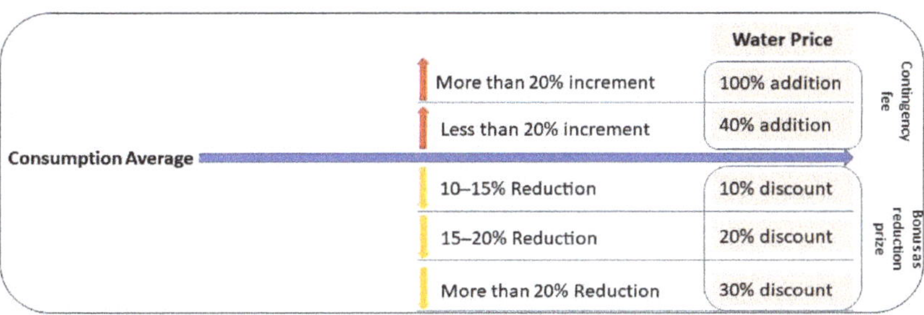

Figure 4. Companhia de Saneamento Básico do Estado de São Paulo S.A (SABESP) program to reduce overconsumption of water [49].

In 2016, the City of Machado evaluated the effectiveness of SABESP's program. The study took the average consumption of water per household during the period of 2013 to 2015 covering 25 districts of the municipality. The dependent variable analyzed was the reduction in water consumption. An increase in average residential consumption in all districts between January 2013 and January 2014 was observed. In contrast, between January 2014 and January 2015, the average reduction in this period was 25.0%, with a median of 25.98% attributed to an "awareness effect" as a function of the serious water crisis experienced. According to Gilbertson et al. [45], attitudes and behaviors in relation to water conservation in households differ between regions and are based geographically [49,50]. However, in the case of São Paulo, even in the described scenario, households received water because of SABESP's program, whereas households of other municipalities in the region were affected by the water crisis [49]. The strategy used by SABESP is similar to what is called Hybrid Policy, through which subsidy and taxation systems are implemented at the same time based on consumption patterns and trends. However, the success of SABESP during the water crisis was not accomplished by the technical characteristic of the water pricing strategy (the Hybrid Policy) but by the complementary strategies applied in conjunction with the Hybrid Policy. This awareness effect was created by the intensive promotion of water conservation that SABESP deployed in parallel with the Hybrid Policy. Therefore, this case study also illustrates that water policy strategies must work holistically, so they become tools that recognize and manage the complexity of social, political, economic, and climatic dimensions [51], which are characteristics of water security under our framework.

Case Study 4: Water Pricing and Conflict—Cochabamba, Bolivia

The privatization of Cochabamba's water utilities was initiated by the World Bank, and the International Monetary Fund (IMF). The privatization agreement was the result of a three-year, USD $138 million national debt relief program from the IMF [52]. Following privatization, a private international consortium (Aguas del Tunari) elevated water premiums by as much as 60% for some consumers [53], resulting in months of protests, hundreds injured, and at least six dead. The protests concluded with the President agreeing to a concession and returning water ownership to the municipal authorities [54–56]

Prior to privatization, Cochabamba's water services were owned and operated by the Municipal Water and Sewage Service of Cochabamba (SEMAPA). Under SEMPAPA's ownership, only 50.20% of residents were connected to municipal water, with many people relying instead on public and private wells, cistern trucks, springs, and neighbors as alternative sources of water [52]. The privatization agreement was believed to address Cochabamba's significant water scarcity issues in addition to alleviating national debt [52]. The privatization guaranteed a return on investment of at least 15%, thereby justifying Bechtel's (the majority stakeholder in the consortium) approach to increasing water price, resulting in a multiple tier block rate structure [57].

It was argued that the price increase was necessary to reflect service costs and was consistent with other utility contracts in high-risk countries. In addition, the policy was meant to be inclusive, as the price increase among the poor was claimed to be no more than 30% [50]. However, as Shultz [58] stated, price increases were actually 41% among the poor and 51% on average for all users. Many residents felt that the new pricing structure was unfair and unaffordable, resulting in the violent protests and eventually termination of the contract. Despite control returning to SEMAPA, issues of water scarcity have not improved in Cochabamba, as fewer people have access to water than prior to the privatization [59,60].

A key characteristic of the IMF and the World Bank's policy intervention was that it only considered water from an economic perspective. Bechtel and the consortium paid no regard to two crucial elements in the city: the overall lack of water resources, and the spatial component of the water scarcity. The lack of a holistic strategy led to a situation in which it was not the water pricing strategy itself that was inapplicable, but the unilateral policy implementation that considered only a return on investment.

Case Study 5: Governance Reforms in a Developed Economy—France

In France, urban water services are publicly owned, but operational responsibilities can be given to private operators if public operators cannot meet certain requirements [61]. Although national legislation has had a history of affecting water pricing and management (e.g., the 1992 law), the current water pricing structure is strongly influenced by France's 2006 water law (Loi sur l'eau et les milieux aquatiques; LEMA). Under this law, specifically under article 57, water service providers are forbidden the use of flat water pricing rates and declining water pricing rates (with exception to small cities and water plenty areas). The 2006 law was meant to emphasize water conservation. However, it should be noted that, overall, France has an abundance of surface and groundwater, although certain parts, mainly the south and east regions, experience dry seasonal weather.

Before LEMA, the water pricing structure was mainly uniform volumetric (applied to 57% of the districts). After the legislation of LEMA, statistics from 2013 show that the number of districts using uniform volumetric pricing systems increased to 61%, and decreasing pricing systems—like flat rate structure—dropped to 4%. Furthermore, the percentage of districts using increasing pricing systems grew from 1% in 2003 to 29% in 2013 [62]. The pricing policy reform led to an increase in water efficiency by 3% from 78% in 2008 to 81% in 2010, and a decline of 4.2% in the number of lead pipes in the country [63]. Finally, France ensures that water prices remain below the national minimum wage and water utilities have provided grants for low-income families that cannot meet their monthly water usage bill [64].

In the case when there is an ideal situation between demand and supply, water pricing strategies should focus on the optimization of the sector, an aspect that national agencies do not consider at first. In the case of the French government, they chose to optimize their water pricing policies by practical analysis and study of policy effect, which is a characteristic of the water security framework (expanded research agenda) [51].

5. Potential of Water Pricing as a Tool for Water Security

As previously described in the case studies, the power of water pricing as a policy of regulation for water consumption is quite relevant, as it implies that, as a water policy, pricing can control the demand sector. Therefore, water pricing is considered a crucial issue for decision makers, water utilities managers, and consumers. Water security can be severely affected if innovations in water pricing are not developed and implemented efficiently to improve and ensure the effectiveness of another strategies already in place. Efforts have commenced to address this issue in some parts of the world. For example, the European Union Water Framework Directive (WFD) created a pricing structure that taxes water users to reflect the scarce value of water [65]. The European Commission report states that water pricing should be used as a key tool to support water management decisions, and that underpriced water may lead to its unsustainable use. The report goes on to state that water pricing should be discussed within social policies to help ensure that they are fair to all sections of society and environmentally sustainable (i.e., ensuring that developing populations do not suffer from high water pricing policies). Applying a participatory approach to water-related decision making is recommended [50]. In contrast, other case examples reflect how water prices could lead to social conflicts and increase government reluctance to adopt this approach, since the different views regarding government's role in water regulation are based on cultures, religions, and political interests [66]. Moreover, negotiation principles demonstrating different and often heavily asymmetric bargaining positions of partners are related to water pricing of existing cross-border utilities [21]. Consequently, to establish whether water pricing could serve as a tool for water security, from the case studies, we concluded that the price of water as a strategy is highly sensitive in societies with problems of scarcity. Therefore, its application must be carefully studied, considering the understanding as well as who is in charge under different scenarios, and commitments of the water allocation, distribution, and overall management.

Yepes [67] sketched a pricing policy as one that distributes the costs of services among users, which must be designed to achieve the objectives of economic efficiency, using multiple investment and operations to provide an additional unit of service; financial sufficiency. The policy should have the necessary resources for the provision and maintenance of the service in an efficient and sustainable manner, along with the possibility of growing and improving infrastructural investments. The policy should provide universal access for all to the water service with simplicity and transparency [67]. The pricing policy must be understood by all persons involved: users and officials. Regarding transparency, rules must be clear with respect to the allocation of user costs so that they are accepted by society. Hughes [68] explained how revenue sufficiency is the primary financial objective for water utility systems that operate as enterprises and considers the other aspects (stated as below) as secondary objectives.

(1) Affordability: Ensuring that water is affordable to a community for basic services is a priority of many utilities and their governing boards. Maintaining affordable rates should almost never take precedence over charging rates that are necessary to recover the full costs of service.
(2) Conservation promotion: The amount that customers pay for water service acts as a price signal, often encouraging the customers to decrease consumption.
(3) Economic development: Utilities may strive to attract new or maintain existing commercial customers through water rates to foster greater community benefit.
(4) Short-term revenue stability: Year to year, most water utilities rely on revenue from water consumption charges to cover the predominantly fixed costs of the utility. Yet, water consumption can vary and is on the decline for many utilities, undermining water utility revenue stability, which some are calling the new normal.

Considering that the amount of water is finite, and the marginal cost of water can be very high in droughts or when a reservoir runs dry at the end of the season, and once a person (or a crop) receives sufficient enough water to alleviate physical stress and strain, the utility of additional units rapidly falls and becomes negative. To evaluate potential drivers of an efficient water pricing policy as to tool for water security realization, different perspectives—illustrated using the selected case studies—present a fair argument that water pricing can be used either effectively or ineffectively to promote water security. For instance, in the agricultural sector, water pricing draws special interest from the national governments due to the link between production and food security [8,9]. There is consensus that water pricing in agro-production and subsistence livelihood sectors report negative trends, referring to impacts like change in cropping composition or introduction of less profitable crops [33].

Kanakoudis et al., Elnaboulsi, and Aidam [22,34,36] outlined a base-price design through mathematical models. The Ghanaian and Spanish case studies exemplify the use of this concept, and modeled evidence to establish a tariff that would not negatively affect the farmers' income that would help encourage saving water through consumption management using the MATA and LP approaches. Whereas models are only reflective of scenarios specific to different contexts and situations, researchers often point towards data gaps [69], showing that you cannot methodologically cover the entire context. Therefore, water pricing policy should not be outlined as a standalone tool but as part of overall water governance measures (i.e., irrigation technology policies) to achieve effective results.

In the municipal context, planning water supply employs a base-price design agreement and embeds scenarios of droughts, water stress, or the incremental cost of operations, making it hard to propose reforms without polemic consequences or physical conflicts [37]. For many municipalities, water accessibility is a persistent issue, as illustrated in the Cochabamba case study. The privatization contract was intended to increase water accessibility in this city, but the consortium failed to assess the stakeholders' economic and social needs (a multidisciplinary framework showed in water security framework, not applied in this case).

However, increasing water rates does not always lead to social unrest, as illustrated in the French case study. Although water pricing structures for many French municipalities changed from flat and decreasing pricing structures to volumetric and increasing ones, the average rate of water remained below the national minimum wage. In theory, this approach would allow affordability while promoting water conservation, aligning with national goals of water security. In addition, it allowed for the provision of water subsidies for low-income families, which addressed issues of income inequality.

The new water management policies need to factor in multiple aspects highlighted in the water security nexus. Many regions worldwide are facing natural- and anthropogenic-driven water scarcity, and therefore require smart thinking (nonlinear) in valuing and pricing provisioning services. The case of study in São Paulo presents a good example of a practical strategy to manage demand and consumption that can in turn promote water conservation. Figure 5 illustrates how a smart water pricing policy based on the presented arguments could enable prioritizing policies and approaches dealing with positive and negative spillovers while ensuring water secure futures.

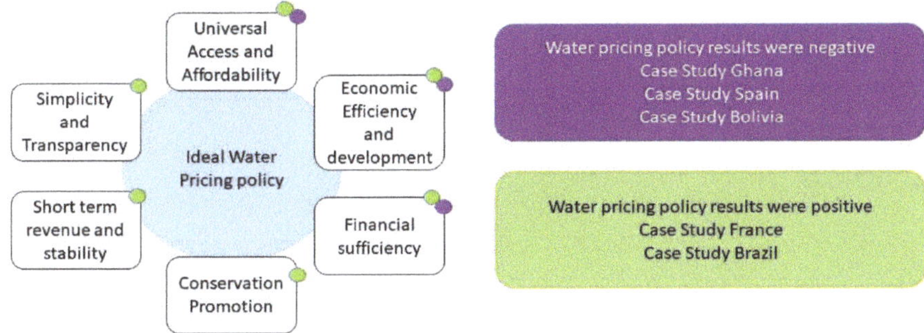

Figure 5. Prioritizing strategies lead to affect positively and negatively water pricing.

Furthermore, Figure 6 exemplifies the water pricing conceptual feedback process at the level and scale analyzed in this study (national and sectorial scale for urban and irrigation sectors) in terms of a water security framework. One of the outcomes may include tackling the risk of inadequate water quality by understanding the importance of ecosystems services as a water source and instruments of water pollution control and restoration. In such a scenario, we not only ensure or secure drinking water and human well-being, but the society at large acknowledges, through conservation, that natural ecosystems can assist in tackling climate change and disaster prevention [70] to provide long-term revenue. In addition, even in the absence of human-induced climate change, more severe drought is likely to occur in the future than has occurred in the past 30 years [71], which leads us to a scenario of reduced water availability. Creating resilient systems is one of the strategies included for short-term revenue generation, as in the case of Brazil. Using these systems leads to creating more sustainable cities and an improvement in water resource management by providing viable, cost-efficient, and effective solutions.

In addition, if solutions need to be developed and financially supported by fair pricing structures, then this kind of setting may provide stability and better acceptance by society to newer policies and arrangements. To provide this stability and ensure transparency of the water supply in urban areas, IBNET (The International Benchmarking Network for Water and Sanitation Utilities) indicators of the World Bank could help [72]. These indicators are based on a framework where the water utility must provide services to all customers at affordable prices, while controlling quality and maintaining financial incentives for its staff [73]. This analysis is based on a compilation of a databases from utilities and companies in charge of distributing water. Special considerations about the data must be evaluated prior the estimation itself, as many of the variables and performance indicators upon which IBNET reports do not fit a normal distribution but rather form a skewed distribution, for which the specific mean and the median differ. For those cases, the Blue Book recommends the median as a better representation of the performance. Some of the IBNET indicators include: water coverage, waste water coverage, nonrevenue water, staff productivity, operating cost coverage ratio, operation and maintenance cost, operating revenues, water consumption, collection period, affordability of water and sewage service, and cross subsidies. IBNET indicators can help to reveal management efficiency and allow for continuous updates to water structures according to their needs.

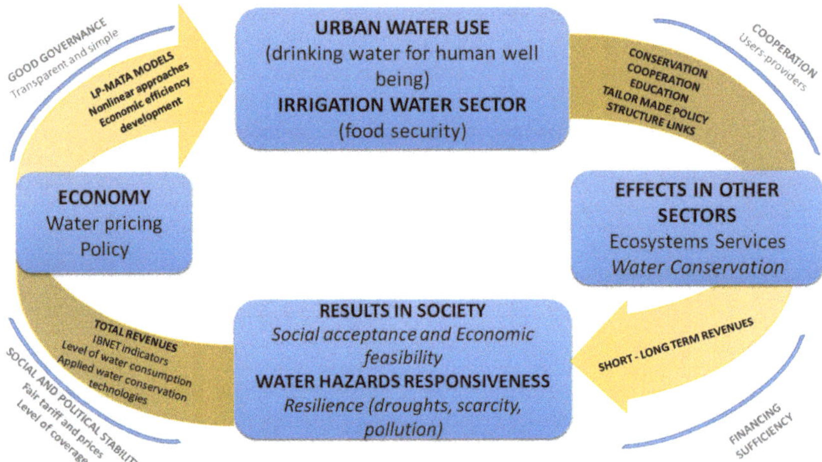

Figure 6. Suggested model for water pricing embedding water security guiding framework.

Mathematical economic models like MATA and LP are used to assess all the potential scenarios that produce negative impacts on society, economy, and the environment in the case of irrigation. It is also essential that the pricing of water services covers the costs of providing service for both operations and maintenance, as well as capital expenses including future, operations, maintenance, and capital costs. In the case of Ghana and Spain, the results of omitting the economic efficiency indicator led to water being distributed under unsustainable growth. The LP and MATA models offer a perspective on how to better implement water pricing changes under reasonable boundaries (case studies 1 and 2). Other characteristics of the suggested model are simplicity and transparency. Learning from Cochabamba (case study 4), wherein the water pricing policy was applied directly without consensus with the population and simplicity in the applied policies led to an unsustainable relationship between the water company and the society. The case of São Paulo (case study 3) showed how the government outlined and applied a set of combined policy strategies for drought (water scarcity) management. It was interesting to note how the population and communities reacted to managing their own consumption post-crisis. The reaction drives us to follow the loop on cooperation (users-providers) to achieve education and conservation as part of the suggested model. This highlights that the complementary policy framework must not only be applied in the face of water crisis but also in both the short and long-term policy trend, simply and transparently leveraging triggers in the policy framework and their impacts. In order to avoid conflicts and organize resilient communities, water pricing can be used as a tool for water security under specific principles and strategies, as shown in Figure 6.

As mentioned above, the water pricing policy is quite sensitive in terms of social response despite the characterization of water as an economic good, which is often considered inelastic. This implies that both simplicity and transparency are required features of the implementation of the policy, without mentioning that it must be managed in a participatory way. In the case of Cochabamba, due to the simple lack of communication between decision makers and society, the policy could not be deployed and social context was not considered (poor sectors in the metropolitan area without access to water and with the same rate as any other another sector). In the case of São Paulo, due to the simple intensive communication between the state government and society, unforeseen positive effects were achieved (water conservation by the population); the context of the drought crisis favored the application of the politics.

The intention of this paper was not to promote flat rates over uniform volumetric rates or privatization over public ownership. Instead, the goal was to illustrate that effective water pricing can be used in a water security framework. Water pricing is complex, and as shown, it needs to be case-specific to consider the ability of a system to respond to changes in prices. Unfortunately, there is no single water pricing structure that can achieve the water security agenda. Therefore, instead of a specific pricing system, we suggest that a water pricing policy based on current and future events is needed that considers seasonal variability, household income, and reality of each sector and ideally the vision of water security that the nations and the communities are planning to adopt.

6. Conclusions

Considering that water insecurity is a looming challenge to address when planning strategies for sustainable development, discussion of re-organizing water pricing, especially in countries with extreme hydro-climatic variability, is crucial. We attempted to present a set of arguments stating how smart planning of water pricing design, structure, and enforcement can serve as potential tools under the water security agenda. The water pricing case studies presented in this paper are not intended to be viewed as best practices, but rather as examples of realities on the ground in different countries. They also serve to illustrate how geographic, economic, and social diversity influence water management planning and decision-making. Water pricing can be considered as a policy tool to address some of the challenges and dimensions of the larger nexus of water security. If countries and regions are able to balance the supply-demand dynamics while assuring a water-secure future for their communities and citizens, pricing models that fit with the socio-economic and socio-political and socio-cultural complex systems will need to be carefully chosen. Recognizing this complexity in the relationship between government and society and their corresponding economic sectors, policies adjusted to the reality of each region can be implemented within the framework of water security, with strategies and tools working on systems in parallel (complementarity). Therefore, water pricing tools must not be implemented in insolation, but governments must execute them within an integral set of policies that have at least one complementary tool of social participation and cooperation (transparency and simplicity).

The critical analysis of a range of case studies makes it increasingly clear that the social parameters, environmental conditions, and cost recovery mechanisms are primary when pricing a natural resource. For example, impoverished population groups or vulnerable economic sectors should not treat water as a public or a pure economic good because of the current and predicted environmental conditions from climate change and other kinds of uncertainties. The contribution of water pricing toward the sustainable management of water resources requires large investments and financial commitments to manage direct and indirect pressures impacting the quantity and quality of water resources. We argue that these decisions should be taken in tandem with additional aspects with respect to innovations in clean water technologies, building capacity of customers/user on larger merit of valuing water, putting in place mechanisms and policies to manage and mitigate conflict, and overall, to set a long-term agenda to develop water resilient communities and nations, with sustainable management of water as normative.

Let us revisit some quick examples to set the final argument. Beginning with the Ghanaian case study, the policy did not associate water price with efficiency indexes of water consumption and therefore, instantly reflected unsustainable growth. In Spain, a fixed cost managed by the communities led to subsidized extensive agriculture with no control over efficient use of water. The São Paolo case study illustrated how urban water security was balanced in the face of the most critical drought in recent history. The Hybrid Policy—balancing the subsidy and taxation systems—demonstrated how the objective of conserving water resources was achieved without much altering the regular water price structures. The case study of drinking water provision in Cochabamba reflected the ineffective process of privatization that resulted in a general negative impression of the privatization approach to service provision. Finally, France's national water policy increased the use of a uniform volumetric pricing system and simultaneously improved infrastructural maintenance.

Overall, water pricing strategies and goals have the potential to tackle both physical and economic water scarcity if they are intelligently planned and implemented. Furthermore, they demonstrate that financial reserves appear to be the valuable aspect of the scenarios that can be achieved through cost-recovery from water users (case studies 3 and 5), leading to a negative response from society (case studies 1, 2 and 4). That being said, water pricing has the potential to positively trigger the water security agenda as managed water extraction, supply, and strata-based management and better investments in infrastructure can help ensure a sustainable water future. The feedback loop of smart water pricing strategies can also have positive spillover effects. For example, managed water withdrawals can help conserve services and benefits from aquatic ecosystems, whereas financial availability can ensure investments in associated aspects such as water-borne diseases and water-related disasters. The result of not considering water pricing as a tool for improved management and behavior change on future water thinking and policies leaves us in a situation where agricultural and urban water demands will continually increase without sustainable structures regulating efficient use (as explained in case study 1: insulated efficient use does not control overall irrigation consumption), conservation (as explained in case study 3: social water conservation promotion did not work on its own), maintenance, utility for expansion, and revenue. All these factors are prerequisites for countries to ensure water security, in the short-, medium-, and long-term.

Author Contributions: Conceptualization, N.N., G.A., T.A.D., and P.C.S.R.; Methodology, N.N., T.A.D., and P.C.S.R; Writing-Original Draft Preparation, P.C.S.R, N.N., T.A.D., and G.A., Writing-Review & Editing, P.C.S.R, N.N., T.A.D., and G.A.; Visualization, T.A.D., and P.C.S.R.; G.A., N.N., Supervision, N.N. and P.C.S.R.; Project Administration, N.N. and P.C.S.R.

Funding: UNU INWEH would like to acknowledge the funding and support of Global Affairs Canada to its research, policy and development mandate, programs and projects.

Acknowledgments: We would like to thank the Ministry of Environment and Water in Bolivia for their support and collaboration.

Conflicts of Interest: The authors declare no conflict of interest.

References

1. Kejser, A. European attitudes to water pricing: Internalizing environmental and resource costs. *J. Environ. Manag.* **2016**, *183*, 453–459. [CrossRef] [PubMed]
2. Cook, C.; Bakker, K. Water security: Debating an emerging paradigm. *Glob. Environ. Chang.* **2012**, *22*, 94–102. [CrossRef]
3. Ministerial Declaration of the Hague on Water Security in the 21st Century. Available online: http://www.worldwatercouncil.org/sites/default/files/World_Water_Forum_02/The_Hague_Declaration.pdf (accessed on 5 August 2018).
4. UN-Water. Analytical Brief on Water Security and the Global Water Agenda: A UN-Water Analytical Brief. 2013. Available online: https://collections.unu.edu/eserv/UNU:2651/Water-Security-and-the-Global-Water-Agenda.pdf (accessed on 8 October 2017).
5. Mehta, P.; Nagabhatla, N. *Without Water, Nothing Is Secure*; UNU-INWEH Policy Brief; United Nations University Institute for Water Environment and Health: Hamilton, ON, Canada, 2017.
6. Reddy, V.R. Water pricing as a demand management option: Potentials, problems, and prospects. In *Strategic Analyses of the National River Linking Project (NRLP) of India, Series 3—Promoting Irrigation Demand Management in India: Potentials, Problems, and Prospects*; Saleth, R.M., Ed.; International Water Management Institute: Colombo, Sri Lanka, 2009; pp. 25–46.
7. Wehn, U.; Montalvo, C. Exploring the dynamics of water innovation: Foundations for water innovation studies. *J. Clean. Prod.* **2018**, *171*, S1–S19. [CrossRef]

8. Frederick, K.D. *Balancing Water Demands with Supplies: The Role of Management*; The World Bank: Washington, DC, USA, 1993; Available online: http://documents.worldbank.org/curated/en/321231468766544380/Balancing-water-demands-with-supplies-the-role-of-management-in-a-world-of-increasing-scarcity (accessed on 1 April 2017).
9. Ahmad, M. Water pricing and markets in the Near East: Policy issues and options. *Water Policy* **2000**, *2*, 229–242. [CrossRef]
10. World Economy Forum (WEF). Global Risks 2017. Available online: http://reports.weforum.org/global-risks-2017/part-1-global-risks-2017/ (accessed on 20 October 2017).
11. UN-Water. What Is Water Security? Infographic. Available online: http://www.unwater.org/publications/water-security-infographic/ (accessed on 10 September 2017).
12. Sokolow, S.; Godwin, H.; Cole, B.L. Impacts of Urban Water Conservation Strategies on Energy, Greenhouse Gas Emissions, and Health: Southern California as a Case Study. *Am. J. Public Health* **2016**, *106*, 941–948. [CrossRef] [PubMed]
13. Evans, R.G.; Sadler, E.J. Methods and technologies to improve efficiency of water use. *Water Resour. Res.* **2008**, *44*. [CrossRef]
14. Environmental Protection Agency (EPA). Cases in Water Conservation: How Efficiency Programs Help Water Utilities Save Water and Avoid Costs. Available online: https://www.epa.gov/sites/production/files/2017-03/documents/ws-cases-in-water-conservation.pdf (accessed on 10 October 2017).
15. Linstead, C.L. The contribution of improvements in irrigation efficiency to environmental flows. *Front. Environ. Sci.* **2018**, *6*. [CrossRef]
16. Shiklomanov, I.A. World Water Resources and Water Use: Modern Assessment and Outlook for the 21st Century. Available online: http://documentos.dga.cl/PHI710.pdf (accessed on 31 August 2018).
17. World Meteorological Organization (WMO). *International Conference on Water and the Environment: Development Issues for the 21st Century*; ICWE Secretariat, World Meteorological Organization: Dublin, Ireland, 1992.
18. McNeill, D. Water as an economic good. *Nat. Resour. Forum* **1998**, *22*, 253–261. [CrossRef]
19. Perry, C.J.; Seckler, D.; Rock, M.T.; Seckler, D.W. *Water as an Economic Good: A Solution, or a Problem?* Research Report 014; International Irrigation Management Institute: Colombo, Sri Lanka, 1997.
20. Briscoe, J. Water as an Economic Good: The Idea and What it Means in Practice. In Proceedings of the World Congress of the International Commission on Irrigation and Drainage, Cairo, Egypt, September 1996; The World Bank: Washington, DC, USA, 1996.
21. Banovec, P.; Domadenik, P. Pricing Approaches in the Case of Cross-Border Water Supply. *Procedia Eng.* **2016**, *162*, 601–610. [CrossRef]
22. Kanakoudis, V.; Papadopoulou, A.; Tsitsifli, S. Domestic water pricing in Greece: mean net consumption cost versus mean payable amount. *Fresenius Environ. Bull.* **2014**, *23*, 2742–2749.
23. Tsitsifli, S.; Gonelas, K.; Papadopoulou, A.; Kanakoudis, V.; Kouziakis, C.; Lappos, S. Socially fair drinking water pricing considering the Full Water Cost recovery principle and the Non-Revenue Water related cost allocation to the end users. *Desalin. Water Treat.* **2017**, *99*, 72–82. [CrossRef]
24. Renzetti, S.; Dupont, D.P. Water pricing in Canada: Recent Developments. In *Water Pricing Experiences and Innovations*; Springer International Publishing: New York, NY, USA, 2015; pp. 63–81.
25. Donnelly, K.; Christian-Smith, J. *An Overview of the "New Normal" and Water Rate Basics*; Pacific Institute: Oakland, CA, USA, 2011; Available online: http://pacinst.org/wp-content/uploads/2013/06/pacinst-new-normal-and-water-rate-basics.pdf (accessed on 1 September 2017).
26. Bogaert, S.; Vandenbroucke, D.; Dworak, T.; Berglund, M.; Interwies, E.; Görlitz, S.; Schmidt, G.; Álvaro, M.H. The Role of Water Pricing and Water Allocation in Agriculture in Delivering Sustainable Water Use in Europe—Final Report; Project Number 11589; Arcadis: Brussels, Belgian. Available online: http://www.enorasis.eu/uploads/files/Water%20Governance/role_water_pricin.pdf (accessed on 10 May 2017).
27. Vander Ploeg, C.G. *Water Pricing: Seizing a Public Policy Dilemma by the Horns: Canadian Water Policy Backgrounders*; Canada West Foundation: Calgary, AB, Canada, 2011; Available online: http://cwf.ca/wp-content/uploads/2015/11/CWF_WaterBackgrounder7_SEP2011.pdf (accessed on 3 May 2017).

28. Mostert, E.; Van Beek, E.; Bouman, N.W.M.; Hey, E.; Savenije, H.H.G.; Thissen, W.A.H. River basin management and planning. In *the International Workshop on River Basin Management*; Technical Documents in Hydrology No. 31; The United Nations Educational, Scientific and Cultural Organization (UNESCO): The Hague, The Netherlands, 1999; pp. 24–55.

29. Green, C.H. *The Handbook of Water Economics: Principles and Practice*; John Wiley & Sons: Chichester, West Sussex, England, 2003; ISBN 0-471-98571-6.

30. Ahlers, R. Fixing and Nixing: The politics of water privatization. *Rev. Radic. Polit. Econ.* **2010**, *42*, 213–230. [CrossRef]

31. Zaag, P.; Savenije, H.H.G. *Water as an Economic Good: The Value of Pricing and the Failure of Markets*; Value of Water Research Report Series No. 19; UNESCO-IHE: Delft, The Netherlands, 2006.

32. Chambers, R.; Pacey, A.; Thrupp, L.A. *Farmer First: Farmer Innovation and Agricultural Research*; Intermediate Technology Publications: London, UK, 1989; ISBN 1853390070.

33. Berbel, J.; Gómez-Limón, J.A. The impact of water-pricing policy in Spain: An analysis of three irrigated areas. *Agric. Water Manag.* **2000**, *43*, 219–238. [CrossRef]

34. Elnaboulsi, J.C. An incentive water pricing policy for sustainable water use. *Environ. Resour. Econ.* **2009**, *42*, 451–469. [CrossRef]

35. Kanakoudis, V.; Tsitsifli, S.; Gonelas, K.; Papadopoulou, A.; Kouziakis, C.; Lappos, S. Determining a Socially Fair Drinking Water Pricing Policy: The Case of Kozani, Greece. *Procedia Eng.* **2016**, *162*, 486–493. [CrossRef]

36. Aidam, P.W. The impact of water-pricing policy on the demand for water resources by farmers in Ghana. *Agric. Water Manag.* **2015**, *158*, 10–16. [CrossRef]

37. Hailu, D.; Osorio, R.G.; Tsukada, R. Privatization and renationalization: What went wrong in Bolivia's water sector? *World Dev.* **2012**, *40*, 2564–2577. [CrossRef]

38. *The State of the Ghanaian Economy Report, 2012*; Institute of Statistical, Social and Economic Research (ISSER): Legon, Ghana, 2013.

39. World Bank. 3RD GHANA ECONOMIC UPDATE Agriculture as an Engine of Growth and Jobs Creation. 2018. Available online: http://documents.worldbank.org/curated/en/113921519661644757/pdf/123707-REVISED-Ghana-Economic-Update-3-13-18-web.pdf (accessed on 31 August 2018).

40. Drechsel, P.; Keraita, B. *Irrigated Urban Vegetable Production in Ghana: Characteristics, Benefits, and Risks*, 2nd ed.; International Water Management Institute: Colombo, Sri Lanka, 2014. [CrossRef]

41. *Food and Agriculture Sector Development Policy (FASDEP)*; The Ministry of Food and Agriculture (MOFA): Accra, Ghana, 2012.

42. Botwe, B.O.; Ntow, W.J.; Kelderman, P.; Drechsel, P.; Carboo, D.; Nartey, V.K.; Gijzen, H.J. Pesticide residues contamination of vegetables and their public health implications in Ghana. *J. Environ. Issues Agric. Dev. Ctries.* **2011**, *3*, 10.

43. *Water Resource Management Policy Paper*; Report Number 12335; The World Bank: Washington, DC, USA, 1993.

44. Molle, F.; Berkoff, J. Water pricing in irrigation: Mapping the debate in the light of experience. In *Irrigation Water Pricing: The Gap between Theory and Practice*; Molle, F., Berkoff, J., Eds.; The Centre for Agriculture and Bioscience International (CABI) Publishing: Wallingford, CT, USA, 2007; pp. 21–93.

45. Obour, P.B.; Dadzie, F.A.; Kristensen, H.L.; Rubæk, G.H.; Kjeldsen, C.; Saba, C.K.S. Assessment of farmers' knowledge on fertilizer usage for peri-urban vegetable production in the Sunyani Municipality, Ghana. *Resour. Conserv. Recycl.* **2015**, *103*, 77–84. [CrossRef]

46. Amponsah, O.; Vigre, H.; Schou, T.W.; Boateng, E.S.; Braimah, I.; Abaidoo, R.C. Assessing low-quality water use policy framework: The Case study from Ghana. *Resour. Conserv. Recycl.* **2015**, *97*, 1–15. [CrossRef]

47. Gerard, F.; Marty, I.; Lancon, F.; Versapuech, M. *Measuring the Effects of Trade Liberalization: The Multilevel Analysis Tool for Agriculture*; Working Papers 32721; Regional Coordination Center for Research and Development of Coarse Grains, Pulses, Roots and Tuber Crops in the Humid Topics of Asia and the Pacific (The CGPRT Center): Bogor, Indonesia, 1998; p. 171.

48. Gilbertson, D.D. Runoff (floodwater) farming and rural water supply in arid lands. *Appl. Geogr.* **1986**, *6*, 5–11. [CrossRef]

49. A SABESP Está Prorrogando o Programa de Incentivo à Redução de Consumo. Available online: http://site.sabesp.com.br/site/interna/Default.aspx?secaoId=544 (accessed on 22 December 2017).

50. Banovec, P.; Domadenik, P. Paying too much or too little? Pricing approaches in the case of cross-border water supply. *Water Sci. Technol. Water Supply* **2017**, *18*, 577–585. [CrossRef]

51. Zeitoun, M.; Lankford, B.; Krueger, T.; Forsyth, T.; Carter, R.; Hoekstra, A.Y.; Taylor, R.; Varis, O.; Cleaver, F.; Boelens, R.; et al. Reductionist and integrative research approaches to complex water security policy challenges. *Glob. Environ. Chang.* **2016**, *39*, 143–154. [CrossRef]

52. Press Release: IMF Approves Three-Year Arrangement under the ESAF for Bolivia. Available online: https://www.imf.org/en/News/Articles/2015/09/14/01/49/pr9841 (accessed on 17 July 2017).

53. Bechtel Perspective on the Aguas del Tunari Water Concession in Cochabamba, Bolivia. Available online: http://www.bechtel.com/files/perspective-aguas-del-tunari-water-concession/ (accessed on 16 May 2017).

54. Nickson, A.; Vargas, C. The limitations of water regulation: The failure of the Cochabamba concession in Bolivia. *Bull. Lat. Am. Res.* **2002**, *21*, 99–120. [CrossRef]

55. Dige, G.; De Paoli, G.; Strosser, P.; Anzaldua, G.; Ayres, A.; Lange, M.; Lago, M.; Oosterhuis, F.H.; Hrabar, M.; Navrud, S. *Assessment of Cost Recovery through Water Pricing*; Technical report No. 16/2013.; European Environment Agency: Copenhagen, Denmark, 2013. [CrossRef]

56. Angel, J. Potential Impacts of Climate Change on Water Availability. Available online: http://www.isws.illinois.edu/iswsdocs/wsp/climate_impacts_012808.pdf (accessed on 20 June 2017).

57. Peredo-Beltran, E. *Water, Privatization and Conflict: Women from the Cochabamba Valley*; Heinrich Boll Foundation: Berlin, Germany, 2004.

58. Shultz, J. The Cochabamba Water Revolt, and Its Aftermath. In *Dignity and Defiance: Stories from Bolivia's Challenge to Globalization*; Shultz, J., Draper, M., Eds.; University of California Press: Oakland, CA, USA, 2008.

59. Marston, A. The scale of informality: Community-run water systems in peri-urban Cochabamba, Bolivia. *Water Altern.* **2014**, *7*, 72–88.

60. *Bolivia Water Management: A Tale of Three Cities*; Report Number 222; The World Bank: Washington, DC, USA, 2002.

61. Chong, E.; Huet, F.; Saussier, S.; Steiner, F. Public-private partnerships and prices: Evidence from water distribution in France. *Rev. Ind. Organ.* **2006**, *29*, 149–169. [CrossRef]

62. Montginoul, M.; Loubier, S.; Barraqué, B.; Agenais, A.L. Water pricing in France: Toward more incentives to conserve water. In *Water Pricing Experiences and Innovations*; Springer International Publishing: New York, NY, USA, 2015; pp. 139–160.

63. Demouliere, R.; Bensaid Schemba, J.; Berger, J.L.; Ait Kaci, A.; Rougier, F. Public water supply and sanitation services in France: Economic, social and environmental data. Available online: /http://www.fp2e.org/userfiles/files/publication/etudes/Etude%20FP2E-BIPE%202012_VA.pdf (accessed on 8 August 2017).

64. Forrer, D.A.; Boudreau, J.; Boudreau, E.; Garcia, S.; Nugent, C.; Allen, D.; Lubin, A.C. The Effects of Water Utility Pricing on Low Income Consumers. *J. Int. Energy Policy* **2016**, *5*, 9–18. [CrossRef]

65. European Commission. Science for Environment Policy: Pricing policies for efficient water management. Available online: http://ec.europa.eu/environment/integration/research/newsalert/pdf/307na6_en.pdf (accessed on 3 July 2017).

66. Dinar, A. *The Political Economy of Water Pricing Reforms*; The World Bank: Washington, DC, USA, 2000. Available online: http://documents.worldbank.org/curated/en/199301468771050868/pdf/multi-page.pdf (accessed on 1 September 2017).

67. Yepes, G. Los Subsidios Cruzados en los Servicios de Agua Potable y Saneamiento. Available online: https://publications.iadb.org/handle/11319/5013 (accessed on 3 July 2017).

68. Tiger, M.; Hughes, J.; Eskaf, S. *Designing Water Rate Structures for Conservation & Revenue Stability*; University of North Carolina Environmental Finance Center and Sierra Club, Lone Star Chapter: Chapel Hill, NC, USA; Austin, TX, USA, 2014.

69. Deybe, D. Can agricultural sector models be a tool for policy analysis? An application for the case of Burkina Faso. *Agric. Syst.* **1998**, *58*, 367–380. [CrossRef]

70. Van den Berg, C.; Danilenko, A. *The IBNET Water Supply and Sanitation Performance Blue Book: The International Benchmarking Network for Water and Sanitation Utilities Databook*; The World Bank: Washington, DC, USA, 2011.

71. Danilenko, A.; Van den Berg, C.; Macheve, B.; Moffitt, L.J. *The IBNET Water Supply and Sanitation Blue Book 2014: The International Benchmarking Network for Water and Sanitation Utilities Databook*; The World Bank: Washington, DC, USA, 2014.

72. Kennemore, A.; Weeks, G. Twenty-First Century Socialism? The Elusive Search for a Post-Neoliberal Development Model in Bolivia and Ecuador. *Bull. Lat. Am. Res.* **2011**, *30*, 267–281. [CrossRef]
73. Spronk, S. Roots of Resistance to Urban Water Privatization in Bolivia: The "New Working Class," the Crisis of Neoliberalism, and Public Services. *Int. Labor Work. Cl. History* **2007**, *71*, 8–28. [CrossRef]

Review

Compound Extremes in Hydroclimatology: A Review

Zengchao Hao [1,*], **Vijay P. Singh** [2] **and Fanghua Hao** [1]

[1] Green Development Institute, College of Water Sciences, Beijing Normal University, Beijing 100875, China; fhhao@bnu.edu.cn

[2] Department of Biological and Agricultural Engineering and Zachry Department of Civil Engineering, Texas A&M University, College Station, TX 77843-2117, USA; vsingh@tamu.edu

* Correspondence: haozc@bnu.edu.cn; Tel.: +86-10-5880-1971

Received: 24 April 2018; Accepted: 30 May 2018; Published: 1 June 2018

Abstract: Extreme events, such as drought, heat wave, cold wave, flood, and extreme rainfall, have received increasing attention in recent decades due to their wide impacts on society and ecosystems. Meanwhile, the compound extremes (i.e., the simultaneous or sequential occurrence of multiple extremes at single or multiple locations) may exert even larger impacts on society or the environment. Thus, the past decade has witnessed an increasing interest in compound extremes. In this study, we review different approaches for the statistical characterization and modeling of compound extremes in hydroclimatology, including the empirical approach, multivariate distribution, the indicator approach, quantile regression, and the Markov Chain model. The limitation in the data availability to represent extremes and lack of flexibility in modeling asymmetric/tail dependences of multiple variables/events are among the challenges in the statistical characterization and modeling of compound extremes. Major future research endeavors include probing compound extremes through both observations with improved data availability (and statistical model development) and model simulations with improved representation of the physical processes to mitigate the impacts of compound extremes.

Keywords: compound extremes; climate change; multivariate distribution; quantile regression; indicator

---

1. Introduction

The climate system has been changing significantly as exhibited by global warning, which is expected to intensify (and accelerate) the hydrologic cycle due to the involvement of certain temperature dependence processes. This would lead to changes in the duration, frequency, spatial extent, and timing of extreme weather and climate events [1–3]. A variety of climate and weather related extremes, such as droughts, floods, heavy rainfalls, heat waves, tornadoes, cyclones or storms, have been shown to change significantly in the past, posing serious challenges to different sectors of the society including water, energy and food and their nexus (i.e., the water-energy-food nexus (WEF)) [4–10]. Moreover, recent studies based on climate projections have revealed a potential increase in these extremes in the future [3,11–15], which calls for an improved understanding of the changes in extremes and their impacts under global warming.

Traditionally, studies on weather and climate extremes have mostly focused on the extremes from a single process or variable, such as heavy precipitation or maximum temperature. For example, a multitude of studies have shown increases in the severity, duration, and frequency of precipitation and temperature extremes [4,16–21]. The extreme value theory (EVT) constitutes the basis for statistical modeling of univariate extremes in this regard, which can generally be achieved with the probability distribution of individual extremes, such as generalized extreme value (GEV) distribution or generalized Pareto distribution (GPD) based on the annual maxima or peak over threshold [22,23]. However, hydroclimatic variables are interconnected and thus focusing on a single variable or extreme may not be sufficient to comprehensively characterize the impact of extremes, which calls for multivariate modeling techniques.

Compound extremes, which are also referred to as simultaneous, concurrent, or coincident extremes (e.g., the concurrence or succession of multiple extremes/events), may exacerbate an adverse impact, leading to larger impacts to human society and the environment than those from individual extremes alone [24–26]. The past decade has witnessed an upsurge in studies of compound extremes, such as drought and heat wave (or low precipitation and high temperature) at different regions including Europe (2003 and 2015), Russia (2010), and California in the U.S. (2014) [4,27–31]. For example, the recent 2012 extreme drought in the central U.S. with record deficit in precipitation was accompanied by high temperatures during the May–August growing season, which significantly affected crop yields [32].

The physical mechanism of a compound extreme (e.g., compound flood) is rather complex depending on a variety of weather and hydrological processes [24,25,33–35]. Studies based on both models and observations have been devoted to the modeling of multiple drivers or processes of compound extremes. Correlations between occurrences of extremes may be induced due to a common external factor (e.g., regional warming), mutual reinforcement of two events (e.g., land surface feedback) or conditional dependence of the occurrence of one event to another event (e.g., extreme precipitation and soil moisture for flood) in the weather or hydroclimatic system [25,36]. Statistical methods have been commonly employed to model the correlation or interaction of multiple variables or processes that may lead to compound extremes. One example is the occurrence of drought and extreme heat in summer, which may be largely due to the land atmosphere feedback. Studies have shown that the number of occurrences of the compound drought and extreme heat increased in different regions [29,37]. In addition, the quantile regression method has been widely used for the assessments of the contribution of the antecedent soil moisture deficit on the occurrence of high temperature [38,39], leading to a compounded dry and hot extreme. In addition, coastal flooding may be caused by large waves combined with a high sea level and its multivariate distribution has been employed to study the compound extreme of wave height and water level at different coastline stretches around the globe [40–46]. These studies advanced our understanding of compound extremes and how to enhance the capacity to cope with the adverse impacts of these climate anomalies. However, a thorough introduction and comparison of statistical methods for assessing compound extremes is lacking.

The aim of this study therefore is to review commonly used statistical methods for the characterization and modeling of compound extremes in hydroclimatology. This paper is organized as follows. The definition and types of compound extremes are introduced in Section 2. An introduction of statistical modeling of compound extremes is provided in Section 3. Section 4 discusses several topics related to compound extremes, followed by conclusions in Section 5.

2. Compound Extremes

2.1. Definition of Compound Extremes

There are different definitions of compound extremes/events. The IPCC Special Report on Managing the Risks of Extreme Events and Disasters to Advance Climate Change Adaptation (IPCC SREX) defines the compound event as follows [25]:

> "(1) two or more extreme events occurring simultaneously or successively, (2) combinations of extreme events with underlying conditions that amplify the impact of the events, or (3) combinations of events that are not themselves extremes but lead to an extreme event or impact when combined."

It is emphasized in this definition that the coincidence of several factors, each of which may not necessary to be extreme, may lead to adverse and extreme impacts [47,48]. By categorizing compound extremes into different classes, this definition is crucial in understanding the phenomenon; however, some events are hard to be classified under this definition [24]. Recently, another definition of compound event has been given as follows [24]:

> "A compound event is an extreme impact that depends on multiple statistically dependent variables or events."

This definition of compound extremes/events emphasizes three aspects, including the impact, presence of multiple variables (or events), and statistical dependence. A common feature of the two definitions is that the compound extreme generally involves the interaction (or dependence) of multiple drivers (or variables/events), either of the same type (e.g., extreme rainfall at both upstream and downstream) or different types (e.g., compound drought and hot extreme). This highlights that the dependence modeling of multivariate random variables plays a central role in the statistical modeling of compound extremes.

2.2. Typical Compound Extremes

The compound extremes refer to a variety of cases in hydrology and climatology. In this study, compound extremes in the bivariate case are classified into different types in which four regions of extremes (I, II, III, and IV) are defined, as illustrated in Figure 1. A summary of typical compound extremes is provided in the Appendix A, such as the drought and hot extreme, precipitation and temperature extreme, and compound flood. Note that there are other types of compound events that may be of particular interest, such as the combined wind speed and storm surges (or precipitation) [36,49,50], combined humidity and temperature extremes [51], and the co-occurrence of particulate matter (PM2.5) and maximum temperature [52].

The compound nonexceedance- nonexceedance extremes are illustrated in region I, in which both variables lower than certain quantiles are of primary interest, such as the deficit in both precipitation and soil moisture. Drought, a typical example in this case, is commonly classified as meteorological, hydrological, agricultural, or socioeconomic drought [53], which may occur separately or simultaneously. Recently, substantial efforts have been devoted to the drought characterization from a multivariate perspective based on the concurrent deficit of multiple variables [54]. The compound exceedance- exceedance extremes type falls in Region III, in which two variables higher than certain values are of primary interest, such as storm surges (or sea level) and high precipitation (river discharge) [35,48,55–60]. For example, in coastal areas, the risk of flood will increase if high coastal water level (or storm surge) occurs simultaneously with high precipitation/runoff, leading to the increase of the river water level [55,56,61,62]. In addition, the combination of high temperature and heavy precipitation in the spring season in certain regions (e.g., Norway) can result in flooding when runoff from snow-melt adds to river discharge due to rainfall [47].

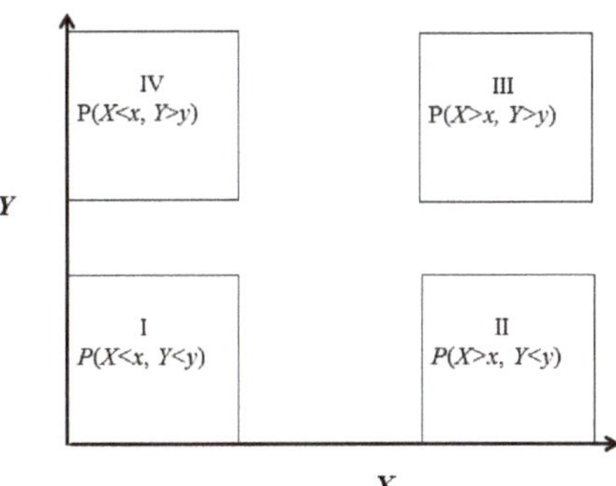

Figure 1. Different types of compound extremes in the bivariate case for two variables X and Y.

Regions II and IV indicate the compound exceedance- nonexceedance extremes, in which one variable lower (or higher) than a threshold with the other variable higher (or lower) than a threshold is of primary interest. Drought and heatwave have been among the most commonly investigated compound extremes of this type [27,37,63–65], which may promote wildfires [66], affect agricultural production [67,68], or reduce the net primary productivity (NPP) [69,70]. Drought and hot extremes are interconnected, which is mostly due to the positive feedback in transitional regions between a wet and dry climate, in that drought condition may be amplified/exacerbated when accompanied by heat wave, while drought may create favorable conditions for hot extreme or heat wave [4,8,13,38,71,72].

Apart from multiple extremes at the same time introduced above, the compound extreme also refers to sequential (or temporal clustering, successive) extremes, such as the temporal clustering of extreme sea level and skew surge events [73]. The compound extremes at different locations (or spatial extremes) may also be of interest, such as simultaneous flood in a wide area or extreme precipitation at multiple stations [74–76]. Understanding these properties of the compound extreme is important for the characterization and modeling to facilitate mitigation measures to reduce its impacts.

3. Statistical Approaches

Both models and observations have been used to explore the relationship between multiple variables/components of compound extremes. Due to limited observations of extremes (or rare events), the statistical inference of compound extremes and the extrapolation beyond observations are of particular interest. In the following, we mainly focus on methods for modeling compound extremes from a statistical perspective. These methods include the empirical approach, multivariate distribution, the indicator approach, quantile regression, and the Markov Chain model.

3.1. Empirical Approach

The empirical approach for the analysis of compound extremes is executed through counting the number of concurrent or consecutive occurrences of multiple extremes [29,51,57,77–80]. This approach mainly characterizes the occurrence or variability of compound extremes. The individual extreme is first defined (e.g., based on a threshold or percentile) [81], which is then used to obtain the quantity of compound extremes based on the co-occurrence of individual extremes. For example, a wide variety of indices of daily precipitation and temperature extremes (e.g., warm days defined as percentage of time when daily max temperature >90th percentile) have been developed by the joint CCI/CLIVAR/JCOMM Expert Team on Climate Change Detection and Indices (ETCCDI) (available at: http://cccma.seos.uvic.ca/ETCCDI/list_27_indices.html), which are also known as the 'ETCCDI' indices [82–85]. Accordingly, a multitude of compound extremes can be defined by counting the concurrence of these individual extremes. Usually the number of compound extremes for each period (i.e., month, year) is first computed and statistical analysis (e.g., trend analysis, change point analysis) is employed to detect associated changes.

The compound precipitation and temperature extreme has been widely explored. The precipitation (and temperature) extreme for a specific period can be defined as dry/wet (and cold/warm) when precipitation (and temperature) is lower/higher than certain thresholds (e.g., 25th/75th percentile). The associated four compound extremes (dry-warm, dry-cold, wet-warm, wet-cold) can then be defined when the two extremes occur concurrently [29,77,80]. The assessment of changes of these four compound extremes generally showed an increase in the warm mode of compound extremes (i.e., dry-warm, wet-warm). For example, it has been found that the occurrence of warm/dry and warm/wet modes in Europe have increased in the 20th century and will continue to increase in the 21th century [77]. In addition, assessments of combined precipitation and temperature modes in Spanish mountains revealed an increase in the frequency of dry-warm and wet-warm days [80].

We use the monthly precipitation and temperature data near Melbourne, Australia (Longitude: 144.9, Latitude: −37.8) to illustrate this approach. The monthly data for the period 1901–2016 were obtained from the Climatic Research Unit (CRU TS3.25) (using the nearest grid). Note that the gridded data should be used with caution in analyzing extreme events due to the reason that extremes may be smoothed during the interpolation process [86–89]. We use these data here and in the following sections mainly for illustrative purposes. Based on the definition above, we computed the number of compound dry-warm extreme occurrences for each year (and 5-year running average) during 1901–2016 for Melbourne, Australia, as shown in Figure 2. A significant increase in the occurrence of the dry-warm extreme is shown during the past century, implying the increased risk of a compound extreme under global warming in this region [90].

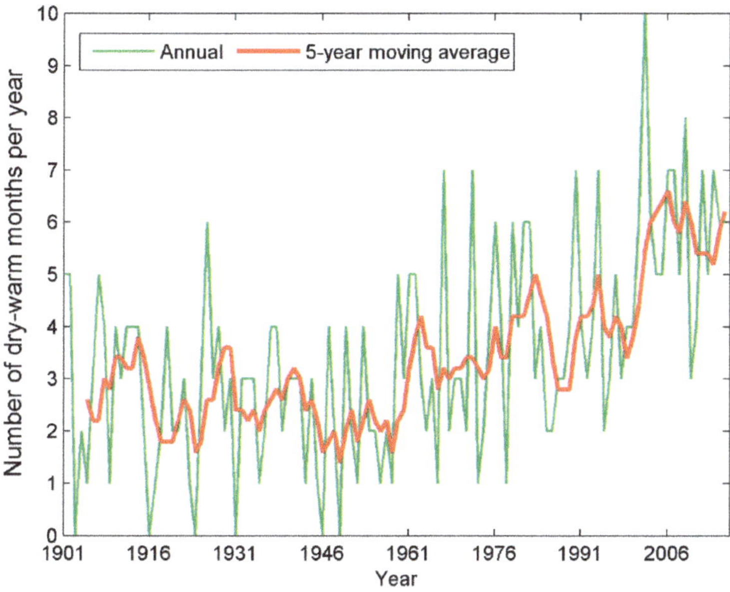

Figure 2. Number of the compound dry-warm extreme for each year (and 5-year running average) during 1901–2016 for Melbourne, Australia.

3.2. Multivariate Distribution

A key property of the compound extreme is that dependence between different contributing variables (or drivers) generally exists. The multivariate distribution plays a critical role in modeling dependence of multiple variables/extremes for a variety of applications (e.g., estimate the combined risk of extremes) [49,91]. For example, the multivariate distribution has been used to explore joint properties of precipitation and temperature (or their extremes) for frequency assessments or statistical simulations [92–97]. There are many ways to construct the multivariate distribution, such as parametric distribution, copula, entropy, and nonparametric models [98]. Copula is among the most recent advances in multivariate dependence modeling for a variety of applications in hydrology and climatology [99–112]. It enables the construction of the joint distribution in a flexible way in which the marginal distribution is independent of the modeling of dependence structure. In the following, we introduce the copula model for constructing the multivariate distribution to model compound extremes.

3.2.1. Copula Approach

For two random variables X and Y with marginal distributions U and V, respectively, the joint distribution can be expressed with a copula C as [113]:

$$F(x,y) = P(X \leq x, Y \leq y) = C(U,V;\theta) \tag{1}$$

where θ is the parameter of the copula.

Several copula families, including elliptical (e.g., Gaussian, Student t), Archimedean (e.g., Frank, Clayton, Gumbel), and extreme-value copula (Gumbel, Galambos, extremal-t, and Hüsler–Reiss), have been commonly used for the construction of multivariate distributions, which show different properties in dependence modeling. Four commonly used 2-parameter copulas are shown in Table 1. Random samples with a sample size of 1000 from these four copulas are illustrated in Figure 3 to show the different properties of these copulas in modeling multivariate variables. Variables drawn from the Gaussian and Frank copula exhibit symmetric dependences. The difference is that the dependence in the Frank copula is weaker in tails and stronger in the center of the distribution compared with the Gaussian copula [114]. Both the Clayton and Gumbel copula exhibit asymmetric dependences. Specifically, the Clayton copula exhibits the lower tail dependence (LTD) while the Gumbel copula shows the upper tail dependence (UTD) [115]. These properties imply that the Clayton copula is best suited for applications with two outcomes likely experiencing low values together while the Gumbel copula is suitable when two outcomes are likely to realize upper tail values simultaneously [114]. The extreme-value copula is commonly used for modeling the dependence structure between a rare event and the Gumbel copula is the only copula that is both an extreme value copula and an Archimedean copula [116–119].

Table 1. Four copulas and their parameter spaces.

Copulas	$C(u,v)$	Parameter
Gaussian	$\Phi_2(\Phi^{-1}(u), \Phi^{-1}(v))$ *	$\theta \in [-1,1]$
Clayton	$\left(u^{-\theta} + v^{-\theta} - 1\right)^{-1/\theta}$	$\theta \in (0,\infty)$
Frank	$-\frac{1}{\theta} \ln\left[1 + \frac{(e^{-\theta u}-1)(e^{-\theta v}-1)}{e^{-\theta}-1}\right]$	$\theta \in (-\infty,\infty)$
Gumbel	$\exp\left\{-\left[(-\log u)^{-\theta} + (-\log v)^{-\theta}\right]^{-1/\theta}\right\}$	$\theta \in [1,\infty)$

* Φ and Φ_2 represent the standard normal distribution in the univariate and bivariate case.

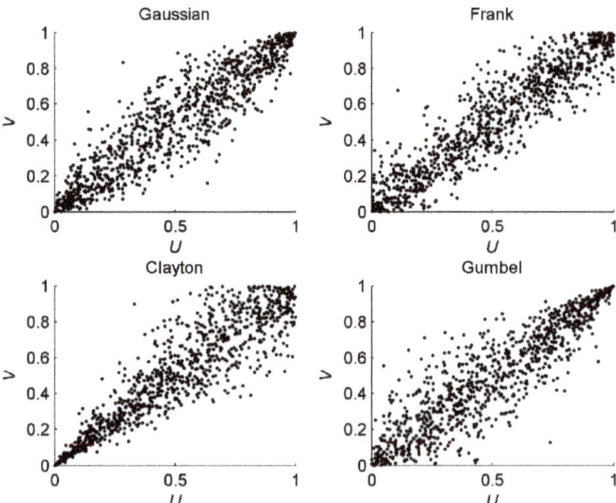

Figure 3. Random samples (sample size 1000) from Gaussian, Frank, Clayton, and Gumbel copulas.

3.2.2. Joint Probability

The multivariate distribution can be employed to model the joint behavior of compound extremes and enables the comparison of individual and compound extremes based on the marginal probability and joint probability (or percentile). We use the compound meteorological and hydrological drought as an example to illustrate the comparison, which may be applied to other compound extremes (e.g., drought and heatwave, storm surge and rainfall). Based on the monthly precipitation and runoff for the period 1932–2011 from climate division 2 in Texas, USA (obtained from the National Climatic Data Center, National Oceanic and Atmospheric Administration), the percentile of precipitation and runoff (6-month time scale) in July is constructed with the Gumbel copula. The 20th percentile is specified as the threshold to define the drought condition, which is shown in Figure 4 (i.e., lines L1 and L2). The 20th percentile of the joint distribution is also specified as a threshold to measure the compound extreme and the joint percentile is shown in Figure 4 (L3 represents the 20th joint percentile). The upper left region (e.g., $P1(0.07, 0.58)$) or the lower right region (e.g., $P3(0.29,0.10)$) is the case with the occurrence of only meteorological drought or hydrological drought. The lower left region (e.g., $P2(0.16,0.07)$) is the case with the concurrence of both meteorological and agricultural drought (i.e., compound drought). Of particular interest is the marked region bounded by L1, L2 and L3 (e.g., $P4(0.32,0.32)$), in which neither the meteorological drought nor the hydrological drought occurs. It can be seen that the joint percentile of this region is lower than the 20th percentile, which indicates the occurrence of the compound meteorological and hydrological drought. This highlights that the compound extreme may occur even when neither of its components are extreme.

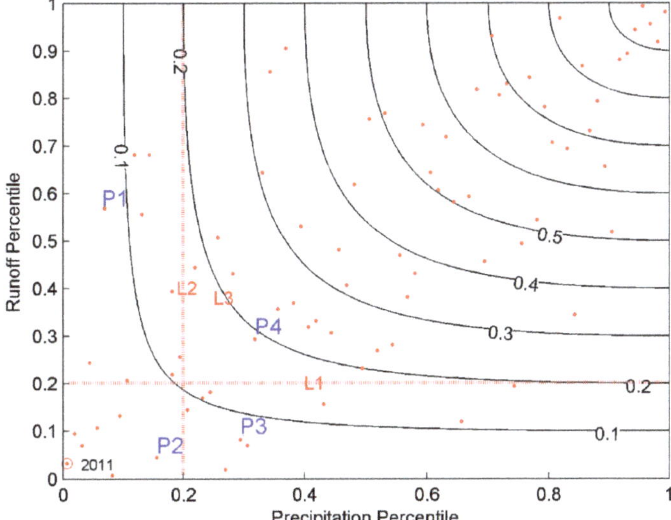

Figure 4. Comparison of individual and compound drought based on the percentile of precipitation and runoff (6-month time scale) for the period 1932–2011 using copula. The circled point represents the data pairs during 2011.

The multivariate distribution has been applied for the frequency analysis of compound extremes by computing the return period. An important way to quantify the risk of extremes is through a frequency analysis of individual extremes entailing a return period (or return level) [120,121]. Usually this return period can be obtained from the probability P with the relationship $T = \mu/(1 - P)$, where μ is the mean interval time and P is the nonexceedance (or nonexceedance) probability of interest [122]. For univariate extremes, efforts are needed in selecting the distribution of extremes either through block maxima or threshold exceedance. In addition, other approaches such as the fractal approach (or power law distribution) have also been discussed for the frequency analysis of rare events in hydrology [123–125]. In the multivariate case, there are different ways to estimate the return period of multiple variables or extremes either based on the joint distribution or the Kendall distribution function [122,126–132]. Traditionally, the commonly used method in the bivariate case refers to the estimation of the joint probability $P(X \leq x$ and $Y \leq y)$, $P(X \leq x$ and $Y > y)$ and $P(X > x$ and $Y > y)$, which can be applied for different cases of the compound extremes in the Appendix A. For example, $P(X \leq x$ and $Y \leq y)$ is of interest for the compound meteorological-agricultural drought while $P(X \leq x$ and $Y > y)$ is of interest for the compound drought-hot extremes.

We use the compound meteorological and agricultural drought to illustrate the application based on the monthly precipitation and runoff for the period 1932–2011 from climate division 2 in Texas, USA. The Standardized Precipitation Index (SPI) [133] and Standardized Runoff Index (SRI) [134] are commonly used drought indicators to track the meteorological drought and hydrological drought and are used in this study. For the computation of these indices, the empirical Gringorten plotting position formula is used to estimate the marginal probability of precipitation and runoff (6-month time scale) [135]. The joint return period of the SPI and SRI can be used for the frequency analysis of compound meteorological and agricultural drought. The empirical return period of the individual meteorological drought and hydrological drought in 2011 is 143 and 32 years, respectively. Based on joint probability, the joint return period of the compound meteorological-hydrological drought is 396 years. These results imply that the drought event in 2011 in this climate division is a 396-year event if both the meteorological and agricultural drought are taken into account. These results are consistent with previous studies, which highlighted that the risk of the compound extreme may be underestimated if the dependence between contributing variables was ignored [48,61,136].

3.2.3. Conditional Probability

The multivariate distribution approach also enables the quantification of the conditional relationship between two or more extremes. For example, the conditional distribution of the maximum temperature given antecedent meteorological drought can be used to quantify the impact of drought on hot extremes, which reflects the land surface feedback that contributes to the compound drought and hot extremes [38,137,138]. Existing approaches for multivariate extreme modeling are generally applicable to the case in which all variables are simultaneously extreme. As stated before, compound extremes may occur when not all variables need to be extreme. In this case, the conditional extreme model [139,140] is an attracting choice since it enables the dependence modeling of a multivariate extreme in which only part of the component is extreme [141,142].

Two types of conditional distributions are of primary interest in studying compound extremes, either conditioned on a specific value (e.g., $u = u_0$) or range (e.g., $u < u_0$). The conditional probability of $V \leq v$ given $U = u_0$ can be expressed with a copula C as [143]:

$$P(V \leq v | U = u_0) = \frac{\partial C(U, V)}{\partial U} | U = u_0 \tag{2}$$

The conditional probability of $V > v$ conditioned on $U \leq u_0$ can be expressed as [128,138]:

$$P(V > v | U \leq u_0) = 1 - \frac{C(u_0, v)}{u_0} \tag{3}$$

Based on Equations (2) and (3), the conditional probability and return period can be derived accordingly [143,144].

We use two examples to illustrate the application of the conditional distribution in a compound extreme analysis. Based on the SPI and SRI of Climate Division 2 in Texas, the conditional distribution of the hydrological drought (SRI) given meteorological drought (SPI) (SPI = −0.84 and 0.84) can be modeled with the Gumbel copula, which is shown in Figure 5. It can be seen that given the meteorological drought (SPI = −0.84) in the antecedent period, the probability of SRI lower than −0.5 is higher than that given the wet condition (SPI = 0.84). As another example, based on monthly precipitation and daily maximum temperature for the station at Dallas Fort Worth, TX for the period 1948–2010 obtained from the Global Historical Climatology Network (GHCN) version 2 (https://www.ncdc.noaa.gov/ghcnm/v2.php), the conditional distribution of maximum temperature given antecedent meteorological drought can be constructed to quantify the impact of drought on hot extremes. For the SPI of July and daily maximum temperature of August (with Pearson correlation −0.33), the joint distribution is constructed using the Frank copula for illustration purposes. The conditional probability of hot extremes higher than the 80th percentile conditioned on the SPI (SPI ≤ −0.84 and SPI > 0.84) in the antecedent period is computed as 0.40 and 0.06, implying the impact of an antecedent drought on the subsequent hot extremes (i.e., drought induced high temperature). The univariate return period of hot extremes higher than the 80th percentile is 5 years, which is longer than the conditional return period given SPI ≤ −0.84 (2.5 years). These results indicate that ignoring the dependences between contributing variables of compound extreme may lead to an underestimation of risk.

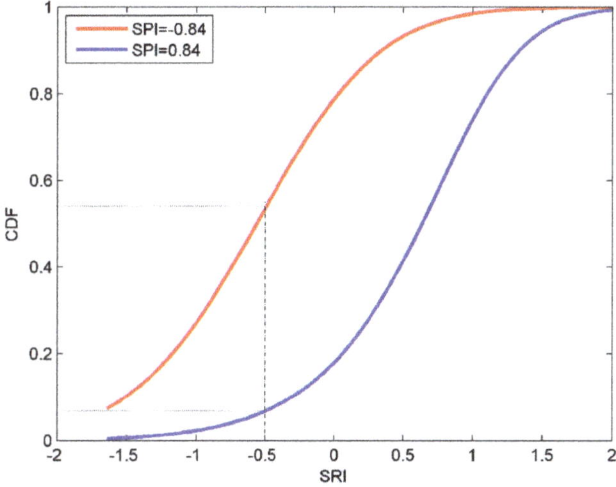

Figure 5. Conditional distribution of hydrological drought (represented with SRI) conditioned on the meteorological drought (represented with SPI).

3.3. Indicator Approach

Different indicators have been defined for the extremes in the univariate case [82,145]. The difference between univariate extremes and compound extremes is that in the multivariate setting, there is no natural order of extremes (or variables) in higher dimensions. As such, a "threshold" that can be used to define extremes in the multivariate setting does not exist [126]. For the indicator approach, information from multiple variables or spatiotemporal fields can be distilled into an indicator (by integrating multiple measures of extremes into one index) to inform users the condition and variability of extremes in a specific area [7]. To summarize, an indicator I can be defined to study compound extremes of multiple variables $X, Y, \ldots, Z$ as follows:

$$I = G(X, Y, \ldots, Z) \tag{4}$$

where the function G can be any form (e.g., a linear combination, maximum, joint distribution) that serves to project the multiple variable into a unique index I. Based on the indicator I, the compound extremes can be characterized flexibly in different aspects including duration, severity, intensity, and spatial extent.

In the compound extreme analysis, the Climate Extremes Index (CEI) [146–148] is such an index defined as the average (or linear combination) of multiple indicators of extremes (including drought, and extremes of precipitation and temperature) in the U.S. [148]. The multiple variables or extremes can also be combined based on the "structure variable", such as $Z = \max (M_x, M_y)$, where M_x and M_y are two extremes (e.g., combined thunderstorm and tornado) [149–152]. Among different ways of developing indicators, the multivariate distribution has been commonly explored for characterizing compound extremes from a multivariate perspective, since it completely describes the joint behavior of two or more variables [54,60,153]. The joint probability or percentile $P_1 = P(X \leq x, Y \leq y)$ can be employed as the measure of the compound extreme of both variable X and Y lower than certain thresholds (e.g., 10th percentile). Similarly, the joint probability $P_2 = P(X \leq x, Y > y)$ can be used as a measure of the compound extreme with X lower than a specific threshold (e.g., 10th percentile) and Y exceeding a higher threshold (e.g., 90th percentile), such as the compound drought and hot extremes.

As an example, the Multivariate Standardized Drought Index (MSDI) can be defined to characterize the compound meteorological and hydrological drought based on the joint probability of SPI and SRI, which can be expressed as [135]:

$$\text{MSDI} = \Phi^{-1}[F(\text{SPI}, \text{SRI})] \tag{5}$$

where Φ is the standard normal distribution. Here the joint distribution is estimated with the empirical Gringorten distribution in the bivariate case.

To illustrate the application of the MSDI, monthly precipitation and runoff data for climate division 2 in Texas, USA were used to compute the MSDI based on the SPI and SRI, as shown in Figure 6. It can be seen that when both meteorological and hydrological droughts show deficit, the concurrent drought is more severe than those from SPI and SRI. The usefulness of MSDI partly resides in that it enables to compare the severity of composite drought condition. For July 1936 and 1940 with SPI and SRI values $(-0.91, -0.27)$ and $(-0.54, -1.39)$, it is not straightforward to define the overall drought severity of these two periods (the SPI is more severe for the first period, while the SRI is more severe for the second period). The MSDI for the two periods are -1.04 and -1.50, respectively, indicating that the compound drought for July 1940 are more severe than that for July 1936.

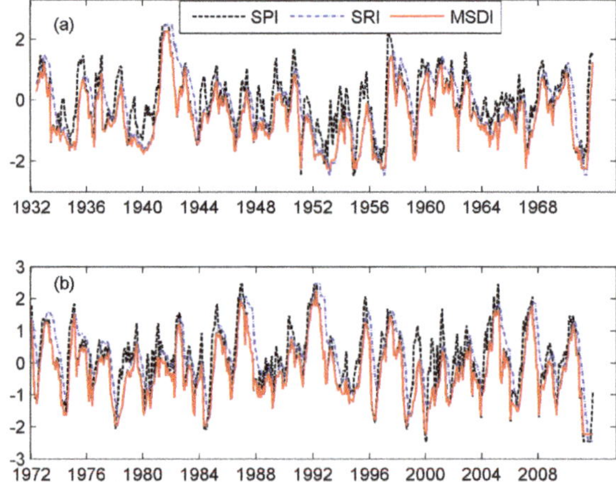

Figure 6. Comparison of drought condition based on SPI, SRI and MSDI in the bivariate case for the period from 1932-2011. (**a**) 1932–1971; (**b**) 1972–2011.

3.4. Quantile Regression

The linear regression has been commonly used to explore the relationship between the response variable and predictors. However, it only estimates the rate of change in the mean of the response variable and thus is generally not suitable for exploring the relationship between extremes. The quantile regression is capable of estimating the functional relationship between the response variable and predictors for all portions of the data and is flexible in modeling data with heterogeneous variance or conditional distribution [154,155].

The quantile regression of a response variable Y as a function of predictor X is introduced as follows [39]. In the traditional linear regression, the conditional mean of Y is related linearly to X as:

$$E(Y|X) = \alpha + \beta X = f(X; \alpha, \beta) \tag{6}$$

where α and β are the intercept and the slope parameters, respectively.

For the quantile regression, quantile τ of the response variable Y conditioned on X is used instead, i.e., $Q_\tau(Y \mid X)$. Specifically, for any quantile τ in (0,1), the quantile regression can be expressed as:

$$Q_\tau(Y|X) = f_\tau(X; \alpha_\tau, \beta_\tau) \tag{7}$$

where α_τ and β_τ are the parameters associated with quantile τ. By changing quantile τ, the relationship between Y and X can be explored.

In studying compound extremes, the regression of high (or low) quantile of Y with respect to X is generally of particular interest. In the past decade, the quantile regression has been commonly used to assess the relationship between two extremes or variables, such as soil moisture deficit and temperature extremes (or drought and hot extremes) [38,39,156,157] or rainfall frequency and hot days [158]. For example, extensive studies have shown that the antecedent dry condition (or low soil moisture/precipitation) may induce or intensify the heat wave or high temperature in different regions including Australia [159], Europe [39], and Oklahoma, USA [157], leading to the compound drought and hot extremes.

As an example, based on monthly precipitation and daily maximum temperature for one grid (99.625 E, 30.875 N) in northeastern China (data obtained from [160]), the hot extremes in terms of the number of hot days (NHD) during summer months of June, July, and August (JJA) and an antecedent drought indicator SPI for May, June, and July (MJJ) were obtained from these data. The quantile regression of NHD with respect to different SPI values during summer for different quantiles (i.e., 95th, 75th, 50th, and 25th percentile) is shown in Figure 7. It can be seen that there is an overall negative relationship between NHD and SPI indicated by the negative regression slope. The negative dependence increases toward higher quantiles of NHD, which is more sensitive to the antecedent drought condition. These results reveal the impact of drought on hot extremes in that the dry surface condition intensifies hot extremes in this grid, from which the occurrence of compound drought and hot extremes is partly explained.

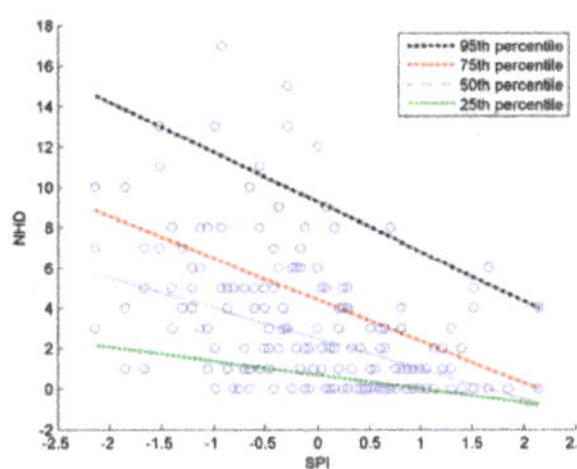

Figure 7. Quantile regression of hot extremes (number of hot days, NHD) with respect to the drought indicator SPI.

3.5. Markov Chain Model

The Markov Chain model is commonly used to describe a sequence of possible events in which the present state only depends on the antecedent state. It has been employed to analyze the variability of the individual variable or extreme, such as a drought [161] or a heat wave [162]. A variety of quantities, such as periodicity, persistence, or recurrence time, of the underlying sequence can be defined for characterizing the variables of interest.

A discrete stochastic process X_t ($t > 0$) with each random variable taking values in the set $S = (1, 2, \ldots, m)$ is a Markov Chain if [161],

$$P(x_t = j | x_{t-1} = i, x_{t-2} = k, \ldots) = P(x_t = j | x_{t-1} = i) \tag{8}$$

where $1 \leq i, j, k \leq m$. Denote p_{ij} the transition probability of the Markov chain. The transition matrix with element p_{ij} can be estimated as:

$$p_{ij} = P(x_t = j | x_{t-1} = i) \tag{9}$$

The transition probability is commonly used for characterizing properties of sequence of X_t (e.g., persistence, recurrence time). For example, the persistence of X_t stays in the state j and will reside in the same state j in the following time step can be expressed as p_{jj}. Recently, the Markov Chain model has been employed to study compound extremes (e.g., heavy precipitation- cold in winter and hot-dry days in summer) for central Europe under changing climate [87].

We use the monthly precipitation and temperature data from 1901–2016 near Melbourne, Australia (Longitude: 144.9, Latitude: −37.8) from CRU (TS3.25) to define the compound event of low precipitation and high temperature based on the 50th percentile (i.e., precipitation < 50th percentile and temperature > 50th percentile). The monthly precipitation and temperature series (detrended) are then partitioned in two states (occurrence of compound extreme or not) and a Markov Chain analysis of the occurrence sequence of compound precipitation and temperature sequence was then performed. For illustration purposes, we computed the transition probability for the sequence of every 60 years (i.e., 1901–1960, \ldots , 1957–2016). The temporal variability of the transition probability to compound extreme was shown in Figure 8. The overall increasing of the transition probability indicates that the occurrence of compound extreme is expected to increase under global warming for this grid, which is consistent with previous examples based on the empirical approach.

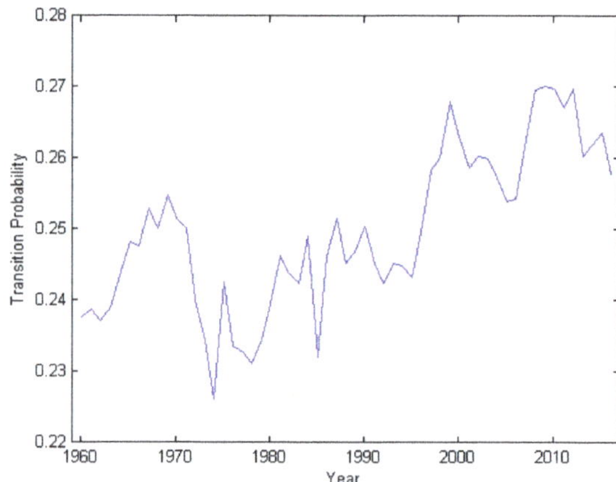

Figure 8. The change of the transition probability to the compound dry and hot extreme during 1960–2016 for Melbourne, Australia.

4. Discussion

4.1. Comparison of Approaches

The approaches introduced above possess different properties in the characterization and modeling of compound extremes. The empirical counting approach and indicator approach have been commonly used to characterize the variability of compound extremes (e.g., occurrence frequency, trend analysis). The empirical approach can be selected when the detection of changes of a compound extreme (e.g., drought and hot extreme) is of primary interest [29]. The indicator approach can be employed to combine different types of extremes for further analysis (e.g., spatial extent [147]). However, they do not describe the interaction between the contributing variables of a compound extreme. The empirical approach usually requires large amounts of data to assess the variation of compound extremes. For 100 data pairs of precipitation and temperature, there would be only 4 co-occurrences of compound dry and hot extremes on average with the 20th and 80th percentile as thresholds of precipitation and temperature if they are assumed to be independent. Observations are usually not abundant enough and the parametric methods, such as the multivariate distribution and quantile regression, can be employed for the statistical extrapolation and inference of compound extremes. The multivariate distribution is capable of modeling the joint behavior of multiple variables or events that lead to the compound extreme. This approach is commonly used for statistical inference and risk assessments of compound extremes based on joint/conditional probability or associated return periods [126,138]. Statistical modeling of multiple extremes through characterizing the dependence among multiple variables or locations is the key in this approach and copula has been among the most commonly used models due to its advantage of flexible dependence modeling. The potential limitation is that most of the commonly used multivariate distributions generally fall short in capturing a complicated dependence in higher dimensions. In an extreme analysis, one is generally interested in the characteristics of a specific quantile (e.g., 80th or 20th percentile) and its relationship with other variables. The quantile regression is advantageous in this case to estimate the relationship between the extreme quantile and the independent variable. However, challenges still remain in the performance of quantile regression in high quantiles (e.g., 99th percentile) due to the limitation of the sample size [163]. The impact of compound extremes depends not only on the number of occurrences but also the sequences of the occurrences [87]. The Markov Chain approach is particularly useful in this case, which enables the characterization of occurrence sequences of different compound extremes.

4.2. High Dimensional Modeling

In this study, most of the compound extremes are introduced in the bivariate case. In reality, there may be multiple variables involved in the occurrence of compound extremes. In this case, one has to model the dependence among a large number of variables (with or without time lags), either at a single location or multiple locations [98]. However, the construction of multivariate distribution in high dimensions remains a challenge (i.e., curse of dimensionality). This calls for suitable tools for dependence modeling in the high dimension to capture a complicated dependence, such as the tail dependence and asymmetric dependence.

The multivariate normal distribution is a traditional model for statistical modeling of multiple random variables even in high dimensions [164]. It is flexible in modeling multiple variables with the dependence structure completely characterized by the variance-covariance matrix, but falls short in modeling complicated dependences (e.g., asymmetric or tail dependence). The multivariate parametric copula is commonly employed for the bivariate case while its extension to higher dimensions (say 4-dimension) is limited in capturing complicated dependence. The vine copula (or the pair-copula construction, PCC), which decomposes the dependence structure into a bivariate dependence that can be modeled with bivariate copulas [165], is among the most recent development in multivariate modeling in a variety of areas [48,166–168]. It is expected that the vine copula may provide useful alternatives in modeling compound extremes due to its advantage of flexible dependence modeling even in high dimensions. In addition, the influence diagram, which provides a graphical representation of the conditional dependence and allows the joint distribution of variables to be factorized according to local conditional relationships, is also a potential method for risk estimations of compound extremes in high dimensions [24].

4.3. Compound Extremes Under Climate Change

An important assumption in statistical approaches to model extremes is the stationary assumption of the underlying system [23,169]. It has been highlighted that the fundamental assumption of stationary has been affected by climate change and anthropogenic effects and is not applicable for water resources planning and management [170]. As such, it is advised to check the non-stationary property based on the univariate or multivariate trend analysis using methods such as the (multivariate) Mann–Kendall and Spearman tests [171,172]. The nonstationary property should also be taken into account in the statistical modeling of compound extremes (e.g., frequency analysis). A commonly used method is to incorporate the non-stationary property as covariates [173–177]. For example, the dynamical copula model has been employed for the compound or multivariate extreme analysis with the copula parameter (or parameters of marginal distributions) varying with time [172,173,177].

Assessment of the climate change impact on compound extremes in a hydroclimatic framework is of particular importance for adaptation measures due to their tendency to have a larger impact than an individual extreme [3]. Efforts in assessing the variation of the compound extreme in the future have been growing based on climate projections [28,77,87]. For example, an overall increase of the number of incidences of compound drought and extreme heat is shown for the projection period (2021–2050) over central Europe [28]. In addition, a substantial increase in US East coast flood hazard is expected to occur over the twenty-first century based on the joint projections of the US East coast sea level and storm surge [178]. Research along this line also includes the climate change impact on the dependence structure of multiple variables or extremes, which may affect the risk of compound extremes [179–181]. An uncertainty quantification of the impact on the compound extreme is needed to assess the reliability of the change signal, which can be addressed based on the ensemble simulation from the General Circulation Models (GCMs) or Regional Climate Models (RCMs) [28].

5. Conclusions

Analyses of compound extremes have received increasing attention in recent decades due to the exaggerated impacts of multiple extremes that may occur concurrently or consecutively. In this study, we review commonly used statistical methods for characterizing and modeling compound extremes, including the empirical approach, multivariate distribution, the indicator approach, quantile regression and the Markov Chain model. The purposes, data requirements, pros and cons of different approaches are also elaborated upon. A discussion of related topics on compound extremes, such as climate change impacts, is also provided. This study introduced commonly used tools for the characterization and modeling of compound extremes; however, other methods may also be applied, such as a complex network analysis or the covariate approach [182,183]. In addition, we mainly focused on the statistical approaches in modeling different properties of compound extremes. The dynamical model simulations have also been employed for analyzing occurrences of compound extremes [58,184], such as storm surge and sea level/precipitation.

Future efforts are needed to understand the physical processes that lead to compound extremes based on observations and models. Data availability is a potential challenge in investigating compound extremes, since the occurrence of compound extremes is rare, which limits the accurate identification of long-term changes. This may be partly bypassed through pooling observations from multiple sites or employing ensembles from physical models (or simulations from statistical models) [35,56,185]. In addition, modeling the dependence of multiple variables based on historical records is of particular importance in investigating the occurrence or risk of compound extremes. The incorporation of asymmetric dependence and tail dependence in the statistical modeling of compound extremes is still challenging, especially in high dimensions. Moreover, a potential limitation of global climate model for studying extremes is the low resolution that hinders the representation of smaller scale features potentially relevant to climate and weather extremes. Modeling compound extremes with high spatial resolution (e.g., through statistical downscaling or RCMs) is of great importance for resolving local scale phenomena (or interactions) [24,25]. The limitation in the parameterization of certain physical processes (e.g., convective parameterization) also needs to be addressed in the accurate simulation of compound extremes across a wide range of temporal and spatial scales.

Human activities may also influence the occurrence of certain compound extremes. For example, the negative impact of flood occurring immediately after a long term drought, during which a reservoir should be kept as full as possible, may be exacerbated [186–188], as shown in the 2011 flooding of Brisbane, Australia that occurred after an exceptional multiyear drought (or "Millennium Drought") [189]. Thus, continuous efforts are needed in incorporating the interaction of human activity to mitigate the potential impacts of compound extremes. At last, a variety of studies have assessed the impacts of individual extremes (e.g., drought or heat wave) on different sectors, such as agricultural production, for the mitigation and resilience under global warming. The even larger impact of the compound extreme than that of the individual extreme calls for enhanced assessments of the changes and impacts to alleviate the potential threat caused by compound extremes, especially under a changing climate that may induce more extremes.

Author Contributions: Writing-Original Draft Preparation, Z.H.; Writing-Review & Editing, Z.H., V.S., and F.H.

Funding: This research was funded by National Natural Science Foundation of China (Grant number 41601014) and the Fundamental Research Funds for the Central Universities (Grant number 2017XTCX01).

Acknowledgments: Climate division data used in this study were obtained from the Climate Prediction Center (CPC), National Oceanic and Atmospheric Administration (NOAA) (ftp://ftp.cpc.ncep.noaa.gov/wd51yf/us). The monthly precipitation and temperature data from Climatic Research Unit (CRU) can be obtained from (http://www.cru.uea.ac.uk/data). We thank the editors and anonymous reviewers for the valuable comments and suggestions in different rounds of submissions and revisions of this manuscript.

Conflicts of Interest: The authors declare no conflict of interest.

Appendix A. Types of Compound Extremes

Table A1. Summary of Different Types of Compound Extremes and Statistical Approaches.

Type of Compound Extremes	Combined Variables/Events/Extremes	Approaches
Compound drought and hot extreme	Drought and heat wave (hot days or months, high temperature) [37–39,65,90,138,156,157,159]	Empirical approach, Quantile regression, Multivariate distribution
Compound precipitation and temperature extreme	Heavy precipitation and cold/warm condition [47,87,184,190]	Empirical approach, Markov Chain approach
	Low precipitation and high temperature [29,179,181,191]	Empirical approach, Multivariate distribution
	Dry-warm/dry-cold/wet-warm/wet-cold condition [29,77,80,192]	Empirical approach
Compound flood	Storm surge and high rainfall [35,57,185,193]	Multivariate distribution
	Storm surge and high discharge/runoff [48,55,56,60,61,136]	Empirical approach, Indicator approach, Multivariate distribution
	Storm surge and sea level [73,178]	Empirical approach, Indicator approach
	Sea levels and rainfall/river flow [59,97,136,194]	Multivariate distribution
Compound drought	Deficit from precipitation, soil moisture, runoff or other variables [54,153,195]	Indicator approach, Multivariate distribution
Combined drought, moisture surplus, precipitation/temperature, and other extremes	Drought indices, precipitation extremes and temperature extremes [146–148,196]	Indicator approach

References

1. Reichstein, M.; Bahn, M.; Ciais, P.; Frank, D.; Mahecha, M.D.; Seneviratne, S.I.; Zscheischler, J.; Beer, C.; Buchmann, N.; Frank, D.C. Climate extremes and the carbon cycle. *Nature* **2013**, *500*, 287–295. [CrossRef] [PubMed]
2. Easterling, D.; Groisman, P.Y.; Karl, T.; Evans, J.; Kunkel, K.; Ambenje, P. Observed variability and trends in extreme climate events: A brief review. *Bull. Am. Meteorol. Soc.* **2000**, *81*, 417–425. [CrossRef]

3. IPCC. *Managing the Risks of Extreme Events and Disasters to Advance Climate Change Adaptation (Srex) a Special Report of Working Groups I And II of the Intergovernmental Panel on Climate Change*; Cambridge University Press: Cambridge, UK; New York, NY, USA, 2012; p. 582.

4. Horton, R.M.; Mankin, J.S.; Lesk, C.; Coffel, E.; Raymond, C. A review of recent advances in research on extreme heat events. *Curr. Clim. Chang. Rep.* **2016**, *2*, 242–259. [CrossRef]

5. Burt, T.P.; Howden, N.J.K.; Worrall, F. The changing water cycle: Hydroclimatic extremes in the british isles. *Wiley Interdiscip. Rev. Water* **2016**, *3*, 854–870. [CrossRef]

6. Mishra, A.K.; Singh, V.P. Changes in extreme precipitation in texas. *J. Geophys. Res. Atmos.* **2010**, *115*, D14106. [CrossRef]

7. Heim, R.R. An overview of weather and climate extremes—Products and trends. *Weather Clim. Extrem* **2015**, *10*, 1–9. [CrossRef]

8. Trenberth, K.E.; Dai, A.; van der Schrier, G.; Jones, P.D.; Barichivich, J.; Briffa, K.R.; Sheffield, J. Global warming and changes in drought. *Nat. Clim. Chang.* **2014**, *4*, 17–22. [CrossRef]

9. Beniston, M.; Stephenson, D.B.; Christensen, O.B.; Ferro, C.A.T.; Frei, C.; Goyette, S.; Halsnaes, K.; Holt, T.; Jylhä, K.; Koffi, B. Future extreme events in european climate: An exploration of regional climate model projections. *Clim. Chang.* **2007**, *81*, 71–95. [CrossRef]

10. Zisopoulou, K.; Karalis, S.; Koulouri, M.-E.; Pouliasis, G.; Korres, E.; Karousis, A.; Triantafilopoulou, E.; Panagoulia, D. Recasting of the wef nexus as an actor with a new economic platform and management model. *Energy Policy* **2018**, *119*, 123–139. [CrossRef]

11. Dosio, A.; Mentaschi, L.; Fischer, E.M.; Klaus, W. Extreme heat waves under 1.5 °C and 2 °C global warming. *Environ. Res. Lett.* **2018**, *13*, 054006. [CrossRef]

12. Naumann, G.; Alfieri, L.; Wyser, K.; Mentaschi, L.; Betts, R.A.; Carrao, H.; Spinoni, J.; Vogt, J.; Feyen, L. Global changes in drought conditions under different levels of warming. *Geophys. Res. Lett.* **2018**, *45*, 3285–3296. [CrossRef]

13. Dai, A. Drought under global warming: A review. *Wiley Interdiscip. Rev. Clim. Chang.* **2011**, *2*, 45–65. [CrossRef]

14. Alfieri, L.; Bisselink, B.; Dottori, F.; Naumann, G.; Roo, A.; Salamon, P.; Wyser, K.; Feyen, L. Global projections of river flood risk in a warmer world. *Earth's Future* **2016**, *5*, 171–182. [CrossRef]

15. Rummukainen, M. Changes in climate and weather extremes in the 21st century. *Wiley Interdiscip. Rev. Clim. Chang.* **2012**, *3*, 115–129. [CrossRef]

16. Stott, P. How climate change affects extreme weather events. *Science* **2016**, *352*, 1517–1518. [CrossRef] [PubMed]

17. Trenberth, K.E.; Fasullo, J.T.; Shepherd, T.G. Attribution of climate extreme events. *Nat. Clim. Chang.* **2015**, *5*, 725–730. [CrossRef]

18. Rahmstorf, S.; Coumou, D. Increase of extreme events in a warming world. *Proc. Natl. Acad. Sci. USA* **2011**, *108*, 17905–17909. [CrossRef] [PubMed]

19. Alexander, L.V. Global observed long-term changes in temperature and precipitation extremes: A review of progress and limitations in ipcc assessments and beyond. *Weather Clim. Extrem* **2016**, *11*, 4–16. [CrossRef]

20. Panagoulia, D.; Vlahogianni, E.I. Nonlinear dynamics and recurrence analysis of extreme precipitation for observed and general circulation model generated climates. *Hydrol. Proccess.* **2014**, *28*, 2281–2292. [CrossRef]

21. Panagoulia, D.; Vlahogianni, E.I. Recurrence quantification analysis of extremes of maximum and minimum temperature patterns for different climate scenarios in the mesochora catchment in central-western greece. *Atmos. Res.* **2018**, *205*, 33–47. [CrossRef]

22. Katz, R.; Parlange, M.; Naveau, P. Statistics of extremes in hydrology. *Adv. Water Resour.* **2002**, *25*, 1287–1304. [CrossRef]

23. Katz, R.W. Statistics of extremes in climate change. *Clim. Chang.* **2010**, *100*, 71–76. [CrossRef]

24. Leonard, M.; Westra, S.; Phatak, A.; Lambert, M.; van den Hurk, B.; McInnes, K.; Risbey, J.; Schuster, S.; Jakob, D.; Stafford-Smith, M. A compound event framework for understanding extreme impacts. *Wiley Interdiscip. Rev. Clim. Chang.* **2014**, *5*, 113–128. [CrossRef]

25. Seneviratne, S.I.; Nicholls, N.; Easterling, D.; Goodess, C.M.; Kanae, S.; Kossin, J.; Luo, Y.; Marengo, J.; McInnes, K.; Rahimi, M. Changes in climate extremes and their impacts on the natural physical environment. In *Managing the Risks of Extreme Events and Disasters to Advance climate Change Adaptation*; Field, C.B., Barros, V., Stocker, T.F., Qin, D., Dokken, D., Ebi, K.L., Mastrandrea, M.D., Mach, K.J., Plattner, G.-K., Allen, S.K., et al., Eds.; A Special Report of Working Groups I And II of the Intergovernmental Panel on Climate Change (IPCC); Cambridge University Press: Cambridge, UK, 2012; pp. 109–230.
26. McPhillips, L.E.; Chang, H.; Chester, M.V.; Depietri, Y.; Friedman, E.; Grimm, N.B.; Kominoski, J.S.; McPhearson, T.; Méndez-Lázaro, P.; Rosi, E.J.; et al. Defining extreme events: A cross-disciplinary review. *Earth's Future* **2018**, *6*, 441–455. [CrossRef]
27. Orth, R.; Zscheischler, J.; Seneviratne, S.I. Record dry summer in 2015 challenges precipitation projections in central Europe. *Sci. Rep.* **2016**, *6*, 28334. [CrossRef] [PubMed]
28. Sedlmeier, K.; Feldmann, H.; Schädler, G. Compound summer temperature and precipitation extremes over central Europe. *Theor. Appl. Climatol.* **2018**, *131*, 1493–1501. [CrossRef]
29. Hao, Z.; AghaKouchak, A.; Phillips, T.J. Changes in concurrent monthly precipitation and temperature extremes. *Environ. Res. Lett.* **2013**, *8*, 034014. [CrossRef]
30. Lyon, B. Southern africa summer drought and heat waves: Observations and coupled model behavior. *J. Clim.* **2009**, *22*, 6033–6046. [CrossRef]
31. Albright, T.P.; Pidgeon, A.M.; Rittenhouse, C.D.; Clayton, M.K.; Wardlow, B.D.; Flather, C.H.; Culbert, P.D.; Radeloff, V.C. Combined effects of heat waves and droughts on avian communities across the conterminous United States. *Ecosphere* **2010**, *1*, 1–12. [CrossRef]
32. Livneh, B.; Hoerling, M.P. The physics of drought in the U.S. Central great plains. *J. Clim.* **2016**, *29*, 6783–6804. [CrossRef]
33. Mazzarella, A.; Rapetti, F. Scale-invariance laws in the recurrence interval of extreme floods: An application to the upper po river valley (Northern Italy). *J. Hydrol.* **2004**, *288*, 264–271. [CrossRef]
34. Merz, B.; Aerts, J.; Arnbjerg-Nielsen, K.; Baldi, M.; Becker, A.; Bichet, A.; Blöschl, G.; Bouwer, L.M.; Brauer, A.; Cioffi, F.; et al. Floods and climate: Emerging perspectives for flood risk assessment and management. *Nat. Hazards Earth Syst. Sci.* **2014**, *14*, 1921–1942. [CrossRef]
35. Van den Hurk, B.; van Meijgaard, E.; de Valk, P.; van Heeringen, K.-J.; Gooijer, J. Analysis of a compounding surge and precipitation event in the Netherlands. *Environ. Res. Lett.* **2015**, *10*, 035001. [CrossRef]
36. Martius, O.; Pfahl, S.; Chevalier, C. A global quantification of compound precipitation and wind extremes. *Geophys. Res. Lett.* **2016**, *43*, 7709–7717. [CrossRef]
37. Mazdiyasni, O.; AghaKouchak, A. Substantial increase in concurrent droughts and heatwaves in the United States. *Proc. Natl. Acad. Sci. USA* **2015**, *112*, 11484–11489. [CrossRef] [PubMed]
38. Mueller, B.; Seneviratne, S.I. Hot days induced by precipitation deficits at the global scale. *Proc. Natl. Acad. Sci. USA* **2012**, *109*, 12398–12403. [CrossRef] [PubMed]
39. Hirschi, M.; Seneviratne, S.I.; Alexandrov, V.; Boberg, F.; Boroneant, C.; Christensen, O.B.; Formayer, H.; Orlowsky, B.; Stepanek, P. Observational evidence for soil-moisture impact on hot extremes in southeastern Europe. *Nat. Geosci.* **2011**, *4*, 17–21. [CrossRef]
40. Wahl, T.; Plant, N.G.; Long, J.W. Probabilistic assessment of erosion and flooding risk in the northern Gulf of Mexico. *J. Geophys. Res. Oceans* **2016**, *121*, 3029–3043. [CrossRef]
41. Serafin, K.A.; Ruggiero, P. Simulating extreme total water levels using a time-dependent, extreme value approach. *J. Geophys. Res. Oceans* **2014**, *119*, 6305–6329. [CrossRef]
42. Li, F.; van Gelder, P.H.A.J.M.; Ranasinghe, R.; Callaghan, D.P.; Jongejan, R.B. Probabilistic modelling of extreme storms along the Dutch coast. *Coast. Eng.* **2014**, *86*, 1–13. [CrossRef]
43. Corbella, S.; Stretch, D.D. Simulating a multivariate sea storm using archimedean copulas. *Coast. Eng.* **2013**, *76*, 68–78. [CrossRef]
44. Svensson, C.; Jones, D.A. Dependence between extreme sea surge, river flow and precipitation in eastern Britain. *Int. J. Climatol.* **2002**, *22*, 1149–1168. [CrossRef]
45. Reed, A.J.; Mann, M.E.; Emanuel, K.A.; Lin, N.; Horton, B.P.; Kemp, A.C.; Donnelly, J.P. Increased threat of tropical cyclones and coastal flooding to new york city during the anthropogenic era. *Proc. Natl. Acad. Sci. USA* **2015**, *112*, 12610–12615. [CrossRef] [PubMed]

46. Hiroaki, I.; Yukiko, H.; Dai, Y.; Sanne, M.; Ward, P.J.; Winsemius, H.C.; Martin, V.; Shinjiro, K. Compound simulation of fluvial floods and storm surges in a global coupled river-coast flood model: Model development and its application to 2007 Cyclone Sidr in Bangladesh. *J. Adv. Model. Earth Syst.* **2017**, *9*, 1847–1862.

47. Benestad, R.E.; Haugen, J.E. On complex extremes: Flood hazards and combined high spring-time precipitation and temperature in Norway. *Clim. Chang.* **2007**, *85*, 381–406. [CrossRef]

48. Bevacqua, E.; Maraun, D.; Hobæk Haff, I.; Widmann, M.; Vrac, M. Multivariate statistical modelling of compound events via pair-copula constructions: Analysis of floods in Ravenna (Italy). *Hydrol. Earth Syst. Sci.* **2017**, *21*, 2701–2723. [CrossRef]

49. Trepanier, J.C.; Yuan, J.; Jagger, T.H. The combined risk of extreme tropical cyclone winds and storm surges along the U.S. Gulf of Mexico Coast. *J. Geophys. Res. Atmos.* **2017**, *122*, 3299–3316. [CrossRef]

50. Petroliagkis, T.I. Estimations of statistical dependence as joint return period modulator of compound events. Part I: Storm surge and wave height. *Nat. Hazards Earth Syst. Sci. Discuss.* **2017**, *2017*, 1–46. [CrossRef]

51. Fischer, E.; Knutti, R. Robust projections of combined humidity and temperature extremes. *Nat. Clim. Chang.* **2012**, *3*, 126–130. [CrossRef]

52. Schnell, J.L.; Prather, M.J. Co-occurrence of extremes in surface ozone, particulate matter, and temperature over eastern north America. *Proc. Natl. Acad. Sci. USA* **2017**, *114*, 2854–2859. [CrossRef] [PubMed]

53. Wilhite, D.; Glantz, M. Understanding: The drought phenomenon: The role of definitions. *Water Int.* **1985**, *10*, 111–120. [CrossRef]

54. Hao, Z.; Singh, V.P. Drought characterization from a multivariate perspective: A review. *J. Hydrol.* **2015**, *527*, 668–678. [CrossRef]

55. Klerk, W.J.; Winsemius, H.C.; van Verseveld, W.J.; Bakker, A.M.R.; Diermanse, F.L.M. The co-incidence of storm surges and extreme discharges within the rhine–meuse delta. *Environ. Res. Lett.* **2015**, *10*, 035005. [CrossRef]

56. Kew, S.F.; Selten, F.M.; Lenderink, G.; Hazeleger, W. The simultaneous occurrence of surge and discharge extremes for the rhine delta. *Nat. Hazards Earth Syst. Sci.* **2013**, *13*, 2017–2029. [CrossRef]

57. Wahl, T.; Jain, S.; Bender, J.; Meyers, S.D.; Luther, M.E. Increasing risk of compound flooding from storm surge and rainfall for major US cities. *Nat. Clim. Chang.* **2015**, *5*, 1093–1097. [CrossRef]

58. Muis, S.; Verlaan, M.; Winsemius, H.C.; Aerts, J.C.; Ward, P.J. A global reanalysis of storm surges and extreme sea levels. *Nat. Commun.* **2016**, *7*, 11969. [CrossRef] [PubMed]

59. Bengtsson, L. Probability of combined high sea levels and large rains in Malmö, Sweden, southern Resund. *Hydrol. Proccess.* **2016**, *30*, 3172–3183. [CrossRef]

60. Paprotny, D.; Vousdoukas, M.I.; Morales-Nápoles, O.; Jonkman, S.N.; Feyen, L. Compound flood potential in Europe. *Hydrol. Earth Syst. Sci. Discuss.* **2018**, *2018*, 1–34. [CrossRef]

61. Khanal, S.; Ridder, N.; de Vries, H.; Terink, W.; van den Hurk, B. Storm surge and extreme river discharge: A compound event analysis using ensemble impact modelling. *Hydrol. Earth Syst. Sci. Discuss.* **2018**, *2018*, 1–25. [CrossRef]

62. Svensson, C.; Jones, D.A. Dependence between sea surge, river flow and precipitation in south and west Britain. *Hydrol. Earth Syst. Sci.* **2004**, *8*, 973–992. [CrossRef]

63. Diffenbaugh, N.S.; Swain, D.L.; Touma, D. Anthropogenic warming has increased drought risk in California. *Proc. Natl. Acad. Sci. USA* **2015**, *112*, 3931–3936. [CrossRef] [PubMed]

64. Otkin, J.A.; Anderson, M.C.; Hain, C.; Mladenova, I.E.; Basara, J.B.; Svoboda, M. Examining rapid onset drought development using the thermal infrared–based evaporative stress index. *J. Hydrometeorol.* **2013**, *14*, 1057–1074. [CrossRef]

65. Sharma, S.; Mujumdar, P. Increasing frequency and spatial extent of concurrent meteorological droughts and heatwaves in India. *Sci. Rep.* **2017**, *7*, 15582. [CrossRef] [PubMed]

66. Witte, J.C.; Douglass, A.R.; Silva, A.D.; Torres, O.; Levy, R.; Duncan, B.N. Nasa a-train and terra observations of the 2010 Russian Wildfires. *Atmos. Chem. Phys.* **2011**, *11*, 19113–19142. [CrossRef]

67. Rizhsky, L.; Liang, H.; Mittler, R. The combined effect of drought stress and heat shock on gene expression in Tobacco. *Plant Physiol.* **2002**, *130*, 1143–1151. [CrossRef] [PubMed]

68. Jiang, Y.; Huang, B. Drought and heat stress injury to two cool-season turfgrasses in relation to antioxidant metabolism and lipid peroxidation. *Crop Sci.* **2001**, *41*, 436–442. [CrossRef]

69. Rammig, A.; Wiedermann, M.; Donges, J.F.; Babst, F.; Von Bloh, W.; Frank, D.; Thonicke, K.; Mahecha, M.D. Coincidences of climate extremes and anomalous vegetation responses: Comparing tree ring patterns to simulated productivity. *Biogeosciences* **2015**, *12*, 373–385. [CrossRef]

70. Ciais, P.; Reichstein, M.; Viovy, N.; Granier, A.; Ogée, J.; Allard, V.; Aubinet, M.; Buchmann, N.; Bernhofer, C.; Carrara, A. Europe-wide reduction in primary productivity caused by the heat and drought in 2003. *Nature* **2005**, *437*, 529–533. [CrossRef] [PubMed]

71. Yuan, W.; Cai, W.; Chen, Y.; Liu, S.; Dong, W.; Zhang, H.; Yu, G.; Chen, Z.; He, H.; Guo, W. Severe summer heatwave and drought strongly reduced carbon uptake in Southern China. *Sci. Rep.* **2016**, *6*, 18813. [CrossRef] [PubMed]

72. Koster, R.; Schubert, S.; Suarez, M. Analyzing the concurrence of meteorological droughts and warm periods, with implications for the determination of evaporative regime. *J. Clim.* **2009**, *22*, 3331–3341. [CrossRef]

73. Haigh, I.D.; Wadey, M.P.; Wahl, T.; Ozsoy, O.; Nicholls, R.J.; Brown, J.M.; Horsburgh, K.; Gouldby, B. Spatial and temporal analysis of extreme sea level and storm surge events around the coastline of the UK. *Sci. Data* **2016**, *3*, 160107. [CrossRef] [PubMed]

74. Davison, A.C.; Gholamrezaee, M.M. Geostatistics of extremes. *Proc. R. Soc. A Math. Phys. Eng. Sci.* **2012**, *468*, 581–608. [CrossRef]

75. Cooley, D.; Cisewski, J.; Erhardt, R.J.; Jeon, S.; Mannshardt, E.; Omolo, B.O.; Sun, Y. A survey of spatial extremes: Measuring spatial dependence and modeling spatial effects. *REVSTAT* **2012**, *10*, 135–165.

76. Hawkes, P.J.; Gonzalez-Marco, D.; Sánchez-Arcilla, A.; Prinos, P. Best practice for the estimation of extremes: A review. *J. Hydraul. Res.* **2008**, *46*, 324–332. [CrossRef]

77. Beniston, M. Trends in joint quantiles of temperature and precipitation in Europe since 1901 and projected for 2100. *Geophys. Res. Lett.* **2009**, *36*, L07707. [CrossRef]

78. Miao, C.; Sun, Q.; Duan, Q.; Wang, Y. Joint analysis of changes in temperature and precipitation on the Loess Plateau during the period 1961–2011. *Clim. Dyn.* **2016**, *47*, 3221–3234. [CrossRef]

79. Leng, G.; Tang, Q.; Huang, S.; Zhang, X.; Cao, J. Assessments of joint hydrological extreme risks in a warming climate in China. *Int. J. Climatol.* **2016**, *36*, 1632–1642. [CrossRef]

80. Morán-Tejeda, E.; Herrera, S.; López-Moreno, J.I.; Revuelto, J.; Lehmann, A.; Beniston, M. Evolution and frequency (1970–2007) of combined temperature–precipitation modes in the Spanish mountains and sensitivity of snow cover. *Reg. Environ. Chang.* **2013**, *13*, 873–885. [CrossRef]

81. Schär, C.; Ban, N.; Fischer, E.M.; Rajczak, J.; Schmidli, J.; Frei, C.; Giorgi, F.; Karl, T.R.; Kendon, E.J.; Tank, A.M.G.K. Percentile indices for assessing changes in heavy precipitation events. *Clim. Chang.* **2016**, *137*, 1–16. [CrossRef]

82. Zhang, X.; Alexander, L.; Hegerl, G.C.; Jones, P.; Tank, A.K.; Peterson, T.C.; Trewin, B.; Zwiers, F.W. Indices for monitoring changes in extremes based on daily temperature and precipitation data. *Wiley Interdiscip. Rev. Clim. Chang.* **2011**, *2*, 851–870. [CrossRef]

83. Frich, P.L.; Lv, A.; Della-Marta, P.; Gleason, B.; Haylock, M.; Klein, T.A.; Peterson, T.C. Observed coherent changes in climatic extremes during 2nd half of the 20th century. *Clim. Res.* **2002**, *19*, 193–212. [CrossRef]

84. Tank, A.M.G.K.; Wijngaard, J.B.; Können, G.P.; Böhm, R.; Demarée, G.; Gocheva, A.A.; Mileta, M.; Pashiardis, S.; Hejkrlik, L.; Kern-Hansen, C. Daily surface air temperature and precipitation dataset 1901–1999 for European Climate Assessment (ECA). *Int. J. Climatol.* **2002**, *22*, 1441–1453.

85. Alexander, L.; Zhang, X.; Peterson, T.; Caesar, J.; Gleason, B.; Klein Tank, A.; Haylock, M.; Collins, D.; Trewin, B.; Rahimzadeh, F. Global observed changes in daily climate extremes of temperature and precipitation. *J. Geophys. Res.* **2006**, *111*, D05109. [CrossRef]

86. Sunyer, M.A.; Sørup, H.J.D.; Christensen, O.B.; Madsen, H.; Rosbjerg, D.; Mikkelsen, P.S.; Arnbjerg-Nielsen, K. On the importance of observational data properties when assessing regional climate model performance of extreme precipitation. *Hydrol. Earth Syst. Sci.* **2013**, *17*, 4323–4337. [CrossRef]

87. Sedlmeier, K.; Mieruch, S.; Schädler, G.; Kottmeier, C. Compound extremes in a changing climate-a markov chain approach. *Nonlinear Process. Geophys.* **2016**, *23*, 375–390. [CrossRef]

88. Donat, M.G.; Sillmann, J.; Wild, S.; Alexander, L.V.; Lippmann, T.; Zwiers, F.W. Consistency of temperature and precipitation extremes across various global gridded in situ and reanalysis datasets. *J. Clim.* **2014**, *27*, 5019–5035. [CrossRef]

89. Hofstra, N.; New, M.; McSweeney, C. The influence of interpolation and station network density on the distributions and trends of climate variables in gridded daily data. *Clim. Dyn.* **2010**, *35*, 841–858. [CrossRef]

90. Kirono, D.G.C.; Hennessy, K.J.; Grose, M. Increasing risk of months with low rainfall and high temperature in southeast Australia for the past 150 years. *Clim. Risk Manag.* **2017**, *16*, 10–21. [CrossRef]

91. Trepanier, J.C.; Needham, H.F.; Elsner, J.B.; Jagger, T.H. Combining surge and wind risk from hurricanes using a copula model: An example from Galveston, texas. *Prof. Geogr.* **2015**, *67*, 52–61. [CrossRef]

92. Tebaldi, C.; Sansó, B. Joint projections of temperature and precipitation change from multiple climate models: A hierarchical Bayesian approach. *J. R. Stat. Soc. Ser. A* **2009**, *172*, 83–106. [CrossRef]

93. Sexton, D.M.H.; Murphy, J.M.; Collins, M.; Webb, M.J. Multivariate probabilistic projections using imperfect climate models part I: Outline of methodology. *Clim. Dyn.* **2011**, *38*, 1–30. [CrossRef]

94. Watterson, I. Calculation of joint pdfs for climate change with properties matching recent Australian projections. *Aust. Meteorol. Oceanogr. J.* **2011**, *61*, 211–219. [CrossRef]

95. Estrella, N.; Menzel, A. Recent and future climate extremes arising from changes to the bivariate distribution of temperature and precipitation in Bavaria, Germany. *Int. J. Climatol.* **2013**, *33*, 1687–1695. [CrossRef]

96. Rodrigo, F. On the covariability of seasonal temperature and precipitation in Spain, 1956–2005. *Int. J. Climatol.* **2015**, *35*, 3362–3370. [CrossRef]

97. Hawkes, P.J. Joint probability analysis for estimation of extremes. *J. Hydraul. Res.* **2008**, *46*, 246–256. [CrossRef]

98. Hao, Z.; Singh, V.P. Review of dependence modeling in hydrology and water resources. *Prog. Phys. Geogr.* **2016**, *40*, 549–578. [CrossRef]

99. Joe, H. *Multivariate Models and Dependence Concepts*; Chapman & Hall: London, UK, 1997.

100. Kao, S.C.; Govindaraju, R.S. Trivariate statistical analysis of extreme rainfall events via the Plackett Family of copulas. *Water Resour. Res.* **2008**, *44*, W02415. [CrossRef]

101. Song, S.; Singh, V. Frequency analysis of droughts using the plackett copula and parameter estimation by genetic algorithm. *Stoch. Environ. Res. Risk Assess.* **2010**, *24*, 783–805. [CrossRef]

102. Chebana, F.; Ouarda, T. Multivariate quantiles in hydrological frequency analysis. *Environmetrics* **2011**, *22*, 63–78. [CrossRef]

103. Mishra, A.K.; Ines, A.V.M.; Das, N.N.; Prakash Khedun, C.; Singh, V.P.; Sivakumar, B.; Hansen, J.W. Anatomy of a local-scale drought: Application of assimilated remote sensing products, crop model, and statistical methods to an agricultural drought study. *J. Hydrol.* **2015**, *526*, 15–29. [CrossRef]

104. Schöelzel, C.; Friederichs, P. Multivariate non-normally distributed random variables in climate research–introduction to the copula approach. *Nonlinear Process. Geophys.* **2008**, *15*, 761–772. [CrossRef]

105. Durante, F.; Salvadori, G. On the construction of multivariate extreme value models via copulas. *Environmetrics* **2010**, *21*, 143–161. [CrossRef]

106. Song, S.; Singh, V.P. Meta-elliptical copulas for drought frequency analysis of periodic hydrologic data. *Stoch. Environ. Res. Risk Assess.* **2010**, *24*, 425–444. [CrossRef]

107. Jonathan, P.; Ewans, K. Statistical modelling of extreme ocean environments for marine design: A review. *Ocean Eng.* **2013**, *62*, 91–109. [CrossRef]

108. Renard, B.; Lang, M. Use of a gaussian copula for multivariate extreme value analysis: Some case studies in hydrology. *Adv. Water Resour.* **2007**, *30*, 897–912. [CrossRef]

109. Rueda, A.; Camus, P.; Tomás, A.; Vitousek, S.; Méndez, F.J. A multivariate extreme wave and storm surge climate emulator based on weather patterns. *Ocean Model.* **2016**, *104*, 242–251. [CrossRef]

110. Bardossy, A. Copula-based geostatistical models for groundwater quality parameters. *Water Resour. Res.* **2006**, *42*. [CrossRef]

111. Bardossy, A.; Li, J. Geostatistical interpolation using copulas. *Water Resour. Res.* **2008**, *44*, W07412. [CrossRef]

112. Kazianka, H.; Pilz, J. Spatial interpolation using copula-based geostatistical models. In *geoENV VII—Geostatistics for Environmental Applications*; Atkinson, P., Lloyd, C., Eds.; Springer: Dordrecht, The Netherlands, 2010; pp. 307–319.

113. Nelsen, R.B. *An Introduction to Copulas*; Springer: New York, NY, USA, 2006.

114. Trivedi, P.K.; Zimmer, D.M. Copula modeling: An introduction for practitioners. *Found. Trends Econom.* **2005**, *1*, 1–111. [CrossRef]

115. Serinaldi, F.; Bárdossy, A.; Kilsby, C.G. Upper tail dependence in rainfall extremes: Would we know it if we saw it? *Stoch. Environ. Res. Risk Assess.* **2015**, *29*, 1211–1233. [CrossRef]

116. Genest, C.; Kojadinovic, I.; Nešlehová, J.; Yan, J. A goodness-of-fit test for bivariate extreme-value copulas. *Bernoulli* **2011**, *17*, 253–275. [CrossRef]

117. Salvadori, G.; de Michele, C.; Kottegoda, N.; Rosso, R. *Extremes in Nature: An. Approach Using Copulas*; Springer: New York, NY, USA, 2007.

118. Cormier, E.; Genest, C.; Nešlehová, J.G. Using b-splines for nonparametric inference on bivariate extreme-value copulas. *Extremes* **2014**, *17*, 633–659. [CrossRef]

119. Genest, C.; Rivest, L.-P. A characterization of gumbel's family of extreme value distributions. *Stat. Probab. Lett.* **1989**, *8*, 207–211. [CrossRef]

120. Singh, V.P.; Jain, S.K.; Tyagi, A. *Risk and Reliability Analysis: A Handbook for Civil and Environmental Engineers*; ASCE Press: Reston, VA, USA, 2007.

121. Singh, V.P. *Handbook of Applied Hydrology*; McGraw Hill Professional: New York, NY, USA, 2016.

122. Serinaldi, F. An uncertain journey around the tails of multivariate hydrological distributions. *Water Resour. Res.* **2013**, *49*, 6527–6547. [CrossRef]

123. Mazzarella, A. A fractal approach to sea-surge occurrences in the Northern Adriatic Sea. *J. Coast. Res.* **1998**, *14*, 1265–1268.

124. Malamud, B.D.; Turcotte, D.L. The applicability of power-law frequency statistics to floods. *J. Hydrol.* **2006**, *322*, 168–180. [CrossRef]

125. Mazzarella, A.; Diodato, N. The alluvial events in the last two centuries at Sarno, Southern Italy: Their classification and power-law time-occurrence. *Theor. Appl. Climatol.* **2002**, *72*, 75–84. [CrossRef]

126. Salvadori, G.; Durante, F.; Michele, C. Multivariate return period calculation via survival functions. *Water Resour. Res.* **2013**, *49*, 2308–2311. [CrossRef]

127. Gräler, B.; Petroselli, A.; Grimaldi, S.; De Baets, B.; Verhoest, N. An update on multivariate return periods in hydrology. *Proc. Int. Assoc. Hydrol. Sci.* **2016**, *373*, 175–178. [CrossRef]

128. Serinaldi, F. Dismissing return periods! *Stoch. Environ. Res. Risk Assess.* **2015**, *29*, 1179–1189. [CrossRef]

129. Brunner, M.I.; Seibert, J.; Favre, A.C. Bivariate return periods and their importance for flood peak and volume estimation. *Wiley Interdiscip. Rev. Water* **2016**, *3*, 819–833. [CrossRef]

130. Gräler, B.; van den Berg, M.; Vandenberghe, S.; Petroselli, A.; Grimaldi, S.; De Baets, B.; Verhoest, N. Multivariate return periods in hydrology: A critical and practical review focusing on synthetic design hydrograph estimation. *Hydrol. Earth Syst. Sci.* **2013**, *17*, 1281–1296. [CrossRef]

131. Hao, Z.; Hao, F.; Singh, V.P.; Wei, O.; Cheng, H. An integrated package for drought monitoring, prediction and analysis to aid drought modeling and assessment. *Environ. Model. Softw.* **2017**, *91*, 199–209. [CrossRef]

132. Salvadori, G.; Durante, F.; de Michele, C.; Bernardi, M.; Petrella, L. A multivariate copula-based framework for dealing with hazard scenarios and failure probabilities. *Water Resour. Res.* **2016**, *52*, 3701–3721. [CrossRef]

133. McKee, T.B.; Doesken, N.J.; Kleist, J. The Relationship of Drought Frequency and Duration to Time Scales. In Proceedings of the Eighth Conference on Applied Climatology, Anaheim, CA, USA, 17–22 January 1993; pp. 179–184.

134. Shukla, S.; Wood, A.W. Use of a standardized runoff index for characterizing hydrologic drought. *Geophys. Res. Lett.* **2008**, *35*, L02405. [CrossRef]

135. Hao, Z.; AghaKouchak, A. Multivariate standardized drought index: A parametric approach for drought analysis. *Adv. Water Resour.* **2013**, *57*, 12–18. [CrossRef]

136. Moftakhari, H.R.; Salvadori, G.; AghaKouchak, A.; Sanders, B.F.; Matthew, R.A. Compounding effects of sea level rise and fluvial flooding. *Proc. Natl. Acad. Sci. USA* **2017**, *114*, 9785–9790. [CrossRef] [PubMed]

137. Koster, R.D.; Wang, H.; Schubert, S.D.; Suarez, M.J.; Mahanama, S. Drought-induced warming in the continental United States under different SST regimes. *J. Clim.* **2009**, *22*, 5385–5400. [CrossRef]

138. Hao, Z.; Hao, F.; Singh, V.P.; Ouyang, W. Quantitative risk assessment of the effects of drought on extreme temperature in Eastern China. *J. Geophys. Res. Atmos.* **2017**, *122*, 9050–9059. [CrossRef]

139. Heffernan, J.E.; Tawn, J.A. A conditional approach for multivariate extreme values (with discussion). *J. R. Stat. Soc. B* **2004**, *66*, 497–546. [CrossRef]

140. Heffernan, J.E.; Resnick, S.I. Limit laws for random vectors with an extreme component. *Ann. Appl. Probab.* **2007**, *17*, 537–571. [CrossRef]

141. Keef, C.; Papastathopoulos, I.; Tawn, J.A. Estimation of the conditional distribution of a multivariate variable given that one of its components is large: Additional constraints for the heffernan and tawn model. *J. Multivar. Anal.* **2013**, *115*, 396–404. [CrossRef]

142. Cheng, L.; Gilleland, E.; Heaton, M.J.; AghaKouchak, A. Empirical bayes estimation for the conditional extreme value model. *Stat* **2014**, *3*, 391–406. [CrossRef]

143. Zhang, L.; Singh, V.P. Bivariate rainfall frequency distributions using archimedean copulas. *J. Hydrol.* **2007**, *332*, 93–109. [CrossRef]

144. Yue, S.; Rasmussen, P. Bivariate frequency analysis: Discussion of some useful concepts in hydrological application. *Hydrol. Proccess.* **2002**, *16*, 2881–2898. [CrossRef]

145. Sillmann, J.; Kharin, V.V.; Zhang, X.; Zwiers, F.W.; Bronaugh, D. Climate extremes indices in the CMIP5 multimodel ensemble: Part 1. Model evaluation in the present climate. *J. Geophys. Res. Atmos.* **2013**, *118*, 1716–1733. [CrossRef]

146. Gallant, A.J.E.; Karoly, D.J. A combined climate extremes index for the Australian Region. *J. Clim.* **2010**, *23*, 6153–6165. [CrossRef]

147. Gallant, A.J.; Karoly, D.J.; Gleason, K.L. Consistent trends in a modified climate extremes index in the United States, Europe, and Australia. *J. Clim.* **2014**, *27*, 1379–1394. [CrossRef]

148. Karl, T.R.; Knight, R.W.; Easterling, D.R.; Quayle, R.G. Indices of climate change for the United States. *Bull. Am. Meteorol. Soc.* **1996**, *77*, 279–292. [CrossRef]

149. Perkins, S.E.; Alexander, L.V. On the measurement of heat waves. *J. Clim.* **2013**, *26*, 4500–4517. [CrossRef]

150. Brooks, H.E.; Lee, J.W.; Craven, J.P. The spatial distribution of severe thunderstorm and tornado environments from global reanalysis data. *Atmos. Res.* **2003**, *67*, 73–94. [CrossRef]

151. Brooks, H.E. Proximity soundings for severe convection for Europe and the United States from reanalysis data. *Atmos. Res.* **2009**, *93*, 546–553. [CrossRef]

152. Coles, S. *An Introduction to Statistical Modeling of Extreme Values*; Springer: London, UK, 2001.

153. Kao, S.C.; Govindaraju, R.S. A copula-based joint deficit index for droughts. *J. Hydrol.* **2010**, *380*, 121–134. [CrossRef]

154. Koenker, R.; Bassett, G., Jr. Regression quantiles. *Econometrica* **1978**, *46*, 33–50. [CrossRef]

155. Koenker, R. *Quantile Regression*; Cambridge University Press: Cambridge, UK, 2005.

156. Meng, L.; Shen, Y. On the relationship of soil moisture and extreme temperatures in East China. *Earth Interact.* **2014**, *18*, 1–20. [CrossRef]

157. Ford, T.W.; Quiring, S.M. In situ soil moisture coupled with extreme temperatures: A study based on the Oklahoma mesonet. *Geophys. Res. Lett.* **2014**, *41*, 4727–4734. [CrossRef]

158. Quesada, B.; Vautard, R.; Yiou, P.; Hirschi, M.; Seneviratne, S.I. Asymmetric European summer heat predictability from wet and dry southern winters and springs. *Nat. Clim. Chang.* **2012**, *2*, 736–741. [CrossRef]

159. Herold, N.; Kala, J.; Alexander, L. The influence of soil moisture deficits on Australian heatwaves. *Environ. Res. Lett* **2016**, *11*, 064003. [CrossRef]

160. Zhang, X.-J.; Tang, Q.; Pan, M.; Tang, Y. A long-term land surface hydrologic fluxes and states dataset for China. *J. Hydrometeorol.* **2014**, *15*, 2067–2084. [CrossRef]

161. Steinemann, A. Drought indicators and triggers: A stochastic approach to evaluation. *J. Am. Water Resour. Assoc.* **2003**, *39*, 1217–1233. [CrossRef]

162. Shaby, B.A.; Reich, B.J.; Cooley, D.; Kaufman, C.G. A markov-switching model for heat waves. *Ann. Appl. Stat.* **2016**, *10*, 74–93. [CrossRef]

163. Friederichs, P.; Hense, A. Statistical downscaling of extreme precipitation events using censored quantile regression. *Mon. Weather Rev.* **2007**, *135*, 2365–2378. [CrossRef]

164. Wilks, D.S. *Statistical Methods in the Atmospheric Sciences*; Academic Press: San Diego, CA, USA, 2011; Volume 100.

165. Aas, K.; Czado, C.; Frigessi, A.; Bakken, H. Pair-copula constructions of multiple dependence. *Insur. Math. Econ.* **2009**, *44*, 182–198. [CrossRef]

166. Liu, Z.; Törnros, T.; Menzel, L. A probabilistic prediction network for hydrological drought identification and environmental flow assessment. *Water Resour. Res.* **2016**, *52*, 6243–6262. [CrossRef]

167. Brechmann, E.C.; Schepsmeier, U. Modeling dependence with C-and D-vine copulas: The R-package cdvine. *J. Stat. Softw.* **2013**, *52*, 1–27. [CrossRef]

168. Liu, Z.; Zhou, P.; Chen, X.; Guan, Y. A multivariate conditional model for streamflow prediction and spatial precipitation refinement. *J. Geophys. Res. Atmos.* **2015**, *120*, 10116–10129. [CrossRef]

169. Milly, P.C.D.; Betancourt, J.; Falkenmark, M.; Hirsch, R.M.; Kundzewicz, Z.W.; Lettenmaier, D.P.; Stouffer, R.J.; Dettinger, M.D.; Krysanova, V. On critiques of "stationarity is dead: Whither water management?". *Water Resour. Res.* **2015**, *51*, 7785–7789. [CrossRef]

170. Milly, P.; Betancourt, J.; Falkenmark, M.; Hirsch, R.; Kundzewicz, Z.; Lettenmaier, D.; Stouffer, R. Climate change. Stationarity is dead: Whither water management? *Science* **2008**, *319*, 573–574. [CrossRef] [PubMed]

171. Chebana, F.; Ouarda, T.B.; Duong, T.C. Testing for multivariate trends in hydrologic frequency analysis. *J. Hydrol.* **2013**, *486*, 519–530. [CrossRef]

172. Sarhadi, A.; Ausín, M.C.; Wiper, M.P. A new time-varying concept of risk in a changing climate. *Sci. Rep.* **2016**, *6*, 35755. [CrossRef] [PubMed]

173. Bender, J.; Wahl, T.; Jensen, J. Multivariate design in the presence of non-stationarity. *J. Hydrol.* **2014**, *514*, 123–130. [CrossRef]

174. Jonathan, P.; Ewans, K.; Randell, D. Non-stationary conditional extremes of northern North Sea storm characteristics. *Environmetrics* **2014**, *25*, 172–188. [CrossRef]

175. Jonathan, P.; Randell, D.; Wu, Y.; Ewans, K. Return level estimation from non-stationary spatial data exhibiting multidimensional covariate effects. *Ocean Eng.* **2014**, *88*, 520–532. [CrossRef]

176. Jiang, C.; Xiong, L.; Xu, C.Y.; Guo, S. Bivariate frequency analysis of nonstationary low-flow series based on the time-varying copula. *Hydrol. Proccess.* **2015**, *29*, 1521–1534. [CrossRef]

177. Sarhadi, A.; Burn, D.H.; Concepción Ausín, M.; Wiper, M.P. Time-varying nonstationary multivariate risk analysis using a dynamic bayesian copula. *Water Resour. Res.* **2016**, *52*, 2327–2349. [CrossRef]

178. Little, C.M.; Horton, R.M.; Kopp, R.E.; Oppenheimer, M.; Vecchi, G.A.; Villarini, G. Joint projections of us east coast sea level and storm surge. *Nat. Clim. Chang.* **2015**, *5*, 1114–1120. [CrossRef]

179. Zscheischler, J.; Seneviratne, S.I. Dependence of drivers affects risks associated with compound events. *Sci. Adv.* **2017**, *3*, e1700263. [CrossRef] [PubMed]

180. Hao, Z.; Singh, V.P. Integrating entropy and copula theories for hydrologic modeling and analysis. *Entropy* **2015**, *17*, 2253–2280. [CrossRef]

181. Serinaldi, F. Can we tell more than we can know? The limits of bivariate drought analyses in the United States. *Stoch. Environ. Res. Risk Assess.* **2016**, *30*, 1691–1704. [CrossRef]

182. Whan, K.; Zscheischler, J.; Orth, R.; Shongwe, M.; Rahimi, M.; Asare, E.O.; Seneviratne, S.I. Impact of soil moisture on extreme maximum temperatures in Europe. *Weather Clim. Extrem* **2015**, *9*, 57–67. [CrossRef]

183. Sun, A.Y.; Xia, Y.; Caldwell, T.G.; Hao, Z. Patterns of precipitation and soil moisture extremes in Texas, US: A complex network analysis. *Adv. Water Resour.* **2018**, *112*, 203–213. [CrossRef]

184. Tencer, B.; Bettolli, M.L.; Rusticucci, M. Compound temperature and precipitation extreme events in southern South America: Associated atmospheric circulation, and simulations by a multi-rcm ensemble. *Clim. Res.* **2016**, *68*, 183–199. [CrossRef]

185. Zheng, F.; Westra, S.; Sisson, S.A. Quantifying the dependence between extreme rainfall and storm surge in the coastal zone. *J. Hydrol.* **2013**, *505*, 172–187. [CrossRef]

186. Di Baldassarre, G.; Martinez, F.; Kalantari, Z.; Viglione, A. Drought and flood in the anthropocene: Feedback mechanisms in reservoir operation. *Earth Syst. Dyn.* **2017**, *8*, 225–233. [CrossRef]

187. Van Loon, A.F.; Stahl, K.; Di Baldassarre, G.; Clark, J.; Rangecroft, S.; Wanders, N.; Gleeson, T.; Van Dijk, A.I.J.M.; Tallaksen, L.M.; Hannaford, J.; et al. Drought in a human-modified world: Reframing drought definitions, understanding, and analysis approaches. *Hydrol. Earth Syst. Sci.* **2016**, *20*, 3631–3650. [CrossRef]

188. Di Baldassarre, G.; Viglione, A.; Carr, G.; Kuil, L.; Yan, K.; Brandimarte, L.; Blöschl, G. Debates—Perspectives on socio-hydrology: Capturing feedbacks between physical and social processes. *Water Resour. Res.* **2015**, *51*, 4770–4781. [CrossRef]

189. Dijk, A.I.; Beck, H.E.; Crosbie, R.S.; Jeu, R.A.; Liu, Y.Y.; Podger, G.M.; Timbal, B.; Viney, N.R. The millennium drought in southeast Australia (2001–2009): Natural and human causes and implications for water resources, ecosystems, economy, and society. *Water Resour. Res.* **2013**, *49*, 1040–1057. [CrossRef]

190. Tencer, B.; Weaver, A.; Zwiers, F. Joint occurrence of daily temperature and precipitation extreme events over Canada. *J. Appl. Meteorol. Climatol.* **2014**, *53*, 2148–2162. [CrossRef]

191. AghaKouchak, A.; Cheng, L.; Mazdiyasni, O.; Farahmand, A. Global warming and changes in risk of concurrent climate extremes: Insights from the 2014 California drought. *Geophys. Res. Lett.* **2014**, *41*, 8847–8852. [CrossRef]

192. Arsenović, P.; Tošić, I.; Unkašević, M. Trends in combined climate indices in Serbia from 1961 to 2010. *Meteorol. Atmos. Phys.* **2015**, *127*, 489–498. [CrossRef]

193. Zheng, F.; Westra, S.; Leonard, M.; Sisson, S.A. Modeling dependence between extreme rainfall and storm surge to estimate coastal flooding risk. *Water Resour. Res.* **2014**, *50*, 2050–2071. [CrossRef]

194. Lian, J.J.; Xu, K.; Ma, C. Joint impact of rainfall and tidal level on flood risk in a coastal city with a complex river network: A case study of Fuzhou city, China. *Hydrol. Earth Syst. Sci.* **2013**, *17*, 679–689. [CrossRef]

195. Beersma, J.J.; Buishand, T.A. Joint probability of precipitation and discharge deficits in the Netherlands. *Water Resour. Res.* **2004**, *40*, W12508. [CrossRef]

196. Gleason, K.L.; Lawrimore, J.H.; Levinson, D.H.; Karl, T.R.; Karoly, D.J. A revised us climate extremes index. *J. Clim.* **2008**, *21*, 2124–2137. [CrossRef]

MDPI
St. Alban-Anlage 66
4052 Basel
Switzerland
Tel. +41 61 683 77 34
Fax +41 61 302 89 18
www.mdpi.com

Water Editorial Office
E-mail: water@mdpi.com
www.mdpi.com/journal/water